ENGINEERING
PLASTICITY

ENGINEERING PLASTICITY

W. JOHNSON *1922 –*

Professor of Mechanical Engineering
University of Manchester Institute of Science and Technology

and

P. B. MELLOR

Professor of Mechanical Engineering
University of Bradford

VAN NOSTRAND REINHOLD COMPANY
LONDON

NEW YORK　　CINCINNATI　　TORONTO　　MELBOURNE

VAN NOSTRAND REINHOLD COMPANY LTD
25-28 Buckingham Gate, London, SWIE 6LQ

INTERNATIONAL OFFICES
New York   Cincinnati   Toronto   Melbourne

Library of Congress Catalog Card No 72-186765
ISBN 0 442 04151 9

First published 1973

Made and Printed in Great Britain by
The Garden City Press Limited
Letchworth, Hertfordshire SG6 1JS

PREFACE

Plasticity for Mechanical Engineers which first appeared in 1962 has enjoyed a steady sale over the intervening years, both in English and in Japanese translation. The present book is based on this previous text but has been considerably expanded to take account of new applications of plasticity theory and of recent experimental work. The original title *Plasticity for Mechanical Engineers* has been criticized over the years, especially by overseas readers, as too restrictive and somewhat misleading. After due consideration we have therefore agreed that this new text should appear under the title *Engineering Plasticity*.

The major change has been the addition of two new chapters, Chapters 15 and 16 by one of the authors (W.J.). These are concerned with 'Load Bounding'—one dealing with the plastic bending of plates and the other with axisymmetric problems. In addition several chapters have been expanded: the basic equations for the yield and flow of anisotropic materials are now included with application to some simple sheet metal forming problems; experimental results are given for instability of tubes subjected to axial tension and internal fluid pressure and also for rotating discs; more work has been included on elastic and plastic bending and on the swaging and up-setting of a cylindrical bar.

In order to accommodate this new material some deletions had to be made and it was decided to omit the chapter on 'Creep' and the appendix on 'Crystal Plasticity' which appeared in the original book.

The book is considered suitable for students of mechanical engineering, production engineering and metallurgy at final year degree level and for post-graduate studies in Universities and Polytechnics. Wherever possible any theoretical work is placed in the context of a practical problem and experimental results are given wherever appropriate.

University of Manchester Institute of Science and Technology　W. JOHNSON
University of Bradford　P. B. MELLOR
1972

NOTES ON S.I. UNITS

Since the publication of *Plasticity for Mechanical Engineers* on which this book is based, S.I. units have been introduced. The question therefore arose as to whether we should convert all units to the S.I. system. Examination of the text, however, showed that this would lead to many absurdities. Where units are used in the text they are in reference to work that has appeared in the literature and since we hope that readers will turn to some of the quoted references it hardly seems helpful to replot a graph in S.I. units. Again it seems reasonable to quote work that has been done for example with sheet metal of thickness 0·036 in in these units and not as 0·8944 mm.

Many first-year textbooks have appeared in S.I. units and they are serving a proselytizing purpose. Readers of this book will, however, have come to realize that they must be familiar with all systems of units and be able to convert from one to another with ease. In order to help such conversion we are setting out the information that will be most commonly required. For general guidance the reader is referred to the booklet 'The Use of S.I. Units' PD 5686: 1972 published by British Standards Institution.

Units of Length
   1 in $= 0{\cdot}0254$ m

Units of Force
   1 tonf $= 9{\cdot}96 \times 10^3$ N
   1 lbf $= 4{\cdot}45$ N
   1 kgf $= 9{\cdot}81$ N

Units of Pressure, Stress
   1 atm $= 101{\cdot}325$ kN/m²
   1 kgf/m² $= 9{\cdot}81$ N/m²
   1 lbf/in² $= 6{\cdot}89$ kN/m²
   1 tonf/in² $= 15{\cdot}444$ MN/m²

CONTENTS

CHAPTER 10   PLASTIC INSTABILITY

Contents

CHAPTER 15   LOAD BOUNDING APPLIED TO THE
PLASTIC BENDING OF PLATES

CHAPTER 16   LOAD BOUNDING APPLIED TO AXISYMMETRIC
INDENTATION AND RELATED PROBLEMS

Contents

Chapter 1

INTRODUCTION

'The tensile test (is) very easily and quickly performed but it is not possible to do much with its results, because one does not know what they really mean. They are the outcome of a number of very complicated physical processes. . . . The extension of a piece of metal (is) in a sense more complicated than the working of a pocket watch and to hope to derive information about its mechanism from two or three data derived from measurement during the tensile test (is) perhaps as optimistic as would be an attempt to learn about the working of a pocket watch by determining its compressive strength.'

E. OROWAN, F.R.S., *Proc. Instn. Mech. Engrs* Vol. 151, p. 133, 1944.

1.1 Phenomenological Nature of Theory of Plasticity

The theories of elasticity and plasticity describe the mechanics of deformation of most engineering solids. Both theories, as applied to metals and alloys, are based on experimental studies of the relations between stress and strain in a polycrystalline aggregate under simple loading conditions. Thus they are of a phenomenological nature on the macroscopic scale and, as yet, owe little to a knowledge of the structure of a metal. However, in order to understand the limitations so imposed on the theories, the engineer, with his main interest in design and manufacture, must have some knowledge of the structure of metals. It is to the metal physicist that he looks for an explanation of fracture and to the metallurgist for supplying him with materials that will withstand increasingly severe conditions. Alternatively, the metallurgist must have some knowledge of the theory of plasticity if he is to understand the requirements of the engineer. In this introductory chapter the macroscopic behaviour of metals will be discussed with particular reference to the effects of hydrostatic pressure, temperature, and strain rate.

1.2 The Load-extension Diagram in Simple Tension

A simple tension test on an annealed mild steel bar is perhaps the most familiar example of elastic and plastic deformation. Consider a bar of mild steel extended in a testing machine at room temperature and at a strain rate of approximately 2×10^{-3} per second, readings of load and extension being recorded. It is essential that the machined specimen be aligned correctly in the testing machine and have a reduced central section to ensure that there is a uniform axial stress distribution throughout the central gauge length. A typical load-extension diagram for mild steel is shown in Fig. 1.1.

Initially the relation between load and extension is essentially linear—i.e. portion OA of the curve, where A defines the limit of proportionality. On further straining, the relation between load and extension is no longer linear but the material is still elastic and upon release of the load the specimen reverts to its original length. The maximum load that can be applied without causing permanent deformation defines the elastic limit. Usually there is little difference between the proportional limit, A, and the elastic limit, B. Both are dependent on the sensitivity of the measuring devices used and on certain details of testing technique. Point B marks the end of purely elastic straining and the point of initiation of plastic deformation. It is known as the upper yield point, and the upper yield stress is defined as the load at this point divided by the original cross-sectional area. The extension per unit length at this point is of the order of 10^{-3}.

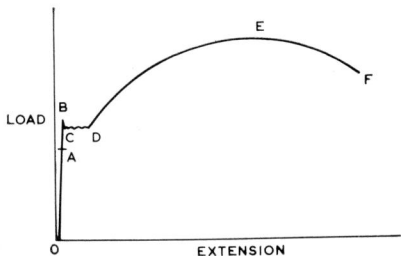

FIG. 1.1 Load-extension diagram for mild steel.

Further straining is accompanied by a sudden drop in load and then extension at approximately constant load. The extension CD is of the order 10^{-2} per unit gauge length and therefore, over an 8-inch gauge length it can easily be observed with the aid of a pair of dividers spanning the gauge length. The lower yield stress is defined as the load at CD divided by the original cross-sectional area of the bar.

After D the load increases with further strain. This effect of the material being able to withstand a greater load despite the uniform reduction in cross-sectional area is known as strain-hardening or work-hardening. The true stress in the bar is the applied load divided by the current cross-sectional area. At E the rate of work-hardening is unable to keep pace with the rate of reduction in the cross-sectional area and a maximum occurs in the load, followed by local straining (necking of the bar) leading to fracture at F. The tensile strength (T.S.) is defined as the maximum load divided by the original cross-sectional area. It is the maximum load that a bar of unit original cross-sectional area will withstand in tension. The T.S. is not a measure of the intrinsic strength of the material; it is indicative only of an instability condition (end of uniform straining and initiation of necking) in the tensile test, and it is discussed at greater length below.

Note that until the T.S. is reached all elements of the bar can be considered to be identically deformed under the load, i.e., until the point E is reached, the whole bar is homogeneously strained.

After the maximum load, a localized 'neck' or waist is formed in which a triaxial tensile stress system develops. The tensile test therefore reaches its limit of simple usefulness not at fracture, but at the maximum load condition. From the point of instability up to fracture, the straining takes place in the 'neck' under a complicated and continuously changing triaxial tensile stress system; the higher the rate of hardening of the metal the greater is the extent of the 'neck' which is formed. Outside the neck, material is unloaded because cross-sections of the bar are called upon to carry a decreasing load. In the neck the mean stress on a cross-section increases as the load decreases, whereas outside the neck the mean stress decreases.

If a tensile specimen, loaded and extended to a degree defined by point A in Fig. 1.2 is completely unloaded, it recovers some of its extension elastically. AB represents the unloading line and it is evident that when the load is

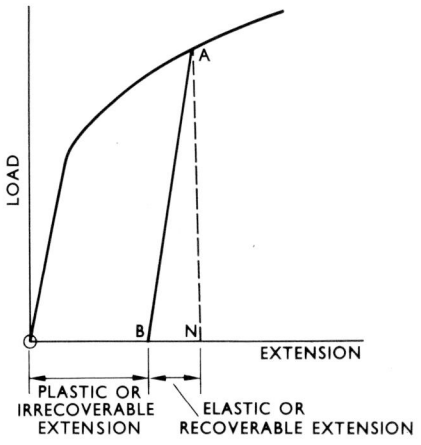

FIG. 1.2 Load-extension diagram for metal or alloy not exhibiting pronounced initial yield.

completely removed, there remains a permanent extension, OB. In unloading, extension BN is elastically recovered whereas by contrast OB represents plastic, irrecoverable extension. Reloading this specimen would cause it to follow the load—elastic extension path BA. On approaching load A further plastic or permanent extension would be induced. Note then, that in the strain-hardening range, elements under load possess a total extension (and also a transverse contraction to increase our generality of statement) which has permanent (i.e., plastic or irrecoverable) and impermanent (i.e., elastic or recoverable) components.

Most metals and alloys do not show the pronounced initial yielding exhibited by annealed mild steel. The change from purely elastic to elastic-plastic deformation is gradual; otherwise the load-extension diagrams have the same form, see Fig. 1.2. In such cases, it is standard practice to define a 'proof stress' instead of a yield stress. A 'proof stress' is defined as the stress (load divided by original area of cross-section) which will cause a permanent extension, i.e., OB, equal to a specified percentage of the original gauge length. (The proof stress finds no application in the mathematical theory of plasticity.)

During the plastic extension of a specimen, it has been found by FARREN and TAYLOR (1925) that only 85–90 per cent of the heat equivalent of the work done reappears as heat. The remaining energy is stored in the deformed lattice of the strain-hardened bar. Recovery of this stored energy is discussed by GORDON (1955).

1.3 The True Stress-Strain Diagram in Simple Tension

Nominal stress is defined as the load divided by the original cross-sectional area of the bar and conventional or engineering strain as the extension per unit original length. A nominal stress-strain curve is therefore simply a load-extension diagram for a bar with unit original cross-sectional area and unit gauge length.

The true stress-strain curve is a more informative diagram for plasticity purposes. The true stress is the load divided by the current cross-sectional area and may be obtained by taking simultaneous measurements of load and current diameter of the bar. However, we may proceed by neglecting the very small changes in volume that occur during a tensile test. If therefore the material is assumed to be incompressible

$$Xl = X_0 l_0 \tag{1.1}$$

where X is the current cross-sectional area and l is the current gauge length. The suffixes on the right-hand side of the equation refer to the original dimensions.

If P is the current load and σ the true stress, then

$$P = \sigma X \tag{1.2}$$

or

$$\sigma = \frac{P}{X_0} \cdot \frac{l}{l_0} = \sigma_0 (1 + e) \tag{1.3}$$

where σ_0 is the nominal stress and $e = (l - l_0)/l_0$, the engineering strain.

At the maximum load $dP = 0$ and thus using equation (1.2)

$$\sigma \, dX + X \, d\sigma = 0 \tag{1.4}$$

Also from the condition of incompressibility (1.1)

$$l \, dX + X \, dl = 0 \tag{1.5}$$

From equations (1.4) and (1.5) the true stress at the instability condition is given by

$$\frac{d\sigma}{\sigma} = \frac{dl}{l} = d\epsilon \tag{1.6}$$

Now the natural or logarithmic strain is defined as

$$\epsilon = \int_{l_0}^{l} \frac{dl}{l} = \ln\frac{l}{l_0} = \ln(1 + e) \tag{1.7}$$

At the maximum load therefore

$$\frac{d\sigma}{d\epsilon} = \frac{\sigma}{1} \quad \text{or} \quad \frac{d\sigma}{de} = \frac{\sigma}{1 + e} = \sigma_0 \tag{1.8}$$

If the true stress, σ, is plotted against the logarithmic strain, ϵ, the true stress and strain at instability will be found by drawing a tangent to the (σ, ϵ) curve such that the length of the sub-tangent measured along the strain axis is unity, Fig. 1.3(a). A similar construction, Fig. 1.3(b), gives the instability strain, e, from the (σ, e) curve.

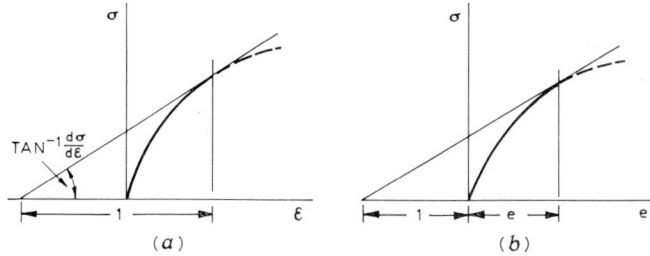

FIG. 1.3 True stress-strain curves in simple tension showing construction for determining point of instability.

The definition of logarithmic strain was suggested by LUDWICK (1909). For small extensions, the engineering strain, e (first defined by Cauchy), is approximately equal to the logarithmic strain ϵ. There are two main advantages in using the logarithmic definition of strain.

1. Logarithmic strains are additive, whereas engineering strains are not. This follows from the definition, but we may illustrate this further by considering the simple example of a bar of material with gauge length l_1, being strained under uniaxial stress to a length l_2. The strain $\epsilon_1 = \ln(l_2/l_1)$. If later the bar is further extended from length l_2 to length l_3, the further strain will be $\epsilon_2 = \ln(l_3/l_2)$ and the total strain will be $\epsilon = \ln(l_2/l_1) + \ln(l_3/l_2) = \ln(l_3/l_1)$.

It will be shown later that a comparison of stress-strain curves obtained under different stress-systems can be made on the basis of logarithmic strain, but not on a basis of engineering strain.

2. It can be shown experimentally that for large plastic strains, such as occur in most metal-forming operations, the material may be considered incompressible. The volume constancy condition is, expressed in principal engineering strains

$$(1 + e_1)(1 + e_2)(1 + e_3) = 1 \tag{1.9}$$

which, in terms of logarithmic strains, is written more simply

$$\epsilon_1 + \epsilon_2 + \epsilon_3 = 0. \tag{1.10.i}$$

Both these equations are true for any value of strain. For infinitesimally small strains such as occur in the elastic deformation of metals, we may neglect the products of the strains and equation (1.9) reduces to

$$e_1 + e_2 + e_3 = 0. \qquad (1.10.\text{ii})$$

In cylindrical co-ordinates (r, θ, z), in place of (1.10.i) we should have,

$$\epsilon_\theta + \epsilon_r + \epsilon_z = 0, \qquad (1.10.\text{iii})$$

where ϵ_θ is a principal hoop strain, ϵ_r the radial strain and ϵ_z the natural axial strain.

In forging, the volume of steel may be reduced by 0·6 per cent and of copper by 1·3 per cent when subject to a hydrostatic pressure of 100 kgf/mm²; the metal caesium at 150 kgf/mm² undergoes a *temporary* change of 30 per cent (UNKSOV (1961)).

In cold-rolling copper with an 80 per cent reduction the weight per unit volume may change from 8·95 to 8·89 gmf/cm². Results and several references to density changes due to cold-working will be found in the paper by SHELTON (1961).

NADAI (1963) gives a number of references to the compressibility of rocks and refers to tests by Hughes and McQueen who support an opinion that iron and nickel in the centre of the earth, at very high temperature and under pressures of about 60×10^6 lbf/in², are elastically compressed to less than one half of their normal (*sic*) values.

Wood and cast iron (fibrous materials) after a modest degree of compression return to their original volume.

1.4 An Example of the Use of the Volume Constancy Equation: Drifting, Low-speed Plate Perforation or Hole Flanging

An example of the use of equation (1.10.i)—the assumption that there is no change in volume of material during a plastic deformation process—is provided by considering a plate of uniform thickness h_0 slowly perforated by a frictionless conical-ended drift of diameter $2b_0$. Such a plate as deformed after the drift has passed through it, is shown in Fig. 1.4. The length of the lip of the cylindrical, deformed portion is denoted by H and its thickness is h at distance z from its head. It is assumed that each element of the cylindrical lip has been deformed under a predominantly uniaxial tensile hoop stress; for a plate in which h_0/b_0 is small this will be nearly true.

For constancy of volume, for each element,

$$2\pi s \, ds \cdot h_0 = 2\pi b_0 \, dz \cdot h \qquad (1.11)$$

and hence,

$$\ln \frac{b_0}{s} + \ln \frac{h}{h_0} + \ln \frac{dz}{ds} = 0$$

or

$$\epsilon_\theta + \epsilon_t + \epsilon_r = 0, \qquad (1.10.\text{iii})$$

where ϵ_θ, ϵ_t and ϵ_r are, respectively, the natural hoop strain, thickness strain and radial strain.

Now, as the element is stretched only in uniaxial tension, then the other two strains, i.e. the thickness strain $\ln h/h_0$ and the radial strain $\ln (dz/ds)$ will be equal and thus, substituting in (1.11) for dz/ds,

$$\frac{s}{b_0} = \left(\frac{h}{h_0}\right)^2 \text{ or } h = h_0 \sqrt{\frac{s}{b_0}}.$$

And alternatively,
$$\frac{ds}{dz} = \sqrt{\frac{b_0}{s}},$$

so that,
$$\int_0^{b_0} \sqrt{s} \cdot ds = \int_0^H \sqrt{b_0} \cdot dz$$

and hence
$$\left[\frac{2}{3} s^{3/2}\right]_0^{b_0} = \sqrt{b_0} \, H.$$

Thus,
$$H = \frac{2}{3} b_0.$$

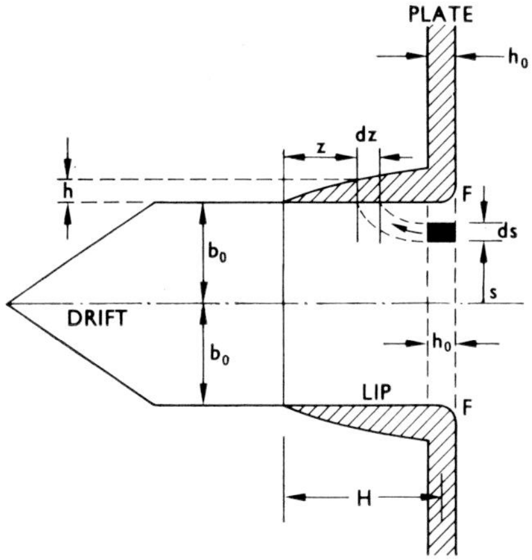

FIG. 1.4 Low speed perforation of plate.

Experiments conducted to test the mode of deformation described show it to be an oversimplification, because

1. the orifice is not sharp but rounded at the base F, due to bending and shear, and
2. the material at the head of the lip E, usually splits and a number of axial fractures are encountered; if no fracturing occurred the hoop strain would be infinite.

In Problems 11 to 16 the perforation process is analysed further.

The approach above is due to TAYLOR (1948). (Taylor gives $H = \frac{3}{4} b_0$, but this appears to be in error.) Dynamic analyses which include the inertia of the plate material have been made by THOMSON (1955), ZAID and PAUL (1957) and BROWN (1964). Detailed experimental results will be found in a paper on 'Hole Flanging', *J. Strain Anal.*, 1973.

A process for producing a pipe tee-junction enlarges a small hole in the side of a cylinder, by pulling a drift or tool radially outwards from the cylinder axis and by applying heat.

1.5 Some Deviations from the Stress-Strain Curves Described Above

So far we have discussed the behaviour of a bar of metal under a continuously and slowly applied load. It is important to remember that the mode of deformation and fracture for any metal in a given metallurgical condition depends, among other things, on the rate of straining, on the temperature and on the stress system to which the material is subjected, see below.

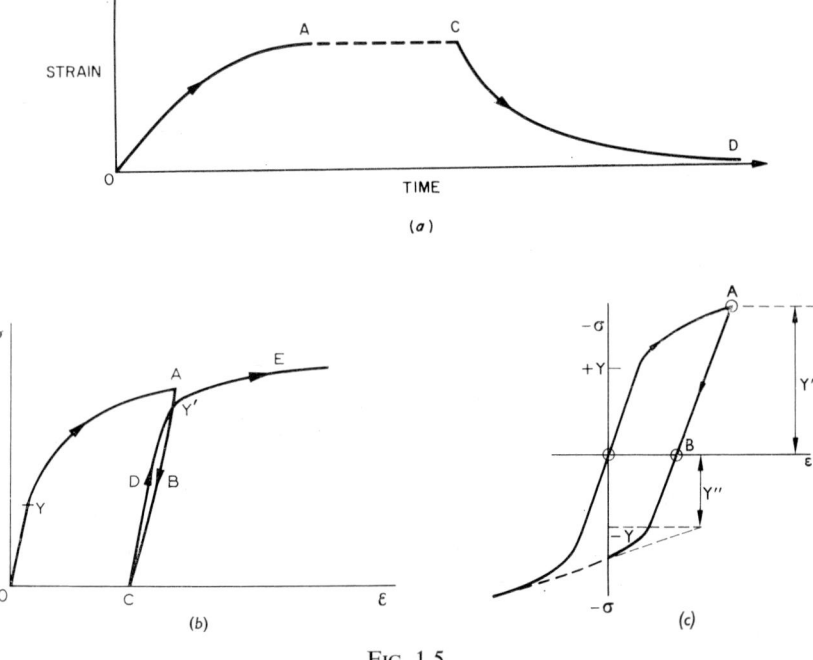

FIG. 1.5

(a) The effect of unloading in a simple tensile test.
(b) Hysteresis effect.
(c) Bauschinger effect.

In the elastic straining of a bar in simple tension, a sensitive extensometer would show that at an instant after load removal, a permanent set is left in the bar, no matter how small the applied load was initially. On applying a load, the strain approaches its final value exponentially along OA, Fig. 1.5(a), and requires a finite time; and when the load is removed the strain recovers exponentially towards zero along CD.

Again, at very low rates of strain, almost whatever the temperature, there will be some inelastic deformation under constant stress. This is the pheno-menon known as creep. For the 'elastic' loading of most engineering materials at room temperatures, the effect of creep is negligible, but at such relatively high temperatures as occur, for example, in steam or gas turbines, a study of creep phenomena becomes of prime importance.

A general discussion of creep does not come within the scope of this book, and for further information, the reader is referred to the texts by FINNIE and HELLER (1959) and PENNY and MARRIOTT (1971). It must be emphasized that the mathematical theory of plasticity does not take into account these small anelastic deformations, and as with the theory of elasticity, the straining of components is considered to be time independent.

If in the simple tension test the load is removed after the yield stress has been exceeded and then reloading occurs, a hysteresis loop is formed. On unloading from a point A, Fig. 1.5(b), and on reloading from point C, the curves ABC and CDE are initially very nearly parallel to the original elastic loading line. The hysteresis loop is greatly exaggerated on the diagram and, excepting under conditions of continuous cyclic loading, the deformation in the region may be assumed to be elastic, the value of Young's modulus being unaltered by the plastic deformation. OC is the permanent (plastic) strain left in the bar on complete unloading from A.

On reloading the bar from point C, the reloading line may now be con-sidered linear, yielding occurs at Y' (a larger value than the original yield stress Y) and the curve $Y'E$ becomes virtually a continuation of the curve YA. Note that every point on the continuous curve beyond Y is a yield point.

Two further phenomena need to be mentioned—anisotropy and the Bauschinger effect.

When a metal is loaded so that plastic deformation occurs, the crystallo-graphic directions in each grain are gradually rotated towards a common axis dictated by the loading, thus creating a preferred orientation. In par-ticular, a metal in which initially the grains are orientated at random (iso-tropy) is rendered anisotropic when plastically deformed. This means that the yield stress and ductility of the metal are dependent on the direction in which they are measured. For example, COOK (1934) has shown that in heavily cold-rolled brass, the tensile yield stress transverse to the direction of rolling may be as much as 10 per cent greater than the tensile yield stress measured in the direction of rolling. (But note that a material in which the tensile yield stress and compressive yield are different for the same direction, may still be isotropic.)

When a metal is plastically deformed and then unloaded, residual stresses, on a microscopic scale, are left in the material. This is mainly due to the different states of stress in the variously orientated crystals before unloading. If a metal is plastically deformed in uniform tension to point A in Fig. 1.5(c), unloaded to point B, and then subjected to uniform compression in the opposite direction, it is found that because of the residual stresses, yielding occurs at a stress Y'' which is less than the beginning stress, Y'. This is the Bauschinger effect, and is present whenever there is a stress reversal.

In plastic deformation processes where there is differential strain and which involve reversal of stress, as in unloading in a bending operation, the Bauschinger effect is important and cannot be indiscriminately ignored. In other cases, including the experimental determination of stress-strain curves where there is in general no more than one stress reversal, the Bauschinger effect can safely be neglected. In practice anisotropy also is often ignored, but it is desirable where possible to obtain some measure of this quantity, both in the original material and as anisotropy further develops during the test.

Anisotropy in rolled sheet for use in deep-drawing operations has assumed major importance in recent years.

The residual stresses, and hence the Bauschinger effect, can be largely removed by a mild annealing, but to change preferred orientation the heat treatment has generally to be carried out above the re-crystallization temperature.

1.6 Frictionless Compression: Homogeneous Compression

If a cylindrical block of isotropic material is compressed so that every small element of it is always deformed to the same degree at the same instant, it is said to be under homogeneous compression, see Figs. 1.6(a) and (b). It is usually supposed that if a cylindrical block of metal, whose height is less than about 1·5 times its diameter, is plastically compressed between rigid, parallel, frictionless platens or dies, that it will be homogeneously compressed. As

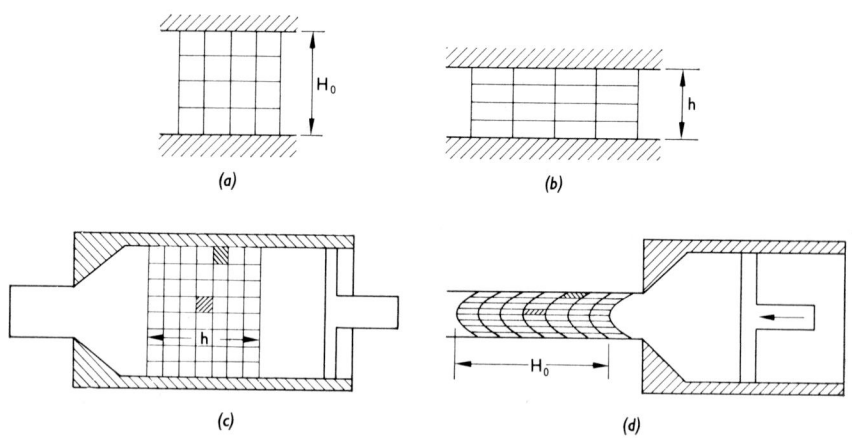

FIG. 1.6

(a) A diametral section of a cylindrical block on which a square grid of lines is marked.

(b) Homogenous deformation: all squares on compression have become identical rectangles.

(c) Block in container before extruding and showing square grid of lines on diametral section.

(d) Inhomogenous deformation: partially extruded block, showing the shapes taken up by the grid squares.

the dies approach one another, it is imagined that every transverse cross-section of the block remains circular and increases in diameter at the same rate as every other cross-section; the block remains prismatic. No bulging of the cylindrical vertical sides occurs—these remain straight and vertical. The original circular ends increase in area by sliding outwards over the dies; each element in the block undergoes, or will have undergone, exactly the same compressive strain, and the same extension hoop and radial strains.

As remarked above, homogeneous compression is usually discussed in terms of frictionless compression, but it should be noted that frictionless compression can (and in plane strain certainly does, see Fig. 13.6(a)) result in inhomogeneous compression. However, for the purposes of discussion in this section, it will be tacitly assumed that frictionless compression brings about homogeneous deformation.

Recall, incidentally, that homogeneous extension has been described in connection with the simple tension test conducted as far as the necking point.

If a cylindrical block of non-hardening material of yield stress Y, initial height H_0 and cross-sectional area A_0, is homogeneously compressed, the force necessary to start plastic compression is $A_0 Y$. When the height of the block is reduced to H, the cross-sectional area $= A_0 H_0 / H$ and the compressive force, F, to enforce deformation is $F = A_0 H_0 Y / H$. The compressive force thus increases hyperbolically.

When the current height of the block is H, let it be considered to be compressed an amount $- dH$. The uniaxial compressive strain increment $d\epsilon_c$, as in the case of tension is then defined as,

$$d\epsilon_c = - dH/H. \tag{1.12}$$

The total compressive strain is reducing the height of the block from H_0 to h is then,

$$\epsilon_c = \int_{H_0}^{h} - \frac{dH}{H} = \ln H_0/h. \tag{1.13}$$

The work that must be done to enforce homogeneous plastic compression on the block, i.e., in reducing its height uniformly from H_0 to h, is found thus. When the current block height is H, to compress the block an amount $-dH$, an amount of plastic work dW must be performed where

$$dW = F \, . \, (-dH) = \frac{A_0 H_0 Y}{H} \cdot (-dH).$$

Thus the total work done in compressing the block from H_0 to h is W and

$$W = A_0 H_0 Y \int_{H_0}^{h} - \frac{dH}{H} = A_0 H_0 Y \ln H_0/h. \tag{1.14}$$

The work done per unit volume of the material is

$$w = \frac{W}{A_0 H_0} = Y \ln H_0/h = Y \, . \, \epsilon_c \tag{1.15}$$

ϵ_c is the compressive natural strain imposed. Of course, if a bar of non-hardening material of length h was homogeneously stretched in tension to length H_0, the work done per unit volume would also be

$$Y\epsilon_t = Y ln H_0/h.$$

Since no energy is used in overcoming friction between the compressing dies and the material, and since every element is identically strained, it follows that the alteration in shape of the block is achieved with maximum efficiency; indeed, homogeneous straining (or shape-changing) is usually taken to be *the* most efficient method of securing deformation, and it is often accepted as the criterion by which all other methods (e.g., drawing) of securing the same final shape are judged. Homogeneous deformation is never realized in practical metal-working operations; it is most nearly found in a long rod subject to simple tension. Real metal-working operations include redundant (or more than is absolutely necessary) deformation and are therefore less than 100 per cent efficient by comparison with frictionless compression or extension. Useful approximate calculations of load or pressure can be made for many working operations based on the assumption that the shape-change sought is brought about by homogeneous deformation. If a bar of material of length h was, say, frictionlessly extruded or drawn so that its final length was H_0, then experiment (and theory, see Chapter 12) would show this change in shape to have been obtained inhomogeneously; if the cylindrical block had been sectioned on a diametral plane, and on one of the plane faces a square grid of lines had been stamped, then, after putting the two halves of the cylinder together and extruding it, the grid would appear as in Fig. 1.6(*d*) each square would have become (nearly) a parallelogram, but the degree of deformation undergone by each of them would vary from the centre-line outwards.

1.7 General Approach to Stress Analysis in Elasticity and Plasticity

1.7.1 *The Theory of Elasticity in Isotropic Solids*
The theory of elasticity deals with the methods of calculating stresses and strains in deformed perfectly elastic solids. A material is perfectly elastic if, when forces are applied to it, the resulting strains appear instantaneously and if, when the forces are removed, the initial shape of the body is instantaneously recovered. In its simplest form, the theory further assumes that the solid is homogeneous and isotropic, that there is a linear relation between stress and strain and that the strains are infinitesimally small. These further conditions mean that (except for a few special cases where the small displacements affect the action of the external forces) the method of superposition of deformations and stresses is legitimate and calculations can be based throughout on the initial dimensions and initial shape of the body.

To obtain a complete solution for the stresses it is necessary to establish:

1. Equilibrium equations between the internal and external forces. This means that the conditions at the boundaries of the solid must be known.

2. A continuity or compatibility equation describing the geometry of deformation.

3. The general relationship between stress and strain.

The first condition is a problem in statics, the second in geometry of strain and the third condition is based, for a specific solution, on experimental observations of the elastic deformation of the given material.

It is not always mathematically possible to obtain a complete solution to a particular problem. In such cases, it is usual to obtain a solution by making some assumption about the compatibility condition and correlating the result with experiment. When this is done, the subject is usually known as Strength of Materials. For example, in calculating the longitudinal and shear stresses in a laterally loaded beam, the assumption is made that transverse sections plane before bending remain plane after bending. The longitudinal and shear stresses are then calculated from the equilibrium equations. Although the solution is not mathematically or physically correct, warping of the cross-sections having been ignored, comparisons with experiment show that the solutions for the stresses and for the deflections of the beam are quite accurate. In all such cases, the accuracy of a solution must be checked by experiment.

The elastic constants are obtained from simple tests. It is usual to determine the values of Young's modulus, E, and the modulus of rigidity, G, experimentally and then derive the bulk modulus, K, and Poisson's ratio, v. In general, any two elastic constants are required for the solution of a particular problem and, although it is usually more convenient to work in terms of E and v, the constants G and K must be regarded as the more fundamental since, in general, any deformation consists of a change in shape (distortion) and a change in volume.

1.7.2 *The Theory of Plasticity*

The theory of plasticity deals with the methods of calculating stresses and strains in a deformed body after part or all of the body has yielded. It is necessary, as for elastic theory, to establish equations of equilibrium and compatibility and to determine the experimental relations between stress and increments of strain.

The most difficult problems to solve in plasticity are those of constrained plastic flow. These are cases where part of the body has yielded and part is still elastic, the plastic strains being of the order of the elastic strains. The compatibility equations and the stress-strain relations are difficult to handle and very few *complete* solutions have been obtained to such problems.

For cases where the plastic strains are large compared with the elastic strains, the change in dimensions of the body is of prime importance, but it is sometimes permissible to neglect the elastic strains altogether, thus simplifying the solution.

The general relation between stress and strain must contain:

1. The elastic stress-strain relations.

2. The stress condition (yield criterion) which indicates onset of plastic flow.

3. The plastic stress-strain or stress-strain increment relations.

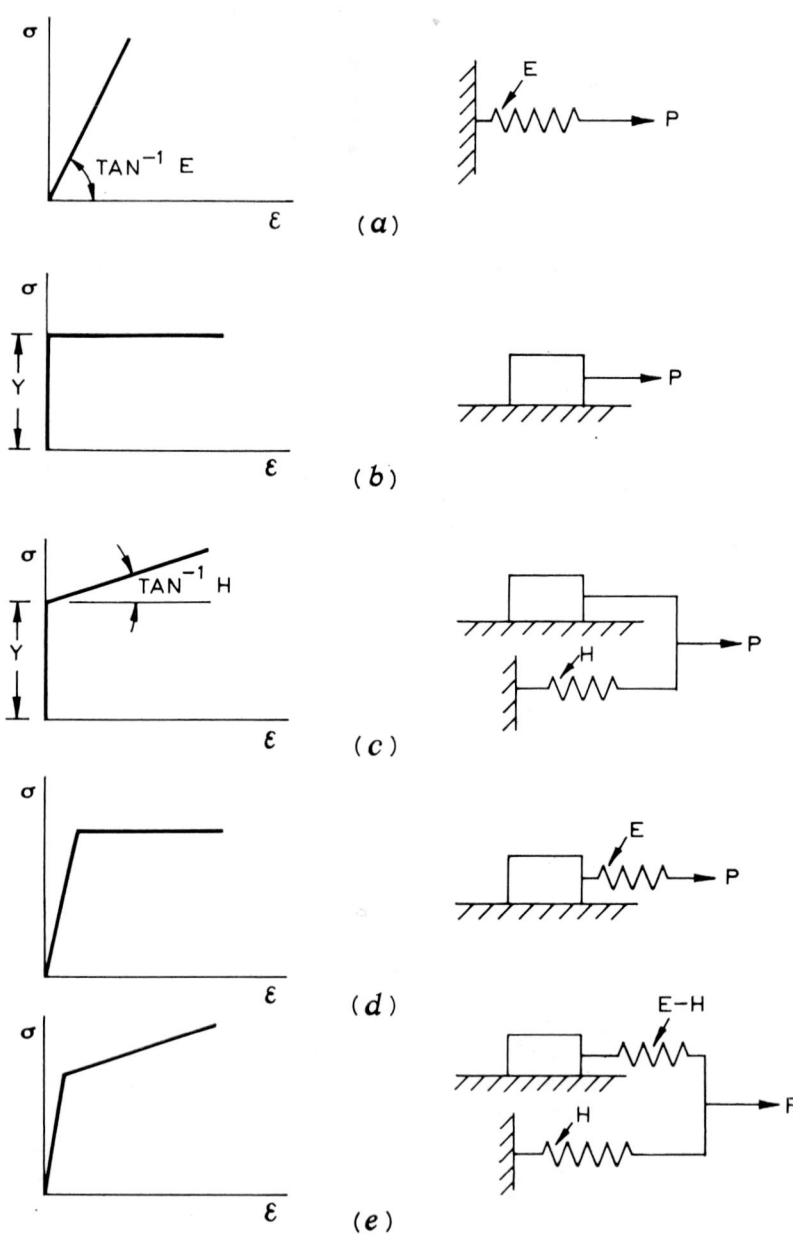

FIG. 1.7 Idealized stress-strain diagrams and dynamic models of mechanical loading behaviour.

(a) Perfectly elastic.
(b) Rigid, perfectly plastic.
(c) Rigid, linear work-hardening.

(d) Elastic, perfectly plastic.
(e) Elastic, linear work-hardening.

The plastic stress-strain relations will include work-hardening in certain cases, though it is not possible to take work-hardening *fully* into account in the general case. The above relations are based on experiments carried out under uniform stress conditions. Their validity when applied to non-uniform stress systems must be checked by experiment. Again, since the *actual* stress-strain relations vary with the speed of deformation and the ambient temperature, it is very important to obtain the experimental stress-strain curve under conditions similar to those existing in the system under consideration.

In order to obtain a solution to a deformation problem, it is necessary to idealize the stress-strain relation. In some instances, it is permissible to neglect the elastic strains and/or the effect of work-hardening. In effecting any idealization of the stress-strain diagram, we must always question the validity of the approximation and be prepared to restrict its field of application. Figure 1.7 shows such idealized stress-strain diagrams under uniaxial stress together with dynamic models of the mechanical behaviour. Considering Fig. 1.7(*a*), the deflection of the spring is proportional to the applied force, thus representing a perfectly elastic material. In Fig. 1.7(*b*), the block is subject to solid friction and can only move if the force exceeds a certain amount. Once movement begins, it continues under constant force. The remaining stress-strain diagrams can be explained in terms of a combination of these two elementary models. But note that some of these models do not represent truly the behaviour of a metal for unloading conditions, i.e., when *P* is reduced from its greatest value to zero.

1.8 Empirical Equations to Stress-Strain Curves

In certain simple problems in plasticity, it is useful to set up empirical equations for the stress-strain curves. In the following equations, the logarithmic definition of strain will be used. It must be remembered that if the equations are used to define an elastic-plastic deformation, where the plastic strains are of the same order as the elastic strains, then the use of the engineering definitions may be preferable.

1. Ludwik's expression

$$\sigma = Y + H\epsilon^n. \tag{1.16}$$

(i) When $n = 1$, the equation represents a material which is rigid up to the yield stress Y, followed by deformation at a constant strain-hardening rate H. It may be applied to cold-worked materials and gives an especially good fit for 'half-hard' aluminium. This is the case shown in Fig. 1.7(*c*).

(ii) The case $0 \leqslant n \leqslant 1$, is shown in Fig. 1.8(*a*). Again this expression is applicable when the elastic strains in an analysis may be safely neglected. When $Y = 0$, the curves are as given in Fig. 1.8(*b*). Note that $d\sigma/d\epsilon$ for $n \neq 1$ is infinite at $\epsilon = 0$ so that this expression should not be used for small strains.

2. Another common method of approximately representing the true stress-strain curve is to use two—bilinear—or more—multi-linear—expressions between stress and strain. This is illustrated in Fig. 1.8(*c*), where $\sigma = E\epsilon$

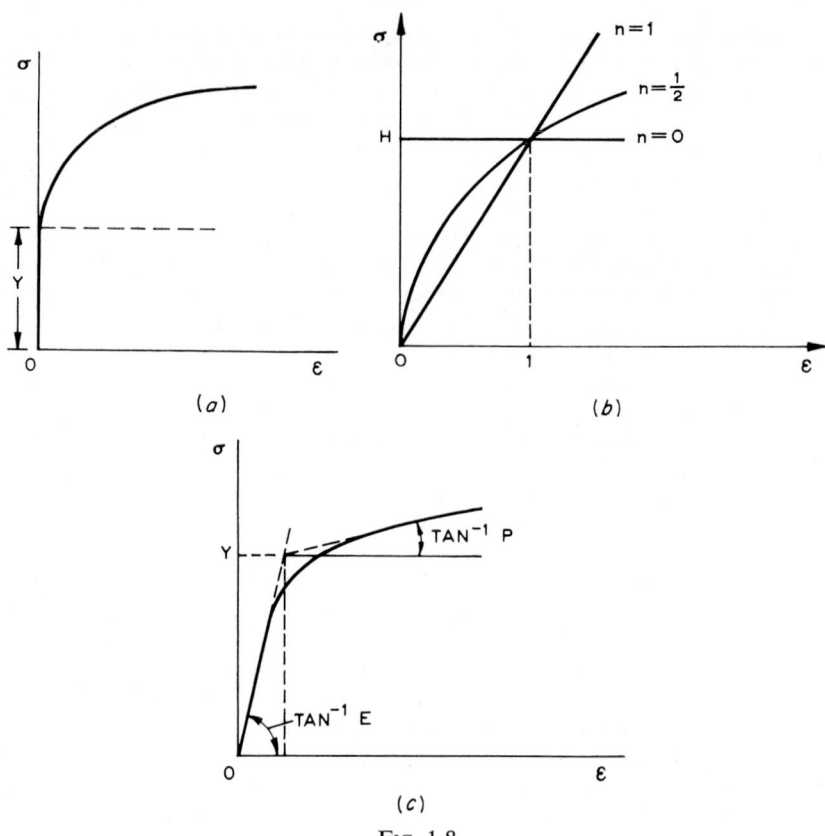

(a)

(b)

(c)

FIG. 1.8

(a) Rigid, work-hardening material.
(b) Empirical curves for $\sigma = H\epsilon^n$.
(c) Approximating the stress-strain curve with two linear expressions:
    a bilinear representation of the stress-strain curve.
$$\sigma = E\epsilon \text{ from 0 to Y.}$$
$$\sigma = P\epsilon \text{ from Y onwards.}$$

from 0 to Y and $\sigma = P\epsilon$ from Y onwards. E may be interpreted as an elastic modulus and P as a plastic modulus.

3. The expression $\sigma = a + (b - a)\{1 - \exp(-n\epsilon)\}$ where a, b and n are constants, has been suggested by Voce (1948). It gives a good fit to a stress-strain curve, but because of its complexity it is seldom suitable for use in a theoretical analysis.

4. An expression due to Swift, here presented in a simplified form is

$$\sigma = c(a + \epsilon)^n \quad \text{where} \quad 0 \leqslant n \leqslant 1 \tag{1.17}$$

and c, a and n are constants for a given material. It is a more realistic equation to employ, in instances where the strains are large, than those in 1, but the algebraic manipulation resulting from such an expression may be difficult.

This last equation can be understood to express the true stress—natural strain relationship for a material which has been cold-worked or strain-hardened in simple tension to a true strain of a from the annealed condition, ϵ being the measure of subsequent strain.

5. A form of expression proposed by Prager is

$$\sigma = Y \tanh (E\epsilon/Y). \tag{1.18}$$

This curve has a tangent modulus of E at zero-strain and approaches Y asymptotically at a fast rate.

1.9 Empirical Equations and the Maximum Load in Simple Tension

At the maximum load in simple tension it was established, see equation (1.8) that

$$\frac{d\sigma}{d\epsilon} = \frac{\sigma}{1} \quad \text{or} \quad \frac{d\sigma}{de} = \frac{\sigma}{1+e}. \tag{1.8}$$

If the empirical equation of the true stress-natural strain curve is $\sigma = A'\epsilon^n$ where A and n are constants, then

$$\frac{d\sigma}{d\epsilon} = A'n^n\epsilon^{-1} \quad \text{and} \quad \sigma = A'\epsilon^n,$$

so that using (1.8), we find that $\epsilon = n$; the logarithmic strain at maximum load is just the exponent n. If, alternatively, the given empirical equation was $\sigma = Be^m$ then it would be found that at maximum load $e = m/(1-m)$.

For a ductile material whose stress-engineering strain curve is given by $\sigma = 32e^{0.22}$, where σ is measured in ton-force per square inch, it may easily be shown that at maximum load in simple tension the engineering strain is 0.282 and the nominal stress 18.8 tonf/in².

Not too much reliance should be placed on the constancy of the coefficients in these empirical expressions; they can be obtained so that the expression fits an experimental curve over a specific strain range excellently, but beyond the range the expression may be a poor fit.

Figure 1.9 shows a two-bar truss which is frictionlessly pin-jointed at C and which hangs in a vertical plane being freely supported from fixed points at A and B distant $2h_0$ apart in a horizontal plane. The bars are identical in all respects and the true stress-engineering strain curve for the material of which they are made is given by $\sigma = Be^m$. Let us find the maximum load, W, the truss can carry.

When the two bars are just at maximum load P we have

$$P = A\,\sigma = A_0\,\sigma_T/(1 + e_T),$$

where A_0 is the initial cross-sectional area of a bar and σ_T and e_T are the true stress and engineering strain at the T.S. It has been shown above

that $\qquad e_T = m/(1 - m) \quad \text{and} \quad \sigma_T = B\,[m/(1 - m)]^m,$

so that $\qquad P = A_0\,Bm^m\,(1 - m)^{1-m}.$

Now, $\qquad W = 2P \cos \theta$

and $$\cos\theta = \sqrt{\frac{l_0}{(1-m)^2} - h_0^2} \Big/ \left(\frac{l_0}{1-m}\right).$$

Thus, $$W = \frac{2A_0\, Bm^m\,(1-m)^{1-m}}{l_0/(1-m)} \cdot \sqrt{\left(\frac{l_0}{1-m}\right)^2 - h_0^2}$$

$$= 2\,A_0\,Bm^m \sqrt{1 - \left(\frac{h_0\,(1-m)}{l_0}\right)^2}.$$

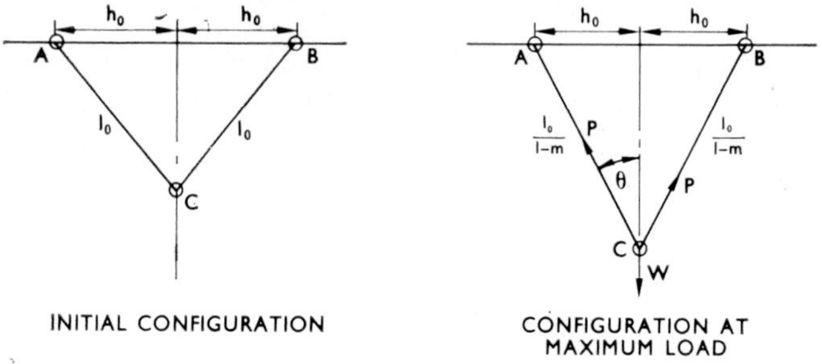
h₀
h₀
A
B
l₀
l₀
C
INITIAL CONFIGURATION
h₀
h₀
A
B
l₀/(1−m)
P
l₀/(1−m)
θ
P
C
W
CONFIGURATION AT MAXIMUM LOAD

FIG. 1.9 Pin-jointed two-bar truss.

1.10 **Compression of a Work-hardening Material: Adiabatic Temperature Rise**

The work done in frictionlessly and homogeneously compressing a cylindrical block from height H_0 to h, when the true stress-natural strain curve is expressed by the empirical equation (see 1.8) $\sigma = A'\epsilon^n$ is given by

$$W = \int_{H_0}^{h} - A\sigma . dH = \int_{H_0}^{h} - A_0\,H_0(A'\epsilon^n)\cdot\frac{dH}{H}$$

where A_0 is the original and A the current cross-sectional area.

But $-dH/H = d\epsilon$, where ϵ is the compressive natural strain and, thus

$$\frac{W}{A_0\,H_0} = A'\left[\frac{\epsilon^{\,n+1}}{n+1}\right]_0^{\epsilon} = A'\,\frac{(\ln H/h)^{n+1}}{n+1}. \tag{1.19}$$

The right-hand side of the last expression is just the area under the appropriate stress-strain curve from zero to the imposed final strain, ϵ.

The energy supplied to effect plastic compression largely reappears as heat and causes a rise in temperature of the material.

An estimate of the greatest possible temperature rise may be arrived at if adiabatic compression is assumed with complete conversion of plastic work to heat, if there is no change in the material properties of the work-piece. For the perfectly plastic material, if ρ denotes density, c specific heat, J the mechanical equivalent of heat and $\Delta\theta$ the uniform temperature rise in the block, then

$$A_0\,H_0\,\rho c.\Delta\theta\,J = A_0\,H_0\,Y\ln H/h$$

or $$\Delta\theta = \frac{Y}{J\rho c}\ln H/h. \tag{1.20}$$

As an example, for super pure aluminium at room temperature, for which the specific weight is $\rho = 0 \cdot 0975$ lbf/in³, $c = 0 \cdot 21$ CHU/in³, $J = 1400$ ft lb/CHU and $Y = 18,000$ lbf/in², if the block is compressed and its height halved, i.e., ln $H/h \approx 0 \cdot 7$ then

$$\Delta\theta = \frac{18,000 \times \ln 2}{(1400 \times 12)\,(0 \cdot 0975 \times 0 \cdot 21)} = \frac{12,600}{345} \simeq 36\,°C \,.$$

This example shows that when a metal is cold-worked—here compressed to a significant degree—the rise in temperature due to the plastic work done on it, is not likely seriously to alter its material properties.

1.11 Brittle and Ductile Materials: Cleavage- and Shear-type Fractures: Transition Temperature

Historically, the engineer's view of what constitutes a brittle or ductile material has been conditioned by the behaviour of a given material in the simple tensile test. In these terms, cast iron is said to be a brittle material since it fractures after very little strain, the fracture surface being perpendicular to the longitudinal axis. On the other hand, the familiar cup-and-cone fracture of a mild steel tensile specimen is said to be symptomatic of a ductile material. At the same time, however, it may be realized that under impact loading, this same mild steel could fracture in a brittle manner.

It seems preferable when discussing metals other than cast iron, to speak of cleavage- and shear-type fractures rather than of brittle and ductile materials. This distinction for cast iron is necessary because the free graphite it contains makes its behaviour more comparable with some non-metallic brittle substances.

It has already been mentioned that after reaching the maximum load condition in the simple tension test, a 'neck' begins to form, giving rise to a triaxial tensile stress system. From this point up to fracture, most of the straining takes place in the 'neck' under a complicated and continuously changing triaxial tensile stress system. BRIDGMAN (1944) and DAVIDENKOV and SPIRIDONOVA (1946) have shown that the maximum values of the triaxial tensions occur on the longitudinal axis and the minimum values on the periphery. Fracture first begins on, and perpendicular to, the longitudinal axis, and on macroscopic examination appears to be a cleavage-type fracture. Towards the periphery, the fracture changes from one of a cleavage type to one of a shear type. We therefore note that a cleavage-type fracture can occur after large plastic strains, and also that in the normal tensile test at room temperature, the fracture is usually a mixture of cleavage and shear type.

Lowering the temperature of testing increases the likelihood of a cleavage-type fracture. PARKER, DAVIS and FLANIGAN (1946) showed that by lowering the temperature sufficiently, the cup-and-cone fracture of mild steel can be changed to one of cleavage, similar to the fracture of a cast-iron bar at room temperature. Research into the effect of temperature on fracture has been intensified since the 1939–45 war, as a result of the failure of some all-welded structures. It is noted that for some materials and specimen sizes, a given stress-system and method of testing, there exists a temperature, the transition temperature, above which the metal is ductile and below which it is brittle. This low temperature brittleness has been reviewed by TIPPER (1957). Most work has been done on mild steel, but the phenomenon of a transition temperature has been shown to exist for some other materials such as molybdenum, chromium and tin. The behaviour of tin at different temperatures in the simple tensile test is shown in Fig. 1.10, MAGNUSSON and BALDWIN (1954). From the graph it is clear that there is a narrow band of temperature below which the reduction in area is small and nearly constant and above which it is fairly large and nearly constant.

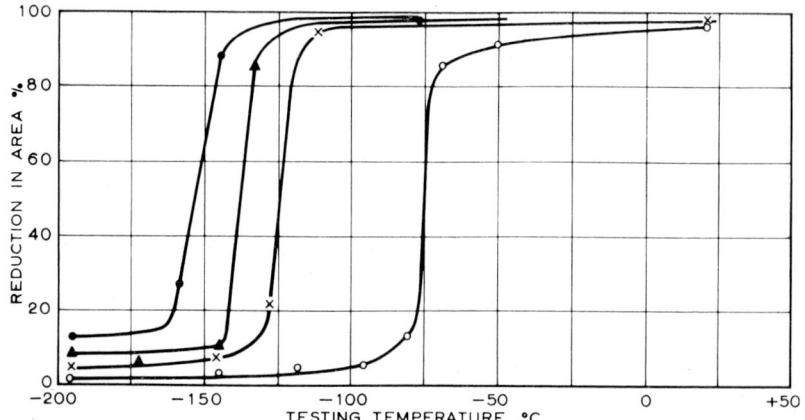

FIG. 1.10 The effect of temperature and strain rate on the ductility of body-centred tetragonal tin in tension.

- 0·05 in/in/min.
▲ 10 in/in/min.
× 100 in/in/min.
○ 19,000 in/in/min.

(*After* Magnusson and Baldwin, *J. Mech. Phys. Solids*)

Metals which show no transition temperature and which are ductile, and hence can be worked at all temperatures, are aluminium, copper, nickel, gold, silver, platinum—and most of their alloys—and the austenitic types of stainless steel.

Most steels and zinc become brittle at temperatures substantially below room temperature, and hence cannot then be formed.

Magnesium and its alloys, tungsten and molybdenum are insufficiently ductile at room temperature to be formed. Tungsten usually requires a temperature of at least 400°C.

1.12 The Effects of Hydrostatic Pressure

Hydrostatic or fluid pressure has long been known to influence the physical behaviour of metals. The major work done on this topic is of sufficient potential practical importance to be worth summarizing. VON KARMAN in 1911 subjected sandstone and marble to compression tests under fluid pressure and showed that materials brittle at atmospheric pressures 'bulged' and deformed in a manner typical of ductile materials. BOKER (1914) investigated the tensile properties of marble and cast zinc under fluid pressure and under a combination of direct stress and torque, the specimens being protected from fluid entering minute cracks formed on the surfaces by encasing them in brass foil. The results for marble as obtained by both workers, demonstrated that ductility increases with the pressure. BRIDGMAN (1947, 1952) has carried out tensile tests on metallic and non-metallic materials, sheathed and unsheathed, under very large pressures (about 25,000 atmospheres). In this work dealing purely with the effects of high pressure, he has shown that some materials are compressible to a significant degree, though much of this is recovered when the load is removed.

But, over the range of pressures important in metal-working, the compressibility is very small. Bridgman also showed, using mild-steel specimens, that the fracture mode changed with increasing hydrostatic pressure, the familiar cup-and-cone fracture changing into one of almost pure shear with a very great percentage reduction in area.

The opinion of Bridgman, as a consequence of nearly 200 papers on the effect of high pressures on materials, seemed to be that the holes and cracks must develop to permit fracture and that the effect of hydrostatic pressure was to keep these closed up or to delay their development.

It is worth noting that whilst in many materials removal of the pressure is accompanied by reversion to its original condition, this is not always so; some materials under pressure undergo allotropic changes. The success of many chemical processes depends on the effect of a high pressure. Considerable interest in the effect of high pressure is evinced by geologists and geophysicists. BREDTHAUER (1956) investigated the behaviour of jacketed rock samples under the influence of moderate hydrostatic pressure (0 to 15,000 lbf/in^2) when loaded in axial compression and verified the brittle-to-ductile transition. This latter kind of work is useful in so far as this range of pressure is equivalent to the pressure exerted by drilling muds upon the face of formations exposed to the cutting action of rock bit teeth in drilling at great depths for oil.

Tests, in compression on cast steel by ROS and EICHINGER (1929), in torsion on copper and steel by COOK (1934) and on several materials including copper and steel by CROSSLAND (1954) agree with Bridgman's work that pressure has relatively little effect on the yield point and flow stress level of 'ductile' materials. In the case of torsion tests on rubber-jacketed grey cast iron, subject to pressures of up to 35 $tonf/in^2$, CROSSLAND and DEARDEN (1958) have shown that the shear-stress-strain diagram is raised, though less than proportionately, as the pressure is increased. At high pressures, the material effectively behaved as a completely 'ductile' material.

Recently, CROSSLAND and MITRA (unpublished, 1968) have confirmed that the level of the flow stress-strain curve for several steels tested under torsion, are not significantly affected when simultaneously subjected to hydrostatic pressures of up to 140,000 lbf/in^2; also, as before, the principal effect of the pressure is to increase torsional ductility, to some degree in proportion to the pressure. They did, however, observe some unusual effects in applying high pressure to the hexagonal metal, magnesium, also when under torsion.

A potentially very useful application of this knowledge is to the extrusion of materials which cannot be satisfactorily extruded at atmospheric pressure. By carrying out the extrusions under high fluid pressure, satisfactory products are obtained (PUGH and GREEN 1958). The product is extruded not into atmospheric pressure as in conventional extrusion but into a container filled with pressurized fluid. The extruded products of materials such as magnesium and bismuth, which cannot be extruded cold into atmospheric pressure without cracking, show great improvements when extruded into a back pressure. The hydrostatic pressure acting on the extruded product inhibits the formation of cracks. Experimental results indicate that a super-imposed hydrostatic pressure of 2 to 3 tonf/in^2 may be adequate to suppress the brittle behaviour of magnesium and bismuth at low reductions. The possibility of cold extruding large metal flanges against fluid pressure to inhibit fracture has been investigated by ALEXANDER and LENGYEL (1964). Aluminium and copper flanges were extruded against 10, 20 and 25 tonf/in^2 fluid pressure. The fracture of the flanges was delayed with increase in fluid pressure.

An interesting account of relatively early work into the effects of high pressure on metals and a statement of early Russian contributions to this field will be found in the book by BERESNEV, VERESHCHAGIN, RYABININ and LIVSHITS (1960).

Industrial applications of hydrostatic pressures to metal forming were discussed at the High Pressure Engineering Conference (1967) and the International Conference on Manufacturing Technology (1967). Recent publications are two books edited by PUGH (1970) and (1971).

In summary, the conditions for plastic flow are well understood, but not those for fracture; we can only describe qualitatively the factors that delay fracture and promote flow. The superposition of a hydrostatic pressure on a simple tensile test can counteract the tensile stresses set up in the 'neck' of the specimen and promote greater plastic flow. On the other hand, the presence of flaws, cracks or notches in a member under load may lead to very high triaxial tensile stresses being set up and under certain conditions a catastrophic brittle fracture may result. It is usual to think of a cleavage-type fracture as depending on a certain function of the tensile principal stresses or strains, and flow, as will be clear from Chapter 4, on some function of the principal stress differences.

1.13 Strain Rate in Simple Uniaxial Compression and Tension

Suppose a hammer, rigid platen or die whose speed v at time t after first making contact with the circular end of a cylindrical block of initial height H_0, and current height H, which rests on a rigid foundation, enforces homogeneous (frictionless) compression. Then the increment of logarithmic compressive strain, $d\epsilon$, enforced in time dt is $-dH/H$. Thus the rate of compressive strain is $\dot{\epsilon}$ or $d\dot{\epsilon}/dt$, and

$$\dot\epsilon = \frac{d\epsilon}{dt} = -\frac{dH/H}{dt} = -\frac{1}{H}\cdot\frac{dH}{dt} = -\frac{v}{H}. \tag{1.21}$$

The engineering strain rate is $\dot{e}$ and

$$\dot{e} = -\frac{dH/H_0}{dt} = -\frac{1}{H_0}\frac{dH}{dt} = -\frac{v}{H_0}. \tag{1.22}$$

Note that strain rate has the dimension of (time)$^{-1}$.

If the hammer impinges on the block initially with speed v_0 and all its kinetic energy is dissipated in doing plastic work on the block, but such that the compression operation is terminated at time T, when a strain of ϵ_0 or e_0 has been enforced and the block height has been reduced from H_0 to h, then

1. A simple mean logarithmic strain rate is

$$\bar\epsilon = \frac{\epsilon_0}{T} = \frac{\epsilon_0}{(H_0 - h)/(v_0/2)} = \frac{\epsilon_0\,v_0}{2(H_0 - h)} \equiv \frac{v_0}{2}\frac{\ln H_0/h}{(H_0 - h)}. \tag{1.23}$$

This assumes that the hammer speed decreases in direct proportion to the amount of compression applied $(H_0 - h)$.

2. A simple mean engineering rate is,

$$\bar{\dot{e}} = \frac{e_0}{T} = \frac{e_0}{(H_0 - h)/(v_0/2)} = \frac{(H_0 - h)/H_0}{(H_0 - h)/(v_0/2)} = \frac{v_0}{2H_0}. \tag{1.24}$$

The simple mean engineering strain rate is thus one half the initial strain rate. In fairly representative circumstances where a cylindrical block is upset under a drop hammer, the mean engineering strain rate is more nearly two thirds of the initial strain rate.

It will be obvious that in a tension test the logarithmic strain rate is $\dot\epsilon = v/H$, using the same notation as above except that v denotes the current speed of extension of a current gauge length H. Similarly $\dot{e} = v/H_0$. Note that in a typical tensile test on, say, an initial 8-inch gauge length of annealed mild steel which is stretched to give an engineering strain of 0·25 in ten minutes, by a crosshead moving at about 0·2 in/min the average engineering strain rate for the test is less than 10^{-3}/s.

1.14 Cold- and Hot-working: Recrystallization and Homologous Temperature

COLD-WORKING

Conventionally, cold-forming or cold-working refers to room temperature forming. More exactly, metals are said to be cold-worked if they become permanently harder during the working process (e.g., the common tensile test). Strain-hardening occurs at room temperature with most metals. Pure lead, tin and cadmium only permanently strain-harden below room temperature; zinc is nearly in this group. If these latter metals are strained and left at room temperature they again become soft, or self anneal, after a period of time.

HOT-WORKING

During plastic deformation there is a generation, movement and interlocking of dislocations. The greater the degree of deformation the larger the number of dislocations produced, and because of their mutual interaction and obstruction, larger stresses are required to enforce their movement and hence cause further plastic flow; broadly, this explains work-hardening.

Above a certain slightly indefinite temperature—the recrystallization temperature T_r—a metal may be worked and after a certain small strain it becomes no harder as the amount of strain imposed is increased; the yield stress during, say compression, would remain constant, see Fig. 1.11. A balance would have been established between the tendency to work-harden and the tendency to soften; these two competing tendencies are concurrent. The concurrency first seriously occurs over a narrow range of temperature near T_r; below this range, as has been remarked, the material perceptibly hardens during straining, but above it the yield stress is constant, though highly strain-rate dependent. The softening, or work-hardening removal processes, are recovery, recrystallization and grain growth and they are all thermally activated ones.

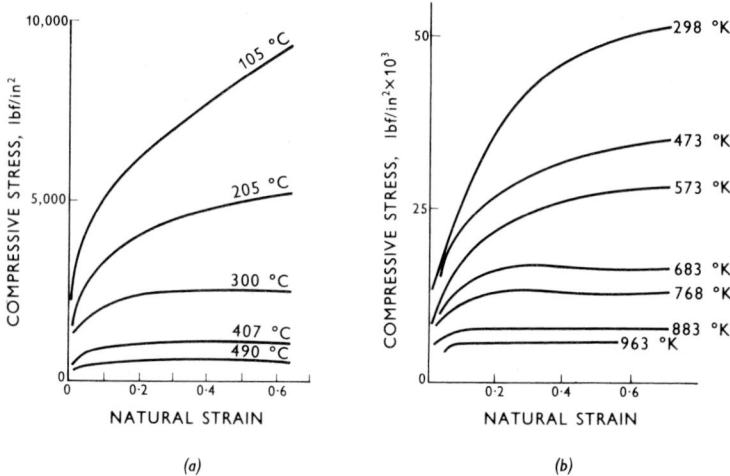

(a)   (b)

FIG. 1.11 Quasi-static compressive stress-strain curves at various temperatures ($\dot{\epsilon} \simeq 10^{-3}/\text{s}$).

((a) For super-pure aluminium *after* Baraya, Johnson and Slater, *Int. J. mech. Sci.* (1965).

(b) For annealed copper, BS. 1433: *after* Mahtab, Johnson and Slater, *Proc. Instn Mech. Engrs* (1965)).

Figures 1.11(a) and (b) show sets of curves for metals compressed at slow strain rate, in a compression testing machine ($\sim 10^{-3}/\text{sec}$). Those compressed at high strain rates, in a cam plastometer, ($\sim 10^1 - 10^2/\text{s}$) at temperatures above and below the recrystallization temperature are shown in Fig. 1.12.

RECRYSTALLIZATION AND HOMOLOGOUS TEMPERATURE

A very useful criterion for interpreting the stress–strain–strain-rate—temperature behaviour of metals is that of the homologous temperature, T_m; this is defined as

$$T_m = \frac{\text{Testing temperature (K)}}{\text{Melting point temperature (K)}}$$

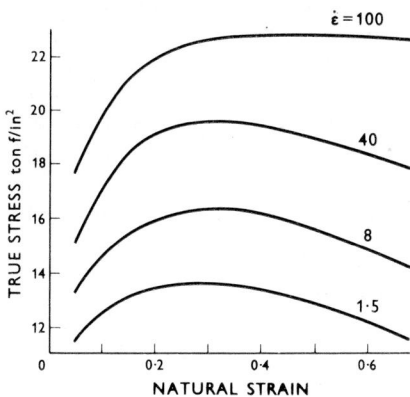

FIG. 1.12 Compressive stress-strain curves for En 52 steel at 1000 °C ($\epsilon \simeq 10^1$ to 10^2/s) (*After* Cook, *Proc. I. mech. Engrs.*)

It is a means of reducing to a 'common scale' and thus facilitating comparisons between metals which have different melting point temperatures.

The recrystallization temperature expressed in this non-dimensionalized way, is usually found to be between 0·4 and 0·5 K. However, the value of the recrystallization temperature is very dependent on the strain and strain rate and the above figures are only given as a rough guide.

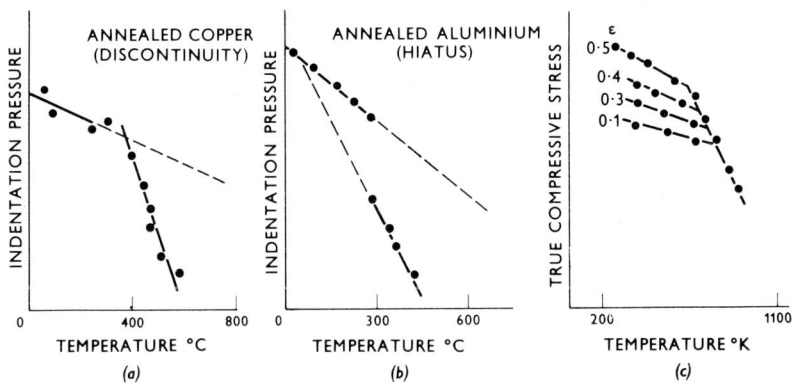

FIG. 1.13 Change in mechanical properties at the recrystallization temperature. (*After* Mahtab, Johnson and Slater, *Proc. I. mech. Engrs.*)

Hardness testing is fundamentally related to the stress-strain properties of a material, and thus indentation tests may be conducted for the purpose of illuminating material behaviour generally. Indentation pressure is, in effect, a mean yield stress for a given mean strain. In Figs. 1.13(a) and (b) the static indentation pressure, using a lubricated cone of 90° angle, is plotted against temperature and it is observed that in (a) there is a discontinuity of slope, and (b) a hiatus. The pivotal role thus played by the recrystallization temperature is clearly evident. The compressive stress after various amounts of compressive strain at different absolute temperatures is shown in Fig. 1.13(c) and again the recrystallization temperature region is well defined by the discontinuity.

A distinction between cold- and hot-working may be usefully based on the recrystallization temperature, T_r. We may understand cold-working to occur if the temperature at which it starts is less than T_r; if the temperature of working exceeds T_r we may call this hot-working. The term 'warm-working' is sometimes used to refer to working at a temperature which is intermediate between typical room temperatures and a recrystallization temperature.

An account of the mechanical behaviour of metals at less than room temperature will be found in the paper by COFFIN and CONRAD (1965).

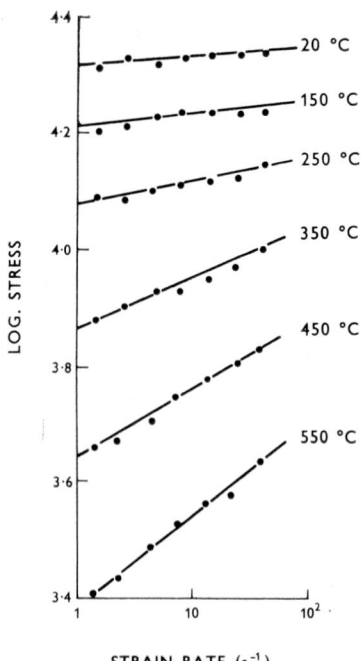

FIG. 1.14 Effect of strain rate on the stress required to compress aluminium to 40 per cent reduction at various temperatures. (*After* Alder and Philips, *J. Inst. Metals*).

1.15 Strain Rate in Relation to Recrystallization Temperature

Figure 1.14 was given by ALDER and PHILIPS (1954) and shows how the logarithm of compressive stress Y (for 40 per cent compression) varies with logarithmic strain rate at various temperatures, for aluminium. If it is assumed that

$$Y = Y_0 \, \dot\epsilon^n \tag{1.25}$$

then the slope of the straight lines in Fig. 1.14 provides the value of the coefficient n. Numerous other workers have demonstrated similar relationships.

It may be noted:

1. that equation (1.25) fits experimental data well. Typical results for n are given in Fig. 1.15 for strain rates between 1 and 40 per second.

Metal	Temp. °C	Value of n for a compression of 10%	30%	50%
Aluminium	18	0·013	0·018	0·020
	350	0·055	0·073	0·088
	550	0·130	0·141	0·155
Copper	18	0·001	0·002	0·010
	450	0·001	0·008	0·031
	900	0·134	0·154	0·190
Mild steel	930	0·088	0·094	0·105
	1200	0·116	0·141	0·196

FIG. 1.15

2. that at the recrystallization temperature, experimental results take on a different character. A study of how exponent n varies with homologous temperature using the data of Alder and Philips on compressive strains up to 0·5 shows that,
(a) for $T_r < 0.55$, $n \simeq 0.055$ and
(b) for $T_r > 0.55$, $n \simeq 0.43$.
These values apply for copper and mild steel as well as aluminium.

3. that if the bottom three lines of Fig. 1.14, i.e., where the temperature is greater than T_r, are extrapolated they are found to be nearly concurrent.

1.16 A Point of Reversal in Compression

Below the recrystallization temperature the softening rate is small and metals work-harden, but above it a substantially constant yield stress (but one highly dependent on strain rate) is obtained during working because an equilibrium between the hardening and the softening rates is established.

This view overlooks the fact that during compression work is being performed on the specimen which to some extent manifests itself as a temperature increase. Over a substantial range of temperature, any temperature increase due to working seems not to be important, but it is evident that a point will be reached when its contribution to the softening tendency will be decisive. With this in mind it now seems reasonable therefore to expect that a region will exist in which the 'integrated' softening rate will predominate over the hardening rate. Experiments using a cam plastometer show that after sufficient high constant rate of strain, and elevated temperature, the yield stress of metal may indeed be found to decrease with increasing strain, see Fig. 1.12. This point of reversal is probably known in practice because it is reported that some metals are easier to forge the more they are worked.

1.17 The Ratio of Dynamic to Static Yield Stress

Many methods have been used for finding the dynamic and static flow stress of metals at temperatures above and below T_r, such as torsion, indentation, simple metal ball bouncing, tensile and compression tests. However, few investigations show how the ratio of dynamic, σ_D, to static, σ_S, flow stress (i.e., the yield stress at a specific strain) varies with temperature. Static yield stress is taken to be the flow stress at about 10^{-3}/s, the strain rate which is typical of a slow speed compression testing machine. Dynamic flow stress is the flow stress associated with an impact test, e.g., that due to a falling hammer, which is likely to be about 10^3/s. Thus σ_D/σ_S refers to a strain rate ratio of about 10^6. Figures 1.16(a) and (b), show how σ_D/σ_S varies with temperature for steel and copper. The static experimental results were obtained from slow

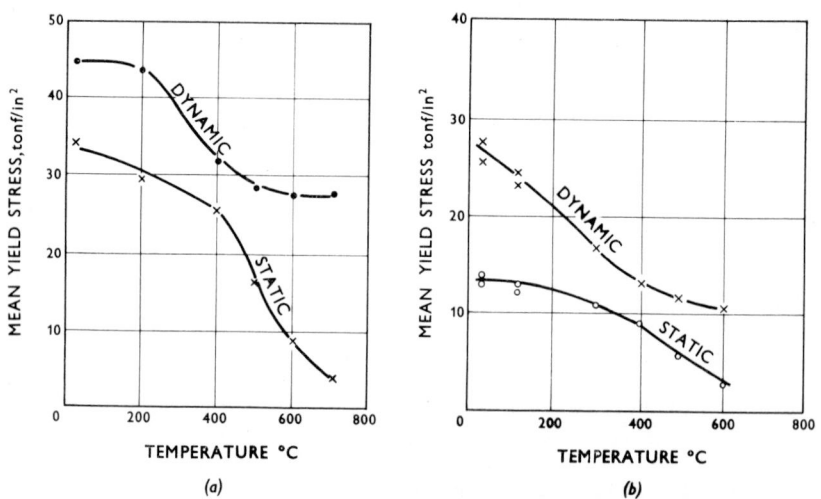

Fig. 1.16 Variation of mean dynamic yield stress σ_D and mean static yield stress σ_S with temperature for (a) annealed mild steel En 2, and (b) cold drawn, as received, copper. (*After* Hawkyard, Eaton and Johnson, *Int. J. mech. Sci.*)

speed compression tests and the dynamic results by firing hot and cold flat-ended projectiles at a rigid anvil (HAWKYARD, EATON and JOHNSON, 1968.)

The results bear out what has previously been found, namely that, from the metal-working point of view where the strains imposed are large,

1. strain rate effects below the recrystallization temperature are not pronounced and only lead to $1 < \sigma_D/\sigma_S < 2$;
2. above T_r, σ_D/σ_S is very sensitive to strain rate. When $T_H \simeq 0.6$ or 0.7, for many metals the ratio is 10.

These simple facts are not widely recognized. (2) is very important because it implies that when working hot metal at temperatures exceeding T_r, at a fast rate, their 'stiffness' is very greatly increased. As Figs. 1.16(a) and (b) show, fast-worked copper and mild steel at temperatures slightly in excess of $T_H = 0.5$, are nearly as strong as they would be if slowly worked at room temperature.

Point (2) may be emphasized by the following.

By compressing a block of super pure aluminium in a simple compression testing machine, SPEAKMAN (1966), at 490 °C it was observed that the yield stress was 500 lbf/in². By dropping a weight of 15.7 lbf to impinge on a specimen 1 in diameter and 1 in high at 31.1 ft/s (from a height of 16 ft) and thus generating a material strain of 0.64 at a mean natural strain rate of 263/s, the mean yield stress may be calculated to be 6000 lbf/in². Using equation (1.15), and equating the potential energy in the weight to the plastic work done,

$$Y = \frac{4.15 \cdot 7 \cdot 16 \cdot 12}{\pi \cdot 0 \cdot 64}.$$

The ratio of the two yield stresses is $\simeq 12$. By firing a 90° cone at about 200 ft/s into hot aluminium (at 500 °C) and copper at 600 °C, the ratios of the mean effective pressure during each of these processes to that which prevails when the operation is carried out statically is about 32 in the first case and 16 in the second (MAHTAB, JOHNSON and SLATER, 1965).

A striking representation of the manner in which stress-strain rate (over three decades of strain rate) and temperature are interrelated is shown in Fig. 1.17; these results have been given by HOCKETT (1966). The work of SUZUKI *et al.* (1968) and SAMANTA (1968) and (1969) should be consulted for further results and references.

Early and notable work on tension testing at elevated temperatures and high speed was that carried out by MANJOINE and NADAI (1940). In later investigations by MANJOINE (1944), tests were carried out over the range of strain rates from $10^{-6}/s$ to $10^3/s$ and for temperatures from about 25 to 600 °C. At room temperature, the tensile strength was found to decrease slightly at very low rates and then increase with strain rate, showing a 40 per cent increase at the highest strain rate. The lower yield point was found to increase throughout the range of strain rates with an overall increase of 170 per cent. The tensile strength and yield point values were noted to be practically constant for strain rates in the region in which ordinary short time tensile tests are made.

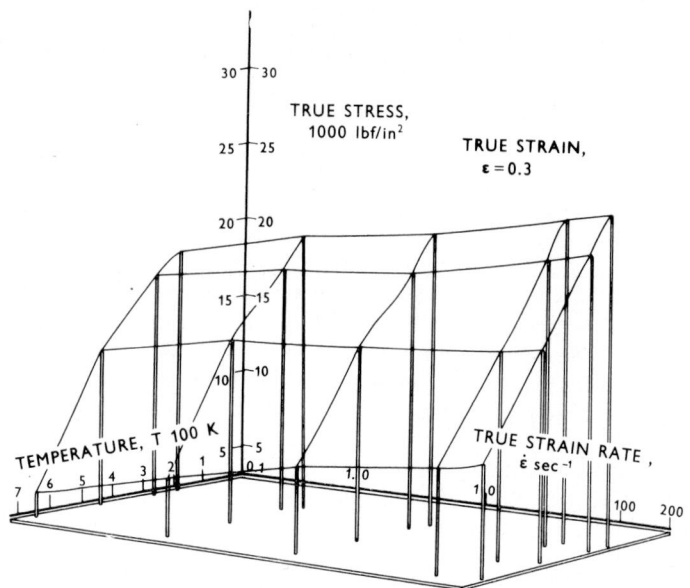

FIG. 1.17 True stress versus log true-strain rate versus temperature.
1100–0 aluminium. True strain, $\epsilon = 0\cdot3$. (*After* Hockett, Los Alamos
Scientific Laboratory of the University of California.)

1.18 Some Test Techniques for Providing Uniaxial Compression and Simple Stress-Strain Data with Special Reference to Rate Effects

1. COMPRESSION TESTING MACHINE

There is little need to dwell on the use of this machine for obtaining stress-strain curves at room and elevated temperatures. The maximum strain rate that can be achieved by most machines is about 10^{-2}/s.

2. THE HYDRAULIC FORGING PRESS

This was used for tests on hot steel by LUEG, MULLER and KRAUSE (1957) and covered the range 10^{-2}/s to 10^{1}/s.

3. THE CAM PLASTOMETER

This is a device in which a cam actuates a bottom platen and causes it to compress a specimen (perhaps at high temperature) against a fixed upper platen; the cam is designed to ensure that the compression occurs at a constant rate of strain. The plastometer introduced by OROWAN (1950) and designed by Los was specifically intended to provided constant strain rate data such as would be useful for rolling mill calculations. It operates in the strain rate range 10/s to 10^{2}/s.

4. The Drop Hammer

The kinetic energy of a falling weight is used to effect the compression of metal. The nominal input energy is easily ascertained and the final nominal compressive strain (for a prismatic block) readily obtained. The useful strain range is small since the impinging mass must exceed about 6 ft/s to give consistent results, but the greatest impact speed is limited by the height of fall, and if this is, say, 49 ft, then 56 ft/s is the greatest impact velocity. Thus the strain rate range is only about one magnitude at 10^2/s.

Load cells may be introduced to record the dynamic compressive stress during compression and displacement transducers introduced to help in composing a stress-strain record during impact.

The drop hammer test is scientifically weak because it provides a record in which the strain rate varies throughout the compression (SLATER, AKU and JOHNSON, 1968). See however HAWKYARD and POTTER (1971).

5. Air and Cartridge Actuated Guns

The speed limitation of the drop hammer may be overcome by driving a projectile or cylindrical specimen with compressed air or a low explosive cartridge against a solid anvil or pressure bar. Air will usually enable speeds of a few hundred feet per second to be attained, and cartridges can be employed for speeds up to three or five thousand feet per second (WHIFFIN, 1948). The strain rates may then reach 10^4/s. Light gas guns will drive projectiles up to about 10,000 ft/s.

Strain rate effects at 10^4/s can also be studied by having similar blocks of materials impinge on one another in free flight. With the help of diffraction gratings this system has been greatly used by BELL (1968) to assess the validity of the strain rate independent theory of plastic wave propagation.

6. Other Tests

1. *Tensile and compression* tests using the Hopkinson-type pressure bar techniques have frequently been used to provide basic high strain rate data (MALVERN, 1965 and RIPPERBERGER, 1965). However, the amount of strain attainable in these circumstances is often too small for use as forging data. Rotary impact and pendulum impact machines tend to suffer from the same limitations, and difficulties due to stress wave propagation, arise.

2. Dynamic *torsion* tests have been performed, but if solid bars are used then a graphical construction must be carried out to derive torsional stress-strain data. If thin-walled tubes are used to avoid the construction, large strains cannot be attained without the tube first failing due to compression stress instability or buckling.

3. During the *orthogonal machining* process strain rates are very high; for quite modest cutting speeds the strain rate is high because the deformation of the material, i.e., the taking of a cut, occurs largely during the length of time taken by the material to cross the narrow shear plane and become the chip. Certainly, strain rates of 10^4/s and even 10^6/s can be achieved in machining. However, difficulty attaches to the task of assessing the operative shear stress

mainly because the magnitude of the frictional force between the chip and tool face is unknown, but see OXLEY (1968).

4. By detonating a *high explosive* against the surface of a metal block and measuring the amplitude of the elastic wave transmitted into the block, the stress at strain rates of 10^6/s may be obtained.

The pressure generated at the explosive-metal interface are several million lbf/in^2, and this is so high that to some significant degree the volume of the metal may be reduced, DUVALL (1962).

7. MULTI-AXIAL STRESS STATE METHODS

All the testing techniques described above are primarily, and nominally, concerned with uniaxial or simple stress states. Strain rate effects in two- and three-dimensional stress states have been little studied. Dynamic biaxial stressing has been considered by a few authors in which thin sheets of metal are subject to lateral impact loads (GERARD and PAPIRNO, 1957).

1.19 Some Proposed Formulae for Correlating Stress, Strain, Strain Rate and Temperature

1. LUDWIK (1909) proposed a semi-logarithmic dependence of tensile yield strength on strain rate, and subsequent work for certain metals in certain ranges of strain have satisfied it. The form of the equation is,

$$\sigma = \sigma_1 + \sigma_0 \ln \dot{\epsilon}/\dot{\epsilon}_0$$

where σ_0, σ_1 and $\dot{\epsilon}_0$ are constants. Constant strain rate tests were used to give σ/ϵ curves for given $\dot{\epsilon}$ and T.

The above equation seems to be generally satisfactory provided that σ does not increase rapidly with $\dot{\epsilon}$.

MANJOINE and NADAI (1940) found that tensile strength at elevated temperature varied nearly linearly with the logarithm of strain rate.

2. ALDER and PHILIPS (1954) in work on copper at up to 600 °C, on aluminium at up to 500 °C and on steel at up to 930 to 1200 °C, were best able to summarize their results in the strain rate range 1 to 40/s. by

$$\sigma = \sigma_0 \dot{\epsilon}^n$$

where σ_0 and n are constants. These authors found that the power law provided 'a slightly better interpretation' of their data than did the Ludwik-type of equation. Sokolov, from earlier tests in compression over the range 20 °C to melting temperature, has advanced the same opinion as Alder and Philips.

3. MACGREGOR and FISHER (1946) introduced the idea of a 'velocity modified temperature', T_m so that

$$T_m = T(1 - m \ln \dot{\epsilon}/\dot{\epsilon}_0).$$

T is the test temperature (absolute) and m and $\dot{\epsilon}_0$ are constants. This form gives the expected qualitative result that an increase in strain rate is equivalent to a decrease in temperature. This equation is only acceptable as an approximation at less than the recrystallization temperature.

4. INOUYE (1955) has used the expression

$$\sigma = \sigma_0 \epsilon^n \dot{\epsilon}^m \exp{(A/Tk)}$$

where σ_0, n, m, A and k are constants.

5. MALVERN (1965) introduced an equation of the form

$$\dot{\epsilon} = \frac{\dot{\sigma}}{E} + F\,[\sigma - \sigma(e)]$$

if $\sigma > \sigma(e)$. $\sigma = \sigma(e)$ is the 'static' stress–strain curve.
Special forms of Malvern's equation are

$$\dot{\epsilon} = \frac{\dot{\sigma}}{E} + D\left(\frac{\sigma}{\sigma_0} - 1\right)p$$

and

$$\dot{\epsilon} = \frac{\sigma}{E} + A\left[\exp\left(\frac{\sigma}{\sigma_0} - 1\right)^q - 1\right]$$

where A, D, p and q are empirical constants and σ_0 is the 'static' yield stress.

The above equation is satisfactory in that it gives an apparent increase in initial yield stress at high rates of strain and it allows plastic strain increments to be propagated at elastic wave speeds along bars initially stressed into the plastic range. Both these features are experimental facts.

6. RIPPERBERGER (1965) has maintained and shown that his experimental data in plastic wave propagation work can be best summarized by adapting Malvern's equation to

$$\dot{\epsilon} = \frac{1}{\tau}\left[\frac{\sigma - \sigma_0(\epsilon)}{\sigma_0(\epsilon)}\right]^m .$$

τ is the relaxation time of the material, $\sigma_0(\epsilon)$ the 'static stress' at strain ϵ, and thus $\sigma - \sigma_0(\epsilon)$ is the 'over stress', or the 'excess stress' to which the material is subject at the given strain rate.

7. Other formulae will be found and are discussed in the book by THOMSEN, YANG and KOBAYASHI (1965).

1.20 The Yield Stress of Steel at about 15 °C

1. YIELD STRESS: TENSION AND COMPRESSION

An extensive summary of work on this topic will be found in the book by GOLDSMITH (1960). This work is worthy of special mention as it bears heavily on the design of structures to withstand impact loading at everyday temperature levels.

From the first work by Hopkinson in 1905 to that by Taylor, Campbell and others in the 1950s, and including the findings of Russian workers, the conclusion seems to emerge fairly clearly from quasi-static and impact tension tests (strain rates 10^{-3} to 10^2 or 10^3/s) that the ratio of dynamic to static yield stress in mild steel is in excess of unity, usually is about two and sometimes as high as three. This ratio decreases with increasing plastic strain, where flow stress instead of initial yield stress is used for the comparison.

Flow stress for a given strain is, generally, the larger the more rapid is the load application. For example, the above ratio for yield in mild steel can fall from about 2·5 to 1·3 when the time for loading reduces from 10^{-5} to 10^{-2} s. It also appears that this ratio is the smaller the higher the absolute value of the static yield stress.

2. DELAY TIMES

There is a clear delay time between load application and the onset of plastic flow in low carbon steels; it is evidently associated with the existence of a distinct yield point and yield mechanism. The time is the shorter the higher the load; the maximum delay period at 25 °C is of the order of one second; for a delay period of only 10^{-6} s the yield stress may have to be 2·5 times as great. At −60 °C, with this ratio unaltered the delay time may increase to more than 10 s; however, at 121 °C the time may be reduced to 10^{-2} to 10^{-3} s. At high temperatures (1600 °F) the delay time does not decrease rapidly with necessary tensile stress, e.g., by a ratio of 1·5 over the range 10^{-2} to 10^{-1} s. The recent work of LEBLOIS and MASSONNET (1972) should be consulted.

3. REPEATED LOADING

Low carbon steel subjected to identical low speed repeated impacts sufficient to induce plastic strain can be made to give rise to two curves. These are obtained by plotting the greatest and the terminal impact stress against total permanent strain.

See problems 1-16

REFERENCES

ALDER, J. F. and PHILLIPS, K. A.   1954   'The Effect of Strain Rate and Temperature on the Resistance of Aluminium, Copper and Steel to Compression'
J. Inst. Metals **83**, 80

ALEXANDER, J. M. and LENGYEL, B.   1964   'On the Cold Extrusion of Flanges against High Hydrostatic Pressure'
J. Inst. Metals **93**, 137

BARAYA, G. L., JOHNSON, W. and SLATER, R. A. C.   1965   'The Dynamic Compression of Circular Cylinders of Super-Pure Aluminium at Elevated Temperatures'
Int. J. mech. Sci. **7**, 621

BELL, J. F.   1968   *The Physics of Large Deformation of Crystalline Solids*
Springer-Verlag, New York Inc., 253 pp.

BERESNEV, B. I., VERESHCHAGIN, L. F., RYABININ, YU. N., and LIVSHITS, L. D.   1960   *Some Problems of Large Plastic Deformation of Metals at High Pressure*
Translated from the Russian and available from Pergamon Press, 1963, p. 79

BOKER, R.   1914   *The Mechanics of Plastic Deformation in Crystalline Bodies*
Dissertation, Technische Hochschule, Aachen

BREDTHAUER, R. D.   1956 'Strength Characteristics of Rock Samples under Hydrostatic Pressure' *Trans A.S.M.E.* 56–PET–23

BRIDGMAN, P. W.   1944 *Trans Am. Soc. Metals* **32**, 553
1947 *The Effect of Hydrostatic Pressure on the Fracture of Brittle Substances* *J. Appl. Phys.* **18**, 246
1952 *Studies in Large Plastic Flow and Fracture with Special Emphasis on the Effects of Hydrostatic Pressure* McGraw-Hill, New York

BROWN, A.   1964 'A Quasi-Dynamic Theory of Containment' *Int. J. mech. Sci.* **6**, 257

COFFIN, L. F. and CONRAD, H.   1965 *The Cryogenic Properties of Metals in High Strength Materials* Wiley, 436 pp.

COOK, G.   1934 *The Effect of Fluid Pressure on the Permanent Deformation of Metals by Shear* Instn Civ. Engrs, Selected Paper No. 170

COOK, P. M.   1957 'True stress-strain curves for steel in compression at high temperatures and strain rates' *Proc. Conf. Properties of Materials at High Rates of Strain*, Instn Mech. Engrs, London, 86

CROSSLAND, B.   1954 'The Effect of Fluid Pressure on the Shear Properties of Metals' *Proc. Instn mech. Engrs* **169**, 935

CROSSLAND, B. and DEARDEN, W. H.   1958 'The Plastic Flow and Fracture of a "Brittle" Material (Grey Cast Iron) with Particular Reference to the Effect of Fluid Pressure' *Proc. Instn mech. Engrs* **172**

CROSSLAND, B. and MITRA, A. K.   1968 'Effect of Hydrostatic Pressure on Metals in Torsion' *Thesis* A. K. Mitra for the M.Sc. degree, Queen's University of Belfast

DAVIDENKOV, N. N. and SPIRIDONOVA, N. I.   1946 'Analysis of Tensile Stress in the Neck of an Elongated Test Specimen' *Proc. A.S.T.M.* **46**, 1147

DUVALL, G. E.   1962 'Some Properties and Applications of Shock Waves' *Response of Metals to High Velocity Deformation*, Interscience Publishers, **9**, 165

FARREN, W. S. and TAYLOR, G. I.   1925 The Heat Developed during Plastic Extension of Metals *Proc. R. Soc., Lond. Ser. A.*, **107**, 422

FINNIE, I. and HELLER, W. R.   1959 *Creep of Engineering Materials* McGraw-Hill, New York

GENSAMER, M. 1940 'Strength of Metals under Combined Stresses'
Trans Am. Soc. Metals **28**, 54

GERARD, G. and 1957 'Dynamic Biaxial Stress-Strain Characteristics
PAPIRNO, R. of Aluminium and Mild Steel'
Trans A.S.M. **49**, 132

GOLDSMITH, W. 1960 *Impact* Edward Arnold, London, 379 pp.

GORDON, P. 1955 'Micro-calorimetric Investigation of
Recrystallisation of Copper'
Trans A.I.M.E. **203**, 1043

HAWKYARD, J. B., 1968 'Dynamic Yield Strength of Copper and Low
EATON, D. and Carbon Steel at Elevated Temperatures'
JOHNSON, W. *Int. J. mech. Sci.* **10**, 929

HAWKYARD, J. B. 1971 'A Novel Form of Drop Hammer Pressure Bar'
and POTTER, T. B. *Int. J. mech. Sci.* **14**, 95

HOCKETT, J. E. 1966 'On Relating the Flow Stress of Aluminium
to Strain, Strain Rate and Temperature'
Los Alamos Scientific Laboratory of Univ.
of California, Report LA-3544

INOUYE, K. 1955 'Studies on the Hot-Working Strength of
Steels' (In Japanese)
Tetsu to Hagane **41**, 593

KARMAN, TH. V. 1911 'Festigkeitsversuche unter allseitigem Druck'
Z. Verb. dt. Ing. **55**, 1749

LEBLOIS, C. R. and 1972 'Influence of the Upper Yield Stress on the Be-
MASSONNET, CH. haviour of Mild Steel in Bending and Torsion'
Int. J. mech. Sci. **14**, 95

LUDWIK, P. 1909 *Elemente der technologischen Mechanik*
Springer, Berlin

LUEG, W., 1957 *Arch. Eisenhutt Wes.* **28**, (8), 505
MULLER, H. G. and
KRAUSE, U.

MCADAM, D. J., 1947 'Influence of Plastic Extension and
GEIL, G. W. and Compression on the Fracture Stress of Metals'
JENKINS, W. D. *Proc. A.S.T.M.* **47**, 554

MACGREGOR, C. W. 1940 'The Tension Test'
Proc. A.S.T.M. **40**, 508

MACGREGOR, C. W. 1946 'A Velocity-Modified Temperature for
and FISHER, J. C. Plastic Flow of Metals'
Trans A.S.M.E., J. Appl. Mech. A–12: A–11

MAGNUSSON, A. W. 1957 'Low Temperature Brittleness'
and BALDWIN, W. M. *J. Mech. Phys. Solids* **5**, 172

MAHTAB, F. U., 1965 'The Dynamic Indentation of Copper and an
JOHNSON, W. and Aluminium Alloy with a Conical Projectile at
SLATER, R. A. C. Elevated Temperature'
Proc. Instn Mech. Engrs **180**, 285

MALVERN, L. E. 1965 'Experimental Studies of Strain Rate Effects
and Plastic Wave Propagation in Annealed
Aluminium'

		Proc. A.S.M.E. Coll. on Behaviour of Materials under Dynamic Loading, 81
MANJOINE, M. J. and NADAI, A.	1940	'High Speed Tension Tests at Elevated Temperatures' Proc. Am. Soc. Testing Materials **40**, 822
MANJOINE, M. J.	1944	'Influence of Rate of Strain and Temperature on Yield Stresses of Mild Steel' J. Appl. Mech. **11**, A–211
MORRISON, J. L. M.	1934	'The Influence of Rate of Strain in Tension Tests' The Engineer **158**, 183
	1939	'The Yield of Mild Steel with Particular Reference to the Effect of Size of Specimen' Proc. Instn mech. Engrs **142**, 193
NADAI, A.	1963	Theory of Flow and Fracture of Solids Vol. 2, McGraw-Hill, New York
OROWAN, E.	1950	'The Cam Plastometer' B.I.S.R.A. Report MW/F/22/50
OXLEY, P. L. B.	1963	'Rate of Strain Effect in Metal Cutting' Trans. A.S.M.E., **85**, Series B, 335.
PARKER, E. R., DAVIS, H. E. and FLANIGAN, A. E.	1946	'A Study of the Tension Test' Proc. A.S.T.M. **46**, 1159
PENNY, R. K. and MARRIOTT, D. L.	1971	Design for Creep McGraw-Hill, New York
PUGH, H. Ll. D. and GREEN, D.	1958	'The Behaviour of Metals under High Hydrostatic Pressure' M.E.R.L. Plasticity Report No. 147.
PUGH, H. LL. D. (Ed.)	1970	Mechanical Behaviour of Materials under Pressure Elsevier, Amsterdam, 785 pp.
	1971	Engineering Solids under Pressure Instn of Mech. Engrs, London, 191 pp.
RIPPERBERGER, E. A.	1965	'Experimental Studies of Strain Rate Effects and Plastic Wave Propagation in Annealed Aluminium' Proc. A.S.M.E., Coll. on Behaviour of Materials under Dynamic Loading, Chicago, p. 62
ROS, M. and EICHINGER, A.	1929	'Versuche zur Klärung der Frage der Bruchgefahr III Metalle' Metalle Diskussionsbericht No. 34 Eidg. Materialprüfungsanstalt, Zurich
SHELTON, A.	1961	'On the ratio of Transverse to Axial Strain and Other Tensile Properties of a Cold Rolled Steel Alloy' J. Mech. Eng. Sci. **3**, 89

SAMANTA, S. K.   1968 'Resistance to Dynamic Compression of Low-Carbon Steel and Alloy Steels at Elevated Temperatures and at High Strain Rates' *Int. J. mech. Sci.* **10**, 613

     1969 'On Relating the Flow Stress of Aluminium and Copper to Strain, Strain-rate and Temperature' *Int. J. mech. Sci.* **11**, 433

SLATER, R. A. C., AKU, S. Y. and JOHNSON, W.   1968 'Experiments in the Fast Upsetting of Short Pure Lead Cylinders and a Tentative Analysis' *Int. J. mech. Sci.*, **10**, 143

SPEAKMAN, T. N.   1966 'Dynamic Compression of Pure Lead and Super-Pure Aluminium at Elevated Temperatures' *M.Sc. Thesis*, University of Manchester

SUZUKI, H. et al   1968 'Studies on the Flow Stress of Metals and Alloys' Report of the Inst. of Industrial Science, University of Tokyo, **18**, 141

TAYLOR, G. I.   1948 'Formation and Enlargement of Circular Hole in Thin Plastic Sheet' *Quart. Jl mech. appl. Math.* **1**, 103

THOMSON, W. T.   1955 'Approximate Theory of Armor Penetration' *J. appl. Physics*, **26**, 80

THOMSEN, E. G., YANG, C. T. and KOBAYASHI, S.   1965 *Plastic Deformation in Metal Processing* Macmillan, New York, 486 pp.

TIPPER, C. F.   1957 'The Brittle Fracture of Metals at Atmospheric and Sub-zero Temperatures' *Metall. Rev., Inst. of Metals* **2**, 195

     1948 *Fracturing of Metals* Amer. Soc. for Metals, Cleveland, Ohio

UNKSOV, E. P.   1961 *An Engineering Theory of Plasticity* Butterworths, London

VOCE, E.   1948 'The Relationship between Stress and Strain for Homogeneous Deformation' *J. Inst. Metals* **74**, 537. Also correspondence in same volume, p. 760

WHIFFIN, A. C.   1948 'The Use of Flat-ended Projectiles for Determining Dynamic Yield Stress. II Tests on Various Metallic Materials' *Proc. R. Soc.* **A.194**, 300

ZAID, M. and PAUL, B.   1957 'Mechanics of High Speed Projectile Perforation' *J. Franklin Inst.* **264**, 117

Chapter 2

STRESS

2.1 Definitions

Consider a given point P in a solid body which is transmitting or sustaining load and through P describe a small plane area δA, see Fig. 2.1. Across δA a force δF, say, will be transmitted and the limit of the ratio $\delta F/\delta A$ as δA

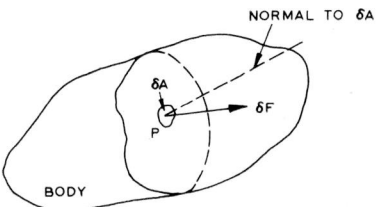

FIG. 2.1 Internal forces in solid body.

approaches zero is a certain quantity σ. σ is the force per unit area or stress across δA at P in the direction of δF at P if δA is small. Instead of referring to σ, other equivalent statements may be made. Resolve δF perpendicular and parallel to δA, into, say, δN and δS and obtain $\delta N/\delta A$ and $\delta S/\delta A$.

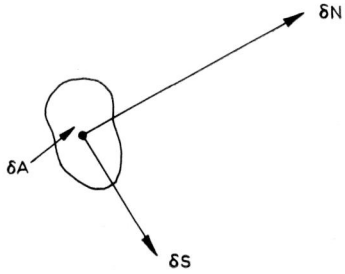

FIG. 2.2 Normal and shear forces on small area.

$\delta N/\delta A$ is the normal stress on δA and $\delta S/\delta A$ the shear stress across δA, see Fig. 2.2. Further, suppose δF had been resolved in three perpendicular directions Ox, Oy, Oz, Fig. 2.3. Ox is the direction of the normal to δA. Oy and Oz are chosen arbitrarily in the plane of δA. Let the components of the force δF along these axes be δF_x, δF_y and δF_z. Three stresses may now be identified:

$$\underset{\delta A \to 0}{\text{Lt}} \frac{\delta F_x}{\delta A} = \sigma_{xx}; \qquad \underset{\delta A \to 0}{\text{Lt}} \frac{\delta F_y}{\delta A} = \sigma_{xy}; \qquad \underset{\delta A \to 0}{\text{Lt}} \frac{\delta F_z}{\delta A} = \sigma_{xz}.$$

A double suffix notation is employed to define the direction of each of the stresses. σ_{xy} means the stress on a plane perpendicular to Ox and parallel to axis Oy. The first suffix gives the direction of the normal to the plane on which a stress acts, and the second one its direction with respect to that plane.

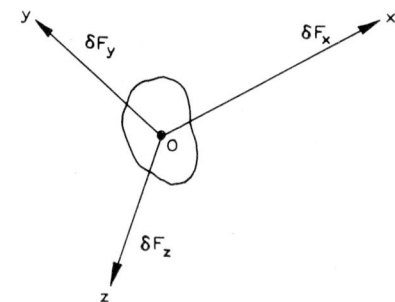

FIG. 2.3 Resolution of force δF along Cartesian axes, where OX is normal to the surface.

If the examination had started by first establishing the direction of the axes Ox, Oy, Oz in space, the discussion would have proceeded by considering the elementary force δF whose components parallel to the axes are δF_x, δF_y and δF_z acting on an elementary area δA_x which is perpendicular to Ox; then σ_{xx}, σ_{xy}, σ_{xz} would have been arrived at as above. If an area δA_y, i.e., a small area at P perpendicular to direction Oy across which the net force is δP, acts, having components δP_x, δP_y, δP_z, then the notation for the stresses across δA_y would be σ_{yz}, σ_{yy}, σ_{yz}. Yet a third set of stresses across area δA_z would be σ_{zx}, σ_{zy}, σ_{zz}; see Fig. 2.4(a).

If instead of Cartesian co-ordinates, cylindrical co-ordinates had been used, Fig. 2.5(a), the equivalent nine components would be

$$\sigma_{rr}, \sigma_{r\theta}, \sigma_{rz}, \sigma_{\theta r}, \sigma_{\theta\theta}, \sigma_{\theta z}, \sigma_{zr}, \sigma_{z\theta}, \sigma_{zz}.$$

Both these notations could be simply summarized by writing σ_{ij} and remembering that for i and j may be substituted any of three specified perpendicular directions at a point. In particular, it is to be noted that when $i = j$, there results σ_{xx}, σ_{yy} and σ_{zz}, or σ_{rr}, $\sigma_{\theta\theta}$ and σ_{zz}, i.e., three direct or normal stresses; the other six quantities associated with σ_{ij} are shear stresses.

It is more usual, however, when not using the suffix notation, to write the normal stresses simply as σ_x, σ_y, σ_z and the shear stresses as τ_{xy}, τ_{yz}, τ_{zx}. The equivalence of the two notations is illustrated in Fig. 2.4 and Fig. 2.5.

σ_{ij} is usually known as the stress tensor at a point. As far as this book is concerned, it may be considered as a shorthand method of referring to an array of nine quantities at a point.

Suffix notation is very useful in proving theorems or making general statements, but it is of no value for specific problems.

2.2 The Equations of Force Equilibrium

Circumscribe symmetrically the point P in a stressed body, with a rectangular parallelepiped whose length of sides are δx, δy and δz. Employing

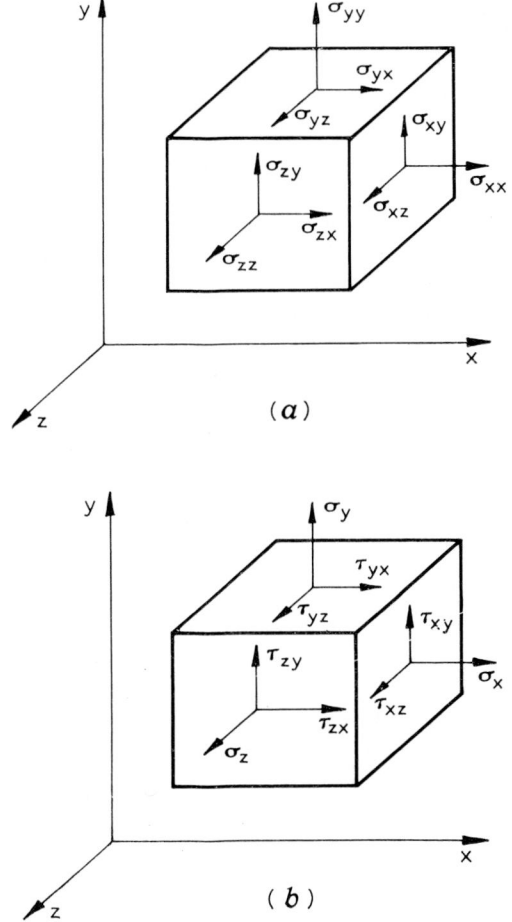

FIG. 2.4 Stress components referred to Cartesian co-ordinates.

the first notation, the *rates* of change of the stress components with increase in x, on area δA_x, when y and z are kept constant are

$$\frac{\partial \sigma_{xx}}{\partial x}, \quad \frac{\partial \sigma_{xy}}{\partial x} \quad \text{and} \quad \frac{\partial \sigma_{xz}}{\partial x}.$$

Similar expressions for directions y and z are:

$$\frac{\partial \sigma_{yx}}{\partial y}, \quad \frac{\partial \sigma_{yy}}{\partial y}, \quad \frac{\partial \sigma_{yz}}{\partial y}, \quad \frac{\partial \sigma_{zx}}{\partial z}, \quad \frac{\partial \sigma_{zy}}{\partial z}, \quad \frac{\partial \sigma_{zz}}{\partial z},$$

see Fig. 2.6. There are thus eighteen forces on the six sides of the block and the conditions for equilibrium may be found by considering the forces in

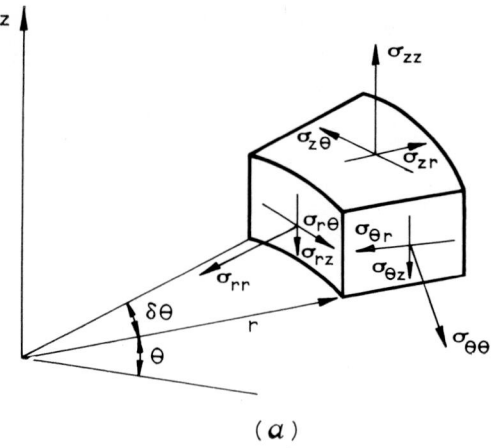

(a)

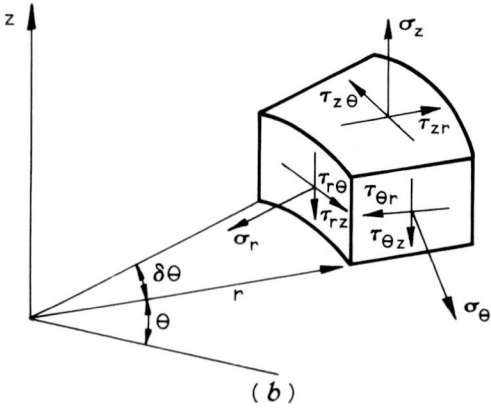

(b)

FIG. 2.5 Stress components referred to cylindrical co-ordinates.

each of the directions parallel to the axes. Examining conditions for equilibrium in direction Ox,

$$\left(\sigma_{xx} - \frac{1}{2} \cdot \frac{\partial \sigma_{xx}}{\partial x} \cdot \delta x \right) \delta z \, \delta y + \left(\sigma_{yx} - \frac{1}{2} \cdot \frac{\partial \sigma_{yx}}{\partial y} \cdot \delta y \right) \delta x \, \delta z + \left(\sigma_{zx} - \frac{1}{2} \frac{\partial \sigma_{zx}}{\partial z} \cdot \delta z \right) \delta x \, \delta y$$

$$= \left(\sigma_{xx} + \frac{1}{2} \frac{\partial \sigma_{xx}}{\partial x} \cdot \delta x \right) \delta z \, \delta y + \left(\sigma_{yx} + \frac{1}{2} \frac{\partial \sigma_{yx}}{\partial y} \cdot \delta y \right) \delta x \, \delta z + \left(\sigma_{zx} + \frac{1}{2} \frac{\partial \sigma_{zx}}{\partial z} \cdot \delta z \right) \delta x \, \delta y.$$

And thus on simplifying

$$\frac{\partial \sigma_{xx}}{\partial x} + \frac{\partial \sigma_{yx}}{\partial y} + \frac{\partial \sigma_{zx}}{\partial z} = 0, \tag{2.1}$$

assuming that no body forces are operative. Again in the shorthand notation,

$$\frac{\partial \sigma_{ij}}{\partial x_i} = 0 \tag{2.2}$$

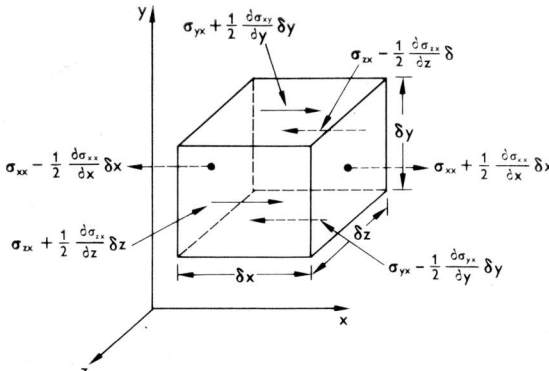

FIG. 2.6 Force equilibrium in direction OX.

if it be agreed that on deciding upon and fixing a specific direction for j, the other suffix i is then given all possible values and all such terms are then summed. For example, if x is substituted for j and kept fixed whilst putting x, y, z for i in turn and the three terms are added together, then equation (2.1) is recovered. The remaining two equations of equilibrium, i.e., for directions Oy and Oz, are

$$\frac{\partial \sigma_{xy}}{\partial x} + \frac{\partial \sigma_{yy}}{\partial y} + \frac{\partial \sigma_{zy}}{\partial z} = 0 \tag{2.3}$$

$$\frac{\partial \sigma_{xz}}{\partial x} + \frac{\partial \sigma_{yz}}{\partial y} + \frac{\partial \sigma_{zz}}{\partial z} = 0. \tag{2.4}$$

It is evident that equation (2.2) also stands for equations (2.3) and (2.4); thus equation (2.2) represents three equations containing three terms when written out in full. If it was desired to include body forces of f_j per unit mass acting in the direction j, (2.2) would be written

$$\frac{\partial \sigma_{ij}}{\partial x_i} + \rho f_j = 0$$

where ρ is the density.

2.3 Couple Equilibrium

The condition for moment or couple equilibrium is next determined. In Fig. 2.7 are shown the stresses acting so as to tend to rotate the block about a line through P parallel to Oz. If these are in equilibrium,

$$\left(\sigma_{yx} + \frac{1}{2}\frac{\partial \sigma_{yx}}{\partial y}.\delta y\right)\delta x\,\delta z\,\frac{\delta y}{2} + \left(\sigma_{yz} - \frac{1}{2}\frac{\partial \sigma_{yx}}{\partial y}.\delta y\right)\delta x\,\delta z\,\frac{\delta y}{2}$$

$$= \left(\sigma_{xy} + \frac{1}{2}\frac{\partial \sigma_{xy}}{\partial x}.\delta x\right)\delta y\,\delta z\,\frac{\delta x}{2} + \left(\sigma_{xy} - \frac{1}{2}\frac{\partial \sigma_{xy}}{\partial x}.\delta x\right)\delta y\,\delta z\,\frac{\delta x}{2}.$$

On simplifying,

$$\sigma_{yx} = \sigma_{xy}. \tag{2.5}$$

Similarly, for the other two directions

$$\sigma_{zy} = \sigma_{yz} \text{ and } \sigma_{xz} = \sigma_{zx}. \tag{2.6}$$

These equations may again be summarized as

$$\sigma_{ij} = \sigma_{ji}. \tag{2.7}$$

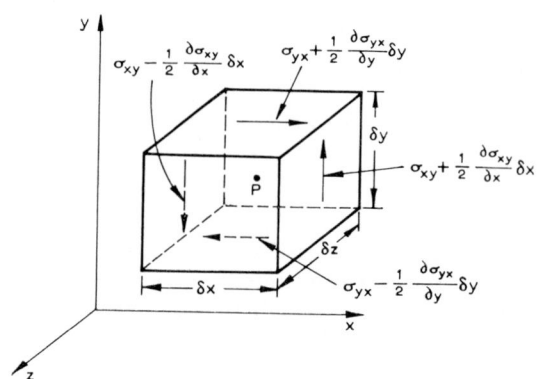

FIG. 2.7 Couple equilibrium in plane XY.

The three equations (2.5) and (2.6) or (2.7) express the well-known idea of 'complementary shear stress'. The requirement for moment or couple equilibrium on any block is expressed by saying that σ_{ij} is symmetrical.

Thus the actual stress at a point is known if the six independent magnitudes and directions of σ_{ij} are specified, i.e., three normal stresses and three shear stresses; in effect, because of the requirement of couple equilibrium three pairs of shear stresses are automatically required.

2.4 Three-dimensional Stress Systems

The six components of stress at a point being given in magnitude and direction as in Fig. 2.8, suppose it is desired to evaluate the normal and shear stress on any doubly oblique plane, or in particular the principal stresses and their directions at that point. Principal stresses are stresses acting normal to a plane across which there is no shear stress. Let the doubly oblique plane be represented by triangle ABC whose normal ON has direction cosines (l, m, n). Denote by s the resultant stress on triangle ABC, which has direct stress component s_n normal to plane ABC and a shear stress component s_s in the plane ABC. s_x, s_y, s_z are components of s parallel to Ox, Oy, Oz, respectively. Δ denotes the area of triangle ABC and Δ_x, Δ_y, Δ_z, the areas of triangles OBC, OAC, OAB. For equilibrium of the forces on the tetrahedron $OABC$ in direction Ox,

$$\Delta s_x = \Delta_x \sigma_x + \Delta_y \tau_{yx} + \Delta_z \tau_{zx},$$

$$\therefore s_x = l\sigma_x + m\tau_{yx} + n\tau_{zx} \qquad (2.8)$$

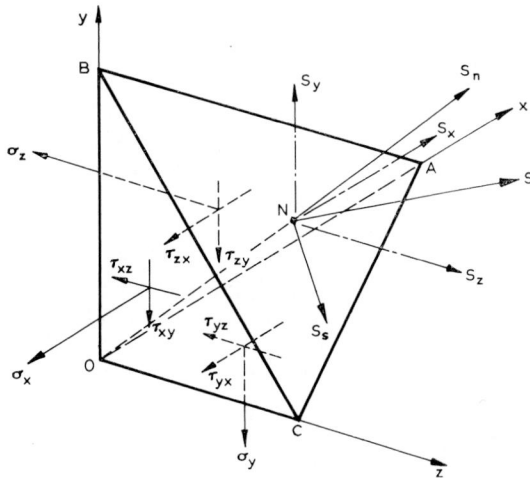

FIG. 2.8 Stresses on a doubly oblique plane.

because $\Delta_x . OA = \Delta . ON$ and therefore $\Delta_x / \Delta = ON/OA = \cos \alpha = l$. α is the angle between ON and Ox.
Similarly,

$$s_y = l\tau_{xy} + m\sigma_y + n\tau_{zy} \qquad (2.9)$$

and

$$s_z = l\tau_{xz} + m\tau_{xz} + n\sigma_z. \qquad (2.10)$$

Now

$$s_n = ls_x + ms_y + ns_z$$

$$= l^2\sigma_x + m^2\sigma_y + n^2\sigma_z + 2(lm\tau_{xy} + mn\tau_{yz} + nl\tau_{zx}) \qquad (2.11)$$

utilizing equations (2.8), (2.9) and (2.10).

Further,

$$s_s^2 = s^2 - s_n^2$$

$$= s_x^2 + s_y^2 + s_z^2 - s_n^2. \qquad (2.12)$$

Equations (2.11) and (2.12) yield the normal and shear stress on any defined plane when the general state of stress at a point is given.

If s_n is a principal stress, say s, i.e. $s_s = 0$, and if its direction is (l, m, n), then by resolving in the directions Ox, Oy and Oz, respectively

$$l(s - \sigma_x) - m\tau_{xy} - n\tau_{xz} = 0 \tag{2.13}$$

$$-l\tau_{xy} + m(s - \sigma_y) - n\tau_{yz} = 0 \tag{2.14}$$

$$-l\tau_{xz} - m\tau_{yz} + n(s - \sigma_z) = 0. \tag{2.15}$$

If (2.13) and (2.14) are solved for m/l and n/l and these values substituted into (2.15), a relation between the coefficients of l, m, n in these equations results. Only if this relation is fulfilled will equations (2.13), 2.14) and (2.15) be satisfied, and then it is found that

$$s^3 - (\sigma_x + \sigma_y + \sigma_z)s^2 - (\tau_{xy}^2 + \tau_{yz}^2 + \tau_{zx}^2 - \sigma_x\sigma_y - \sigma_y\sigma_z - \sigma_z\sigma_x)s$$
$$- (\sigma_x\sigma_y\sigma_z + 2\tau_{xy}\tau_{yz}\tau_{zx} - \sigma_x\tau_{yz}^2 - \sigma_y\tau_{zx}^2 - \sigma_z\tau_{xy}^2) = 0. \tag{2.16}$$

This cubic equation in a real physical situation has three real solutions which are the principal stresses, say, σ_1, σ_2, σ_3 of this particular stress system.

Suppose that with the same stress system acting, the stress components are, initially, given with respect to a set of three perpendicular axes other than Ox, Oy, Oz as before, say Ox', Oy', Oz',

i.e. $\qquad \sigma_{x'}$, $\sigma_{y'}$, $\sigma_{z'}$, $\tau_{x'y'}$, $\tau_{y'z'}$, $\tau_{z'x'}$.

Now the principal stresses at P would be given by a cubic equation that could be derived in exactly the same way as was equation (2.16); it would be

$$s^3 - (\sigma_{x'} + \sigma_{y'} + \sigma_{z'})s^2 - (\tau_{x'y'}^2 + \tau_{y'z'}^2 + \tau_{z'x'}^2 - \sigma_{x'}\sigma_{y'} - \sigma_{y'}\sigma_{z'} - \sigma_{z'}\sigma_{x'})s$$
$$- (\sigma_{x'}\sigma_{y'}\sigma_{z'} + 2\tau_{x'y'}\tau_{y'z'}\tau_{z'x'} - \sigma_{x'}\tau_{y'z'}^2 - \sigma_{y'}\tau_{z'x'}^2 - \sigma_{z'}\tau_{x'y'}^2) = 0. \tag{2.17}$$

Clearly the principal stresses remain the same, regardless of the axes chosen, and hence the coefficients of the unknown s in equations (2.16) and (2.17) must be the same—in order that they will give the same values of principal stresses σ_1, σ_2, σ_3. Thus equations (2.16) and (2.17) may be written in the form

$$s^3 - I_1 s^2 - I_2 s - I_3 = 0 \tag{2.18}$$

where

$$I_1 = \sigma_x + \sigma_y + \sigma_z = \sigma_{x'} + \sigma_{y'} + \sigma_{z'} \tag{2.19}$$

$$I_2 = -(\sigma_x\sigma_y + \sigma_y\sigma_z + \sigma_z\sigma_x) + \tau_{xy}^2 + \tau_{yz}^2 + \tau_{zx}^2$$
$$= -(\sigma_{x'}\sigma_{y'} + \sigma_{y'}\sigma_{z'} + \sigma_{z'}\sigma_{x'}) + \tau_{x'y'}^2 + \tau_{y'z'}^2 + \tau_{z'x'}^2 \tag{2.20}$$

$$I_3 = \sigma_x\sigma_y\sigma_z + 2\tau_{xy}\tau_{yz}\tau_{zx} - \sigma_x\tau_{yz}^2 - \sigma_y\tau_{zx}^2 - \sigma_z\tau_{xy}^2$$
$$= \sigma_{x'}\sigma_{y'}\sigma_{z'} + 2\tau_{x'y'}\tau_{y'z'}\tau_{z'x'} - \sigma_{x'}\tau_{y'z'}^2 - \sigma_{y'}\tau_{z'x'}^2 - \sigma_{z'}\tau_{x'y'}^2 \tag{2.21}$$

I_1, I_2, I_3 are evidently quantities independent of the direction of the axes chosen; they are called the three invariants (or non-varying quantities) of the stress at P. The first and second invariants have particular physical significance for the theory of plasticity, see p. 63.

The values of I_1, I_2 and I_3 in terms of the principal stresses are

$$I_1 = \sigma_1 + \sigma_2 + \sigma_3$$

$$I_2 = -(\sigma_1\sigma_2 + \sigma_2\sigma_3 + \sigma_3\sigma_1)$$

$$I_3 = \sigma_1\sigma_2\sigma_3 \tag{2.22}$$

The roots of the cubic equation (2.18) are the three principal stresses σ_1, σ_2, σ_3. This equation is solved either by guessing or, directly, by using the goniometric method. For example, suppose the stress system indicated in Fig. 2.9 is given. It is readily computed that

$$I_1 = 15, \quad I_2 = -60 \quad \text{and} \quad I_3 = 54$$

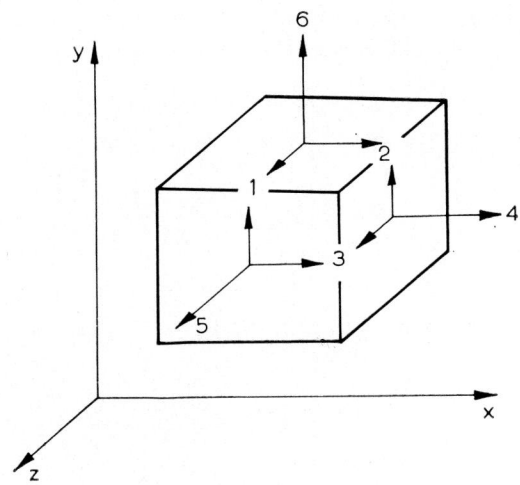

Fig. 2.9 Particular stress system.

Equation (2.18) becomes

$$s^3 - 15s^2 + 60s - 54 = 0 \tag{2.23}$$

Write $s = (\sigma' + 5)$ in order to remove the s^2 term, and equation (2.23) becomes

$$\sigma'^3 - 15\sigma' - 4 = 0. \tag{2.24}$$

Now recall the identity

$$\cos 3\theta = 4\cos^3\theta - 3\cos\theta$$

in the form $\cos^3\theta - \tfrac{3}{4}\cos\theta - \tfrac{1}{4}\cos 3\theta = 0.$ \hfill (2.25)

Put $\sigma' = r\cos\theta$ and equation (2.24) becomes

$$\cos^3\theta - \frac{15}{r^2}\cos\theta - \frac{4}{r^3} = 0. \tag{2.26}$$

Equations (2.25) and (2.26) are identical if

$$\frac{15}{r^2} = \frac{3}{4}, \text{ i.e. } r = \sqrt{20} = 4.47$$

and
$$\frac{4}{r^3} = \frac{\cos 3\theta}{4} \text{ or } \cos 3\theta = \frac{16}{89.4} = 0.179.$$

The principal values are $\theta_1 = 26.6°$, $\theta_2 = 93.4°$ and $\theta_3 = 146.6°$. Thus $r_1 \cos \theta_1 = 4.00$, $r_2 \cos \theta_2 = -0.27$, $r_3 \cos \theta_3 = -3.73$. The principal stresses are obtained by adding 5 to these quantities and they become 9.00, 4.73 and 1.27.

Equation (2.23) could have been factorized immediately, if it had been noticed that $s = 9$ was a solution, as $(s - 9)(s^2 - 6s + 6) = 0$ and the remaining two roots of s would be $(3 + \sqrt{3})$ and $(3 - \sqrt{3})$.

In general, the principal stresses from (2.18) are s and

$$s = r \cos \theta + I_1/3.$$

θ and hence $\cos \theta$ has the three principal values given by

$$\cos 3\theta = \frac{2I_1^3 + 9I_1I_2 + 27I_3}{2(I_1^2 + 3I_2)^{3/2}}.$$

Also,
$$r = 2(I_1^2 + 3I_2)^{\frac{1}{2}}/3.$$

If in equation (2.18) for s is substituted $(\sigma' + I_1/3)$, then the resulting equation is

$$\sigma'^3 - \left(\frac{I_1^2 + 3I_2}{3}\right)\sigma' - \left(\frac{2I_1^3 + 9I_1I_2 + 27I_3}{27}\right) = 0. \tag{2.27}$$

σ' or $(s - I_1/3)$ is known as a reduced stress; in effect, a hydrostatic stress of $I_1/3$ is 'taken away' from the original stress system. Equation (2.27) can also be written

$$\sigma'^3 - J_1\sigma'^2 - J_2\sigma' - J_3 = 0 \tag{2.28}$$

where
$$J_1 = 0$$

$$J_2 = (I_1^2 + 3I_2)/3$$

$$J_3 = (2I_1^3 + 9I_1I_2 + 27I_3)/27.$$

J_1, J_2 and J_3 are described as the first, second and third invariants of the reduced stresses and they have special importance when considering the yielding of metals.

The cubic equation formed from a knowledge of its roots, or of the reduced principal stresses σ_1', σ_2', σ_3' is

$$\sigma'^3 - (\sigma_1' + \sigma_2' + \sigma_3')\sigma'^2 + (\sigma_1'\sigma_2' + \sigma_2'\sigma_3' + \sigma_3'\sigma_1')\sigma' - \sigma_1'\sigma_2'\sigma_3' = 0$$

or
$$\sigma'^3 + \left(\sum \sigma_1'\sigma_2'\right)\sigma' - \sigma_1'\sigma_2'\sigma_3' = 0.$$

The values of J_1, J_2 and J_3 in terms of the reduced principal stresses are

$$J_1 = \sum \sigma_1' = 0$$
$$J_2 = -\sum \sigma_1' \sigma_2' = \tfrac{1}{2} \sum \sigma_1'^2$$
$$J_3 = \sigma_1' \sigma_2' \sigma_3' = \tfrac{1}{3} \sum \sigma_1'^3. \tag{2.29}$$

2.5 Mohr's Circles for Three-dimensional Stress Systems

If the principal stresses σ_1, σ_2, σ_3 are given, then the normal stress, σ, on a doubly oblique plane whose direction cosines are l, m, n, is, using equation (2.11)

$$\sigma = l^2 \sigma_1 + m^2 \sigma_2 + n^2 \sigma_3. \tag{2.30}$$

The total shear stress, τ, on this plane is, using equation (2.12),

$$\tau^2 = l^2 \sigma_1^2 + m^2 \sigma_2^2 + n^2 \sigma_3^2 - (l^2 \sigma_1 + m^2 \sigma_2 + n^2 \sigma_3)^2. \tag{2.31}$$

The identity

$$l^2 + m^2 + n^2 = 1 \tag{2.32}$$

is also available.

σ and τ can be calculated from equations (2.30) and (2.31) or, alternatively, a graphical construction, originally due to Mohr, may be used.

First, find an expression for l^2. From equation (2.32), $m^2 = 1 - l^2 - n^2$; substitute for m^2 in equation (2.30) to obtain,

$$\sigma = l^2 \sigma_1 + \sigma_2 (1 - l^2 - n^2) + n^2 \sigma_3 \tag{2.33}$$

and therefore

$$n^2 = \frac{\sigma + l^2(\sigma_2 - \sigma_1) - \sigma_2}{(\sigma_3 - \sigma_2)}. \tag{2.34}$$

And substituting in equation (2.31) for m^2 we obtain,

$$\tau^2 = l^2(\sigma_1^2 - \sigma_2^2) + \sigma_2^2 + n^2(\sigma_3^2 - \sigma_2^2) - [l^2(\sigma_1 - \sigma_2) + \sigma_2 + n^2(\sigma_3 - \sigma_2)]^2 \tag{2.35}$$

Substitute for n^2 in equation (2.35) from equation (2.34) and simplify to find

$$l^2 = \frac{\tau^2 + (\sigma_2 - \sigma)(\sigma_3 - \sigma)}{(\sigma_2 - \sigma_1)(\sigma_3 - \sigma_1)}. \tag{2.36}$$

Similarly,

$$m^2 = \frac{\tau^2 + (\sigma_3 - \sigma)(\sigma_1 - \sigma)}{(\sigma_3 - \sigma_2)(\sigma_1 - \sigma_2)} \tag{2.37}$$

and

$$n^2 = \frac{\tau^2 + (\sigma_1 - \sigma)(\sigma_2 - \sigma)}{(\sigma_1 - \sigma_3)(\sigma_2 - \sigma_3)}. \tag{2.38}$$

Rewrite equation (2.36) as

$$(\sigma - \sigma_2)(\sigma - \sigma_3) + \tau^2 = l^2(\sigma_1 - \sigma_2)(\sigma_1 - \sigma_3)$$

or

$$\left(\sigma - \frac{\sigma_2 + \sigma_3}{2}\right)^2 + \tau^2 = l^2(\sigma_1 - \sigma_2)(\sigma_1 - \sigma_3) + \left(\frac{\sigma_2 - \sigma_3}{2}\right)^2. \tag{2.39}$$

Thus if l, m, n are given, then σ and τ lie on the circle specified by equation (2.39) when σ are the abscissae and τ the ordinates, the circle having a centre at $(\sigma_2 + \sigma_3)/2$, 0 and a radius

$$\sqrt{l^2(\sigma_1 - \sigma_2)(\sigma_1 - \sigma_3) + [(\sigma_2 - \sigma_3)/2]^2}.$$

Mark the points P_1, P_2 and P_3 on the σ-axis where $OP_1 = \sigma_1$, $OP_2 = \sigma_2$ and $OP_3 = \sigma_3$. Draw circles on diameters P_1P_2, P_2P_3, P_3P_1 having their centres C_1, C_2 and C_3 at

$$\left(\frac{\sigma_1 + \sigma_2}{2}, 0\right), \quad \left(\frac{\sigma_2 + \sigma_3}{2}, 0\right) \text{ and } \left(\frac{\sigma_3 + \sigma_1}{2}, 0\right).$$

Through P_1, P_2, P_3 draw lines parallel to the τ-axis, say P_1T_1, P_2T_2, P_3T_3; see Fig. 2.10. From P_1 draw a line at angle α to P_1T_1 where $l = \cos\alpha$, to cut the circles on P_1P_2 and P_1P_3 at Q_3 and Q_2 respectively. Calculate length C_2Q_3. The co-ordinates of Q_3 are

$$\sigma_2 + (\sigma_1 - \sigma_2)\cos^2\alpha, \quad (\sigma_1 - \sigma_2)\cos\alpha\sin\alpha.$$

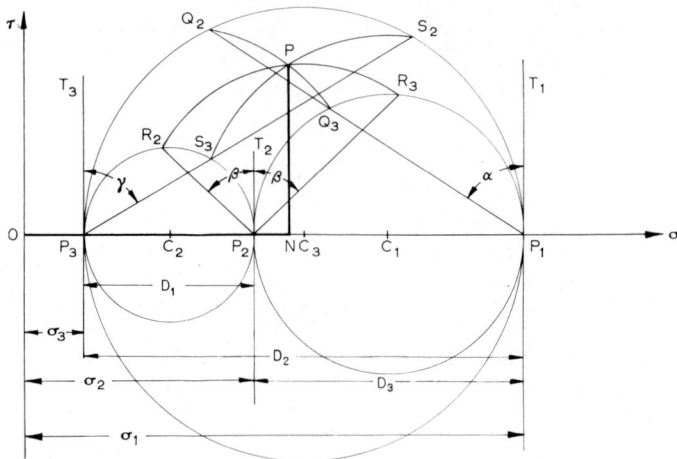

FIG. 2.10 Mohr's circles for three-dimensional stress system.

Hence,

$$(C_2Q_3)^2 = \left[\frac{(\sigma_2 - \sigma_3) + 2(\sigma_1 - \sigma_2)l^2}{2}\right]^2 + \left[(\sigma_1 - \sigma_2)l\sqrt{1 - l^2}\right]^2$$

$$= \left[\left(\frac{\sigma_2 - \sigma_3}{2}\right)^2 + l^2(\sigma_1 - \sigma_2)(\sigma_1 - \sigma_3)\right],$$

i.e., the radius of the circle defined by equation (2.39) is C_2Q_3. Similarly, it follows that $C_2Q_2 = C_2Q_3$.

In the same way, starting from the expressions for m^2 and n^2, σ and τ lie on

1. A circle centre C_3 and radius C_3R_2 or C_3R_3 where R_2 and R_3 are the points of intersection of lines drawn from P_2 at angle β (where $\cos \beta = m$) to P_2T_2 with the circles on P_2P_3 and P_2P_1.

2. A circle centre C_1 and radius C_1S_3 or C_1S_2, where S_2 and S_3 are the points of intersection of a line drawn from P_3 at angle γ (where $\cos \gamma = n$) to P_3T_3, with the circles on P_3P_2 and P_3P_1.

The stresses σ and τ are thus uniquely identified as the point of intersection, P, in Fig. 2.10 and the construction for graphically determining the normal and shear stresses on a plane whose direction cosines are (l, m, n), with respect to the directions of the principal stresses $\sigma_1, \sigma_2, \sigma_3$, is arrived at, thus

1. Plot on the σ-axis, the points P_1, P_2, P_3 such that

$$OP_1 = \sigma_1, \quad OP_2 = \sigma_2, \quad OP_3 = \sigma_3.$$

2. On P_1P_2, P_2P_3, P_3P_1 as diameters, draw circles with centres C_1, C_2, C_3, respectively.

3. At P_1, P_2, P_3 erect perpendiculars P_1T_1, P_2T_2, P_3T_3 to the σ-axis.

4. Draw Q_2P_1 at angle α (where $\cos \alpha = l$) to P_1T_1; draw S_2P_3 at angle γ (where $\cos \gamma = n$), to P_3T_3. Mark the point of intersection of these lines with the circles which pass through the point from which they originate, i.e. mark Q_2 and Q_3 and S_2 and S_3 (see Fig. 2.10).

5. With centre C_1 and radius C_1S_2 draw an arc S_2S_3; with centre C_2 and radius C_2Q_2 draw an arc Q_2Q_3.

6. From P_1 the point of intersection of arcs S_2S_3 and Q_2Q_3, draw PN perpendicular to the σ-axis. Then the normal stress on the plane is represented by ON and the total shear stress by PN.

Suppose it is required to find the normal stress on a plane whose direction cosines are $(0{\cdot}530, 0{\cdot}695, 0{\cdot}500)$, the principal stresses being $\sigma_1 = 5{\cdot}4$ (say, tonf/in²) parallel to Ox, $\sigma_2 = 2{\cdot}45$ parallel to Oy, and $\sigma_3 = 0{\cdot}65$ parallel to Oz. The construction described above as applied to this problem is displayed in Fig. 2.10. It is found that $\sigma = 2{\cdot}80$ and $\tau = 1{\cdot}77$.

The value of the shear stress τ, on the oblique plane is determined in magnitude but not in direction, by the Mohr circle construction, SWIFT (1946). In Fig. 2.11, the force S_1 due to principal stress σ_1 is

$$s_1 = \sigma_1 . \varDelta OBC . \sin R\hat{O}A.$$

R is the foot of the perpendicular from O on to the plane ABC and thus R is also the orthocentre of triangle ABC. Similar expressions for S_2 and S_3 may be written. Figure 2.12 shows the true shape of the oblique triangle ABC; BM is perpendicular to CA. Resolve S_1, S_2 and S_3, parallel and perpendicular to BM. Let the direction of the resultant force, and hence shear stress, define a line BD to which it is perpendicular. After some manipulation, it can be shown that the position of D is such that

$$CD/CA = (\sigma_3 - \sigma_2)/(\sigma_3 - \sigma_1).$$

Thus to decide the direction of the shear stress in plane ABC, the following construction is arrived at:

1. Draw a true shape triangle ABC; the ratio of the lengths of its sides are determined by the direction cosines of the normal from O, i.e., $l: m: n$.

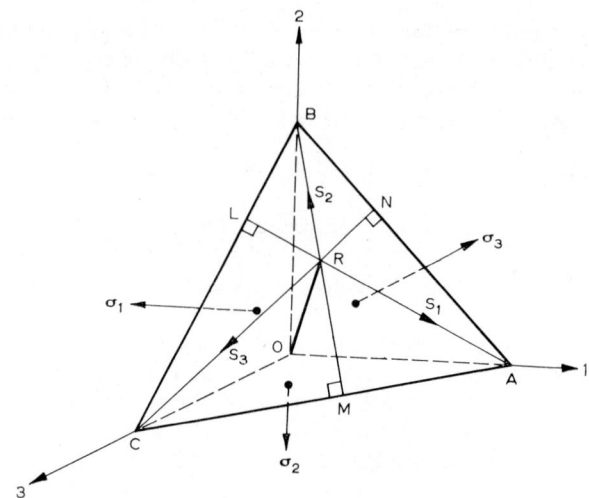

FIG. 2.11 Shear forces on an oblique plane.

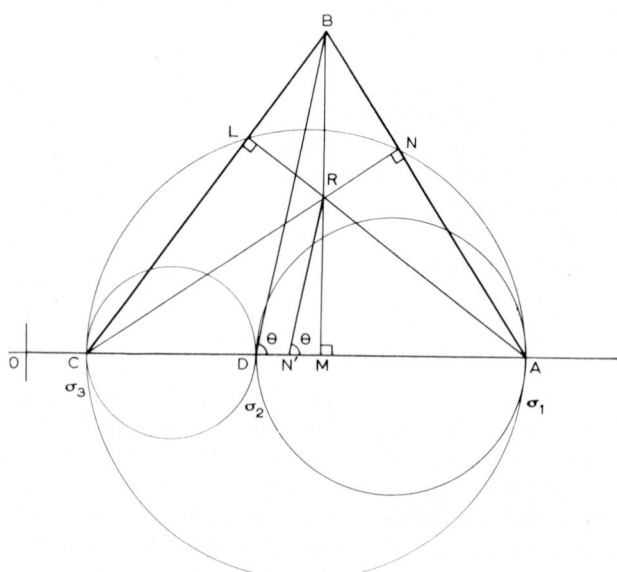

FIG. 2.12 True shape of the triangle ABC, shown in Fig. 2.11, and the construction for determining the direction of shear stress in its plane.

2. On base AC produced, mark off point O such that lengths OC, OD, OA represent the principal stresses σ_3, σ_2, σ_1, to some scale.

3. The direction of the shear stress is perpendicular to BD.

It is worth noting that a line drawn from orthocentre R, parallel to BD, cuts CA in N' and that ON', to the appropriate scale, is the normal stress on plane ABC.

2.6 The Shortcoming of Mohr's Circle for Three-dimensional Stress Systems

If in the two-dimensional stress state shown in Fig. 2.13, σ_x, σ_y and σ_{xy}, are given and if σ denotes a principal stress which is sought, we have directly, by resolving the forces due to the stresses which act on a small element, first in the x-direction and then in the y-direction,

$$(\sigma - \sigma_x).\sin\theta = \sigma_{xy}.\cos\theta$$

and $$(\sigma - \sigma_y).\cos\theta = \sigma_{xy}.\sin\theta.$$

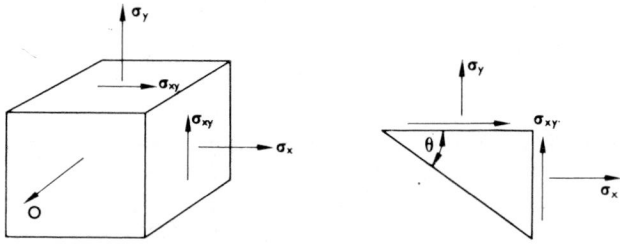

FIG. 2.13 State of plane stress.

Eliminating θ,

$$(\sigma - \sigma_x)(\sigma - \sigma_y) = \sigma_{xy}^2$$

or $$\sigma^2 - \sigma(\sigma_x + \sigma_y) + \sigma_x\sigma_y - \sigma_{xy}^2 = 0. \qquad (2.40)$$

This equation can, of course, be obtained immediately from equation (2.16) by putting σ_z and σ_{yz}, σ_{zx} equal to zero. Thus, to find the principal stresses, the roots of this equation must be found. This can be done in the classical way using rule and compasses. In Fig. 2.14, along Ox, mark point C where $OC = (\sigma_x + \sigma_y)/2$ and then describe a circle on OC as diameter. Further, from centre O describe a circle of radius $(\sigma_x\sigma_y - \sigma_{xy})^{\frac{1}{2}}$ to intersect the first circle in P_1 and P_2. Draw a circle centre C and radius CP_1 to cut OC produced at S_1 and S_2. Then the principal stresses are $\sigma_1 = OS_1$ and $\sigma_2 = OS_2$. This is a simple graphical construction for finding the roots of the quadratic equation.

It is not possible simply to extend this method and develop an approach for finding the three principal stresses when all the six elements of stress at a point are given (none zero). As shown earlier in the chapter, to find the three principal stresses a cubic equation must be solved. However, as the Greek geometers well knew, using only 'rule and compasses', this is not possible. (See, DERRINGTON and JOHNSON, 1960).

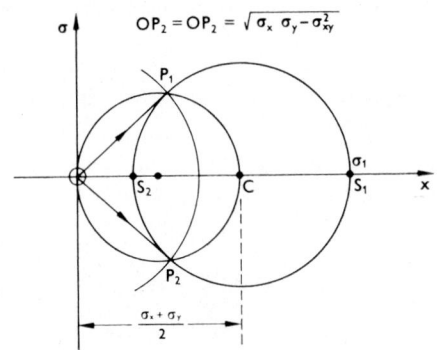

FIG. 2.14 Graphical construction for determining the principal stresses.

REFERENCES

DERRINGTON, M. G. and JOHNSON, W. 1960 'The Shortcoming of Mohr's Circle for Three-dimensional Stress' *Bull. mech. Engng Educ.* No. 18 (July), 1960. (University of Manchester Institute of Science and Technology publication)

SWIFT, H. W. 1946 'Plastic Strain in an Isotropic Strain Hardening Material' *Engineering*, **162**, 381

Chapter 3

STRAIN

The study of strain is the study of the displacement of points in a body relative to one another when the body is deformed. It is not concerned with rigid body movements.

Infinitesimal strain components will be considered first. Such components are applicable to the elastic strains in engineering materials and also to strain-increments in the general case of finite straining. In the latter type of straining, lines which are initially straight may become curved, and the equations relating the strains are so complicated as to be of little practical use. A simple case of finite straining will be considered. This is the homogeneous straining of a straight line. Strain is defined as being homogeneous if all lines which are initially straight remain straight after straining, and lines which are initially parallel remain parallel.

3.1 Infinitesimal Strains as Functions of Displacement

Figure 3.1 shows two neighbouring points $A(x, y, z)$ and $B(x + \delta x, y + \delta y, z + \delta z)$, in a body which is to be deformed so that A is displaced to $A'(x + u, y + v, z + w)$ and B to $B'(x + \delta x + u + \delta u, y + \delta y + v + \delta v, z + \delta z + w + \delta w)$. u, v, w are the projections on the axes of the displacement of A and $u + \delta u, v + \delta v, w + \delta w$ of B.

Assuming that u is a continuous function of (x, y, z), i.e.,

$$u = u(x, y, z) \tag{3.1}$$

then

$$\delta u = \frac{\partial u}{\partial x} . \delta x + \frac{\partial u}{\partial y} . \delta y + \frac{\partial u}{\partial z} . \delta z \tag{3.2}$$

if the powers of the coefficients of δx, δy, δz higher than the first can be neglected; this is the case for small displacements. $\partial u/\partial x . \delta x$ is the alteration in the original length δx or AC; hence $\partial u/\partial x$ is the direct strain in the direction Ox at A which is denoted by e_{xx}. This nomenclature may be interpreted as the rate of movement in the Ox direction (first suffix to e) of a point on a line parallel to Ox (which gives the second suffix to e) at A. $\partial u/\partial y$ is the rotation of CD, in plane yOx due to movement u; or, alternatively, the rate of slide or shear, of planes perpendicular to Oy parallel to Ox and which can be denoted by e_{xy}. $\partial u/\partial y$ may also be said to be the angular strain of CD, the rate of movement in the Ox direction (first suffix to e) of a point on a line parallel to Oy (second suffix to e) at A. Similarly $\partial u/\partial z$ is the rotation or angular strain of BD in a plane parallel to zOx, which may be denoted by e_{xz}.

Similar expressions and meanings may be obtained in connection with δv and δw.

It is necessary at this point to define a convention for deciding when a rotation shall be regarded as positive or negative. We adopt a right-hand screw rule. Looking out from the origin along a given axis, the direction in

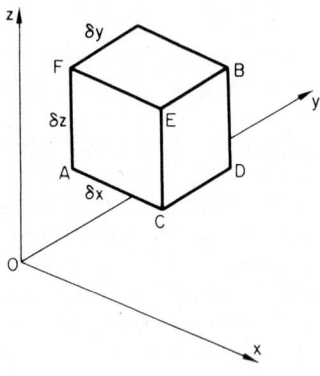

Fig. 3.1

which a right-handed screw would rotate is counted as positive, e.g. in plane yOx, counter-clockwise, or from Ox to Oy, if looking from the origin along Oz.

The change in right angle FAC, see Fig. 3.2(a), i.e. in planes perpendicular to Oy, is called the engineering shear strain and designated ϕ_{zx}, the two suffixes referring to the plane concerned. It follows that

$$\phi_{zx} = \frac{\partial u}{\partial z} + \frac{\partial w}{\partial x}. \tag{3.3}$$

Clearly this expression is composed of two slides; it consists of the relative sliding, parallel to the axes x and z of planes perpendicular to the other; in the notation above

$$e_{xz} + e_{zx} = \phi_{zx}.$$

The small strain of the block of Fig. 3.1 is thus defined by three direct strains and three shear strains:

$$e_{xx} = \frac{\partial u}{\partial x}; \qquad e_{yy} = \frac{\partial v}{\partial y}; \qquad e_{zz} = \frac{\partial w}{\partial z};$$

$$\phi_{yz} = \frac{\partial w}{\partial y} + \frac{\partial v}{\partial z}; \qquad \phi_{zx} = \frac{\partial u}{\partial z} + \frac{\partial w}{\partial x}; \qquad \phi_{xy} = \frac{\partial v}{\partial x} + \frac{\partial u}{\partial y}. \qquad (3.4)$$

In Fig. 3.2(a) the line AE where EAC is 45° is rotated through angle ω_y to take up direction $A'E'$ and thus

$$\tan\left(\frac{\pi}{4} - \omega_y\right) = \left(1 + \frac{\partial w}{\partial x}\right)\Big/\left(1 + \frac{\partial u}{\partial z}\right) \simeq \left(1 + \frac{\partial w}{\partial x}\right)\left(1 - \frac{\partial u}{\partial z}\right)$$

$$\frac{1 - \omega_y}{1 + \omega_y} \simeq \left(1 + \frac{\partial w}{\partial x} - \frac{\partial u}{\partial z}\right) \quad \text{or} \quad \omega_y \simeq \tfrac{1}{2}\left(\frac{\partial u}{\partial z} - \frac{\partial w}{\partial x}\right) = \tfrac{1}{2}(e_{xz} - e_{zx}).$$

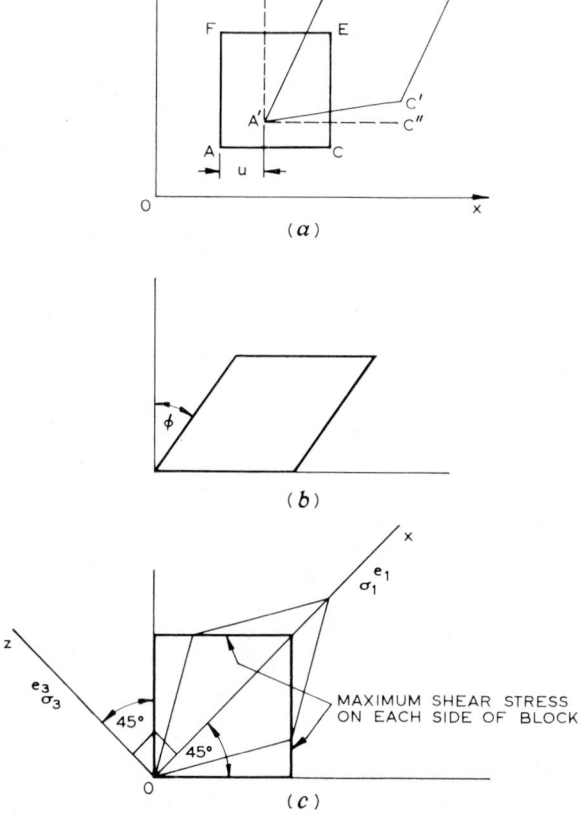

(a)

(b)

MAXIMUM SHEAR STRESS
ON EACH SIDE OF BLOCK

(c)

FIG. 3.2 Direct and shear strains in plane xz

The quantity ω_y is thus to a first approximation the mean angle through which the volume is rotated in an anti-clockwise (i.e. positive) sense about the Oy axis. AC rotates anti-clockwise through angle $C''A'C'$, i.e. $-\partial w/\partial x$, and AF through angle $F''A'F'$, i.e. $+ \partial u/\partial z$. Similar expressions for ω_x and ω_z are

$$\omega_x = \tfrac{1}{2}\left(\frac{\partial w}{\partial y} - \frac{\partial v}{\partial z}\right) \quad \text{and} \quad \omega_z = \tfrac{1}{2}\left(\frac{\partial v}{\partial x} - \frac{\partial u}{\partial y}\right)$$
$$= \tfrac{1}{2}(e_{zy} - e_{yz}) \qquad\qquad = \tfrac{1}{2}(e_{yz} - e_{xy})$$

$$\left.\phantom{\begin{array}{c}a\\a\end{array}}\right\} \quad (3.5)$$

When $\omega_x = \omega_y = \omega_z = 0$, the strain is irrotational, and then $e_{zx} = e_{xz}$, etc. Also in this case, $e_{xz} = \phi_{xz}/2$.

It will be observed that the three component displacements serve to define the six strain components in equation (3.4) and hence the latter cannot be independent. It may be verified that the strain components are related through six equations, three of the type

$$\frac{\partial^2 e_{xx}}{\partial y^2} + \frac{\partial^2 e_{yy}}{\partial x^2} = \frac{\partial^2 \phi_{xy}}{\partial x\, \partial y} \tag{3.6}$$

and three of the type

$$2\frac{\partial^2 e_{zz}}{\partial x\, \partial y} = \frac{\partial}{\partial z}\left(\frac{\partial e_{yz}}{\partial x} + \frac{\partial \phi_{zx}}{\partial y} - \frac{\partial \phi_{xy}}{\partial z}\right). \tag{3.7}$$

These are the conditions of compatibility for strain.

In irrotational cases, in Fig. 3.2(a), $A'E'$ remains parallel to AE. The state of strain in which there is only change of shape is one of pure shear. If the whole block was rotated clockwise through angle e_{zx}, the familiar engineering manner of depicting shear strain is arrived at, see Fig. 3.2(b). Then the angle of shear, ϕ or $\phi_{xz} = e_{zx} + e_{xz} = 2e_{zx}$. The common engineering picture for defining shear strain thus implicitly depicts an anti-clockwise rigid-body rotation of angle equal to half the shearing strain ϕ. ϕ is, in effect, the sum of the angular strains of two lines lying initially in the directions of the principal strains. See also Fig. 3.2(c).

The following relations may be obtained as exercises,

1. that to a second approximation

$$e_{xx} = \frac{\partial u}{\partial x} + \frac{1}{2}\left\{\left(\frac{\partial u}{\partial x}\right)^2 + \left(\frac{\partial v}{\partial x}\right)^2 + \left(\frac{\partial w}{\partial x}\right)^2\right\}$$

and

$$\phi_{xy} = \frac{\partial v}{\partial x} + \frac{\partial u}{\partial y} + \frac{\partial u}{\partial x}\cdot\frac{\partial u}{\partial y} + \frac{\partial v}{\partial x}\cdot\frac{\partial v}{\partial y} + \frac{\partial w}{\partial x}\cdot\frac{\partial w}{\partial y},$$

2. the cubical dilation Δ (alteration in volume divided by original volume) is

$$\Delta = \frac{\partial u}{\partial x} + \frac{\partial v}{\partial y} + \frac{\partial w}{\partial z}$$

3. $du = e_{xx}dx + e_{yy}dy/2 + e_{zz}dz/2 - \omega_x dy + \omega_y dz$, with corresponding values for dv and dw.

3.2 A Note on the Strain Tensor

Reference was made to the use of the suffix notation and the stress tensor in the previous chapter. The same suffix notation is applied to the strain tensor.

The symbol e_{ij} is introduced and used to define the nine components of strain, e_{xx}, e_{yy}, e_{zz}, e_{yz}, e_{zx}, e_{xy}, derived in Section 3.1. For example, $e_{xx} = \partial u/\partial x$ and $e_{xy} = \partial u/\partial y$ so that, in general, $e_{ij} = \partial u_i/\partial x_j$; where u_i denotes displacement in the 'i' direction (i.e., x, y or z) and x_j refers to an axis (Ox, Oy or Oz). Write,

$$e_{ij} = \epsilon_{ij} + \omega_{ij} \tag{3.8}$$

where $\epsilon_{ij} = (e_{ij} + e_{ji})/2$ and $\omega_{ij} = (e_{ij} - e_{ji})/2$. (See equations (3.4) and (3.5).)

In particular, $e_{yx} = (e_{yx} + e_{xy})/2 + (e_{yx} - e_{xy})/2$

$$= \left(\frac{\partial v}{\partial x} + \frac{\partial u}{\partial y}\right)\Big/2 + \left(\frac{\partial v}{\partial x} - \frac{\partial u}{\partial y}\right)\Big/2$$

$$= \epsilon_{yx} + \omega_{yx} \equiv \epsilon_{yx} + \omega_z$$

and, simply, $e_{xx} = \left(\dfrac{\partial u}{\partial x} + \dfrac{\partial u}{\partial x}\right)\Big/2 + \left(\dfrac{\partial u}{\partial x} - \dfrac{\partial u}{\partial x}\right)\Big/2$

$$= \epsilon_{xx}$$

For convenience, when $i \neq j$, instead of ϵ_{ij} we may sometimes write γ_{ij}.

It is clear that the ϵ_{ij} portion of e_{ij} in equation (3.8) provides a statement about the strain at a point and ω_{ij} a statement about rotation. ϵ_{ij} is the strain tensor and, because $\epsilon_{ij} = \epsilon_{ji}$, it is said to be symmetrical. On the other hand, $\omega_{ij} = -\omega_{ji}$ and hence is said to be anti-symmetrical. Thus we may write out the array for the strain tensor as

$$
\begin{array}{ccc}
\epsilon_{xx} & \gamma_{xy} & \gamma_{xz} \\
\gamma_{yx} & \epsilon_{yy} & \gamma_{yz} \\
\gamma_{zx} & \gamma_{zy} & \epsilon_{zz}
\end{array}
\tag{3.9}
$$

where, it will be remembered, γ is equal to half the conventional engineering shear strain ϕ.

It is instructive to represent the above symbols diagrammatically. This is done for a two-dimensional case in Fig. 3.3. A special case to which attention may be drawn concerns the definitions of simple shear and pure shear: these will be obvious from Fig. 3.4.

3.3 The Geometry of Large and Small Strains

Consider a line OP of unit length L_0, Fig. 3.5, having direction cosines l, m, n relative to three mutually perpendicular axes, $O1$, $O2$ and $O3$, in a material which is subject to a system of forces which causes lineal strain, e_1, e_2 and e_3 in the directions of the axes so that P moves to P'. The length of the unstrained unit line L_0 as projected on to the axes is l, m, n and the latter become, on straining $l(1 + e_1)$, $m(1 + e_2)$, $n(1 + e_3)$; e_1, e_2, e_3 need not be small, and e

means the increase in length per unit original length. The strained unit length denoted by L is given by

$$L^2 = l^2(1 + e_1)^2 + m^2(1 + e_2)^2 + n^2(1 + e_3)^2 \qquad (3.10)$$

$$= l^2 l_1^2 + m^2 m_1^2 + n^2 n_1^2 \qquad (3.11)$$

where, $\qquad l_1 = 1 + e_1, \qquad m_1 = 1 + e_2, \qquad n_1 = 1 + e_3.$

Projecting the strained length OP' along the original direction OP and calling it L_0', we have

$$L'_0 = l^2 l_1 + m^2 m_1 + n^2 n_1 \qquad (3.12)$$

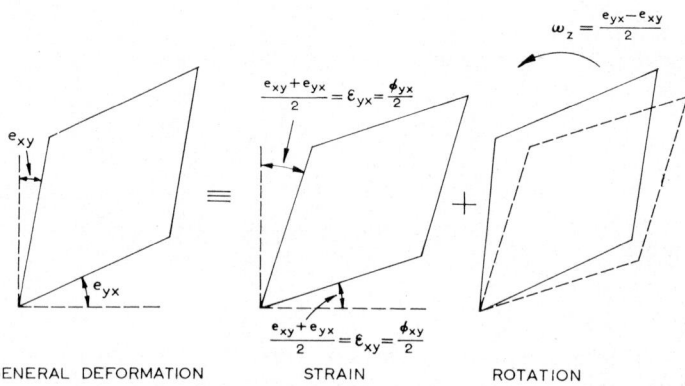

GENERAL DEFORMATION   STRAIN   ROTATION

FIG. 3.3 General deformation consists of shear strain and a rigid body rotation.

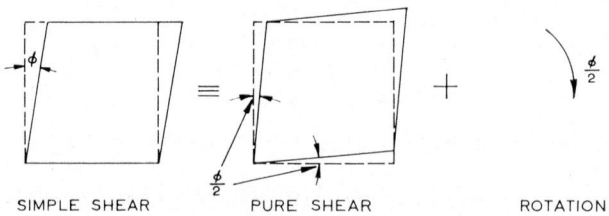

SIMPLE SHEAR   PURE SHEAR   ROTATION

FIG. 3.4 Difference between simple and pure shear.

By reference to Section 2.5 we see that if l_1, m_1, n_1, i.e., strained unit lengths parallel to $O1$, $O2$ and $O3$, are identified with principal stresses σ_1, σ_2, σ_3, then on the doubly oblique plane whose direction cosines are l, m, n, with resultant stress S, where $S^2 = \sigma^2 + \tau^2$, is to be identified with strained length L and normal stress σ with the strained length L'_0. *Thus equations* (3.11) *and* (3.12) *show that the Mohr circle diagram may be applied generally to the comparison of strained lengths in material. This is a method of dealing with finite strains using Mohr circles.*

The linear strain in the original direction of the unit line is, from equation (3.12)

$$e = L'_0 - 1 = e_1 l^2 + e_2 m^2 + e_3 n^2 \tag{3.13}$$

If θ is small the rotation of the unit line, see Fig. 3.6, is given by

$$e^2 + \theta^2 = (PP')^2 = e_1^2 l^2 + e_2^2 m^2 + e_3^2 n^2 \tag{3.14}$$

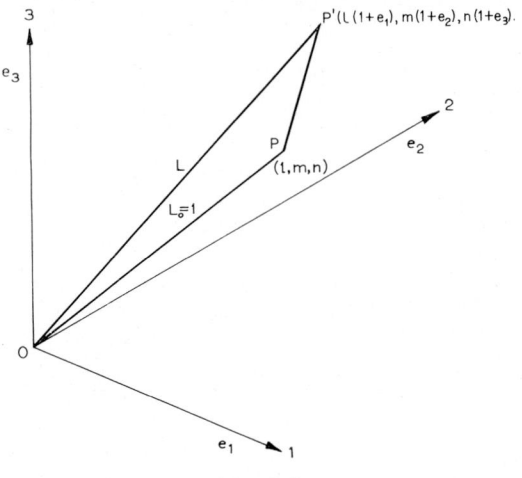

FIG. 3.5

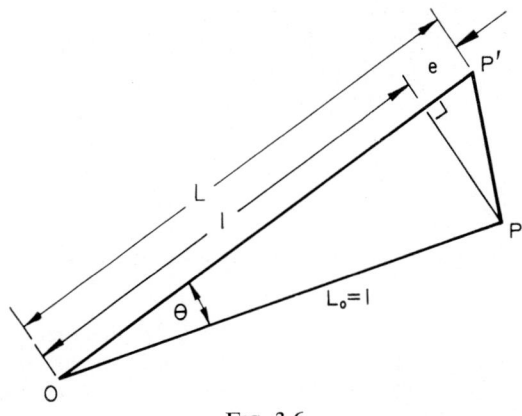

FIG. 3.6

The expressions for e and θ are analogous to those for normal stress, σ, and shear stress, τ, in the stress equations and *show that the Mohr circle diagram can be applied to strains, as distinct from strained lengths, provided the strains are small.* In Fig. 3.7, if OP_1, OP_2 and OP_3 represent strains e_1, e_2 and e_3, then for a unit length whose direction cosines are initially l, m, n, its linear strain is e, represented by $O'N$ and its rotation, θ, by NP.

By pursuing the analogy, see Fig. 2.12, with respect to shear stress on a doubly oblique plane, it may be shown that the resultant rotation or angular strain of the initial unit length line will be in a direction perpendicular to BD where $CD/CA = (e_2 - e_3)/(e_1 - e_2)$. There is thus a complete analogy between the circle diagrams for stress and for small strains, lineal strain corresponding to normal stress and angular strain to tangential stress.

It is possible to show that one set of circles suffices to facilitate the determination of stress and strain on any doubly oblique plane. From elementary elastic theory,

$$Ee_1 = \sigma_1 - v(\sigma_2 + \sigma_3)$$
$$= \sigma_1(1 + v) - 3v\sigma_m$$

where

$$3\sigma_m = \sigma_1 + \sigma_2 + \sigma_3.$$

Hence,

$$e_1 = \frac{\sigma_1}{2G} - \frac{3v\sigma_m}{E}$$

$$= \left[\sigma_1 - \frac{3K - 2G}{3K} \sigma_m \right] \bigg/ 2G.$$

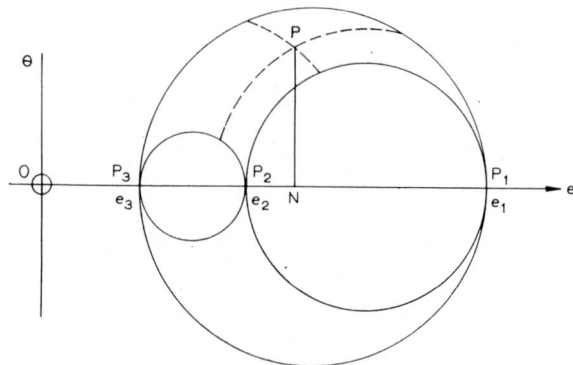

Fɪɢ. 3.7 Mohr's circles for small strains.

Hence, take a new origin O', at $(3K - 2G)\sigma_m/3K$ distance from the origin for stresses and use it as the new origin for strains. The lineal strains are then obtained by multiplying the normal stresses by $1/2G$. And since

$$\sigma_1 - \sigma_2 = 2G(e_1 - e_2)$$

the angular strains corresponding to τ are also obtained by multiplying by $1/2G$.

Chapter 4

THE YIELD CRITERIA OF METALS

4.1 General Considerations

A yield criterion is a hypothesis concerning the limit of elasticity *under any possible combination of stresses*. The suitability of any proposed yield criterion must be checked by experiment.

Consider an element of ductile material subject to the principal stresses σ_1, σ_2, σ_3 and represent them by a Mohr circle diagram, as in Fig. 2.10. If the principal stresses had been $(\sigma_1 + \sigma_m)$, $(\sigma_2 + \sigma_m)$ and $(\sigma_3 + \sigma_m)$, the Mohr circle system would have consisted of circles of the same size as in the previous case, the two differing only in that the circle system of the latter had been shifted a distance σ_m further along the σ-axis; additional stresses σ_m make up a hydrostatic (tensile or compressive) stress system. It is found that the absolute sizes of the Mohr circles alone determine the limit of elasticity at ordinary industrial presures and temperatures and these are independent of their position along the σ-axis. This is to say that under common working conditions, the yield criterion is a function of $(\sigma_1 - \sigma_2)$, $(\sigma_2 - \sigma_3)$ and $(\sigma_3 - \sigma_1)$, or equivalently—$(\sigma_1 + \sigma_m - \sigma_2 + \sigma_m)$, etc. The yield criterion is independent of the hydrostatic stress component, $(\sigma_1 + \sigma_2 + \sigma_3)/3$ or $\sum \sigma_1/3$ (The symbol $\sum$ is used to mean the sum of all similar terms, i.e. $\sum \sigma_1 \equiv \sigma_1 + \sigma_2 + \sigma_3$ and $\sum (\sigma_1 - \sigma_2)^2 \equiv (\sigma_1 - \sigma_2)^2 + (\sigma_2 - \sigma_3)^2 + (\sigma_3 - \sigma_1)^2.$)

Thus yielding occurs when some scalar function of the principal stress differences reaches a critical magnitude. Generally this may be written as

$$f(\overline{\sigma_1 - \sigma_2}, \overline{\sigma_2 - \sigma_3}, \overline{\sigma_3 - \sigma_1}) = \text{constant}. \qquad (4.1)$$

Since only isotropic materials are to be considered, no more 'weight' should be given to, say, $(\sigma_1 - \sigma_2)$ than $(\sigma_2 - \sigma_3)$ or $(\sigma_3 - \sigma_1)$. By a scalar function we mean here a function which, when particular magnitudes are given to the component variables, enables a quantity to be calculated. When this quantity attains a certain magnitude, the stress condition is such that yielding will occur.

Two approaches to the interpretation of a yield criterion are possible; a purely mathematical (or statistical) approach and one which seeks to provide a physical justification.

Perhaps the simplest function imaginable which satisfies equation (4.1) is one of the form

$$|\sigma_3 - \sigma_1| = \text{constant}. \qquad (4.2)$$

Of the three magnitudes $|\sigma_1 - \sigma_2|$, $|\sigma_2 - \sigma_3|$, $|\sigma_3 - \sigma_1|$ yielding must occur when the largest one of them attains the critical, and—for a given material—constant value.

Another admissible function that may come to mind is

$$(\sigma_1 - \sigma_2)^2 + (\sigma_2 - \sigma_3)^2 + (\sigma_3 - \sigma_1)^2 = \text{constant}. \tag{4.3}$$

The incorporation of $(\sigma_1 - \sigma_2)$, etc. as quadratic terms, dispenses with the need to use the modulus sign. As in statistical work, 'squares' are easier to deal with than algebraic quantities. In this type of function, each of the principal stresses contributes to yielding.

On the purely mathematical level, that hydrostatic or volumetric stress does not affect yielding is equivalent to saying that I_1 is of no physical importance as regards yield, whilst $J_2 = \text{constant}$ is a possible yield criterion, since

$$\sum(\sigma_1 - \sigma_2)^2 = 3\sum(\sigma_1 - I_1/3)^2 = 3\sum\sigma_1'^2$$

$$= 3\left[(\sum\sigma_1')^2 - 2\sum\sigma_1'\sigma_2'\right] = 6J_2.$$

See equations (2.22) and (2.29).

Equation (4.2) is TRESCA's criterion given in 1864. Equation (4.3) was proposed by HUBER (1904), VON MISES (1913) and by J. C. Maxwell in a letter to Kelvin in 1856! Tresca's criterion is usually stated as: 'yielding occurs when the greatest absolute value of any one of the three maximum shear stresses in the material reaches a certain value'. The Huber–Mises criterion was interpreted by HENCKY to mean that yielding began when the shear strain energy reached a critical value.

Other forms of yield criterion are perfectly possible; the two presented are the simplest and most easily applied.

A yield criterion is held to be correct for any combination of stresses and therefore the constants in equations (4.2) and (4.3) may be determined from simple stress states. It is usual to identify the constants with the tensile yield stress Y, or the yield shear stress k for a state of pure shear.

At yielding in simple tension $\sigma_1 = Y$, $\sigma_2 = \sigma_3 = 0$ and the constants in equations (4.2) and (4.3) are therefore Y and $2Y^2$, respectively. For pure shear $\sigma_1 = -\sigma_3 = k$ and the intermediate principal stress $\sigma_2 = 0$, giving constants as $2k$ and $6k^2$. Equations (4.2) and (4.3) may then be written as

$$|\sigma_3 - \sigma_1| = Y = 2k. \tag{4.4}$$

$$(\sigma_1 - \sigma_2)^2 + (\sigma_2 - \sigma_3)^2 + (\sigma_3 - \sigma_1)^2 = 2Y^2 = 6k^2. \tag{4.5.i}$$

It will be noticed from the Mises criterion, equation (4.5.i), that

$$k = \frac{2}{\sqrt{3}}\left(\frac{Y}{2}\right).$$

This predicts that the yield shear stress in pure torsion, k, is greater than the maximum elastic shear stress in simple tension, $Y/2$, by a factor 1·155, whereas the Tresca criterion, equation (4.4), predicts that these two quantities are equal.

4.2 Stress Space Representations of Yield Criteria

In Fig. 4.1(a), are shown three mutually perpendicular axes, $O\sigma_1$, $O\sigma_2$ and $O\sigma_3$. Each of the three axes is devoted to one particular principal stress. If the principal stresses at a point in a body are (σ_1, σ_2, σ_3), this state or stress system is represented by a point P in this 'stress space', the co-ordinates of P being σ_1, σ_2, σ_3. The state may be written as the sum of the three vectors $OP_1(=\sigma_1)$, $P_1M(=\sigma_2)$ and $MP(=\sigma_3)$.

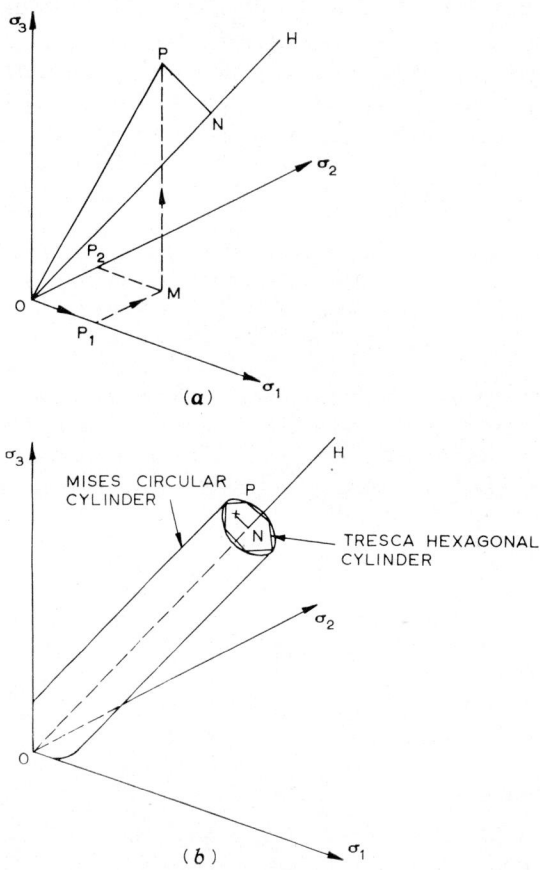

FIG. 4.1 Stress space representations of yield criteria.

Consider a line OH, equally inclined to each of the three axes, i.e., the direction cosines of OH are

$$(1/\sqrt{3},\quad 1/\sqrt{3},\quad 1/\sqrt{3}).$$

Project OP on this line to become ON. Then

$$ON = \sigma_1 \cdot \frac{1}{\sqrt{3}} + \sigma_2 \cdot \frac{1}{\sqrt{3}} + \sigma_3 \cdot \frac{1}{\sqrt{3}},$$

and $\qquad PN^2 = OP^2 - ON^2$

$$= \sum \sigma_1^2 - (\sum \sigma_1)^2/3$$

$$= \sum \sigma_1^2 - (\sum \sigma_1^2 + 2\sum \sigma_1 \sigma_2)/3$$

$$= \tfrac{1}{3}\sum(\sigma_1 - \sigma_2)^2.$$

Now, the von Mises criterion states that yielding occurs when $\sum(\sigma_1 - \sigma_2)^2$ attains a certain value, $2Y^2$. Thus yielding, or the yield locus as it is most frequently referred to, may be represented by a circle of radius $PN = Y\sqrt{2/3}$. It follows that the von Mises yield criterion may be represented in principal stress space by a circular cylinder of radius $Y\sqrt{2/3}$ whose axis is the line through the origin equally inclined to the axes of co-ordinates, see Fig. 4.1(b). States of stress which plot as points within the cylinder are completely elastic states of the material element. 'Nearness' to yielding is thus determined by the distance of the stress point from the line OH, e.g., PN. Yielding is un-affected by the magnitude of ON which represents a hydrostatic state of stress. ON is the vector sum of the spherical components and NP is the vector sum of the deviatoric components of the stress. The intention of the second term is clear—the amount by which the 'whole' stress 'deviates' from the mean normal stress.

To show directly that the Mises criterion as represented in principal stress space is a circular cylinder about OH as axis, choose a new set of three mutually perpendicular axes, $O\sigma_1''$, $O\sigma_2''$, $O\sigma_3''$ their direction cosines with respect to the original axes being $(1/\sqrt{3}, 1/\sqrt{3}, 1/\sqrt{3})$, (l_1, m_1, n_1), (l_2, m_2, n_2).

Then $\qquad\qquad \sigma_1 = \dfrac{\sigma_1''}{\sqrt{3}} + \sigma_2'' l_1 + \sigma_3'' l_2,$

$$\sigma_2 = \dfrac{\sigma_1''}{\sqrt{3}} + \sigma_2'' m_1 + \sigma_3'' m_2,$$

and $\sigma_3 = \dfrac{\sigma_1''}{\sqrt{3}} + \sigma_2'' n_1 + \sigma_3'' n_2.$

Thus substituting in equation (4.5), i.e.

$$2Y^2 = (\sigma_1 - \sigma_2)^2 + (\sigma_2 - \sigma_3)^2 + (\sigma_3 - \sigma_1)^2$$

we obtain $\quad 2Y^2 = [\sigma_2''(l_1 - m_1) + \sigma_3''(l_2 - m_2)]^2$
$$+ [\sigma_2''(m_1 - n_1) + \sigma_3''(m_2 - n_2)]^2$$
$$+ [\sigma_3''(n_1 - l_1) + \sigma_2''(n_2 - l_2)]^2 = 3(\sigma_2''^2 + \sigma_3''^2),$$

the coefficient of σ_2'', σ_3'' in the expansion being zero. Thus by the aid of a change of axes, the Mises criterion is demonstrated directly to be a circular cylinder of constant cross-section and radius $Y\sqrt{2/3}$.

The Tresca criterion may be written $|\sigma_3 - \sigma_1| = Y$ and when drawn in principal stress space it is represented by a regular hexagonal cylinder which is conventionally inscribed within the Mises cylinder, see Fig. 4.1(b). This is best demonstrated by considering the projection of the stress states, such as P

in Fig. 4.1(a) on to a plane through the origin and perpendicular to OH. This plane is called the π-plane or the synoptic plane and in the original principal stress space has the equation

$$\sigma_1 + \sigma_2 + \sigma_3 = 0.$$

This means that hydrostatic stresses are neglected, a fact that is immaterial here since it has been assumed that hydrostatic states of stress do not enter into the yield criteria. The positive principal axes $O\sigma_1$, $O\sigma_2$, $O\sigma_3$ appear in the π-plane, Fig. 4.2(a), inclined at 120° to each other. When the negative axes of principal stress are included, as broken lines in Fig. 4.2(a), the whole area or plane is divided into six equal sectors.

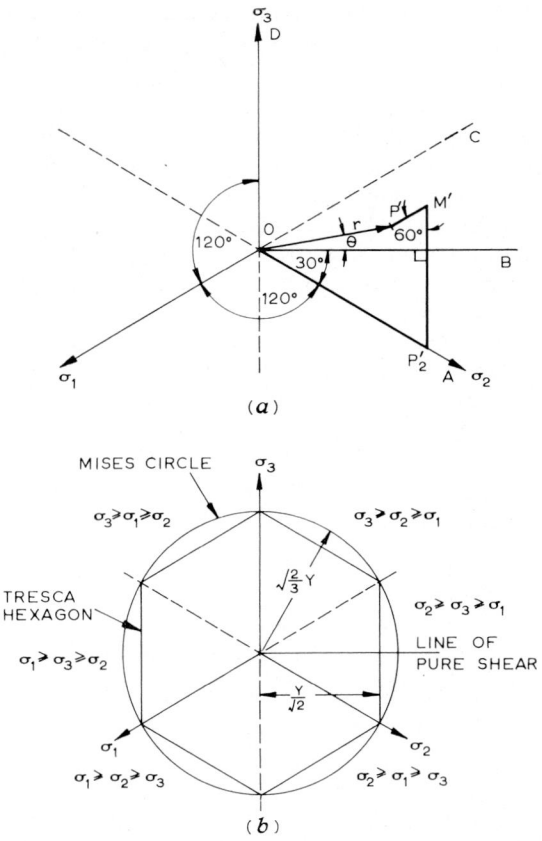

FIG. 4.2 Representation of yield criteria on π-plane.

The co-ordinates of P in Fig. 4.1(a) appear in Fig. 4.2(a) as the projected lengths $OP_2' = \sqrt{(2/3)}\sigma_2$, $P_2'M' = \sqrt{(2/3)}\sigma_3$, $M'P' = \sqrt{(2/3)}\sigma_1$. If the Cartesian co-ordinates of P' with respect to OB and OD are (x, y), then from Fig. 4.2(a)

$$x = \frac{\sigma_2 - \sigma_1}{\sqrt{2}}, \quad y = \frac{2\sigma_3 - \sigma_2 - \sigma_1}{\sqrt{6}}. \tag{4.6}$$

Alternatively, if the polar co-ordinates of P' are (r, θ), then

$$
\left.
\begin{aligned}
r &= \sqrt{x^2 + y^2} = \frac{1}{\sqrt{3}} \sqrt{(\sigma_1 - \sigma_2)^2 + (\sigma_2 - \sigma_3)^2 + (\sigma_3 - \sigma_1)^2} \\
\text{and} \quad \theta &= \tan^{-1}\left[\frac{1}{\sqrt{3}}\left(\frac{2\sigma_3 - \sigma_2 - \sigma_1}{\sigma_2 - \sigma_1}\right)\right]
\end{aligned}
\right\} \quad (4.7)
$$

Length r of equation (4.7) is immediately identified with the von Mises criterion of yielding, when $r = Y\sqrt{2/3}$, as obtained earlier.

The stress state represented in sector COA, Fig. 4.2(a), is such that $\sigma_2 \geqslant \sigma_3 \geqslant \sigma_1$ and the Tresca criterion of yielding is then $\sigma_2 - \sigma_1 = Y$. Thus the Tresca criterion of yielding in sector COA is represented by a line parallel to the $O\sigma_3$ axis and at a distance $x = Y/\sqrt{2}$ from it. The complete yield locus is a regular hexagon as shown in Fig. 4.2(b). Considering Fig. 4.2(b), it is clear that according to the Mises criterion any point which falls inside the Mises circle represents an elastic state of stress and points on the circumference represent yielding. If the material is a non-hardening one, yielding is always represented by a point on the circumference; the stress point cannot move outside the circle. Unloading is represented by the path of a stress point which moves inside the circle away from the circumference. A similar interpretation is given to the Tresca hexagon.

Note that in obtaining these yield loci both yield criteria have been identified with Y, the tensile yield stress. Thus the Mises circle passes through the corners of the Tresca hexagon. The loci then differ most for pure shear where the Mises criterion gives a yield stress $2/\sqrt{3}$ times that given by the Tresca criterion.

LODE (1926) in his experimental investigations of the yield criterion introduced the parameter

$$
\mu = \frac{2\sigma_3 - \sigma_1 - \sigma_2}{\sigma_1 - \sigma_2} \tag{4.8}
$$

to characterize the influence of the intermediate principal stress. It can be seen from equation (4.7) that on the π-plane

$$
\mu = -\sqrt{3}\tan\theta. \tag{4.9}
$$

When $\mu = 0$ the stresses are σ_1, σ_2 and $\sigma_3 = (\sigma_1 + \sigma_2)/2$ which is a pure shear—deviators $(\sigma_1 - \sigma_2)/2$, $(\sigma_2 - \sigma_1)/2$, 0—and a hydrostatic stress $(\sigma_1 + \sigma_2)/2$. This stress state is represented on the π-plane by putting $\theta = 0$, Fig. 4.2(b) When $\mu = -1$ the stresses are σ_1, $\sigma_2 = \sigma_3$ which is a uniaxial stress $(\sigma_1 - \sigma_2)$ together with a hydrostatic stress σ_2. On the π-plane this stress state gives $\theta = 30°$. In the experimental determination of the yield criterion it is sufficient to investigate stress conditions between $\mu = 0$ and $\mu = -1$, that is for a 30° sector on the π-plane. Lode's parameter is discussed further in Section 5.8.

In a case of plane stress, $\sigma_2 = 0$, the yield criteria are sometimes represented with σ_1 and σ_3 as rectangular co-ordinate axes. The yield loci shown in Fig. 4.3 represent the curves of intersection of the Mises circle and the Tresca regular hexagon with the $\sigma_2 = 0$ plane. The equation of the Mises ellipse, obtained by putting $\sigma_2 = 0$ in equation (4.5) is

$$\sigma_1^2 + \sigma_3^2 - \sigma_1\sigma_3 = Y^2$$

whose major axis is $2Y\sqrt{2}$, minor axis $2Y\sqrt{2/3}$, these axes being the perpendicular bisectors of the angles between the σ_1 and σ_3 axes.

The Tresca hexagonal cylinder, whose boundary of intersection is shown in Fig. 4.3, is plotted from the conditions

(i) $|\sigma_1 - \sigma_3| = Y$ (lines AF and CD)

(ii) $|\sigma_2 - \sigma_3| = Y$ and $\therefore |\sigma_3| = Y$ (lines AB and DE)

(iii) $|\sigma_2 - \sigma_1| = Y$ and $\therefore |\sigma_1| = Y$ (lines BC and EF)

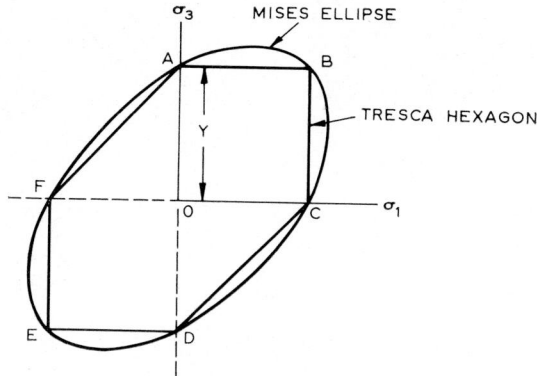

FIG. 4.3 Representation of yield criteria for case of plane stress.

The Tresca criterion can be expressed as a function of the reduced stress invariants, but it is much too complicated to be of practical value.

It is interesting to observe that alternative rectangular Cartesian co-ordinates which could be chosen are τ_1, τ_2 and τ_3 where $2\tau_1 = \sigma_2 - \sigma_3$, $2\tau_2 = \sigma_3 - \sigma_1$ and $2\tau_3 = \sigma_1 - \sigma_2$. The Mises criterion then plots in principal shear stress space, see Fig. 4.4(a) as a sphere of radius $Y/\sqrt{2}$. We have,

$$\tau_1^2 + \tau_2^2 + \tau_3^2 = (Y/\sqrt{2})^2. \tag{4.10}$$

However, τ_1, τ_2, τ_3 are not independent and

$$\tau_1 + \tau_2 + \tau_3 = 0 \tag{4.11}$$

so that the yield locus is restricted to be the circle of intersection of the sphere and the plane (4.11). This plane is, of course, normal to the line which is equally inclined to the three positive and three negative axes of τ_1, τ_2 and τ_3.

Calling this plane the τ-plane, a view towards the origin along the normal to the τ-plane shows the Mises criterion as a circle of radius $Y/\sqrt{2}$, see Fig. 4.4(b).

The Tresca criterion in this stress space is represented by the intersection of a cube whose sides are parallel to the co-ordinate axes, i.e., $\tau_1 = \pm Y/2$, $\tau_2 = \pm Y/2$ and $\tau_3 = \pm Y/2$, and the τ-plane. This cube when projected on to the τ-plane appears as the regular hexagon $ABCDEF$, see Fig. 4.4(b); where cut by the τ-plane it gives the regular hexagon $GHIJKL$. The points common to the Mises circle and the Tresca hexagon represent states of stress concerning which the two criteria make identical predictions—uniaxial stress states.

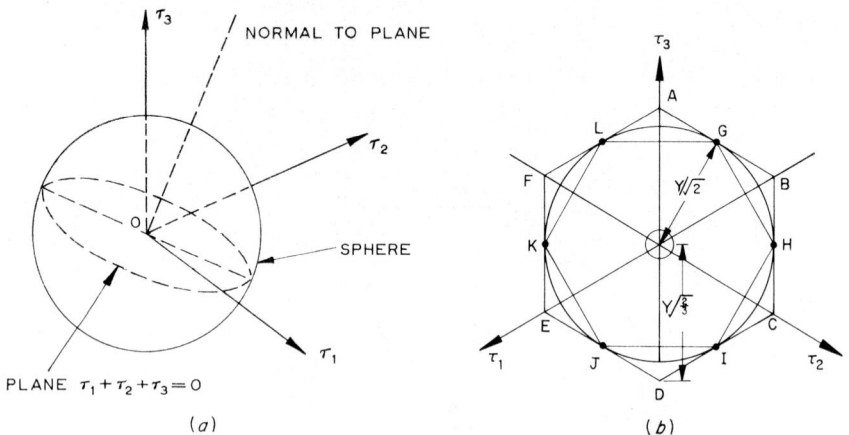

FIG. 4.4 Yield criteria plotted in principal shear stress space.

4.3 Shear and Volumetric Resilience and the Mises Criterion

Let $e_i(i = 1, 2, 3)$ denote principal *elastic* strains and σ_i principal stresses. Then the work done, W, on unit volume of an isotropic block in loading it from zero stress to σ_i is

$$W = \tfrac{1}{2}\sum \sigma_1 e_1 = \frac{1}{2E}\sum \sigma_1(\sigma_1 - v\sigma_2 - v\sigma_3)$$

$$= \frac{1}{2E}\sum \sigma_1^2 - \frac{v}{E}\sum \sigma_1 \sigma_2$$

where E is Young's Modulus and v is Poisson's ratio.

Now $v = \dfrac{3K - 2G}{6K + 2G}$ and $E = \dfrac{9KG}{3K + G}$

where K is the bulk modulus and G the shear modulus. Thus

$$W = \frac{3K + G}{18KG} \sum \sigma_1^2 - \frac{3K - 2G}{6K + 2G} \cdot \frac{3K + G}{9KG} \sum \sigma_1 \sigma_2$$

$$= \left(\frac{1}{6G} + \frac{1}{18K}\right) \sum \sigma_1^2 - \left(\frac{1}{6G} - \frac{1}{9K}\right) \sum \sigma_1 \sigma_2$$

$$= \frac{\sum \sigma_1^2 - \sum \sigma_1 \sigma_2}{6G} + \frac{\sum \sigma_1^2 + 2\sum \sigma_1 \sigma_2}{18K}$$

$$= \frac{\sum \tau_1^2}{3G} + \frac{(\sum \sigma_1)^2}{18K} \quad \text{where} \quad \tau_1 = \frac{(\sigma_2 - \sigma_3)}{2}$$

$$= \frac{\sum \tau_1^2}{3G} + \frac{(\sum \sigma_1/3)^2}{2K}. \tag{4.12}$$

The volumetric resilience V is $(\sum \sigma_1/3)^2/2K$ and the shear resilience, S, is $\sum \tau_1^2/3G$. K and G may be regarded as the fundamental physical elastic constants, since K measures the resistance of the material to change of volume without change of shape and G measures the resistance to change of shape without change of volume. This observation was made by STOKES in 1845. In more general terms, when the six components of the stress tensor σ_{ij} are given, i.e., the principal stresses not given, we may proceed thus

$$\sum (\sigma_1 - \sigma_2)^2 = 2 \left[\sum \sigma_1^2 - \sum \sigma_1 \sigma_2\right].$$

Now
$$\sum \sigma_1^2 = (\sum \sigma_1)^2 - 2\sum \sigma_1 \sigma_2$$

and from equations (2.19) and (2.20)

$$\sum \sigma_1 = \sum \sigma_{xx} \quad \text{and} \quad \sum \sigma_1 \sigma_2 = \sum \sigma_{xx} \sigma_{yy} - \sum \sigma_{xy}^2$$

Thus
$$\sum (\sigma_1 - \sigma_2)^2 = 2 \left[(\sum \sigma_1)^2 - 3\sum \sigma_1 \sigma_2\right]$$

$$= 2 \left[(\sum \sigma_{xx})^2 - 3\sum \sigma_{xx} \sigma_{yy} + 3\sum \sigma_{xy}^2\right]$$

$$= 2 \left[\sum \sigma_{xx}^2 - \sum \sigma_{xx} \sigma_{yy} + 3\sum \sigma_{xy}^2\right]$$

$$= \sum (\sigma_{xx} - \sigma_{yy})^2 + 6\sum \sigma_{xy}^2$$

Alternatively, then, equation (4.12) may be written

$$W = \frac{\sum (\sigma_{xx} - \sigma_{yy})^2 + 6\sum \sigma_{xy}^2}{12G} + \frac{(\sum \sigma_{xx})^2}{18K} = S + V \tag{4.13}$$

In relation to the Mises yield criterion, observe that a hydrostatic stress imposed on the whole system simply increases V. However, for an isotropic incompressible material K is infinite and then, always, $W = S$, i.e., plastic yielding is associated purely with change of shape and shear strain energy. Further, a more general way of stating the Mises yield criteria is,

$$(\sigma_{xx} - \sigma_{yy})^2 + (\sigma_{yy} - \sigma_{zz})^2 + (\sigma_{zz} - \sigma_{xx})^2 + 6(\sigma_{xy}^2 + \sigma_{yz}^2 + \sigma_{zx}^2) = 2Y^2. \tag{4.5.ii}$$

If an aelotropic material—one whose stress-strain relations depend on the directions of the axes of reference and which thus shows no, or only a limited amount of, symmetry—had been considered, it is intuitively obvious that, contrary to what we have just obtained where in a purely hydrostatic stress state the whole of W is attributable to V, a purely hydrostatic stress state will create both S and V components.

4.4 Experimental Evidence for the Tresca and Mises Criteria of Yielding

The most common method of experimentally investigating the yield criterion is to use an apparatus in which thin-walled tubes are subject to combined stresses. For example, Fig. 4.5 shows a tube subject to an elastic torque, T, and then loaded in tension P till yield occurs. Readings of angle of twist and extension of the tube are noted as the tensile load increases and yield

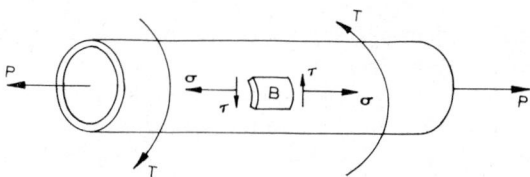

FIG. 4.5 Thin-walled tube subjected to tension P and torque T.

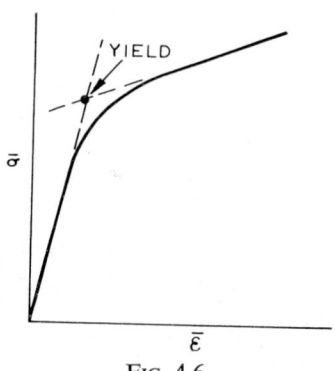

FIG. 4.6

is understood to have occurred soon after the rapid increase in strains occurs and when the material is clearly no longer elastic. Such a curve is shown in Fig. 4.6. (The co-ordinate axes are actually 'representative' or 'generalized' stress and strain, concepts necessitated when dealing with combined stress systems in the elastic-plastic range, see Chapter 5.)

Imagine an element at B cut from the tube wall and let σ be the applied tensile stress, and τ the shear stress which may be assumed constant through the tube wall since it is thin (see Fig. 4.7). Then the principal stresses at a point in the tube wall are

$$\sigma_1 = \tfrac{1}{2}\sigma + (\tfrac{1}{4}\sigma^2 + \tau^2)^{\frac{1}{2}}$$

$$\sigma_2 = \tfrac{1}{2}\sigma - (\tfrac{1}{4}\sigma^2 + \tau^2)^{\frac{1}{2}}$$

$$\sigma_3 = 0.$$

Thus,
$$\sigma_1 - \sigma_2 = 2(\tfrac{1}{4}\sigma^2 + \tau^2)^{\frac{1}{2}}$$

and
$$(\sigma_1 - \sigma_2)^2 + (\sigma_2 - \sigma_3)^2 + (\sigma_3 - \sigma_1)^2 = 2(\sigma^2 + 3\tau^2).$$

If Y is the uniaxial yield stress of the material in tension, then

1. Tresca's criterion gives

$$\frac{\sigma^2}{4} + \tau^2 = \left(\frac{Y}{2}\right)^2 \quad \text{or} \quad \left(\frac{\sigma}{Y}\right)^2 + \left(\frac{\tau}{Y/2}\right)^2 = 1 \qquad (4.14)$$

2. Mises' criterion gives

$$\left(\frac{\sigma}{Y}\right)^2 + \left(\frac{\tau}{Y/\sqrt{3}}\right)^2 = 1 \qquad (4.15)$$

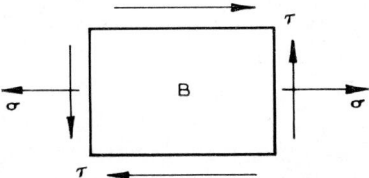

FIG. 4.7 Stresses on element in tube wall.

Equations (4.14) and (4.15) plot as ellipses with rectangular axes of σ/Y and τ/Y, see Fig. 4.8.

For most metals it is generally found that while the experimental points fall between the two ellipses, they incline towards Mises' ellipse. The classic experimental work of this kind is that of TAYLOR and QUINNEY (1931) and many repeat experiments have been performed. Experiments were earlier performed by LODE (1925) on tubes subject to tension and internal pressure.

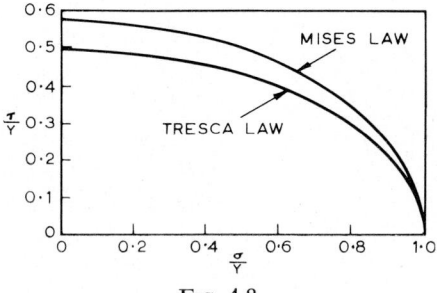

FIG. 4.8

Bending and torsion was used by SIEBEL in 1953. All results confirm the general opinion that the best simple yield criterion for metals is that of Mises. (An exception is the upper yield point of annealed mild steel which is said to follow Tresca's criterion.)

An extensive survey of the experimental and theoretical literature up to 1948 was made by DRUCKER.

Thin-walled tube tests require complicated apparatus and in recent years a new method has been tried following suggestions and a theory of localized necking put forward by HILL (1953). The apparatus is simple and depends on the testing in simple tension of a thin, uniform rectangular strip, in which a groove has been cut, see Fig. 4.9. This idea of using a notched strip was earlier considered and discussed by BILJAARD (1940) and the method has been used by LIANIS and FORD (1957) and PARKER and BASSETT (1964).

After examining many experimental results concerned with attempts to verify the form of the yield locus, or the yield criterion—these remarks also

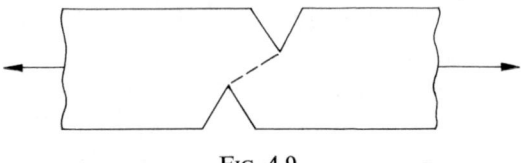

FIG. 4.9

apply to the next section—and from experience of doing this themselves (FIKRI and JOHNSON, 1956), the authors expressed the opinion in *Plasticity for Mechanical Engineers* (1962) that there was a great amount of latitude open to experimenters to decide what they should take to be the yield point load, or stress, in a given combined stress experiment. This experimenters have tended to confirm during the last few years. The difficulty lies in defining yield experimentally; the commonest method is to assume the initial yield point, for the purposes of calculation, to be defined by the intersection of the elastic and the elastic-plastic lines, see Fig. 4.10. With a well-rounded stress-

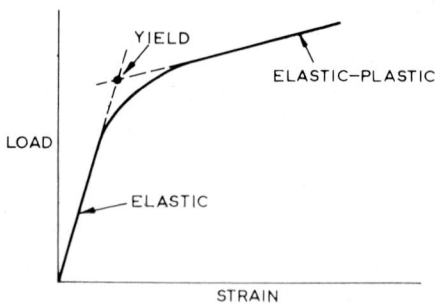

FIG. 4.10

strain or load-deflection curve, it will be appreciated that a wide range of estimates or choices of the yield point are possible. Other definitions of yield point are the stress at a specified offset of strain (IVEY, 1961) and the intersection of the elastic-plastic line with the load or stress axis. Results in accordance with any of these definitions indicate varying degrees of suitability or accuracy of the different yield criteria. As remarked above, this applies especially to the study of the shape of yield surfaces after specified amounts of over-strain, see Section 4.5 (MAIR and PUGH (1963) and PARKER and BASSETT (1964)). It is at points such as this that theories based on high degrees of mathematical exactness are seen to be unrealistic.

4.5 Isotropic Work-hardening Materials

If the material considered is a work-hardening one, and presuming that a Mises criterion is obeyed, then some initial circle in the π-plane represents primary yield. Further plastic straining alters the shape of the yield locus. For instance, if Y_0 is the primary yield stress, then the radius of the Mises circle is $Y_0\sqrt{2/3}$. Suppose that the straining is continued beyond Y_0 to Y_1 and that the material is then completely unloaded; further assume that the straining has not rendered the material anisotropic. This material now possesses a yield locus which is a circle of radius $Y_1\sqrt{2/3}$. In the π-plane this circle of radius $Y_1\sqrt{2/3}$ surrounds but is concentric with the original circle of radius $Y_0\sqrt{2/3}$.

The implication here is that the material is an isotropic work-hardening one, represented by a circular yield locus which expands with strain and stress history retaining the same shape throughout. An isotropic strain-hardening 'Tresca' material would appear as a series of concentric regular hexagons. It is easy to envisage the representation of some imaginary modified Mises material by supposing that the radius of the circle is proportional to the volumetric stress; the yield locus for this material appears as a cone whose axis is the line of hydrostatic stress states.

The concept of an isotropic work-hardening material gives rise to mathematical expressions which are easy to handle, but in reality they may only be accepted as first-degree approximations. Most strikingly the Bauschinger effect intervenes to 'reduce the size' of the locus on one side as that on the 'opposite' side is increased. Thus progressively, there are changes in the shape of the yield locus. The point is well illustrated in work by NAGHDI, ESSENBURG and KOFF (1958); it is an extension of substantially the same experiments as those already alluded to, performed by TAYLOR and QUINNEY. Their tension-torsion experimental results are plotted on orthogonal Cartesian co-ordinates but they none-the-less reflect a decisive change in the shape of a yield locus as it would appear on the π-plane.

Tests were carried out with aluminium alloy tubes 3/4 in internal diameter and 0·075 in thickness. They were initially subjected to tension alone, followed by torsion with the tension held almost constant, the tube remaining elastic but approaching yield. Both tension and torsion were then increased till yield had distinctly occurred.

Several tests of this kind with various ratios of torsion and tension thus provided data for an initial yield surface. The load on each of the tubes was, after complete removal, followed by a re-loading in torsion to a pre-chosen figure, and then full unloading. Each tube was now subject to the same sequence of testing as was previously the case and until yielding was again evident. The data from this set of experiments enabled a first subsequent yield surface to be found. A second subsequent yield surface was calculated from further tests on the tubes after a second re-loading to a pre-chosen size of load in torsion—one which was larger than the first one.

The yield surfaces obtained are shown in Fig. 4.11. Most surprising in these results is the lack of cross-effect that torsion produced; prior over-strain in torsion apparently had no effect whatsoever on the uniaxial tensile yield stress. The change of shape of the initial Mises yield curve with in-creasing pre-torsion, indicates that isotropic strain-hardening theories are somewhat away from reality. Note the Bauschinger effect as reflected in progressively reduced yield stresses in reversed torsion. See also, SHIRATORI and IKEGAMI, (1968).

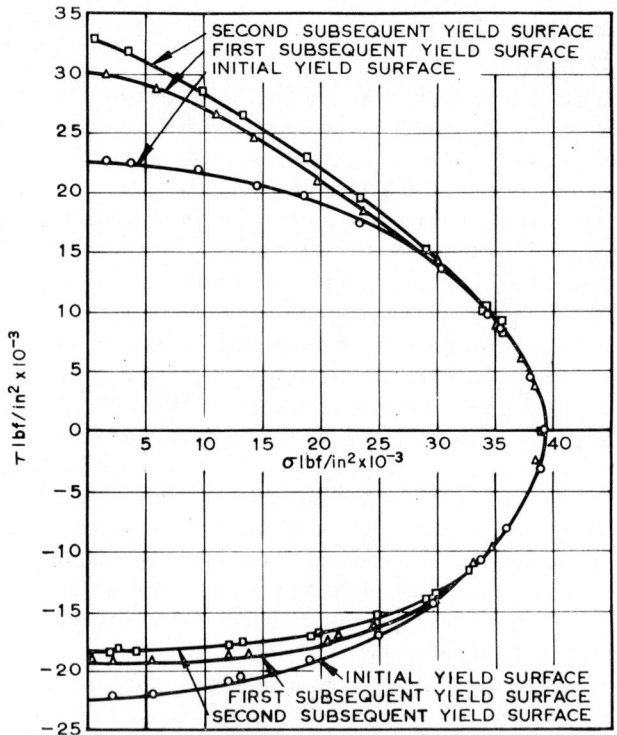

FIG. 4.11 Initial and subsequent yield surfaces. (*After* Naghdi, Essenberg and Koff, *J. appl. Mech., Trans. A.S.M.E.*)

4.6 An Anisotropic Yield Criterion

All metals exhibit anisotropy to a greater or lesser degree when deformed at room temperature, that is, the mechanical properties of the metal vary in different directions, and the amount and type of anisotropy is characteristic of the mechanical and heat treatments to which it has been subjected in previous processing. Practical interest in recent years has centred on anisotropy in sheet metals and this is discussed further in Chapters 6 and 11. In sheet metals it is convenient to define two types of anisotropy. Planar anisotropy describes the variations of mechanical properties in the plane of the sheet and is characterized by the amount of earing that occurs when drawing a cylindrical cup from a flat circular blank. Normal anisotropy describes the strength of the metal through the thickness of the sheet relative to its strength in the plane of the sheet. In sheet metal forming it is usually desirable to approach an isotropic condition in the plane of the sheet and to have increased relative strength through the thickness.

Theories describing anisotropic behaviour have been proposed by JACKSON, SMITH and LANKFORD (1948), HILL (1948) and DORN (1949). Much of this work was done during the 1939–1945 War and released for publication afterwards. The theory of Hill (1948–1950) will be used here to describe a state of simple orthotropic anisotropy, that is, there are three mutually orthogonal planes of symmetry at every point. The intersections of these planes are known as the principal axes of anisotropy. The yield criterion proposed by Hill when referred to these axes has the form

$$2f(\sigma_{ij}) \equiv F(\sigma_y - \sigma_z)^2 + G(\sigma_z - \sigma_x)^2 + H(\sigma_x - \sigma_y)^2$$
$$+ 2L\tau_{yz}^2 + 2M\tau_{zx}^2 + 2N\tau_{xy}^2 = 1 \quad (4.16)$$

where F, G, H, L, M, N are parameters characteristic of the current state of anisotropy. It is assumed that there is no Bauschinger effect and that a hydrostatic stress does not influence yielding. Hence, linear terms are not included and only differences between normal stress components appear in the yield criterion. The conditions for planar isotropy (rotational symmetry about the z-axis) are determined by noting that equation (4.16) must remain invariant for arbitrary (x, y) axes of reference. It can then be shown that

$$N = F + 2H = G + 2H, \qquad L = M. \quad (4.17)$$

For complete isotropy

$$L = M = N = 3F = 3G = 3H. \quad (4.18)$$

The dimensions of the parameters can be seen by considering the tensile yield stresses X, Y, Z in the principal directions of anisotropy. Thus

$$\left.\begin{aligned} \frac{1}{X^2} &= G + H \\[4pt] \frac{1}{Y^2} &= H + F \\[4pt] \frac{1}{Z^2} &= F + G \end{aligned}\right\} \quad (4.19)$$

When anisotropy is vanishingly small the expression (4.16) reduces to the von Mises criterion. Substituting from (4.18) into (4.16) the yield criterion then becomes

$$(\sigma_y - \sigma_z)^2 + (\sigma_z - \sigma_x)^2 + (\sigma_x - \sigma_y)^2 + 6\tau_{yz}^2 + 6\tau_{zx}^2 + 6\tau_{xy}^2 = \frac{1}{F} = 6k^2$$

$$(4.20)$$

since for pure shear $\sigma_x = -\sigma_y = k, \sigma_z = 0$

and $\qquad\qquad\qquad\qquad \tau_{yz} = \tau_{zx} = \tau_{xy} = 0$

giving $\qquad\qquad\qquad\qquad \dfrac{1}{F} = 6k^2.$

See Problems 18-20

REFERENCES

BILJAARD, P. P.   1940 *Publs int. Ass. Bridge struct. Engng* **6**, 27

DORN, J. E.   1949 'Stress strain relations for anisotropic plastic flow' *J. appl. Phys.* **20**, 15

DRUCKER, D. C.   1950 'Stress-strain Relations in the Plastic Range— a Survey of Theory and Experiment' O.N.R. Report, NR–041–032

FIKRI, K. and JOHNSON, W.   1955 'The Effect of Tensile Pre-strain on the Plastic Distortion of Metals' B.I.S.R.A. Report, MW/E/59/55

HENCKY, H.   1924 *Z. angew. Math. Mech.* **4**, 323

HILL, R.   1950 *The Mathematical Theory of Plasticity* Chap. 12, O.U.P.

   1953 'On Discontinuous Plastic States with Special Reference to Localized Necking in Thin Sheets' *J. Mech. Phys. Solids* **1**, 19

HUBER, M. T.   1904 *Czasopismo techniczne* 22, 81, Lemberg

IVEY, J. H.   1961 'Plastic Stress-Strain Relations and Yield Surfaces for Aluminium Alloys' *J. mech. Engng Sci.* **3**, 15

JACKSON, L. R., SMITH, K. F. and LANKFORD, W. T.   1948 'Plastic flow in anisotropic sheet metal' *Metals Technology* Tech. Pub. No. 2440., and *J. Metals* **1**, 323, (1949)

LIANIS, G. and FORD, H.   1957 'An Experimental Investigation of the Yield Criterion and the Stress-Strain Law' *J. Mech. Phys. Solids* **5**, 215

LODE, W.   1925 *Z. angew. Math. Mech.* **5**, 142

MAIR, W. N. and PUGH, H. Ll. D.   1963 N.E.L. Report No. 90

MISES, R. VON   1913 *Göttinger Nachrichten, math.–phys. K1*, 582

NAGHDI, P. M., ESSENBURG, F. and KOFF, W.   1958 'An Experimental Study of Initial and Subsequent Yield Surfaces in Plasticity' *Trans. A.S.M.E.* **80**; *J. appl. Mech.* 201

PARKER, J. and 1964 'Plastic Strain Relationships—Some Experi-
BASSETT, M. B. ments to Derive Subsequent Yield Surface'
Trans. A.S.M.E. Series E, **31**, 676

SHIRATORI, E. and 1968 'Experimental Study of the Subsequent
IKEGAMI, K. Yield Surface by Using Cross-Shaped Specimens' *J. Mech. Phys. Solids* **16**, 373

SIEBEL, M. P. L. 1953 'The Combined Bending and Twisting of
Thin Cylinders in the Plastic Range'
J. Mech. Phys. Solids **1**, 189

STOKES, G. G. 1845 *On the Theories of the Internal Friction
of Fluids in Motion, and of the Equilibrium
and Motion of Elastic Solids*
Cambridge Philosophical Society

TAYLOR, G. I. and 1931 'The Plastic Distortion of Metals'
QUINNEY, H. *Phil. Trans. R. Soc. A.*, **230**, 323

TRESCA, H. 1864 *C. r. Acad. Sci., Paris* **59**, 754

Chapter 5

STRESS-STRAIN RELATIONS

The complete stress-strain relations describe the elastic and plastic deformation of a solid. In the following, the effects of time and temperature are not considered, but work-hardening is taken into account. It will also be assumed that the solid is isotropic and the Bauschinger effect negligible.

It should be noted that the shearing strains, $\gamma_{xy}, \gamma_{yz}, \gamma_{zx}$ are the shear components of the strain tensor and therefore have values equal to half the corresponding values of engineering shear strain.

5.1 The Elastic Stress-Strain Relations

For an isotropic solid, the elastic stress-strain relations are usually written in the form

$$
\left.
\begin{aligned}
Ee_x &= \sigma_x - v(\sigma_y + \sigma_z) \\
Ee_y &= \sigma_y - v(\sigma_z + \sigma_x) \\
Ee_z &= \sigma_z - v(\sigma_x + \sigma_y) \\
2G\gamma_{yz} &= \tau_{yz} \\
2G\gamma_{zx} &= \tau_{zx} \\
2G\gamma_{xy} &= \tau_{xy}
\end{aligned}
\right\}
\tag{5.1}
$$

E is Young's modulus, v Poisson's ratio and G the torsion modulus.

In the present context, it is convenient to distinguish between strains describing a change in shape and those describing a change in volume. If σ_m is the hydrostatic stress and e_m the corresponding volumetric strain, then, employing the relations between the elastic constants, equations (5.1) may be written in the form

$$
e_x = \frac{1}{2G}(\sigma_x - \sigma_m) + \frac{(1 - 2v)}{E}\sigma_m
$$

$$
e_y = \frac{1}{2G}(\sigma_y - \sigma_m) + \frac{(1 - 2v)}{E}\sigma_m
$$

$$
e_z = \frac{1}{2G}(\sigma_z - \sigma_m) + \frac{(1 - 2v)}{E}\sigma_m
$$

$$\gamma_{yz} = \tau_{yz}/2G$$
$$\gamma_{zx} = \tau_{zx}/2G$$
$$\gamma_{xy} = \tau_{xy}/2G$$

where $\quad 3\sigma_m = \sigma_x + \sigma_y + \sigma_z \quad$ and $\quad e_m = e_x + e_y + e_z = \dfrac{\sigma_m}{K}.$ (5.2)

$(\sigma_x - \sigma_m)$, etc. are reduced or deviatoric components and are written in the form σ_x'. The complete elastic stress-strain relations may therefore be written, when using an obvious double suffix notation, as

$$\left.\begin{aligned} e_{ij} &= \frac{\sigma_{ij}'}{2G} + \frac{(1 - 2v)}{E}\,\delta_{ij}\sigma_m \\[2mm] \sigma_m &= \tfrac{1}{3}\sigma_{ii} \end{aligned}\right\} \tag{5.3}$$

The delta symbol, δ_{ij}, is equal to unity when $i = j$ and to zero when $i \neq j$.

5.2 The Prandtl-Reuss Equations

The stress-strain relations for an elastic-perfectly plastic solid were first proposed by PRANDTL (1924) for the case of plane-strain deformation. The general form of the equations was given by REUSS (1930). Prandtl's equations are an extension of the earlier Lévy-Mises equations which are discussed in the next section.

Reuss assumed that the *plastic* strain *increment*—denoted by a superfix p in the equations—is at any instant proportional to the instantaneous stress deviation and the shear stresses, thus

$$\frac{d\epsilon_x^p}{\sigma_x'} = \frac{d\epsilon_y^p}{\sigma_y'} = \frac{d\epsilon_z^p}{\sigma_z'} = \frac{d\gamma_{yz}^p}{\tau_{yz}} = \frac{d\gamma_{zx}^p}{\tau_{zx}} = \frac{d\gamma_{xy}^p}{\tau_{xy}} = d\lambda$$

or $\qquad\qquad\qquad d\epsilon_{ij}^p = \sigma_{ij}'d\lambda.$ (5.4)

$d\lambda$ is an instantaneous non-negative constant of proportionality which may vary throughout a straining programme.

The equations state that a small increment of plastic strain depends on the current deviatoric stress, not on the stress increment which is required to bring it about. Also, the principal axes of stress and plastic strain increment coincide. The equation is only a statement about the *ratio* of the plastic strain increments in the various x, y, z directions; it gives no direct information about their absolute magnitude. Experimental verification of these statements is discussed in Section 5.7.

The total strain increment is the sum of the elastic strain increment (henceforth defined as $d\epsilon^e$ instead of de) and the plastic strain increment.

Thus, $\qquad\quad d\epsilon_{ij} = d\epsilon_{ij}^p + d\epsilon_{ij}^e$

$$= \sigma_{ij}'\,d\lambda + \frac{d\sigma_{ij}'}{2G} + \frac{(1 - 2v)}{E}\,\delta_{ij}\,d\sigma_m \tag{5.5}$$

from equations (5.3) and (5.4).

Since plastic straining causes no change of plastic volume, we may write the condition of incompressibility, in terms of the principal or normal strains, as

$$d\epsilon_1^p + d\epsilon_2^p + d\epsilon_3^p = d\epsilon_x^p + d\epsilon_y^p + d\epsilon_z^p = 0$$

or
$$d\epsilon_{ii}^p = 0. \tag{5.6}$$

Equation (5.4), considering principal stress directions, gives

$$\frac{d\epsilon_1^p - d\epsilon_2^p}{\sigma_1 - \sigma_2} = \frac{d\epsilon_2^p - d\epsilon_3^p}{\sigma_2 - \sigma_3} = \frac{d\epsilon_3^p - d\epsilon_1^p}{\sigma_3 - \sigma_1} = d\lambda. \tag{5.7}$$

Equation (5.7) states that the Mohr circles of stress and plastic strain increment are similar, Fig. 5.1.

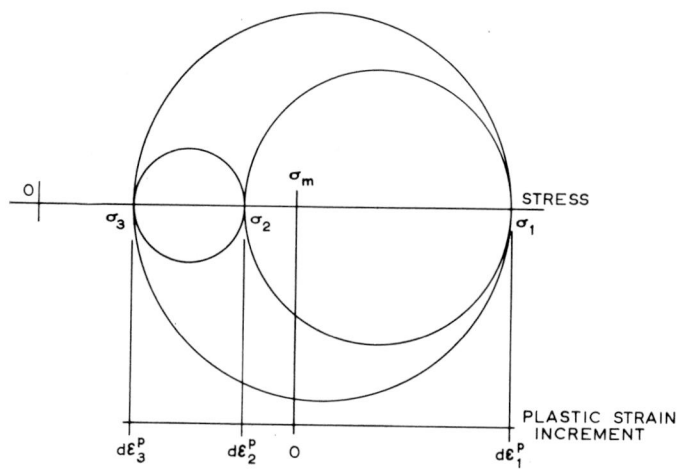

FIG. 5.1 Mohr's circles for stress and plastic strain increment.

Equation (5.4) may, of course, be rewritten in terms of the normal stresses giving rise to equations of the form

$$d\epsilon_x^p = \tfrac{2}{3} d\lambda [\sigma_x - \tfrac{1}{2} (\sigma_y + \sigma_z)].$$

Equation (5.5) thus consists of three equations of the type

$$d\epsilon_x = \tfrac{2}{3} d\lambda [\sigma_x - \tfrac{1}{2}(\sigma_y + \sigma_z)] + [d\sigma_x - \nu(d\sigma_y + d\sigma_z)]/E$$

and three of the type

$$d\gamma_{yz} = \tau_{yz} d\lambda + d\tau_{yz}/2G. \tag{5.8}$$

Finally, on examining equation (5.5), it will be seen that the volumetric and deviatoric strain increments can be separated in the expression for the total strain increment. Including the Mises yield criterion, the Prandtl-Reuss equations may then be written as

$$d\epsilon'_{ij} = \sigma'_{ij}d\lambda + d\sigma'_{ij}/2G$$

$$d\epsilon_{ii} = \frac{(1 - 2v)}{E} d\sigma_{ii} \qquad\qquad (5.9)$$

$$\sigma'_{ij}\sigma'_{ij} = 2k^2.$$

These equations for an elastic-plastic solid are usually difficult to handle in a real problem, and, in consequence, there are relatively few solutions which employ them.

In problems of large plastic flow, the elastic strain may often be neglected altogether. The material is then considered as being a rigid-perfectly plastic solid. When the stresses are below the yield point, no straining takes place, and the total strain increment and the plastic strain increment are identical. Stress-strain relations for such a material were proposed by Lévy and von Mises.

5.3 The Lévy-Mises Equations

In presenting the relations between stress and strain, we have not followed the historical development of the subject. At the present time, it seems more logical to consider the Lévy-Mises equations as a special form of the Prandtl-Reuss equations. However, it was SAINT VENANT (1870) who first proposed that the principal axes of strain *increment* coincided with the axes of principal stress. The general relationship between strain increment and the reduced stresses was first introduced by LÉVY (1871) and independently by VON MISES (1913). These equations are now known as the Lévy-Mises equations and may be written as

$$\frac{d\epsilon_x}{\sigma'_x} = \frac{d\epsilon_y}{\sigma'_y} = \frac{d\epsilon_z}{\sigma'_z} = \frac{d\gamma_{yz}}{\tau_{yz}} = \frac{d\gamma_{zx}}{\tau_{zx}} = \frac{d\gamma_{xy}}{\tau_{xy}} = d\lambda. \qquad (5.10)$$

The superfix, p, of equation (5.4) may be dropped, since the total strain increment and the plastic strain increment are now identical. Further, the Mohr circles of stress and strain increment are identical. Written in terms of total stresses, the Lévy-Mises relation has three equations of the type

$$d\epsilon_x = \tfrac{2}{3} d\lambda \left[\sigma_x - \tfrac{1}{2}(\sigma_y + \sigma_z) \right]$$

and three of the type

$$d\gamma_{yz} = \tau_{yz} d\lambda. \qquad (5.11)$$

Since the elastic strains are not taken into account, the Lévy-Mises relations obviously cannot be used to obtain information about 'elastic springback' or residual stresses. The more complex Prandtl-Reuss equation must then be used.

The Lévy-Mises equations will be used extensively throughout the book where cases of unrestricted plastic flow arise.

5.4 Work-hardening

When a real material is cold formed, it 'work-hardens', that is, as the material deforms, its resistance to further deformation increases.

One hypothesis is that the degree of hardening is a function only of the total plastic work and is otherwise independent of the strain path. This is sometimes known as the equivalence of plastic work. In other words, the resistance to further distortion depends only on the amount of work which has been done on the material since it was in its initial annealed state. This 'resistance to distortion' is assessed via the yield criterion.

It has been established by careful tests that the von Mises yield criterion is the simplest relation which fits experimental data reasonably well, no matter what the degree of pre-strain. According to this criterion, the final yield locus is independent of the strain path and also independent of the hydrostatic component of stress. It may be expressed in terms of principal stresses σ_1, σ_2, σ_3 as

$$(\sigma_1 - \sigma_2)^2 + (\sigma_2 - \sigma_3)^2 + (\sigma_3 - \sigma_1)^2 = 6k^2$$

where k is a parameter depending on the amount of pre-strain.

HILL (1950) has used this isotropic hardening rule. It assumes that the yield surface so expands during plastic flow that its original shape and position with respect to the line of hydrostatic stress states remains fixed (see p. 65). PRAGER (1955) has proposed a hardening rule such that the yield surface is of fixed size, but moves in stress space in the direction of the strain increment. This accounts for the Bauschinger effect, which the first does not, but it is much more difficult to handle mathematically.

For the sole reason of familiarity, it is convenient to write the criterion as

$$\bar{\sigma} = \sqrt{\tfrac{1}{2}\{(\sigma_1 - \sigma_2)^2 + (\sigma_2 - \sigma_3)^2 + (\sigma_3 - \sigma_1)^2\}}$$
$$= \sqrt{\tfrac{3}{2}(\sigma_1'^2 + \sigma_2'^2 + \sigma_3'^2)} \tag{5.12}$$

where $\bar{\sigma}$ is known as the effective, generalized or equivalent stress. The numerical factor has been chosen so that $\bar{\sigma} = Y$ for simple tension. The effective stress $\bar{\sigma}$ is, according to the above hypothesis, a function of the total plastic work, W_p.

$$\bar{\sigma} = F(W_p) \tag{5.13}$$

It will be realized that these expressions are strictly valid only for an idealized metal where the deformation is isotropic and there is no Bauschinger effect. Also, the yield criterion implies that no plastic work is done by the hydrostatic components of stress and therefore that there is no permanent change in volume. This is a close approximation to the experimental data.

The increment of plastic work per unit volume is

$$dW_p = \sigma_1'\, d\epsilon_1^p + \sigma_2'\, d\epsilon_2^p + \sigma_3'\, d\epsilon_3^p$$

and it may be found in terms of geometrical quantities in the following manner. Since $d\epsilon_1^p + d\epsilon_2^p + d\epsilon_3^p = 0$, the plastic strain increment can be represented by a vector in the π-plane. If the factor $2G$ is introduced, to obtain the dimensions of stress, the plastic strain increment vector may be drawn on the same diagram as the deviatoric stress vector. Further, since it is assumed that the principal axes of plastic strain increment coincide with the axes of principal stress, then in Fig. 5.2 the stress vector $\overrightarrow{OP}$ is drawn parallel to the plastic-strain increment vector $\overrightarrow{RQ}$, and

$$dW_p = \frac{\overrightarrow{OP}.\overrightarrow{RQ}}{2G}$$

where $\quad |OP| = \sqrt{(\sigma_1'^2 + \sigma_2'^2 + \sigma_3'^2)} = \sqrt{\tfrac{2}{3}}\bar{\sigma}$

and $\quad |RQ| = 2G\sqrt{(d\epsilon_1^{p2} + d\epsilon_2^{p2} + d\epsilon_3^{p2})} = 2G\sqrt{\dfrac{3}{2}}\,d\bar{\epsilon}^p.$

(5.14)

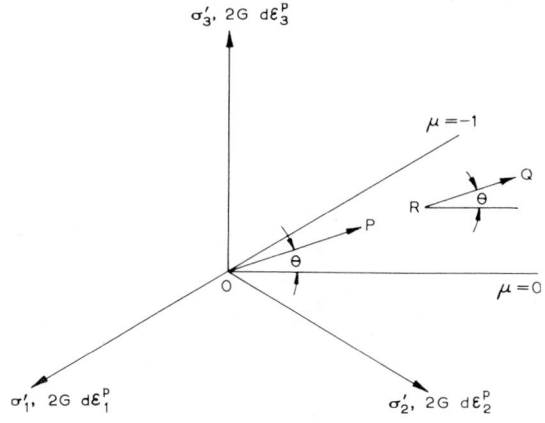

FIG. 5.2 Plastic strain increment vector $\overrightarrow{RQ}$ is assumed parallel to the stress vector $\overrightarrow{OP}$.

$d\bar{\epsilon}^p$ is the same invariant function of the components of the plastic strain increment as $\bar{\sigma}$ is of the components of the deviatoric stress. It may be written

$$d\bar{\epsilon}^p = \sqrt{\tfrac{2}{9}\{(d\epsilon_1^p - d\epsilon_2^p)^2 + (d\epsilon_2^p - d\epsilon_3^p)^2 + (d\epsilon_3^p - d\epsilon_1^p)^2\}} \tag{5.15}$$

Thus, $\qquad\qquad\qquad dW_p = \bar{\sigma}\,d\bar{\epsilon}^p. \tag{5.16}$

The hypothesis for strain-hardening (5.13) can now be written as

$$\bar{\sigma} = F\left(\int \bar{\sigma}\,d\bar{\epsilon}^p\right)$$

and it follows that $\bar{\sigma}$ is a function only of $\int d\bar{\epsilon}^p$ where the integral is taken over the strain path. Therefore

$$\bar{\sigma} = H\int d\bar{\epsilon}^p. \tag{5.17}$$

This expression is usually more convenient to use than equation (5.13). It must be stressed that equations (5.13) and (5.17) may not lead to identical results when applied to a real material because of anisotropy and the Bauschinger effect.

If we are dealing with a material that is rigid up to yield, then we can write equation (5.15) in the form

$$d\bar{\epsilon} = \sqrt{\tfrac{2}{9}\{(d\epsilon_1 - d\epsilon_2)^2 + (d\epsilon_2 - d\epsilon_3)^2 + (d\epsilon_3 - d\epsilon_1)^2\}} \qquad (5.18)$$

since the plastic strain increment is identical with the total strain increment. Further, in the special case where the principal axes of successive strain increment do not rotate relative to the element being strained and where the strain increment components stand in a constant ratio one to another, then putting

$$\frac{d\epsilon_2}{d\epsilon_1} = x, \quad \frac{d\epsilon_3}{d\epsilon_1} = y$$

and since

$$d\epsilon_1 + d\epsilon_2 + d\epsilon_3 = 0$$

$$x + y + 1 = 0 \quad \text{or} \quad y = -(1 + x).$$

Thus

$$d\bar{\epsilon}^2 = \tfrac{2}{3} d\epsilon_1^2 [1 + x^2 + y^2]$$

or

$$d\bar{\epsilon} = \frac{2}{\sqrt{3}} (1 + x + x^2)^{\frac{1}{2}}.d\epsilon_1.$$

This may be integrated to give

$$\bar{\epsilon} = \frac{2}{\sqrt{3}} (1 + x + x^2)^{\frac{1}{2}}.\epsilon_1. \qquad (5.19)$$

Hence, for this special case, equation (5.17) can be written as

$$\bar{\sigma} = H(\bar{\epsilon}) \qquad (5.20)$$

and the equation of incompressibility has the integrated form

$$\epsilon_1 + \epsilon_2 + \epsilon_3 = 0.$$

5.5 The Complete Stress-Strain Relations

From equations (5.7), (5.9), (5.12) and (5.15) the complete stress-strain relations for a work-hardening material can be written as

$$\left. \begin{array}{l} d\epsilon'_{ij} = \tfrac{3}{2}\sigma'_{ij} \dfrac{d\bar{\epsilon}^p}{\bar{\sigma}} + \dfrac{d\sigma'_{ij}}{2G} \\[2ex] d\epsilon_{ii} = \dfrac{(1 - 2v)}{E}.d\sigma_{ii} \\[2ex] \sigma'_{ij}\sigma'_{ij} = 2k^2. \end{array} \right\} \qquad (5.21)$$

Equations (5.21) include such equations as

$$d\epsilon_x = [\sigma_x - \tfrac{1}{2}(\sigma_y + \sigma_z)]\frac{d\bar{\epsilon}^p}{\bar{\sigma}} + \frac{1}{E}[d\sigma_x - v(d\sigma_y + d\sigma_z)]$$

$$d\gamma_{yz} = \tfrac{3}{2}\tau_{yz}\frac{d\bar{\epsilon}^p}{\bar{\sigma}} + \frac{d\tau_{yz}}{2G}. \tag{5.22}$$

Equations (5.21) were developed in a more general manner by HILL (1950).

If work-hardening is neglected altogether, some constant value of yield stress being assumed for the material, the equations for the plastic strain increment are

$$d\epsilon_{ij}^p = \tfrac{3}{2}\sigma_{ij}'\frac{d\bar{\epsilon}^p}{Y} \tag{5.23}$$

since $\bar{\sigma} = Y$ from equation (5.12). Equations (5.23) were first given by HILL and are a re-statement of the Prandtl-Reuss equations in a more useful form.

When the elastic strains can be neglected, then for a Lévy-Mises material, the stress-strain relations have the form

$$d\epsilon_{ij} = \tfrac{3}{2}\sigma_{ij}'\frac{d\bar{\epsilon}}{\bar{\sigma}}. \tag{5.24}$$

Equations (5.24) consist of three equations of the type

$$d\epsilon_x = [\sigma_x - \tfrac{1}{2}(\sigma_y + \sigma_z)]\frac{d\bar{\epsilon}}{\bar{\sigma}}$$

and three of the type

$$d\gamma_{yz} = \tfrac{3}{2}\tau_{yz}\frac{d\bar{\epsilon}}{\bar{\sigma}}. \tag{5.25}$$

5.6 Total Strain Theory

HENCKY'S STRESS-STRAIN EQUATIONS

It appears that only over the past three decades has it become generally recognized that plasticity problems are all incremental in nature. HENCKY'S equations (1924) were and are seemingly an attempt to extend the total strain theory of elasticity to plasticity. They state, in effect, that

$$\epsilon_{ij}^p = \phi\sigma_{ij}', \tag{5.26}$$

i.e. that the components of *total* plastic strain—in contradistinction to the Reuss equations which treat of components of plastic strain *increment*—are proportional to the deviatoric stress components. The Hencky equations may be written as

$$\left.\begin{aligned} \epsilon_{ij}' &= \left(\phi + \frac{1}{2G}\right)\sigma_{ij}' \\ \epsilon_{ii} &= \frac{(1 - 2v)}{E}\sigma_{ii} \end{aligned}\right\} \tag{5.27}$$

where ϕ is a scalar quantity which is positive throughout loading and zero during unloading. It is obvious that (5.27) implies that if a value of the stress at a point is given, the total strain is immediately defined. This is clearly untrue except in the special cases when the stress and strain ratios are maintained constant. In this circumstance, the Hencky equations follow from a straightforward integration of the Reuss equations. HILL's remarks (1950) are very much to the point on this topic:

'It is very easy to show that the Hencky equations are unsuitable to describe the *complete* plastic behaviour of a metal. Suppose that after a certain plastic deformation, the element is unloaded, partially or completely, and then reloaded to a different stress state on the same yield locus. While the stress-point lies inside the yield locus, only elastic changes of strain can occur, and the total plastic strain is unchanged. According to (5.26), however, the plastic strain ratios are now entirely different since the state of stress has changed. This means that the plastic strain itself has altered during the unloading and reloading, which is absurd.'

5.7 The Lévy-Lode Variables

The first experiments to investigate the plastic stress-strain relationships were carried out by LODE (1926). He introduced the two parameters

$$\mu = \frac{(\sigma_3 - \sigma_1) - (\sigma_2 - \sigma_3)}{\sigma_1 - \sigma_2} \tag{5.28}$$

$$v = \frac{(d\epsilon_3^p - d\epsilon_1^p) - (d\epsilon_2^p - d\epsilon_3^p)}{d\epsilon_1^p - d\epsilon_2^p}. \tag{5.29}$$

It can be seen that if the theoretically proposed equations (5.7) are correct, then μ should be equal to v. (Care should be taken not to confuse Lode's parameter v with Poisson's ratio.)

If $\mu = -1$, $\sigma_2 = \sigma_3$ and this is the same as a uniaxial stress of $(\sigma_1 - \sigma_2)$ together with a hydrostatic stress of σ_2. When $\mu = 0$, $\sigma_3 = (\sigma_1 + \sigma_2)/2$ and this is the same as a state of pure shear $[(\sigma_1 - \sigma_2)/2, (\sigma_2 - \sigma_1)/2, 0]$ with a hydrostatic stress of $(\sigma_1 + \sigma_2)/2$. A plot of $\mu = v$ representing various states of stress between the extremes of pure shear and uniaxial tension is the straight line OA, Fig. 5.3.

Experimental method is much simplified if in any test the stress-ratios are kept constant throughout the straining. This reduces the elastic component of strain to a minimum and for a slightly pre-strained material, it can be neglected altogether. When this is the case, v can be written in terms of the total strain,

$$v = \frac{(\epsilon_3 - \epsilon_1) - (\epsilon_2 - \epsilon_3)}{\epsilon_1 - \epsilon_2}. \tag{5.30}$$

All experiments to evaluate Lode's variables have been carried out under conditions of plane stress. LODE (1926), in his experiments, stressed thin-walled tubes of iron, copper and nickel in combined tension and internal pressure. The ratio between axial and circumferential stress was kept approximately constant throughout each test. Despite considerable scatter, due to anisotropy in the tubes, the results indicated a deviation from the line OA,

Fig. 5.3. A systematic deviation was noted by TAYLOR and QUINNEY (1931) who stressed thin-walled tubes of aluminium, copper and mild steel in combined tension and torsion. They took considerable care to measure and limit the degree of anisotropy in the tubes. In these latter tests, the axial load was held constant while the torque was increased—hence the stress ratios were not constant and equation (5.30) does not apply.

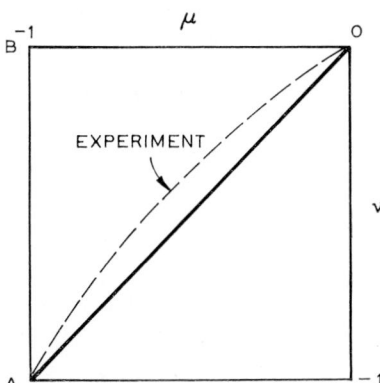

FIG. 5.3 Lévy-Lode variables.

PUGH (1953) has shown that it is impossible to be absolutely certain that a thin-walled tube is isotropic. To overcome this difficulty, HILL (1953) has proposed that the experimental determination of the plastic stress-strain relations should be carried out on a notched strip specimen, since a greater control can be exercised on the degree of anisotropy. This idea has been put into practice by HUNDY and GREEN (1954) and they obtained results which were in very good agreement with the Lévy-Mises relation.

Further experiments by HOHENEMSER (1931) and MORRISON and SHEPHERD (1950) whereby thin-walled tubes were subjected to various combinations of torsion and tension, demonstrate the approximate validity of the Prandtl-Reuss equations. In these experiments, the elastic and plastic strains were of the same order—a condition not usually satisfied in verifying the Lode variables.

5.8 The Plastic Potential and Flow Rules

So far we have discussed only the Mises yield criterion and the Lévy-Mises flow rule, since these hypotheses are in greatest accord with present experimental evidence. We will now look at the problem of yield and plastic flow in a more general manner, using the concept of a plastic potential. The hypothesis is made that the plastic potential is that scalar function of the stress, say $g(\sigma_{ij})$, from which the ratios of the components of the plastic strain increments, $d\epsilon_{ij}^p$, are derivable by partially differentiating $g(\sigma_{ij})$ with respect to σ_{ij}. Thus

$$d\epsilon_{ij}^p = \frac{\partial g\,(\sigma_{ij})}{\partial \sigma_{ij}}\,d\lambda'$$

or
$$\frac{d\epsilon_{ij}^{p}}{\partial g/\partial \sigma_{ij}} = \ldots = d\lambda', \qquad (5.31)$$

$d\lambda'$ being a non-negative constant. (If $d\lambda'$ *was* negative it would mean that a negative strain would be associated with a positive stress, which is absurd.) This set of equations becomes identical with the Lévy-Mises equations if for g the Mises yield function f is substituted. Thus, in terms of principal stresses,

$$g(\sigma_{ij}) = f(\sigma_{ij}) = \sum(\sigma_1 - \sigma_2)^2.$$

Then
$$\frac{\partial f}{\partial \sigma_1} = 2(\sigma_1 - \sigma_2) - 2(\sigma_3 - \sigma_1)$$

$$= 4\sigma_1 - 2\sigma_2 - 2\sigma_3$$

$$= 6(\sigma_1 - \sigma_m) \quad \text{where} \quad \sigma_m = \tfrac{1}{3}\sum\sigma_1.$$

Thus, in accordance with (5.31) above,

$$d\epsilon_1^p = \frac{\partial g}{\partial \sigma_{ij}} d\lambda' = \frac{\partial f}{\partial \sigma_1} d\lambda' = 6(\sigma_1 - \sigma_m) d\lambda' = 6\sigma_1' d\lambda'$$

and hence
$$\frac{d\epsilon_1^p}{\sigma_1'} = \ldots = d\lambda, \qquad (5.32)$$

$d\lambda$ being another non-negative constant of proportionality.

A function $g(\sigma_{ij})$ which is to serve as a yield criterion as well as a plastic potential, must always so be chosen that it is a symmetric function of the three stress invariants. That is to say that g does not depend on the system of co-ordinates chosen—or indeed the principal directions—and by symmetric it is meant that each of the principal stresses is of equal 'weight'.

If along the σ_1-axis in three-dimensional stress space, see Fig. 5.4, the principal plastic increment $d\epsilon_1^p$ is marked off, and similarly $d\epsilon_2^p$ and $d\epsilon_3^p$ are treated with respect to the σ_2 and σ_3 axes, being also components of the plastic strain increment vector due to yielding under stress $(\sigma_1, \sigma_2, \sigma_3)$, then we may conveniently place this vector at the point $\sigma_1, \sigma_2, \sigma_3$ on the yield cylinder. Now, the direction of the strain increment vector is the same as the direction of the outward drawn normal at $(\sigma_1, \sigma_2, \sigma_3)$ on the $g(\sigma_{ij})$ surface. This follows because the ratios of the direction-cosines of the outward drawn normal to the surface $g(\sigma_{ij})$ at $(\sigma_1, \sigma_2, \sigma_3)$ are by the methods of ordinary three-dimensional Cartesian co-ordinate geometry,

$$\frac{\partial g}{\partial \sigma_1} \quad : \quad \frac{\partial g}{\partial \sigma_2} \quad : \quad \frac{\partial g}{\partial \sigma_3}.$$

This is the geometrical way of thinking about the way in which a *flow rule* may be obtained from a plastic potential; thus

$$d\epsilon_1^p \quad : \quad d\epsilon_2^p \quad : \quad d\epsilon_3^p = \frac{\partial g}{\partial \sigma_1} \quad : \quad \frac{\partial g}{\partial \sigma_2} \quad : \quad \frac{\partial g}{\partial \sigma_3}. \qquad (5.33)$$

A restriction that must, however, be placed on g, follows from the usual assumption about plastic incompressibility. We require g to be such that

$$d\epsilon_1^p + d\epsilon_2^p + d\epsilon_3^p = \frac{\partial g}{\partial \sigma_1} + \frac{\partial g}{\partial \sigma_2} + \frac{\partial g}{\partial \sigma_3} = 0.$$

The Mises yield criteria obviously conforms to this restriction.

It will now be seen that the assumption concerning the normality of the plastic strain increment vector to the yield surface, leads to another stress-strain increment relation for the Tresca yield criterion; it means that, corners excepted, there are only six differently directed normals associated with it, one for each of the six planes defining the cylinder. Only the maximum shear stress that has the greatest absolute value is applicable for a given $(\sigma_1, \sigma_2, \sigma_3)$, say $(\sigma_1 - \sigma_3)$. Then $g(\sigma_{ij}) = (\sigma_1 - \sigma_3)$ and using the rule in (5.31),

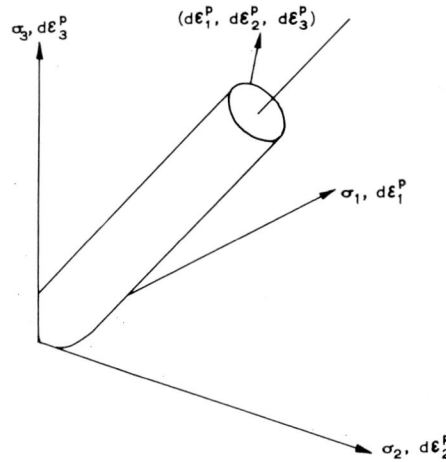

FIG. 5.4 Components of plastic strain increment shown in three-dimensional stress space.

$$d\epsilon_1^p \;:\; d\epsilon_2^p \;:\; d\epsilon_3^p = \frac{\partial(\sigma_1 - \sigma_3)}{\partial \sigma_1} \;:\; \frac{\partial(\sigma_1 - \sigma_3)}{\partial \sigma_2} \;:\; \frac{\partial(\sigma_1 - \sigma_3)}{\partial \sigma_3}$$

$$= 1 \;:\; 0 \;:\; -1. \tag{5.34}$$

This merely states that plastic straining takes place only in the plane of σ_1 and σ_3 and that the plastic strain increments are equal in magnitude but of opposite sign. The Tresca yield criterion with its associated flow rule are not used in this book, but reference may be made to KOITER (1953), PRAGER (1955) and BLAND (1956) for their application to particular problems.

5.9 The Principle of Maximum Work Dissipation

The energy increment dissipated per unit volume of a plastic-rigid material when the principal stresses in it are $(\sigma_1, \sigma_2, \sigma_3)$ or σ_i are

$$\delta w = \sigma_1\, d\epsilon_1 + \sigma_2\, d\epsilon_2 + \sigma_3\, d\epsilon_3 = \sigma_i\, d\epsilon_i.$$

where $(d\epsilon_1, d\epsilon_2, d\epsilon_3)$ are the principal plastic strain increments. Alternatively,

$\delta w \equiv$ scalar product of the stress deviation

vector $\overrightarrow{OP}$ and the strain increment vector $\left.\begin{array}{c} \\ \\ \\ \end{array}\right\} = \overrightarrow{OP}.\overrightarrow{PQ}$

$(d\epsilon_1, d\epsilon_2, d\epsilon_3), \overrightarrow{PQ}.$

This latter vector $\overrightarrow{PQ}$ is shown in Fig. 5.5, and is normal to the yield locus at

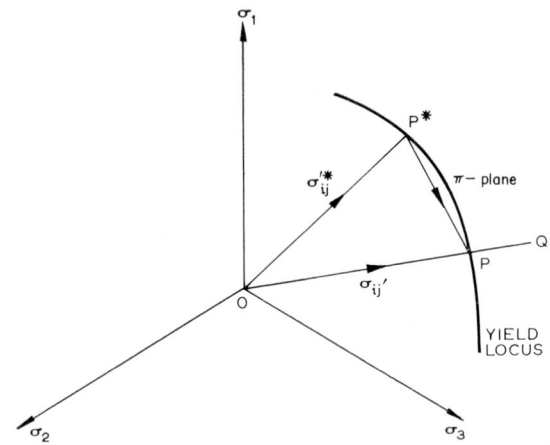

FIG. 5.5 Plastic strain increment vector $\overrightarrow{PQ}$ is normal to the yield locus at P.

P. Now the hydrostatic components of stress do no work and only the stress deviators enter into considerations of the plastic work dissipated. Thus the view shown in Fig. 5.5 may be taken to relate to the π-plane and hence $\delta w = \sigma_i'.d\epsilon_i$.

Now consider,

$$\delta w^* = \sigma_1^{*'} d\epsilon_1 + \sigma_2^{*'} d\epsilon_2 + \sigma_3^{*'} d\epsilon_3 = \sigma_i^{*'}.d\epsilon_i$$

in which $\overrightarrow{OP^*} \equiv (\sigma_1^{*'}, \sigma_2^{*'}, \sigma_3^{*'})$ and which also satisfies the yield criterion, i.e., P^* is on the yield locus as in Fig. 5.5.

Then

$$\delta w - \delta w^* = \overrightarrow{OP}.\overrightarrow{PQ} - \overrightarrow{OP^*}.\overrightarrow{PQ} = (\sigma_i' - \sigma_i^{*'}) d\epsilon_i. \qquad (5.35.i)$$

More generally, for an element of volume dV subject to stresses σ_{ij} which cause strain increments $d\epsilon_{ij}$, and stresses σ_{ij}^*, corresponding to (5.35.i), we will have, for the difference of work increments,

$$(\sigma_{ij}' - \sigma_{ij}^{*'}) d\epsilon_{ij}.dV. \qquad (5.35.ii)$$

Now consider a rigid perfectly plastic body undergoing plastic deformation throughout a volume V, such that the strain increments are $d\epsilon_{ij}$ at each point.

Then the true increment in plastic work δW required to bring about a given set of strain increments, is greater than the increment in plastic work δW^* which would be required to bring about this same strain increment by any other distribution of stresses throughout V which conforms to the same yield criterion. This follows from (5.35.ii) as,

$$\delta W - \delta W^* = \int_V (\sigma_{ij}' - \sigma_{ij}^{*\prime}) \, d\epsilon_{ij} \, dV. \tag{5.35.iii}$$

Instead of treating with strain increments, statements may be made in terms of strain rates and rates of working. With an obvious change of notation, and dots referring to time rates of change,

$$\dot{W} - \dot{W}^* = \int_V (\sigma_{ij}' - \sigma_{ij}^{*\prime}) . \dot{\epsilon}_{ij} \, dV. \tag{5.35.iv}$$

We may restate (5.35.iv) in its more usual form as

$$\int_V (\sigma_{ij}' - \sigma_{ij}^{*\prime}) \, \dot{\epsilon}_{ij} \, dV \geqslant 0, \tag{5.36}$$

for a concave-to-the-origin yield locus.

Inequality (5.36) states that the rigid-perfectly plastic material undergoes distortion or deformation in such a way as to cause a maximum dissipation of energy. This is the Principle of Maximum Work Dissipation. It has a wider application than to plastic deformation only and an interesting example is in relation to ordinary mechanical friction. On a horizontal plane, see Fig. 5.6, the friction force is μP where P is the weight of the body and μ is the

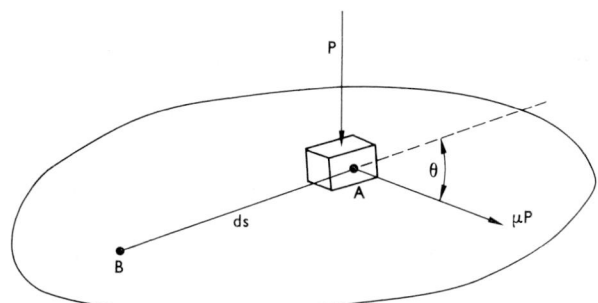

FIG. 5.6 Illustrating the principle of maximum work dissipation.

coefficient of friction. If the body is to be moved from A to B through distance ds, the work W dissipated in overcoming friction is $W = \overrightarrow{\mu P} . \overrightarrow{ds} = \mu . P . ds . \cos \theta$. W is greatest when $\theta = 0$, i.e., the friction force is in the direction opposite to that in which motion occurs; force P acts so as to maximize the work done. This can be thought of as an example of the maximum work principle. This analogy should not, however, be carried too far, see DRUCKER (1953).

5.10 Flow Rule for Anisotropic Material

The hypothesis is made, as for the isotropic material, that $f(\sigma_{ij})$ is the plastic potential. The incremental strains are thus derived by partially differentiating $f(\sigma_{ij})$ with respect to σ_{ij}. Since

$$2f(\sigma_{ij}) = F(\sigma_y - \sigma_z)^2 + G(\sigma_z - \sigma_x)^2$$

$$+ H(\sigma_x - \sigma_y)^2 + 2L\tau_{yz}^2 + 2M\tau_{zx}^2 + 2N\tau_{xy}^2$$

$$\frac{\partial f}{\partial \sigma_x} = G(\sigma_x - \sigma_z) + H(\sigma_x - \sigma_y)$$

and hence

$$\frac{d\epsilon_x{}^p}{G(\sigma_x - \sigma_z) + H(\sigma_x - \sigma_y)} = d\lambda.$$

Similar expressions are obtained for the other components of the strain-increment and can be written down as

$$
\begin{aligned}
d\epsilon_x{}^p &= d\lambda[H(\sigma_x - \sigma_y) + G(\sigma_x - \sigma_z)], & d\gamma_{yz}{}^p &= d\lambda L\tau_{yz} \\
d\epsilon_y{}^p &= d\lambda[F(\sigma_y - \sigma_z) + H(\sigma_y - \sigma_x)], & d\gamma_{zx}{}^p &= d\lambda M\tau_{zx} \qquad (5.37) \\
d\epsilon_z{}^p &= d\lambda[G(\sigma_z - \sigma_x) + F(\sigma_z - \sigma_y)], & d\gamma_{xy}{}^p &= d\lambda N\tau_{xy}.
\end{aligned}
$$

These expressions of course satisfy the incompressibility condition $d\epsilon_x{}^p + d\epsilon_y{}^p + d\epsilon_z{}^p = 0$. When the elastic strain increments are small compared with the plastic strain increments the superfix 'p' in the above expressions can be dropped as for a Lévy-Mises rigid-plastic material.

If a simple tension is applied to a strip lying in the (x,y) plane and cut parallel to the x-axis of anisotropy the incremental strain ratios are

$$d\epsilon_x{}^p \quad : \quad d\epsilon_y{}^p \quad : \quad d\epsilon_z{}^p = G + H \quad : \quad -H \quad : \quad -G.$$

The ratio of thickness to width strain is known as the r-value and therefore

$$r_x = \frac{d\epsilon_y{}^p}{d\epsilon_z{}^p} = \frac{H}{G}$$

where the suffix 'x' denotes that the specimen is oriented along the 'x'-direction. For a strip cut in the y-direction

$$d\epsilon_x{}^p \quad : \quad d\epsilon_y{}^p \quad : \quad d\epsilon_z{}^p = -H \quad : \quad F + G \quad : \quad -F$$

$$\text{and } r_y = \frac{d\epsilon_x{}^p}{d\epsilon_z{}^p} = \frac{H}{F}.$$

So far, we have considered loading to be applied along the axes of anisotropy. In order to derive the required anisotropic parameters for plane-stress deformation of sheet it is necessary to carry out a tensile test in at least one other direction in the plane of the sheet. If anisotropic sheet is subjected to forces in the plane of the sheet, the (x,y) plane, then τ_{yz} and τ_{zy} will be zero. Suppose a strip tensile specimen is cut at an angle to the x-direction then from considerations of equilibrium

$$\sigma_x = \sigma \cos^2 \alpha, \quad \sigma_y = \sigma \sin^2\alpha, \quad \tau_{xy} = \sigma \sin \alpha \cos \alpha$$

where σ is the applied tensile yield stress. Substitution of these values in equation (5.37) gives

$$d\epsilon_x{}^p = [(G + H) \cos^2\alpha - H \sin^2\alpha] \, \sigma d\lambda$$

$$d\epsilon_y{}^p = [(F + H) \sin^2\alpha - H \cos^2\alpha] \, \sigma d\lambda \qquad (5.38)$$

$$d\epsilon_z{}^p = - [F \sin^2\alpha + G \cos^2\alpha] \, \sigma d\lambda$$

$$d\gamma_{xy}{}^p = [N \sin \alpha \cos \alpha] \, \sigma d\lambda.$$

From considerations of the geometry of small strain, the width strain increment $d\epsilon^p_{\alpha + (\pi/2)}$ is given by

$$d\epsilon^p_{\alpha + (\pi/2)} = d\epsilon_x{}^p \sin^2\alpha + d\epsilon_y{}^p \cos^2\alpha - 2 \, d\gamma_{xy}{}^p \sin \alpha \cos \alpha$$

and therefore

$$r_\alpha = \frac{d\epsilon^p_{\alpha + (\pi/2)}}{d\epsilon_z{}^p} = \frac{d\epsilon_x{}^p \sin^2\alpha + d\epsilon_y{}^p \cos^2\alpha - 2 \, d\gamma_{xy}{}^p \sin \alpha \cos \alpha}{d\epsilon_z{}^p}$$

$$r_\alpha = \frac{H + (2N - F - G - 4H) \sin^2\alpha \cos^2\alpha}{F \sin^2\alpha + G \cos^2\alpha}. \qquad (5.39)$$

In sheet metal the rolling direction is usually an axis of anisotropy and the x-direction is then taken to coincide with the rolling direction. The above equation then yields

$$r_x = r_0 = \frac{H}{G},$$

$$r_y = r_{90} = \frac{H}{F}, \qquad (5.40)$$

$$r_{45} = \frac{2N - (F + G)}{F + G} \quad \text{or} \quad \frac{N}{G} = \left(r_{45} + \frac{1}{2}\right)\left(1 + \frac{r_0}{r_{90}}\right).$$

r_0, r_{45} and r_{90} indicate r-values along, at forty-five degrees and ninety degrees respectively to the direction of rolling.

It is of course assumed in this derivation that the ratios of anisotropic parameters remain constant over the range of measurement. This is something that must be checked experimentally for individual materials. This would seem to be a correct assumption for aluminium (KLINGER and SACHS, 1948), for deep-drawing steel and titanium (BRAMLEY and MELLOR, 1966, 1967), but AVERY, HOSFORD and BACKOFEN (1965) found that the r-values for some magnesium alloys varied greatly with strain. It must be stressed that the simple theory cannot be expected to describe the anisotropic behaviour of all metals and alloys and a need exists to define more clearly its range of useful engineering applications. Yielding of anisotropic materials has been studied by LEE and BACKOFEN (1966), MEHAN (1961) and by BABEL, EITMAN and MCIVER (1966).

For metal forming purposes the r-values of sheet materials are usually assessed for longitudinal strains greater than five per cent. It can then reasonably be assumed that the elastic strains are negligible and that therefore volume constancy applies. The longitudinal and width strains are measured and the thickness strain (which is difficult to measure accurately in thin sheet) derived. Thus

$$r = \frac{\ln (w_0/w)}{\ln (t_0/t)} = \frac{\ln (w_0/w)}{\ln (wl/w_0 . l_0)} \tag{5.41}$$

where w, t and l are the current width, thickness and longitudinal gauge lengths. The suffix 0 refers to initial conditions. The measurement of r-value has been considered in detail by ATKINSON (1967).

5.11 Generalized Stress and Strain Relations for Anisotropic Work-hardening Material

When a material is deformed plastically the state of anisotropy changes. However, it will be assumed here that this change in anisotropy is negligible compared with the anisotropy existing in the material at the start of testing. The theory will therefore be expected to give closest agreement with a material that is strongly anisotropic to begin with. A similar argument is of course used in isotropic theory where a material is assumed to be isotropic initially and to remain isotropic as plastic deformation occurs. It is known that this is not strictly correct but experience shows that it is a reasonable approximation for most applications.

If the state of anisotropy remains constant the yield stresses must increase in strict proportion as the material work-hardens and it follows that the anisotropic parameters must decrease in strict proportion. The ratios of the parameters will therefore remain constant and it is the ratios not the absolute values of the individual parameters that are determined by experiment. It has already been stated in the previous section that the strain ratios remain constant for some materials but not for others. Caution should therefore be used in applying anisotropic theory to particular materials.

HILL (1950) has proposed that the equivalent stress should be defined as

$$\bar{\sigma} = \sqrt{\frac{3}{2}} \left[\frac{F(\sigma_y - \sigma_z)^2 + G(\sigma_z - \sigma_x)^2 + H(\sigma_x - \sigma_y)^2 + 2L\tau_{yz}^2 + 2M\tau_{zx}^2 + 2N\tau_{xy}^2}{F + G + H} \right]^{\frac{1}{2}} \tag{5.42}$$

where it is understood that only ratios of the anisotropic parameters, not the absolute values, will be considered. The expression reduces to equation (5.12) when anisotropy is negligible and loading is along the principal axes.

Following JACKSON, SMITH and LANKFORD (1948), HILL (1950) assumed by analogy with isotropic theory that $\bar{\sigma}$ is a function of the plastic work. The increment of plastic work per unit volume, for an assumed rigid plastic material, is

$$dw = \sigma_{ij} \, d\epsilon_{ij} = \sigma_{ij} \frac{\partial f}{\partial \sigma_{ij}} \, d\lambda. \tag{5.43}$$

Using equation (4.16) and Euler's theorem on homogeneous functions (see for example SOKOLNIKOFF (1941), p. 136),

$$dw = \sigma_{ij} \frac{\partial f}{\partial \sigma_{ij}} d\lambda = 2f \, d\lambda = d\lambda. \tag{5.44}$$

From equations (5.37) we obtain, for a rigid plastic material the equations

$$G \, d\epsilon_y - H \, d\epsilon_z = (FG + GH + HF)(\sigma_y - \sigma_z) \, d\lambda$$
$$H \, d\epsilon_z - F \, d\epsilon_x = (FG + GH + HF)(\sigma_z - \sigma_x) \, d\lambda$$
$$F \, d\epsilon_x - G \, d\epsilon_y = (FG + GH + HF)(\sigma_x - \sigma_y) \, d\lambda.$$

The generalized strain increment, $d\bar{\epsilon}$, can therefore be defined from

$$dw = \bar{\sigma} \, d\epsilon = d\lambda$$

as

$$d\bar{\epsilon} = \frac{d\lambda}{\bar{\sigma}} = \sqrt{\frac{3}{2}} \left[F + G + H \right]^{\frac{1}{2}} \left[F \left(\frac{G \, d\epsilon_y - H \, d\epsilon_z}{FG + GH + HF} \right)^2 + \ldots \frac{2 d\gamma_{yz}^2}{L} + \ldots \right]^{\frac{1}{2}} \tag{5.45}$$

For a sheet material subjected to plane stress, with rotational symmetry about the z-axis, so that

$$r = \frac{H}{G} = \frac{H}{F},$$

equations (5.42) and (5.45) reduce to

$$\bar{\sigma} = \sqrt{\frac{3}{2}} \left[\frac{\sigma_x^2 + \sigma_y^2 + r(\sigma_x - \sigma_y)^2}{2 + r} \right]^{\frac{1}{2}}, \text{ and} \tag{5.46}$$

$$d\bar{\epsilon} = \sqrt{\frac{2}{3}} \left[\frac{2 + r}{(1 + 2r)^2} \left\{ (d\epsilon_y - r \, d\epsilon_z)^2 + (d\epsilon_x - r \, d\epsilon_z)^2 + r(d\epsilon_x - d\epsilon_y)^2 \right\} \right]^{\frac{1}{2}} \tag{5.47}$$

The above equations are useful for qualitative discussion of the effect of normal anisotropy in sheet forming.

See Problem 21

REFERENCES

ATKINSON, M. 1967 *Assessing Normal Anisotropy of Sheet Metals* Sheet Metal Industries, March

AVERY, D. H., HOSFORD, W. F. Jr. and BACKOFEN, W. A. 1965 'Plastic Anisotropy in Magnesium Alloy Sheets' *Trans. metall. Soc. A.I.M.E.* **233**, 71

BABEL, H. W., EITMAN, D. A. and MCIVER, R. W. 1966 'The Biaxial Strengthening of Textured Titanium' *A.S.M.E.* Paper 66—Met. 6

BRAMLEY, A. N. and MELLOR, P. B. 1966 'Plastic Flow in Stabilized Sheet Metal' *Int. J. mech. Sci* **8**, 101

1967 'Plastic Anisotropy of Titanium and Zinc Sheet—I. Macroscopic Approach' *Int. J. mech. Sci.* **10**, 211

BLAND, D. R.   1956 'Elastoplastic Thick-walled Tubes of Work-hardening Material Subject to Internal and External Pressures and to Temperature Gradients'
J. Mech. Phys. Solids **4**, 209
     1957 'The Associated Flow Rule of Plasticity'
J. Mech. Phys. Solids **6**, 71

DRUCKER, D. C.   1953 'Coulomb Friction, Plasticity and Limit Loads'
J. appl. mech. A.S.M.E. Paper No. 53-A-57

HENCKY, H.   1924 'Zur Theorie plastischer Deformationen und der hierdurch im Material hervorgerufenen Nachspannungen'
Z. angew. Math. Mech. **4**, 323

HILL, R.   1950 *The Mathematical Theory of Plasticity* O.U.P.
     1953 'A New Method for Determining the Yield Criterion and Plastic Potential of Ductile Metals'
J. Mech. Phys. Solids **1**, 271

HILL, R., LEE, E. H. and TUPPER, S. J.   1947 'The Theory of Combined Plastic and Elastic Deformation'
Proc. R. Soc. A. **191**, 278

HOHENEMSER, K.   1931 'Fliessversuche an Röhren aus Stahl bei kombinierter Zug- und Torsionsbeanspruchung'
Z. angew. Math. Mech. **11**, 15

HUNDY, B. B. and GREEN, A. P.   1954 'A Determination of Plastic Stress-Strain Relations'
J. Mech. Phys. Solids **3**, 16

JACKSON, L. R., SMITH, K. F. and LANKFORD, W. T.   1948 *Plastic Flow in Anisotropic Sheet Metal* Metals Technology Tech. Pub. 2440

KLINGER, L. G. and SACHS, G.   1948 'Dependence of the stress-strain curves of cold worked metals upon the testing direction'
J. aeronaut. Sci. **15**, 599

KOITER, W. T.   1953 *Biezeno Anniversary Volume* Stam, Haarlem, 232

LEE, D. and BACKOFEN, W. A.   1966 'An Experimental Determination of the Yield Locus for Titanium and Titanium Alloy Sheet'
Trans. metall. Soc. A.I.M.E. **236**, 1077

LÉVY, M.   1870 *C. r. Acad. Sci., Paris* **70**, 1323

LODE, W.   1926 'Versuche über den Einfluss der mittleren Hauptspannung auf das Fliessen der Metalle Eisen, Kupfer und Nickel'
Zeitsch. Phys. **36**, 913

MEHAN, R. L.   1961 'Effect of Combined Stress on Yield and Fracture Behaviour of Zircaloy-2'
Trans. A.S.M.E., J. Basic Eng. **83**, 499

MISES, R. VON   1913 *Göttinger Nachrichten, math.-phys. Kl.*, 582
                                1928 'Mechanik der plastischen Formänderung von Kristallen'
                                Z. ang. Math. Mech. **8**, 161

MORRISON, J. L. M.   1950 'An Experimental Investigation of Plastic
and SHEPHERD, W. M.   Stress-Strain Relations'
                                Proc. Instn Mech. Engrs **163**, 1

PRAGER, W.   1955 'The Theory of Plasticity: A Survey of Recent Achievements'
                                Proc. Instn Mech. Engrs **169**, 41

PRANDTL, L.   1924 Proc. 1st Int. Congr. App. Mech., Delft, 43

PUGH, H. Ll. D.   1953 'A Note on a Test of the Plastic Isotropy of Metals'
                                J. Mech. Phys. Solids **1**, 284

REUSS, A.   1930 *Z. ang. Math. Mech.* **10**, 266

SAINT-VENANT, B. DE   1870 *C. r. Acad. Sci., Paris* **70**, 473

SOKOLNIKOFF, I. S.   1941 *Higher Mathematics for Engineers and*
and E. S.   *Physicists*
                                McGraw-Hill, New York

TAYLOR, G. I.   1947 'A Connexion between the Criteria of Yield and the Strain Ratio Relationship in Plastic Solids'
                                Proc. R. Soc., A. **191**, 441

TAYLOR, G. I. and   1931 'The Plastic Distortion of Metals'
QUINNEY, H.   *Phil. Trans. R. Soc.* A. **230**, 323

Chapter 6

METHODS OF DETERMINING
WORK-HARDENING CHARACTERISTICS

6.1 Introduction

At some point in any analysis of the deformation of metals under a given loading system resort must always be made to experimentally derived relations between stress and strain. It is desirable, though not usually practicable, to determine the stress-strain relations under conditions similar to those prescribed in the analysis. When this is not possible, use has to be made of an effective stress-strain relationship and it will be remembered that such relationships cannot, in general, take into account anisotropy or the Bauschinger effect.

In the determination of an experimental stress-strain curve, it will be noted that the results will also depend on the speed of testing and on the temperature. The discussion will be limited to the cold-working of metals so that high-temperature creep is not a factor to be considered, and it will be further assumed that the tests are carried out at approximately the same strain rates as ordinary short-time tensile tests (about 2×10^{-3} per second). It has been shown (MANJOINE, 1944) that in this region the tensile strength and yield point values are practically constant.

In many cases of metal forming, the plastic strains are so large that the elastic strains are negligible by comparison. In obtaining stress-strain curves for such processes, it is permissible to neglect elastic strains altogether. Our idealized material is then one that is perfectly rigid up to yield and the plastic strain is the total strain.

In the previous chapter effective stress and effective strain-increment were derived, equations (5.12) and (5.18), on the basis of equivalence of plastic work. These relationships apply to a Lévy-Mises solid—that is, to one with equal Lode variables, $\mu = \nu$ and zero elastic strains.

The equations have the form

$$\sigma = \sqrt{[(\sigma_1 - \sigma_2)^2 + (\sigma_2 - \sigma_3)^2 + (\sigma_3 - \sigma_1)^2]/2} \qquad (6.1)$$

$$d\bar{\epsilon} = \sqrt{2[(d\epsilon_1 - d\epsilon_2)^2 + (d\epsilon_2 - d\epsilon_3)^2 + (d\epsilon_3 - d\epsilon_1)^2]/3}. \qquad (6.2)$$

In the experimental determination of stress-strain curves, the strain-ratios will usually be constant and there will be no rotation in the element of the principal axes of successive strain-increments. Equation (6.2) can then be used in the integrated form

$$\bar{\epsilon} = \sqrt{2[(\epsilon_1 - \epsilon_2)^2 + (\epsilon_2 - \epsilon_3)^2 + (\epsilon_3 - \epsilon_1)^2]/3}. \tag{6.3}$$

(This was shown to be valid in Section 5.5. Simple torsion is an example where the principal axes *do* rotate relatively to the element).

Equations (6.1) and (6.3) were first suggested in this form by Ros and Eichinger (1929). The constants $1/\sqrt{2}$ and $\sqrt{2/3}$ in the equations are chosen so that in simple tension $\bar{\sigma} = Y$, the uniaxial yield stress, and $\bar{\epsilon} = \epsilon$, the corresponding longitudinal strain. Similar relations, differing only in the values of the numerical constants, have been put forward by Nadai (1937) and Swift (1946).

Nadai introduced the concept of 'octahedral shearing stress' and 'octahedral shearing strain' defined by the equations

$$\tau_{oct} = \sqrt{[(\sigma_1 - \sigma_2)^2 + (\sigma_2 - \sigma_3)^2 + (\sigma_3 - \sigma_1)^2]/3} \tag{6.4}$$

$$\gamma_{oct} = 2\sqrt{[(\epsilon_1 - \epsilon_2)^2 + (\epsilon_2 - \epsilon_3)^2 + (\epsilon_3 - \epsilon_1)^2]/3}. \tag{6.5}$$

'Octahedral shearing stress' is the shearing stress on planes having direction cosines of $1/\sqrt{3}$ with respect to the principal stress axes.

Swift suggested the use of a representative shear stress, q, defined as the root mean square of the maximum shear stresses. The appropriate equations are (ψ being the representative shear strain)

$$q = \sqrt{[(\sigma_1 - \sigma_2)^2 + (\sigma_2 - \sigma_3)^2 + (\sigma_3 - \sigma_1)^2]}/2\sqrt{3} \tag{6.6}$$

$$\psi = \sqrt{[\epsilon_1 - \epsilon_2)^2 + (\epsilon_2 - \epsilon_3)^2 + (\epsilon_3 - \epsilon_1)^2]}/2\sqrt{3}. \tag{6.7}$$

The definition of effective stress, contained in equations (6.1), (6.2) and (6.3) will be used throughout the book.

It is worth while comparing the stress-strain curves for simple tension, simple compression and the torsion of a thin-walled tube, directly, on a basis of equivalent plastic work. Elastic strains can be considered in these cases and the importance of logarithmic strains will be demonstrated.

Consider the stress-strain curve for a circular bar in tension. If the current gauge length is l and current yield stress Y, then the *increase* in total work per unit volume in increasing the length by dl is $Y\,dl/l$. The recoverable elastic *increment* of work constitutes an amount $Y \cdot dY/E$ of this total and thus the increment of plastic work is $Y(dl/l - dY/E)$. The total plastic work is therefore

$$W_p = \left(\int_{l_0}^{l} Y\frac{dl}{l} \right) - \frac{Y^2}{2E}$$

where l_0 is the original gauge length. Thus

$$Y = F\left\{ \left(\int_{l_0}^{l} Y\frac{dl}{l} \right) - \frac{Y^2}{2E} \right\}. \tag{6.8}$$

If Y is plotted against $\ln (l/l_0) - Y/E$ the argument of F is the area under the curve up to the ordinate Y.

In an exactly similar manner, it can be shown that for simple compression, the compressive stress

$$\sigma = F\left\{\left(\int_h^{h_0} \sigma \frac{dh}{h}\right) - \frac{\sigma^2}{2E}\right\} \qquad (6.9)$$

where h_0 and h are the initial and current heights of the specimen. It is seen that the compressive stress, σ, is the same function of $\ln h_0/h$ as Y is of $\ln l/l_0$.

Thus, according to this hypothesis, the stress-strain curves for simple tension and simple compression will coincide if Y is plotted against $\ln l/l_0$ and σ against $\ln h_0/h$. The terms $\ln l/l_0$ and $\ln h_0/h$ are the natural or logarithmic strains. Note therefore that the curves should coincide if the stresses are plotted against the ratios l/l_0 and h_0/h but not if plotted against the conventional engineering strains $(l - l_0)/l_0$ and $(h_0 - h)/h_0$.

Finally, consider an isotropic thin-walled circular tube subjected to a pure torque. On twisting, a line on the tube parallel to the axis becomes a helix making an angle ϕ with the axial direction. For simple torsion $\sigma_1 = -\sigma_2 = \tau$ the maximum shear stress, $\sigma_3 = 0$, and the effective stress $\bar{\sigma} = \sqrt{3}\tau$ from equation (6.1). The work per unit volume done by the shear stress, τ, in a further small twist $d\phi$ is $\tau\, d(\tan \phi)$, and so

$$\bar{\sigma} = \sqrt{3}\tau = F\left\{\left[\int_0^\phi \tau\, d(\tan \phi)\right] - \frac{\tau^2}{2G}\right\}. \qquad (6.10)$$

If $\sqrt{3}\tau$ is plotted against $(\tan \phi - \tau/G)/\sqrt{3}$, the argument of F is the area under the curve up to the ordinate $\sqrt{3}\tau$. In other words, $\sqrt{3}\tau$ is the same function of $(\tan \phi - \tau/G)/\sqrt{3}$ as Y is of $(\ln l/l_0 - \sigma/E)$ and the stress-strain curve is plotted accordingly.

MORRISON (1948) and SHEPHERD (1948) have determined the yield stress in tension experimentally for a chromium molybdenum steel and have shown that the shear stress curve derived from this test is in good agreement with experimental results obtained by twisting a solid bar of the same material. Nadai's construction was used for deriving the stress-strain results for the solid bar. This construction is discussed in section 6.7.

6.2 Simple Tension

The simple tensile test is the most commonly used method of measuring the mechanical properties of metals. A machined circular bar or strip specimen is subjected to an axial tensile load. An increase in the load increases the tensile stress, reduces the cross-sectional area and increases the longitudinal or axial strain. The axial stress $\sigma = P/X$ where P is the load and X the current cross-sectional area. The conventional axial strain is $(l - l_0)/l_0$ and the logarithmic strain is $\ln l/l_0$ where l_0 is the original and l the current gauge length. The current cross-sectional area may be derived from measurements of the bar, but it is usually sufficiently accurate to neglect the small elastic changes in volume and estimate its value from the condition of incompressibility

$$Xl = X_0 l_0$$

where X_0 is the original cross-sectional area. Thus the true stress

$$\sigma = Pl/X_0 l_0.$$

The main disadvantage of the tensile test as a means of obtaining a strain-hardening characteristic is that at the maximum load the specimen begins to neck down at comparatively low strains. The stress distribution in the neck becomes triaxial and cannot be predicted theoretically. BRIDGMAN (1944) and DAVIDENKOV and SPIRIDONOVA (1946) have determined the approximate distribution experimentally in specific cases.

6.3 Balanced Biaxial Tension

Larger strain values under tensile conditions are reached by deforming a circular diaphragm by uniform lateral pressure. This is achieved by clamping the circular diaphragm at its periphery and applying oil pressure on one side, Fig. 6.1. The thickness-to-diameter ratio of the blank must be such that bending and shearing stresses can be neglected. The results to be quoted were obtained using material of thickness 0·036 to 0·039 in and a blank diameter of 10 in (MELLOR, 1956).

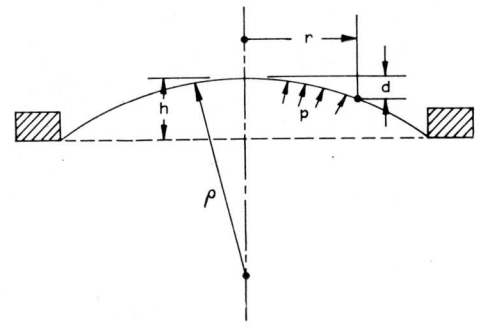

FIG. 6.1 Deforming metal diaphragm by fluid pressure.

The conditions at the pole of the dome are given by

$$p = 2\sigma_h t/\rho \qquad (6.11)$$

where p is the oil pressure, σ_h is the hoop stress, t the current material thickness and ρ the radius of curvature. It is implicit in this equation that there is rotational symmetry about the pole and hence that the deformation is assumed to be isotropic. Strain and other measurements were taken across diameters along and perpendicular to the rolling direction of the original sheet and the results averaged. Otherwise, no assessment was made of anisotropy. For reasons of safety and to reduce creep to a minimum, the pressure was removed before taking measurements across the dome. It was shown by taking readings for both continuous and interrupted tests that this made no appreciable difference in the stress-strain results.

The profile of the dome was measured with a depth micrometer and it was shown to be very nearly spherical within two inches of the pole. To estimate the radius of curvature at the pole, the general method used by BROWN and SACHS (1948) was adopted. Average radii of curvature were measured for various chord lengths and the radius at the pole was obtained by extrapolation. Each of these radii was calculated from the chord length and its corresponding sagitta, neglecting any deviation from the arc of a circle. If d is the vertical distance of a point from the pole and r its horizontal distance from the pole then the radius of the circle passing through the pole and the given point, and having its centre on the axis of symmetry, is $\rho = (r^2 + d^2)/2d$.

To estimate the hoop strain at the pole, concentric ink circles were drawn from the centre of the original blank at radial intervals of 0·2 in and the radial movement with increasing pressure measured with a vernier microscope. If r_0 is the initial radius to the particle and r the current radius, then

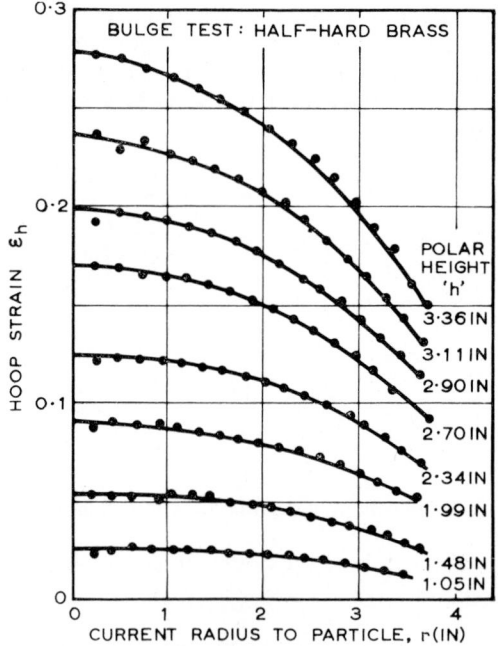

FIG. 6.2 Distribution of hoop stress across metal diaphragm. (*After* Mellor, *J. Mech. Phys. Solids.*)

the logarithmic hoop strain $\epsilon_h = \ln r/r_0$. From symmetry the hoop strain at the pole is equal to the radial strain ϵ_r and the thickness strain ϵ_t is given by the equation of incompressibility

$$\epsilon_h + \epsilon_r + \epsilon_t = 0.$$

Therefore $$\epsilon_t = -2\epsilon_h = -\ln t_0/t \qquad (6.12)$$

where t_0 is the original thickness of the sheet. Hence from a measurement of hoop strain the thickness strain and the current thickness of the material at the pole may be calculated. The distribution of hoop strain for a particular case is shown in Fig. 6.2.

Now consider the stress system at the pole. For a thin membrane the stress normal to the sheet is negligible and therefore the material at the pole is subjected to a balanced bi-axial tension, σ. Since a hydrostatic pressure has no effect on yielding, the system is therefore equivalent to a simple compressive stress, σ, normal to the sheet, and a plot of σ against ϵ_t gives the effective stress-strain curve. This can also be derived from equations (6.1) and (6.3).

Typical stress-strain curves derived in this manner are shown in Figs. 6.3

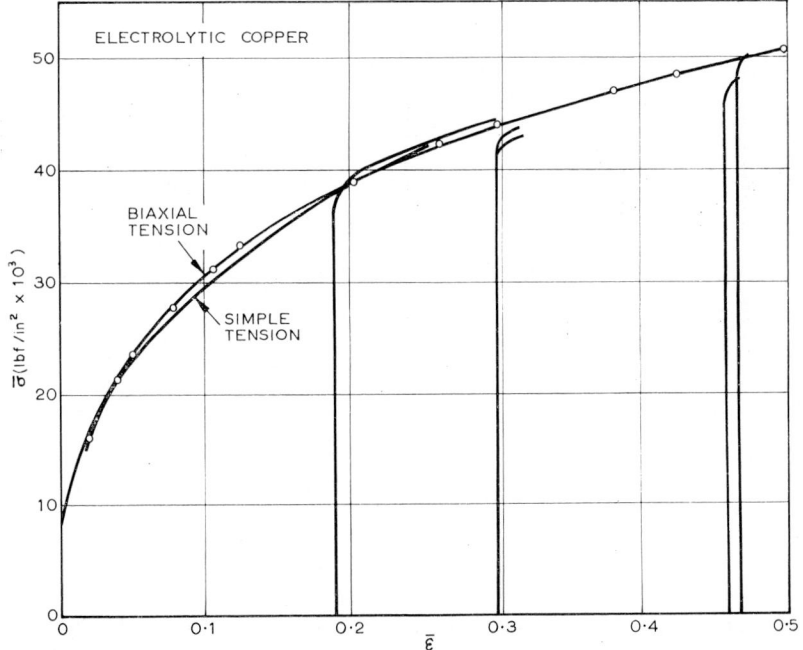

FIG. 6.3 Effective stress-strain curve for copper.
(Curves not starting at $\bar{\epsilon} = 0$ were obtained by rolling.) (*After* Mellor, *J. Mech. Phys. Solids.*)

and 6.4. Also shown are stress-strain values obtained on the same material by rolling and testing in tension (see next section).

Manual and automated biaxial test extensometers have been described by DUNCAN and JOHNSON (1965) and BELL, DUNCAN and JOHNSON (1965) respectively.

Diaphragm tests on anisotropic materials were first carried out by JACKSON, SMITH and LANKFORD (1948). When considering conditions at the pole of the diaphragm they assumed that the loading was along the anisotropic axes of the material. They pointed out that this assumption was incorrect but in fact when they predicted the biaxial stress-strain curves from simple tension there was fairly good correlation for deep-drawing steel. At the pole of a diaphragm the applied forces in the plane of the sheet act in all directions not just along the x-, y-axes and BRAMLEY and MELLOR (1966 a) therefore

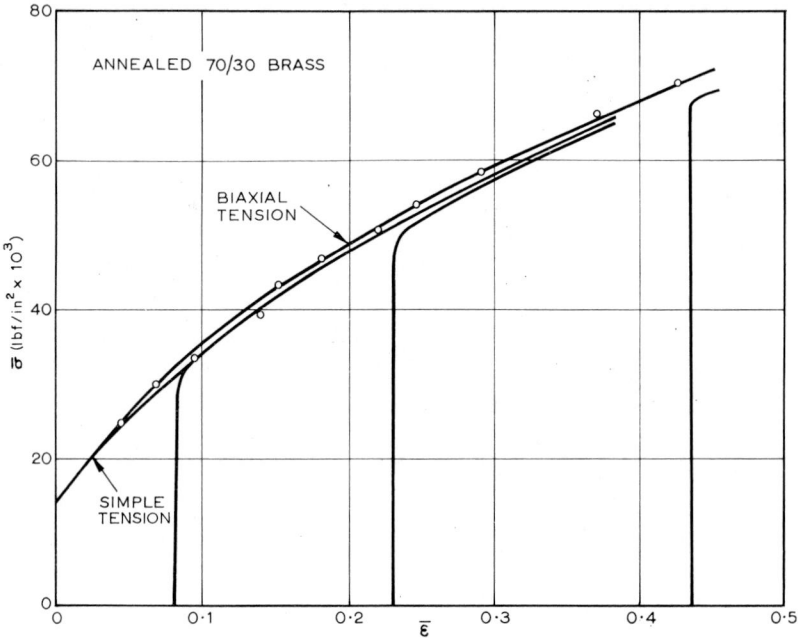

FIG. 6.4 Effective stress-strain curve for brass.
(Curves not starting at $\bar{\varepsilon} = 0$ obtained by rolling.) (*After* Mellor, *J. Mech. Phys. Solids.*)

suggested that it would be better to average the planar properties of the material thus imposing rotational symmetry about the z-axis. Measurements on aluminium killed steel diaphragms showed that for these materials there was very little angular variation in polar strains or in polar radii of curvature and that therefore little error would be introduced by the assumption of rotational symmetry. As a basis for calculation the r-values were determined from tensile specimens cut every ten degrees to the direction of rolling, Fig. 6.5, and the average r-value, $\bar{r}$, was calculated from the area under the curve. The stress-strain curve corresponding to the $\bar{r}$-value was also derived from the experimental data.

The ratios of the anisotropic parameters for the four killed steels tested are given in Fig. 6.6. According to the theory there will be rotational symmetry when

$$N = F + 2H = G + 2H. \qquad (4.17)$$

On the basis of this equation Steel D would be expected to show least planar anisotropy and this was borne out by the small spread of the tensile stress-strain curves compared with the other steels. With the assumption of rotational symmetry the stresses at the pole in the plane of the sheet are

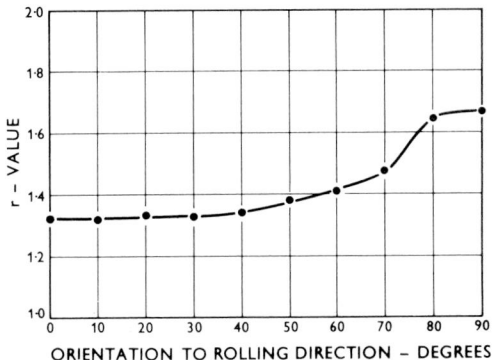

FIG. 6.5 Variation of r-value with orientation: steel D. (*After* Bramley and Mellor, *Int. J. mech. Sci.*)

Material	F	:	G	:	H	:	N	: $(F + 2H)$	:	$(G + 2H)$
Steel A	1·00	:	1·07	:	1·91	:	3·31	: 4·82	:	4·89
Steel B	1·00	:	1·11	:	1·91	:	3·69	: 4·82	:	4·93
Steel C	1·00	:	1·15	:	1·79	:	3·38	: 4·58	:	4·73
Steel D	1·00	:	1·27	:	1·68	:	4·20	: 4·36	:	4·63

FIG. 6.6.   Ratio of anisotropic parameters for four killed steels. (F has arbitrarily been taken as unity) (*After* Bramley and Mellor, *Int. J. mech. Sci.*)

equal and since a superimposed hydrostatic stress does not affect yielding the system is equivalent to a simple compressive stress, σ_z, normal to the sheet. For copper and brass it has been shown, Figs. 6.3 and 6.4, that plotting σ_z against the thickness strain ϵ_z gives a curve that is practically identical to the simple tension curve. If we repeat this procedure for killed steel the two curves are not coincident, Fig. 6.7. In this case, the stress-strain curve from the diaphragm test can be predicted from the simple tensile results as follows. The normal stress σ_z can be related to the average tensile stress, σ_{av}, by equation (5.46).

$$\sqrt{\frac{3}{2}} \left[\frac{2}{2 + r} \right]^{\frac{1}{2}} \sigma_z = \bar{\sigma} = \sqrt{\frac{3}{2}} \left[\frac{1 + r}{2 + r} \right]^{\frac{1}{2}} \sigma_{av}$$

or

$$\sigma_z = \sigma_{av} \left(\frac{1 + r}{2} \right)^{\frac{1}{2}}. \tag{6.13}$$

This equation has often been used incorrectly to predict the change in stress from one system to another without any reference to strain. Although this will predict the correct stress level for non-work-hardening materials it leads to large errors as the rate of work-hardening increases.

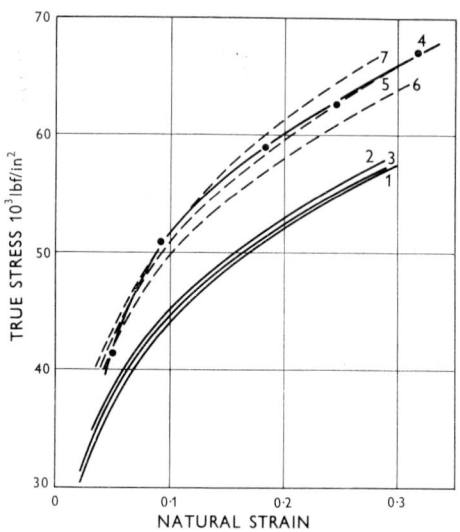

FIG. 6.7 Work-hardening characteristics: steel D. (1) Experimenta_ curve, simple tension 0° to direction of rolling (2) Experimental curve, simple tension 45° to direction of rolling (3) Experimental curve, simple tension 90° to direction of rolling (4) Experimental curve, diaphragm test (5) Theoretical curve based on average r-value and corresponding work-hardening characteristic (6) Theoretical curve based on 90° tensile curve and on an assumption of loading along the anisotropic axes (7) Theoretical curve based on 0° tensile curve and on an assumption of loading along the anisotropic axes. (*After* Bramley and Mellor, *Int. J. mech. Sci.*)

The accompanying strain change can be derived from equations (5.37) and (5.47). Since there is no change in the strain ratios during straining total strains can be substituted for incremental strains and then from equation (5.37).

$$\epsilon_x : \epsilon_y : \epsilon_z = [(r+1)\,\sigma_x - r\sigma_y - \sigma_z] : [(r+1)\,\sigma_y - r\sigma_x - \sigma_z] : [2\,\sigma_z - \sigma_x - \sigma_y]$$

In simple tension

$$\sigma_x = \sigma_{av}, \, \sigma_y = \sigma_z = 0$$

and therefore

$$\epsilon_x \; : \; \epsilon_y \; : \; \epsilon_z = (r+1) \; : \; -r \; : \; -1.$$

At the pole of the diaphragm

$$\epsilon_x \; : \; \epsilon_y \; : \; \epsilon_z = 1 \; : \; 1 \; : \; -2.$$

Hence relating the average longitudinal strain $\epsilon_x = \epsilon_{av}$ from the experimental tensile data to the polar thickness strain ϵ_z in the diaphragm through equation (5.47).

$$\sqrt{\frac{2}{3}}\left[\frac{2+r}{2}\right]^{\frac{1}{2}}\epsilon_z = \bar{\epsilon} = \sqrt{\frac{2}{3}}\left[\frac{2+r}{1+r}\right]^{\frac{1}{2}}\epsilon_{av}$$

or
$$\epsilon_z = \epsilon_{av}\left(\frac{2}{1+r}\right)^{\frac{1}{2}}. \tag{6.14}$$

Finally, therefore, to convert the average tensile curve to the stress-strain curve from the diaphragm we use

$$\sigma_z = \sigma_{av}\left(\frac{1+\bar{r}}{2}\right)^{\frac{1}{2}}, \; \epsilon_z = \epsilon_{av}\left(\frac{2}{1+\bar{r}}\right)^{\frac{1}{2}}, \tag{6.15}$$

where $\bar{r}$ is the average of the experimentally determined r-values. (For Steel D, $\bar{r} = 1\cdot42$). It will be seen from Fig. 6.7 that there is good agreement between the experimental and predicted curves and similar agreement was obtained for Steels A, B, and C.

BRAMLEY and MELLOR (1967) have repeated the same experiments for titanium and zinc. There was reasonable agreement for titanium but not for zinc. The same two materials have been studied crystallographically by ROGERS and ROBERTS (1967) and the reader is referred to their paper for a detailed discussion of the slip and twinning systems that occur. PEARCE (1968) has studied the behaviour of steel and aluminium diaphragms where the r-values were less than unity and has reported that equations (6.15) could not be correlated with experiment.

The diaphragm test has also been used to investigate strain-rate effects in sheet metals, BRAMLEY and MELLOR (1966b).

6.4 Rolling and Simple Tension

While results from balanced bi-axial tension tests are suitable in investigating stretch-forming processes, the strain values reached are still too low for application to forming processes under compressive stresses. To obtain strain-hardening characteristics to very high strain values, some form of compression test must be used. One method applicable to sheet and strip material consists of work-hardening a number of specimens by different amounts in cold-rolling followed by tests in simple tension. This technique has been thoroughly investigated by FORD (1948).

To obtain the results presented in Figs. 6.3 and 6.4, 3-inch wide strips of the material were cut from the parent sheet in the direction of rolling, and rolled successively, without tension, in approximately 10 per cent reductions, one strip being taken out of the batch after every reduction. (It was ascertained that the lateral spread of the strip was negligible.) Tensile specimens were then cut from the centre of the strip and tested in the normal manner, at least two specimens being tested for each temper. During rolling with inhibited spread, the longitudinal strain ϵ, is equal and opposite to the thickness strain. The effective strain at the beginning of the tensile tests is therefore, from equation (6.3), $\bar{\epsilon} = 2\epsilon/\sqrt{3}$.

It will be noticed in Figs. 6.3 and 6.4 that the stress-strain curves obtained by rolling and tension generally lie below the stress-strain curves from the diaphragm. The effect of any redundant shearing in the rolling process would have the effect of raising the stress-strain curve. It is thought that the ratio of roll diameter to sheet thickness was so great that redundant shearing was negligible. Accepting this, the fact that the rolling and tension curves fall below the balanced bi-axial tension curves is probably due to the fact that the tension tests were carried out in the rolling direction only. For most practical purposes, the agreement between the work-hardening characteristics is good. The plane strain compression of anisotropic materials is discussed in Section 6.6.

6.5 Simple Compression

At first sight, it would seem to be an easy matter to obtain a strain-hardening characteristic in uniaxial compression by compressing a short circular cylinder of metal. However, homogeneous compression is difficult to achieve because of friction between the ends of the specimen and the deforming tools. This causes barrelling of the specimen with increasing load (Fig. 6.8). It has been shown experimentally that, because of the restriction in flow at the ends of the specimen, cone-shaped zones of relatively undeformed material form at each end of the specimen. This is indicated by the shaded areas in Fig. 6.8. SIEBEL and POMP (1927) suggested that a uniform distribution of stress could be achieved if a specimen was compressed between cones. However, this introduces further difficulties and uncertainties in the choice of cone angle and in the machining of the ends of the specimen.

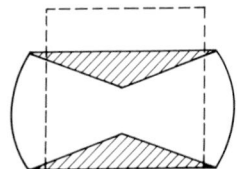

FIG. 6.8 Barrelling of specimen in simple compression.

An approximation to homogeneous compression may be obtained by ensuring good lubrication between deforming platens and the ends of the specimen. This entails the use of incremental loading with further lubrication between each loading, and when barrelling does become evident it is necessary to reduce the diameter of the specimen before further compression. The yield stress is taken to be the pressure on the deforming platens and the strain is best measured under no-load conditions. Such techniques have been used by TAYLOR and QUINNEY (1934) and JOHNSON (1956) up to logarithmic strains of about 4. A similar technique was used by LOIZOU and SIMS (1953) to measure the yield stress of pure lead. In this last case, concentric grooves turned in the ends of the specimen entrapped the lubricant, producing more efficient lubrication and virtually eliminating barrelling.

The extrapolation method of SACHS (1924) and COOK and LARKE (1945) has been used successfully for determining the stress-strain curve in uniaxial compression. This is based on the fact that if cylinders of equal diameter but different heights are compressed, the degree of barrelling depends on the original length of the cylinder and is least for the longest one. Theoretically, for a cylinder of infinite length the end effects would be negligible, the barrelling effect would therefore be absent, and the mean compressive pressure could be taken as the true stress in axial compression.

To obtain a stress-strain curve in axial compression for copper, COOK and LARKE used four cylinders of equal diameter but different heights such that the ratios of initial diameter d_0 to initial height h_0 were 0·5, 1·0, 2·0 and 3·0.

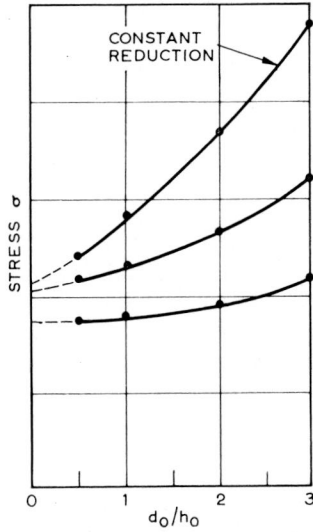

FIG. 6.9 Extrapolation to zero d_0/h_0 for true axial compressive stress.

A cylinder of infinite height then corresponds to a ratio d_0/h_0 of zero. Each cylinder was compressed in turn and the loads corresponding to various percentage reductions in height, $(1 - h/h_0) \times 100$ where h is the current height, were noted. The stress was calculated as though the deformation was homogeneous, that is stress $\sigma = 4P/\pi d^2$ where P is the load and d is the current diameter of the specimen. Assuming constancy of volume, $h_0 d_0^2 = h d^2$ and the stress $\sigma = 4Ph/\pi d^2 h_0$ is estimated from the measurement of load and current height. Graphs of stress against percentage reduction in height were plotted and from these were abstracted values of stress at 5 per cent, 10 per cent, 20 per cent, etc. reductions in height. The method of plotting and extrapolating for zero d_0/h_0 is shown in Fig. 6.9. Extrapolating to zero means in effect extrapolating to the true axial compressive stress since this is the condition for no barrelling. Finally, values of true stress against percentage reduction in height were plotted.

WATTS and FORD (1955) have used this method successfully on copper and have suggested simplifications and improvements in the technique. They suggested the use of incremental loading and lubrication, measurement of the current cylinder height under no-load conditions and extrapolation on the basis of equal load instead of equal percentage reduction. Extrapolation then gives the percentage reduction which would have occurred with an infinitely long specimen. They also demonstrated that extrapolation with high friction between tools and specimen leads to a stress-strain curve which flattens too rapidly above approximately 40 per cent reduction.

Perhaps the most accurate and nowadays certainly the simplest way of obtaining a true stress-strain curve in simple compression is to place P.T.F.E. sheet between the ends of the cylindrical specimen and the compressing platens. The thickness of the P.T.F.E. sheet should be between 0·002 and 0·005 in. The effect of the P.T.F.E. is not just to provide good lubrication between specimen and platens. As the compression load is increased the P.T.F.E. deforms the ends of the specimen and a raised peripheral rim is formed. The mechanism has been discussed by Hsu (1967) and it results in a changeover from 'barrelling' to 'bollarding' in which the diameters at the ends of the specimen became greater than the diameter at the centre of the specimen. For a particular case there is an optimum thickness of P.T.F.E. which will suppress 'barrelling' and keep 'bollarding' to a minimum. The test is best performed incrementally, renewing the P.T.F.E. after each loading. Measurements of load and diameter, at the centre of the specimen, are the only recordings necessary. The true stress is then the load divided by the current cross-sectional area at the centre of the specimen and the true strain is $\ln (d/d_0)^2$ where d_0 is the original diameter and d is the current diameter at the centre of the specimen. A true strain of approximately 0·8 can usually be attained before it becomes necessary to re-machine the specimen to a right cylindrical shape ready for further straining.

SIEBEL (1923) has derived an approximate expression for the mean pressure on a cylinder in terms of the axial yield stress, assuming Coulomb friction between compressing platens and the specimen. The expression is only true for small values of friction and provided there is no appreciable barrelling of the cylinder. It is assumed that the normal pressure, p, varies with radius r, but is constant through the height, h, of the cylinder. The radial equilibrium equation of a small element of the cylinder is, neglecting shear stresses $\sigma_{\theta r}$, from Fig. 6.10,

$$\sigma_r rh \, \delta\theta + 2\sigma_\theta h \, \delta r \frac{\delta\theta}{2} = 2\mu pr \, \delta r \, \delta\theta + h(\sigma_r + \delta\sigma_r)(r + \delta r) \, \delta\theta$$

where μ is the coefficient of friction and compressive stresses are written as positive. The radial equilibrium equation for the cylinder is then

$$\frac{d\sigma_r}{dr} + \frac{\sigma_r - \sigma_\theta}{r} = \frac{-2\mu p}{h}. \tag{6.16}$$

It is further assumed that the radial and circumferential stress components, σ_r and σ_θ, are equal and that both are principal stresses. This can be justified, in part, by using the Lévy-Mises equation $\epsilon_r/\sigma_r' = \epsilon_\theta/\sigma_\theta' = \epsilon_z/\sigma_z'$. Using the fact of volumetric constancy $\pi r^2 h = $ constant, hence $2\delta r/r + \delta h/h = 0$ or

$2\epsilon_\theta + \epsilon_z = 0$ and with $\epsilon_\theta + \epsilon_r + \epsilon_z = 0$, $\epsilon_r = \epsilon_\theta$ so that $\sigma_r = \sigma_\theta$. This assumption means that $p - \sigma_r = Y$, the yield stress in frictionless uniaxial compression, since a hydrostatic pressure does not affect yielding. Substituting this yield criterion into equation (6.16)

$$\frac{dp}{dr} = \frac{-2\mu p}{h}. \tag{6.17}$$

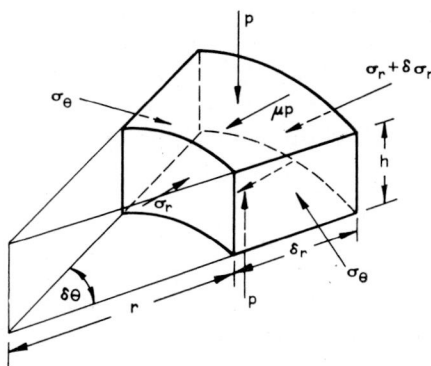

FIG. 6.10 Stresses on element in cylinder.

Integrating and inserting the boundary conditions $p = Y$ at $r = a$, the outside radius, where $\sigma_r = 0$

$$p = Y \exp 2\mu(a - r)/h. \tag{6.18}$$

The maximum pressure occurs at the centre line of the cylinder, $r = 0$, and the mean pressure is given by

$$\bar{p} = \frac{\int_0^a 2\pi r p \, dr}{\pi a^2} \simeq \left(1 + \frac{2\mu a}{3h}\right) Y. \tag{6.19}$$

For example, if $\mu = 0\cdot1$, $a/h = 1/2$, then $\bar{p} = 1\cdot03\,Y$.

Further work has been done on this topic by SHIELD (1955) and BISHOP (1958). SCHROEDER and WEBSTER (1949) have examined the forging of thin circular discs in a similar manner. HAWKYARD and JOHNSON (1966) have analysed the compression of short circular hollow cylinders.

When a circular disc is compressed by overlapping platens, the effect of friction on the straining is isotropic. This is not so if a long, narrow and thin rectangular piece of metal is compressed in the same manner. Such a lamina deforms after a finite compression to a 'cigar' shape. This indicates that away from the ends of the lamina, frictional effects inhibit longitudinal strain. This test was used by WATTS and FORD to secure plane strain conditions over the greater part of the specimen. The problem has been discussed by HILL (1950a) where he demonstrated that the test may be used for estimating coefficients of friction between a plastically deforming metal and elastically deformed tools.

WATTS and FORD (1952) compressed a copper strip of 0·100 in thickness, 1/4 in wide and 2 5/8 in long, using incremental loading with graphite grease as lubricant. They found, after small increments of compression, that at no part of the specimen was there complete plane strain and that their observations meant that the value of μ was not greater than 0·015. This last figure is of importance, since the above test was carried out under the same frictional conditions as the tests on the yielding of strip in plane strain compression (WATTS and FORD, 1952 and 1955).

6.6 Plane Strain Compression

The plane compression test is one of the most exact methods of obtaining a compressive stress-strain curve. A diagram of the apparatus is shown in Fig. 6.11. The indenting dies overlap the strip in its width direction and are

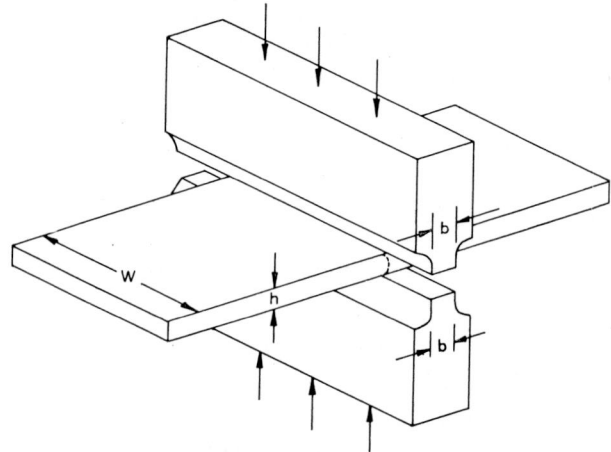

FIG. 6.11 Dies for plane strain compression.

narrow in breadth, b, so that there is unstressed material on each side. This unstressed material prevents strain in the direction of the width, w, of the strip, giving a state of plane strain where, from a consideration of volume constancy, the thickness strain is equal to the longitudinal strain. The pressure surfaces of the dies are polished and incremental loading with repeated lubrication ensures that conditions are as near frictionless as possible. This arrangement was suggested by NADAI (1931) and put forward by Orowan as a suitable test for determining the yield stress characteristic in the cold-rolling of strip. The test was first used by FORD (1948) and was followed by an intensive investigation by WATTS and FORD. High conductivity copper was used throughout their investigations because of its reliable and repeatable stress-strain curve.

The values of mean pressure on the strip were assumed to be the yield stress in plane compression, but early investigations showed that pressure values varied with the ratio of the thickness of the strip to the die breadth h/b.

This had been predicted theoretically by GREEN (1951) for the compression of an ideally plastic material between smooth flat plates. He predicted that the mean pressure would only be equal to $2k$, where k is the yield stress in pure shear, for integral values of h/b. (See also Upper Bound Solution, p. 424.) Experimental results on copper showed good correlation with the theoretical predictions and demonstrated the sensitivity of the test.

After further tests, it was concluded that an accurate stress-strain curve in plane strain compression could be obtained for large deformations if the following techniques and restrictions were adhered to:

1. Incremental loading in the same indentation, with lubrication and measurement of the strain at each increment. The increment to be 1/2–2 per cent, or sufficient to ensure the material is fully plastic.

2. The ratio die-breadth/strip-thickness to be maintained between 2 and 4, the dies being changed appropriately during the test.

3. To reduce the time taken, deformations of 10–15 per cent reduction can be made by a continuous application of load, but the stress and strain measurements must be made by incremental technique between each major deformation.

4. The width of the strip to be at least six times the die breadth and preferably more.

Point (1) may be explained by reference to Fig. 6.12, which shows a plane strain compression curve OA. The ordinate is mean pressure over the dies,

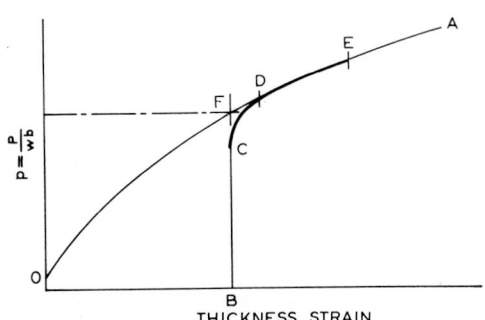

FIG. 6.12 Estimation of yield stress.

$p = P/wb$, where P is the load, and the abscissa is a measure of the reduction in thickness of the strip under the dies. Suppose that a particular test starts with an initial strain B. For this test, point C on the curve is the actual yield point, and D the point where it first coincides with the enveloping curve OA. The yield stress is defined by extrapolating the curve ED to zero deformation, B, for that particular test. This is the point F on the diagram. The part of the curve DE is obtained by incremental loading as described in (1), the increments being sufficient in each case to ensure that the strain to reach such a point D is exceeded.

In the plane compression test, the lateral strain is zero, and assuming volume constancy, the thickness strain ϵ_t is numerically equal to the strain in the longitudinal direction of the strip. For narrow dies, the stress in the longitudinal direction of the strip can be considered zero and it follows from the Lévy-Mises relationship that the normal stress, p, is then twice the stress in the lateral direction. Substituting these conditions in equations (6.1) and (6.3) the effective stress $\bar{\sigma} = Y = p\sqrt{3/2}$, and the effective strain $= 2\epsilon_t/\sqrt{3}$. WATTS and FORD (1955) compared the stress-strain curves in uniaxial compression and plane strain compression by plotting Y and p against an effective strain base. According to the simple theory the ratio p/Y should be equal to $2/\sqrt{3}$. This ratio had a value $2/\sqrt{3}$ for zero reduction, but, owing to the development of anisotropy, this fell gradually to a value just below 1.1.

ALEXANDER (1955), using slip-lines, examined the ratio $p/2k$ versus b/h in the plane compression test for different states of Coulomb friction. For $\mu = 0.015$, which is a reasonable value in the above tests, and a ratio $b/h = 3$,

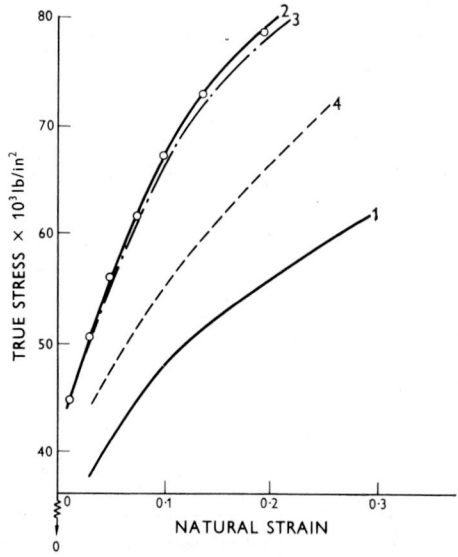

FIG. 6.13 Work-hardening characteristic: steel *C*. (1) Experimental curve, simple tension, specimen cut parallel to rolling direction. (2) Experimental curve –o–o– for plane strain compression. Pressure, *p*, is plotted against thickness strain. Zero strain in rolling direction. (3) Plane strain compression curve predicted from curve (1) using Hill's anisotropic theory. (4) Plane strain compression curve predicted from curve (1) on the assumption that the material is isotropic. (*After* Taghvaipour and Mellor, *Proc. Inst. mech. Engrs.*)

he found that $p/2k = 1.024$. Thus, in the plane compression test, the normal stress will be overestimated by about 2·4 per cent—an error which can be considered small for most practical cases.

Some sheet materials nowadays are processed specifically to have a certain amount of normal anisotropy. Killed steel and titanium have already been mentioned as examples of materials having an average r-value greater than unity, a factor which is beneficial in deep-drawing processes. The plane strain compression test can be used to predict the resistance such materials would have to further cold-rolling. The techniques and restrictions suggested by Watts and Ford for isotropic materials were found to be satisfactory for anisotropic materials. The amount by which a particular killed steel is strengthened in the through thickness direction is illustrated in Fig. 6.13 (TAGHVAIPOUR and MELLOR (1970)). Curve 1 is the stress-strain curve obtained from a tensile specimen cut in the rolling direction. Curve 4, is the predicted variation of the normal stress, p, with natural thickness strain in plane strain compression, which results from assuming that the material is isotropic and has a generalized stress-strain characteristics represented by Curve 1. Curve 2 gives the actual variation of p with natural thickness strain as measured by the plane strain compression test when the indenter is aligned along the rolling direction. The steel used for this work has been denoted as Steel C and the anisotropic parameters of this material are given in Fig. 6.6. The average r-value is 1·41. We can now predict the plane strain compression curve using these values in Hill's theory of anisotropy.

If x is the rolling direction then in the plane strain compression test $\epsilon_x = 0$ and $\epsilon_y = -\epsilon_z$ from the incompressibility condition. Also, since the loading is parallel to the x-axis, $\gamma_{xy} = 0$. Assuming that there is no friction between the compressing platens and the sheet $\tau_{yz} = \tau_{zx} = 0$ and since there is no tension in the y-direction $\sigma_y = 0$. Equations (5.37) then give

$$\sigma_x = \frac{G}{H + G}\sigma_z$$

or
$$\sigma_z = p = (1 + r_0)\sigma_x \qquad (6.20)$$

from equation (5.40), where r_0 is the r-value obtained from a tensile specimen aligned in the rolling direction. Note that as r_0 increases, the induced stress σ_x becomes a smaller and smaller fraction of the normal stress p. Substituting equation (6.20) into equation (5.42) and noting that

$$\frac{G}{F} = \frac{r_{90}}{r_0}, \quad \frac{H}{G} = r_0, \quad \frac{H}{F} = r_{90},$$

the generalized stress is

$$\bar{\sigma} = \sqrt{\frac{3}{2}}\left[\frac{1 + r_0 + r_{90}}{(1 + r_0)\left(1 + r_{90} + \dfrac{r_{90}}{r_0}\right)}\right]^{\frac{1}{2}} p. \qquad (6.21)$$

Also substituting the strain conditions into equation (5.45),

$$\bar{\epsilon} = \sqrt{\frac{2}{3}}\left[\frac{F + G + H}{(FG + GH + HF)^2}\left\{F(G + H)^2 + GH^2 + HG^2\right\}\right]^{\frac{1}{2}}\epsilon_z$$

or
$$\bar{\epsilon} = \sqrt{\frac{2}{3}}\left[\frac{(1 + r_0)\left(1 + r_{90} + \dfrac{r_{90}}{r_0}\right)}{1 + r_0 + r_{90}}\right]^{\frac{1}{2}}\epsilon_z \qquad (6.22)$$

The variation of p with ϵ_z can now be related to simple tension in the rolling direction through expressions (6.21) and (6.22) noting that for simple tension in the x-direction

$$\bar{\sigma} = \sqrt{\frac{3}{2}} \left[\frac{1 + r_0}{1 + r_0 + \frac{r_0}{r_{t0}}} \right]^{\frac{1}{2}} X$$

$$\text{and } \bar{\epsilon} = \sqrt{\frac{2}{3}} \left[\frac{1 + r_0 + \frac{r_0}{r_{90}}}{1 + r_0} \right]^{\frac{1}{2}} \epsilon_x. \tag{6.23}$$

This has been done in Fig. 6.13 where Curve 3 is predicted from Curve 1. It is seen that there is good correlation between theory and experiment for this material.

It is useful to discuss these relationships for the case when $r_0 = r_{90} = r$. Then equations (6.21) and (6.22) reduce to

$$\bar{\sigma} = \sqrt{\frac{3}{2}} \left[\frac{(1 + 2r)}{(1 + r)(2 + r)} \right]^{\frac{1}{2}} p$$

$$\text{and } \bar{\epsilon} = \sqrt{\frac{2}{3}} \left[\frac{(1 + r)(2 + r)}{(1 + 2r)} \right]^{\frac{1}{2}} \epsilon_z \tag{6.24}$$

and the equations for simple tension reduce to

$$\bar{\sigma} = \sqrt{\frac{3}{2}} \left[\frac{1 + r}{2 + r} \right]^{\frac{1}{2}} X$$

$$\text{and } \bar{\epsilon} = \sqrt{\frac{2}{3}} \left[\frac{2 + r}{1 + r} \right]^{\frac{1}{2}} \epsilon_x. \tag{6.25}$$

Therefore, to convert from simple tension to plane strain compression the following equations apply.

$$p = \frac{(1 + r)^{\frac{1}{2}}}{(1 + 2r)} X \tag{6.26}$$

$$\text{and } \epsilon_z = \frac{(1 + 2r)^{\frac{1}{2}}}{(1 + r)} \epsilon_x.$$

It is seen that when $r = 0$ the plane strain compression curve and the simple tension curve should coincide. For all other positive values of r the plane strain compression curve will always be higher than the simple tension curve. Taking a fairly high value for r, say $r = 5$, we see that $p = 1.82 X$ and $\epsilon_z = 0.55 \epsilon_x$. This should be compared with the values $p = 1.55 X$ and $\epsilon_z = 0.866 \epsilon_x$ for an isotropic material.

6.7 Simple Torsion

Torsion of a thin-walled tube has been used as a means of determining work-hardening characteristics. (TAYLOR and QUINNEY, (1931), ZENER and

HOLLOMAN (1946); SWIFT (1947).) The test requires an accurately machined specimen and measurements are made of torque and angle of twist. Neglecting the elastic part of the curve, the effective stress is $\sqrt{3}\,\tau$, where τ, the shear stress, is assumed constant through the wall thickness, and the effective strain is $\tan\phi/\sqrt{3}$. This test is not suitable for measuring high values of strain since the tube buckles and 'wrings' out of true.

It is much more convenient experimentally to carry out a test on a solid bar of circular cross-section. However, owing to the inequality of stress distribution across the bar, some assumptions have to be made in order to interpret the results measured in terms of torque and twist. NADAI has suggested a construction for deriving the stress-strain curve from the torque-twist curve which is based on the assumptions that an isotropic material remains isotropic during the straining, that transverse cross-sections of the twisted shaft continue to remain plane and their radii remain straight. These assumptions are identical to those made in the elastic torsion of a round bar and must be justified by experiment.

Consider an isotropic cylinder, of unit gauge length and outside radius r_0, subjected to a plastic torsional strain. For an isotropic metal there will be no change in the gauge length and no change in volume.

At any radius r, the shear strain $\gamma = \tan\phi$ and for a given angle of twist per unit length θ (Fig. 6.14), $r\theta = \tan\phi$ or

$$\theta = \frac{\gamma}{r} = \frac{\gamma_0}{r_0} \tag{6.27}$$

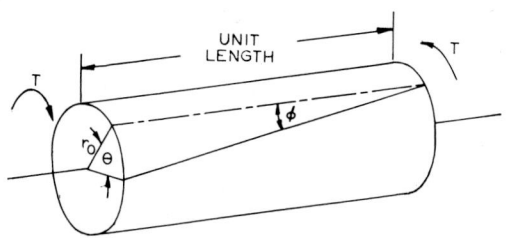

FIG. 6.14 Torsion of solid circular shaft.

where γ_0 is the shear strain at the outside radius. The resisting torque T is therefore

$$T = \int_0^{r_0} \tau(2\pi r\, dr)r$$

and from equation (6.27)

$$r = \frac{\gamma}{\theta}, \qquad dr = \frac{d\gamma}{\theta}.$$

Therefore,

$$T = \frac{2\pi}{\theta^3} \int_0^{r_0} \tau\gamma^2\, d\gamma \tag{6.28}$$

Thus, if we know the shear stress-strain curve $\tau = f(\gamma)$ we can determine the $T - \theta$ curve. Rearranging equation (6.28) and differentiating with respect to θ

$$d\left(\frac{T\theta^3}{2\pi}\right) = \tau_0 \gamma_0^2 d\gamma_0 = \tau_0 . r_0^2 \theta^2 . r_0 d\theta$$

or

$$\frac{d}{d\theta}\left(\frac{T\theta^3}{2\pi}\right) = r_0^3 \theta^2 \tau_0$$

that is,

$$\frac{dT}{d\theta} . \theta^3 + T . 3\theta^2 = 2\pi r_0^3 \theta^2 \tau_0$$

and

$$\tau_0 = \frac{1}{2\pi r_0^3}\left\{3T + \theta\frac{dT}{d\theta}\right\}. \qquad (6.29)$$

If the $T-\theta$ curve is obtained, the τ/γ curve can be derived. Note that $\tau_0 = 3T/2\pi r_0^3$ is the usual simple expression for torque due to a maximum yield shear stress of τ_0 in the fully plastic condition for a non-work-hardening material. The second term represents a correction for work-hardening.

From Fig. 6.15, the slope at B on the $T-\theta$ curve is given by

$$\frac{dT}{d\theta} = \frac{BC}{DC} \quad \text{or} \quad BC = \theta\frac{dT}{d\theta} \quad \text{since} \quad DC = \theta$$

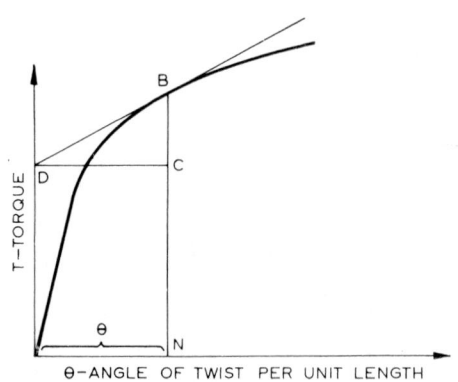

FIG. 6.15 $T-\theta$ curve.

Hence the shear stress at the outside surface, τ_0, corresponding to a twist θ is

$$\tau_0 = \frac{1}{2\pi r_0^3}\{3BN + BC\}.$$

Therefore by drawing tangents to points on the $T-\theta$ diagram, the $\tau-\tan\phi$ diagram can be derived.

This method is not very accurate for the early part of the $T-\theta$ curve where $dT/d\theta$ is decreasing rapidly. The accuracy is improved by writing equation (6.29) as

$$\tau_0 = \frac{1}{2\pi r_0^3}\left\{4T + \theta^2\frac{d}{d\theta}\left(\frac{T}{\theta}\right)\right\}, \qquad (6.30)$$

(HILL, (1950b)). T/θ is constant over the elastic range and then decreases steadily. Therefore the derivative is zero in the elastic range and is small compared with the other term when the rate of work-hardening is high. Equation (6.19) gives the greater accuracy for larger strains.

SHEPHERD (1948) has used Nadai's construction for small strains and SWIFT (1947) using Nadai's construction for large plastic strains has shown that there is good correlation of the stress-strain curves for hollow and solid bars of mild steel. Swift also noted the permanent increase in length of the specimens which is due to anisotropy developed during straining (See also p. 195).

See Problems 22-25

REFERENCES

ALEXANDER, J. M.	1955	'Plane Strain Compression of a Short Block' *J. Mech. Phys. Solids* **3**, 233
BELL, R. and DUNCAN, J. L.	1965	The Evolution of a Prototype Machine for Automatically Recording the True Stress–Strain Curve for Sheet Metal Using the Hydrostatic Bulge Test *Proc. 5th Conf. M.T.D.R.*, Pergamon
BISHOP, J. F. W.	1958	'On the Effect of Friction on Compression and Indentation Between Flat Dies' *J. Mech. Phys. Solids* **6**, 132
BRAMLEY, A. N., and MELLOR, P. B.	1966a	'Plastic Flow in Stabilized Sheet Steel' *Int. J. mech. Sci.* **8**, 101
	1966b	'The Effect of Strain Rate on the Plastic Flow Characteristics of Steel and Aluminium Sheet' *J. Strain Analysis* **1**, 439
	1968	'Plastic Anisotropy of Titanium and Zinc Sheet—I' Macroscopic approach *Int. J. mech. Sci.* **10**, 211
BRIDGMAN, P. W.	1944c	'The Stress Distribution at the Neck of a Tension Specimen' *Trans. A.S.M.E.* **32**, 553
BROWN, W. F. and SACHS, G.	1948	'Strength and Failure Characteristics of Thin Circular Membranes' *Trans. A.S.M.E.* **70**, 241
COOK, M. and LARKE, E. C.	1945	'Resistance of Copper and Copper Alloys to Homogeneous Deformation in Compression' *J. Inst. Metals* **71**, 371
DAVIDENKOV, N. N. and SPIRIDONOVA N. I.	1946	'Analysis of Tensile Stress in the Neck of an Elongated Test Specimen' *Proc. A.S.T.M.* **46**, 1146

DAVIS, E. A.   1943 'Yielding and Fracture of Medium-carbon
                                     Steel Under Combined Stress'
                                     J. appl. Mech. **67**, A–13
DUNCAN, J. L. and   1965 'The Use of a Biaxial Test Extensometer'
  JOHNSON, W.   *Sheet Metal Industries*, 271
FORD, H.   1948 'Researches into the Deformation of Metals
                                     by Cold Rolling'
                                     Proc. Instn mech. Engrs **159**, 115
GREEN, A. P.   1951 'A Theoretical Investigation of the
                                     Compression of a Ductile Material Between
                                     Smooth Flat Dies'
                                     Phil. Mag. 7 Ser. **42**, 900
HAWKYARD, J. B. and   1966 'An Analysis of the Changes in Geometry of
  JOHNSON, W.   a Short Hollow Cylinder During Axial Com-
                                     pression'
                                     Int. J. mech. Sci. **9**, 163
HILL, R.   1950a 'On the Inhomogeneous Deformation of a
                                     Plastic Lamina in a Compression Test'
                                     Phil. Mag. **41**, 733
                              1950b *The Mathematical Theory of Plasticity* O.U.P.
HSU, T. C.   1967 'A Study of the Compression Test for
                                     Ductile Materials'
                                     A.S.M.E. 67—W.A./Met 11
JACKSON, L. R.,   1948 'Plastic Flow in Anisotropic Sheet Steel'
  SMITH, K. F. and   *Metals Technology Tech. Pub.* 2440, and
  LANKFORD, W. T.   *J. Metals* **1**, 323, 1949
JOHNSON, W.   1956 'Experiments in Plane Strain Extrusion'
                                     J. Mech. Phys. Solids **4**, 269
LOIZOU, N. and   1953 'Yield Stress of Pure Lead in Compression'
  SIMS, R. B.   *J. Mech. Phys. Solids* **1**, 234
MACGREGOR, C. W.   1940 'The Tension Test'
                                     Proc. A.S.T.M. **40**, 508
MANJOINE, M.   1944 'Influence of Rate of Strain and Temperature
                                     on Yield Stresses of Mild Steel'
                                     J. appl. mech. **11**, 211
MELLOR, P. B.   1956 'Stretch-forming under Fluid Pressure'
                                     J. Mech. Phys. Solids **5**, 41
MORRISON, J. L. M.   1948 'The Criterion of "Yield" of Gun Steels'
                                     Proc. Instn mech. Engrs **159**, 81
NADAI, A.   1931 *Plasticity* McGraw-Hill, New York
                               1937 'Plastic Behaviour of Metals in the Strain-
                                     hardening Range'
                                     J. appl. Phys. **8**, 205
                               1947 'The Flow of Metals Under Various Stress
                                     Conditions'
                                     Proc. Instn mech. Engrs **157**, 121
                               1950 *Theory of Flow and Fracture of Solids*
                                     McGraw-Hill, New York

PEARCE, R.   1968 'Some Aspects of Anisotropic Plasticity in Sheet Metals' *Int. J. mech. Sci.* **10**, 995

ROGERS, D. H. and ROBERTS, W. T.   1968 'Plastic Anisotropy of Titanium and Zinc Sheet—II. Crystallographic Approach *Int. J. mech. Sci.* **10**, 221

ROS, M. and EICHINGER, A.   1929 Metalle Diskussionsbericht No. 34 der Eidg. Materialprüfungsanstalt, Zürich

SACHS, G.   1924 *Z. Metallk.* **16**, 55

SCHROEDER, W. and WEBSTER, D. A.   1949 'Press-forging Thin Sections' *J. appl. mech.* **16**, 289

SHIELD, R. T.   1955 'On the Plastic Flow of Metals Under Conditions of Axial Symmetry' *Proc. R. Soc. A.*, **233**, 267

SHEPHERD, W. M.   1948 'Plastic Stress-strain Curves' *Proc. Instn mech. Engrs* **159**, 95

SIEBEL, E.   1923 Stahl und Eisen, Düsseldorf **43**, 1295

SIEBEL, E. and POMP, A.   1927 'Die Ermittlung der Formänderungsfestigkeit von Metallen durch den Stauchversuch' *Mitt. K.-Wilhelm-Inst. Eisenforsch.* **9**, 157

SWIFT, H. W.   1946 'Plastic Strain in an Isotropic Strain-hardening Material' *Engineering* **162**, 381

1947 'Length Changes in Metals Under Torsional Overstrain' *Engineering* **163**, 253

TAGHVAIPOUR, M and MELLOR, P. B.   1970 –71 'Plane Strain Compression of Anisotropic Sheet Metal' *Proc. Instn mech. Engrs* **185**, 593

TAYLOR, G. I. and QUINNEY, H.   1931 'The Plastic Distortion of Metals' *Trans. R. Soc. A*, **230**, 323

1934 'The Latent Energy Remaining in a Metal After Cold-working' *Proc. R. Soc. A*, **143**, 307

WATTS, A. B. and FORD, H.   1952 'An Experimental Investigation of the Yielding of Strip Between Smooth Dies' *Proc. Instn mech. Engrs* (B), **1B**, 448

1955 'On the Basic Yield Stress Curve for a Metal' *Proc. Instn mech. Engrs* **169**, 1141

ZENER, C. and HOLLOMAN, J. H.   1946 'Problems in Non-elastic Deformation of Metals' *J. appl. Phys.* **17**, 69

Chapter 7

ELEMENTARY ANALYSES OF THE ELASTIC-PLASTIC BENDING OF BEAMS, RINGS AND PLATES

7.1 Introduction

The plastic bending of a bar would appear at first sight to be easily amenable to mathematical analysis. In fact, the complete solution is as yet unknown and all existing theories make assumptions which lead to varying degrees of approximation to the correct result. These approximate theories are very useful, provided they are applied within the limitations of the basic assumptions and provided they have been checked experimentally for a particular application.

The difficulties involved in analysing plastic bending can be appreciated by considering the bending of a straight bar of rectangular cross-section in which the breadth and depth are of about the same magnitude. If such a bar is subjected to pure bending, Fig. 7.1 (i.e., a constant bending moment

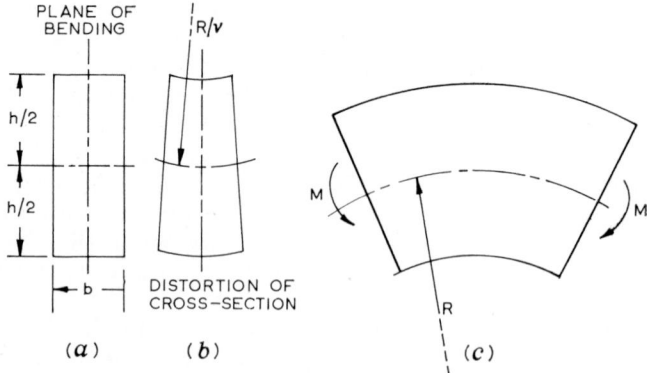

PLANE OF
BENDING R/v

$h/2$

$h/2$

b

DISTORTION OF
CROSS–SECTION

(a) (b)

M M

R

(c)

Fig. 7.1 Bar of rectangular cross-section subjected to pure bending.

and therefore no shearing force), then provided the elastic limit is not reached at any point, transverse sections which are plane before bending remain plane after bending. If R is the radius of curvature of the neutral axis in the plane of bending, then the anticlastic curvature is $-v/R$ where v is

Poisson's ratio. As bending continues, yielding begins at the outer fibres and gradually spreads towards the centre of the bar but the strains are still largely controlled by the central elastic core. However, the Poisson's ratio effect now varies over the cross-section, v being about 0·3 for elastic strains and 0·5 for plastic strains. To attain the necessary continuity of strain across the elastic-plastic boundary, some transverse stresses must be present. To overcome this difficulty, some writers assume that the material dealt with is incompressible, that is $v = 0·5$ in both elastic and plastic strain. Whilst this facilitates a mathematically correct analysis, if it is applied to a real material, it is tantamount to neglecting the transverse stresses. (A survey of the literature on anticlastic curvature in plastic bending is given in a paper by HORROCKS and JOHNSON, 1967.)

Again in elastic bending, the same compatibility conditions which hold for pure bending are assumed to hold for laterally loaded beams, i.e., plane sections remain plane on bending, provided the shear stresses are small compared with the longitudinal bending stresses. This is so if the longitudinal dimensions are large compared with the dimensions of the cross-section. This same approximation is sometimes made in plastic bending and is additional to the neglect of the transverse stresses mentioned above.

7.2 Simple Theory of Plastic Bending

7.2.1 Straight Rectangular Section Beams

From the elementary theory of bending, the elastic stress distribution over, say, a bar of rectangular cross-section, Fig. 7.1, subject to a *pure* bending moment M is linear, as shown in Fig. 7.2(*b*). The maximum stresses occur at distances from the neutral axis of $\pm h/2$ and are given by

$$\sigma_{\max} = \frac{M}{I} \cdot \frac{h}{2}$$

where I is the second moment of area of the section about the neutral axis, NN. Suppose the beam is made of a material whose stress-strain curve is the same in pure tension and compression and is as shown in Fig. 7.3. As M increases, the stress distribution remains linear until $\sigma_{\max} = Y$, the yield stress, Fig. 7.2(*c*). With further increase in M, the linear distribution ceases. The strain at a distance y from the neutral axis, provided the strains be small is y/R, where R is the radius of curvature of the neutral fibre of the bent beam. For a given M, R is constant (being very large) across the section and hence the strain is directly proportional to y, provided sections originally plane remain plane. If the strain in the outermost fibres, i.e., at $y = \pm h/2$, is ϵ_A, the corresponding stress, σ_A, may be obtained directly from the stress-strain diagram. In fact the stress distribution at each point on the cross-section is obtained simply by re-drawing on the base or zero stress line, the stress-strain diagram, see Fig. 7.2(*d*), one portion being positive (tensile stresses), the other negative (compressive stresses), the origin of the σ/ϵ curve being placed on the neutral axis. Equality of these positive and negative areas

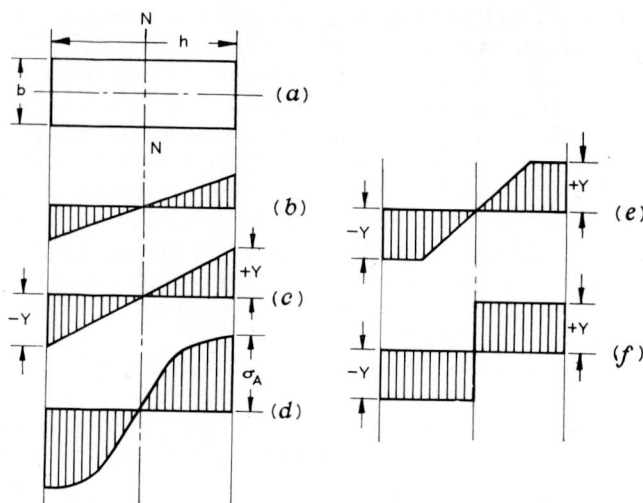

FIG. 7.2 Stress distribution over a bar of rectangular cross-section.

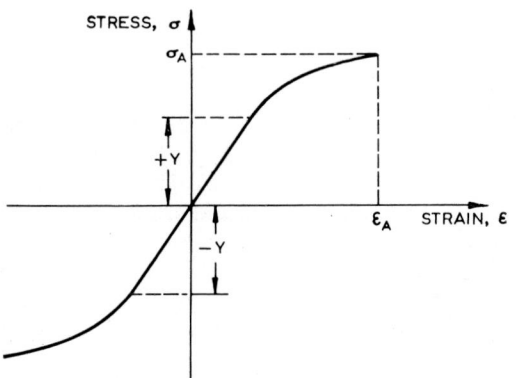

FIG. 7.3 Stress-strain curve.

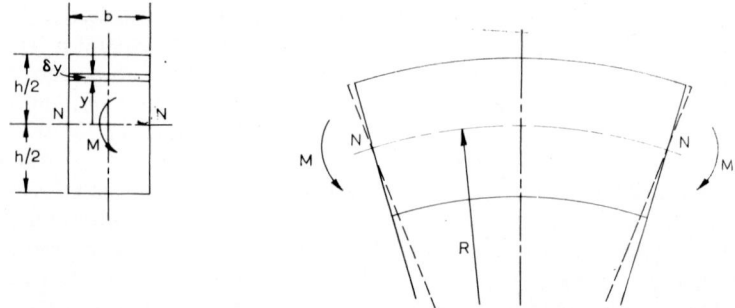

FIG. 7.4 Bar of rectangular cross-section subjected to pure bending.

ensures zero total force on a section as long as the section is symmetrical about NN. The external moment is given by

$$M = \int_{-h/2}^{+h/2} \sigma b y \, dy$$

7.2.2 Expressions for M, *Using a Non-linear Stress-Strain Law*

Assuming that the stress-strain relation is

$$\sigma = Ee + Fe^n \tag{7.1}$$

and since $e = y/R$, then, with a notation which is clear from Fig. 7.4

$$M = \int_{-h/2}^{+h/2} \left(\frac{Eby^2}{R} + \frac{Fby^{n+1}}{R^n} \right) dy.$$

$$= \frac{E}{R} I_1 + \frac{F}{R^n} I_n \tag{7.2}$$

where $\quad I_1 = \int_{-h/2}^{+h/2} by^2 \, dy \quad$ and $\quad I_n = \int_{-h/2}^{+h/2} by^{n+1} \, dy.$

I_1 is the expression for the second moment of area commonly used in Strength of Materials. I_n is the integral arising from the non-linear behaviour of the material.

Putting $E = 0$ in the stress-strain relation (7.1)

$$\frac{M}{I_n} = \frac{F}{R^n} = \frac{\sigma}{y^n} \tag{7.3}$$

follows, being a little more general than the conventional trinity of equations used by engineers in the elastic case, i.e. when $n = 1$. The value of I_n is then, if b is constant

$$I_n = \frac{bh^{n+2}}{2^{n+1}(n+2)}.$$

7.2.3 Expression For Deflection

A very interesting treatment of elementary non-linear bending theory for straight beams, when the material of the beam has different stress-strain properties in tension and compression was given by St. Venant and account of this is given in the book by TIMOSHENKO (1953).

Since R is large, then as in elementary beam theory dy/dx is small and negligible in the expression for curvature

$$\frac{1}{R} = \frac{d^2y/dx^2}{\{1 + (dy/dx)^2\}^{3/2}}.$$

Thus the expression for the deflection of a point in a beam can be calculated by substituting for $1/R$, d^2y/dx^2 and using equation (7.3). Thus

$$\frac{d^2y}{dx^2} = \left(\frac{M}{FI_n} \right)^{1/n}.$$

Applied to the simple cantilever of length L carrying an end load W, see Fig. 7.5, $M = W(L-x)$ and

$$\frac{dy}{dx} = \frac{-C(L-x)^{1/n+1}}{(1+1/n)} + B.$$

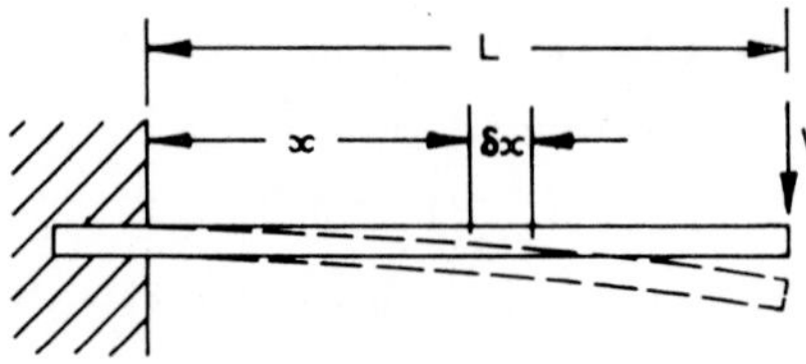

FIG. 7.5 Cantilever with end load W.

$dy/dx = 0$ at $x = 0$, and thus integration constant $B = CL^{1/n+1}/(1/n + 1)$, where $C = (W/FI_n)^{1/n}$. Integrating again,

$$y = \frac{C(L-x)^{1/n+2}}{(1/n+1)(1/n+2)} + Bx + D.$$

When $x = 0$, $y = 0$ and $D = -CL^{1/n+2}/(1/n+1)(1/n+2)$.

Thus, $\quad y = \dfrac{CL^{1/n+2}}{(1/n+1)(1/n+2)}\left[\left(1-\dfrac{x}{L}\right)^{1/n+2} + \dfrac{x}{L}\left(\dfrac{1}{n}+2\right) - 1\right]$

The deflection under the load is δ_w and

$$\delta_w = \frac{CL^{1/n+2}}{(1/n+1)(1/n+2)}(1/n+1) = W^{1/n}f, \text{ say.}$$

In the elastic case, $n = 1$ and

$$\delta_w = \frac{WL^3}{3FI}.$$

This example illustrates the obviousness of approach for this and similar types of problem. Many results for y, dy/dx, I_n and δ for complicated cases are given in the book by PHILLIPS (1956).

When a load W_1 other than W is applied to the end of the cantilever then

$$\delta_{w_1} = W_1^{1/n}f.$$

The deflection due to the simultaneous action of W and W_1 is

$$\delta = (W + W_1)^{1/n}f.$$

Note that
$$\delta \neq \delta_{w_1} + \delta_{w_2}, \quad \text{i.e.} \quad (W + W_1)^{1/n} \neq W_1^{1/n} + W_2^{1/n}$$

unless $n = 1$, and hence the principle of superposition is not permissible except when $n = 1$.

7.2.4 Shear Stress Distribution

A very simple application of the use of this particular stress-strain law is to find the shear stress distribution in a rectangular beam subject to simple bending. Referring to Fig. 7.6, which shows a portion δx of the beam in Fig. 7.5, the shear stress, τ on plane $ABCD$ is, following the usual elementary methods of strength of materials,

$$\tau b \, \delta x = \int_y^{h/2} b \frac{\partial \sigma}{\partial x} \, \delta x \, dy$$

Now from equation (7.3), $M = \sigma I_n / y^n$ and thus

$$\frac{\partial M}{\partial x} = \frac{I_n \, d\sigma}{y^n \, dx} \quad \text{but} \quad \frac{\partial M}{\partial x} = -W \quad \text{and so} \quad \frac{\partial \sigma}{\partial x} = \frac{-W y^n}{I_n}.$$

Hence, $$\tau = \int_y^{h/2} \frac{-W}{I_n} y^n \, dy = \frac{-W}{I_n} \frac{[(h/2)^{n+1} - y^{n+1}]}{(n+1)}.$$

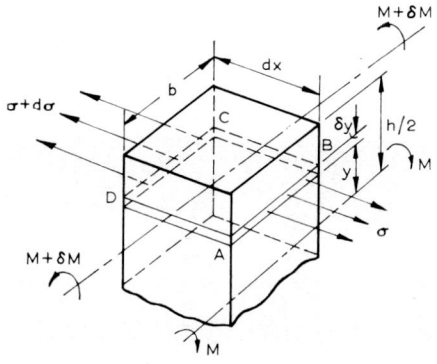

FIG. 7.6 Equilibrium of element in beam subjected to lateral loading.

The ratio of the maximum shear stress on the neutral axis for this material to that for a linear elastic material is $2(n+2)/3(n+1)$.

7.2.5 *Idealized Materials in Bending*

These idealized materials have been introduced in Fig. 1.7. The moment, M_E, required to bring the outermost fibre of a rectangular section beam of perfectly plastic-elastic material, to its yield stress is

$$M_E = bh^2 Y/6. \tag{7.4}$$

When the applied moment $M > M_E$, the stress distribution across the section is as in Fig. 7.2(e). As the radius of curvature of the beam decreases the thickness of the layer of yielded material increases and the elastic-plastic boundaries approach the neutral axis; when this boundary is distant y from the neutral axis, the moment is given by,

$$M = Yb(3h^2 - 4y^2)/12 \tag{7.5.i}$$

(In mild steel the yielding does not proceed in this fashion; the yield in beams wide relative to their thickness—approaching a plane strain situation—is discontinuous and dark bands of plastically deformed material at $\pm 45°$ to the surfaces are visible. In between the bands, light-coloured elastic material is visible; the apices of the bands suggest an advancing plastic-elastic interface). As the bending moment increases, this latter interface advances towards the neutral axis, the quantity of unyielded material nearer the neutral surface being progressively reduced, see Figs. 7.7(a) and (b). Clearly, there always remains a residual layer of elastic material, no matter what the size of M,

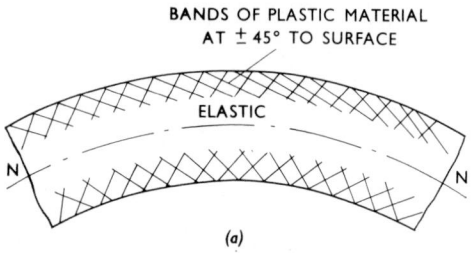

FIG. 7.7 (a) and (b). Plastic yielding of a mild steel plate due to a couple.

but through it there may be a very large gradient of stress. The thinner this elastic core, the less it will be in error to assume that the material is finally plastic across the whole section though the radius of curvature progressively decreases; two plastic stress blocks thus represent this limiting case, see Fig. 7.2(f). The more closely this case is approached, the smaller does the radius of curvature of the beam become and the more in error the initial assumptions of the analysis. (This ultimate case also displays a stress discontinuity, i.e., sudden jump from $+Y$ to $-Y$, across the neutral axis.) However, M_P, the moment required to make the section fully plastic is found from equation (7.5.i) by putting $y = 0$, which gives,

$$M_P = bh^2 Y/4. \tag{7.5.ii}$$

The result in equation (7.5.ii) is also obtained if the beam is considered to be made of rigid-perfectly plastic material. Whilst this idealized type of material is less realistic than the former, it enables a self-consistent argument to be maintained. Since elastic strains in a rigid-plastic material are zero, there can be no deformation or deflection of a rigid-plastic beam until the moment is so large that the *whole* of the material of the section has become plastic. The existence of a residual elastic core at any time implies complete rigidity; when once the whole section is plastic, the beam will 'collapse' or bend without hardening. This collapse load, M_P, is thus half as large again as M_E. It demonstrates that in elastically designed members (buckling excepted) there is a considerable reserve of strength due to the possibility of utilizing their plasticity. It also leads to a new method of designing certain structures which frequently entails much less effort than does an elastic analysis. The method will be applied, in Section 7.3, to some simple and straightforward cases of loaded beams and portal frames, and to links. In the book by HODGE (1959), a detailed discussion of this and many other points touched on in this chapter, will be found.

7.2.6 *Shape Factor*

For the case of the rectangular section beam of ideal elastic-perfectly plastic material above, the ratio $M_P/M_E = 1.5$; this ratio of the bending moment required to make a section fully plastic, to that carried by the same section elastically, with the extreme fibre stress just attaining the yield stress, is known as the shape factor.

For a circular section beam, of radius a, it is easily shown that,

$$M_P = \frac{4}{3} a^3 Y \quad \text{and} \quad M_E = \frac{\pi}{4} a^3 Y.$$

Thus its shape factor is $16/3\pi \simeq 1.7$. Evidently, then, the full load carrying capacity of a circular section bar (based on the completely plastic section) is underestimated by about 41 per cent, by the elastic moment with the yield stress reached in the outermost fibres.

For beam sections which are not symmetrical about any plane perpendicular to the plane of bending, the neutral plane must first be found; for purely linear elastic bending the neutral plane passes through the centroid of the section, but when the section is partly plastic the neutral plane no longer coincides with the elastic neutral plane. Indeed the neutral plane shifts as the depth of the plastic layer in the beam increases. The position of the neutral plane in a section of a beam which is wholly plastic is, however, easily found from the same force equilibrium considerations as prevail for the elastic case. For a beam subject only to a bending moment, the net tensile forces on one side of the neutral plane must equal the net compressive forces on the other. Since the normal bending stress on the section is everywhere constant and equal to Y, the equilibrium condition simply requires the neutral plane to be placed in such a position that the area of the section above it equals the area of the section below it.

An example is that of a beam section which is an isosceles triangle of height H and base length B, any neutral plane in bending being parallel to the base, see Figs. 7.8 (*a*) and (*b*).

(i) For purely elastic bending the neutral axis is distant $H/3$ above the base and it is easily shown that when the bending stress just reaches the yield stress at the apex of the triangle, $M_E = YBH^2/24$.

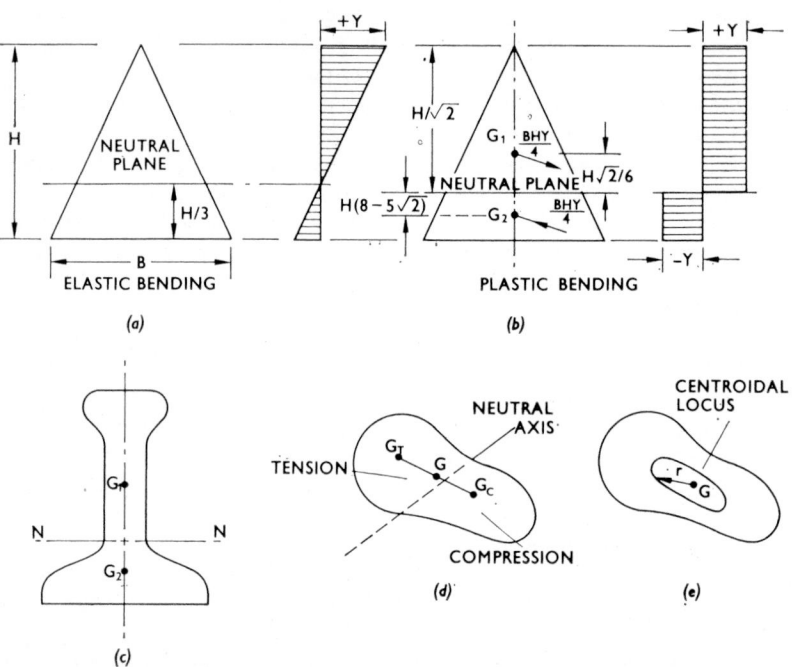

FIG. 7.8 Bending of beam having non-symmetrical cross-sections.

(ii) For full plastic bending the neutral axis is distant $H/\sqrt{2}$ below the apex; the net tensile force on the section may be assumed to act through the centroid of the triangular portion of the section which is above the neutral plane, i.e., at G_1 and the net compressive through G_2 the centroid of the trapezium below the neutral plane. It may be shown that, $M_P = (2 - \sqrt{2}) YBH^2/6$.

The shape factor for this section is $4(2 - \sqrt{2}) \simeq 2\cdot34$.

Certain sections of beam which possess only one axis of symmetry, like the triangular one just discussed, may possess a boundary which can only be treated numerically, e.g., the bulb section of Fig. 7.8(*c*). For most practical purposes the fully plastic bending moment can be obtained by simple experiment. A thin metal sheet or cardboard lamina cut-out of the section shape is taken and, with the help of a planimeter, its area is found. The position of the neutral plane for plastic bending NN can be found; the area above NN must equal that below it. The lamina is then cut along NN and by balancing each of the two portions in turn on a knife edge, the centroid of each is found, i.e., G_1 and G_2. Immediately we have $M_P/Y = 1/2$ area of lamina times the distance $G_1 G_2$.

For beams of strain-hardening material graphical and numerical procedure must be used, see Chapter 22 of Vol. 1 in the book by NADAI (1950).

7.2.7 *Plastic Asymmetrical Bending*

Asymmetrical bending occurs whenever the plane of an applied bending moment is not parallel or perpendicular to an axis of symmetry of the cross-section. Plastic asymmetrical bending has been considered by Johansen, Harrison, Barrett and Brown.

Figure 7.8(d) shows a general cross-section of a beam of cross-sectional area A with an arbitrarily chosen neutral plane. If the section is fully plastic then for force equilibrium due to bending stresses normal to the section, the area of the section in tension on one side of the neutral axis, A_T, must equal the area of the section in compression on the other side of the neutral axis, A_C. We must have,

$$Y A_T - Y A_C = 0.$$

Also then, $A_T = A_C = A/2$. The position of the neutral axis *for each given direction* is thus provided from the family of area bisectors of the section.

The net tensile force $A_T . Y$ acts through the centroid G_T of A_T and the net compressive force $A_C . Y$ through G_C, the centroid of A_C. Hence the full plastic moment of the section is M_P and

$$M_P = r . A Y$$

the direction of the axis of M_P being perpendicular to $G_T G_C$. $2r$ is the distance $G_T G_C$ and the mid point of $G_T G_C$ is obviously the centroid of the whole section G.

The locus of G_T and G_C is called the centroidal locus, see Fig. 7.8(c); when this is known, M_P about any axis is easily found, r being the distance from G to the locus in a direction perpendicular to the moment axis. Further, the tangent to the centroidal locus at the point giving r, provides the direction of the neutral axis. This follows from the fact that a given diameter of the centroidal locus is perpendicular to the moment axis; in the limit an 'adjacent' diameter is also perpendicular to the same moment axis and thus the lines joining the points at the ends of the two diameters—tangents—must also be parallel to the two moment axes and hence the neutral plane. This topic is discussed in detail by BROWN (1967).

7.3 **Plastic Bending Followed by Elastic Unloading**

7.3 *Residual Stress Distribution*

The beam of non-hardening material, of rectangular cross-section, Fig. 7.1, is loaded so that the whole of the section becomes elastic-plastic, Fig. 7.2(e).

The moment required to bring this about is $bh^2 Y/4$. If this bending moment is removed, it is equivalent to adding a negative bending moment of amount $bh^2 Y/4$. The beam springs back or recovers elastically; it is supposed that the recovery in each layer of the beam is elastic. This being so, the elastic strain of recovery varies linearly with the distance from the neutral axis, the

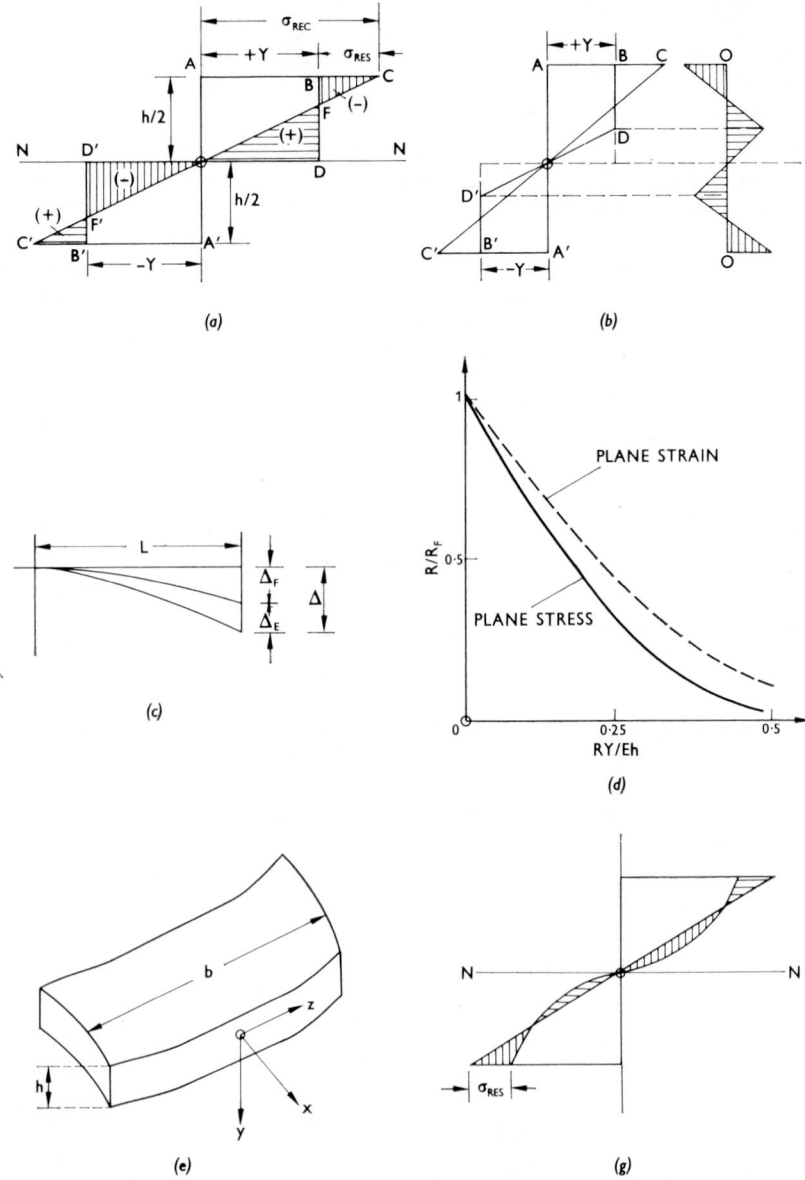

FIG. 7.9 Residual stress distribution and springback in beams of elastic-perfectly plastic material.

elastic recovery stresses being the greater the further away the layer is from the neutral axis. The elastic recovery near the axis will be comparatively small. This recovery is represented in Fig. 7.9(a) by CC'. The moment of the area of the triangle OAC about NN must be the same as that of $OABD$. The value of σ_{REC}. is easily shown to be $3Y/2$ and the actual residual stress distribution is shown shaded in Fig. 7.9(a). From the Fig. 7.9(a) it is apparent that there will thus be compressive residual stresses in the fibres of the beam between F and B and F' and D', and tensile residual stresses between F and D, and F' and B'; $BF/FD = 1/2$ or BF and $B'F'$ are the outer $1/6$ of the whole beam thickness.

7.3.2 Springback Calculations

(i) PLANE STRESS

The determination of the residual stresses in a beam of ideal elastic-perfectly plastic material when $M_{\text{E}} < M < M_{\text{P}}$, is in principle identical with that just discussed for the case when $M = M_{\text{P}}$. In Fig. 7.9(b), $ABDOD'B'A'A$ represents the elastic-plastic stress block due to M; $ACOC'A'A$ represents the elastic recovery stresses and its moment is also M. Thus, the net residual stress distribution is as shown in Fig. 7.9(c).

If the radius of curvature of the neutral axis for the beam in this elastic-plastic condition is R,

$$\frac{y}{R} = \frac{Y}{E} ;$$

hence, substituting for y in equation (7.5.i),

$$M = b \, [3 \, h^2 \, Y - 4 \, Y^3 \, R^2/E^2]/12. \tag{7.5.iii}$$

Now the *change* in the radius of curvature due to springback is R_{E} and

$$M = EI/R = E \, b \, h^2/12 \, R_{\text{E}} \tag{7.5.iv}$$

Figure 7.9(c) shows a beam which has undergone a deflection due to M of $\varDelta$; when unloading occurs the elastic springback is $\varDelta_{\text{E}}$ so that the final deflection is $\varDelta_{\text{F}}$. We have,

$$\varDelta_{\text{F}} = \varDelta - \varDelta_{\text{E}}$$

and because $\varDelta . 2R \simeq L^2$,

$$\frac{1}{R_{\text{F}}} = \frac{1}{R} - \frac{1}{R_{\text{E}}} \quad \text{or} \quad \frac{R}{R_{\text{F}}} = 1 - \frac{R}{R_{\text{E}}}. \tag{7.5.v}$$

R_{F} is the final radius of curvature.

Equating (7.5.iii) and (7.5.iv) and substituting in (7.5.v), we find

$$\frac{R}{R_{\text{F}}} = 1 - 3 \left(\frac{YR}{Eh} \right) + 4 \left(\frac{YR}{hE} \right)^3 \tag{7.5.vi}$$

$$= \left(\frac{YR}{Eh} + 1 \right) \left(2 . \frac{YR}{Eh} - 1 \right)^2 \tag{7.5.vii}$$

When $R/R_F = 0$, there is complete springback, i.e., the bending is wholly elastic; if, at the other extreme, $R/R_F = 1$, there is no springback at all. Figure 7.9(d) shows how R/R_F varies with (YR/Eh).

(ii) PLANE STRAIN

In the sections immediately above, the bending discussed is plane stress bending; dimensions b and h were about equal and thus there was no significant stress in the plane perpendicular to the bending plane. If b is very much greater than h, anticlastic curvature is suppressed apart from small regions near the two sides of the beam; the greater portion of the beam, see Fig. 7.9(e), is flat and thus strain e_z in the direction of b, or z, is zero. For purely elastic strains,

$$e_z = 0 = (\sigma_z - \nu\,\sigma_x)/E; \qquad (7.5.\text{viii})$$

σ_y is assumed everywhere to be zero and ν is Poisson's ratio.

Also, $e_x = (\sigma_x - \nu\,\sigma_z)/E = y/R,$

and hence, $\sigma_x = Ey/R + \nu.\sigma_z. \qquad (7.5.\text{ix})$

Substituting in (7.5.ix) from (7.5.viii) for σ_z and simplifying,

$$\sigma_z = \frac{Ey}{(1 - \nu^2)\,R}.$$

Hence,

$$M_{\text{E}} = \int_{-h/2}^{h/2} \sigma_x\, y\, dy = Eh^3/12R\,(1 - \nu^2) = E'h^3/12R \qquad (7.5.\text{x})$$

From (7.5.x) it is seen that when plane strain bending occurs we must use $E/(1 - \nu^2)$ or E' rather than just E as in plane stress bending. Thus the expression equivalent to (7.5.vii) for plane strain bending is

$$\frac{R}{R_F} = 1 - 3\left(\frac{YR}{Eh}\right)(1 - \nu^2) + 4\left(\frac{YR}{Eh}(1 - \nu^2)\right)^3. \qquad (7.5.\text{xi})$$

This last expression is plotted in Fig. 7.9(d).

It may be observed, however, that the direct yield stress in plane strain bending for a Mises material is $(2/\sqrt{3})\,Y$ but for a Tresca material is just Y; from experimental results the latter is most appropriate for mild steel in simple bending.

Residual stresses in beams of real materials may be estimated, in principle, in the same way, using the real stress-strain curve of the material though the computation is tedious, see Fig. 7.9(g). Such a procedure is valid, provided the residual stresses are everywhere less than the yield stress of the material in *reversed* loading. This means not merely the primary yield stress, but something less than it; this is to say that the Bauschinger effect must be kept in mind. In principle, also, it is no more difficult to deal with non-rectangular sections.

7.4 The Collapse Load in Simple Structures: Plastic Hinges

7.4.1 *The Built-in Beam*

A uniform beam built in at both ends carries a concentrated load, W, at its mid-point, Fig. 7.10. The elastic bending moment diagram is as shown,

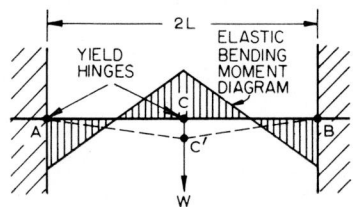

FIG. 7.10 Mode of collapse of built-in beam.

being greatest at the two built-in ends and at the centre. Increasing W eventually causes yielding at each of these three points, and when the material is plastic across the whole of these three sections, the deflections of the beam will become large for further small increases in load. In terms of the rigid-plastic material, no deflection would be possible until the three plastic regions had fully formed. Between these regions the two portions of the beam would be rigid. The real situation may thus be thought of as consisting of three plastic hinges (or four, there being two contiguous hinges under the central load) about which rotation of the two rigid links occurs. This notion is widely used in conjunction with the principle of virtual work to furnish an estimate of the load-carrying capacity of the beam. At the instant of collapse of the beam, the rate at which the external force W is doing work is, if it is moving downwards with unit speed, $W \times 1$. (This work is dissipated as heat as a result of the plastic work done at the hinges.) The rate of rotation of each rigid half of the beam is $1/L$; if M_P denotes the moment required to make the section fully plastic at a hinge, then at A and B the rate of dissipation of energy is M_P/L and at C it is twice this. Thus

$$W \,.\, 1 = 4M_P/L \qquad (7.6)$$

This value of W applies only at the instant at which collapse starts. Immediately W descends below C, to C' say, the situation is complicated by the necessity to stretch AC and BC. In the strict sense in which a mechanical engineer uses the term, this structure is never a mechanism because the structural configuration cannot alter and the links yet remain rigid. But refer to the remarks on Fig. 7.16 below. It will be noted later (Chapter 12) that the method of yield-hinges in fact furnishes an Upper Bound or an over-estimate to the load required to start collapse.

7.4.2 *Portal Frames*

In solving problems in more complicated structures, it is the practice to guess the position of sufficient yield-hinges to enable the structure to become a mechanism and to compute the associated collapse load using the principle of virtual work; the best or most correct collapse load is then the smallest.

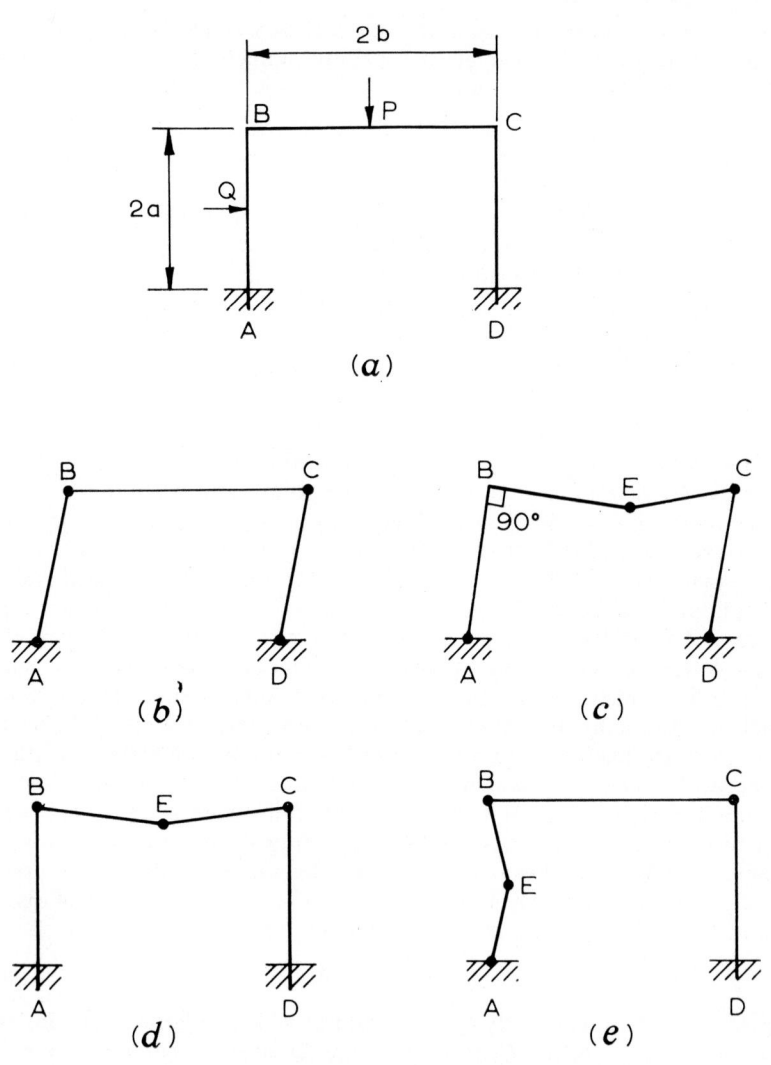

FIG. 7.11 Modes of collapse of portal frame.

To exemplify the procedure, consider a portal frame of which the horizontal beam requires a moment C times as large as does the two identical vertical members to make it fully plastic across a section. It is loaded and of the dimensions shown in Fig. 7.11(*a*). Several guessed modes of deformation or possible mechanisms are proposed in Fig. 7.11(b) to 7.11(e). The associated collapse loads are calculated thus:

(a) For the mechanism in Fig. 7.11(b), Q moving horizontally with unit speed

$$Q.1 = \frac{M_A}{a} + \frac{M_B}{a} + \frac{M_C}{a} + \frac{M_D}{a} = \frac{4M}{a}.$$

It is clear that $M_A = M_B = M_C = M_D$. Note that at B and C, the hinge moments are always M.

(b) For the mechanism in Fig. 7.11(c), where $M_E = CM$,

$$Q.1 + P.\frac{b}{a} = \frac{M_A}{a} + M_E.\frac{2}{a} + \frac{2M_C}{a} + \frac{M_D}{a} = \frac{2(C+2)M}{a}.$$

(c) For the mechanism in Fig. 7.11(d), P moving vertically with unit speed,

$$P.1 = \frac{M_B}{b} + \frac{2M_E}{b} + \frac{M_C}{b} = \frac{2(1+C)M}{b}.$$

(d) For the mechanism in Fig. 7.11(e)

$$Q.1 = \frac{M_A}{a} + \frac{2M_C}{a} + \frac{M_B}{a} = \frac{4M}{a}.$$

These four estimates are now compared and, supposing $Q = nP$, then that value of P is chosen which is least. The mechanism giving this least value of P is not necessarily indicative of the real mode of collapse; an even lower value of P might be ascertained by employing a collapse mechanism which is less obvious than those depicted.

In the above example, if $Q = 2P$, $C = 1.5$, and $b/a = 3$, then the values of P are, for

(a) $2M/a$ (b) $1.4M/a$ (c) $1.67M/a$ (d) $2M/a$

The best value of P is then $1.4M/a$. It tells one that collapse will certainly occur for a value $P = 1.4M/a$.

The plastic analysis of steel structures dates back to KAZINCZY (1914) and KIST (1917) (see PRAGER, 1955). More recently, the plastic analysis of steel-framed structures has been carried very much further by BAKER, HEYMAN, HORNE, RODERICK and others. (See *The Steel Skeleton*, Vols. I and II, by J. F. BAKER et al., 1956).

7.4.3 Oval Links and Circular Rings

Figure 7.12(a) shows a link symmetrical about centre-line AB. To find the collapse load due purely to bending, and neglecting shear forces, assume plastic hinges at A, B, C_1 and C_2. Consider A to remain stationary and B to move along AB with unit speed when collapse occurs. Sections AC_1, C_1B, BC_2 and C_2A may be considered to constitute a '4-bar chain', with I the instantaneous centre of 'link' BC_1. The 'links' AC_1 and C_1B rotate with angular velocity Ω and ω, respectively. (C_1 and C_2 need not be equidistant from A and B as shown in Fig. 7.12a.)

Then $\qquad IB.\omega = 1 \quad$ and $\therefore \quad \omega = 1/IB$

$$IC_1.\omega = AC_1.\Omega, \quad \therefore \quad \Omega = \frac{IC_1}{AC_1}\omega.$$

Thus, if M denotes the fully plastic bending moment

$$P.1/2 = 2M\,(\omega + \Omega).$$

Therefore $\qquad\quad P/4M = \dfrac{IC_1 + AC_1}{AC_1}.\dfrac{1}{IB} = \dfrac{1}{C_1N}.$

Thus $\qquad\qquad \left(\dfrac{P}{4M}\right)_{min} = \left(\dfrac{1}{C_1N}\right)_{max}$ $\qquad$ (7.7)

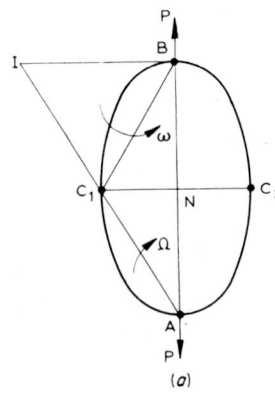
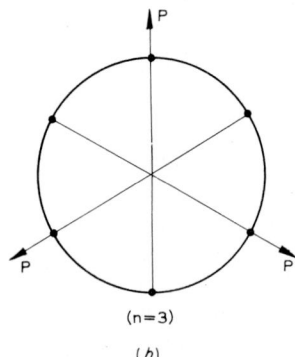

Fig. 7.12
(a) Collapse mechanism of oval link.
(b) Collapse mechanism of circular link carrying n radial loads.

The hinge at C_1 occurs (approximately) at the section furthest from AB.

If a link was welded-in as AB and the load required to cause it to yield in tension was F, then, to find the value of P to ensure collapse, $(P + F)$ would be substituted for P in equation (7.7).

A circular ring carrying n loads, P, acting radially outwards and spaced uniformly at intervals of $2\pi/n$ radians, Fig. 7.12(b), requires

$$P = 4M \cot(\pi/2n)/R$$

if collapse is assumed to occur simultaneously at the $2n$ plastic hinges. R is the radius of the ring. The approach described here is originally due to GREENBERG and PRAGER (1951). Experimental work to test the theory has been reported by JOHNSON (1956) and by SOWERBY, JOHNSON and SAMANTA (1968). The latter authors have also re-examined the problem theoretically using slip-line field theory and taking account of the shear force under each load. They have derived the shape of the four hinges formed. The elementary theory was found to well predict the collapse of a ring, even for fairly thick rings. Experimental and theoretical results based on an elementary stress-block method, for half-rings of trapezoidal cross-section loaded by a concentrated force across a diameter are given by JOHNSON and SENIOR (1957).

7.4.4 Stud Link

This is a case similar to the oval link described above, the stud being a member transverse to the line of action of the load, see Fig. 7.13(a). Assume the stud is rigid, that no buckling is possible in it. This means that for each half of the link to become a mechanism, four new hinges must be postulated between the quarter points, e.g., one at C between A and B (see Fig. 7.13(b)). Proceeding as before,

$$P = 4M(\Omega + \omega)/V_A.$$

Ω, ω, V_A having obvious meaning from Fig. 7.13(b). I is the instantaneous centre for AC.

Now $\qquad \Omega = V_A/IA, \quad \omega = V_A IC/BC.IA$

and hence $\qquad P/4M = IB/BC.IA.$

But $\qquad y/IA = BC/BI$

and thus $\qquad P/4M = 1/y$

where $\qquad y = CQ.$

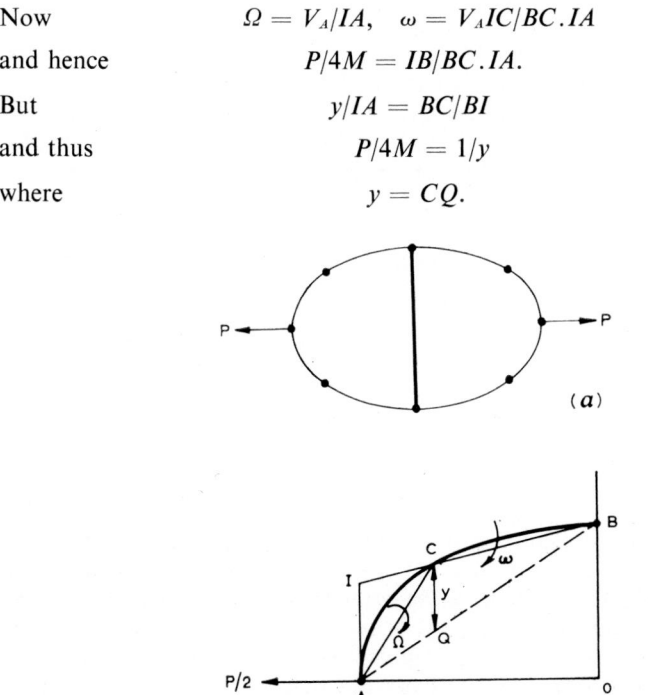

FIG. 7.13 (a) and (b) Collapse mechanism of stud link.

P is least when y is greatest, which defines the position of hinge C. This is the point on the link at which the tangent is parallel to AB.

For an elliptical shackle or link whose Cartesian equation is $x^2/a^2 + y^2/b^2 = 1$, (origin O as in Fig. 7.13(b) with $OB = b$ and $OA = a$) it is easy to show that $CQ = b(\sqrt{2}-1)$ and hence that $P/4M = (1 + \sqrt{2})/b$.

If the stud is insufficiently rigid it may undergo compressive yielding (buckling not being envisaged) and give rise to a transverse force AY where A is its cross-sectional area. The plastic hinges at the ends of the stud do not then form and the link behaves as a 6-bar linkage. The collapse load can be shown to be $P/4M = (1 + 2a/c)^{\frac{1}{2}}/b$ for an elliptical link, where $c = 4M/AY$; the off-the-centre line hinges form at a distance $a^2/(a+c)$ from the transverse stud, at a location which makes P a minimum.

7.5 The Diagram of Angular Velocities

In estimating the collapse load for the simple structures above, we saw that calculations of the internal energy dissipation depended on ascertaining the rate of change of angle at a hinge between two rigid links. We now introduce the diagram of angular velocity vectors in connection with the simple beam-type structures shown in Figs. 7.10 and 7.12(a) and (b). We do this for two reasons, (a) it will, for some readers, render load calculations easier in this type of problem, but largely we do it (b) in anticipation of their extended and much more valuable use in connection with the study of the plastic bending of plates, with which we shall deal in Chapter 14.

In Fig. 7.10 we analysed the case of the plastic collapse of a beam built-in at both ends. Let the two portions of the beams each rotate with angular velocity ω; they will do so in opposite senses, see Fig. 7.14(a). This fact can be represented in a diagram, Fig. 7.14(b), in which $\overrightarrow{oa}$, of a length propor-tional to ω, represents the rate of rotation of AC about A; similarly $\overrightarrow{ob}$ describes the rate of rotation of CB about B. It follows that the rate of change of angle at C as between AC and BC, i.e., $\overrightarrow{ab}$ in Fig. 7.14(b), is 2ω. As previously then,

$$P.\omega L = M_{\text{P}}.\omega_{AC} + M_{\text{P}}\,\omega_{BC} + M_{\text{P}}.2\omega$$

or
$$P = 4\,M_{\text{P}}/L.$$

ω or $\overrightarrow{oa}$ or ω_{AC} is the magnitude of an angular velocity discontinuity at A; the wall part of the structure does not rotate and then suddenly at A, the structure, i.e., link AC, is found to be rotating. 2ω is the angular velocity discontinuity, or jump, i.e., from, say, $+\omega$ to $-\omega$, that takes place at C as between AC and BC; and ω_{BC} is the angular velocity discontinuity at B.

The diagram of angular velocities for the rigid portions of the circular ring of Fig. 7.12(b), see Fig. 7.14(c), which is undergoing plastic collapse because of the three equal outward acting radial forces, is shown as Fig. 7.13(d). We have,

$$3P.v = 3M_{\text{P}}.(2\omega) + 3M_{\text{P}}.(2\omega)$$
$$\text{(i)} \qquad\qquad \text{(ii)}$$

(Lower case letters in Fig. 7.14(d) refer to the rigid links in Fig. 7.14(c)).

Term (i) is the rate of energy dissipation at points under a load. 2ω, the distance from f to a in Fig. 7.14(d), is the magnitude of the discontinuity in angular velocity at these points; it is the change from $+\omega$ to $-\omega$ as between a rigid link on one side of a load point hinge and the rigid link on the other.

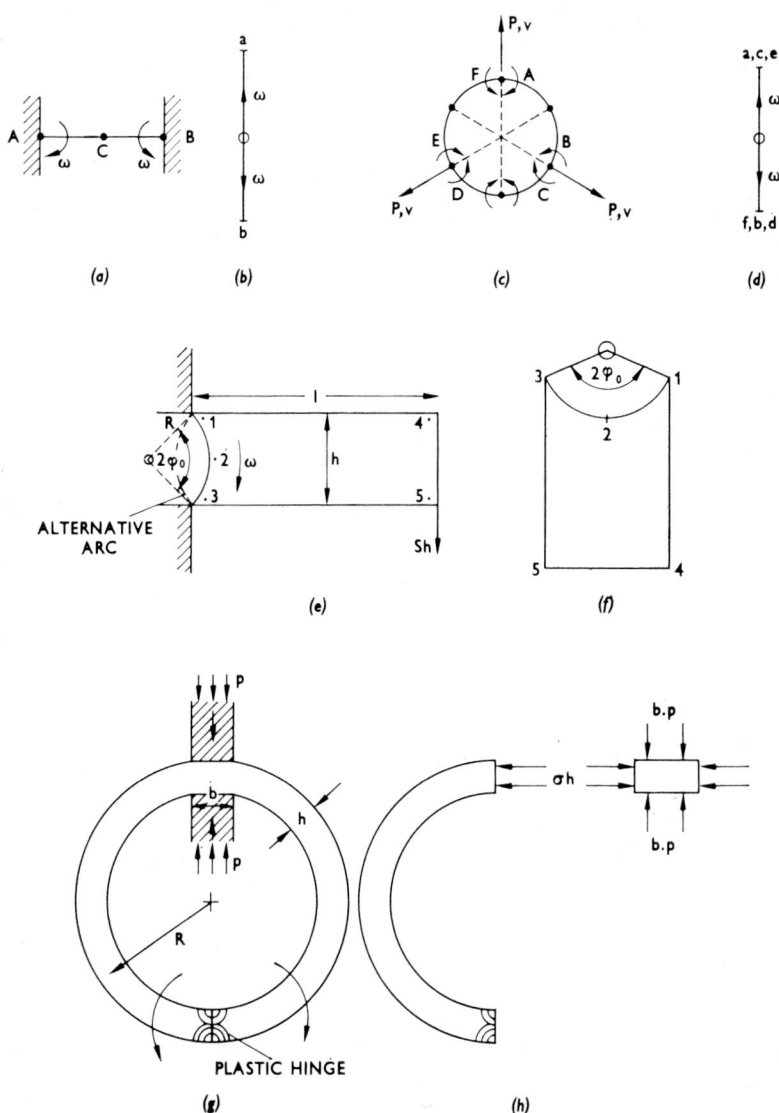

FIG. 7.14

There are identical angular velocity discontinuities or jumps at the hinges formed in the ring exactly between load points.

    The use of angular velocity diagrams as just described for structures

6—EP * *

whose elements are rotating in one plane only, and which therefore give rise to vectors all of which lie along one straight line, does not provide a method of load calculation much more straightforward than that described in earlier sections. However, we repeat, when the diagram of angular velocities is used in the analysis of plate bending problems later, it makes for great simplicity as compared with the more common approaches.

7.6 The Plane Strain Bending of Cantilevers by a Transverse Shear Force

Figure 12.29 shows a short wide cantilever carrying an end load, which just causes plastic collapse, sectioned and etched to reveal the geometrical structure of the plastic hinge at the cantilever root. Apart from the regions at the top and bottom of the beam, it is clear that a circular arc is present in which the material is entirely plastic; both in the body of the beam and in that portion which is firmly fixed into the wall, there is no yielding. Evidently when collapse is about to take place, the beam rotates over the arc as a hinge and the dark arc is a thin plastic region in which shearing occurs. With this idea in mind we may make a simple analysis of the plastic bending of wide beams; this is predominantly plane strain bending because nearly all the material in a given vertical plane remains in that plane when bending occurs. At the instant the bending takes place let the yield shear stress in the circular arc be k, which we imagine to span the entire depth of the beam, see Fig. 7.14(e). Let the radius of the arc be R, the angle it subtends at its centre $2\phi_0$ and denote by Sh, the shear force which causes the bending. Fig. 7.14(f) is a velocity vector diagram showing the speed at all points in the rigid rotating beam.

When plastic collapse occurs, because the beam is rotating about 0, the work rate per unit width is $(Sh)\,(l + R \cos \phi_0)\,.\omega$; and this is dissipated in doing plastic work in the circular arc because the shear force $(R.2\phi_0)k$ is being moved at a speed of $R\omega$. Hence, equating these two quantities,

$$Sh.(l + R \cos \phi_0)\,\omega = (R.2\,\phi_0.k).R\omega$$

Thus,
$$\frac{S}{k} = \frac{2.R^2\,\phi_0}{h\,(l + R \cos \phi_0)}$$

Substituting for R because $h = 2R \sin \phi_0$, and simplifying,

$$\frac{S}{k} = \frac{\phi_0}{\tfrac{1}{2} \sin 2\,\phi_0 + \dfrac{l}{h}\,(1 - \cos 2\,\phi_0)}.$$

It is sensible to choose that value of ϕ_0 which makes S/k a minimum. Differentiating S/k with respect to ϕ_0 and equating to zero gives,

$$2\frac{l}{h} = \frac{\sin 2\,\phi_0 - 2\,\phi_0 \cos 2\,\phi_0}{2\,\phi_0 \sin 2\,\phi_0 + \cos 2\,\phi_0 - 1}.$$

Using the last two equations, the table below may be drawn up.

l/h	0·42	0·875	3·08	12·9	∞
S/k	0·965	0·571	0·208	0·052	0
$\phi_0{}^\circ$	30	45	60	65·2	$\sim 67^\circ$
$S'k$	1·19	0·577	0·16	0·039	0

When $2\phi_0 = \tan \phi_0$, $l/h \to \infty$ and $S/k \to 0$.

If the elementary theory earlier described was used, we should have, using equation (7.5.ii) and replacing Y by $2k$,

$$\frac{h^2 . 2k}{4} = Sh . l$$

that is
$$\frac{S}{k} = \frac{1}{2} . \frac{1}{l/h} \equiv \frac{S'}{k}$$

The fourth line in the table, denoted S'/k, is compiled using the above equation. The arc type of deformation just described, because it gives the smallest value for S'/k for values of $l/h \leqslant 1$, is evidently the more realistic one to choose when dealing with the loading of short stubby cantilevers such as spur gear teeth. More refined analyses justify this conclusion. Appropriate references to the pioneer work of GREEN in this field will be found in JOHNSON and SOWERBY (1967). In this latter paper the arc type of approach just described is used for investigating other modes of deformation [for instance, where the arc is contained in the portion of the beam which is embedded in the wall, see the broken curve in Fig. 7.14(c)] and for curved bars, wedge-shaped bars and notched bars, etc. The reader will be interested to note that much of the work in this section is formally identical with the notched bar problem dealt with in Chapter 13.

7.7 Bending in Wide Ring Cogging

An interesting and unexpected circumstance in which a plastic hinge is found is in considering cogging, which is represented in Fig. 7.14(g). A circular ring of rigid-perfectly plastic material of breadth w, in a plane perpendicular to the plane of the paper, and thickness h, is forged or indented by a pair of rigid dies, of width b. For simplicity we suppose the ring has been prepared with flats to receive the dies. If w/h is large then the mode of deformation as the dies approach will, in general, possess two major features. (a) When the material between the dies is wholly plastic it will, in the main, flow laterally (in the plane of the page) as the dies approach; little or no flow out of the plane of the paper is envisaged. (b) This kind of flow thus causes the ring to open, so that we must suppose that a plastic hinge is formed in the ring diametrically opposite the dies. Between the dies and the hinge the material is rigid and it rotates as a part of one of the two rigid links about the hinge.

This situation described is one encountered in forging practice though many of the assumptions described apply only in part.

If the ratio of the ring thickness, h, to the mean radius, R, of the ring is small it is reasonable to suppose that the lateral normal stress, σ, through the ring at the dies is uniform. Now the force σhw must be great enough to create a moment at the bottom of the ring which is equal to the full plastic bending moment $wh^2 Y/4$, where Y is the appropriate yield stress. Thus,

$$\sigma . hw . 2R = wh^2 Y/4 \quad \text{or} \quad \sigma = hY/8R;$$

i.e., when σ attains this value the two rigid halves of the ring may rotate about the plastic hinge.

The forging or indenting pressure p and σ may now be related through the Tresca yield criterion, thus

$$(-\sigma) - (-p) = Y$$

and substituting for σ in the previous equation,

$$p/Y = 1 + h/8R.$$

The second term, $h/8R$ above will, in most practical cases, be small enough to be ignored.

Experiments testify well, however, to the existence of a plastic hinge in situations such as the one described.

7.8 Combined Bending and Tension: Example of Use of a Yield Inequality

The method of dealing with the plastic design of structures or machine elements in which both tension and bending are combined is well illustrated by the following adaptation of a problem originally analysed by Hodge (1955). The aim is to determine certain critical angular velocities for a non-symmetric rotating ray. In Fig. 7.15(a) and (b), the cylindrical disc and the plane radial ray are shown in elevation and plan. Due to the centrifugal force acting on the ray, the ray will tend to straighten itself and hence on any section there will be acting both a radial tensile force and a bending moment. There will also be a small shearing force due to the weight of the section, but this will be taken to be negligible. For simplicity the analysis of a flat-bottomed ray only is considered, Fig. 7.15(c).

7.8.1 Elastic Analysis

The total tensile or radial force N acting on the section of height h_2 is

$$N = \int_r^b \rho hts\omega^2 \, ds = \frac{\rho t\omega^2 h_0}{6L}[b^3 - r^2 (3b - 2r)] \tag{7.8}$$

noting that $h/h_0 = (b - s)/L$. Thus the average tensile stress over the section is

$$\sigma_t = \frac{N}{h_2 t} = \frac{\rho \omega^2}{6(b - r)}[b^3 - r^2 (3b - 2r)] = \frac{\rho \omega^2}{L}[b (b + r) - 2r^2]. \tag{7.8.i}$$

The total bending moment, M, also acting on the section h_2, about its centre

$$M = \int_r^b \rho h t s \omega^2 \left(\frac{h_2 - h}{2}\right) ds$$

$$= \int_r^b \frac{\rho t \omega^2 h_0^2 s}{2L^2} (b - s)(s - r) \, ds$$

$$= \frac{\rho t \omega^2 h_0^2}{24L^2} (b - r)^2 (b^2 - r^2). \qquad (7.8.\text{ii})$$

The maximum bending stress σ_m occurs in the outermost fibres of the ray, and hence

$$\sigma_{m,max} = (b^2 - r^2) \rho \omega^2 / 4. \qquad (7.9)$$

The greatest (tensile) stress σ_m on section h_2 occurs then in the bottom-most fibre, and combining equations (7.8.i) and (7.9),

$$\sigma_m = \frac{\rho \omega^2}{12} [3(b^2 - r^2) + 2b(b + r) - 4r^2]$$

$$= \frac{\rho \omega^2}{12} [5b^2 + 2br - 7r^2]. \qquad (7.10)$$

Thus for rays whose disc diameter, $a < b/7$, the greatest stress occurs in the ray at $r = b/7$, and is of magnitude $\sigma_m = 3\rho\omega^2 b^2/7$. If $a > b/7$, σ_m occurs where the ray joins the hub and is of magnitude $\rho\omega^2 (b - a)(5b + 7a)/12$. Hence for full elastic loading such that $\sigma_m = Y$, the tensile yield stress,

$$\left.\begin{array}{ll}
(i) & \dfrac{b}{7} > a, \qquad \omega^2 \leqslant \dfrac{7Y}{3\rho b^2} \\[4mm]
(ii) & \dfrac{b}{7} < a, \qquad \omega^2 \leqslant \dfrac{12Y}{\rho(b - a)(5b + 7a)}.
\end{array}\right\} \qquad (7.11)$$

7.8.2 *Plastic Analysis*

When the speed of rotation exceeds that given by equations (7.11), part of the section on the lower side of the ray will yield. Further increase in ω will cause the yielded region to grow, and eventually plasticity will set in along the top edge of the ray. Supposing that the material is perfectly plastic, the yield stress in the plastic regions, once reached, will not be exceeded. The two zones of plasticity will be separated by an elastic zone. No attempt will be made to trace the growth of these zones with increase in ω, but attention will be restricted to finding across which section of the ray the material first becomes wholly plastic. The speed at which this occurs will be critical for the ray, because immediately large deflections become possible. When the tension and bending have combined, the fully plastic stress distribution across the critical section is as shown in Fig. 7.15(d), η being the section height at which the stress discontinuity occurs. The total tensile force on the section

$$N = 2tY\left(\eta - \frac{h}{2}\right)$$

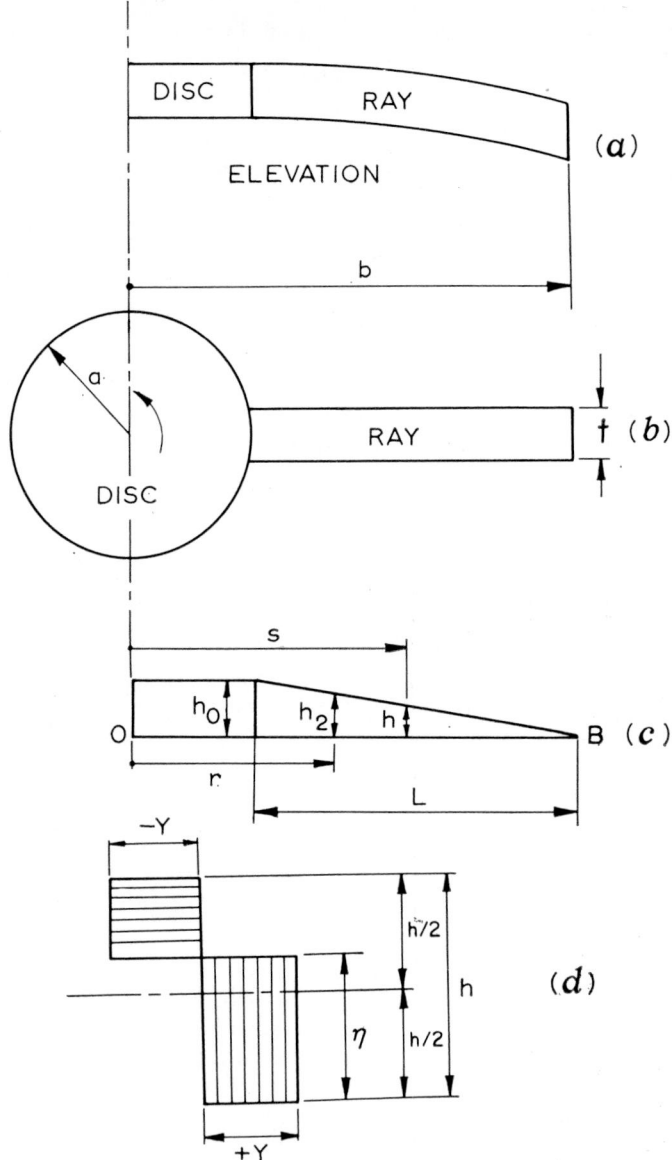

Fig. 7.15 Non-symmetric rotating ray.

and the total moment, $M = tY\eta(h - \eta)$. When $\eta = h$ the stress distribution is one of pure tension, the tensile force being $N_0 = tYh$. When $\eta = h/2$, the stress distribution is one of pure bending and the bending moment, $M_0 = tYh^2/4$. Thus

$$\left(\frac{N}{N_0}\right)^2 = \left(\frac{\eta - h/2}{h/2}\right)^2 \quad \text{and} \quad \frac{M}{M_0} = \frac{\eta (h - \eta)}{h^2/4}.$$

Eliminating η between these two equations

$$\left(\frac{N}{N_0}\right)^2 + \frac{M}{M_0} = 1.$$

It is evident that since a perfectly plastic cross-section *cannot* carry stresses larger than Y, for all sections, we can write down a *yield inequality* as follows

$$\left(\frac{N}{N_0}\right)^2 + \frac{M}{M_0} < 1$$

We now substitute for M and N from equations (7.8) and (7.8.ii) in the above yield inequality to obtain

$$\left(\frac{N}{N_0}\right)^2 = \left(\frac{\rho\omega^2}{6Y}\right)[b(b+r)-2r^2]^2 \quad \text{and} \quad \frac{M}{M_0} = \left(\frac{\rho\omega^2}{6Y}\right)(b^2-r^2).$$

Putting $\rho\omega^2 b^2/6Y = P$ and writing x for $(1-r/b)$, the inequality gives

$$P^2x^2(3-2x)^2 + Px(2-x) - 1 \leqslant 0.$$

Give x values between 1 and 0·6 and calculate the corresponding positive value of P. The results are

x	1·0	0·9	0·8	0·6
P	0·615	0·594	0·592	0·63

Evidently P is least and $\simeq 0.59$ when $x \simeq 0.82$. This means that the inequality is satisfied at each particular section—as indicated by x—provided P (and hence ω) is less than the calculated value. Obviously the maximum permissible angular speed of the ray is associated with the smallest admissible, greatest value of P.

Hence $0.59 = \rho\omega^2 b^2/6Y$

or $\omega = 1.88\sqrt{Y/\rho b^2}.$

This solution applies only if the critical section is on the ray, i.e. if $a < 0.18b$ or $6.5a < b$. Thus when,

(i) $a < 0.18b,$ $\omega_P = 1.88\sqrt{Y/\rho b^2}$

and (ii) $a < b/7,$ $\omega_E = 1.53\sqrt{Y/\rho b^2},$ using (7.11),

ω_E being the angular speed at which yielding starts.

A simple means of displaying results is shown in Fig. 7.16. We plot (i) the yield inequality, and (ii) that relationship between N and M which is such that if they operate together, they cause the yield stress just to be reached. For (ii) let σ denote a uniform tensile stress on a section; then,

$$N/N_0 = \sigma/Y$$

and thus $\dfrac{M}{M_0} = \dfrac{bh^2(Y-\sigma)}{6} \cdot \dfrac{4}{bh^2Y} = \dfrac{2}{3}\left(1-\dfrac{\sigma}{Y}\right) = \dfrac{2}{3}\left(1-\dfrac{N}{N_0}\right),$

which appears as a straight line in Fig. 7.16. All combinations of M and N which involve some plastic yielding, plot between these two curves. Note that small values of N have little effect on the required moment for yield.

The vertical deflection of the 'mechanism' of Fig. 7.10 should cause the 'links' to extend and work to be dissipated. However, the parabola of Fig. 7.16 suggests that this is extremely small, and hence may be ignored.

7.9 The Elastic–Plastic Bending of Wide Plates having an Initial Curvature

Many aspects of this particular problem have been thoroughly analysed by SHAFFER et al. (1955, 1957) and here only a résumé of that work will be given.

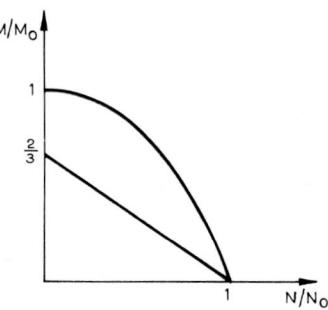

FIG. 7.16 Yielding and collapse relationships for a rotating ray.

Figure 7.17(a) shows a section of a wide plate of constant curvature. The bending moment M per unit length perpendicular to the plane of the paper, except near the point of its application at the ends, will produce the same stress distribution across each radial plane. Fig. 7.17(b) shows the stresses

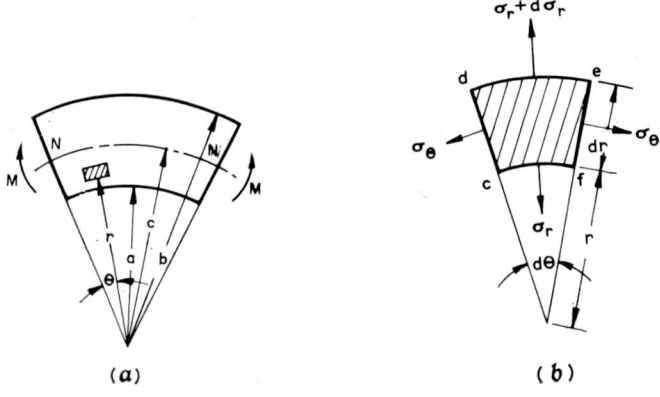

(a)   (b)

FIG. 7.17

(a) Geometry of the curved bar.
(b) Stresses on a typical element.

on an element of the beam and the equation of equilibrium in the radial direction is

$$\frac{d\sigma_r}{dr} = \frac{\sigma_\theta - \sigma_r}{r}. \tag{7.12}$$

7.9.1 *Purely Elastic Case*

The solution in this instance is well known; it is a hyperbolic stress distribution for σ_θ. (See Figs. 7.18(*b*), (*c*).) The neutral axis does not pass through the centroids of the cross-sections.

By introducing the Airy stress function ϕ,

$$\sigma_r = \frac{1}{r}\frac{d\phi}{dr} \quad \text{and} \quad \sigma_\theta = \frac{d^2\phi}{dr^2}, \text{and} \tag{7.13}$$

equilibrium equation (7.12) is satisfied. The compatibility equation is, (TIMOSHENKO and GOODIER, 1951)

$$\left(\frac{\partial}{\partial r^2} + \frac{1}{r}\frac{\partial}{\partial r} + \frac{1}{r^2}\frac{\partial^2}{\partial \theta^2}\right)\left(\frac{\partial^2\phi}{\partial r^2} + \frac{1}{r}\frac{\partial\phi}{\partial r} + \frac{1}{r^2}\frac{\partial^2\phi}{\partial \theta^2}\right) = 0 \tag{7.14}$$

but here $\partial^2/\partial\theta^2 = 0$ and thus (7.14) becomes

$$\frac{d^4\phi}{dr^4} + \frac{2}{r}\frac{d^3\phi}{dr^3} - \frac{1}{r^2}\frac{d^2\phi}{dr^2} + \frac{1}{r^3}\frac{d\phi}{dr} = 0. \tag{7.15}$$

The solution to (7.15) is

$$\phi = A \ln r + Br^2 \ln r + Cr^2 + D.$$

For this particular case, in which the surfaces at $r = a$ and b are stress free,

$$\sigma_r = \frac{-4M}{N}\left(\frac{a^2b^2}{r^2}\ln\frac{b}{a} + b^2\ln\frac{r}{b} + a^2\ln\frac{a}{r}\right) \tag{7.16}$$

$$\sigma_\theta = \frac{-4M}{N}\left(\frac{-a^2b^2}{r^2}\ln\frac{b}{a} + b^2\ln\frac{r}{b} + a^2\ln\frac{a}{r} + b^2 - a^2\right) \tag{7.17}$$

where $\qquad\qquad N = (b^2 - a^2)^2 - 4a^2b^2(\ln b/a)^2.$

Yielding commences first on the inner concave surface at $r = a$, and using the Tresca criterion, $\sigma_\theta - \sigma_r = 2k$,

$$M = \frac{k[(b^2 - a^2)^2 - 4a^2b^2\,(\ln b/a)^2]}{2(a^2 - b^2 + 2b^2 \ln b/a)}. \tag{7.18}$$

7.9.2 *Fully Plastic Case*

If the Tresca criterion is used

$$\sigma_\theta - \sigma_r = 2k. \tag{7.19}$$

On substituting and integrating in (7.12), it is found that

$$\sigma_r = 2k \ln r + C. \tag{7.20}$$

Noting that $\sigma_r = 0$ where $r = a$ and b, it is found that

$$\sigma_r = 2k \ln r/a \quad \text{when} \quad a \leqslant r \leqslant c \tag{7.21}$$

because in this region $\sigma_\theta > \sigma_r$,

and $\qquad\qquad\qquad \sigma_r = 2k \ln b/r \quad \text{when} \quad b \geqslant r \geqslant c \tag{7.22}$

because in this region $\sigma_r > \sigma_\theta$.

c is the radius of the neutral axis. Hence, since σ_r must be continuous across $r = c$, equating (7.21) and (7.22), it is found that $c = \sqrt{ab}$. It follows that for $a \leqslant r \leqslant c, \sigma_\theta = 2k(1 + \ln r/a)$ and $c \leqslant r \leqslant b, \sigma_\theta = -2k(1 - \ln b/r)$. The bending moment, M_P, for this fully plastic state is

$$M_\mathrm{P} = k(b - a)^2/2.$$

7.9.3 Elastic–Plastic Case

As M is increased in excess of that given by (7.18), the yielding initiated on the concave layer at $r = a$, spreads outwards to a radius ρ_1, see Figs. 7.18(a) and (b). For $r > \rho_1$ the stress distribution remains elastic. Eventually

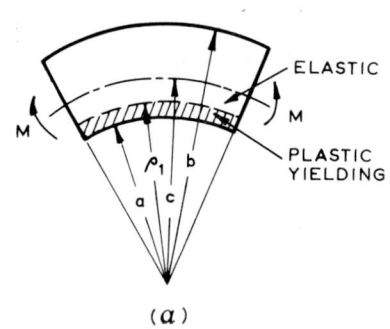

(a)

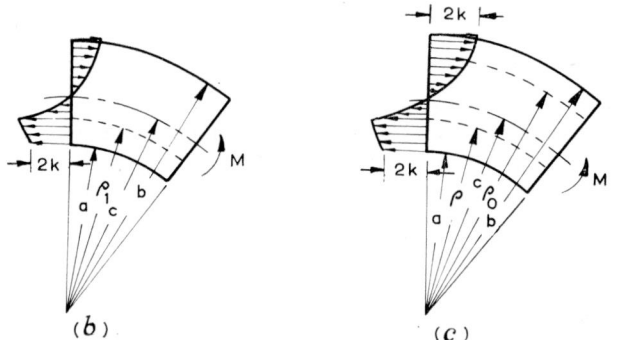

(b)   (c)

FIG. 7.18

(a) First stage of the elastic-plastic solution.
(b) Circumferential stress distribution σ_θ on a sector of the plate during the first stage of the elastic-plastic solution.
(c) Circumferential stress distribution σ_θ on a sector of the plate at the beginning of the second stage of the elastic-plastic solution.

a value of ρ_1 is attained for which yielding in compression commences on the outermost fibre, i.e. $r = b$, see Fig. 7.18(c). Further increase in M causes a progressive extension of the zones of yielding, the elastic layer between them reducing in thickness. The position of the neutral axis alters with the loading.

Fig. 7.19 is a reproduction of diagrams given by SHAFFER and HOUSE (1955). For a plate having $b/a = 2$, yielding first starts when $M = 0.516M_P$ where M_P is the bending moment required to bring about plasticity through the whole plate. Figures 7.19 show how the distributions σ_r/k and σ_θ/k through the plate thickness alter as M/M_P approaches 1 from 0.516. The oscillation in the position of the neutral axis is also given.

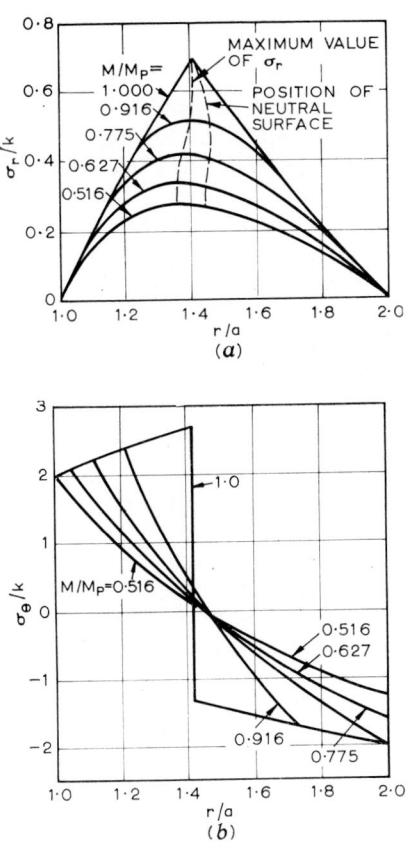

FIG. 7.19
(a) Radial stress distribution σ_r for $b/a = 2.0$.
(b) Circumferential stress distribution σ_θ for $b/a = 2.0$.

The same authors have also examined the strain and displacements which occur in bending. They showed that the change in radii on bending a given wide curved plate is small (less than 2 per cent) if M/M_P is less than 0.95. Thus for all likely practical cases, neglect of changes in geometry throughout the analysis is justified.

7.9.4 *Residual Stresses*

Upon unloading a wide plate in which yielding has occurred, some residual stress distribution will occur. Generally these are distributed entirely elastically. However, under certain conditions, for wide curved plates, yielding may occur on unloading, both in the interior of the plate or/and at its concave boundary. SHAFFER and HOUSE (1957) have discussed the various cases at great length and have identified nine non-trivial cases. They have also provided data on the amount of springback, or elastic recovery displacements, that occur.

7.9.5 *Additional References*

The SHAFFER and HOUSE analysis just described, relates to the elastic-plastic bending of a wide curved bar of incompressible material under conditions of plane strain; for the case when the material is compressible, the solution has been given by EASON (1960a). The corresponding problem for conditions of plane stress has been investigated by SHEPHERD and GAYDON (1957). The elastic-plastic bending of a curved bar by end couples in plane stress has also been examined by EASON (1960b).

The article by PROKSA (1959) makes a valuable contribution to the plate bending theory of strain-hardening materials and also provides a useful source of references to bending theory articles in German.

7.10 **Sheet and Plate Bending as a Forming Operation**

The successful bending of plate or sheet is an everyday press shop operation and many methods for doing this are available.

The particular machine or method used for the bending operation needs to be studied when calculations are to be made, connecting, say, the bending force or moment which the tools must apply, to the curvature or the deflections sought in the product, or concerning the permissible degree of bending and the expected springback. The degree of bending which may be required industrially is often very severe by the criteria encountered in the context of bending in structural elements; and further, the complexity and restraints imposed by press tool shape and friction between materials and tools give rise to great complications.

7.10.1 *Press Brake Forming of Straight Edges*

In forming sheet, say for automobile parts, the bending operation indicated in Fig. 7.20(*a*) is often performed; a shaped punch forces sheet metal to be bent to take up (often allowing for elastic springback) a pre-determined shape. In achieving small bend radii the maximum tensile hoop strain imposed on the sheet may exceed $1/4$ and then transverse contraction is to be expected on the tensile side and thickening on the compression side; the net result is that the neutral axis in the sheet may move from the original section centroid towards the side in compression by as much as 5 per cent of the sheet thickness. Cracks are likely to occur, too, in the centre of the sheet in these circumstances.

To predict the minimum satisfactory bend radius R for a plate thickness T which may be obtained, an empirical relationship often used for sheet when the width to thickness ratio exceeds 10 and when the angle of bend exceeds 70° is

$$\epsilon \text{ bend} = \tfrac{3}{4} \, \epsilon \text{ tension} = (1 + 1\cdot8 \, R/T)^{-1};$$

ϵ bend denotes the bending strain permissible, which is often taken to be three-quarters of the fracture strain in a tension test.

Figure 7.20(*b*) indicates the extent and shape of the plastic deformation zone under the bending tool; this zone extends outside the region of contact with the tool and, as would be expected, the zone is of greatest extent in the outermost fibres where the bending stresses are largest. (Problem 61 on p. 578 also discusses this situation.)

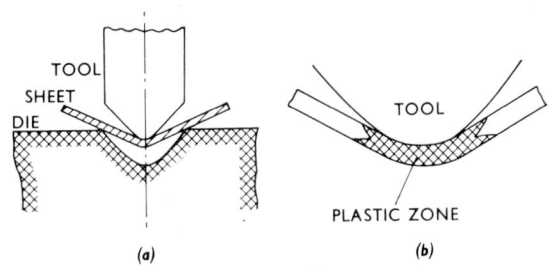

(*a*)   (*b*)

FIG. 7.20 Press brake bending.

Springback considerations are obviously of primary importance in any industrial operation. The approach described above may be used to calculate this and, in particular, the paper by BOTROS (1967) may be consulted.

An excellent detailed discussion of industrial sheet bending practice and a resumé of very useful research and empirical data is to be found in the book *Principles and Methods of Sheet Metal Fabricating* by Sachs as updated by Voegeli; see also the article by PERRY (1956) and the detailed analyses and experiments of WOLTER (1952).

7.10.2 *Plate Bending Using Bending Machines*

There are two traditional types of plate bending machine, see Fig. 7.21, developed originally for forming the cylindrical plates of boilers. The bending of plates is accomplished by rotating rolls so that analyses are difficult due to the introduction of rolling friction phenomena, among other things. A discussion, some analysis and many experimental results as well as references to other works on this topic will be found in a paper by BASSETT and JOHNSON (1966).

Two articles well worthy of reference are those by ALEXANDER (1957) and MASUDA and TOZAWA (1963). Forrest showed that a moderate degree of stretching of material in the same direction as that in which residual stresses act, reduces them very significantly. Alexander has shown theoretically that transverse stretching is nearly as effective. Masuda and Tozawa have discussed the theoretical principles of the suppression of springback and suggest a practical method for doing so. They showed experimentally that springback decreases with increase in compression force applied to the bent part in its lateral direction (the z-direction in Fig. 7.9(f)), and that when this force attains a certain magnitude the springback becomes very small.

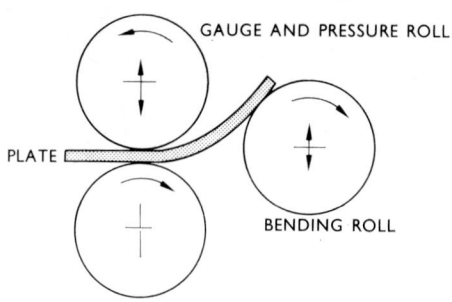

PINCH – TYPE ROLLS

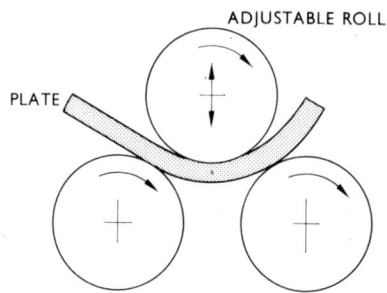

PYRAMID – TYPE ROLL

FIG. 7.21 The two types of plate bending machines.

See Problems 26-38, 58 and 78

REFERENCES

ALEXANDER, J. M.   1957 'An Analysis of the Plastic Bending of Wide Plate, and the Effect of Stretching on Transverse Residual Stresses' *Proc. Instn mech. Engrs* **173**, 73

BAKER, J. F., HORNE, M. R. and HEYMAN, J.   1956 *The Steel Skeleton* C.U.P.

BASSETT, M. B. and JOHNSON, W.   1966 'The Bending of Plate Using a Three-Roll Pyramid Type Plate Bending Machine' *J. Strain Analysis* **1**, 398

BOTROS, B. M.   1967 'Springback in Sheet Metal Forming after Bending' *A.S.M.E.* 67–WA/PROD.–17

BROWN, E. H.   1967 'Plastic Asymmetrical Bending of Beams'
Int. J. mech. Sci., **9**, 77

EASON, G.   1960a 'The Elastic-Plastic Bending of a Compressible Curved Bar'
Appl. Sci. Res. Series A, **9**, 53

EASON, G.   1960b 'The Elastic-Plastic Bending of a Curved Bar by End Couples in Plane Stress'
Q. J. Mech. Appl. Math. XIII, 334

GREENBERG, H. J. and PRAGER, W.   1951 'Limit Design of Beams and Frames'
Proc. Am. Soc. civ. Engrs **77**, 1

HODGE, P. G.   1955 'Rotating Rays'
J. appl. mech. **24**, Paper No. 54–A–96

HODGE, P. G.   1959 Plastic Analysis of Structures,
McGraw-Hill, London

HORROCKS, D. and JOHNSON, W.   1967 'On Anticlastic Curvature with Special Reference to Plastic Bending: A Literature Survey and Some Experimental Results'
Int. J. mech. Sci. **9**, 835

JOHNSON, W.   1956 'The Compression of Circular Rings'
J. R. aeronaut. Soc. **60**, 484

JOHNSON, W. and SENIOR, B. W.   1957 'The Plastic Bending of Heavily Curved Beams'
J. R. aeronaut. Soc., **61**, 824

JOHNSON, W. and SOWERBY, R.   1967 'On the Collapse Load of Some Simple Structures'
Int. J. mech. Sci. **9**, 433

MASUDA, M. and TOZAWA, Y.   1963 'Compression Bending'
Bull. Jap. Soc. precision Engng **1**, 33

NADAI, A.   1950 *Theory of Flow and Fracture of Solids*, Vol. 1
McGraw-Hill, London,

PERRY, T. G.   1956 'Bending and Allied Forming Operations'
Institute of Metals Monograph and Report Series No. 20, p. 91

PHILLIPS, A.   1956 *Introduction to Plasticity*
The Ronald-Press Co., New York

PRAGER, W.   1955 'The Theory of Plasticity: A Survey of Recent Achievements'
Proc. Instn mech. Engrs **169**, 41

PROKSA, F.   1959 'Plastisches Biegen von Blechen'
Der Stahlbau **2**, 29

SACHS, G.   1966 *Principles and Methods of Sheet Metal Fabricating*
Revised and enlarged by H. E. Voegeli, Reinhold, New York

SHAFFER, B. W. and HOUSE, R. N.   1955 'The Elastic-Plastic Stress Distribution Within a Wide Curved Bar Subjected to Pure Bending'
J. appl. Mech. **24**, 305

1957 'Displacements in a Wide Curved Bar Subject to Pure Elastic-Plastic Bending' *J. appl. mech.* **26**, 447

SHAFFER, B. W. and   1957 *Residual Stresses and Displacements in Wide*
UNGAR, E. E.   *Curved Bars Subject to Pure Bending*
                         Office of Ordnance Research

SHEPHERD, W. M. and   1957 'Plastic Bending of a Ring Sector by End
GAYDON, F. A.   Couples'
                         J. Mech. Phys. Solids **5**, 296

SOWERBY, R.,   1968 'The Diametral Compression of Circular
JOHNSON, W. and   Rings by "Point" Loads'
SAMANTA, S. K.   *Int. J. mech. Sci.* **10**, 369

TIMOSHENKO, S.   1953 *History of Strength of Materials*
                         McGraw-Hill, London, 137 pp.

TIMOSHENKO, S. and   1951 *Theory of Elasticity*
GOODIER, J. N.   McGraw-Hill, London

WOLTER, K. H.   1952 *Freies Biegen von Blechen*
                         V.D.I.—Forschungsheft 435, Deutsche
                         Ingenieur Verlag, Düsseldorf

Chapter 8

TORSION OF PRISMATIC BARS OF CIRCULAR AND NON-CIRCULAR SECTION

8.1 Elastic Analyses

8.1.1 *Introduction*

The plastic analysis of prismatic bars in torsion is most usefully approached by first presenting the standard methods and results given by elastic analyses. The plastic analysis, for an elastic-perfectly plastic solid, follows from it quite naturally and easily. The elementary theory of the torsion of prismatic bars is usually restricted to bars of circular cross-section. This simple theory was first given by Coulomb in 1784 and results in the equations

$$\frac{T}{J} = \frac{\tau}{r} = G\theta \tag{8.1}$$

where T is the applied torque, J the polar moment of area, τ the shear stress on a transverse section at a point distance r from the axis of the bar, G the shear modulus and θ the angle of twist per unit length of the bar. The derivation of equation (8.1) proceeds from the assumption that initially plane sections of the bar remain plane after application of the torque. It follows that the magnitude of the shear stress at a point varies directly as its distance from the axis of the bar and this fact sometimes leads students of engineering to suppose, quite erroneously, that the same feature applies in considering bars of other cross-sectional shape. Indeed, NAVIER (1864) proceeding on the assumption that plane sections remain plane after twisting, deduced an incorrect relationship of this kind. (The reader interested in the development of the study of torsion is referred to the work of TODHUNTER and PEARSON (1886) and the book by TIMOSHENKO (1953).) Experiments in the twisting of non-circular prismatic bars show that initially plane sections, or two-dimensional sheets, on twisting become deformed or warped into three-dimensional sheets which are generally of a symmetrical character; Fig. 8.1 illustrates this point. The mathematical analysis of the problem on the assumption that uniform warping will occur was first performed by SAINT-VENANT (1855) and in particular, usually (but not always) leads to the conclusion that the largest shearing stress on a section occurs on the periphery at the point or points nearest to the bar axis.

8.1.2 *The Analysis of Torsion Following Saint-Venant*

The analysis immediately below follows that originally given by Saint-Venant and is usually referred to as his *semi-inverse solution* to the torsion problem. It is best to endeavour to clarify the meaning of the latter term after the details of the analysis have been presented along with some examples of the procedure.

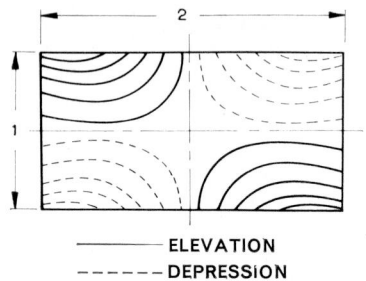

——————— ELEVATION

— — — — — DEPRESSION

FIG. 8.1 Twisting of rectangular bar indicating warping of cross-section.

Figure 8.2 shows a portion of a prismatic bar subject to a torque T. Oz is the axis of the bar parallel to a generator and Ox and Oy, mutually perpendicular, lie in the plane of the cross-section; O is in the end section. Body forces due to gravitational weight, etc. are taken to be of no importance and it is further supposed that Saint-Venant's principle is adhered to, which here means that local stress effects in the material, immediately about the point of application of T, are quickly redistributed to give a stress distribution across each section which does not vary with z.

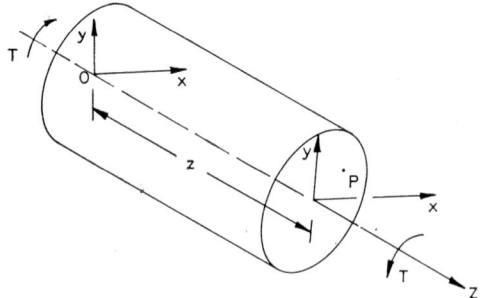

FIG. 8.2 Prismatic bar subject to torque T.

On deforming the bar each section rotates and also warps out of its initial plane.

In Fig. 8.3, P is a point whose initial co-ordinates are (x, y, z) and on twisting positively, P changes to position P', co-ordinates $(x + u, v + y, z + w)$, where u, v, w are small displacements. The angle of rotation of a cross-section at distance z from the origin is $z\theta$, where θ is the angle of twist per unit length. For elastic deformation θ is small and constant along the length of the bar. Then from Fig. 8.3

$$x = r \cos \alpha \quad \text{and} \quad y = r \sin \alpha.$$

Therefore, $\qquad \delta x = -r \sin \alpha \, \delta\alpha \quad \text{and} \quad \delta y = r \cos \alpha \, \delta\alpha$

and hence $\qquad u = \delta x = -y \, \delta\alpha \quad \text{and} \quad v = \delta y = x \, \delta\alpha.$

Thus $\qquad\qquad u = -y.z\theta \quad \text{and} \quad v = x.z\theta. \qquad\qquad (8.2)$

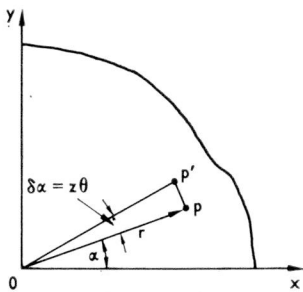

FIG. 8.3 Co-ordinates of point P in section of prismatic bar.

P' is taken to have moved out of the original plane of P by amount w; this amount varies from point to point, i.e. with x and y over the section, and for a given point is also proportional to θ; this is expressed as

$$w = \theta f(x, y). \qquad (8.3)$$

$f(x, y)$ is called the warping function. Note that w is assumed to be independent of z so that warping takes place freely at all sections, including the ends of the bar

From equations (8.2), (8.3) and (3.4) it is seen that

$$\left.\begin{aligned}
\epsilon_x &= \frac{\partial u}{\partial x} = 0 \quad ; \quad \epsilon_y = \frac{\partial v}{\partial y} = 0 \quad ; \quad \epsilon_z = \frac{\partial w}{\partial z} = 0 \\[2mm]
\phi_{xy} &= \frac{\partial u}{\partial y} + \frac{\partial v}{\partial x} = -z\theta + z\theta = 0 \\[2mm]
\phi_{yz} &= \frac{\partial v}{\partial z} + \frac{\partial w}{\partial y} = x\theta + \theta \frac{\partial f}{\partial y} \\[2mm]
\phi_{xz} &= \frac{\partial u}{\partial z} + \frac{\partial w}{\partial x} = -y\theta + \theta \frac{\partial f}{\partial x}.
\end{aligned}\right\} \qquad (8.4)$$

It follows from Hooke's law and equations (8.4) that

$$\sigma_x = \sigma_y = \sigma_z = \tau_{xy} = 0.$$

Also
$$\tau_{yz} = G\phi_{yz} = G\theta\left(x + \frac{\partial f}{\partial y}\right)$$

and
$$\tau_{xz} = G\phi_{xz} = G\theta\left(-y + \frac{\partial f}{\partial x}\right). \tag{8.5}$$

Differentiating and subtracting equations (8.5) to eliminate the warping function

$$\frac{\partial \tau_{xz}}{\partial y} - \frac{\partial \tau_{yz}}{\partial x} = -2G\theta. \tag{8.6}$$

Now the equation of equilibrium for the shear forces acting on a small element at P in direction z (see Fig. 8.4) is given by

$$\frac{\partial \tau_{yz}}{\partial y} \delta x \, \delta z \, \delta y + \frac{\partial \tau_{zx}}{\partial x} \delta y \, \delta z \, \delta x = 0$$

or
$$\frac{\partial \tau_{yz}}{\partial y} + \frac{\partial \tau_{zx}}{\partial x} = 0. \tag{8.7}$$

The next step, which is least familiar to engineers, consists in supposing that a function $\psi(x, y)$ exists which is capable of yielding the shear stresses and satisfying the equilibrium equation (8.7). By taking

$$\tau_{xz} = \frac{\partial \psi}{\partial y} \quad \text{and} \quad \tau_{yz} = -\frac{\partial \psi}{\partial x} \tag{8.8}$$

equation (8.7) is seen to be satisfied and equation (8.6) becomes

$$\frac{\partial^2 \psi}{\partial x^2} + \frac{\partial^2 \psi}{\partial y^2} = -2G\theta$$

or
$$\nabla^2 \psi = -2G\theta. \tag{8.9}$$

ψ is termed the stress function. Equation (8.9), Poisson's equation, is applicable to many problems in engineering science. Another better-known equation, Laplace's equation, concerns the warping function. Differentiating equations (8.5) it is found that

$$\frac{\partial \tau_{xz}}{\partial x} = G\theta \frac{\partial^2 f}{\partial x^2} \quad \text{and} \quad \frac{\partial \tau_{yz}}{\partial y} = G\theta \frac{\partial^2 f}{\partial y^2}$$

and substituting these two equations in (8.7)

$$\frac{\partial^2 f}{\partial x^2} + \frac{\partial^2 f}{\partial y^2} = \nabla^2 f = 0. \tag{8.10}$$

Suppose that the boundary of the section is defined by points (x, y) such that $x = x(s)$ and $y = y(s)$; that is x and y are some different functions of the length of the boundary from a specified point. Figure 8.5 shows part of a section of the bar perpendicular to the axis, where s is increasing as x is decreasing. Consider the triangular element shown shaded at the boundary. The shear stresses over an end face of the triangle, area dA, are τ_{zx} and τ_{zy};

together these must be such that they produce no resultant force normal to ds because there is no normal stress over the curved surface of the bar. Thus

$$dA\,\tau_{zx}\sin\theta + dA\,\tau_{zy}\cos\theta = 0$$

i.e.,

$$\frac{\tau_{zy}}{\tau_{zx}} = \frac{dy/ds}{dx/ds}.$$ (8.11)

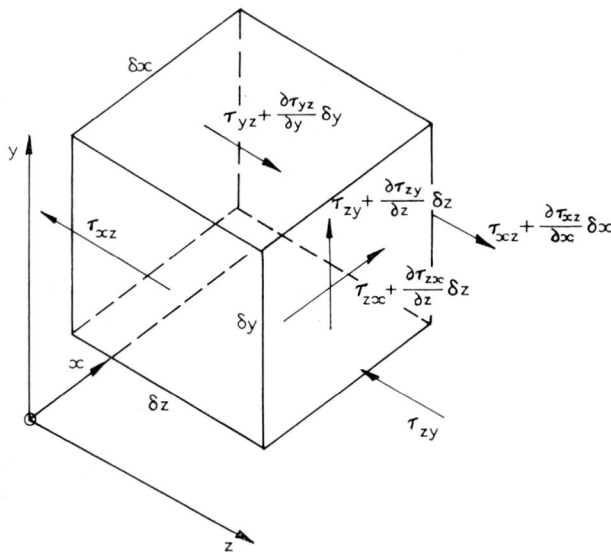

FIG. 8.4 Stresses at point P.

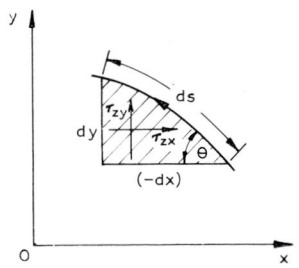

FIG. 8.5 Boundary conditions.

Substituting from equation (8.8) into equation (8.11)

$$\frac{\partial\psi}{\partial x}\cdot\frac{dx}{ds} + \frac{\partial\psi}{\partial y}\cdot\frac{dy}{ds} = 0$$

and thus

$$\frac{d\psi}{ds} = 0,$$ (8.12)

or $\psi = $ a constant along the boundary of the cross-section.

This is true for any boundary, whether it be an external one as in the solid section considered, or an internal one as for the case of a hollow cylinder. Since interest attaches only to the first derivatives of ψ, which give the shear stress components, it is usual to take ψ to be zero over the boundary.

The torque, T, required to give a twist θ is obviously

$$T = \int (x\tau_{zy} - y\tau_{zx})\, dA$$

$$= -\int\int \left(x\frac{\partial\psi}{\partial x} + y\frac{\partial\psi}{\partial y} \right) dx\, dy \quad \text{using equation (8.8)}$$

$$= -\int dy \int x\frac{\partial\psi}{\partial x} dx - \int dx \int y\frac{\partial\psi}{\partial y} dy$$

If the section is not a hollow one, then integrating by parts,

$$T = -\int dy \left[x\psi - \int \psi\, dx \right]_{x_1}^{x_2} - \int dx \left[y\psi - \int \psi\, dy \right]_{y_3}^{y_4} \tag{8.13}$$

where (x_1, y_1), (x_2, y_1), (x_3, y_3), (x_3, y_3) are points on the boundary, see Fig. 8.6.

Thus,

$$T = -\int dy \left[x_2\psi_2 - x_1\psi_1 - \int_{x_1}^{x_2} \psi\, dx \right] - \int dx \left[y_4\psi_4 - y_3\psi_3 - \int_{y_3}^{y_4} \psi\, dy \right]$$

$$= 2\int\int_A \psi\, dx\, dy \tag{8.14}$$

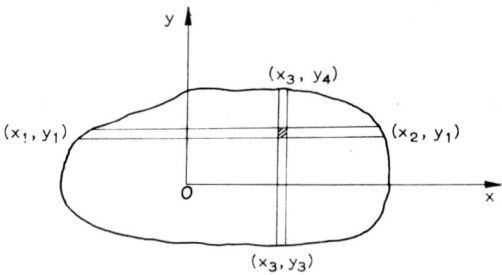

FIG. 8.6 Derivation of torque: points on the boundary of solid section.

since $\psi_1 = \psi_2 = \psi_3 = \psi_4 = 0$ on the boundary, as originally assumed.

Recapitulating the above results, if a function $\psi(x, y)$ can be found which is zero around a boundary which is the shape of the shaft being twisted, and if on calculating $\nabla^2\psi$ it is found to be a constant, the equation $\nabla^2\psi + 2G\theta = 0$ can be satisfied. Hence, the shear stress distribution throughout the section can be found and also the torque T. The magnitude of the resultant shear stress at a point is

$$\sqrt{\left(\frac{\partial\psi}{\partial x}\right)^2 + \left(\frac{\partial\psi}{\partial y}\right)^2} \quad \text{or} \quad \text{grad } \psi.$$

8.1.3 *Elliptical Cross-section Bar*

To illustrate the method of approach, examine the well-known case of a prism of elliptical cross-section whose major and minor axes are $2a$ and $2b$. Make use of the Cartesian equation for its shape, i.e.,

$$\frac{x^2}{a^2} + \frac{y^2}{b^2} = 1$$

to suggest the function ψ which is zero along the boundary. Select

$$\psi = m \left(\frac{x^2}{a^2} + \frac{y^2}{b^2} - 1 \right)$$

where m is a constant whose value will be determined. Then

$$\frac{\partial^2 \psi}{\partial x^2} = \frac{2m}{a^2} \quad \text{and} \quad \frac{\partial^2 \psi}{\partial y^2} = \frac{2m}{b^2}$$

and on using equation (8.9), it is found that

$$m = -\frac{a^2 b^2}{a^2 + b^2} G\theta.$$

The torque is calculated from (8.14) in which ψ is substituted to yield

$$T = 2 \int \int m \left(\frac{x^2}{a^2} + \frac{y^2}{b^2} - 1 \right) dx\, dy,$$

and $\dfrac{T}{8m} = \dfrac{1}{a^2} \int_0^b \left(\int_0^x x^2\, dx \right) dy + \dfrac{1}{b^2} \int_0^a \left(\int_0^y y^2\, dy \right) dx - \int_0^b \left(\int_0^x dx \right) dy$

$$= \frac{1}{a^2} \cdot \frac{\pi a^3 b}{16} + \frac{1}{b^2} \cdot \frac{\pi a b^3}{16} - \frac{\pi a b}{4} = -\frac{\pi a b}{8}.$$

Thus $\qquad T = \dfrac{\pi a^3 b^3}{a^2 + b^2} G\theta.$

To calculate the shear stress at a point

$$\tau_{zx} = \frac{\partial \psi}{\partial y} = -\frac{2a^2}{a^2 + b^2} G\theta y$$

and

$$\tau_{zy} = +\frac{2b^2}{a^2 + b^2} \cdot G\theta x.$$

Thus $\qquad \tau = \sqrt{\tau_{zx}^2 + \tau_{zy}^2} = \dfrac{\pm 2G\theta}{a^2 + b^2} \sqrt{a^4 y^2 + b^4 x^2}.$

The greatest stresses occur at the boundary of the bar and the equation then becomes

$$\tau = \frac{\pm 2G\theta}{a^2 + b^2} \sqrt{a^2 y^2 (a^2 - b^2) + a^2 b^4}$$

thence $\qquad \tau_{\max} = \tau_{y=b} = \dfrac{\pm 2G\theta a^2 b}{(a^2 + b^2)}$

and plastic yielding starts on the periphery at the point nearest to the axis. Note that the direction of the shear stress at any point on a straight line from the origin to a point on the periphery, is the same and is parallel to the tangent at the periphery, because $\tau_{zx}/\tau_{zy} = -y/x$.

For a circular section, $a = b$, the Coulomb result follows and

$$\tau_{\max} = G\theta a.$$

To investigate the warping which occurs, start with equation (8.5)

$$\tau_{zx} = G\theta \left(\frac{\partial f}{\partial x} - y \right) = -\frac{2a^2}{a^2 + b^2} \cdot G\theta y.$$

Therefore

$$\frac{\partial f}{\partial x} - y = -\frac{2a^2}{a^2 + b^2} \cdot y$$

$$\frac{\partial f}{\partial x} = -\frac{a^2 - b^2}{a^2 + b^2} \cdot y.$$

Integrating

$$f = -\frac{a^2 - b^2}{a^2 + b^2} xy + C.$$

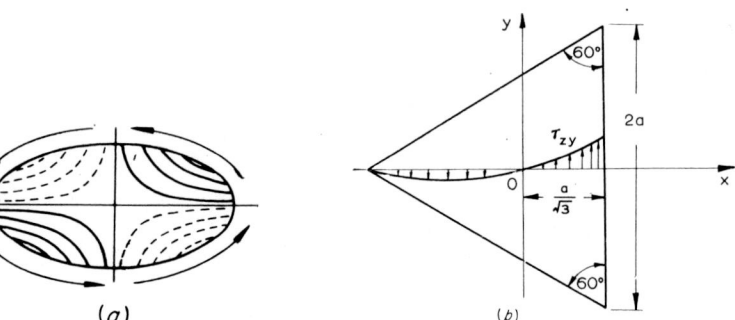

(a)   (b)

FIG. 8.7

(a) Twisting of bar with elliptical cross-section indicating lines of constant warping.

——— depression.

– – – – elevation.

(b) Shear stress distribution across axis of symmetry of triangular cross-section.

Obviously at $x = y = 0, f = 0$ and hence $C = 0$.

Thus

$$f = -\frac{a^2 - b^2}{a^2 + b^2} xy.$$

The contours for warped sections are rectangular hyperbolas whose asymptotes are the axes of the ellipse, see Fig. 8.7(a). In the first and third quadrants for a positive twist, [anticlockwise rotation in Fig. 8.7(a)] the elements of a cross-section are depressed and in the second and fourth quadrants the elements are elevated.

8.1.4 *Equilateral Triangle*

Let the triangle be of side length $2a$, see Fig. 8.7(b), and if the stress function is chosen to be

$$\psi = m(\sqrt{3}x - a)(\sqrt{3}x - 3y + 2a)(\sqrt{3}x + 3y + 2a)$$

it will be found to be zero on its perimeter. (The factors on the right-hand side of the equation are the Cartesian equations of the lines forming the triangle.) It can be confirmed that

$$\frac{\partial^2\psi}{\partial x^2} = 18m(\sqrt{3}x + a) \quad \text{and} \quad \frac{\partial^2\psi}{\partial y^2} = 18m(a - \sqrt{3}x)$$

and hence that Poisson's equation is satisfied if

$$m = -\frac{G\theta}{18a}.$$

Also, $$\tau_{zx} = \frac{\partial\psi}{\partial y} = \frac{G\theta y}{a}(x\sqrt{3} - a) \quad \text{and}$$

$$\tau_{zy} = -\frac{\partial\psi}{\partial x} = \frac{\sqrt{3}G\theta}{2a}\left(x^2 + \frac{2a}{\sqrt{3}}x - y^2\right).$$

When $y = 0$, $\tau_{zx} = 0$ and $\tau_{zy} = \frac{G\theta\sqrt{3}}{2a}\left(x + \frac{2a}{\sqrt{3}}\right)x.$

The shear stress distribution across this axis of symmetry of the bar is indicated in Fig. 8.7(b); τ_{zy} is greatest when $x = a\sqrt{3}/3$ and then

$$\tau_{zy} = \frac{\sqrt{3}}{2}aG\theta$$

when $x = -2a\sqrt{3}/3$, $\tau_{zy} = 0$.

Thus the shear stress at an apex of the triangle is zero and at the middle of each of the three sides, i.e. nearest to the bar axis, is greatest. The torque is easily shown to be

$$T = G\theta a^4\sqrt{3}/5.$$

8.1.5 *The Straightforward Nature of the above Two Examples*

These last two examples might give the reader the impression that a valid stress function ψ which satisfies equation (8.9), is always obtained by simply making use of the equation to the boundary of the cross-section concerned. For instance, for a rectangular cross-section, see Fig. 8.8(a), we might be tempted to believe that

$$\psi = m(x^2 - a^2)(y^2 - b^2) \tag{8.15}$$

would be valid stress function. However, it is soon found that this equation does not give a constant and satisfy (8.9) as is necessary. Further, just as use was made of the boundary for the equilateral triangle above, for suggesting the function ψ, so we might imagine that *any* triangular section could similarly be treated, see Fig. 8.8(*b*). Generally, we might try to use

$$\psi = m(x - a)(a_1 x + b_1 y + c_1)(a_2 x + b_2 y + c_2). \qquad (8.16)$$

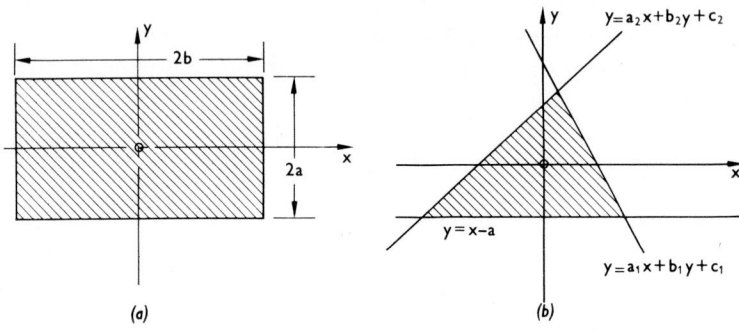

(a)   (b)

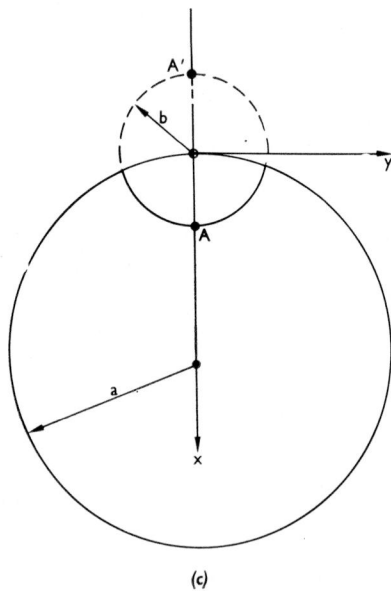

(c)

FIG. 8.8 Twisting of (*a*) a bar with a rectangular cross-section, (*b*) a bar with a triangular cross-section and (*c*) a shaft containing a circular keyway.

Using the above expression for ψ in equation (8.9), $\nabla^2\psi$ is found to be a constant and acceptable only when $a_1/b_1 = -1/\sqrt{3}$ and $a_2/b_2 = 1/\sqrt{3}$; these last two equations imply that (8.16) must define an equilateral triangle.

8.1.6 *Rectangular and Square Sections*

The treatment of solid sections, other than those described above, usually requires more complicated methods than we have employed in the previous sections and some of these are described in the book by WANG (1953). A brief description of the important and interesting case associated with rectangular sections may be referred to. Figure 8.9 shows contours of con-

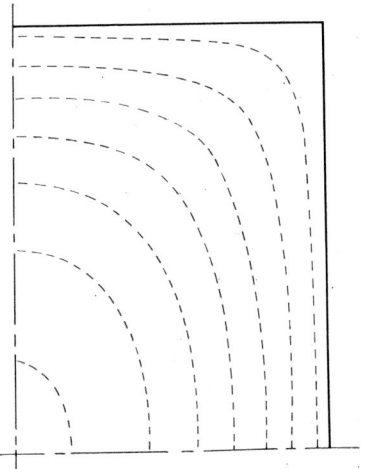

FIG. 8.9 Contours of constant shear stress in the section of a 4 × 3 rectangle.

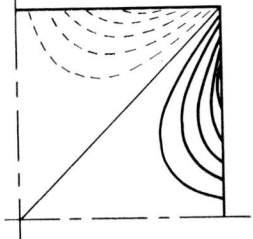

FIG. 8.10 Positive and negative warping contours for one quarter of a square.

stant shear stress magnitude in the section of a 4 × 3 rectangle. Figures 8.10 and 8.1 show the positive and negative warping contours for a square and a 2 × 1 rectangle. In the former there are four positive warping regions as against two in the latter. This difference in physical behaviour is certainly not what would be expected. In an interesting sequence of calculations, the degeneration from 4 to 2 such regions is demonstrated graphically by SOUTH-WELL (1954). The changeover is shown to be complete for a 1·42 × 1 rectangle. The results given by Southwell were obtained using equation (8.9) in finite difference form; the solution was a demonstration of Relaxation techniques.

8.1.7 A Circular Shaft with a Circular Keyway

The case described by the above title is naturally of engineering design interest. Consider a stress function defined by,

$$\psi = m(x^2 + y^2 - b^2)[(x - a)^2 + y^2 - a^2]/(x^2 + y^2) \qquad (8.17.i)$$

where a, n and m are numerical constants.

It may be verified that $\nabla^2\psi = 4m$ and hence that equation (8.9) is satisfied if $m = G\theta/2$. Equation (8.17.i) when $\psi = 0$, describes the boundary of a circular shaft containing a circular keyway, see Fig. 8.8(c). The greatest shear stress arises at A, at the bottom of the keyway; we have,

$$\tau_{xz} = \partial\psi/\partial y = 0 \text{ at } A \text{ or } (b, 0) \text{ and}$$
$$\tau_{yz} = -\partial\psi/dx = G\theta(2a - b).$$

Note, that when $b/a \to 0$, i.e., when the keyway becomes an infinitely small groove or just an axial scratch, $\tau = 2G\theta a$; the stress at the scratch in the shaft boundary is twice what it would be if smooth, or without the scratch.

A circular protuberance, shown by the broken line in Fig. 8.8(c), would contain a shear stress at A' of magnitude $G(2\theta a + b)$.

It is found to be much less tedious to test and show that (8.17.i) satisfies (8.9) if the work is carried out in terms of polar rather than Cartesian co-ordinates. In polar co-ordinates,

$$\nabla^2\psi = \frac{\partial^2\psi}{\partial r^2} + \frac{1}{r}\frac{\partial\psi}{\partial r} + \frac{1}{r^2}\frac{\partial^2\psi}{\partial r^2} \qquad (8.9.ii)$$

and equation (8.17.i) so converted by putting $r\cos\theta$ for x and $r\sin\theta$ for y, and simplifying gives

$$\psi = m(r^2 - b^2)(r - 2a\cos\theta)/r. \qquad (8.17.ii)$$

Thus:

$$\frac{\partial\psi}{\partial r} = m\left(\frac{2rB + A}{r} - \frac{AB}{r^2}\right)$$

$$\frac{\partial^2\psi}{\partial r^2} = m\left(\frac{2B + 2r + 2r}{r} - \frac{2rB + A}{r^2} + \frac{2AB}{r^3} - \frac{2rB}{r^2} - \frac{A}{r^2}\right) \qquad (a)$$

$$\frac{1}{r}\frac{\partial\psi}{\partial r} = m\left(\frac{2rB + A}{r^2}\right) - m\frac{AB}{r^3} \qquad (b)$$

$$\frac{1}{r^2}\frac{\partial^2\psi}{\partial\theta^2} = \frac{m}{r^3}A\, 2a\cos\theta \qquad (c)$$

and

$$\nabla^2\psi = (a) + (b) + (c).$$
$$= 4m.$$

8.1.8 *The Semi-inverse Solution to Saint-Venant's Torsion Problem*

We now endeavour to explain why the semi-inverse solution of Saint-Venant is so called.

In designing and analysing engineering structures the straightforward, direct approach is for applied loads to be given, and from these, for a given structure, we require stresses, displacements and deflections to be calculated. It may happen, however, that this procedure or order of working cannot be followed because of the mathematical difficulties which present themselves. It sometimes transpires, however, that we possess (or can easily obtain), for one reason or another, solutions to a problem and that we are then capable of finding out precisely what useful problem we have solved; in this case an *inverse* method of analysis will have been followed.

As has been remarked on p. 160, Saint-Venant's researches in torsion, which first appeared in his classic mémoire of 1855, led him to develop his well-known *semi-inverse* method. Starting with certain assumptions about the displacements in a bar under torsion (those embodied in (8.2) and (8.3) that warping (or longitudinal strain) of cross-sections occurs which is independent of the axial co-ordinate, that there are no body forces present and that no forces are applied to the curved surface of the shaft, Saint-Venant was able, with the aid of expressions for strain in terms of the displacements (e.g., equations (8.4)), Hooke's law (8.5) and the equations of equilibrium, to formulate a certain fundamental differential equation, (8.9). This equation, together with the requirement that no stresses act on the curved surface, set the situation for solving a number of cases of the torsion of prismatic bodies which are of some engineering interest. This procedure is evidently semi-inverse because it contains elements of what are usually thought of as being sought, namely displacements, whereas the introduction of restrictions about certain surface stresses being zero is usually conceived as a properly and initially formulated requirement.

8.1.9 *Point Matching*

We have remarked that the solution of the elastic torsion problem for some cross-sectional shapes demands the use of methods associated with the names Rayleigh-Ritz, Trefftz and Galerkin and numerical, finite difference methods; see TIMOSHENKO and GOODIER (1951). A technique that is simple is that of point matching and we mention it because it is comparatively new. The method is demonstrated by applying it to two examples.

The Poisson equation (8.9), written in terms of polar co-ordinates r and θ, is

$$\nabla^2 \psi = \frac{\partial^2 \psi}{\partial r^2} + \frac{1}{r} \frac{\partial \psi}{\partial r} + \frac{1}{r^2} \frac{\partial^2 \psi}{\partial \theta^2} = -2G\theta_0. \qquad (8.9.ii)$$

(θ_0 is here the angle of twist per unit length of the bar and should not be confused with angular co-ordinate θ). A general solution to (8.9.*ii*) is

$$\psi = -\frac{G\theta_0}{2} r^2 + A_0 + \sum_{n=1,2\ldots} (A_n r^n \cos n\theta + B_n r^n \sin n\theta). \qquad (8.18)$$

This series in truncated form may be used to indicate the shear stress distribution in shafts of special cross-sectional shapes. For example, consider a form generated by swinging circular arcs of radius $2a$ from the vertices of an equilateral triangle, see Fig. 8.11(a).

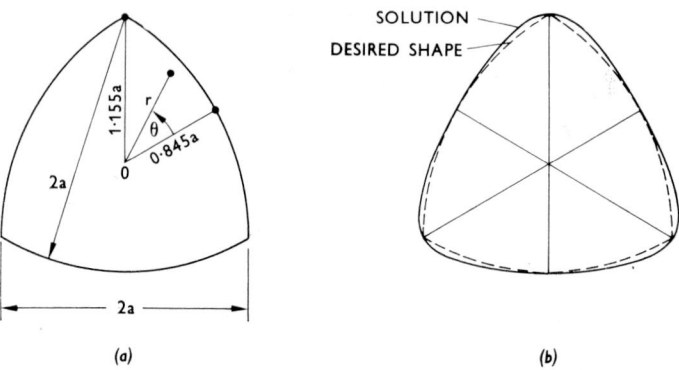

(a)   (b)

FIG. 8.11 Twisting of a shaft having cross-section generated by swinging circular arcs from the vertices of an equilateral triangle.

Now, because we ensure that this figure has three-fold symmetry we choose terms from (8.18) as far as $\cos 3\theta$, i.e., we assume

$$\psi = -\frac{G\theta_0}{2} r^2 + A_0 + A_3 r^3 \cos 3\theta. \tag{8.19}$$

That this expression for ψ is a valid one for the torsion problem may easily be verified by differentiating and showing that it fits equation (8.9). And since ψ must be zero on the boundary, with the origin of the co-ordinate system at the form centroid, we ensure that

$$-\frac{G\theta_0}{2} r^2 + A_0 + A_3 r^3 \cos 3\theta = 0 \tag{8.20}$$

passes through the special points 1 (the mid points of the sides), and 2 (the vertices of the section) of Fig. 8.11(a), thus determining A_0 and A_3. We are then hopeful that the resulting polar equation will closely describe the boundary of some initially prescribed shaft shape. It is not difficult to verify that for the section shape to be made to pass through $(0.845a, 0°)$ and $(1.155a, 60°)$ that $A_0 = 0.444\, G\theta_0\, a^2$ and $A_3 = -0.144\, G\theta_0/a$.

The section shape actually generated with these values of A_0 and A_3 by (8.20), is shown in Fig. 8.11(b) and it evidently gives rise to corners which are slightly more rounded than the initially prescribed shape. The maximum shear stress $\tau_{\max}$ in this section, occurs at point 1 when

$$\tau_{\max} = -\partial\psi/\partial r = 1.15\, G\theta_0 a.$$

This same technique may be applied to endeavour to find a satisfactory solution for a square section—one having fourfold symmetry. We assume

$$\psi = -\frac{G\theta_0 r^2}{2} + A_0 + A_4 r^4 \cos 4\theta \tag{8.21}$$

and arrange for $\psi = 0$ to pass through the mid point of a side at $\theta = 0°$ and through a diagonal at 45°. It may be verified that the shape generated is given by

$$1\cdot 2\, a^2 - r^2 - \frac{0\cdot 2\, r^4 \cos 4\theta}{a^2} = 0; \tag{8.22}$$

this curve passes through $(a, 0°)$ and $(1\cdot55a, 45°)$ instead of $(a, 0°)$ and $(1\cdot41a, 45°)$.

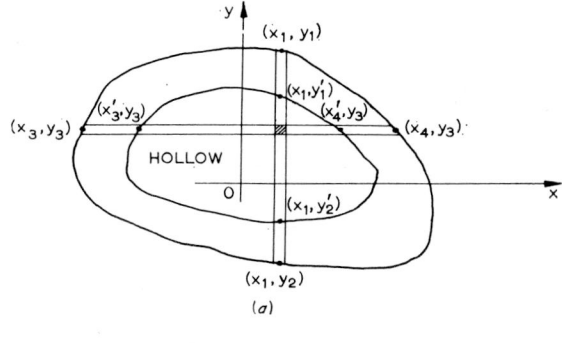

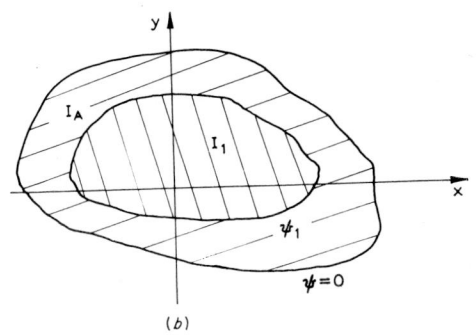

FIG. 8.12
(a) Derivation of torque: points on the boundary of hollow section.
(b) Derivation of torque: integrals of areas of hollow section.

This technique of point matching has certain limitations, and for a discussion of some of these the paper by LEISSA and BRANN (1964) should be read.

8.1.10 *The Torsion of Hollow Cylinders*

Mention may now be made of the torsion of a prismatic cylinder of cross-sectional area A containing a hole of cross-sectional area A_1, see Fig. 8.12(a).

The appropriate expression for the torque on reconsidering equation (8.13) is

$$T = -\int dy \left[x\psi - \int \psi \, dx \right]_{x_3'}^{x_3} - \int dy \left[x\psi - \int \psi \, dx \right]_{x_4'}^{x_4}$$

$$- \int dx \left[y\psi - \int \psi \, dy \right]_{y_1'}^{y_1} - \int dx \left[y\psi - \int \psi \, dy \right]_{y_2'}^{y_2}$$

$$= -\int (x_3\psi_3 - x_3'\psi_3') \, dy + \int\int_{x_3'}^{x_3} \psi \, dx \, dy$$

$$- \int (x_4\psi_4 - x_4'\psi_4') \, dy + \int\int_{x_4'}^{x_4} \psi \, dx \, dy$$

$$- \int (y_1\psi_1 - y_1'\psi_1') \, dx + \int\int_{y_1'}^{y_1} \psi \, dx \, dy$$

$$- \int (y_2\psi_2 - y_2'\psi_2') \, dx + \int\int_{y_2'}^{y_2} \psi \, dx \, dy$$

But $\psi_1 = \psi_2 = \psi_3 = \psi_4 = 0$ and $\psi_1' = \psi_2' = \psi_3' = \psi_4' = \psi_1$ say, bearing in mind equation (8.12).

Thus
$$\frac{T}{2} = \psi_1 I_1 + I_A. \tag{8.23}$$

I_A refers to the annulus over which $\int\int \psi \, dx \, dy$ is taken and I_1 is the area of the 'hollow', ψ_1 being a constant value of the stress function along the internal boundary, see Fig. 8.12(b).

Equation (8.23) may be written in a form which is sometimes more useful

$$\frac{T}{2} = (I_A + \psi_1 I_1 + I_H) - I_H$$

$$= I_0 - I_H$$

where $I_0 = \int\int \psi \, dx \, dy$ is taken over the *whole* section as if it were solid and

$I_H = \int\int \psi \, dx \, dy$ is taken over the 'hollow', with $\psi = 0$ on the inner boundary.

Applying (8.23) to the case of the prismatic elliptical section bar which contains a hole of the same shape as the external boundary, its major and minor axes being one-half those of the perimeter

$$T = \frac{\pi a^3 b^3}{a^2 + b^2} G\theta (1 - (\tfrac{1}{2})^4)$$

$$= \frac{15\pi}{16} \frac{a^3 b^3}{a^2 + b^2} G\theta$$

8.1.11 *The Membrane Analogy*

In Figs. 8.13 and 8.14 a thin biaxially-tensioned membrane is depicted in plan and elevation, the initial position of the membrane being in the (x, y) plane. It may be supposed that a small pressure p is now applied causing the membrane to rise slightly out of the $z = 0^*$ plane and such that the tension

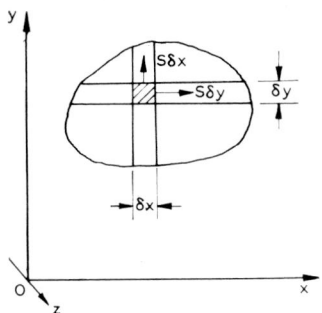

FIG. 8.13 Membrane in plan.

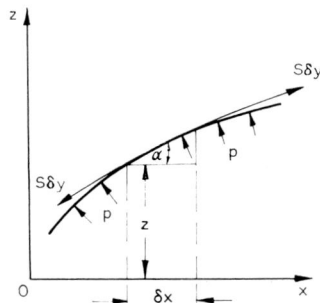

FIG. 8.14 Membrane in elevation.

everywhere remains constant at S per unit length. The boundary of the membrane remains in the $z = 0$ plane. Considering the vertical equilibrium of the membrane indicated, and remembering that angles such as α are small, we then resolve vertically, so that

$$\frac{\partial}{\partial x}\left(S\frac{\partial z}{\partial x}\cdot \delta y\right)\delta x + \frac{\partial}{\partial y}\left(S\frac{\partial z}{\partial y}\delta x\right)\delta y + p\,\delta x\,\delta y = 0$$

which on simplifying reduces to

$$S\frac{\partial^2 z}{\partial x^2} + S\frac{\partial^2 z}{\partial y^2} + p = 0$$

or

$$\nabla^2 z = -\frac{p}{S} \tag{8.24}$$

---

* z, as used in this section should not be confused with the z-axis of the prismatic shaft.

7—EP * *

This is a Poisson equation and it is formally identical with the torsion equation (8.9).

In 1903 PRANDTL suggested that use could be made of this formal similarity for the purpose of solving torsion problems in an analogy, where the stress function ψ is identified with the transverse displacement z, and $2G\theta$ with p/S. If the edge support of the membrane has the same shape as the section of the twisted bar, clearly the surface taken up by the membrane in an equilibrium position is representative of the variation of the stress function over the section of the bar. The volume under the membrane, V, is $\int\int z\,dx\,dy$, Fig. 8.15, which is analogous to $\int\int \psi\,dx\,dy$, i.e., the volume between the membrane and the $z = 0$ plane is proportional to the torque required to cause a twist of θ per unit length of bar.

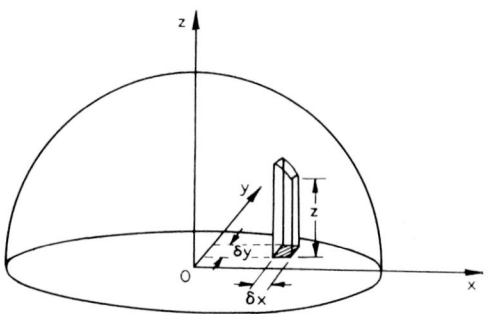

FIG. 8.15 Volume under the membrane.

A comprehensive account of the various experimental techniques for utilizing the membrane analogy is to be found under Analogues, Chapter 16, in the *Handbook of Experimental Stress Analysis*, edited by M. Hetenyi.

The first membrane analogy experiments were apparently carried out by Anthes in 1906 using a soap film under pressure. The apparatus developed by GRIFFITH and TAYLOR (1917) is, however, now widely used in experiments. In their method two soap films or rubber sheets in the lid of an air-tight box are blown up side by side, one being of circular form, the other of the form to be investigated. p/S is the same for both shapes and corresponds to the same value of $2G\theta$ for the two bars under torque; the torques required to give these twists are as the ratio of the volumes between the membranes and the $z = 0$ plane. The volume enclosed by the circular membrane which is a paraboloid of revolution is found, say V_1, and if the other membrane be flattened using a plate, the 'circular' membrane will expand to enclose a new paraboloidal segment of volume V_2. Thus for the same value of $2G\theta$, the ratio of the torques in the bars is $V_1/(V_2 - V_1) = h_1/(h_2 - h_1)$, where h_1 is the initial membrane height and h_2 the final membrane height.

The inclination of the film surface to the $z = 0$ plane at any point may be determined using an optical reflection method and is simply related to $|\,\text{grad }\psi\,|$, the total shear stress at any point.

A qualitative examination of the slope of a membrane is usually sufficient to indicate points in a section which will act as stress-raisers.

The methods involved may be exemplified by reference to the trivial case of the torsion of a uniform circular section, see Fig. 8.16. According to

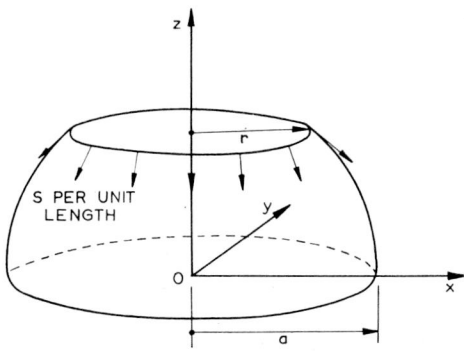

FIG. 8.16 Torsion of uniform circular section: membrane analogy.

equation (8.24) a plane circular extended membrane anchored around a circle of radius a and subjected to a transverse pressure p will be distended to become a segment of a paraboloid. Equating vertical forces,

$$p\pi r^2 = -S\frac{\partial z}{\partial r}\,.\,2\pi r \quad \text{or} \quad \frac{\partial z}{\partial r} = -\frac{p}{2S}\,r.$$

Hence
$$z = -\frac{pr^2}{4S} + C$$

and if $z = 0$ when $r = a$,

$$z = \frac{p}{4S}(a^2 - r^2).$$

Thus,
$$\psi = \frac{G\theta}{2}(a^2 - r^2)$$

The maximum shear stress at a point is $|\,\text{grad }\psi\,|$, i.e. $-\partial\psi/\partial r$; thus $\tau = G\theta r$. $\tau_{\max}$ occurs when $r = a$; hence $\tau_{\max} = G\theta a$.

The volume, V, between $z = 0$ and the membrane is given by

$$V = \int_A z\,dA = \int_0^a \frac{p}{4S}(a^2 - r^2)\,2\pi r\,dr = \frac{p\pi a^4}{8S}$$

Hence,   Torque $T(\equiv 2V) = \dfrac{G\theta\pi a^4}{2} = G\theta J$ where $J = \dfrac{\pi a^4}{2}$ is

the polar moment of the cross-sectional area about the centroidal axis.

Collecting these results together, the well-known trinity of equations is obtained

$$\frac{T}{J} = \frac{\tau}{r} = G\theta \tag{8.1}$$

The method is further exemplified by considering a thin rectangular section in which the breadth b is considerably greater than the thickness, t, see Fig. 8.17. The membrane shape is constant, or independent of x except near the

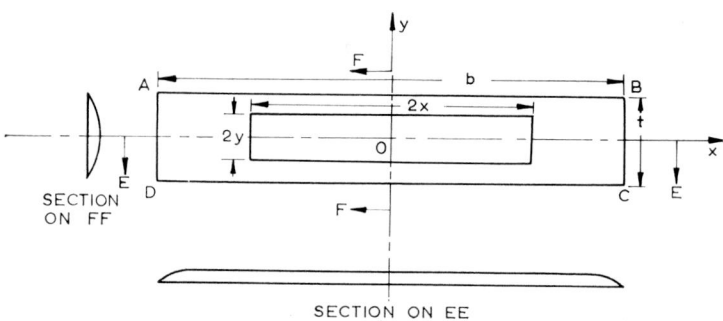

FIG. 8.17 Torsion of thin rectangular section.

ends BC and DA. From vertical equilibrium requirements, on the portion indicated

$$4x\left(-\frac{\partial z}{\partial y}\right)S = p.2x.2y;$$

therefore

$$\frac{\partial z}{\partial y} = -\frac{p}{S}y.$$

Thus,

$$z = -\frac{p}{2S}y^2 + C.$$

Since $z = 0$ when $y = t/2$, then

$$z = \frac{p}{2S}\left(\frac{t^2}{4} - y^2\right) \quad \text{and} \quad \psi = G\theta\left(\frac{t^2}{4} - y^2\right)$$

Now

$$\tau_{zx} = \frac{\partial \psi}{\partial y} = -2G\theta y \quad ; \quad \tau_{yz} = -\frac{\partial \psi}{\partial x} = 0.$$

$$\tau^2 = \tau_{xz}^2 + \tau_{yz}^2 \quad \text{and therefore} \quad |\tau| = \frac{p}{S}y = 2G\theta y.$$

Clearly, τ is greatest when $y = \pm t/2$ and

$$\tau_{max} = G\theta t.$$

The volume under the membrane is

$$V = \int_A z \, dA \doteq \frac{pb}{S} \int_0^{t/2} \left(\frac{t^2}{4} - y^2 \right) dy = \frac{pb}{S} \cdot \frac{t^3}{12}.$$

$$\text{Torque, } T \, (\equiv 2V) = \frac{bt^3 G\theta}{3}.$$

These results may be re-written

$$\frac{T}{bt^3/3} = G\theta = \frac{\tau_{max}}{t}.$$

Whilst points of maximum elastic shear stress are frequently on the periphery *nearest* the centroidal axis, there are, however, unusual shapes of section for which this is not so. (See TIMOSHENKO and GOODIER, 1951, p. 267.)

On examining the nature of the shear stress components, i.e., τ_{xz} and τ_{yz} around the periphery of the rectangle, it is seen that one of them is always zero (because there is zero normal stress along each side or surface) and in particular, at each of the four corners, they are both zero. Thus at each corner of this rectangle, e.g., point α in Fig. 8.18(a), the total shear stress is

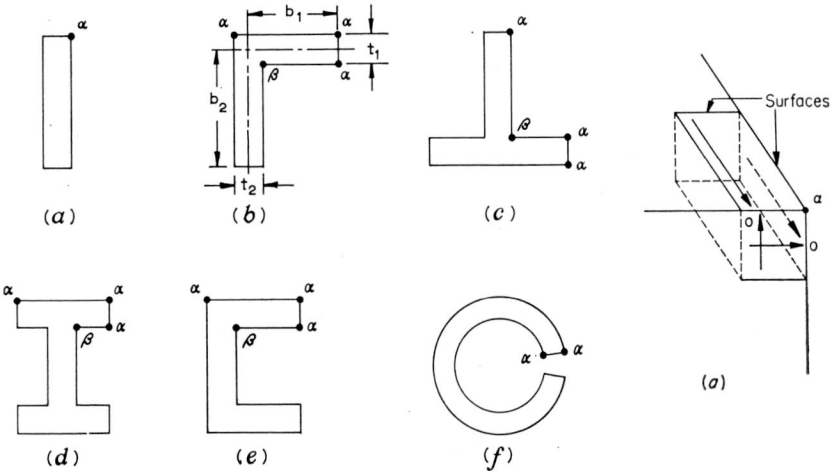

(a)   (b)   (c)

(d)   (e)   (f)

FIG. 8.18 Thin-walled open sections.

zero. This means also that the membrane remains tangential to the $z = 0$ plane, at all the corners. The above result is of considerable help in relation to approximate calculations for the twist of comparatively thin-walled open-sections such as those shown in Fig. 8.18. It will be easily appreciated that these sections may be treated by imagining them to be composed of thin rectangles. Fig. 8.18(b) is equivalent to two rectangles ($b_1 \times t_1$) and ($b_2 \times t_2$). Thus for the twisting of this section, we have

$$T = G\theta \left(\frac{b_1 t_1^3}{3} + \frac{b_2 t_2^3}{3} \right)$$

The 'interaction' of the membrane film at the junction of the two portions of the section is neglected. It has been observed that there is zero shear stress at the external corners labelled α, but at the internal corners labelled β, it is 'infinite', or yielding takes place at sharp corners such as these, for any torque.

Reverting to the result for the torsion of elliptical cross-section bars, the expression for the torque can be put in terms of the cross-sectional area, $A(=\pi ab)$ and the polar moment of area

$$J = \frac{\pi ab}{4}\,(a^2 + b^2);$$

thus, $$\frac{T}{G\theta} = \frac{A^4}{4\pi^2 J} \simeq \frac{A^4}{40J}.$$

This formula is found to be very useful and correct to better than 10 per cent for convex squat sections of the type shown in Fig. 8.19. The contours indicated in Fig. 8.19 are curves of constant ψ; the tangent at any point on the curve gives the shear stress direction at that point.

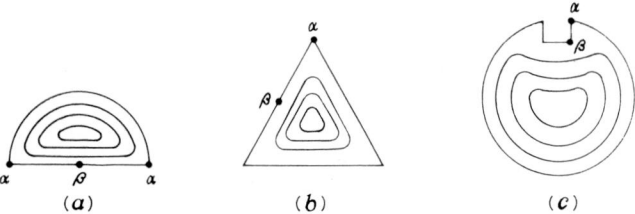

FIG. 8.19 Convex squat sections showing lines of shear stress.

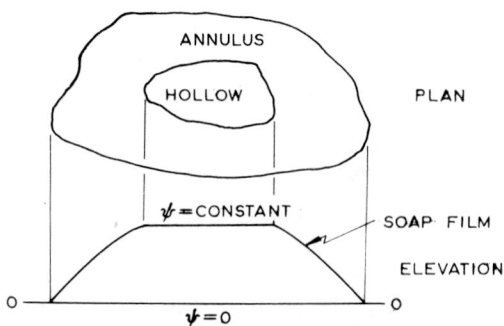

FIG. 8.20 Membrane analogy applied to hollow sections.

Again, in these figures at the corners marked α, the shear stress is zero and at the mid-points of sides marked β, it is greatest. In particular, for sharp corners as at the root of the keyway, the material yields.

The membrane analogy may be applied to hollow sections with some slight modification, as Fig. 8.20 illustrates, in plan and elevation.

The first term of the right-hand side of equation (8.23) is clearly represented by the cylindrical prism formed beneath the 'hollow' and the second term is simply the volume directly under the film. Thus in an experiment, a light rigid plate having the shape of the inner boundary is constrained to move vertically by any amount, the soap film between the plate and the outer boundary being stretched and the plate finding its own height due to the air pressure beneath. A typical mechanism for effecting a displacement of the plate and rendering it, in effect, weightless, so that the air pressure does not have to support it, is shown in Fig. 8.21.

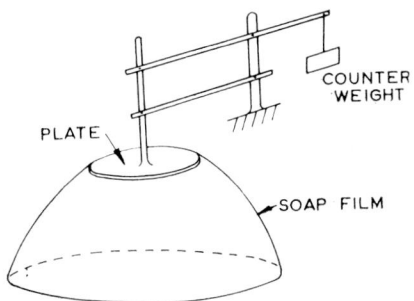

FIG. 8.21 Mechanism for effecting displacement of plate.

A very simple example of the application of this method is available in determining the torque required to twist a rectangular annulus of given dimensions, see Fig. 8.22. The slope of the film is assumed to be constant (the shearing stress is thus taken to be constant across the section). Thus if the central plate rises a distance d, then $2V \doteq 2b_1b_2d$

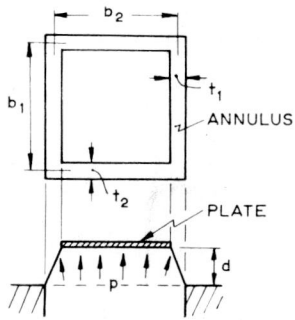

FIG. 8.22 Torsion of a rectangular annulus.

But $\tau_1 t_1 = d, \quad \tau_2 t_2 = d \quad$ and $\quad pb_1b_2 = 2S(b_1/t_1 + b_2/t_2)d$

and thus $T = 2b_1b_2\tau_1t_1 = 2b_1b_2\tau_2t_2 = 2t_1t_2b_1^2b_2^2G\theta/(b_1t_2 + b_2t_1),$

τ denoting the shear stress. It is obvious how, in contrast, squat-sectioned prisms containing a regular symmetrically situated hole would be approached, using the above analogies.

8.1.12 *Other Analogies*

Other analogies have been used for investigating elastic torsion. First, the hydrodynamic analogy utilizes the fact that the elastic twisting of bars of uniform section is mathematically identical with the motion of a frictionless fluid having uniform angular speed inside a tube of the same section; the speed of the circulatory fluid at a point is taken to represent the shearing stress at that point in the twisted bar.

The effect of a hole in a twisted bar on the stress distribution is reflected in the disturbance, caused by an object of the same shape, to the circulating fluid. For example, in Fig. 8.23, which shows a small hole in a circular section

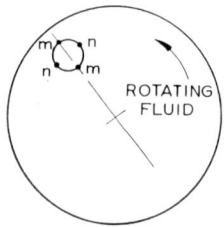

FIG. 8.23 Hydrodynamic analogy.

shaft, the speed at points n is zero and at points m is doubled. The effect of this hole is thus to double the usual shear stress at m in the shaft. More generally, if the hole in the shaft is elliptical (see Fig. 8.24), then if the fluid flows from left to right with unit speed, at n the speed is zero and at m it is $(1 + a/b)$. Thus the stress at m is increased $(1 + a/b)$-fold.

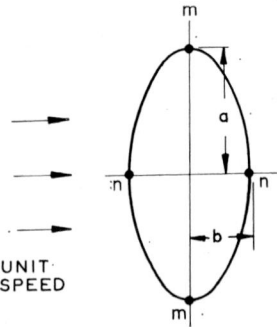

FIG. 8.24 Hydrodynamic analogy: small elliptical hole in shaft.

A second analogy is that between cylindrical torsion and the potential of a plane electric field. This method is applicable for studying stress concentrations at fillets and grooves.

A detailed discussion of the various experimental analogies which have been used for investigating torsion as well as a very extensive bibliography is to be found in a paper by HIGGINS (1944).

8.2 Plastic Analyses

8.2.1 *Plastic Yielding in a Prismatic Bar and the Sand Heap Analogy*

It has been shown that $|\text{grad } \psi|$ represents the total shearing stress at a point and is physically interpreted to be the maximum slope at a point in the membrane surface. If k denotes the yield shear stress of the bar, then for yielding at a point

$$|\text{grad } \psi| = k$$

i.e. a limit is set upon the value that the gradient at a point may attain. Once this constant gradient is attained, further torque so acts as to spread the area over which yielding occurs. Constancy of gradient in the membrane analogy may be simulated by erecting a constant slope roof over a bar's section: see Fig. 8.25.

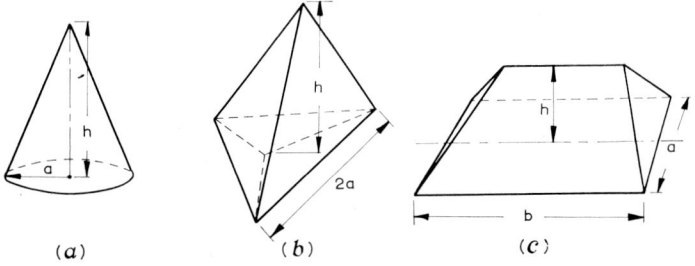

(a)   (b)   (c)

FIG. 8.25 Constant slope roofs indicating full plastic yielding across each section.

It will be easily imagined how in a typical example the membrane under increasing pressure is raised out of the $z = 0$ plane, maintains its largest gradient on the boundary and, having attained the critical slope, proceeds to have increasing contact with the underside of the roof as the pressure increases. In this manner a physical and intuitive conception of the growth

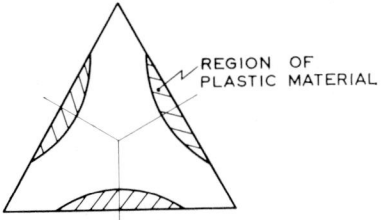

REGION OF
PLASTIC MATERIAL

FIG. 8.26 Torsion of triangular section showing growth of plastic regions.

of the zones of plastic yielding under increasing torque may be had. In Fig. 8.26 the shaded regions about the mid-points of the sides of an equilateral triangle indicate regions over which plastic yielding has occurred, or in plan the regions of contact between the membrane and the pyramidal roof erected over it. This extension of the membrane analogy is due to NADAI (1923 and 1950). From this analogy it will be clear that, in general, yielding usually proceeds from the boundary inwards towards the centroidal axis; throughout the outer plastic zones, the slope or yield stress is constant and within the section, the slope varies or the total shear stress decreases from k, (assuming the material to be perfectly plastic), at the elastic-plastic boundary to zero on the axis of twist. Note that the shear stress is continuous across the boundary. In summary, therefore

$$\text{in the plastic region } \left(\frac{\partial \psi}{\partial x}\right)^2 + \left(\frac{\partial \psi}{\partial y}\right)^2 = k^2$$

$$\text{and in the elastic region } \left(\frac{\partial^2 \psi}{\partial x^2}\right) + \left(\frac{\partial^2 \psi}{\partial y^2}\right) = -2G\theta \tag{8.25}$$

Clearly, problems involving elastic-plastic yielding using the membrane analogy will not be calculable without considerable labour. If one proceeds to consider full or complete plastic yielding, calculations of the torque required to achieve this will be much simpler. The case requires the membrane to be in contact with the whole of the roof; the whole of the material up to the axis is plastic; no elastic core remains. A simple effective method of obtaining a roof of constant slope is to heap dry sand over a plate whose section is similar to that of the prismatic bar, see Fig. 8.25, thus utilizing the constancy of the angle of repose of the sand. Sand heaps are best employed when section shapes are difficult to handle mathematically.

We calculate by this method the torques required to bring about full plastic yielding of bars whose section is (a) a circle, (b) an equilateral triangle, (c) a rectangle. Corresponding sand heap diagrams appear in Fig. 8.25, and from which the meaning of the symbols is evident.

(*a*) Circle of radius *a*

$$\text{The slope is } h/a \equiv k$$
$$\text{The volume } = \tfrac{1}{3}\pi a^2 h$$
$$\text{Torque } (\equiv 2V) = \tfrac{2}{3}\pi a^3 k$$

(b) Equilateral triangle of side 2a

$$\text{The slope is } h/\tfrac{1}{3}a\sqrt{3} \equiv k \tag{8.26}$$
$$\text{Torque } = 2a^3 k/3$$

(c) Rectangle, $a \times b$

$$\text{The slope } h/(a/2) \equiv k$$
$$\text{Volume } = (b - a)ah/2 + 2(\tfrac{1}{3}a^2 h/2)$$
$$\text{Torque } = a^2 k(3b - a)/6$$

When $a = b$, Torque $= a^3 k/3$

As an alternative to calculating the volume of the sand heap the moment or torque about the centroidal axis may be obtained directly. The ridges in the

sand heap indicate lines of shear stress discontinuity and hence, in Fig. 8.27, the direction of the yield shear stress at each point in the section of the rectangular bar will be as shown.
Thus

$$\frac{T}{2} = (\text{Area I}.k) \times OG_1 + (\text{Area II}.k) \times OG_2$$
$$\qquad\qquad \underset{\text{i.e. force}}{} \qquad\qquad\qquad \underset{\text{i.e. force}}{}$$

$$\frac{T}{2k} = \left[2 \cdot \frac{a^2}{8} \cdot \frac{a}{3} \cdot + (b-a) \cdot \frac{a}{2} \cdot \frac{a}{4}\right] + \frac{a^2}{4}\left(\frac{b}{2} - \frac{a}{6}\right)$$

Therefore $\qquad T = \dfrac{a^2 k (3b - a)}{6}$

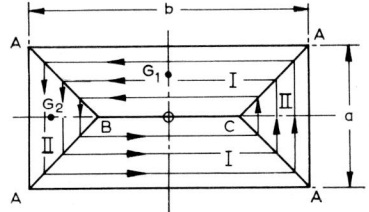

FIG. 8.27 Yield shear stress distribution for a rectangular section.

(Note that in calculating the torque for a long thin section (i.e., $b \ggg a$) the moments of the shear force over the end triangles may *not* be neglected by simply extending areas I to each cover one half of the section; to do this would yield a torque too small by one half. The temptation to omit these moments may arise from recalling the soap or sand-heap analogy where the volume over the whole section is calculated by reference to the whole of extended areas I.)

The problem of finding the volume of the sand heap for a section which is not simple may be avoided by employing a weighing procedure. Suppose, for instance, that sand has been heaped over the section to be investigated and its weight then determined as W. The weight of the sand heap covering a circle of radius a is next found, say W_1. Then the ratio of the torques to cause complete plastic yielding of both the circular and the given shaft is the same as the ratio of W_1 to W.

When a section having a symmetrically situated prismatic hole is considered, the sand heap analogy may be extended by heaping the sand around the section of a fixed cylinder which has the shape of the hole and which stands perpendicular to the section from the hole in the plate. It may easily be verified that for a cylinder of radius a, containing a concentric hole of radius b (see Fig. 8.28),

$$V = \frac{\pi a^2 h}{3} - \frac{\pi b^2 h'}{3} \equiv \frac{\pi}{3} k(a^3 - b^3)$$

because $\qquad k = \dfrac{h'}{b} = \dfrac{h}{a},\quad$ and therefore $\quad T = \dfrac{2\pi k}{3}(a^3 - b^3).$

This formula is easily substantiated by working from first principles.

It is worth noting that corresponding to discontinuities or ridges on the sand heap surface, there are corresponding physical discontinuities in the section of the twisted bar. For instance, for the equilateral triangle in Fig. 8.29(a), there are three regions of uniformly directed shear stress with abrupt changes in direction along the three lines GA, GB, GC; and for the rectangle there are five such lines AB, AC and BC, see Fig. 8.27. For an

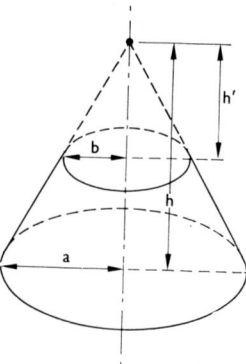

FIG. 8.28 Sand heap analogy for a circular section with a concentric hole.

elliptical cross-section the ridge or discontinuity covers a length less than that of the major axis, see Fig. 8.29(b). The end points of this length are at the centre of curvature for the ellipse at the ends of the major axis; the ridge is defined by the length along the major axis in which normals at points on the boundary of the ellipse intersect.

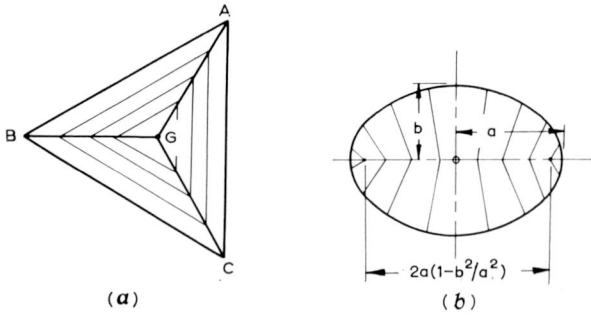

(a)   (b)

FIG. 8.29 Sand heap analogy (a) Triangular section (b) Elliptical section.

Three further interesting examples are shown in Fig. 8.30, concerning bars containing cracks and re-entrant sections which have been twisted to full plasticity. The ridges or lines of discontinuity are the loci of points equidistant from the nearest pair of sides; re-entrant sharp corners generate parabolas to join up the straight lines.

Of the three cases shown Fig. 8.30(a) depicts a square bar having a slit or crack in the middle and at right angles to one of its sides. The 'crater-ridge' is made up of

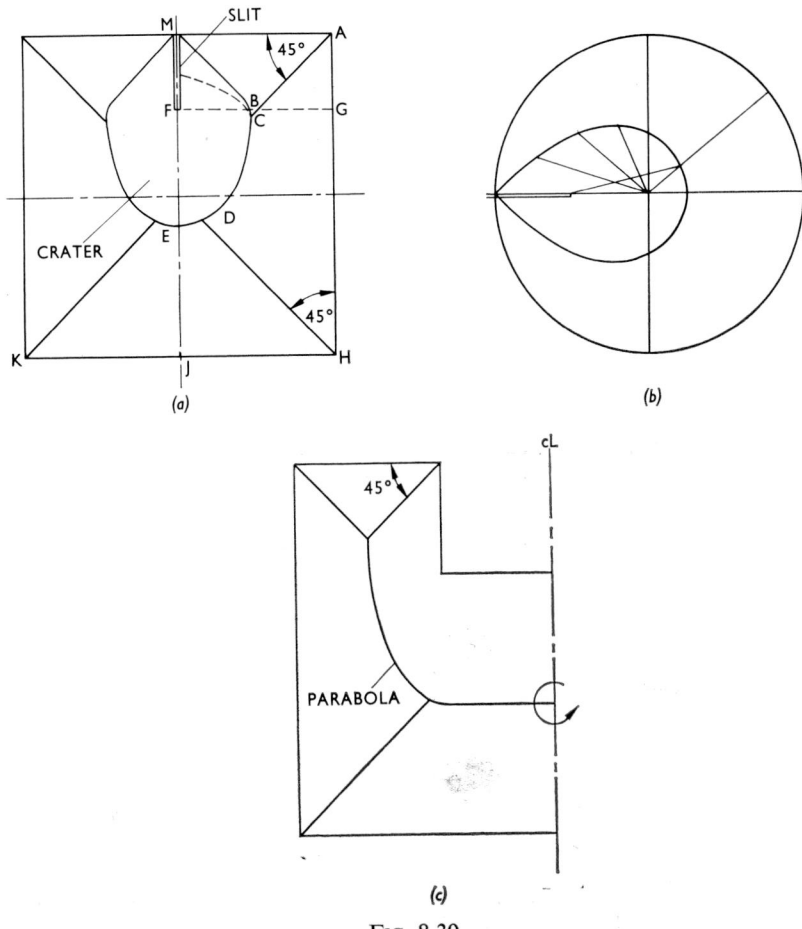

(a)

(b)

(c)

FIG. 8.30

(a) Square bar with slit at the middle and at right angles to one of the sides.
(b) Circular bar with a radial slit.
(c) Rectangular bar containing large rectangular slot: half section.

(i) a straight line MB at 45° to the slit MF and adjacent side, AM

(ii) a parabola from B to C, focus F, vertex the mid-point of MF and directrix MA

(iii) a parabola from C to D, focus F, vertex the mid-point of FG and directrix AGH

(iv) a parabola from *D* to *E*, focus *F*, vertex the mid-point of *FJ* and directrix *KH*.

The effect of a slit in reducing the torque-carrying capacity of the bar (i.e., its creation of a 'crater' in what would otherwise be a solid 'hill') is immediately made obvious.

Figure 8.30(*b*) shows in plan the line of discontinuity or 'ridge', which arises for the case of a circular bar having a radial crack which penetrates half way towards the section centre, along a radius. Figure 8.30(*c*) pertains to a square section bar containing a rectangular slot.

Figure 8.31 shows sand heap models pertaining to shafts of rectangular

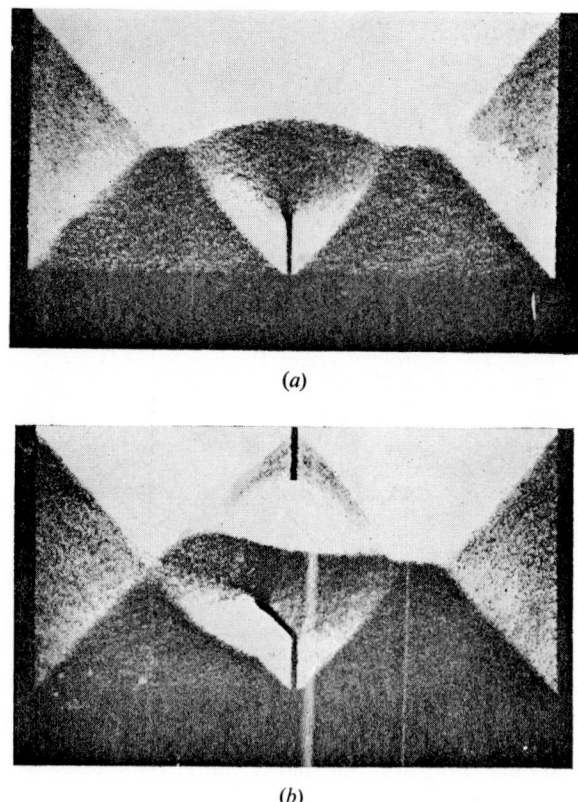

(*a*)

(*b*)

FIG. 8.31

(*a*) Crack in a broad face.
(*b*) Opposing cracks in broad faces.

and circular cross-section which contain cracks and which carry a torque which makes the section fully plastic. These photographs are taken from a paper by McCLINTOCK (1956) in which he successfully used the sand heap analogy to discuss the growth of fatigue cracks in shafts subject to full plastic torsion by directing attention to the strain distribution around the crack.

Attention must be drawn particularly to the fact that sand heap calculations are of limited value. This is because,

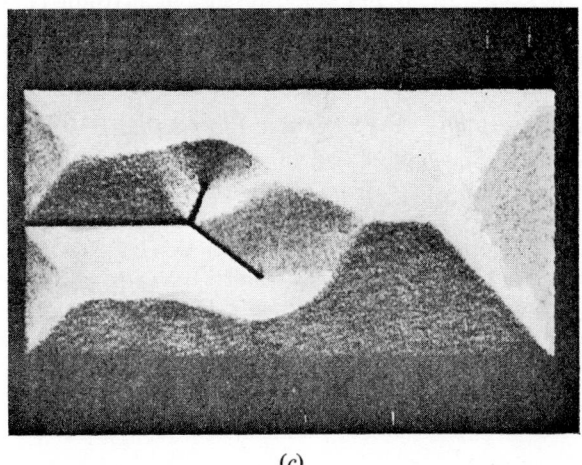

(c)

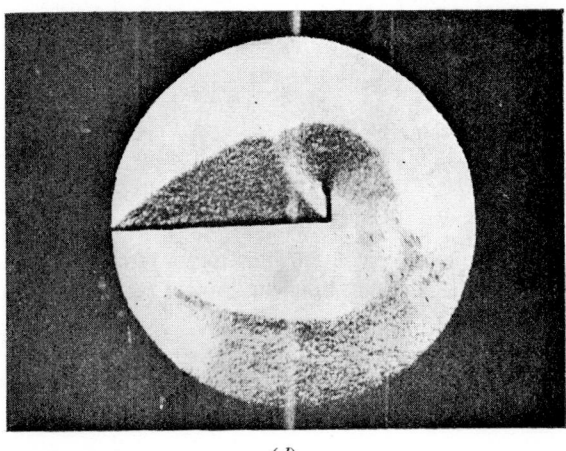

(d)

FIG. 8.31

(c) Forked crack from a narrow face.
(d) Crack in circular cross-section.

(*After* McClintock, *Proc. Instn. mech. Engrs.*)

(i) there is always some work-hardening with real metals and also no section can be rendered completely plastic;

(ii) bars become non-prismatic long before the whole of the section is plastic; this can apply elastically as well as plastically. Changes in the external boundary of a cross-section occur and these are not negligible.

(iii) There are also length changes in twisted shafts, due to the neglect of secondary effects, see particularly the papers by OLSZAK (1965), KJAR (1967) and RONAY (1968).

In spite of these limitations and in the absence of anything better, the sand heap analogy is useful, at least as an approximate quantitative guide to behaviour in plastic torsion.

8.2.2. *Unloading in Hollow Bars Following Elastic–Plastic Monotonic Torsion*

The torsion of a long hollow bar of elastic-perfectly plastic material is well appreciated, but apart from the circular annulus, no analytical solutions are available. Several authors have provided numerical solutions for annuli of different shape but a recent paper by HERAKOVICH and HODGE (1968) has brought to light an unusual and surprising phenomenon. They have shown numerically for some hollow cylinders which have developed plastic enclaves at the inner boundary of the section with monotonically increasing torque, that at some particular angle of twist elastic unloading with further increase of twist takes place; yet further increase of torque leads to plastic reloading. This behaviour does not occur for solid sections.

8.2.3. *Elastic–Plastic Torsion with Work-hardening*

MENDELSON (1968) has outlined a numerical method for handling the analysis of the elastic-plastic torsion of prismatic beams with strain hardening. He gives results of calculations for a rectangular cross-section with linear work-hardening.

8.3 Residual Stresses in Plastically Twisted Shafts

8.3.1. *Circular Shafts*

Let $+T_1$ be the torque which just brings the outermost fibre to yield in a hollow prismatic circular shaft of inner and outer radii b and a respectively. Then denoting b/a by c

$$T_1 = \frac{\pi}{2} a^3 k (1 - c^4)$$

where k is the yield shear stress. The torque $+T_2$ which brings the whole shaft to full plasticity—assuming the material of the shaft is non-strain hardening—is

$$T_2 = \tfrac{2}{3} \pi a^3 k (1 - c^3).$$

The torque $+T$, which causes yielding into radius ρ, is, see Fig. 8.32(c),

$$T = \frac{2}{3} \pi a^3 k \left[1 - \left(\frac{\rho}{a}\right)^3 + \frac{3}{4} \left(\frac{\rho}{a}\right)^3 \left\{ 1 - \left(\frac{b}{\rho}\right)^4 \right\} \right]$$

the associated angle being

$$\theta_0 = \frac{k}{\rho G}.$$

These three cases are represented in Fig. 8.32(*a*), (*b*) and (*c*). Now consider an 'elastic' torque $-T$ to act on the shaft, thus nullifying $+T$ and so finally completely unloading the shaft. Then

$$-\frac{T}{J} = \frac{q}{r} = G\theta.$$

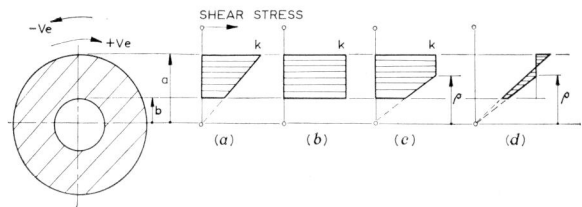

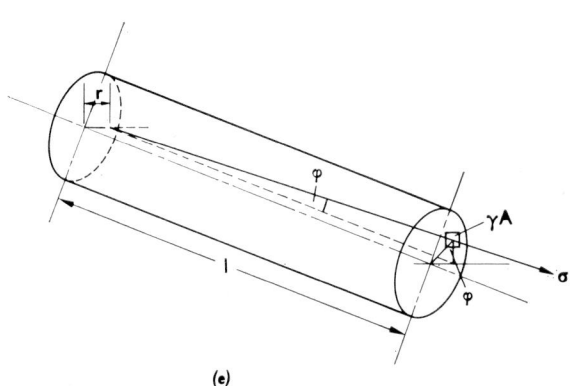

(*e*)

FIG. 8.32 Stress distribution in plastically loaded and unloaded circular hollow shaft.

Thus $-T$ added to the original stress distribution gives the kind of residual shear stress pattern indicated in Fig. 8.32(*d*).

The residual shear stress distribution in a hollow shaft which has been brought to yield to a depth of $(a - \rho)$ is given by

$$-\tau = q - k = \frac{Tr}{J} - k$$

$$= k \left\{ \frac{4r}{3a(1 - c^4)} \left[1 - \frac{1}{4}\left(\frac{\rho}{a}\right)^3 - \frac{3}{4}c^4\left(\frac{a}{\rho}\right) \right] - 1 \right\} \qquad (8.27.\text{i})$$

when $\rho \leqslant r \leqslant a$, and

$$\tau = -\frac{Tr}{J} + q_1$$

$$= r\left\{\frac{k}{\rho} - \frac{4k}{3a(1-c)^4}\left[1 - \frac{1}{4}\left(\frac{\rho}{a}\right)^3 - \frac{3}{4}\left(\frac{a}{\rho}\right)c^4\right]\right\}$$

$$= \frac{rk}{\rho}\left\{1 - \frac{4\rho}{3a} \cdot \frac{1 - \frac{1}{4}\left(\frac{\rho}{a}\right)^3 - \frac{3}{4}\left(\frac{a}{\rho}\right)c^4}{1 - c^4}\right\} \qquad (8.27.\text{ii})$$

when $b \leqslant r \leqslant \rho$. q_1 is the magnitude (positive) of the original elastic shear stress due to the initial torque $+T$.

Equations (8.27) show that the residual stresses are negative in the outer part of the annulus and positive over the remaining inner portion.

If θ_0 is the angle of twist ($= k/G\rho$) due to $+T$, then the angle of untwist θ'_0 is given by

$$T = \tfrac{1}{2}\pi G\theta'_0 a^4(1 - c^4)$$

$$\text{and } \theta'_0 = \frac{4k}{3aG}\left(\frac{1 - \frac{1}{4}\left(\frac{\rho}{a}\right)^3 - \frac{3}{4}\left(\frac{a}{\rho}\right)c^4}{(1 - c^4)}\right)$$

$$= \frac{4\theta_0}{3}\left(\frac{\rho}{a}\right)\left[\frac{1 - \frac{1}{4}\left(\frac{\rho}{a}\right)^3 - \frac{3}{4}\left(\frac{a}{\rho}\right)c^4}{(1 - c^4)}\right].$$

Thus the residual angle of twist is

$$\theta_R = \theta_0 - \theta'_0$$

$$= \theta_0\left\{1 - \frac{\frac{4}{3}\left(\frac{\rho}{a}\right)\left[1 - \frac{1}{4}\left(\frac{\rho}{a}\right)^3 - \frac{3}{4}\left(\frac{a}{\rho}\right)c^4\right]}{(1 - c^4)}\right\}. \qquad (8.28)$$

8.3.2. The Equilateral Triangle

As a further example, consider a solid shaft whose section is an equilateral triangle loaded by torque $+T_2$, sufficient just to make the whole of it plastic.

Then from equation (8.26), $+T_2 = 2a^3k/3$. If θ denotes the angle of untwist (elastic) then $-T_2 = G\theta a \sqrt{3}/5$.

Hence
$$\theta = \frac{2a^3k}{3} \cdot \frac{5}{a^4\sqrt{3G}} = \frac{10\sqrt{3k}}{9Ga}.$$

Now the 'elastic' stress distribution due to $-T_2$, using the above value of θ, is given by expressions

$$\tau_{xz} = \frac{10\sqrt{3k}}{9a^2}y(x\sqrt{3} - a)$$

$$\tau_{zy} = \frac{5k}{3a^2}\left(x^2 + \frac{2a}{\sqrt{3}}x - y^2\right).$$

But in the fully plastic state, $\tau_{xz} = 0$ and $\tau_{zy} = k$.
Hence the residual stress distribution is given by,

$$\tau_{xz} = \frac{10\sqrt{3}k}{9a^2}\,y(x\sqrt{3} - a)$$

$$\tau_{zy} = k\left\{1 - \frac{5}{3a^2}\left(x^2 + \frac{2ax}{\sqrt{3}} - y^2\right)\right\}.$$
(8.29)

This distribution has, of course, a three-fold symmetry, there being discontinuities along the internal bisectors of the angles of the triangle, see Fig. 8.29(a).

The determination of residual stress patterns and angles of twist in the case of *circular* shafts need not be confined to materials which are elastic-perfectly plastic. Strain-hardening materials may be taken into account in the manner used in Chapter 6, for calculating torques.

8.3.3. Using the Sand Heap and Membrane Analogies

To obtain a physical impression of any residual stress distribution simultaneous use may be made of both the soap film and sand heap analogy. Consider for instance the case of a shaft of circular cross-section, twisted to full plasticity. The stress function sand heap analogue is a right circular cone; if the *unloading* is entirely elastic, then it may be represented by the membrane

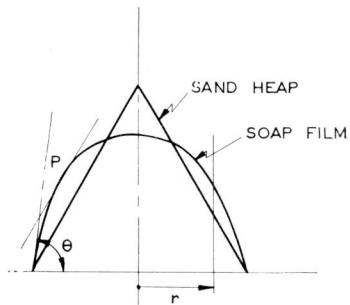

FIG. 8.33 Plastic loading and unloading of circular shaft.

analogue, the volume under the membrane being equal to that of the cone, since the loading and unloading torques are equal. This situation is depicted in Fig. 8.33. The maximum slope at a point at radius r on the sand heap is m_p, and on the membrane is proportional to the unloading elastic stress of the material, say, m_e. Hence the difference between these slopes $(m_e - m_p)$ is a measure of the residual stress along the circle of radius r. A limit is set to this difference, however, because it must nowhere exceed the gradient of a sand heap, i.e., $|m_e - m_p| \leqslant m_p$. This restriction means that the residual stress cannot exceed the yield shear stress of the material. In the case considered, it will be clear that

(i) since the soap film has zero slope on the centre-line (i.e., zero elastic stress), there is a positive residual stress of magnitude k at the centre of the cross-section. The residual stress, though positive, reduces as r increases until

(ii) at P, the tangent to the membrane is the same as that of the sand heap. Then at P the residual stress is zero. Further increase in r leads to negative residual stresses which increase to a maximum on the boundary. Thus,

(iii) unless the membrane gradient on the boundary, i.e., $\tan \theta$, is less than $2m_p$, recovery in unloading will not be entirely elastic. The residual shear stress τ_R in the outer layers is given by equation (8.27.i), as

$$\tau_R/k = 1 - 4r/3a$$

and at $r = a$, by $\qquad \tau_r/k = -1/3.$

$$(8.30)$$

The membrane slope along the boundary is thus less than twice the analogous sand heap gradient.

(iv) This analogy can be carried further. If a negative torque be applied after the unloading, it may be increased until at the outside the membrane touches a second roof, defined by a second sand heap which has a slope twice that of the first; when this occurs, yielding again starts but the stress will be in the opposite sense to that in which it first occurred. This membrane will enclose a volume which is in effect represented by the sum of two volumes, that due to the unloading torque and that due to the further negative torque imposed (see Fig. 8.34).

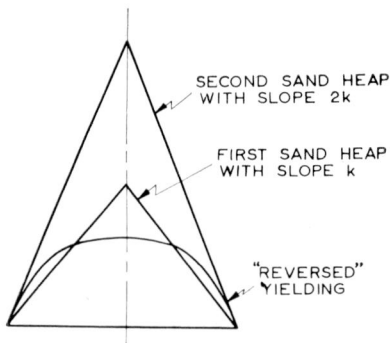

SECOND SAND HEAP
WITH SLOPE 2k

FIRST SAND HEAP
WITH SLOPE k

"REVERSED"
YIELDING

FIG. 8.34 Loading, unloading and reversed re-loading of circular shaft.

It is easy to appreciate how the two analogues may be used in the manner described above when considering, say, the unloading and possibly reversed loading from the fully plastic state, of a shaft having a cross-section which is an equilateral triangle. The determination of the residual stresses is not so straightforward as in the case of the circular shaft, since, for a particular point in the shaft section, they will be given by the vector difference of the maximum gradients of the membrane and the sand heap which lie in non-radial planes.

8.4 The Elastic Shortening of Twisted Bars

This section is included to indicate that length changes due to plastic twisting are to be expected. An elastic analysis is given in order to suggest an approach and to present the ideas associated; the elastic analysis is chosen for ease of presentation.

If a prismatic bar of circular cross-section is subject to torsion by couples applied at its ends, in planes normal to the bar axis, then classically, for small angles of twist per unit length, generating lines remain unextended and the distance between specified sections remains constant. The larger the angle of twist per unit length the less true are the above assumptions and indeed the distance between the end cross-sections reduces so that shortening increases with angle of twist. The inclination of a given fibre to the axis of the prismatic bar is in direct proportion to its distance from the axis (which always remains straight) so that exterior fibres will tend to be shortened most; and since interior fibres and exterior fibres do not strain independently it follows that exterior fibres will tend to be subject to axial tension and interior ones to compression.

If the distance between two sections is l and one is then rotated with respect to the other through angle θ about the axis, a fibre at distance r from the axis takes on a length increase to $(l + dl)$ if the axial distance between the end sections remains constant. We have

$$\frac{dl}{l} = \sqrt{1 + \left(\frac{r\theta}{l}\right)^2} - 1$$

or a tensile strain of

$$e = \frac{dl}{l} \doteq \frac{1}{2} \cdot \frac{r^2\theta^2}{l^2}.$$

The tensile stress set up is,

$$\sigma = E \cdot \frac{dl}{l} = E \frac{r^2\theta^2}{2l^2}.$$

Since the bar as a whole shortens, we may suppose all fibres to have further imposed on them the same amount of compressive strain e_c, so that the *net* axial strain in each fibre is $(e - e_c)$. Now the net normal force on any section of the bar is, however, zero, so that we must have

$$\int E(e - e_c)\,dA = 0 \quad \text{or} \quad e_c\,A = \int e\,dA$$

that is

$$e_c = \frac{\theta^2}{2l^2 A} \cdot \int_A r^2\,dA. \qquad (8.31)$$

Thus for a solid circular bar of radius R,

$$e_c = \frac{\theta^2}{2l^2} \cdot \frac{R^2}{2}. \qquad (8.32)$$

The longitudinal stress, σ_z, which varies from fibre to fibre, for a bar subject to a torque is

$$E\left(e - e_c\right) = \sigma_z = \frac{E\theta^2}{2l^2}\left(r^2 - \frac{R^2}{2}\right). \tag{8.33}$$

The position of the neutral fibre, i.e., that for which $\sigma_z = 0$ is

$$r_n = R/\sqrt{2}.$$

Thus a solid circular bar which is not allowed to change its length requires a torque ΔT in addition to the classical torque $T = JG\theta/l$ for small angles of twist, given by

$$\Delta T = \int(\sigma dA).\sin\phi.r$$

$$= \int \frac{E\theta^2}{2l^2}.r^2.2\pi r dr.\frac{r\theta}{l}.r$$

$$= \frac{E\theta^3\pi}{l^3}\int_0^R r^5.dr$$

$$= \frac{\pi R^6}{6}.E.\left(\frac{\theta}{l}\right)^3. \tag{8.34}$$

Note that,

$$\frac{\Delta T}{T} = \frac{(\pi R^6/6).E.(\theta/l)^3}{(\pi R^4/2).G.(\theta/l)} = \frac{2}{3}.(1 + v).R^2.\left(\frac{\theta}{l}\right)^2. \tag{8.35}$$

For metals the quantity ΔT is usually small because θ/l is small; this is not true of course for a material like rubber. More generally, tension uniformly distributed over an entire cross-section obviously influences the angle of twist of a bar and stiffens it against torsion.

A circular bar which is *not* axially loaded in order to maintain its original length, requires the torque $\Delta T'$, in addition to the 'classical' torque, necessary to give a rotation of θ so that using (8.33),

$$\Delta T' = \frac{\pi E\theta^3}{l^3}.\int_0^R\left(r^2 - \frac{R^2}{2}\right)r^3.dr$$

$$= \frac{\pi R^6}{24}.E.\left(\frac{\theta}{l}\right)^3.$$

The argument developed above with respect to circular section bars may obviously be repeated for prismatic bars of non-circular section. It is easy to show that for rectangular bars of dimensions $b \times h$ that,

(i)
$$e_c = \frac{1}{2}.\left(\frac{\theta}{l}\right)^2.\frac{b^2 + h^2}{12} \tag{8.36}$$

(ii)
$$\sigma_z = \frac{1}{2}.E.\left(\frac{\theta}{l}\right)^2\left(r^2 - \frac{b^2 + h^2}{12}\right) \tag{8.37}$$

(iii)
$$r_n = [(b^2 + h^2)/12]^{\frac{1}{2}} \tag{8.38}$$

(iv)

$$\Delta T = \frac{1}{2}. E. \left(\frac{\theta}{l}\right)^3 \int_A r^4. dA \qquad (8.39)$$

$$= \frac{3b^5h + 3h^5b + 10h^3b^3}{240}. E. \left(\frac{\theta}{l}\right)^3$$

(v)

$$\Delta T' = \frac{1}{2}. E. \left(\frac{\theta}{l}\right)^3 \int_A r^2\left(r^2 - \frac{b^2 + h^2}{12}\right) dA$$

$$= \frac{bh (b^4 + h^4)}{360}. E. \left(\frac{\theta}{l}\right)^3. \qquad (8.40)$$

Note that for the rectangular section bar, the greatest 'secondary' longitudinal stresses arise at the corners of the section and are a quadratic function of θ.

8.5 The Plastic-Torsion of I-Sections with Warping Restraint

In all the cases concerning torsion discussed above, every cross-section of a given shaft has been assumed to be free to warp. Practical situations do not always allow this and, in structural engineering particularly, the ends of I-section members subject to torsion may well be prevented from warping by adjacent structural members. It is, in fact, known from experiment that sand heap values for the plastic torque for I-section members when warping is restrained, are too low. Several investigators have discussed this topic and all the theories proposed tend to underestimate experimentally determined values, see DINNO and GILL (1964).

An expression for the torque, T, to cause full plastic yield under warping restraint conditions is

$$T = T_N + M_P h/l. \qquad (8.41)$$

It was originally proposed by DINNO and MERCHANT (1965) as an upper bound to T, though partly embodied in a paper by BOULTON (1962). In the above expression, T_N is the sand heap torque for the whole section, M_D the full plastic bending moment in one flange about Ox, h the distance between centroids of the I-section flanges and l is the length of the beam. The above equation was arrived at by assuming that the torque on the free end of the cantilever I-section is the sum of (i) a sand heap torque T_N, see Fig. 8.35(c) and (ii) a torque which arises due to a differential bending mechanism, the associated shear stress distribution being shown in Fig. 8.35(d); the net force in each flange is supposed to cause a plastic hinge at the restrained end.

The cantilever, firmly restrained at its built-in end but free to warp at the opposite end, is shown in Fig. 8.35(a) undeformed and in (b) deformed. If the effect of the web is neglected, each shear force across the flange Q is assumed to be capable of causing the flange to become fully plastic at the root, by bending about axis Ox; hence $Q.l = M_P$ where $M_P = Yb^2t/4$.

Thus

$$T = T_N + M_P.h/l.$$

Evidently the sand heap approach developed earlier is always adequate for a relatively long shaft restrained at one end, since $T \rightarrow T_N$ as $l \rightarrow \infty$.

AUGUSTI (1966) has proved rigorously the validity of the above expression, for an ideally plastic material.

In all experimental investigations the presence of strain-hardening makes it difficult to interpret results.

The torque carrying capacity of a shaft is also increased by the second order phenomenon of the helix effect (this effect due to change of geometry is, of course, wholly neglected in the usual theory); the generator of a section assumes the shape of a helix after twist and this is accompanied by normal stresses along it. The components of the latter stresses normal to the bar axis produce a torque and this has been estimated to be as much as 7 per cent of the sand heap value.

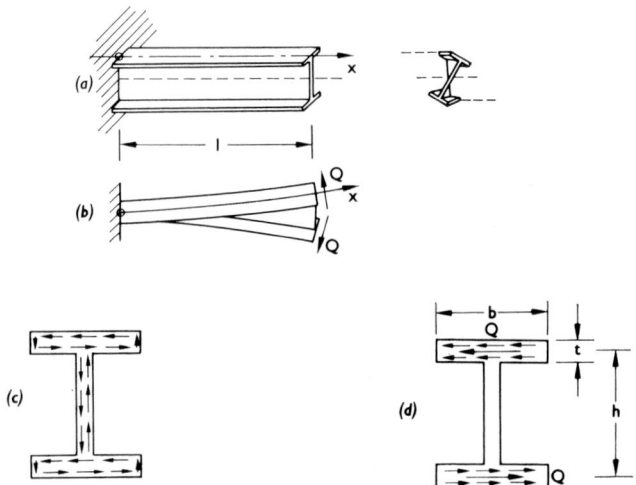

FIG. 8.35 Torsion of I-section.

More general results on this topic can be found in a recent paper by AUGUSTI (1968), but the article by DINNO and GILL is especially valuable for its discussion of the whole subject. This was foreshadowed in Section 8.4.

An experimental investigation into the plastic collapse of structural members under combined bending and torsion is described in a paper by GILL and BOUCHER (1964).

See Problems 31-35

REFERENCES

AUGUSTI, G.   1966 'Full Plastic Torque of I-Beams'
                        Int. J. mech. Sci. **8**, 641
                  1968 'On the Limit Analysis of I-Beams with
                        Warping Restraint'
                        Engineering Plasticity C.U.P., p. 41

BOULTON, N. S. 1962 'Plastic Bending and Twisting of an I-Beam'
Int. J. mech. Sci. **4**, 491

DINNO, K. S. and 1964 'The Plastic Torsion of I-Sections with
GILL, S. S. Warping Restraint'
Int. J. mech. Sci. **6**, 127

DINNO, K. S. and 1965 'A Procedure for Calculating the Plastic
MERCHANT, W. Collapse of I-Sections Under Bending and
Torsion'
Struct. Engng. **43**, 219

GILL, S. S. and 1964 'An Experimental Investigation of Plastic
BOUCHER, J. K. G. Collapse of Structural Members Under
Combined Bending and Torsion'
Struct. Engng. December

GRIFFITH, A. A. and 1917 'The Use of Soap Films in Solving Torsion
TAYLOR, G. I. Problems'
Proc. Instn mech. Engrs. Oct.–Dec

HERAKOVICH, C. T. 1969 'Elastic-Plastic Torsion of Hollow Bars
and HODGE, P. G. by Quadratic Programming'
Int. J. mech. Sci. **11**, 53

HETENYI, M. (ed). 1950 *Handbook of Experimental Stress Analysis.*
Wiley, New York.

HIGGINS, T. J. 1944 'Analogic Experimental Methods in Stress
Analysis as Exemplified by Saint-Venant's
Torsion Problem'
Soc. for expl Stress Analysis **2**, 17

KJAR, A. R. 1967 'The Axis of Distortion'
Int. J. mech. Sci. **9**, 873

LEISSA, A. W. and 1964 'On the Torsion of Bars having Symmetry
BRANN, J. H. Axes'
Int. J. mech. Sci. **6**, 45

MENDELSON, A. 1968 *Plasticity: Theory and Applications*
(Chapter 11)
Macmillan, New York

McCLINTOCK, F. A. 1956 'The Growth of Fatigue Cracks under Plastic
Torsion'
*Conf. on Fatigue of Metals, Instn of mech.
Engrs,* Session 6, Paper 6

NADAI, A. 1923 'Der Beginn des Fliessvorganges in einem
tortierten Stab'
Z. angew. Math. Mechanik **3**, 442
1950 *Theory of Flow and Fracture of Solids*
Vol. 1, McGraw-Hill, New York

NAVIER, M. 1864 '*Résumé des leçons sur l'application de la
mécanique*', 3rd edn, Paris, edited by
Saint-Venant

OLSZAK, W. 1965 'On Anisotropic Twisted Bars.'
Acta techn. hung. **50**, 263

PRAGER, W. and HODGE, P. G.   1951 *Theory of Perfectly Plastic Solids* Wiley, New York

PRANDTL, L.   1903 'Zur Torsion von prismatischen Staeben'. *Physik Z* **4**, 758

RONAY, M.   1968 'Second Order Elongation of Metal Tubes in Cyclic Torsion' *Int. J. Solids Structures*, **4**, 509

SAINT-VENANT, B.   1855 'Mémoire sur la torsion des prismes'. Mem. prés. par div. sav. a l'Ac.Sci., *Sci. math. et phys.* **14**, 233

SOUTHWELL, R.   1954 'Relaxation Methods: A Retrospect'. *Proc. Instn mech. Engrs*, **168**, 7

TIMOSHENKO, S.   1953 *A History of Strength of Materials* McGraw-Hill, N.Y.

TIMOSHENKO, S. and GOODIER, J. N.   1951 *Theory of Elasticity*, McGraw-Hill, London

TODHUNTER, I., and PEARSON, K.   1886 *A History of the Elasticity and Strength of Materials*

WANG, C. T.   1953 *Applied Elasticity* McGraw-Hill, New York

Chapter 9

ELASTIC–PLASTIC PROBLEMS WITH SPHERICAL OR CYLINDRICAL SYMMETRY

9.1 Introduction

In this chapter we shall examine the pressure and steady state thermal loading of thick-walled spherical shells and cylinders, followed by orthodox treatments of the thermal loading of discs and rotating discs.

The design of thick spheres and cylinders for use as testing chambers or for the containment of fluids at high pressures becomes more complicated when high temperatures and temperature gradients are involved. The use of very high temperatures and pressures, particularly in chemical plant, has led to new considerations in the design of such containers.

Most industrial design problems involving temperature and pressure are tractable when the resultant stresses involved are everywhere elastic. The wholly elastic situation is one in which the usual providential boon is available; the total elastic stress is simply the algebraic sum of the elastic stresses due separately to temperature and pressure; as such, these elastic problems are, in principle, straightforward. We shall consider only symmetrical components. Elastic stress distributions will first be examined, followed by some cases of elastic-plastic stress distribution, and finally fully plastic cases. It will be seen that with the introduction of plasticity, the algebraic manipulation becomes complicated and tedious. The analyses become increasingly inexact because of creep as the degree of plasticity increases with temperature.

The more complicated problems, particularly concerning strains and displacements, are not fully treated here, but reference is made to some of the original sources in which they are discussed—if, indeed, they have been discussed at all. It will become apparent that there are many particular problems which are unsolved.

The discussion of the stresses in loaded pressure vessels, given below, is a simplified one; for full and detailed discussions the reader should consult the references at the end of the chapter. To provide a fairly rounded picture, both the elastic and plastic analyses will be included and referred to.

The books by BOLEY and WIENER (1960) and by NOWACKI (1965) are particularly useful for their treatments of thermal stress.

Reference to the heat transfer phenomenon must first be made. It will be assumed that the coefficient of conductivity, k, the coefficient of linear thermal expansion, α, and other physical constants such as Young's Modulus, E, and Poisson's ratio, v, are all unaffected by temperature. It may be shown that the temperature of any element of a body under elastic stress is, by virtue of the first and second laws of thermodynamics, subject to the following transient heat conduction equation, exactly, (BIOT, 1958),

$$\frac{\partial^2 \theta}{\partial x^2} + \frac{\partial^2 \theta}{\partial y^2} + \frac{\partial^2 \theta}{\partial z^2} = \frac{\rho c}{k} \frac{\partial \theta}{\partial t} + \frac{T'\beta}{k} \frac{\partial \epsilon}{\partial t}. \qquad (9.1)$$

θ is the rise in temperature at a point in the body in three-dimensional space referred to Cartesian co-ordinates x, y, z; t is time, $k/\rho c$ is the thermal diffusivity, β is a material constant and the dilatation $\epsilon = \epsilon_{xx} + \epsilon_{yy} + \epsilon_{zz}$; T' is the absolute temperature of the body. Strictly, the second term on the right of equation (9.1) should appear in all transient heat conduction problems; it is never large enough, however, in the cases usually considered by engineers, to be of importance. The cases to be examined below are all ones of steady heat flow, and hence the right-hand side of (9.1) is zero; all temperature gradients used are purely radial.

It may be noted that ROSENFIELD and AVERBACH (1956) have shown how the coefficient of linear expansion, α, is affected by stress. At the yield stress in tension, α for steel is about 10 per cent greater than it is when the specimen is not loaded. The general shape of the α versus tensile stress curve σ is shown in Fig. 9.1. It may easily be shown that $d\alpha/d\sigma$ at a given temperature T is $(dE/dT)/E^2$ for a given elastic stress; E is Young's Modulus.

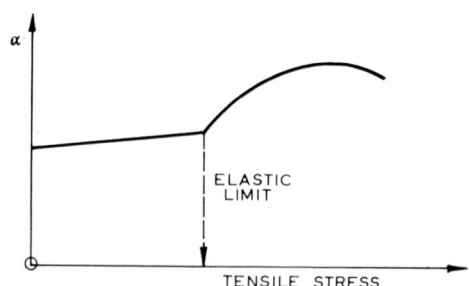

FIG. 9.1 Showing variation of α with tensile stress σ, as experimentally determined by ROSENFIELD and AVERBACH.

9.2 Thick Hollow Spheres

9.2.1 *Elastic Stress Distribution: Steady State Temperature Gradient Only*

In the solution for the elastic stress distribution, we shall follow the treatment of TIMOSHENKO and GOODIER (1951).

The sphere has an internal radius a and an external radius b in its initial unloaded state. The stress on a small element at radius r is as shown in Fig. 9.2. For radial force equilibrium of this element

$$2\sigma_\theta r \, dr \, d\theta \, d\theta = d(\sigma_r r^2 \, d\theta \, d\theta)$$

and thus

$$d\sigma_r/dr = 2(\sigma_\theta - \sigma_r)/r. \qquad (9.2)$$

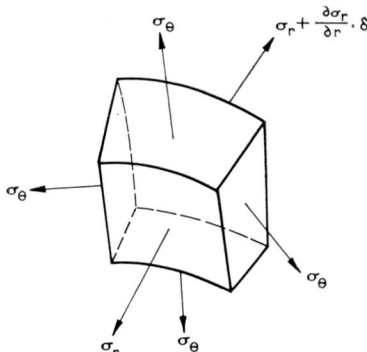

FIG. 9.2 Stresses acting on an element at radius r taken from wall of spherical shell.

The total elastic strain at a point is the sum of that due to stress and that due to a temperature rise T, thus

radial strain,

$$e_r = \frac{\sigma_r - 2v\sigma_\theta}{E} + \alpha T$$

$$\left. \right\} \qquad (9.3)$$

and hoop strain,

$$e_\theta = \frac{\sigma_\theta - v(\sigma_r + \sigma_\theta)}{E} + \alpha T,$$

where α is the linear coefficient of thermal expansion.

Hence,

$$\sigma_\theta = \frac{E}{(1 + v)(1 - 2v)}[e_\theta + v e_r - (1 + v)\alpha T]$$

$$\left. \right\} \qquad (9.4)$$

and

$$\sigma_r = \frac{E}{(1 + v)(1 - 2v)}[(1 - v)e_r + 2v e_\theta - (1 + v)\alpha T].$$

Substituting these last expressions into the equilibrium equation (9.2), noting that

$$e_r = du/dr \quad \text{and} \quad e_\theta = u/r \qquad (9.5)$$

where u is the small radial displacement of a point at radius r, and simplifying, it follows that

$$\frac{d^2u}{dr^2} + \frac{2}{r}\frac{du}{dr} - \frac{2u}{r^2} = \left(\frac{1 + v}{1 - v}\right)\alpha\frac{dT}{dr}$$

which, after some integration, gives

$$u = \left(\frac{1 + v}{1 - v}\right)\frac{\alpha}{r^2}\int_a^r Tr^2 dr + C_1 r + \frac{C_2}{r^2} \qquad (9.6)$$

where C_1 and C_2 are constants of integration. Utilizing the expression for u from equation (9.6) in equation (9.5) to give others for e_r and e_θ and putting these into equation (9.4), it is found that

$$
\sigma_r = \frac{2\alpha E}{(1-\nu)}\left[\frac{r^3-a^3}{(b^3-a^3)r^3}\int_a^b Tr^2\,dr - \frac{1}{r^3}\int_a^r Tr^2\,dr\right],
$$

$$
\sigma_\theta = \frac{2\alpha E}{(1-\nu)}\left[\frac{2r^3+a^3}{2(b^3-a^3)r^3}\int_a^b Tr^2\,dr + \frac{1}{2r^3}\int_a^r Tr^2\,dr - \frac{T}{2}\right].
$$

(9.7)

Assuming that the temperature of the sphere on its inner surface $r = a$ is T_i, and on its outer surface $r = b$ is 0, the temperature distribution through the sphere wall in the steady state is given by

$$
T = \frac{T_i a}{(b-a)}\left(\frac{b}{r} - 1\right)
$$

(9.8)

which substituted into equation (9.7), gives

$$
\sigma_r = \frac{\alpha E T_i}{1-\nu}\cdot\frac{ab}{b^3-a^3}\left[\overparen{a+b} - \frac{b^2+ab+a^2}{r} + \frac{a^2b^2}{r^3}\right]
$$

$$
\sigma_\theta = \frac{\alpha E T_i}{1-\nu}\cdot\frac{ab}{b^3-a^3}\left[\overparen{a+b} - \frac{b^2+ab+a^2}{2r} - \frac{a^2b^2}{2r^3}\right].
$$

(9.9)

9.2.2 *Elastic Stress Distribution: Internal Pressure*, p, *only*

From the analysis immediately above, equations (9.2), (9.3), (9.4) and (9.5) apply if $T = 0$ for all values of r and then $u = C_1 r + C_2/r^2$. Hence

$$
e_r = C_1 - 2C_2/r^3 \quad \text{and} \quad e_\theta = C_1 + C_2/r^3
$$

and thus

$$
\sigma_\theta = A + \frac{B}{2r^3}
$$

and

$$
\sigma_r = A - \frac{B}{r^3},
$$

(9.10)

where A and B are constants fitted to the given pressure or stress boundary conditions.

If at $r = a$, $\sigma_r = -p$, and at $r = b$, $\sigma_r = 0$, it is easily verified that

$$
\sigma_r = \frac{pa^3(b^3 - r^3)}{r^3(a^3 - b^3)}
$$

and

$$
\sigma_\theta = \frac{pa^3(2r^3 + b^3)}{2r^3(b^3 - a^3)}.
$$

(9.11)

9.2.3 *Elastic Stress Distribution: Internal Pressure and a Steady State Temperature Gradient: Onset of Yielding*

The expressions for σ_r and σ_θ in this instance are simply the sum of the stresses independently brought about by the temperature and the pressure. Denoting $\alpha E T_i/(1 - v)$ by β, b/a by m and r/a by R, then

$$\sigma_r/\beta = [m^3(1 - p/\beta) - mR^2(m^2 + m + 1)$$
$$+ R^3(p/\beta + m^2 + m)]/R^3(m^3 - 1)$$

and

$$\sigma_\theta/\beta = [m^3(p/\beta - 1) - mR^2(m^2 + m + 1)$$
$$+ 2R^3(p/\beta + m^2 + m)]/2R^3(m^3 - 1).$$

$$(9.12)$$

Denoting by τ_r the maximum shear stress at a point at radius r

$$\tau_r/\beta = (\sigma_\theta - \sigma_r)/2\beta = [3m^3(p/\beta - 1)$$
$$+ mR^2(m^2 + m + 1)]/4R^3(m^3 - 1). \qquad (9.13)$$

In particular

$$\tau_a/\beta = m[3m^2(p/\beta - 1) + m^2 + m + 1]/4(m^3 - 1) \qquad (9.14)$$

and

$$\tau_b/\beta = [3(p/\beta - 1) + m^2 + m + 1]/4(m^3 - 1). \qquad (9.15)$$

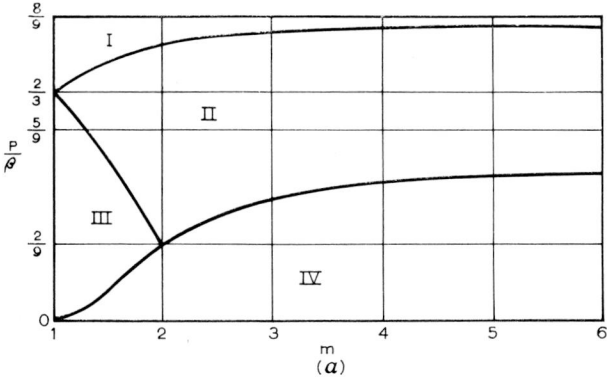

FIG. 9.3

(a) Graph to identify the position of onset of yield, or greatest maximum shear stress, in a sphere, in terms of $m(= b/a)$ for various combinations of pressure p and internal temperature, T_i

$$\beta = \frac{\alpha E T_i}{(1 - v)}$$

Region I: Yield first occurs on inner surface, $R = 1$, $r = a$.
Region II: Yield first occurs in the shell wall, at
$R = 3m\sqrt{(1 - p/\beta)/(m^2 + m + 1)}$
Region III: Yield first occurs on outer surface $R = m$, $r = b$.
Region IV: Yield first occurs on inner surface, $R = 1$, $r = a$.

Yielding will begin at the radius where τ_r reaches its absolutely greatest value whether the criterion be that of Mises or Tresca. According to the combination of p and β chosen, so the radius at which yielding starts may be anywhere in the shell wall; Fig. 9.3(a) identifies the position of this critical radius (see DERRINGTON and JOHNSON, (1958) and COWPER (1958) and (1960)).

For comparison, in Fig. 9.4 the separate elastic stress distributions due to pressure and temperature through the shell thickness, of spheres having outside diameter/thickness ratios of 4 and 2·22 (or $m = 2$ and 10, respectively) are shown. The prime ordinates are ratios of the stress produced in the material to the applied internal pressure; the ratios for the temperature stress are

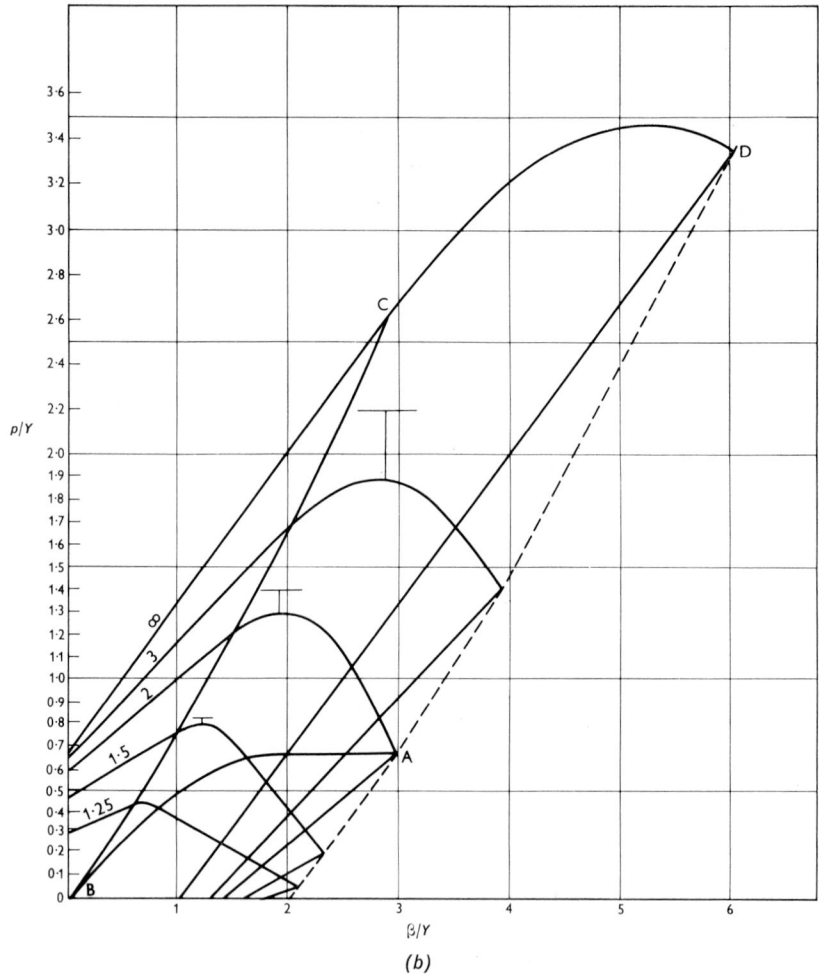

p/Y

β/Y

(b)

FIG. 9.3

(b) Showing relation between p/Y and β/Y for bringing about first yield for $m = 1.25$, 1.5, 2, 3 and ∞.

evaluated by setting $\beta = p$. An immediate inference from these curves is that an outward heat flow induces shear stresses opposite in sign to those caused by the pressure. This leads to the idea of 'stress-saving'; by introducing a suitably chosen temperature gradient, the stresses due to pressure may be reduced. This becomes apparent if, once the critical radius for onset of yield is known, the particular separate values of p/Y and β/Y which apply, are found and the graph shown in Fig. 9.3(b) (JOHNSON 1961, unpublished) is constructed. For particular values of m, curves for first yield are shown. Each curve consists of three parts if $m > 2$ and four parts if $1 \leqslant m \leqslant 2$, thus:

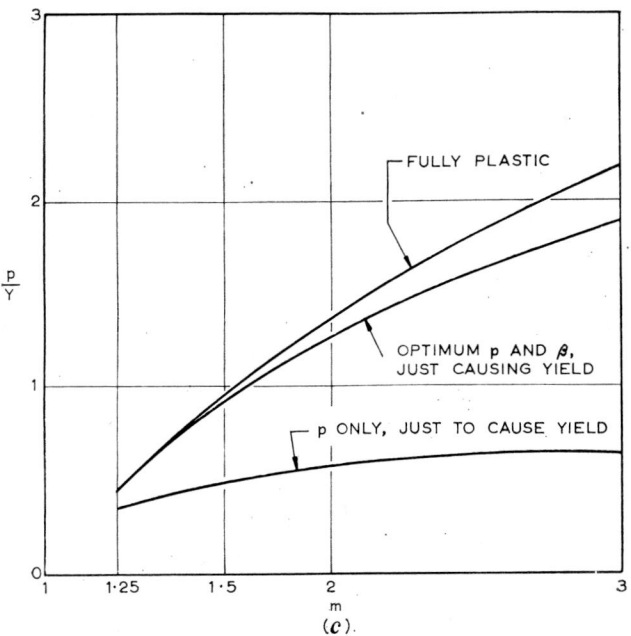

FIG. 9.3

(c) Comparing the values of p/Y to cause first yield without any temperature gradient, with that applying when the internal temperature is optimized.

(i) with points on a curve above BC is to be associated yield on the inside surface.

(ii) With points on a curve within $CBAD$—curved lines—is to be associated yield within the wall.

(iii) With points on a curve below $D1$—on the straight lines where p/Y increases with increasing β/Y—is to be associated yield on the inside surface.

(iv) With points on a curve below BAO, where p/Y increases with decreasing β/Y is to be associated yield on the outside surface.

It is noted that for $m > 2$ first yield on the outside surface cannot occur. Also any point within the curve, for a given m, represents an elastic state if it has been arrived at by 'travelling' to it from the origin and staying wholly within the curve. These curves assist then in the selection of permissible 'routes' to a required (p/Y, β/Y) first yield value. In particular, optimum

'stress saving' values of p/Y are immediately obvious—the maximum on each
m-curve.

Line OAD in Fig. 9.3(b) is the locus of points for a given m that define
permissible maximum β/Y, or temperature loading, for avoiding yield by
employing a suitably chosen value of p/Y.

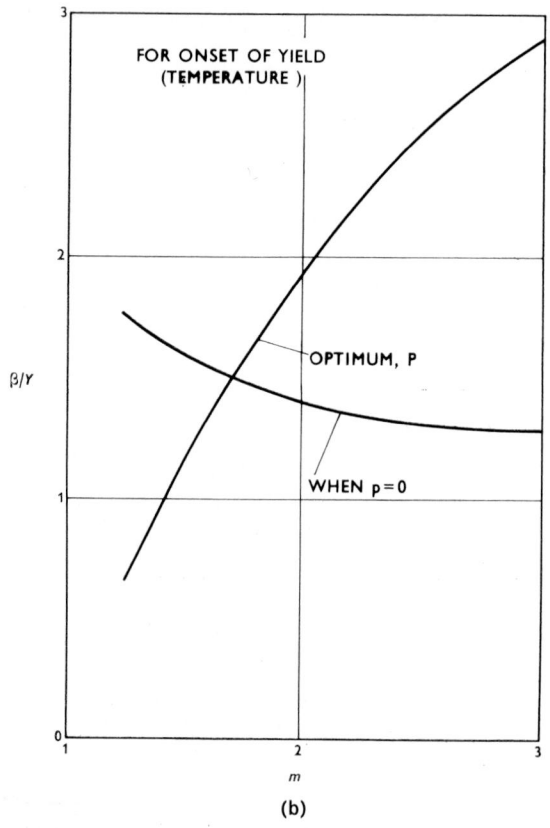

FIG. 9.3

(d) Comparing the values of β/Y to cause first yield with no internal
pressure, with that applying when p is optimized. (*After* Derrington
and Johnson, *Appl. Sci. Res.*)

For $m = 1.5$, 2 and 3, in Fig. 9.3(b), is shown the value of p/Y, marked **T**,
required to cause the whole shell to become completely plastic. (See sub-
section 9.2.5 below.) On Fig. 9.3(c), values required to cause this state of
complete plasticity are compared with

(i) that value of p/Y required to initiate yielding;
(ii) the largest value of p/Y with a suitably chosen β/Y at which yielding
starts.

Similarly, in Fig, 9.3(*d*), the value of β/Y to cause first yield when $p/Y = 0$ is compared with the highest permissible value of β/Y, when p/Y is suitably chosen.

9.2.4 *Partly Plastic Shell; Internal Pressure only*

For simplicity, first suppose that no temperature stresses are operative, i.e. $\beta = 0$. Yielding will commence at the inner surface of the shell, at pressure p^*, such that

$$\frac{Y}{2} = \frac{3m^3 p^*}{4(m^3 - 1)}$$

or $$p^* = \frac{2Y(m^3 - 1)}{3m^3}. \tag{9.16}$$

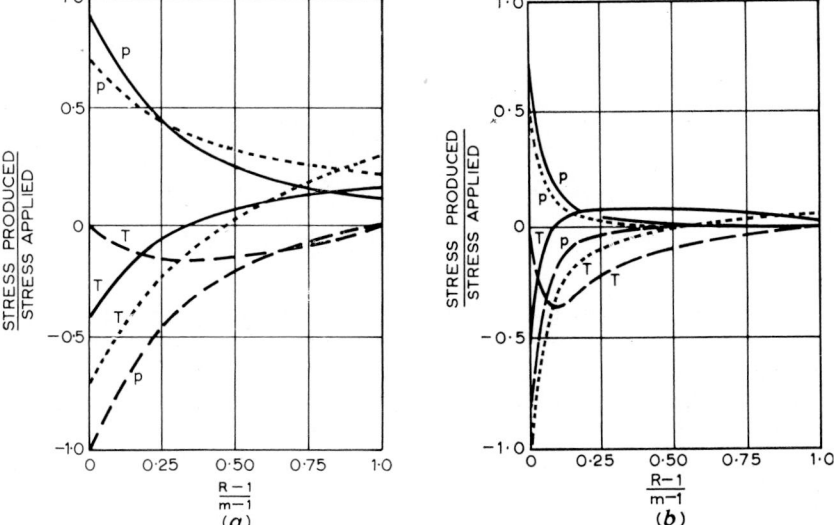

FIG. 9.4 Distribution of stresses due separately to temperature and pressure through wall of shell.

σ_θ – – – – –, σ_r – – – – –, $\tau_r = (\sigma_\theta - \sigma_r)/2$

p refers to pressure, T to temperature

(*a*) $m = 2$ (*b*) $m = 10$.

(*After* Derrington and Johnson, *Appl. Sci. Res.*)

Increasing p causes the spherical zone of yielding to expand. Suppose the radius to which it proceeds is c, then for $r > c$, the sphere is still elastically stressed; thus

$$\frac{Y}{2} = \frac{3z^3 \sigma_c}{4(1 - z^3)} \tag{9.17}$$

where $z = b/c$ and σ_c is the radial stress at $r = c$.

Now at each point in the plastic region, i.e. $r < c$, the yield criterion is

$$\sigma_\theta - \sigma_r = Y$$

and thus substituting in the equilibrium equation (9.2) and integrating

$$\sigma_r = 2Y \ln r + A.$$

From the boundary condition $\sigma_r = -p$ at $r = a$,

$$\sigma_r = 2Y \ln (r/a) - p \qquad (9.18)$$

and hence $\qquad \sigma_\theta = Y(1 + 2 \ln (r/a)) - p. \qquad (9.19)$

Since the radial stress is continuous across the elastic-plastic boundary, the internal pressure to cause yielding to radius c is, from equations (9.17) and (9.18)

$$p/2Y = \ln (c/a) + (1 - c^3/b^3)/3 \qquad (9.20)$$

Note that this plastic stress distribution was obtained without using any stress-strain relation. The various expressions concerning cases of yielding assume, of course, that a and b have not been significantly altered by yielding having penetrated to depth c.

The stresses in the elastic region, $r > c$, are

$$\left. \begin{array}{l} \sigma_r/2Y = -c^3(b^3 - r^3)/3b^3r^3 \\[2mm] \text{and } \sigma_\theta/2Y = c^3(b^3 + 2r^3)/6b^3r^3. \end{array} \right\} \qquad (9.21)$$

In particular yielding through the whole shell occurs when

$$p^{**} = 2Y \ln m. \qquad (9.22)$$

The residual stresses in the sphere, σ_r^R and σ_θ^R, after unloading it from its partly plastic state are obtained by superposing on the stress system due to internal pressure, p, a wholly elastic stress distribution due to $-p$.

One advantage of pressurizing a vessel so as to cause plastic deformation, is that of inducing favourable compressive stresses in the inner layers. By thus inducing residual compressive hoop stresses, the shell may be subsequently used to withstand a greater pressure and yet behave without further plasticity than it would if it had not been overstrained in the first instance. This topic is discussed at length in the book by HILL (1950).

9.2.5 *Partly Plastic Shell: Temperature Gradient Only*

An expression defining the radius at which first yield occurs when the shell is subject to steady heat flow only, is obtained from equation (9.9), thus

$$|\sigma_r - \sigma_\theta| = Y = \frac{\alpha E T_i}{2(1 - \nu)} \cdot \frac{ab}{b^3 - a^3} \left[\frac{3a^2b^2}{r^3} - \frac{b^2 + ab + a^2}{r} \right]$$

It can easily be shown that yield first occurs at the innermost surface of the sphere, i.e., $r = a$. The temperature difference between the walls, T_i^*, to cause yield is

$$T_i^* = \frac{2Y(1 - \nu)}{\alpha E} \cdot \frac{m^2 + m + 1}{m(2m + 1)},$$

where $m = b/a$.

For mild steel, typical values for the constants are $E = 3.10^7$ lbf/in², $Y = 3.10^4$ lbf/in², $v = 0.3$ and $\alpha = 7.5 \cdot 10^{-6}$ per °F and hence values of T_i^* are

m	$\simeq 1$	2	∞
T_i^* °F	187	130	93

Although there is not a great deal of difference between these values of T_i^*, the temperature gradients involved vary considerably.

If T_i is caused to exceed T_i^*, the plastic zone must spread outwards, spherically, to some new radius c, $b > c > a$. Within the region in which the sphere is plastic, i.e. $c \geqslant r \geqslant a$, the yield criterion applies with $\sigma_0 - \sigma_r = -Y$;

$$\left.\begin{array}{l} \sigma_r = -2Y \ln r/a \\[2mm] \text{and } \sigma_0 = -Y(1 + 2 \ln r/a), \end{array}\right\} \qquad (9.23)$$

since $\sigma_r = 0$ at $r = a$. Again note that in this region, the stress distribution is independent of T_i. For the outer elastic region $b \geqslant r \geqslant c$, we have (i) $\sigma_r = 0$ at $r = b$, (ii) σ_r is given by (9.23) at $r = c$ and (iii) also at $r = c$ yielding is taking place. This outer elastic shell is now examined with the help of the results in sub-section 9.2.3. At $r = c$ there is a temperature T_c (say) and a radial pressure given by equation (9.23). From equation (9.8)

$$T_c = T_i \cdot \frac{m - c/a}{(c/a)(m - 1)} = T_i \cdot \frac{m_1 - 1}{m - 1} \qquad (9.24)$$

where $m_1 = b/c$, $m \geqslant m_1 \geqslant 1$, and $T_i > T_i^*$.

Denoting $\alpha E T_c/(1 - v)$ by β' then

$$\beta' = \alpha E T_i (m_1 - 1)/(1 - v)(m - 1) \qquad (9.25)$$

The radial pressure at $r = c$ is from equation (9.23)

$$p' = -2Y \ln m/m_1 \qquad (9.26)$$

Since the inner surface is yielding, from equation (9.14)

$$\frac{-Y}{\beta'} = \frac{3m_1^3(p'/\beta' - 1) + m_1(m_1^2 + m_1 + 1)}{2(m_1^3 - 1)} \qquad (9.27)$$

Substituting p' of equation (9.26) in equation (9.27)

$$\frac{\beta'}{2Y} = -\frac{m_1^3(3 \ln m/m_1 + 1) - 1}{m_1(2m_1^2 - m_1 - 1)} \equiv \frac{\beta'}{p'} \cdot \frac{p'}{2Y} \qquad (9.28)$$

Denoting $\ln m/m_1$ by x, the ratio p'/β' is, from equation (9.28)

$$\frac{p'}{\beta'} = \frac{m_1(2m_1^2 - m_1 - 1)x}{m_1^3(3x + 1) - 1}. \qquad (9.29)$$

Figure 9.5 shows the variation of p'/β' with m_1 for $m = 2$, 4 and 6. These are the curves AB, CD and EF, respectively. When the magnitude of m_1 (for a given value of m) reaches a value on the limiting curve $OBGDF$ any further increase in T_i will cause a second zone of yielding to start in the outer elastic shell. The initial radius of the second zone of yielding, r, is given by

$$r/c = 3m_1 \sqrt{(1 - p'/\beta')/(m_1^2 + m_1 + 1)}.$$

The limits to the values of m_1 are

$$(i) \quad m_1 \leqslant 2, \qquad \frac{2(m_1 - 1)^2}{3(m_1^2 - m_1 + 1)} = \frac{p'}{\beta'} \qquad (9.30)$$

and $\quad (ii) \quad m_1 \geqslant 2, \qquad \dfrac{5 - 4/m_1 - 4/m_1^2}{9} = \dfrac{p'}{\beta'}. \qquad (9.31)$

Equations (9.30) and (9.31) define the two continuous curves, i.e. OBG and GDF, respectively at which the second separate zone of yielding will occur, see Fig. 9.5. We will examine the implications of these two curves separately by considering special cases.

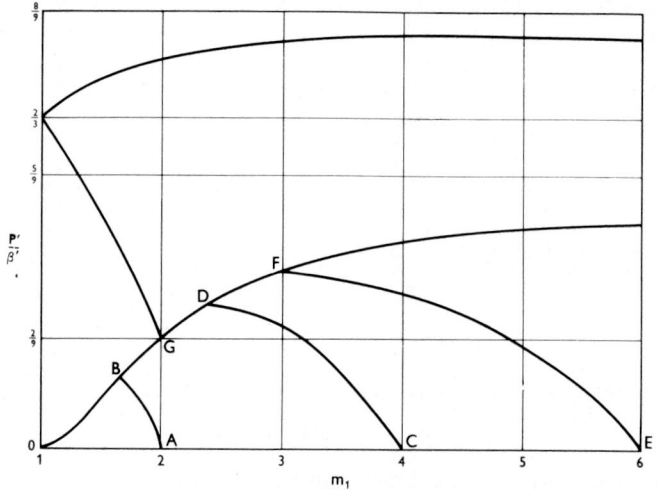

FIG. 9.5 For values of $m = 2, 4, 6$ respectively, AB, CD, and EF show the progress of yield with increasing temperature from first yield, until second yield starts at a new radius which is unconnected with the first yield zone, i.e., B, D, F respectively. The internal pressure is zero.

(a) $m = 2$

For values of $T_i > T_i^*$, yielding will have penetrated to a radius c. Thus selecting values of m_1, p'/β' values may be obtained from equation (9.29). The curve AB shows a sequence of values in Fig. 9.5 and it terminates at B for the value of p'/β' given by equation (9.30) above. In this instance, $c = 1.2a$, at which point the surface of the outer wall starts to yield. Increasing T_i further, the inner plastic zone, corresponding to c, increases in size, and a plastic shell develops from the outer surface and progresses inwards, the two plastic zones being separated by an elastically stressed shell.

For all values of m up to 2·79, second yielding will commence on the outer surface. Figure 9.6 shows the variation of the critical radius m_1 with m for the second zone of yielding to start first at radius b, when $m < 2·79$.

(b) $m = 4$

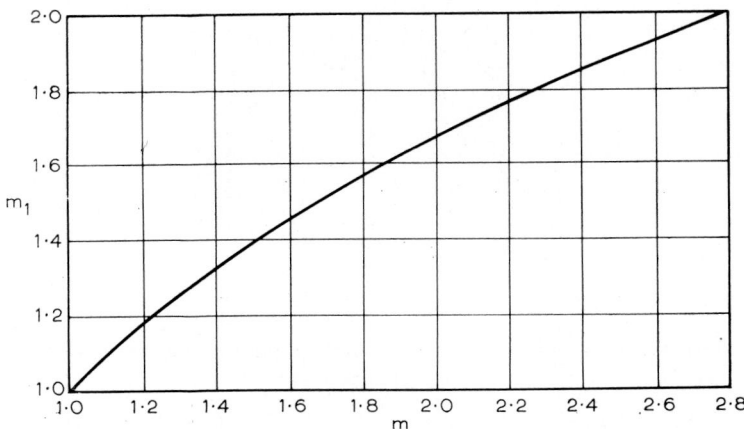

FIG. 9.6 Showing variation of critical radius m_1 with m for yielding to start first on outer surface; $m \leqslant 2·791$.

The sequence of states, i.e. the variation of p'/β' with m_1 for this initial value of m, is found by using equation (9.29) and is shown by curve CD in Fig. 9.5. When $c = 1·69a$, point D is reached and yielding will then start elsewhere in the sphere; in this case at $3·34a$.

For a sphere in which $m = 6$, when $c = 2·03a$, second stage yielding commences at $4·06a$; the sequence of stages up to this point is indicated by EF in Fig. 9.5. For $m = 8$, $c = 2·29a$ when second stage yielding starts at $4·59a$ (not shown in Fig. 9.5).

A further increase in T_i would cause a second plastic shell to develop, so that the situation becomes as shown diagrammatically in Fig. 9.7 with two plastic and elastic shells alternately situated.

In Fig. 9.8 are shown the critical values of m_1, and the radius of onset of second stage yielding, associated with each value of m.

The temperatures at which second stage yielding commences in mild steel shells are, (i) for $m = 4$ and 6, 740° and 953 °F when $T = 0$ °F at $r = b$. (At such temperatures, because of creep, calculations of the above kind become grossly unrealistic.) (ii) When $m = 2$, $T_i = 359$ °F and (iii) when $m = 2·79$, $T_i = 430$ °F. Residual stress distributions on unloading from second stage yielding are discussed by JOHNSON and MELLOR (1962).

9.2.6 Partly Plastic Shell: Steady State Radial Temperature Gradient and an Internal or External Pressure

This particular situation in which a thick-walled shell is simultaneously subject to a temperature and a pressure loading has been treated at length by

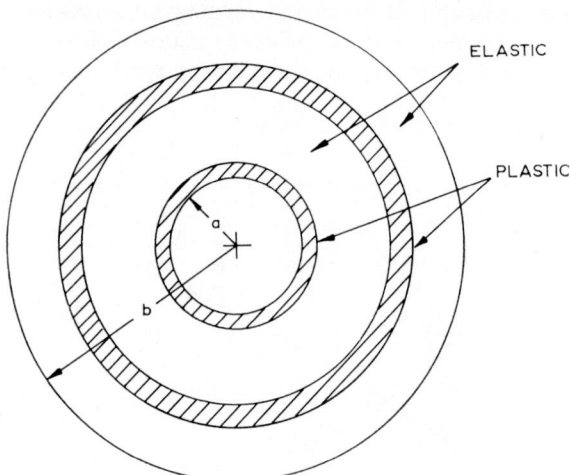

FIG. 9.7 Diagrammatic representation of plastic and elastic shells
alternately situated.

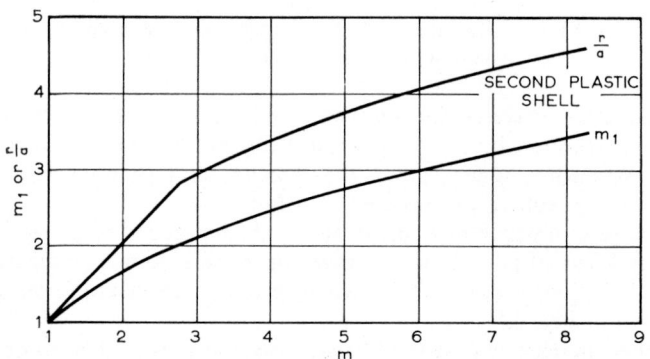

FIG. 9.8 Critical values of m_1 against m, at onset of *second stage*
yielding and showing radius of onset of *second stage* yielding for a
given value of m.

DRABBLE and JOHNSON (1964). Sub-section 9.2.5 is just a special case of this
sub-section. The analyses and results are too tedious and lengthy to give here.

9.2.7 *The Influence of Displacements: Internal Pressure Only*

All the previous results assume that displacements of elements in the
spherical shell are small, that their geometric effect is negligible, and in the
plastic analysis that the radius ratio m remains constant throughout at its
initial value.

We now recalculate the internal pressure required to make the whole
spherical shell just plastic when subject to internal pressure only, if the elastic

modulus is finite, on the assumption that the material is incompressible both elastically (i.e., $v = \frac{1}{2}$) and plastically.

Now the outermost layer of the shell is the last layer to become plastic and since $\sigma_r = 0$ at $r = b^*$ and $\sigma_\theta = \sigma_\phi = Y$, then the hoop strain at yield on the outside of the sphere is,

$$e_{\theta,b^*} = \frac{Y}{E}(1 - v) = \frac{Y}{2E}, \text{ if } v = \frac{1}{2}.$$

b^* is the external radius when yield has occurred.

Hence,

$$\frac{b^* - b}{b} = e_{\theta,b} = \frac{Y}{2E},$$

and thus,

$$\frac{b^*}{b} = 1 + \frac{Y}{2E}. \tag{9.32}$$

At this final stage the internal radius has attained a value a^*, so that,

$$\frac{4}{3}\pi(b^{*3} - a^{*3}) = \frac{4}{3}\pi(b^3 - a^3)$$

and thus

$$\left(\frac{a^*}{a}\right)^3 = \left[\left(\frac{b^*}{b}\right)^3 - 1\right]\left(\frac{b}{a}\right)^3 + 1.$$

Hence,

$$\left(\frac{b^*}{a^*}\right)^3 = \left(\frac{b^*}{b}\right)^3 \cdot \left(\frac{b}{a}\right)^3 \cdot \left(\frac{a}{a^*}\right)^3 = \frac{(b^*/b)^3 m^3}{m^3[(b^*/b)^3 - 1] + 1}$$

$$= \frac{[1 + (Y/2E)]^3 m^3}{m^3\{[1 + (Y/2E)]^3 - 1\} + 1}$$

Since, $Y/E \simeq 10^{-3}$, and writing $x = Y/2E$,

$$\left(\frac{b^*}{a^*}\right)^3 \simeq \frac{m^3(1 + 3x)}{m^3[(1 + 3x) - 1] + 1} \simeq \frac{m^3}{1 + 3m^3 x}.$$

It follows that the pressure to cause full plastic yield p_1^* is,

$$p_1^* = 2Y \ln \frac{b^*}{a^*} = \frac{2}{3} Y \ln\left[\frac{m^3}{1 + (3m^3 Y/2E)}\right] \tag{9.33}$$

In sub-section (9.2.4) we obtained, in effect,

$$p^{**} = \frac{2Y}{3} \ln(m^3). \tag{9.22}$$

How the effective radius ratio b^*/a^* varies with initial radius ratio and how p^* and p^{**} compare is evident from Fig. 9.9 for $Y/E = 10^{-3}$.

$m = b/a$	1·5	2	4	6	8	10
$p^{**}/2Y$	0·405	0·693	1·39	1·79	2·08	2·30
b^*/a^*	1·50	1·99	3·87	5·45	6·57	7·37
$p_1^*/2Y$	0·40	0·689	1·355	1·70	1·89	1·99

FIG. 9.9

The error in calculating the true value p_1^*, by using p^{**} increases as m increases. In the extreme when $m \to \infty$, $\dfrac{p_1^*}{2Y} = \dfrac{1}{3}\ln\dfrac{2E}{3Y}$ and $p^{**}/2Y \to \infty$. To expand a sphere of infinite radius ratio, for a typical metal where $Y/E \simeq 10^{-3}$, $p_1^*/2Y \simeq \ln 8\cdot7 = 2\cdot16$ or $p_1^* \simeq 4\cdot3Y$.

9.2.8 Expanding an Infinitely Small Cavity by Internal Pressure, p^{***}

An infinitely small cavity of radius a in a very large body may be continuously expanded by internal pressure of constant magnitude, and to show this we again assume that the material is incompressible throughout. In this steady state, let the current radius of the elastic-plastic interface be c^*. For the outer elastic sphere where $\sigma_r = 0$ at radius b^*, because the body is infinitely extensive, $b^*/c \to \infty$ and thus from above, $\sigma_r = \frac{2}{3}Y$ at $r = c^*$.

However, we may regard the inside plastic shell as having a shell of infinite radius ratio initially, i.e., $c/a \to \infty$, so that the results of the previous section can be applied, allowing for the external compressive stress of $\frac{2}{3}Y$; then p^* of the previous section will be increased by the hydrostatic stress $2Y/3$.

Thus, $p^{***} = \dfrac{2Y}{3}\ln\left(\dfrac{2E}{3Y}\right) + \dfrac{2Y}{3}$,   or $\dfrac{p^{***}}{2Y} = \dfrac{1 + \ln(2E/3Y)}{3}$ (9.34)

Again, with $Y/E \simeq 1/1000$, $p^{***} \simeq 4\cdot98Y$; with a low value of $Y/E = 1/300$, $p^{***} = 4\cdot19Y$.

Using (9·18) we may calculate c^*/a^*. Substituting,

$$2Y/3 = 2Y\ln c^*/a^* - p^{***} \quad \text{or} \quad c^*/a^* = \exp(2 + \ln 2E/3Y)/3;$$

and hence for $Y/E = 10^{-3}$, $c^*/a^* = 6\cdot25$ and for $Y/E = 1/300$, $c^*/a^* = 4\cdot19$.

9.2.9 The Work-Hardening Material

Earlier we have dealt with spherical shells of elastic-perfectly plastic incompressible material. We now discuss the inclusion of a work-hardening representative stress-representative strain law in the analysis; let this be described by,

$$\bar{\sigma} = Y + P\,\bar{\epsilon}^n,$$

where $\bar{\sigma}$ and $\bar{\epsilon}$ denote representative stress and strain, respectively.

Now, for the spherical shell at any current radius r for an element which has become plastic,

$$2\bar{\sigma}^2 = (\sigma_\theta - \sigma_\phi)^2 + (\sigma_\phi - \sigma_r)^2 + (\sigma_r - \sigma_\theta)^2$$

or, $\qquad\qquad \bar{\sigma} = \sigma_\theta - \sigma_r,$

and $\qquad\qquad d\bar{\epsilon} = 2d\epsilon_\theta = 2dr/r.$

And since the loading path is always 'radial', we may integrate the last equation and write

$$\bar{\epsilon} = 2 \ln \frac{r}{r_0}.$$

r_0 is the original radius. Now the equilibrium equation for an element of the shell is,

$$\frac{d\sigma_r}{dr} = \frac{2(\sigma_\theta - \sigma_r)}{r} = \frac{2\bar{\sigma}}{r}$$

$$= \frac{2[Y + P\bar{\epsilon}^n]}{r}$$

Hence, $\qquad \displaystyle\int_{-p}^{-\sigma_r} d\sigma_r = 2Y \int_a^c \frac{dr}{r} + 2P \int_a^c \frac{(2\ln - r/r_0)^n \, dr}{r}$ (9.35)

For real materials $n < 1$ and thus for a given value of P the strain-hardening effect—evident as the second term in equation (9.35)—will be most pronounced if we take $n = 1$. Also $-\sigma_r$, the upper limit to the integral on the left is such that yield is just occurring at c (or the sphere is elastic for $r > c$) so that from equation (9.16), $-\sigma_r = 2Y(1 - c^3/b^3)$.

Equation (9.35) becomes then,

$$\frac{p}{2Y} = \left(1 - \frac{c^3}{b^3}\right) + \ln \frac{c}{a} + \frac{2P}{Y} \int_a^c \frac{\ln r/r_0}{r} dr$$ (9.36)

Now, $\qquad\qquad r_0^3 = r^3 - a^3 - a_0^3$

and thus, $\qquad \displaystyle\int_a^c \frac{\ln r/r_0}{r} dr = \frac{1}{3} \int_a^c \ln \frac{r^3}{r^3 - a^3 + a_0^3} \frac{dr}{r}$ (9.37)

If a is specified, $a_0 < a < a^*$, and the integral to limit c may be found.

Write x for r/a and consider the case of the complete plastic yielding of a spherical shell of infinite radius ratio; by comparison with a, a_0 is negligible and hence (9.37) becomes

$$\frac{1}{3} \int_1^{c/a} \ln \frac{x^3}{x^3 - 1} \frac{dx}{x} = -\frac{1}{3} \int_1^{m^*} \left[\ln\left(1 - \frac{1}{x^3}\right)\right] \frac{dx}{x}$$

$$= -\frac{1}{3} \int_1^{m^*} \left(\sum \frac{1}{nx^{3n+1}}\right) dx,$$ (9.38)

expanding $\ln\left(1 - \dfrac{1}{x^3}\right)$ in terms of x.

Thus equation (9.38) becomes, on inserting the limits and observing that the insertion of m^* for x will be a negligibly small quantity,

$$\frac{1}{9} \sum_{1}^{n=\infty} \frac{1}{n^2} = \frac{\pi^2}{9.6}.$$

Reverting to (9.36),

$$\frac{p^*}{2Y} = \ln \frac{b^*}{a^*} + \frac{2P}{9Y} \frac{\pi^2}{6}$$

or

$$p^* = \frac{2}{3} Y \ln \frac{2E}{3Y} + \frac{2\pi^2}{27} P,$$

substituting for b^*/a^* using (9.33).

The internal pressure needed to expand an infinitely small cavity in a large spherical shell, is then as in sub-section 9.2.8.

$$\frac{p^*}{2Y} = \frac{1}{3} + \frac{1}{3} \ln \frac{2E}{3Y} + \frac{\pi^2}{27} \cdot \frac{P}{Y} \tag{9.39}$$

For a typical value of $E/Y = 1/1000$ and $P/Y = 1/3$,

$$p^* = 5 \cdot 2 \ Y$$

CHADWICK (1959) discusses the distribution of residual stresses in this latter shell when it is unloaded and demonstrates that the cavity radius is only reduced to a very small extent.

9.2.10 *The Metallic Spherical Shell: Other References*

HWANG (1960) has examined the case of the sphere of elastic- work-hardening material subject to transient thermal loading. An alternative approach to that given above in which the complete elastic-plastic stress-history of a structure (the spherical shell) is determined for design purposes, is one which makes use of a bounding technique; the method is discussed by LECKIE, PAYNE and PENNY (1966). Other useful and interesting references are HOPKINS (1960), CHADWICK (1963) KRAUSE and SHAFFER (1962), and PHILLIPS and YILDIZ (1962).

The plastic design of spherical and other pressure vessels when fitted with nozzles is discussed at length in the book by GILL et al. (1970).

9.2.11 *Stress in a Sphere Due to a Uniform Heat Source*, Q *per Unit Volume*

In a spherical moderator or reflector the rate of generation of heat is a fixed amount per unit volume, Q. Thus if in a solid sphere the temperature on the outside surface of radius a is zero, the temperature T at radius r is given by

$$- 4 \pi r^2 k \frac{dT}{dr} = \frac{4}{3} \pi r^3 \ Q;$$

therefore,

$$T = Q(a^2 - r^2)/6k,$$

where k is the thermal conductivity. Substituting this expression for T in equations (9.7), it is found that

$$\sigma_r = A(r^2 - a^2)$$

and
$$\sigma_\theta = A(2r^2 - a^2),$$

where
$$A = \alpha EQ/15k(1 - \nu).$$

Hence, $\sigma_\theta - \sigma_r = Ar^2$ and the critical radius or size c, when yielding starts in the outside surface is

$$c^2 = \frac{15\,Yk\,(1 - \nu)}{EQ}.$$

Spheres whose radius a exceeds c will consist of a central solid sphere of radius c elastically stressed, and an outer plastic spherical shell of thickness $(a - c)$. Thus we have

$$\sigma_r = 2\,Y \ln r/a, \quad r > c$$

and
$$\sigma_r = A(r^2 - c^2) + 2\,Y \ln c/a, \quad r < c.$$

Also,
$$\sigma_\theta = Y(1 + 2 \ln r/a), \quad r > c$$

and
$$\sigma_\theta = A(2r^2 - c^2) + 2\,Y \ln c/a, \quad r < c.$$

Further details concerning this and other similar cases will be found in the book by GLASSTONE (1956).

9.3 Thick Circular Cylinders: Plane Strain

9.3.1 *Elastic Stress Distribution: General Equations*

The notation to be used will be obvious from Fig. 9.10; in particular the axis of the cylinder is taken as the z-axis.

For radial force equilibrium, the equation is

$$\frac{d\sigma_r}{dr} + \frac{\sigma_r - \sigma_\theta}{r} = 0 \tag{9.40}$$

The elastic strains at a point in the cylinder wall, e_r, e_θ and e_z, are given by

$$\left. \begin{aligned} E(e_r - \alpha T) &= \sigma_r - \nu(\sigma_\theta + \sigma_z) \\ E(e_\theta - \alpha T) &= \sigma_\theta - \nu(\sigma_r + \sigma_z) \\ \text{and } E(e_z - \alpha T) &= \sigma_z - \nu(\sigma_r + \sigma_\theta). \end{aligned} \right\} \tag{9.41}$$

T and α have the same meanings as before.

The cases we shall consider are those in which it may be assumed that the axial displacement of each section is given by $w = 0$, i.e., $e_z = 0$, which is the case of plane strain.

Then
$$\sigma_z = \nu(\sigma_r + \sigma_\theta) - \alpha ET$$

and thus from equation (9.41),

$$e_r = (1 + \nu)\alpha T + \left(\frac{1 - \nu^2}{E}\right)\left(\sigma_r - \frac{\nu}{1 - \nu}\sigma_\theta\right)$$

and
$$e_\theta = (1 + v)\alpha T + \frac{(1 - v^2)}{E}\left(\sigma_\theta - \frac{v}{1 - v}\sigma_r\right).$$

Hence from equation (9.40),

$$\sigma_r - \sigma_\theta = E\frac{(e_r - e_\theta)}{1 + v}$$

and
$$\sigma_r = \frac{E(1 - v)}{(1 - 2v)(1 + v)}\left[\left(e_r + \frac{v}{1 - r}e_\theta\right) - \alpha T\left(\frac{1 + v}{1 - v}\right)\right] \qquad (9.42)$$

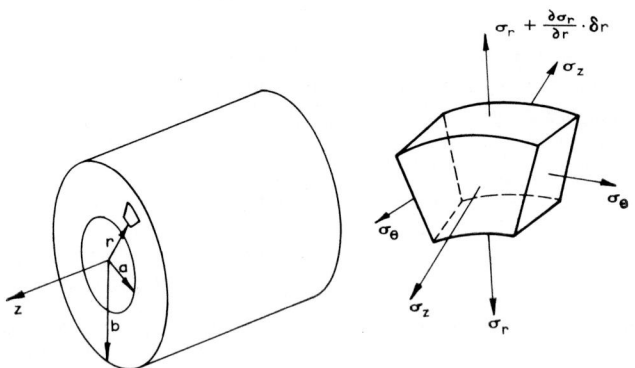

FIG. 9.10 Stresses acting on an element at radius r, taken from wall of circular cylinder.

Substituting for $(\sigma_r - \sigma_\theta)$ and σ_r in the equilibrium equation (9.40) from equation (9.42)

$$\left(\frac{1 - v}{1 - 2v}\right)\left[\frac{de_r}{dr} + \frac{v}{1 - v}\frac{de_\theta}{dr} - \frac{1 + v}{1 - v}\alpha\frac{dT}{dr}\right] + \frac{e_r - e_\theta}{r} = 0. \qquad (9.43)$$

Noting that $e_r = du/dr$ and $e_\theta = u/r$ equation (9.43) reduces to

$$\frac{d^2u}{dr^2} + \frac{1}{r}\frac{du}{dr} - \frac{u}{r^2} = \frac{1 + v}{1 - v}\alpha\frac{dT}{dr}$$

or
$$\frac{d}{dr}\left[\frac{1}{r}\frac{d(ru)}{dr}\right] = \frac{1 + v}{1 - v}\alpha\frac{dT}{dr}.$$

Integrating,

$$\frac{1}{r}\frac{d(ru)}{dr} = \frac{u}{r} + \frac{du}{dr} = \frac{1 + v}{1 - v}\alpha T + C_1$$

and
$$ru = \frac{1 + v}{1 - v}\alpha\int Tr\,dr + \int C_1 r\,dr,$$

or
$$u = \frac{\alpha}{r}\left(\frac{1 + v}{1 - v}\right)\int_a^r Tr\,dr + \frac{C_1 r}{2} + \frac{C_2}{r}.$$

Thus,
$$e_\theta = \frac{u}{r} = \frac{\alpha}{r^2}\left(\frac{1+v}{1-v}\right)\int_a^r Tr\,dr + \frac{C_1}{2} + \frac{C_2}{r^2}$$

and
$$e_\theta + e_r = \frac{1+v}{1-v}\alpha T + C_1.$$

Substituting in equation (9.42), we obtain,

$$\left.\begin{array}{l}
\sigma_r = \dfrac{E}{1+v}\left[\dfrac{C_1}{2(1-2v)} - \dfrac{C_2}{r^2}\right] - \dfrac{\alpha E}{(1-v)r^2}\int_a^r Tr\,dr \\[3mm]
\sigma_r - \sigma_\theta = \dfrac{E\alpha T}{1-v} - \dfrac{2\alpha E}{r^2(1-v}\int_a^r Tr\,dr - \dfrac{2C_2 E}{r^2(1+v)} \\[3mm]
\sigma_\theta = \dfrac{E}{(1+v)}\left[\dfrac{C_1}{2(1-2v)} + \dfrac{C^2}{r^2}\right] - \dfrac{\alpha ET}{(1-v)} + \dfrac{\alpha E}{r^2(1-v)}\int_a^r Tr\,dr \\[3mm]
\text{and } \sigma_z = \dfrac{vEC_1}{(1+v)(1-2v)} - \dfrac{\alpha ET}{(1-v)}.
\end{array}\right\} \quad (9.44)$$

9.3.2 *Elastic Stress Distribution: Any Steady State Temperature Distribution*

For this case the constants C_1 and C_2 are determined from the condition that σ_r is zero on the inside and outside surfaces, i.e., when $r = a$ and $r = b$. Using equation (9.44)

$$\left[\frac{C_1}{2(1-2v)} - \frac{C_2}{a^2}\right] = \left(\frac{1+v}{1-v}\right)\frac{\alpha}{a^2}\int_a^a Tr\,dr = 0$$

and
$$\left[\frac{C_1}{2(1-2v)} - \frac{C_2}{b^2}\right] = \left(\frac{1+v}{1-v}\right)\frac{\alpha}{b^2}\int_a^b Tr\,dr.$$

Thus
$$\frac{C_1}{2(1-2v)} = \frac{1+v}{1-v}\cdot\frac{\alpha}{b^2-a^2}\int_a^b Tr\,dr \qquad (9.45)$$

and
$$C_2 = \frac{\alpha a^2}{b^2-a^2}\left(\frac{1+v}{1-v}\right)\int_a^b Tr\,dr. \qquad (9.46)$$

Hence,
$$\sigma_r = \frac{\alpha E}{1-v}\cdot\frac{1}{r^2}\left[\frac{r^2-a^2}{b^2-a^2}\int_a^b Tr\,dr - \int_a^r Tr\,dr\right] \qquad (9.47)$$

$$\sigma_r - \sigma_\theta = \frac{\alpha E}{1-v}\cdot\frac{1}{r^2}\left[Tr^2 - \frac{2a^2}{b^2-a^2}\int_a^b Tr\,dr - 2\int_a^r Tr\,dr\right] \qquad (9.48)$$

$$\sigma_\theta = \frac{\alpha E}{1-v}\cdot\frac{1}{r^2}\left[\frac{r^2+a^2}{b^2-a^2}\int_a^b Tr\,dr + \int_a^r Tr\,dr - Tr^2\right] \qquad (9.49)$$

$$\text{and } \sigma_z = \frac{\alpha E}{1-v}\left[\frac{2v}{b^2-a^2}\int_a^b Tr\,dr - T\right]. \qquad (9.50)$$

9.3.3 Elastic Stress Distribution: Steady State Heat Flow: Constant Temperature Difference Between the Cylinder Walls Only

If the temperature at $r = a$ is T_i and at $r = b$ is T_0, then if there is a steady state heat flow, the temperature T, at any radius r is

$$T = \lambda \ln b/r \quad \text{where} \quad \lambda = \frac{(T_i - T_0)}{\ln b/a}.$$

Substituting this expression for T into equations (9.47), (9.48), (9.49) and (9.50), it is found that since

$$\int_a^r r \ln b/r \, dr = \frac{r^2}{2} \ln \frac{b}{r} - \frac{a^2}{2} \ln \frac{b}{a} + \frac{(r^2 - a^2)}{4}$$

that

$$\sigma_r = \frac{\alpha E \lambda}{2(1 - \nu)} \left[- \ln \frac{b}{r} - \frac{a^2}{(b^2 - a^2)} \left(1 - \frac{b^2}{r^2} \right) \ln \frac{b}{a} \right] \tag{9.51}$$

$$\sigma_\theta = \frac{\alpha E \lambda}{2(1 - \nu)} \left[1 - \ln \frac{b}{r} - \frac{a^2}{b^2 - a^2} \left(1 + \frac{b^2}{r^2} \right) \ln \frac{b}{a} \right] \tag{9.52}$$

$$\sigma_z = \frac{\alpha E \lambda}{2(1 - \nu)} \left[\nu - 2 \ln \frac{b}{r} - \frac{2\nu a^2}{b^2 - a^2} \ln \frac{b}{a} \right] \tag{9.53}$$

$$\sigma_r - \sigma_\theta = \frac{\alpha E \lambda}{2(1 - \nu)} \left[- 1 + \frac{a^2}{(b^2 - a^2)} \cdot \ln \frac{b}{a} \cdot \frac{2b^2}{r^2} \right] \tag{9.54}$$

9.3.4 Elastic Stress Distribution: Internal Pressure Only

With $\sigma_r = - p_i$ at $r = a$ and $\sigma_r = -p_0$ at $r = b$, (9.44) gives with $T = 0$ everywhere,

$$c_1 = \frac{2(1 + \nu)(1 - 2\nu)}{E (b^2 - a^2)} (p_i a^2 - p_0 b^2)$$

$$c_2 = \frac{(1 + \nu)(p_i - p_0) a^2 b^2}{E (b^2 - a^2)}$$

and hence,

$$\sigma_r = \frac{a^2 b^2 (p_0 - p_i)}{b^2 - a^2} \cdot \frac{1}{r^2} + \frac{p_i a^2 - p_0 b^2}{b^2 - a^2}$$

$$\sigma_\theta = - \frac{a^2 b^2 (p_0 - p_i)}{b^2 - a^2} \cdot \frac{1}{r^2} + \frac{p_i a^2 - p_0 b^2}{b^2 - a^2} \tag{9.55}$$

$$\text{and } \sigma_z = \frac{p_i a^2 - p_0 b^2}{b^2 - a^2} \cdot 2\nu.$$

These expressions are derived on the assumption that there is a plane strain situation or the axial strain is everywhere zero. Note that $\sigma_z = (\sigma_r + \sigma_\theta)/2$ for the incompressible elastic solid, i.e., $\nu = \frac{1}{2}$ and hence that the principal stress system at a point then consists of a 'spherical component' or hydrostatic stress system $\frac{1}{2}(\sigma_\theta + \sigma_r)$ together with a deviatoric or pure shear stress system of $(\sigma_\theta - \sigma_r)/2$, $(\sigma_r - \sigma_\theta)/2$, 0. This last statement is also true even when $\nu \neq \frac{1}{2}$, if it is assumed that under load, plane sections of the cylinder remain plane.

9.3.5 *Elastic Stress Distribution: Steady State Temperature Gradient and Internal Pressure*

The expressions for σ_θ, σ_r, and σ_z, in this case are obtained by using the principle of superposition, e.g., σ_θ is the sum of the circumferential stress due separately to pressure and temperature. It simply requires us to add together the corresponding stresses given in sub-sections 9.3.3 and 9.3.4.

9.3.6 *Brittle Failure*

The simplest, but not necessarily the most correct, criterion of brittle failure usually accepted is that one of the principal tensile stresses reaches a certain critical magnitude. Clearly, it is possible for this to be achieved by any of σ_θ, σ_r or σ_z. No general investigation of this point has to date been reported, but in principle for a specific case, the matter is easily decided.

It is possible for the hoop stress at the bore, due purely to internal pressure, to be reduced by thermal stresses, exactly as in the case of the thick spherical shell analysis above. It is not wise, however, to depend upon such 'stress-saving', since the two stresses may not always act simultaneously.

It can be shown that for a given internal pressure and steady-state transfer of heat, there is a value of b/a which gives a minimum value of the greatest principal stress at the bore.

9.3.7 *Elastic–Plastic Tube: Internal Pressure Only*

The theoretical investigation of the elastic-plastic behaviour of thick-walled tubes, subject to various end conditions, has been treated at great length in many papers and books: the subject is thoroughly discussed in the books by HILL (1950) and by PRAGER and HODGE (1951). Accordingly, only the briefest mention will be made of this topic.

Using expressions (9.55) and putting $p_0 = 0$, it is found that

$$\left.\begin{aligned}
\sigma_\theta - \sigma_r &= \frac{p_i}{m^2 - 1}\left(\frac{2b^2}{r^2}\right), \\[2mm]
\sigma_r - \sigma_z &= \frac{p_i}{m^2 - 1} \cdot (1 - 2v) - \frac{p_i}{m^2 - 1} \cdot \frac{b^2}{r^2}, \\[2mm]
\text{and,}\quad \sigma_z - \sigma_\theta &= \frac{p_i}{m^2 - 1} \cdot (2v - 1) - \frac{p_i}{m^2 - 1} \cdot \frac{b^2}{r^2}.
\end{aligned}\right\} \tag{9.56}$$

(a) It is easily shown that $(\sigma_\theta - \sigma_r)$ is the numerically largest of the above expressions and hence using Tresca's yield criterion, that yielding must first occur on the cylinder bore. Thus,

$$\left.\begin{aligned}
Y &= p_i \frac{2m^2}{m^2 - 1} \equiv p_i \frac{2b^2}{b^2 - a^2}. \\[2mm]
\text{or}\quad \frac{p_i}{Y} &= \frac{m^2 - 1}{2m^2}.
\end{aligned}\right\} \tag{9.57}$$

(b) Using equation (9.56) it may be shown that,

$$(\sigma_\theta - \sigma_r)^2 + (\sigma_r - \sigma_z)^2 + (\sigma_z - \sigma_\theta)^2$$

$$= \left(\frac{p_i}{m^2 - 1}\right)^2 \left[\frac{6b^4}{r^4} + 2(1 - 2\nu)^2\right].$$

Thus, applying the Mises criterion, yield first occurs at $r = a$, again on the cylinder bore, and it is found that,

$$\frac{p_i}{Y} = \frac{(m^2 - 1)}{[3m^4 + (1 - 2\nu)^2]^{\frac{1}{2}}}$$ (9.58)

The progress of yield in the cylinder may be followed using equilibrium equation (9.40),

$$\frac{d\sigma_r}{dr} + \frac{\sigma_r - \sigma_\theta}{r} = 0.$$

Introducing the Tresca criterion on the assumption that $(\sigma_\theta - \sigma_r)$ is the greatest principal stress difference—this is true of course for other conditions besides those of plane strain—we have,

$$\frac{d\sigma_r}{dr} - \frac{Y}{r} = 0.$$

Let the internal pressure p_i be large enough to cause plastic yield out to radius c, see Fig. 9.11(a). Then, integrating,

$$\sigma_r = Y \ln r + C.$$

where C is some constant.
Since $\sigma_r = -p_i$ at $r = a$

$$\sigma_r = Y \ln r/a - p_i$$ (9.59)

and in particular $\sigma_c = Y \ln c/a - p_i.$ (9.60)

Zone B is elastic except for yielding at $r = c$ and thus employing equation (9.57).

$$Y = -\sigma_c \frac{2b^2}{b^2 - c^2}.$$ (9.61)

Eliminating σ_c from equations (9.60) and (9.61), since there must obviously be continuity in the radial pressure at $r = c$,

$$\frac{p_i}{Y} = \ln \frac{c}{a} + \frac{1}{2}\left(1 - \frac{c^2}{b^2}\right).$$ (9.62)

Observe that, as in the case of the sphere, the extent of the plastic deformation is obtained without recourse to equations involving strains, i.e. the problem is a statically determinate one, from the plasticity aspect.

9.3.8 *Onset of Ductile Yielding: Internal Pressure and Temperature Gradient*

A very comprehensive analysis concerning the onset of yield for a variety of end conditions for a thick-walled cylinder subject to an internal pressure and a steady state radial temperature distribution has been given by DERRINGTON (1962). The sequence of events seems to follow that outlined in the section dealing with spherical shells, at least in principle, and yielding may commence anywhere in the tube wall.

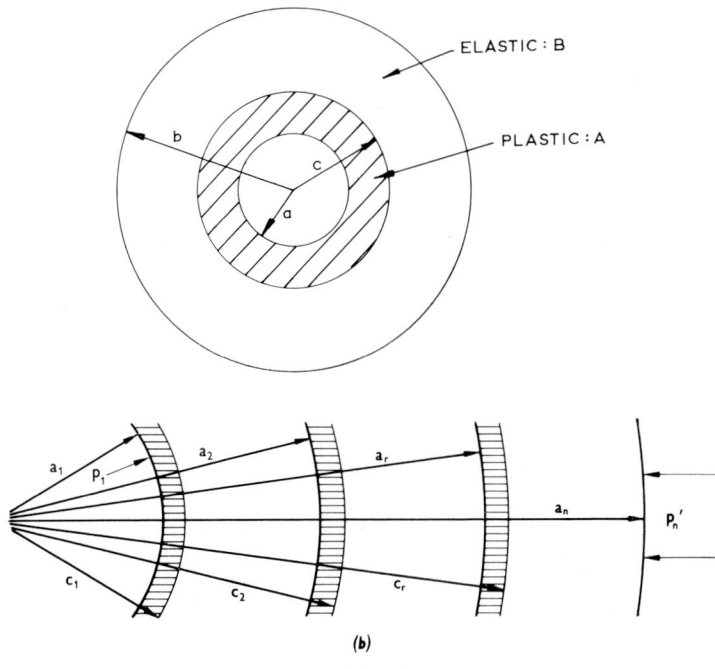

FIG. 9.11

(*a*) Cross-section of thick circular cylinder showing elastic and plastic zones.

(*b*) Compound cylinder in which each cylinder is stressed into the elastic-plastic range.

Derrington has pointed to the very remarkable possibility that if $p_i = E\alpha T/2 (1 - \nu)$, the yield shear stress is uniform through the thickness of the tube and the whole of it may pass instantaneously from the totally elastic to the totally plastic condition; T is the steady state temperature difference between the inner and outer cylindrical surfaces of the tube.

9.3.9 *Determination of the Pressure–Expansion Curve in a Thick-walled Tube which is Closed at its Ends*

Figure 9.12 shows a typical pressure-expansion curve for a thick-walled cylinder. From O to A the cylinder responds in a linear-elastic manner to increase in internal pressure. At A, plastic yield starts and from A to B as the pressure increases, the tube wall becomes progressively more plastic, expands and strain-hardens; the precise shape of the curve depends on the ratio m and on the stress-strain properties of the tube material. At B a point of instability is reached and local bulging begins. The pressure at B is the ultimate pressure for the tube. From B to C, the tube expands and at C rupture occurs.

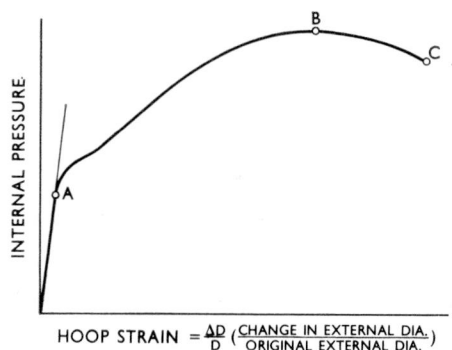

HOOP STRAIN $= \frac{\Delta D}{D}$ (CHANGE IN EXTERNAL DIA. / ORIGINAL EXTERNAL DIA.)

FIG. 9.12 Pressure–expansion curve for thick-walled cylinder.

It is obviously of great importance to engineers that they should be able to calculate the ultimate pressure, p_{ULT}. One approach is to employ the formula given by CROSSLAND and BONES (1958).

$$p_{ULT} = \sigma_{ULT} \frac{2(m - 1)}{m + 1}. \tag{9.63.i}$$

σ_{ULT} is the tensile strength of the tube material and m denotes the *mean* diameter of the cylinder; the authors refer to (9.63.i) as a 'mean diameter formula' and remark that comparison with experimental results shows that it underestimates the ultimate pressure. The underestimate would appear to be less than 10 per cent even for $m = 8.05$. Equation (9.63.i) is similar to that of (10.31), which applies for a thin tube; the expected error in (9.63.i) on a basis of (10.31) is about ± 14 per cent.

CROSSLAND and BONES have also compared the ultimate pressure as calculated following a procedure described by MANNING (1945), which makes use of a shear stress–shear strain curve for the tube material, with experimental results and found very good agreement.

In this particular instance it is assumed that axial strain is negligible and that there is no change in volume so that the state of stress at a point in the tube wall is that of a shear stress $k = (\sigma_\theta - \sigma_r)/2$ plus a hydrostatic stress $\sigma = (\sigma_\theta + \sigma_r)/2$; the axial stress is the mean of the radial and hoop stresses, see Fig. 9.13.

The radial equilibrium equation and the Tresca yield criterion provide the usual equation

$$r \frac{d\sigma_r}{dr} = \sigma_\theta - \sigma_r = 2k,$$

where k is the yield shear stress of the material in plane strain at radius r, which is some function of the shear strain. Hence

$$\int_{-p_i}^{0} d\sigma_r = 2 \int_{a}^{r} \frac{k\,dr}{r}, \tag{9.63.ii}$$

with k some function of r.

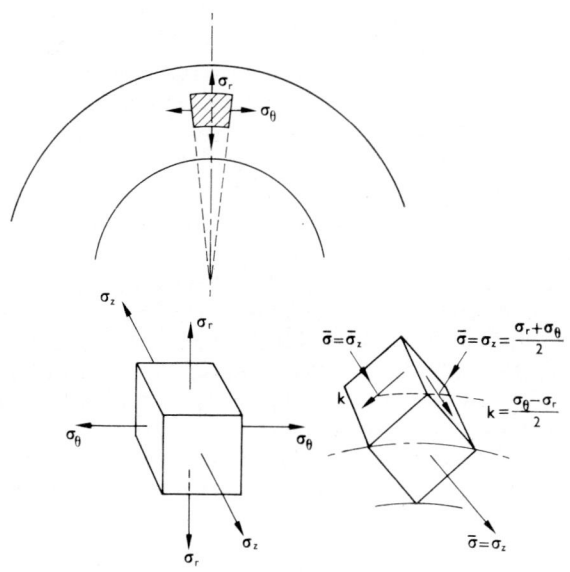

FIG. 9.13 State of stress in thick-walled tube with closed ends.

Letting the alteration in inner and outer radii be u_a and u_b, assuming volumetric constancy of the material and supposing the radial displacement at r is u, then

$$r^2 - a^2 = (r + u)^2 - (a + u_a)^2$$

and

$$r + u = (r^2 + 2au_a + u_a^2)^{\frac{1}{2}}, \tag{9.64.i}$$

i.e., for a chosen value of r, u can be deduced once u_a is measured.

Now

$$d\epsilon_\theta = \frac{d(r + u)}{r + u}$$

therefore

$$\epsilon_\theta = \ln (r + u)/r = -\epsilon_r$$

Hence the engineering shear strain at radius r is $\phi = 2 \ln (r + u)/r$. Assuming that the results of a torsion test on the cylinder material are available, which relates shear stress and strain, i.e., k and ϕ, we find that the shear stress through the cylinder wall can now be obtained for a given displacement of the cylinder bore, u_a. If u_a is given and a is known, for a chosen value of r, using (9.64.i), $(r + u)$, can be found. Hence ϕ can be calculated and the corresponding value of k taken from the given shear stress–shear strain curve for the tube material. The internal pressure p_i to cause displacement u_a is now found numerically or by taking the area under the curve k/r versus r, over the range $(b - a)$, as (9.63.ii) indicates.

When large displacements are used, then in place of equation (9.63) there must be substituted

$$\int d\sigma_r = 2 \int_{a+u_a}^{b+u_b} \frac{k}{(r+u)} d(r+u)$$

MANNING (1945) shows in his original paper how the ultimate pressure-diameter relation for a given tube and material may be derived using this theory. The work of NADAI (1931) and FRANKLIN and MORRISON (1960) should also be consulted.

9.3.10 *Elastic Thermal Shock*

Consider an unpressurized shell entirely at some arbitrary temperature zero. Suppose it is suddenly subjected to an external temperature T on the outer or inner surface at time $t = 0$. 'Instantaneously' hoop stresses will be induced at these surfaces; the radial stresses will obviously be zero. (Strictly, inertia forces will be created which will give rise to stress waves, but these are neglected, CRISTESCU (1967).) Let the response of the shell be elastic.

Consider a very thin annular element in the surface which will have been raised to T; the remainder of the cylinder will still be at zero temperature and, because of this, the heated layer is prevented from altering to its natural unstrained size both radially and axially. Let the surface hoop stress be σ_θ; then

$$e_\theta = \frac{\sigma_\theta - \nu\sigma_z}{E} = \alpha T, \qquad (9.64.\text{ii})$$

with a similar expression for e_z and thus $\sigma_\theta = E\alpha T/(1 - \nu) = \sigma_z$.

If some initial temperature gradient and/or pressure is present, σ_θ would be added to that caused by them. For mild steel $\sigma_\theta/T = 320$ lbf/in² per °F.

Expression (9.64.ii) applies also for a spherical shell.

Further results on transient thermal stress distribution in cylinders will be found in an article by STRUB (1961).

9.3.11 *Special References*

Many long papers have appeared in the last ten years or so on this subject, and space only permits us to refer to them briefly here. Summaries taken from some are quoted to indicate their scope and content.

BLAND's work (1956) on pressure and temperature effects is especially noteworthy.

'Using Tresca's yield criterion and its associated flow rule solutions are obtained for the stresses, the elastic and plastic strains and the displacements when a thick-walled tube of work-hardening material is subject to internal and external pressures and its surfaces are maintained at different temperatures. In general, a numerical integration is necessary, but the solutions can be expressed explicitly when the work-hardening law is linear. On removal of the pressures and the temperature difference, plastic flow may or may not occur. In both cases the residual stresses can be calculated. Specific examples are given.'

Five papers by WHALLEY, *et al.* (1956 and 1960) deal with the design of pressure vessels subjected to thermal stress.

A survey and source of historical references concerning the design of high pressure cylinders will be found in a paper by WILSON and SKELTON (1967). This paper among other things reviews, autofrettage cylinder design, compound cylinder design, the use of tungsten carbide liners, wire and strip wound cylinders, sectored vessel design and cascaded vessel design. Another very useful work with experimental results is that of PARSONS and COLE (1967) on the optimum design of short composite cylinders. Two very useful sources of reference are to be found in the Conferences,

(i) Thermal Loading and Creep in Structures and Components, 1964 and
(ii) High Pressure Engineering, 1967.

The books edited by PUGH (1970), (1971) should be consulted, see references on p. 37.

Some other papers on high pressure cylinders for facilitating extrusion are LENGYEL, BURNS and PRASAD, (1966), SUMNER and MEREDITH, (1966), and FIORENTINO, VAGINS, SABROFF and BOULGER (1966).

Further useful references are, PRZEMIENIECKI (1960), SCHMIDT and SONNEMAN (1960), RICHARDS (1965), SCHWIEBERT, (1965), MARCAL, (1965), KRAUS, (1962), BHARGAVA and SHARMA, (1963), BERMAN, (1960), KAMMASH, MURCH and NAGHDI, (1960) and HORNE, (1955).

STEELE and EICHBERGER (1956) reported work concerning the non-homogeneous yielding of mild steel cylinders. By etching, they showed how yield starts in two or three Lüder lines and propagates through the wall thickness along spirals of maximum shear stress, see Fig. 9.14. The number of these lines increases with increasing pressure. The effect of non-homogeneous yielding does not appear to have been considered theoretically. See problem 39.

9.4 Compound Circular Cylinders and Spherical Shells

9.4.1 *The Compound Cylinder Subjected to Internal Pressure Only*

Figure 9.10(*b*) shows a compound cylinder consisting of *n* cylinders of the same material each of which is stressed into the elastic-plastic range due to an internal pressure. For each cylinder we have, if we assume plane strain conditions to apply at all times,

$$
\left.
\begin{aligned}
p_1 - p_1' &= 2k \ln \frac{c_1}{a_1} + k \left(1 - \frac{c_1^2}{a_2^2} \right) \\[6pt]
p_r - p_r' &= 2k \ln \frac{c_r}{a_r} + k \left(1 - \frac{c_r^2}{a_{r+1}^2} \right) \\[6pt]
&\cdots\cdots\cdots\cdots\cdots\cdots\cdots \\[6pt]
p_n - p_n' &= 2k \ln \frac{c_n}{a_n} + k \left(1 - \frac{c_n^2}{a_{n+1}^2} \right)
\end{aligned}
\right\}
\qquad (9.65.\text{i})
$$

and

where c_r is the radius of the elastic-plastic interface in the rth cylinder, a_r its internal radius and a_{r+1} its external radius, the internal and external pressure being p_r and p_r', respectively.

Summing the n equations of (9.65.i) and because $p_r = p_{r-1}'$

$$p_1 - p_n' = 2k \ln \frac{c_1}{a_1} + \ldots + 2k \ln \frac{c_n}{a_n}$$

$$+ k \left(1 - \frac{c_1^2}{a_2^2}\right) + \ldots + k \left(1 - \frac{c_n^2}{a_{n+1}^2}\right). \qquad (9.65.\text{ii})$$

(a)

Fig. 9.14

(a) Photograph of full yield Lüder's line patterns on the top face of mild steel cylinder—compare with (b).

Suppose that plastic yielding has proceeded to the same degree through each cylinder wall, i.e., $(c_r - a_r)/(a_{r+1} - a_r) = c$ and we wish to find the individual cylinder radii (or radii ratio) for a given p_1, p_n' and a_{n+1} in order to maximize $(p_1 - p_n')$.

Then from equation (9.65.ii),

$$p_1 - p_n' = \sum_1^n 2k \ln \frac{c_r}{a_r} + \sum_1^n k \left[1 - \left(\frac{c_r}{a_r}\right)^2 \cdot \frac{1}{m_r^2}\right]$$

$$= \sum_1^n 2k \ln \left[c(m_r - 1) + 1\right] + \sum_1^n k \left\{1 - \left[\frac{c(m_r - 1) + 1}{m_r}\right]^2\right\}. \qquad (9.65.\text{iii})$$

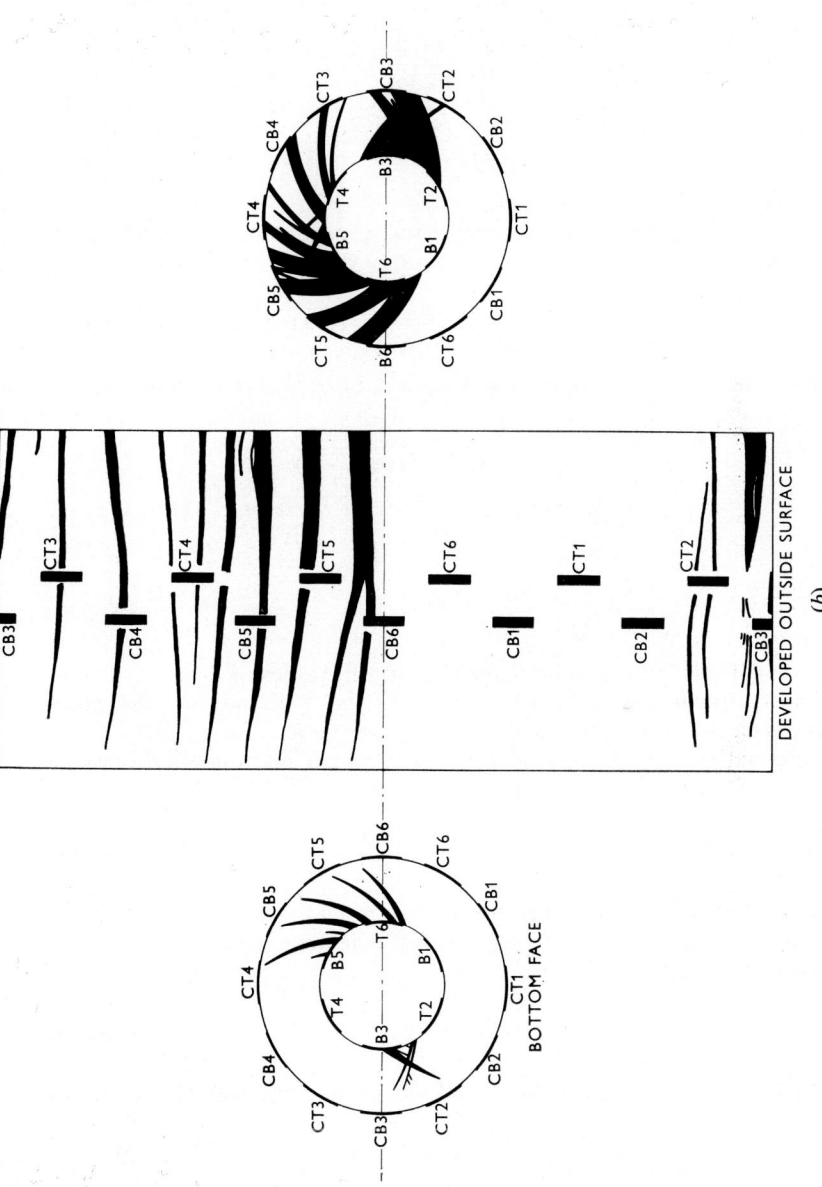

Fig. 9.14

(b) Lüder's line patterns on end faces and outside diameter at full yield pressure (17,310 lbf/in²) for a cylinder of mild steel. (*After* Steele and Eichberger, *Trans, A.S.M.E.*)

Hence,

$$\frac{1}{2k}\frac{\partial}{\partial m_r}(p_1 - p_n') = \frac{c}{c(m_r - 1) + 1} + \frac{c}{c(m_n - 1) + 1}\cdot\frac{\partial m_n}{\partial m_r}$$

$$+ \frac{c(m_r - 1) + 1}{m_r}\cdot\frac{1 - c}{m_r^2}$$

$$+ \frac{c(m_n - 1) + 1}{m_n}\cdot\frac{1 - c}{m_n^2}\cdot\frac{\partial m_n}{\partial m_r} = 0, \qquad (9.65.\text{iv})$$

for a maximum.

The n terms within each of the summation terms in equation (9.65.iii) are not independent. We have

$$m_1.m_2 \ldots m_r \ldots m_n = \frac{a_{n+1}}{a_1} = \text{constant}, K \qquad (9.65.\text{v})$$

and hence $\partial m_n/\partial m_r = -m_n/m_r$. Thus equation (9.65.iv) reduces to

$$\frac{cm_r}{c(m_r - 1) + 1} + \frac{c(m_r - 1) + 1}{m_r}\cdot\frac{1 - c}{m_r} = \frac{c\, m_n}{c(m_n - 1) + 1}$$

$$+ \frac{c(m_n - 1) + 1}{m_n}\cdot\frac{1 - c}{m_n}$$

or

$$\frac{cm_r}{c(m_r - 1)} + \frac{c(m_r - 1) + 1}{m_r}\cdot\frac{1 - c}{m_r} = \lambda \qquad (9.65.\text{vi})$$

which is a constant of the same magnitude for all values of r.

However, from equation (9.65.v) $\Pi m_r = a_{n+1}/a_1 = K$ and thus, the radius ratio for each cylinder is the same and $(a_{n+1}/a_1)^{1/n} = K$.

From equation (9.65.iii) the non-dimensional maximum internal pressure p if p_n', the external pressure, is zero is given by

$$\frac{p}{k} = n\left\{2\ln\left[c(K^{1/n} - 1) + 1\right] + 1 - \left[\frac{c(K^{1/n} - 1) + 1}{K^{1/n}}\right]^2\right\}. \qquad (9.65.\text{vii})$$

This result has several times previously been given for the special case when $c = 0$, i.e., for purely elastic stresses and with yielding just about to start at the inside bore of each cylinder as

$$\frac{p}{k} = n\left[1 - \frac{1}{K^{2/n}}\right]. \qquad (9.65.\text{viii})$$

When $c = 1$, i.e., each cylinder is fully plastic, the well-known result

$$\frac{p}{k} = \ln K, \qquad (9.65.\text{ix})$$

is found.

Strictly, equation (9.65.vii) only applies for the radius ratio in the cylinder shell when under stress, but for all practical purposes this is not significantly different from that which prevails when the cylinder is unstressed.

The interesting result we have derived applies to a nested set of circular cylinders originally unstressed, subject to a given internal pressure which renders the system partly plastic. If the overall internal and external radii of the group is fixed as, say, a_1 and a_n and it is specified that each and every cylinder must yield through the same fraction of wall thickness, then the compound set will sustain the greatest internal pressure, if there is a *geometrical similarity* as between the different cylinders, i.e.,

$$a_{r+1}/a_r = (a_{n+1}/a_1)^{1/n}.$$

9.4.2. *Spherical Shells*

It is easy to see that for an internally pressurized nested set of *spherical shells* (a compound sphere) a similar result will hold.

Further, it can be appreciated intuitively that for compound cylinders or spheres of identical work-hardening material the same condition or conceptions concerning geometrical similarity will apply, for units to sustain a maximum load. The results given in this sub-section are taken from a paper by JOHNSON, MALHERBE and VENTER (1970).

9.5 **Rings Subject to a Steady State Radial Temperature Gradient Only**

The results quoted in this section are taken from a paper by WILHOIT (1958).

The case investigated is that of the elastic-plastic states of stress which arise in a ring subject to a steady state radial temperature distribution, the temperature difference between the inner and outer radii being $|T|$. Essentially, because the ring is thin, it can be taken to be in a state of plane stress, i.e., $\sigma_z = 0$.

The assumptions, as regards α, T, etc., of the analysis made by WILHOIT, are the same as those made in previous sections. The conclusions reached by WILHOIT, and given below, are remarkably similar to those drawn for thick spheres. They are:

(a) A thin ring subjected to a temperature difference $|T|$ begins to yield at the inner surface and this zone of yielding moves outwards with increasing $|T|$.

(b) At a sufficiently large value of $|T|$ a second plastic zone begins to form at the outer surface and moves inwards with increasing $|T|$.

(c) The larger the ratio of outer to inner radius, the lower the $|T|$ for yielding at the inner surface and the higher the value of $|T|$ for yielding at the outer surface.

(d) The ring, theoretically, only becomes fully plastic as $|T| \to \infty$.

The Table below gives the values of $|T|$ required for the existence of two plastic regions:

$\alpha E|T|/Y$	2·25	3·5	
b/a	1·50	4·0	
Killed carbon steel	285	440	Temperature $|T|$
18–8 Cr-Ni Steel	225	350	in °F
Inconel X	750	1170	

9.6 Rotating Discs

In this section a straightforward investigation is made of the distribution of elastic and elastic-plastic stresses in a uniformly thin, flat circular disc rotating with uniform angular speed about an axis through its centroid. For the treatment of circularly symmetric discs of non-uniform thickness, reference should be made to the papers by HEYMAN (1958) and KOBAYASHI and TRUMPLER (1960).

9.6.1. *Elastic Analysis*

The radial equilibrium equation for the forces on an element of a spinning disc, see Fig. 9.15, is

$$r \frac{d\sigma_r}{dr} = (\sigma_\theta - \sigma_r) - \rho\omega^2 r^2 \tag{9.66}$$

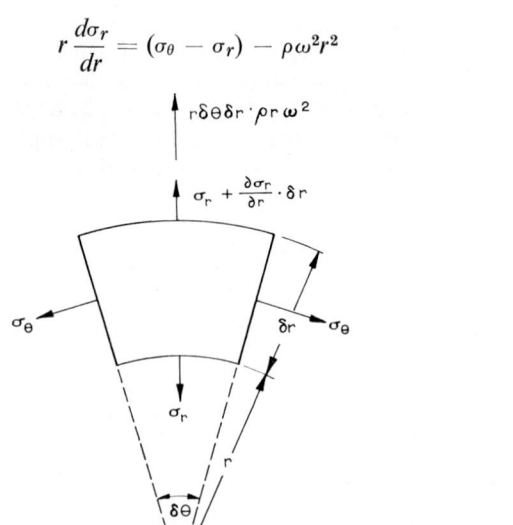

FIG. 9.15 Stresses on an element of a rotating disc.

where ω is the uniform angular speed of the disc and ρ the mass per unit volume. The principal elastic-strain components are

$$e_\theta = u/r \quad \text{and} \quad e_r = du/dr.$$

Hooke's law provides equations

$$Ee_\theta = \sigma_\theta - \nu\sigma_r \quad \text{and} \quad Ee_r = \sigma_r - \nu\sigma_\theta. \tag{9.67}$$

σ_z, the axial stress, regarded as constant through the disc thickness, is zero. Hence from equation (9.67)

$$\sigma_\theta = \frac{E}{1 - \nu^2}(e_\theta + \nu e_r) = \frac{E}{1 - \nu^2}\left(\frac{u}{r} + \nu\frac{du}{dr}\right)$$

and

$$\sigma_r = \frac{E}{1 - \nu^2}(\nu e_\theta + e_r) = \frac{E}{1 - \nu^2}\left(\nu\frac{u}{r} + \frac{du}{dr}\right). \tag{9.68}$$

Substituting from equation (9.68) for σ_θ and σ_r in equation (9.66) and simplifying, it is found that

$$r^2\frac{d^2u}{dr^2} + r\frac{du}{dr} - u + Xr^3 = 0 \qquad (9.69)$$

where $X = \rho\omega^2(1 - v^2)/E$.

Equation (9.69) may be solved by introducing a stress function ψ such that

$$\psi = r\sigma_r \quad \text{and} \quad d\psi/dr = \sigma_\theta - Xr^2$$

or more directly by verifying that

$$u = Ar - \frac{X}{8}r^3 + \frac{C}{r} \qquad (9.70)$$

satisfies equation (9.69) where A and C are constants that are required to fulfil various boundary conditions. Putting this expression for u into that for σ_r in equation (9.68)

$$\sigma_r = \frac{E}{(1 - v^2)}\left[A(1 + v) - \frac{X}{8}r^2(3 + v) + (v - 1)\frac{C}{r^2}\right] \qquad (9.71)$$

Assume that there is no radial stress on the inner and outer surfaces of the disc, i.e., $\sigma_r = 0$ at $r = a$ and b. These conditions determine

$$\left.\begin{array}{l} A = \dfrac{3 + v}{8(1 + v)}(b^2 + a^2)X \\[3mm] \text{and} \qquad C = \dfrac{(3 + v)a^2b^2}{8(1 - v)}X. \end{array}\right\} \qquad (9.72)$$

Thus for an annular disc, we have

$$u = \frac{(3 + v)\rho\omega^2(1 - v^2)r}{8E}\left[a^2 + b^2 - \left(\frac{1 + v}{3 + v}\right)r^2 + \left(\frac{1 + v}{1 - v}\right)\frac{a^2b^2}{r^2}\right]$$

$$\sigma_\theta = \frac{(3 + v)\rho\omega^2}{8}\left[a^2 + b^2 - \left(\frac{1 + 3v}{3 + v}\right)r^2 + \frac{a^2b^2}{r^2}\right] \qquad (9.73)$$

$$\sigma_r = \frac{(3 + v)\rho\omega^2}{8}\left[a^2 + b^2 - r^2 - \frac{a^2b^2}{r^2}\right].$$

If a solid disc is considered, then C in equation (9.70) must be taken as zero, otherwise u cannot be zero at $r = 0$. Thus for $\sigma_r = 0$ at $r = b$, equation (9.71) gives

$$A = \frac{Xb^2}{8}\left(\frac{3 + v}{1 + v}\right) \qquad (9.74)$$

and hence

$$u = \frac{\rho\omega^2(1 - v^2)}{8E} r \left(\frac{3 + v}{1 + v} b^2 - r^2 \right)$$

$$\sigma_r = \frac{\rho\omega^2}{8}(3 + v)(b^2 - r^2)$$

$$\sigma_\theta = \frac{\rho\omega^2(3 + v)}{8} \left(b^2 - \frac{r^2(1 + 3v)}{(3 + v)} \right). \tag{9.75}$$

Note that σ_r and σ_θ are greatest, and equal, when $r = 0$.

i.e. $$\sigma_{r\max} = \sigma_{\theta\max} = \frac{\rho\omega^2(3 + v)b^2}{8}. \tag{9.76}$$

For the annular disc, σ_r is greatest when $r = \sqrt{ab}$ and thus

$$\sigma_{r\max} = \frac{3 + v}{8} \rho\omega^2 (b - a)^2. \tag{9.77}$$

σ_θ is greatest when r is least, i.e. at $r = a$ and then

$$\sigma_{\theta\max} = \frac{3 + v}{4} \cdot \rho\omega^2 \cdot \left(b^2 + \frac{1 - v}{3 + v}a^2 \right). \tag{9.78}$$

It is obvious for both solid and annular discs that σ_θ is always larger than σ_r and thus that

$$\sigma_\theta \geqslant \sigma_r \geqslant \sigma_z = 0.$$

Applying the Tresca criterion, yield starts where σ_θ is greatest and when it equals Y. This occurs at the inner surface and thus for $\sigma_{\theta\max}$ on the left-hand side of equation (9.75), Y is substituted to find the corresponding critical value of ω_1.

For the annular disc, when

$$v = \frac{1}{3}, \qquad \frac{Y}{\rho\omega_1^2} = \left(\frac{5b^2 + a^2}{6} \right).$$

For a solid disc, the critical angular speed for the onset of yield is

$$\omega_1 = \frac{1}{b}\sqrt{\frac{Y}{\rho} \cdot \frac{8}{3 + v}} = \frac{1 \cdot 55}{b}\sqrt{\frac{Y}{\rho}}.$$

As the speed further increases, the disc becomes partially plastic and in discussing this case, the analysis is simplified by assuming that $v = 1/2$ in both the elastic and plastic regions. For such an incompressible material, initial yielding will occur when

$$\omega_1 = \frac{1 \cdot 51}{b}\sqrt{\frac{Y}{\rho}}.$$

9.6.2 *Elastic–Plastic Analysis*

For a velocity $\omega > \omega_1$, the disc will become plastic out to some radius c. The disc now consists of an inner plastic region and an outer elastic one.

In the plastic zone, in the radial equilibrium equation, we proceed by substituting Y for σ_θ; thus equation (9.66) becomes

$$r\frac{d\sigma_r}{dr} = Y - \sigma_r - \rho\omega^2 r^2 \tag{9.79}$$

hence,

$$\frac{d}{dr}(r\sigma_r) = Y - \rho\omega^2 r^2$$

and

$$r\sigma_r = Yr - \frac{\rho\omega^2 r^3}{3} + C. \tag{9.80}$$

(a) Annular Disc

C in equation (9.80) is determined from the condition $\sigma_r = 0$ at $r = a$, and thus

$$r\sigma_r = \frac{\rho\omega^2}{3}(a^3 - r^3) + Y(r - a). \tag{9.81}$$

Supposing the inner plastic zone to extend to $r = c$, we have for $\sigma_{r=c}$ or σ_c

$$\sigma_c = \frac{\rho\omega^2}{3}\left(\frac{a^3 - c^3}{c}\right) + \frac{Y(c - a)}{c}. \tag{9.82}$$

Now σ_c must also be the radial stress at the inner surface of the outer elastic section of the disc; this outer zone must in effect be treated as an elastic rotating disc of outer radius b and inner radius c in which yield is just commencing at the inner radius. Further, there is a radial pressure at c of amount given by equation (9.82). To find the stress distribution in this outer zone, the elastic disc analysis and results must be re-determined with this new boundary condition. The constants A and C of equation (9.70) will be found to be

$$A = -\frac{\sigma_c c^2(1-v)}{E(b^2-c^2)} + \frac{X}{8}\frac{3+v}{1+v}(b^2+c^2); \quad C = -\frac{\sigma_c b^2 c^2(1+v)}{E(b^2-c^2)} + \frac{X}{8}\frac{3+v}{1-v}b^2 c^2. \tag{9.83}$$

(With $\sigma_c = 0$, A and C reduce of course to the expressions in equation (9.72).) Now insert these values of A and C into the expression for the displacement equation (9.70) and using equation (9.68), obtain

$$\sigma_r = \sigma_c \frac{c^2}{b^2 - c^2}\left(-1 + \frac{b^2}{r^2}\right) + \frac{\rho\omega^2}{8}(3 + v)\left(b^2 + c^2 - \frac{b^2 c^2}{r^2} - r^2\right)$$

and

$$\sigma_\theta = -\sigma_c \frac{c^2}{b^2 - c^2}\left(1 + \frac{b^2}{r^2}\right) + \frac{\rho\omega^2}{8}(3 + v)\left(b^2 + c^2 - \frac{1+3v}{3+v}r^2 + \frac{b^2 c^2}{r^2}\right).$$

Substituting in the above equation, Y for σ_θ and for σ_c from equation (9.82), an expression for ω which causes yielding out to radius c is found; this is, where $y = c/b$ and $m = a/b$

$$\frac{\rho\omega^2}{Y/b^2} = \frac{1 + [(1+y^2)/(1-y^2)].[(y-m)/y]}{[(3+v)/4].[1+y^2(1-v)/(3+v)] + [(1+y^2)/(1-y^2)].[(y^3 - m^3)/3m]}.$$

(b) Solid Disc

$r = 0$ implies that constant C in equation (9.80) must be zero. Hence in the plastic zone

$$\sigma_r = Y - \frac{\rho\omega^2 r^2}{3}, \qquad \sigma_\theta = Y.$$

When the plastic zone has extended to radius c, $\sigma_c = Y - \rho\omega^2 c^2/3$. For the outer elastic zone—an annular disc yielding at the inner radius c—where the radial stress is σ_c, it follows, by proceeding as described in the previous section for the annular disc, that the speed to cause yielding to radius c, is given by

$$\omega = \frac{1}{bM}\sqrt{\frac{Y}{\rho}}.$$

where, with $y = c/b$,

$$M^2 = \frac{8 + (1 + 3v)(y^2 - 1)^2}{24}. \tag{9.84}$$

Thus having found the value of c for a given value of ω, it is not difficult to substitute in the various equations above and so find the complete expressions for σ_r and σ_θ in the elastic regions. These are

$$\left. \begin{aligned} \sigma_\theta &= \frac{Y}{24M^2}\left[\frac{c^4}{b^4}\left(1 + \frac{b^2}{r^2}\right)(1 + 3v) + 3(3 + v) - 3(1 + 3v)\frac{r^2}{b^2}\right] \\ \sigma_r &= \frac{Y}{24M^2}\left[3(3 + v) - (1 + 3v)\frac{c^4}{r^2 b^2}\right]\left(1 - \frac{r^2}{b^2}\right). \end{aligned} \right\} \tag{9.85}$$

The value of ω, say ω_2, at which the whole disc becomes plastic is

$$\omega_2 = \frac{1}{b\sqrt{\frac{1}{3}}}\sqrt{\frac{Y}{\rho}} .$$

Hence, $\omega_1/\omega_2 = \sqrt{8/3(3 + v)}$

and with $v = \frac{1}{3}, \omega_1/\omega_2 = 0\cdot89.$ $\left. \right\} \tag{9.86}$

A very useful paper is that by WALDREN, PERCY and MELLOR, (1965–66) which discusses the bursting strength of rotating discs and compares experimental results against certain theoretical predictions, see Chapter 10.

9.7 Elastic–Plastic Analysis of a Rotating Cylinder

This problem has been intensively analysed by HODGE and BALABAN (1962) and references to previous work on this topic will be found therein.

See Problems 36-39

REFERENCES

BERMAN, I.   1960 'Expansion of Thick Walled Cylinders
                                   Fabricated from Cold Bent Plates'
                                   J. appl. Mech. **47**, 505
BHARGAVA, R. D. and   1964 'Elastic-Plastic Medium Containing a
SHARMA, C. B.   Cylindrical Cavity Under Uniform Internal
                                   Pressure'
                                   J. Franklin Inst. **277**, 422
BIOT, M. A.   1958 'Linear Thermodynamics and the Mechanics
                                   of Solids'
                                   Proc. 3rd U.S. Cong. appl. Mech.
BLAND, D. R.   1956 'Elasto-Plastic Thick-walled Tubes of
                                   Work-hardening Material Subject to Internal
                                   and External Pressures and to Temperature
                                   Gradients'
                                   J. Mech. Phys. Solids **4**, 209
BOLEY, B. A. and   1960 *Theory of Thermal Stresses*
WEINER, J. H.   Wiley, New York
COWPER, G. R.   1958 Tech. Report No. 9, 562(20).
                                   Division of Engineering, Brown University
                             1960 'The Elastoplastic Thick-walled Sphere
                                   Subjected to a Radial Temperature Gradient'
                                   Trans. A.S.M.E., J. appl. Mech. **47**, 496
CHADWICK, P.   1959 'The Quasi-static Expansion of a Spherical
                                   Cavity in Metals and Ideal Soils'
                                   Q. J. Mech. and appl. Math XII, 52
                             1963 'Compression of a Spherical Shell of
                                   Work-hardening Material'
                                   Int. J. mech. Sci. **5**, 65
CRISTESCU, N.   1967 *Dynamic Plasticity*,
                                   North-Holland Publishing Co., Amsterdam,
                                   98pp.
CROSSLAND, B. and   1958 'Behaviour of Thick-walled Steel Cylinders
BONES, J. A.   Subjected to Internal Pressure'
                                   Proc. Instn mech. Engrs, **172**, 777
DERRINGTON, M. G.   1958 'The Onset of Yield in a Thick Spherical
and JOHNSON, W.   Shell Subject to Internal Pressure and a
                                   Uniform Heat Flow'
                                   Appl. Sci. Research Series A, **7**, 408
DERRINGTON, M. G.   1962 'The Onset of Yield in a Thick-walled
                                   Cylinder Subject to a Uniform Internal or
                                   External Pressure and Steady Heat Flow'
                                   Int. J. mech. Sci. **4**, 83
DRABBLE, F. and   1964 'The Development of the Zones of Yielding in
JOHNSON, W.   Thick-walled Spherical Shells of Non-work-
                                   hardening Material Subjected to a Steady

State Radial Temperature Gradient and on Internal or External Pressure' Conf. on Thermal Loading and Creep, Paper 19, Instn. Mech. Engrs.

FIORENTINO, R. J.   1966 'An Extrusion Container for Hydrostatic
VAGINS, M.,   Pressures up to 17 Kilobars'
SABROFF, A. M. and   *Annals of the C.I.R.P.* Vol. XIII, 169
BOULGER, F. W.

FRANKLIN, G. J. and   1960 'Autofrettage of Cylinders: Prediction of
MORRISON, J. L. M.   Pressure/External Expansion Curves and
                                 Calculation of Residual Stresses'
                                 Proc. Instn mech. Engrs **174**, 947

GILL, S. S.   1970 *The Stress Analysis of Pressure Vessels
                                 and Pressure Vessel Components*
                                 Pergamon Press, Oxford.

GLASSTONE, S.   1956 *Nuclear Reactor Engineering.*
                                 Macmillan, London

HEYMAN, J.   1958 'Plastic Design of Rotating Discs'
                                 Proc. Instn mech. Engrs **172**, 531

HILL, R.   1950 *The Mathematical Theory of Plasticity.*
                                 O.U.P.

HODGE, P. G. and   1962 'Elastic-Plastic Analysis of a Rotating Cylinder'
BALABAN, M.   *Int. J. mech. Sci.* **4**, 465

HOPKINS, H. G.   1960 'The Dynamic Expansion of Cavities in Metals'
                                 Progress in Solid Mechanics, North-Holland,
                                 Groningen.

HORNE, M. R.   1955 'The Elastic-plastic Theory of Containers
                                 and Liners for Extrusion Presses'
                                 Proc. Instn mech. Engrs **169**, 107

HWANG, C.   1960 'Thermal Stresses in an Elastic Work-harden-
                                 ing Sphere'
                                 J. appl. Mech. **47**, 629

JOHNSON, W.,   1971 'On Geometrical Similarity in Compound
MALHERBE, de M. C.   Circular Cylinders and Spherical Shells
and VENTER, R.   Under Internal Pressure'
                                 Annals of the C.I.R.P., **19**, 653

JOHNSON, W. and   1962 'Elastic-plastic Behaviour of Thick-walled
MELLOR, P. B.   Spheres of Work-hardening Material Subject
                                 to Steady State Radial Temperature Gradient'
                                 Int. J. mech. Sci. **4**, 147

KAMMASH, T. B.,   1960 'The Elastic-plastic Cylinder Subjected to
MURCH, S. A. and   Radially Distributed Heat Source, Lateral
NAGHDI, P. M.   Pressure and Axial Force with Applications
                                 to Nuclear Reactor Fuel Elements'
                                 J. Mech. Phys. Solids, **8**, 1

KOBAYASHI, A. S. and   1960 'Elastic Stresses in a Rotating Disc of
TRUMPLER, P. R.   General Profile'
                                 Int. J. mech. Sci. **2**, 13

KRAUS, H.   1962 'Pressure Stresses in Multi-bore Bodies'
             Int. J. mech. Sci. **4**, 187

KRAUSE, J. and   1962 'Thermal Stresses in Spherical Case-bounded
SHAFFER, B. W.   Propellant Grains'
                       J. Engng Ind. A.S.M.E., 144

LECKIE, F. A.,   1967 'Elliptical Discontinuities in Spherical Shells',
PAYNE, D. J. and   *J. Strain Analysis*, **2**, 34
PENNY, R. K.

LENGYEL, B.,   1966 'Design of Containers for a Semi-continuous
BURNS, D. J. and   Hydrostatic Extrusion Production Machine'
PRASAD, L. V.   *7th Int. M.T.D.R. Conf.*, Pergamon Press, 319

MARCAL, P. V.   1965 'A Note on the Elastic-plastic Thick Cylinder
                      with Internal Pressure in the Open and
                      Closed-end Condition'
                      Int. J. mech. Sci. **7**, 841

MANNING, W. R. D.   1945 'The Overstrain of Tubes by Internal Pressure
                         Engineering, **159**, 101 and 183

NADAI, A.   1931 *Plasticity*
                  McGraw-Hill, New York

NOWACKI, W.   1965 *Thermo-elasticity*
                    Pergamon Press, Oxford, 628 pp.

PARSONS, B. and   1967 'A Generalised Approach to the Optimum
COLE, B. N.   Design of Short Composite Cylinders'
                        Paper No. 20, p. 157, *High Pressure Eng.
                        Conf., Instn mech. Engrs*

PHILLIPS, A, and   1962 'Thick-walled Hollow Sphere of Elastic-locking
YILDIZ, A.   Material',
                        Oesterreichisches Ingenieur-Archiv, **16**, 313

PRAGER, W. and   1951 *The Theory of Perfectly Plastic Solids.*
HODGE, P. G.   Wiley, New York

PRZEMIENIECKI, J. S.   1960 'Design Charts for Transient Temperature and
                            Thermal Stress Distributions in Thermally
                            Thick Plates'
                            Aeronaut. Q. **11**, 269

RICHARDS, T. H.   1965 'Thermal Stresses in a Thick-walled
                        (Rectangular Section) Tube: An Experimental
                        Study by Electrical Analogy'
                        Int. J. mech. Sci. **7**, 103

ROSENFIELD, A. R. and 1956 'Effect of Stress on the Expansion Coefficient'
AVERBACH, B. L.   *J. appl. Physics* (U.S.), **27**, 154–6

TIMOSHENKO, S. and   1951 *Theory of Elasticity.*
GOODIER, J. N.   McGraw-Hill, New York

SCHMIDT, J. E. and   1960 'Transient Temperatures and Thermal Stresses
SONNEMANN, G.   in Hollow Cylinders due to Heat Generation
                          J. Heat Transfer C **82**, 273

SCHWIEBERT, P. D.   1965 'Elastic, Plastic and Creep Deformations in
                         Long, Thick-walled Cylinders of Work
                         Hardening Material Subjected to Transient

Thermal and Mechanical Loading'
Int. J. mech. Sci. **7**, 115

STEELE, M. C. and   1957 'Non-homogeneous Yielding of Steel
EICHBERGER, L. C.   Cylinders: 1—Mild Steel'
                             Trans. A.S.M.E. **79**, 1608

STRUB, R. A.   1961 'Transient Temperature and Thermal Stress
                             Conditions in a Cylinder'
                             Sulzer Research, 46

SUMNER, P. J. C. and   1966 'Some Experiments on Containers for Metal
MEREDITH, K. E. G.   Forming Processes'
                             7th Int. M.T.D.R. Conf., Pergamon Press, 647

WALDREN, N. E.,   1965 'Burst Strength of Rotating Discs'
PERCY, M. J. and   –66 *Proc. Instn mech. Engrs*, **180**, 111
MELLOR, P. B.

WHALLEY, E.   1965a 'General Theory for Monoblock Vessels'
                             Can. J. Technol. **34**, 268

                      1956b 'Steady State Temperature Distribution'.
                             Can. J. Technol. **34**, 291

                      1960c 'The Design of Pressure Vessels Subjected
                             to Thermal Stress: A Review'.
                             Int. J. mech. Sci., **1**, 379

WHALLEY, E. and   1960a 'Thermal Shock'.
MACKINNON, R. F.   *Int. J. mech. Sci.*, **1**, 301

WHALLEY, E. and   1960b 'Multi-layer Vessels'
MORRIS, S.   *Int. J. mech. Sci.* **1**, 369

WILHOIT, J. C.   1958 'Elastic-plastic Stresses in Rings Under
                             Steady State Radial Temperature Variation'
                             Proc. 3rd U.S. Cong. App. Mech. 693

WILSON, W. R. D. and 1967 'Design of High Pressure Cylinders'
SKELTON, W. J.   Paper No. 5, p. 32. *High Pressure Eng. Conf.*,
                             Instn mech. Engrs

                        1964 'Conference on Thermal Loading and Creep
                             in Structures and Components'
                             Instn mech. Engrs. London

Chapter 10

PLASTIC INSTABILITY

10.1 General Considerations of the Buckling of an Ideal Column

When a column is subjected to an increasing compressive load, failure occurs by buckling sideways at a much smaller load than that required to crush the material. Even in the most careful laboratory tests, the applied load will be slightly eccentric and the material of the column inhomogeneous. These two factors give rise to initially small bending stresses in the column, which eventually lead to the instability condition. In practice, eccentric loading may also occur because of initial curvature of the column.

At buckling, the stress in the column may be less or greater than the stress at the elastic limit, depending on the geometry of the column. The first published analysis of the former case was given by EULER in 1744. The solution to the problem of elastic instability is based on the assumption that the equilibrium of the straight column becomes unstable when there are equilibrium positions infinitesimally near to the straight equilibrium position under the same axial load. In formulating the problem it is assumed that the material is perfectly homogeneous, the column perfectly straight and the applied load perfectly axial. Under these idealized conditions, buckling will not occur unless an infinitesimally small disturbance is applied to the column. The buckling load, P, is then that load which will maintain the bent form of equilibrium when the disturbance is removed.

Consider the case of a straight column with frictionless pin-jointed or spherical ends. The small disturbance may be thought of as a lateral force acting at the mid-point of the column, removed when the bent equilibrium position is attained. Since deflections are assumed small, the basic differential equation is (see Fig. 10.1)

$$\frac{d^2y}{dx^2} + \frac{Py}{EI} = 0$$

where P is the critical or buckling load, E the modulus of elasticity and I the least value of the second moment of area of the cross-section. The solution of this equation is

$$y = A \sin \sqrt{\frac{P}{EI}} x + B \cos \sqrt{\frac{P}{EI}} x$$

and the well-known Euler expression for the buckling load follows, i.e.,

$$P = \pi^2 EI/l^2$$

where l is the effective length of the column. This equation may be written in the form

$$\frac{P}{A} = \frac{\pi^2 E}{(l/k)^2}$$

where A is the area of cross-section, k the least radius of gyration and (l/k) the slenderness ratio; it is valid only as long as the deformation is defined by the modulus of elasticity. P/A must be less than the stress at the elastic limit. The equation is therefore only valid for slenderness ratios greater than a certain limiting value which depends on the value of the modulus of elasticity.

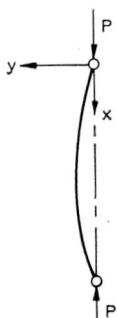

FIG. 10.1 Ideal column with axial load.

For small values of l/k the compressive stress in the column may reach the compressive elastic limit before buckling occurs. In this case, the buckling load can be defined as the smallest axial load at which the bent form of equilibrium of an originally straight and centrally loaded column, resulting from the action of a small disturbance, becomes stable. The problem may be investigated in a similar manner to the elastic case, note being taken of the fact that the deformation is no longer a reversible process. If, after exceeding the elastic limit, there is a stress reversal at any section the unloading line will be parallel to the elastic loading line. The theoretical instability load will therefore depend on the method of applying the infinitesimally small lateral force to initiate buckling.

The first attempt to solve the problem of the plastic buckling of a straight column was made by Engesser in 1889. He suggested that the modulus of elasticity in Euler's equation should simply be replaced by the tangent-modulus, E_t, at the point on the compressive stress-strain curve corresponding to the instability stress (Fig. 10.2). This was criticized by Jasinski on the grounds that during bending, under constant axial load, a portion of the cross-section on the convex side of the column would be subjected to decreasing stress. This meant that the modulus of elasticity, which applies during unloading, should be brought into the solution. Jasinski pointed out that this had been suggested by Considère in 1891. Engesser acknowledged the error in his original theory and showed how the 'double-modulus', a function of E and E_t, could be calculated in the general form. The critical load predicted by this theory is $P = \pi^2 \bar{E} I / l^2$ where $\bar{E}$ is the double-modulus. KARMAN, in 1910, presented the theory again and calculated the value of the double-modulus for a rectangular cross-section and an idealized H section. He also carried out a series of tests designed to show that the assumptions of the double-modulus theory were correct.

The main assumption made in the double-modulus theory is that the equilibrium of the straight column becomes unstable when there are equilibrium positions infinitesimally near to the straight equilibrium position under the same axial load. In formulating this problem, therefore, it is assumed that the buckling load is reached while the column is still straight and then a small disturbance causes the transition to the bent position under constant buckling load. The buckling load holds the column in the bent equilibrium position when the disturbance is removed. This hypothesis is exactly similar to the Euler hypothesis of elastic buckling.

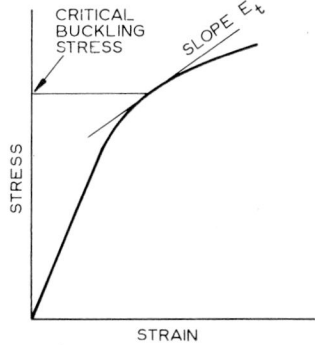

FIG. 10.2 Tangent-modulus and critical buckling stress.

The double-modulus theory was accepted as the only solution until 1946, when SHANLEY suggested that it is possible for the column to bend simultaneously with increasing axial load. If this is so, then it is possible for bending to occur without any strain reversal. In formulating this problem, it is assumed that the transition from the straight to the infinitesimally near bent equilibrium position, requires an increase in axial load. The transition takes place when a small lateral force is applied simultaneously with the final increment of axial load. If it were assumed that bending of the column began with initiation of loading, then the relevant theory would have to assume either a certain initial eccentricity of loading or a certain initial curvature of the column.

This approach by Shanley means in effect that the original tangent-modulus formula, first suggested by Engesser on the incorrect basis of constant axial loading during the transition to the bent position, gives the smallest value of the axial load at which a bifurcation of the equilibrium positions can occur. The maximum buckling load is given by the double-modulus theory. These two loads therefore give the theoretical lower and upper limits for the instability of a perfect column, and experimental results on carefully machined columns should lie between these limits. It has been found by careful tests that the experimental results give better agreement with the tangent-modulus theory except for very short columns (Fig. 10.3). The explanation of this fact would seem to be that short columns are less sensitive than long columns to the small disturbances that occur during an actual test and therefore the experimental buckling loads for short columns tend to approach the upper limit of the critical load.

HILL and SEWELL (1959) have shown that the critical load lies above the bifurcation value suggested by Shanley and coincides with the reduced modulus value, provided this is modified to allow for the effect of shear deformation.

The tangent-modulus formula has always been used in design, in preference to the double-modulus formula, because it is easier to manipulate and gives the more conservative value of the critical load. Since we shall be dealing with small displacements, it will be a good approximation to assume that initially plane sections remain plane after bending.

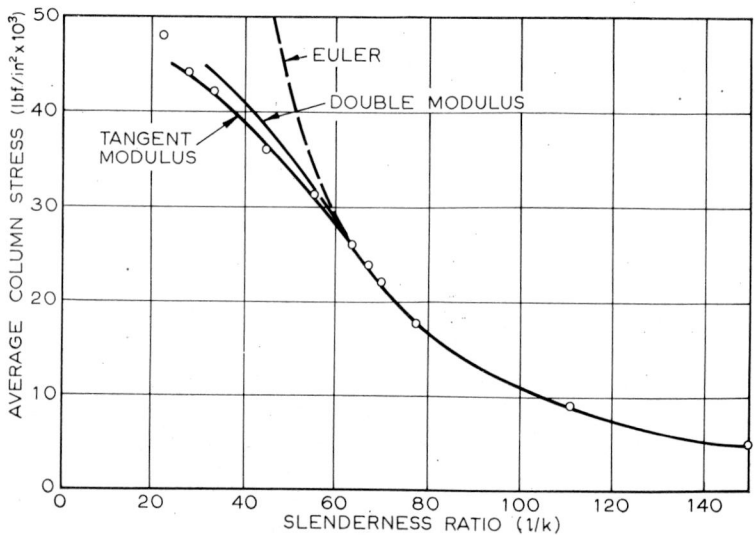

FIG. 10.3 Correlation of experimental results with theories of buckling.

o − Experimental Points (VAN DEN BROEK)
24 ST. 5/8″ SQUARE BAR, SPHERICAL ENDS.

10.2 The Double-modulus Formula

Consider a column of rectangular cross-section, dimensions h and b, subjected to an increasing load, P. The slenderness ratio (l/k) of the column is assumed sufficiently small to prevent elastic buckling. Up to the buckling load, the stress will be uniform and equal to P/bh. If now the transition to the bent position takes place under constant load, there will be a further increase of stress on the concave side and a decrease in stress on the convex side of the column. The unloading depends on the modulus of elasticity and since deflections are assumed small, the loading depends on the value of the tangent-modulus at the point of buckling.

The stress distribution across the column section and the corresponding strain distribution are shown in Fig. 10.4. $\Delta\sigma_1$ is the decrease in compressive stress on the convex outer surface and $\Delta\sigma_2$ is the increase in stress on the concave inner surface. For equilibrium of bending stresses

$$\tfrac{1}{2}\Delta\sigma_1 h_1 b + \tfrac{1}{2}\Delta\sigma_2 h_2 b = 0$$

or

$$\Delta\sigma_1 h_1 + \Delta\sigma_2 h_2 = 0. \tag{10.1}$$

The bending moment at the centre of the column is Py where y is the central deflection and for equilibrium this applied moment is equal to the resisting moment. That is

$$Py = \tfrac{2}{3}h_1(\tfrac{1}{2}\Delta\sigma_1 h_1 b) - \tfrac{2}{3}h_2(\tfrac{1}{2}\Delta\sigma_2 h_2 b). \tag{10.2}$$

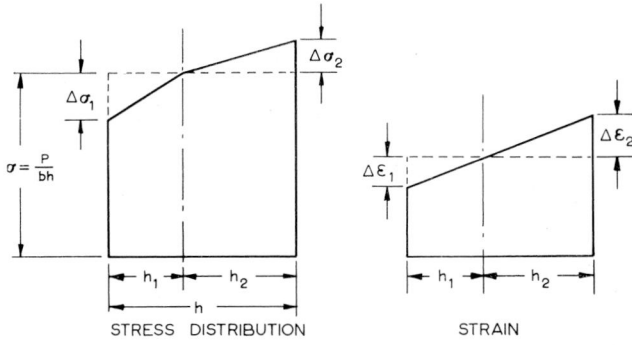

STRESS DISTRIBUTION   STRAIN

FIG. 10.4 Stress and strain distribution across column section according to double-modulus theory.

Since small strains are assumed, the relationships between stress and strain are

$$\Delta\sigma_1 = E\Delta\epsilon_1 \quad \text{and} \quad \Delta\sigma_2 = E_t\Delta\epsilon_2 \tag{10.3}$$

and assuming initially plane sections remain plane

$$\Delta\epsilon_1 = h_1/R \quad \text{and} \quad \Delta\epsilon_2 = -h_2/R \tag{10.4}$$

where R is the radius of curvature of the slightly bent column. Therefore

$$\Delta\sigma_1 = E\frac{h_1}{R} \quad \text{and} \quad \Delta\sigma_2 = -E_t\frac{h_2}{R}. \tag{10.5}$$

From equations (10.1) and (10.5)

$$Eh_1^2 = E_t h_2^2 \quad \text{and since} \quad h_1 + h_2 = h$$

it is easily shown that

$$h_1 = \frac{h\sqrt{E_t}}{\sqrt{E} + \sqrt{E_t}} \quad \text{and} \quad h_2 = \frac{h\sqrt{E}}{\sqrt{E} + \sqrt{E_t}}.$$

Eliminating $\Delta\sigma_1$, $\Delta\sigma_2$, h_1 and h_2 from equation (10.2)

$$Py = \frac{bh^3}{12R}\left[\frac{4EE_t}{(\sqrt{E} + \sqrt{E_t})^2}\right]. \qquad (10.6)$$

The term in brackets is the double-modulus $\bar{E}$, and equation (10.6) may be re-written

$$Py = \bar{E}\frac{I}{R}$$

where I is the least value of the second moment of area of the rectangular cross-section. For small deflections $1/R \simeq -d^2y/dx^2$ and the equation becomes

$$\frac{d^2y}{dx^2} + \frac{Py}{\bar{E}I} = 0$$

which is the same basic differential equation as for elastic buckling.
The buckling load is therefore

$$P = \pi^2\frac{\bar{E}I}{l^2}. \qquad (10.7)$$

Note that, in general, the values of $\bar{E}$ and I depend on the cross-section of the column. The values derived here only apply to a rectangular cross-section. The value of $\bar{E}$ is always greater than E_t since the latter quantity is always less than E.

10.3 The Tangent-modulus Formula

Assuming the same conditions as above, the hypothesis is, now, that the transition to the bent position takes place under increasing axial load. There is no stress and strain reversal and a constant value of E_t is assumed to apply across the entire cross-section. Therefore, the basic differential equation is simply

$$\frac{d^2y}{dx^2} + \frac{Py}{E_tI} = 0$$

and the buckling load is

$$P = \pi^2\frac{E_tI}{l^2}. \qquad (10.8)$$

This value of the buckling load is always less than that predicted by the double-modulus formula and E_t is not affected by the shape of the column section.

10.4 A Comparison Between the two Solutions for the Plastic Buckling of a Column with a Rectangular Cross-section

Figure 10.5 shows the stress-strain curve for an aluminium alloy and the graphically derived curve of tangent-modulus plotted against stress. If now the tangent-modulus formula is written as $l/k = \pi\sqrt{E_t/\sigma}$ and the double-modulus formula as $l/k = \pi\sqrt{\bar{E}/\sigma}$, values of critical buckling stress can be plotted against slenderness ratio for a column of rectangular cross-section made of the above material. The values of $\bar{E}$ are calculated from the bracketed expression in equation (10.6). This has been done in Fig. 10.6, together with a plot of the Euler curve. It should be noted that for all values of slenderness ratio shown, the total strains are of the order of the elastic strains, i.e. not exceeding a value of 0·005.

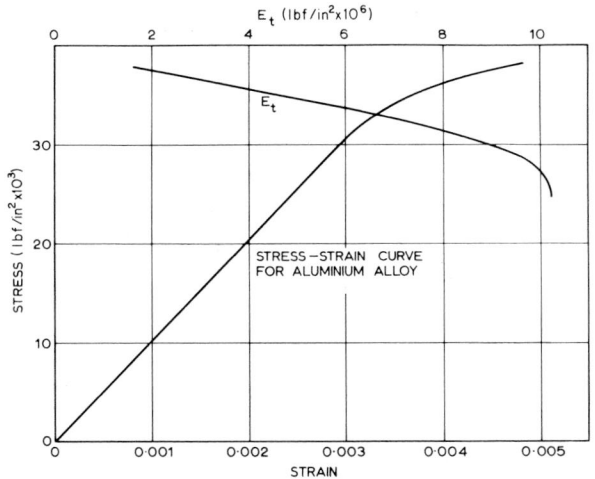

FIG. 10.5 Stress-strain curve for aluminium alloy and graphically derived curve of tangent-modulus.

The above work has dealt with a work-hardening material. In the case of a material that yields at constant stress, the initial buckling stress will be the yield stress of the material.

For a more complete understanding of buckling, the bending of columns with eccentric loading should be studied. This is beyond the scope of this book and the reader is referred to BLEICH (1952) where the whole subject is reviewed and further references given.

10.5 General Considerations of Plastic Instability in Tension

The case of instability in simple tension has been considered in Chapter 1. The solution for that system can now be extended to include the effect of elastic strains. If $d\epsilon_1$ is the total strain increment in the longitudinal direction, then the stress-strain relation for that direction is

$$d\epsilon_1^p = d\epsilon_1 - \frac{d\sigma_1}{E}. \tag{10.9}$$

For the transverse directions,

$$d\epsilon_2^p = d\epsilon_2 + v\frac{d\sigma_1}{E} \tag{10.10}$$

$$\text{and } d\epsilon_3^p = d\epsilon_3 + v\frac{d\sigma_1}{E} \tag{10.11}$$

where equations (10.10) and (10.11) will be identical for an isotropic material. Also

$$d\epsilon_1^p + d\epsilon_2^p + d\epsilon_3^p = 0.$$

Therefore

$$d\epsilon_2 = d\epsilon_3 = -\frac{1}{2}\left[d\epsilon_1 - \frac{d\sigma_1}{E}(1 - 2v)\right].$$

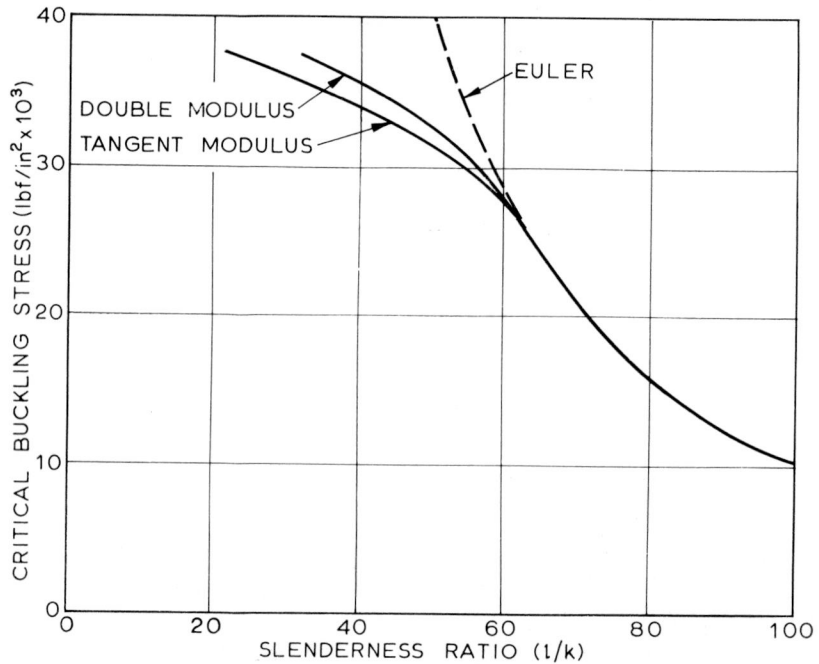

FIG. 10.6 Critical buckling stress based on tangent-modulus and double modulus theories.

Since there is no rotation of the principal axes, we can write

$$\epsilon_2 = \epsilon_3 = -\frac{1}{2}\left[\epsilon_1 - \frac{\sigma_1}{E}(1 - 2v)\right]. \tag{10.12}$$

The tensile load, P, at any time before instability, is

$$P = X_0\sigma_1 \exp 2\epsilon_2$$

$$= X_0\sigma_1 \exp\left\{-\left[\epsilon_1 - \frac{\sigma_1}{E}(1 - 2v)\right]\right\} \tag{10.13}$$

where X_0 is the original cross-sectional area.

If there is no elastic change in volume (i.e., $v = 1/2$), then the condition for instability is $d\sigma_1/d\epsilon_1 = \sigma_1$ as before.

When elastic volume changes are considered, then at instability, from equation (10.13),

$$\frac{dP}{d\epsilon_1} = 0 = -\sigma_1 + \frac{d\sigma_1}{d\epsilon_1} + \frac{(1-2v)}{E} \cdot \sigma_1 \cdot \frac{d\sigma_1}{d\epsilon_1}$$

or

$$\frac{d\sigma_1}{d\epsilon_1} = \frac{\sigma_1}{1 + \dfrac{(1-2v)}{E}\sigma_1} = \frac{\sigma_1}{1 + \dfrac{\sigma_1}{3K}} \tag{10.14}$$

where σ_1 is the stress at instability. The term $\sigma_1/3K$ will usually be very small compared with unity and can be neglected.

Conditions similar to the instability phenomenon in simple tension occur under complex stresses and are of importance in the stretch-forming of sheet metals. In the remainder of this chapter, we shall consider cases of the straining of sheet metal under biaxial tension, the third stress (normal to the sheet) being either zero or negligible in comparison with the other two stresses. Elastic strains will be neglected and the Lévy-Mises equations assumed to apply. Further, in all the cases to be considered, the principal axes do not rotate relative to the element being strained and the principal stress ratios are constant throughout the straining.

For bi-axial tension, we can write $\sigma_2 = x\sigma_1$, $\sigma_3 = 0$ where x, the stress ratio is a proper fraction. Then the Lévy-Mises equations (5.10), for the principal directions, have the form

$$\frac{d\epsilon_1}{2-x} = \frac{d\epsilon_2}{2x-1} = -\frac{d\epsilon_3}{1+x}. \tag{10.15}$$

Similarly the equations for effective stress and effective strain increment, (6.1) and (6.2) may be written

$$\bar{\sigma} = \sigma_1(1 - x + x^2)^{\frac{1}{2}} \tag{10.16}$$

$$d\bar{\epsilon} = d\epsilon_1 \cdot \frac{2}{(2-x)} \cdot (1 - x + x^2)^{\frac{1}{2}} \tag{10.17}$$

Equations (10.15) and (10.17) together give, in integrated form,

$$\frac{\bar{\epsilon}}{2(1 - x + x^2)^{\frac{1}{2}}} = \frac{\epsilon_1}{2-x} = \frac{\epsilon_2}{2x-1} = \frac{-\epsilon_3}{1+x}. \tag{10.18}$$

In order to give the work generality, it is useful to employ an empirical equation for the strain-hardening characteristic. The one to be used here is $\bar{\sigma} = A(B + \bar{\epsilon})^n$ suggested by SWIFT. A is some measure of the basic strength of the material, independent of its initial state, B gives an indication of the initial state of the material and n is a measure of the rate of hardening with strain.

The fit of this empirical relation to two sheet materials, 70/30 brass and commercially pure aluminium, is shown in Figs. 10.7 and 10.8. Formulae for other sheet materials are set down in Fig. 10.9. The value of the strain-hardening index, n, for commercially used materials lies between 0·2 and 0·5.

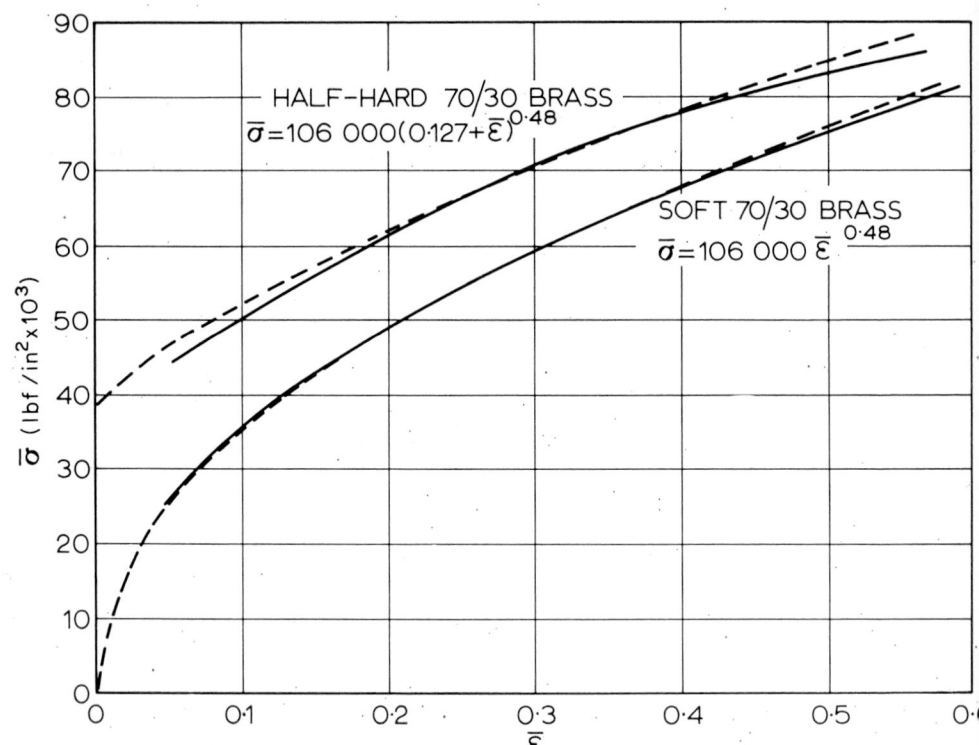

FIG. 10.7 Empirical relation to stress-strain curves of brass. Experimental results denoted by full line. (*After* Mellor, *J. Mech. Phys. Solids.*)

These strain-hardening characteristics were obtained in balanced bi-axial tension (see Section 6.3).

Differentiating the empirical equation, $\bar{\sigma} = A(B + \bar{\epsilon})^n$,

$$\frac{d\bar{\sigma}}{d\bar{\epsilon}} = \frac{n\bar{\sigma}}{(B + \bar{\epsilon})} = \frac{\bar{\sigma}}{z} \qquad (10.19)$$

where the sub-tangent $z = (B + \bar{\epsilon})/n$, see Fig. 10.10.

For simple tension, $z = 1$, as has already been shown in sub-section 1.3.

The value of A does not affect the amount of strain that a material can withstand, but it does, of course, affect the stress required to attain any given value of strain.

10.6 Plastic Instability of a Closed-ended Thin-walled Pipe or Cylinder Subjected to Internal Pressure

If p is the internal pressure, the hoop stress $\sigma_1 = pr/t$ and the longitudinal stress $\sigma_2 = pr/2t$ where r and t are the current values of cylinder radius and

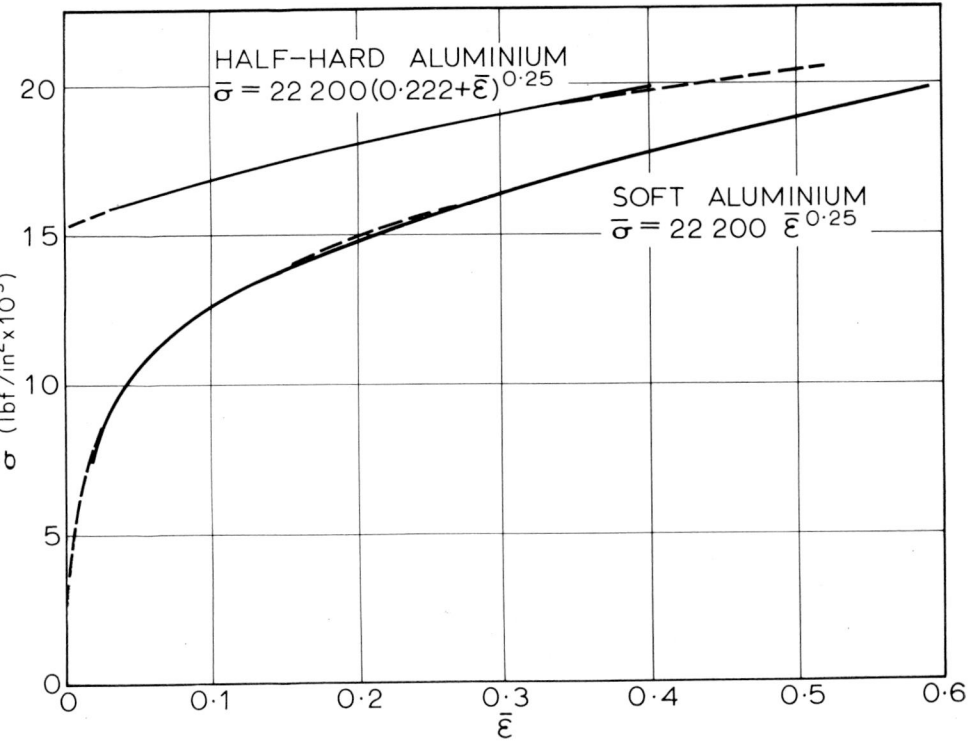

FIG. 10.8 Empirical relation to stress-strain curves of aluminium. Experimental results denoted by full line. (*After* Mellor, unpublished work.)

Metal	$\bar{\sigma} = A(B + \bar{\epsilon})^n \; (lbf/in^2)$	Experimental thickness ratio		Theoretical thickness (*Hill*) t/t_0
		Instability t/t_0	Fracture t/t_0	
Copper, soft	$\bar{\sigma} = 62{,}200(0{\cdot}016 + \bar{\epsilon})^{0{\cdot}3}$	0·555	0·50	0·563
Copper, half-hard	$\bar{\sigma} = 62{,}200(0{\cdot}114 + \bar{\epsilon})^{0{\cdot}3}$	0·59	0·58	0·581
Brass, soft	$\bar{\sigma} = 106{,}000\bar{\epsilon}^{0{\cdot}48}$	No instability	0·50	0·498
Brass, half-hard	$\bar{\sigma} = 106{,}000(0{\cdot}127 + \bar{\epsilon})^{0{\cdot}48}$	No instability	0·51	0·518
Aluminium, soft	$\bar{\sigma} = 22{,}200\bar{\epsilon}^{0{\cdot}25}$	0·565	0·48	0·580
Aluminium, half-hard	$\bar{\sigma} = 22{,}200(0{\cdot}222 + \bar{\epsilon})^{0{\cdot}25}$	No instability	0·65	0·613
Killed steel	$\bar{\sigma} = 91{,}000\bar{\epsilon}^{0{\cdot}2}$	0·57	0·48	0·603
Stainless steel	$\bar{\sigma} = 222{,}000(0{\cdot}016 + \bar{\epsilon})^{0{\cdot}5}$	0·52	0·50	0·487

FIG. 10.9 Experimental and theoretical instability strains for metal diaphragm. (*After* Mellor, *J. Mech. Phys. Solids.*)

wall thickness, respectively. The stress ratio $x = 1/2$, and at instability, $dp = 0$, therefore

$$\frac{d\sigma_1}{\sigma_1} = \frac{d\sigma_2}{\sigma_2} = \frac{dr}{r} - \frac{dt}{t} \tag{10.20}$$

where dr/r is the hoop strain increment $d\epsilon_1$ and dt/t is the incremental thickness strain $d\epsilon_3$.

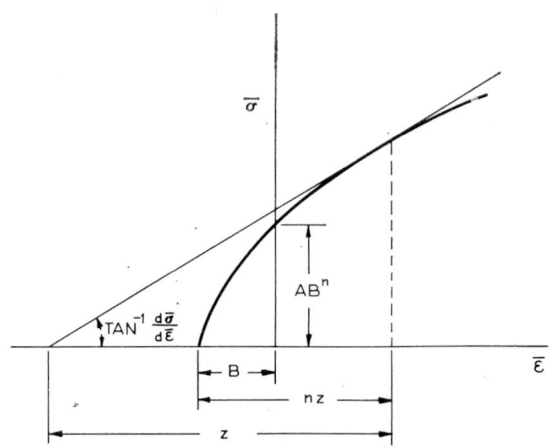

FIG. 10.10 Generalized instability strain.

The values of generalized stress and strain are, from equations (10.16) and (10.17),

$$\left. \begin{array}{l} \bar{\sigma} = (\sqrt{3}/2)\sigma_1 \\[2ex] d\bar{\epsilon} = 2d\epsilon_1/\sqrt{3} = -2d\epsilon_3/\sqrt{3} \quad \text{and} \quad d\epsilon_2 = 0 \end{array} \right\} \tag{10.21}$$

This is a case of plane strain, the strain in the longitudinal direction being zero. It is also assumed, of course, that the end effects are local and can be neglected in the discussion.

From equation (10.21) $d\bar{\sigma} = (\sqrt{3}/2)d\sigma_1$, and the condition at instability may be written

$$d\bar{\sigma} = \sqrt{3}\sigma_1(d\epsilon_1 - d\epsilon_3)/2$$

which in terms of generalized stress and strain, reduces to

$$d\bar{\sigma}/d\bar{\epsilon} = \sqrt{3}\bar{\sigma} \tag{10.22}$$

giving as the value of the sub-tangent, $z = 1/\sqrt{3}$.

For a material with the strain-hardening characteristic $\bar{\sigma} = A(B + \bar{\epsilon})^n$ the generalized strain at instability $\bar{\epsilon} = (n/\sqrt{3}) - B$. The values of the principal stresses and strains are shown in Fig. 10.11 and values of maximum principal strains are compared with the longitudinal strains in uniaxial tension in Fig. 10.12. It can be seen from the latter diagram that the maximum strains

Loading system	$\bar{\epsilon}$	ϵ_1	ϵ_2	$\epsilon_3 = \ln t/t_0$	$\bar{\sigma}/A$	σ_1/A	σ_2/A
Simple tension	$(n - B)$	$(n - B)$	$-\frac{1}{2}(n - B)$	$-\frac{1}{2}(n - B)$	$(n)^n$	$(n)^n$	0
Cylindrical shell	$\left(\dfrac{n}{\sqrt{3}} - B\right)$	$\dfrac{\sqrt{3}}{2}\left(\dfrac{n}{\sqrt{3}} - B\right)$	0	$-\dfrac{\sqrt{3}}{2}\left(\dfrac{n}{\sqrt{3}} - B\right)$	$\left(\dfrac{n}{\sqrt{3}}\right)^n$	$\dfrac{2}{\sqrt{3}}\left(\dfrac{n}{\sqrt{3}}\right)^n$	$\dfrac{1}{\sqrt{3}}\left(\dfrac{n}{\sqrt{3}}\right)^n$
Spherical shell	$\left(\dfrac{2n}{3} - B\right)$	$\dfrac{1}{2}\left(\dfrac{2n}{3} - B\right)$	$\dfrac{1}{2}\left(\dfrac{2n}{3} - B\right)$	$-\left(\dfrac{2n}{3} - B\right)$	$\left(\dfrac{2n}{3}\right)^n$	$\left(\dfrac{2n}{3}\right)^n$	$\left(\dfrac{2n}{3}\right)^n$

Fig. 10.11 Stress and strain and instability.

at instability for the case of the cylinder are very much less than the corresponding longitudinal strains in simple tension. Since it is very probable that rupture of the metal of a thin-walled pressure vessel follows very closely after instability, there will be an apparent lack of ductility of the material.

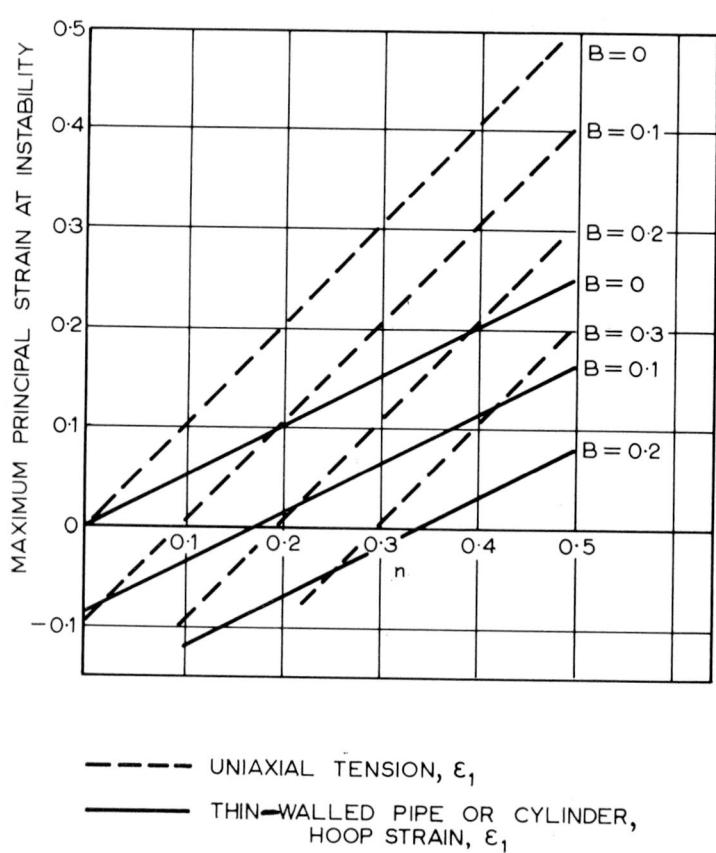

- - - - UNIAXIAL TENSION, ε_1

———— THIN-WALLED PIPE OR CYLINDER, HOOP STRAIN, ε_1

FIG. 10.12 Instability strains for thin-walled pipe or cylinder.

10.7 Plastic Instability of a Spherical Shell Subjected to Internal Pressure

For a spherical shell under internal pressure $\sigma_1 = \sigma_2 = pr/2t$; the stress ratio $x = 1$ and the strain increment $d\epsilon_1 = d\epsilon_2 = -d\epsilon_3/2$. The instability condition is $dp = 0$, so that

$$\therefore \quad \frac{d\sigma_1}{\sigma_1} = \frac{dr}{r} - \frac{dt}{t} = d\epsilon_1 - d\epsilon_3. \tag{10.23}$$

From equations (10.16) and (10.17) for $x = 1$

$$\bar{\sigma} = \sigma_1$$

and $d\bar{\epsilon} = 2d\epsilon_1 = 2d\epsilon_2 = -d\epsilon_3$

At instability, therefore

$$d\bar{\sigma}/d\bar{\varepsilon} = \tfrac{3}{2}\bar{\sigma} \qquad (10.24)$$

and the value of the sub-tangent $z = 2/3$ (see the summary of results in Fig. 10.11). The maximum principal strains at instability are greater than for the case of the thin-walled cylinder but less than for uni-axial tension, Fig. 10.13.

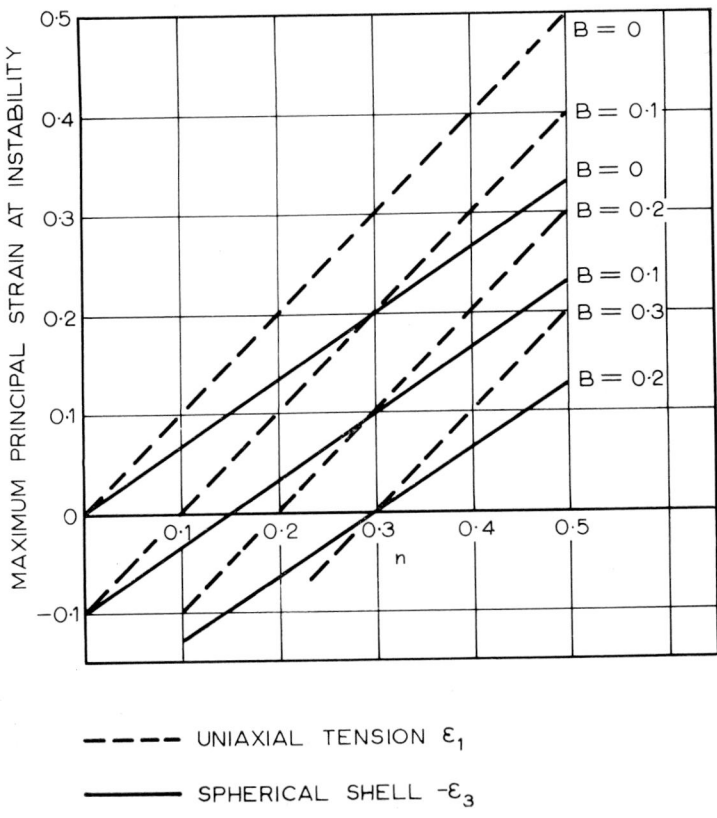

———— UNIAXIAL TENSION ε_1

——— SPHERICAL SHELL $-\varepsilon_3$

FIG. 10.13 Instability strains for spherical shell.

10.8 Plastic Instability of a Thin-walled Pipe or Cylinder Subjected to an Internal Pressure and Independent Axial Load

A more general case of instability in a thin-walled cylinder can now be discussed. The principal stress components σ_θ and σ_z are, from considerations of equilibrium,

$$\sigma_z . 2\pi r t = P + \pi r^2 p$$

$$\sigma_\theta = \frac{pr}{t},$$

where p is the internal pressure and P the independent axial load, r is the current mean radius of the cylinder and t the current wall thickness. It will be assumed here that the loads are applied in such a manner that the ratio of hoop to longitudinal stress remains constant throughout the deformation. Instability will then occur when there is a maximum in either the total axial load or in the internal pressure. This criterion was put forward by LANKFORD and SAIBEL (1947) and has been discussed by MELLOR (1962). The mode of necking will be either a tensile neck or a local radial bulge. If the total axial load is maximum at instability then the sub-tangent is

$$z_1 = \frac{2\left[(\sigma_\theta/\sigma_z)^2 - (\sigma_\theta/\sigma_z) + 1\right]^{\frac{1}{2}}}{2 - (\sigma_\theta/\sigma_z)} \tag{10.25}$$

and if the internal pressure is a maximum then the sub-tangent is

$$z_2 = \tfrac{2}{3}\left[(\sigma_z/\sigma_\theta)^2 - (\sigma_z/\sigma_\theta) + 1\right]^{\frac{1}{2}}. \tag{10.26}$$

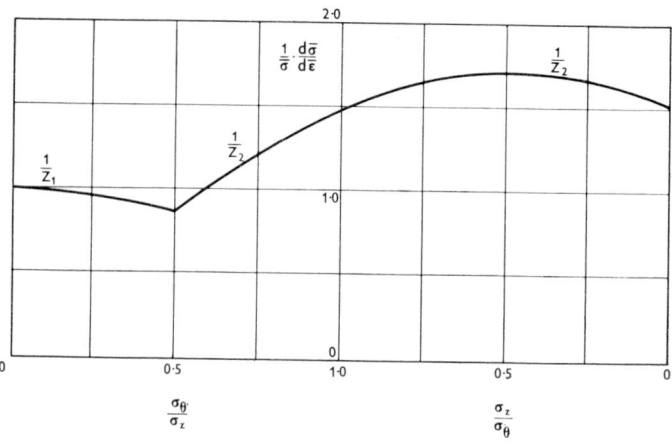

FIG. 10.14 Variation of reciprocal of sub-tangents z with stress ratio for thin-walled tubes.

The variation of the reciprocal of the sub-tangent with stress ratio is shown in Fig. 10.14. Experiments have been carried out by JONES and MELLOR (1967) in which great care was taken to deform nickel-chrome steel cylinders under constant true stress ratio conditions. The types of failure are shown in Fig. 10.15 and the experimental results verified both the assumed modes of instability and the strain attained at instability.

Theories have been put forward by SWIFT (1952), MARCINIAK (1958), HILLIER (1965) and STORAKERS (1968) for more general loading where the true stress ratio is allowed to vary.

DAVIS (1945) carried out tests on mild steel tubes in which the ratio of applied axial load to internal pressure was held constant. However, interest at that time centred on fracture and instability strains were therefore not quoted.

10.9 Instability of a Circular Metal Diaphragm

In the processes considered previously, the stress distribution was uniform up to instability. If now a circular sheet of metal is clamped at its periphery and subjected on one side to fluid pressure the shape of the deforming metal is approximately spherical, except in the region near to the clamped edge, but its divergence from a true sphere is reflected in a changing stress-ratio

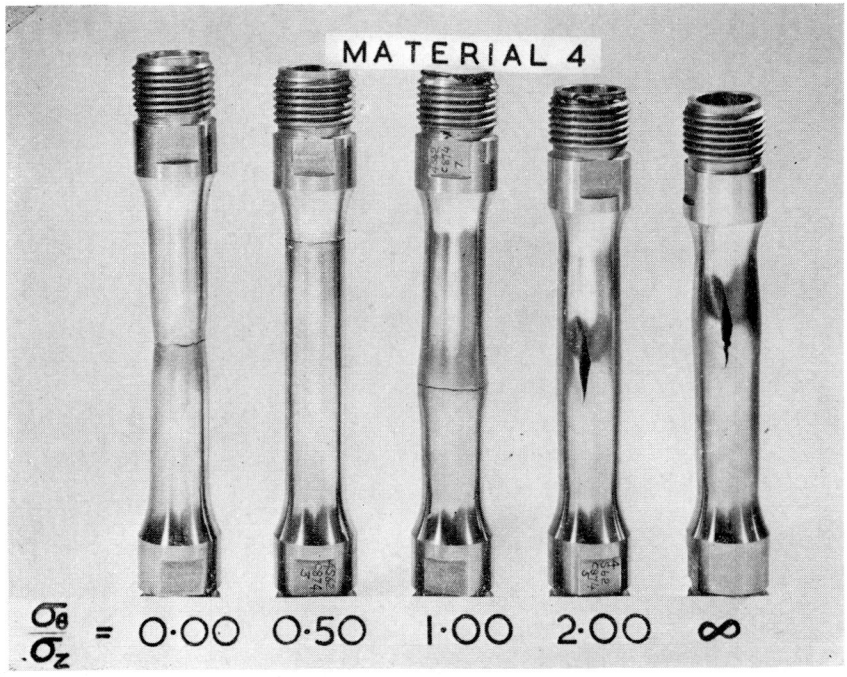

FIG. 10.15 Modes of instability in thin-walled tubes. (*After* Jones and Mellor, *J. Strain Analysis.*)

except near to the crown or pole where, for an isotropic material, $x = 1$ throughout. Fig. 10.16 shows how the stress ratio varies across a steel diaphragm of 5 in radius. The values of stress ratio were derived from strain measurements, across the dome, assuming the Lévy-Mises relationship between stress and strain increment. It is seen that, for most of the straining, the dome is spherical within a radius of 2 in from the pole, and gradually this region extends outwards as the height increases.

The distribution of stress and strain in a plastically deforming circular diaphragm is now well understood. HILL (1950) put forward a theory for a material having a linear strain-hardening characteristic, and this was shown to be in good agreement with experimental results on half-hard aluminium (MELLOR, 1956). Following the work of Hill, THOMAS (1954) took the particular case of soft copper and obtained a solution by a method of successive

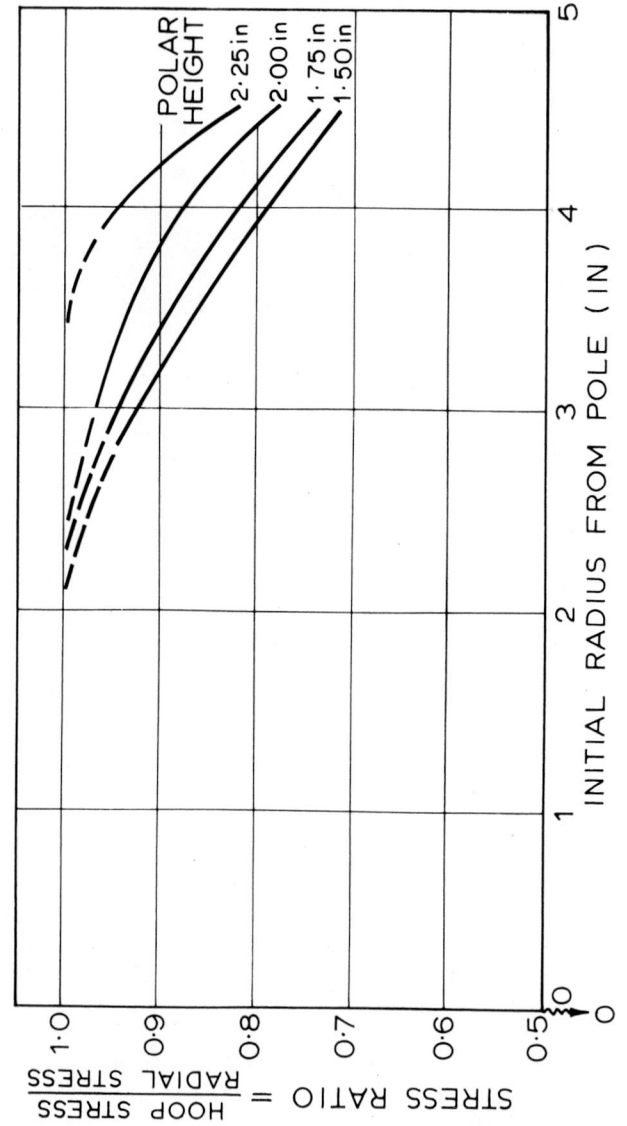

Fig. 10.16 Variation of stress ratio across steel diaphragm.

approximations which was in fairly good agreement with experimental work. However, the solution is lengthy, and if the maximum pressure and instability strain only are required, the special solution by HILL (1950) is worth investigating.

It is known from experiment that with an annealed metal, a maximum pressure occurs and fracture follows under decreasing pressure in the region of the pole. The relation between pressure and polar curvature, ρ, is $p = 2\sigma_1 t/\rho$ where $\sigma_1 = \sigma_2$, the membrane stresses at the pole, and t is the current sheet thickness. The normal stress σ_3 is negligible.

At instability $dp = 0$ and

$$\frac{d\sigma_1}{\sigma_1} = \frac{d\rho}{\rho} - \frac{dt}{t}$$

or

$$\frac{1}{\sigma_1}\frac{d\sigma_1}{d\epsilon_3} = 1 + \frac{1}{\rho}\frac{d\rho}{d\epsilon_3}, \tag{10.27}$$

where the through-thickness strain ϵ_3 is defined as the positive quantity $\ln(t_0/t)$, t_0 being the initial thickness of the sheet.

A simple geometric relationship for spherical bulging is $h(2\rho - h) = a^2$ or $\rho = (a^2 + h^2)/2h$; (10.28.i).
h is the polar height and a the die radius or initial radius of the blank, see Fig. 10.17. Assuming that particles in the membrane near the pole describe paths orthogonal to the momentary profile, $d\epsilon_\theta = dh/\rho$, so that using (10.28.i) for ρ and integrating,

$$\epsilon_3 = 2\epsilon_1 = 2\ln\left(1 + \frac{h^2}{a^2}\right) \tag{10.28.ii}$$

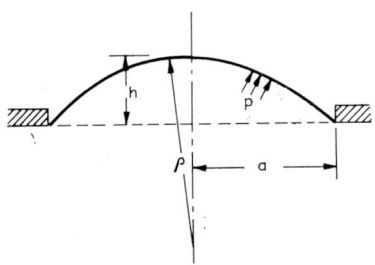

FIG. 10.17 Geometry of diaphragm.

Relations (10.28) are independent of the properties of the metal and are a better approximation for the more work-hardened materials (Fig. 10.18).

From equations (10.27) and (10.28), the point of instability occurs when

$$\frac{1}{\sigma_1}\cdot\frac{d\sigma_1}{d\epsilon_3} = \frac{3}{2} - \frac{\rho}{2h}.$$

On expanding in powers of ϵ_3 this reduces to

$$\frac{1}{\sigma_1}\cdot\frac{d\sigma_1}{d\epsilon_3} \simeq \frac{11}{8} - \frac{1}{2\epsilon_3}. \tag{10.29}$$

If the stress-strain curve is fitted by the relation $\bar{\sigma} = A(B + \bar{\epsilon})^n$, the polar thickness strain at maximum pressure is given by the quadratic equation

$$11\epsilon_3^2 + \epsilon_3(11B - 8n - 4) - 4B = 0. \qquad (10.30)$$

For a fully annealed material, B is zero and

$$\epsilon_3 = \tfrac{4}{11}(2n + 1).$$

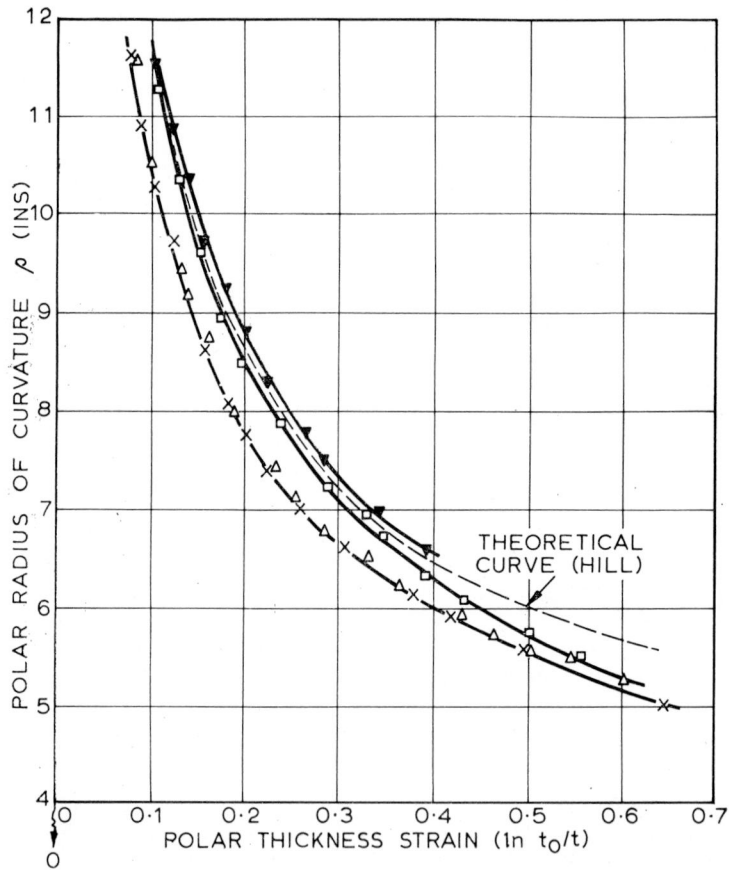

FIG. 10.18 Variation of polar thickness strain with polar radius of curvature.

× Soft copper
▲ Soft aluminium
■ Half-hard copper
▼ Half-hard aluminium

A simpler, though less exact analysis assumes parabolic bulging (JOHNSON *et al*, 1963), for which $\rho = a^2/2h$; with $\sigma = A(B + \epsilon)^n$, $\epsilon_3 = (n + \tfrac{1}{2} - B)$ or with $\sigma = Y + P\epsilon^{m/n}$, $\epsilon_3 = z = (Y/P)[(1 + 2z)/(1 + 2z + m/n)]^{-m/n}$.

The most striking feature about the hydrostatic bulging of a circular metal membrane is that no matter how heavily work-hardened the material is initially, the theoretical strain at fracture is always greater than 4/11. This fact makes the 'bulge test' a very suitable method of obtaining stress-strain characteristics in bi-axial tension. The variation of instability strain with B and n is shown in Fig. 10.19.

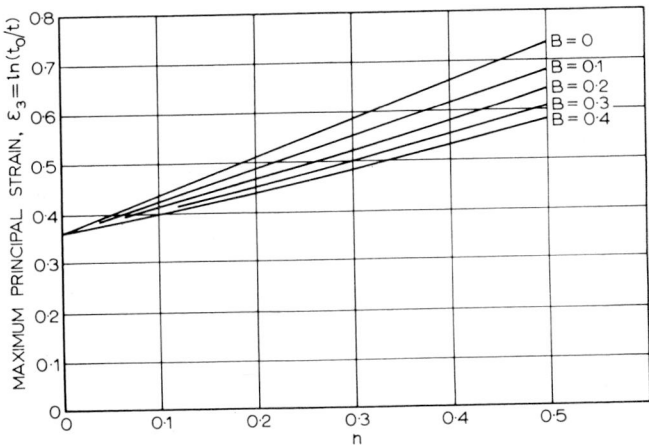

FIG. 10.19 Theoretical instability strains for metal diaphragm.

Theoretical values of thickness ratio at instability obtained from Hill's solution are compared in Fig. 10.9 with experimental values (MELLOR, 1956) for a wide range of materials.

Soft brass, half-hard brass and half-hard aluminium fractured under almost constant pressure, and it can be assumed, when comparing theoretical and experimental values, that instability and fracture occurred simultaneously. It can be seen that the theoretical values give a good prediction of the instability strains.

The case of hydrostatic bulging is much simplified by the fact that there is no friction between the pressure medium and the blank. In stretch-forming with a solid punch, the amount and locality of straining is very dependent on the degree of lubrication between punch and material .The deep-drawing of a circular blank is even more complex, for though instability usually occurs over the punch-head, it is affected by friction and the pull-back forces due to radial drawing and bending over the die. (See sub-section 11.7).

An interesting general approach to the instability of sheet metal subjected to biaxial tension has been put forward by MARCINIAK and KUCZYNSKI (1967). Their hypothesis associates instability with pre-existing inhomogeneities in the sheet metal. This original theory has been extended by SOWERBY and DUNCAN (1971) and some experimental results have been given by VENTER, JOHNSON and DE MALHERBE (1971).

10.10 **The Strength of Thin-walled Shells and Circular Diaphragms subjected to Hydrostatic Pressure**

The tensile strength (T.S.) corresponds to the maximum load attained in simple tension, which, in turn, marks the end of uniform straining. Similar maximum load conditions occur in shells subjected to slowly increasing internal pressure, and it is of interest for design purposes to determine what errors are involved if the maximum pressure in these cases are calculated on a basis of T.S. alone. For this purpose, the conventional instability stress, that is stress based on the initial dimensions of the structure, will be determined for each system and compared with the T.S. The equation $\bar{\sigma} = A(B + \bar{\epsilon})^n$ will be used to define the strain-hardening characteristics of the metals.

In these terms, the T.S. itself has, from equation (10.13) and Fig. 10.11, the form

$$\text{T.S.} = \frac{P_m}{X_0} = An^n \exp(B - n) \tag{10.31}$$

where P_m is the maximum load and X_0 original cross-sectional area.

10.10.1 *Thin-walled Pipe or Cylinder*

The hoop stress at instability is

$$\sigma_1 = \frac{pr}{t} = A \frac{2}{\sqrt{3}} \left(\frac{n}{\sqrt{3}}\right)^n,$$

see Fig. 10.11, where r and t are the mean cylinder radius and wall thickness at instability. It is now necessary to relate the maximum internal pressure to the initial wall thickness t_0 and the initial mean radius of the cylinder r_0. The strains are

$$\epsilon_1 = \ln \frac{r}{r_0} = \frac{\sqrt{3}}{2} \left(\frac{n}{\sqrt{3}} - B\right) \text{ and } -\epsilon_1 = \ln \frac{t_0}{t} = \frac{\sqrt{3}}{2} \left(\frac{n}{\sqrt{3}} - B\right)$$

giving

$$r = r_0 \exp \tfrac{1}{2}(n - \sqrt{3}B) \text{ and } t = t_0 \exp \tfrac{1}{2}(\sqrt{3}B - n).$$

The pressure at instability is

$$p = \frac{\sigma_1 t}{r} = \frac{t_0}{r_0} A \frac{2}{\sqrt{3}} \left(\frac{n}{\sqrt{3}}\right)^n \exp(\sqrt{3}B - n). \tag{10.32}$$

The term

$$A \frac{2}{\sqrt{3}} \left(\frac{n}{\sqrt{3}}\right)^n \exp(\sqrt{3}B - n)$$

is the conventional instability stress, i.e. the stress based on the initial dimensions of the cylinder. Let this quantity be denoted by S_1. What error is involved if the T.S. (the conventional instability stress in tension) is used in equation (10.32) instead of S_1? The ratio

$$\frac{S_1}{(\text{T.S.})} = \frac{2}{(\sqrt{3})^{n+1}} \exp B(\sqrt{3} - 1) = K_1 \quad (\text{say})$$

and the pressure at instability can be written

$$p = K_1(\text{T.S.}) \, t_0/r_0. \tag{10.33}$$

The variation of K_1 with n and B is shown in Fig. 10.20. The condition $n = 0$ refers to an ideally plastic material and instability occurs at yielding, giving $K_1 = 1 \cdot 155$. For cases where K_1 is greater than unity, basing the instability pressure on the T.S. alone would underestimate the strength of the material. The reverse is true for cases where K_1 is less than unity. Depending on the value of n, it is possible to underestimate the maximum pressure by 13 per cent or overestimate it by 15 per cent. The effects of initial work-hardening (increasing B) is to increase the value of K_1, but, of course, the value of the instability strain decreases.

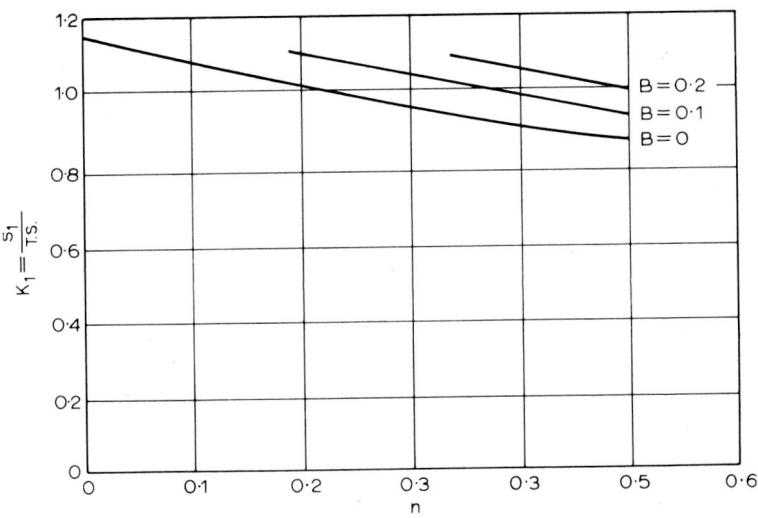

FIG. 10.20 Effect of using tensile strength as a basis for design: thin-walled pipe or cylinder.

As an example, consider a thin-walled cylinder constructed in aluminium (see Fig. 10.9). For soft aluminium, $B = 0$ and $n = 0 \cdot 25$, giving, from Fig. 10.8, $K_1 \simeq 1$. Therefore, in this case, basing the design on the T.S. alone would give the correct maximum pressure. However, considering the half-hard aluminium, where $B = 0 \cdot 222$ and $n = 0 \cdot 25$, instability arises before any general elongation has taken place.

10.10.2 Spherical Shell

The procedure is exactly the same as for the thin-walled cylinder. The principal stresses at instability have the value

$$\sigma_1 = \sigma_2 = \frac{pr}{2t} = A \left(\frac{2n}{3} \right)^n.$$

The strains $\epsilon_1 = \epsilon_2$ and ϵ_3 in terms of the dimensions of the shells are

$$\epsilon_1 = \epsilon_2 = \ln \frac{r}{r_0} = \frac{1}{2} \left(\frac{2n}{3} - B \right) \text{ and } -\epsilon_3 = \ln \frac{t_0}{t} = \left(\frac{2n}{3} - B \right)$$

giving

$$r = r_0 \exp \frac{1}{2} \left(\frac{2n}{3} - B \right) \text{ and } t = t_0 \exp \left(B - \frac{2n}{3} \right).$$

The pressure at instability is

$$p = \frac{2\sigma_1 t}{r} = \frac{2t_0}{r_0} A \left(\frac{2n}{3} \right)^n \exp \left(\frac{3}{2} B - n \right). \tag{10.34}$$

The term

$$A \left(\frac{2n}{3} \right)^n \exp \left(\frac{3}{2} B - n \right)$$

is therefore the conventional instability stress. Let this be denoted by S_2.

The ratio
$$\frac{S_2}{\text{(T.S.)}} = \left(\frac{2}{3} \right)^n \exp \frac{B}{2} = K_2 \text{ (say)}$$

and the pressure at instability can be written

$$p = 2K_2 (\text{T.S.}) \, t_0 / r_0. \tag{10.35}$$

The variation of K_2 with n and B is shown in Fig. 10.21. It is seen that K_2 is always less than unity. This means that if the maximum attainable pressure is calculated on the basis of T.S. alone, it will always be overestimated. For a material with $n = 0.5$ and $B = 0$, this would lead to an overestimation of approximately 22 per cent in the maximum pressure.

10.10.3 Circular Diaphragm

With an annealed metal, a maximum occurs in the pressure and fracture follows, under decreasing pressure, in the region of the pole. The instability strain, $\epsilon_3 = \ln t_0/t$, has been given in equation (10.30) and it follows that the polar stress at instability $\sigma_1 = A(B + \epsilon_3)^n$. Knowing ϵ_3 and σ_1, the polar height and polar radius of curvature are determined from equation (10.28). Hence the maximum pressure, $p = 2\sigma_1 t/\rho$ is determined.

Theoretical and experimental results are compared in Fig. 10.22. Some materials fractured without a decisive maximum in oil pressure, and in these cases experimental results for fracture are presented. It is seen that the theory underestimates the polar height and overestimates the radius of curvature at instability. The maximum oil pressure attained is underestimated for all materials and there is better correlation for the more work-hardened materials. For the annealed materials, the pressure at instability is underestimated by between 5 and 8·5 per cent. This is a consequence of equation (10.28) which is not such a good approximation for the annealed materials as it is for the work-hardened materials. All theoretical results are, of course, dependent on the accuracy with which the empirical law $\bar{\sigma} = A(B + \bar{\epsilon})^n$ can be fitted to the experimentally determined strain-hardening characteristics.

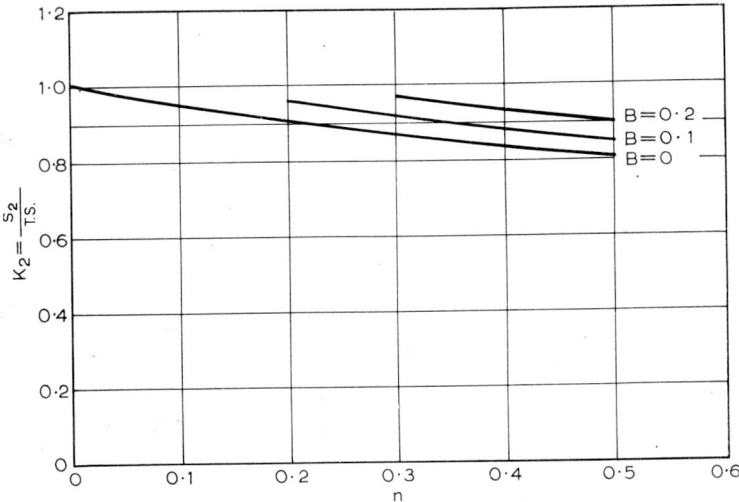

FIG. 10.21 Effect of using tensile strength as basis for design: spherical shell.

10.11 Empirical Equations to represent the Stress-Strain Curve

The most widely used equation to represent a stress-strain curve is of the form $\bar{\sigma} = C\bar{\epsilon}^m$ where C and m are constants for a particular material. The value of the constant m depends on the degree of initial work-hardening, small values indicating a cold-worked material.

WEIL (1958) and SVENSSON (1958) have investigated the bursting pressures of thin-walled cylinders and spherical shells using this equation, and have obtained equations (10.32) and (10.34) for the case $B = 0$, when $n = m$. Curves $B = 0$, Figs. 10.20 and 10.21, may be discussed for the equation $\bar{\sigma} = C\bar{\epsilon}^m$ where work-hardened materials having low values of $m = n$ have the higher K values.

The equation $\bar{\sigma} = A(B + \bar{\epsilon})^n$ has the advantage in a general discussion that the forming capability and the difference between, for example, soft brass and soft aluminium in this respect is emphasized. (See Figs. 10.9 and 10.11) Once the best metal (high value of n) has been chosen, the effect of initial work-hardening on forming capability can be studied. A further advantage of the equation is that it provides a much better fit to the stress-strain curves

Instability Conditions in Circular Diaphragms

Metal	t/t_0 Instability		Oil Pressure lbf/in².		Polar Height (in.)		Polar Radius of Curvature (in.)		Polar Stress lbf/in².	
	Expt.	Theory	Expt.	Theory	Expt.	Theory	Expt.	Theory	Expt.	Theory
Soft copper	0.555	0.563	408	378	3·30	2·88	5·25	5·78	53,000	53,000
Half-hard copper	0·59	0·581	404	396	2·85	2·79	5·65	5·86	52,900	54,800
Soft brass	0·50F	0·498	610F	590	4·18F	3·22	5·05F	5·50	84,900F	89,100
Half-hard brass	0·51F	0·518	650F	633	3·66F	3·12	5·10F	5·57	90,500F	94,300
Soft aluminium	0·565	0·580	148	138	3·20	2·80	5·35	5·86	19,200	19,100
Half-hard aluminium	0·65F	0·613	150F	148	2·43F	2·63	6·35F	6·06	20,400F	20,400
Killed steel	0·56	0·603	602	555	3·23	2·68	5·20	6·00	80,500	79,500
Stainless steel	0·52	0·487	1260	1240	3·90	3·28	5·00	5·45	168,500	190,300

(i) F denotes that the material fractured under increasing pressure.
(ii) The radius of the die was 5 in. and the initial nominal thickness of the metal 0·036 in.

Fig. 10.22

(After Mellor, unpublished work.)

of work-hardened materials. This is illustrated in Fig. 10.23 which shows the stress-strain curves for two tempers of aluminium. The experimental results to high strain values were obtained by taking readings from a plastically deforming circular diaphragm. A stress-strain curve for the half-hard aluminium up to the necking-point in simple tension is also included in the

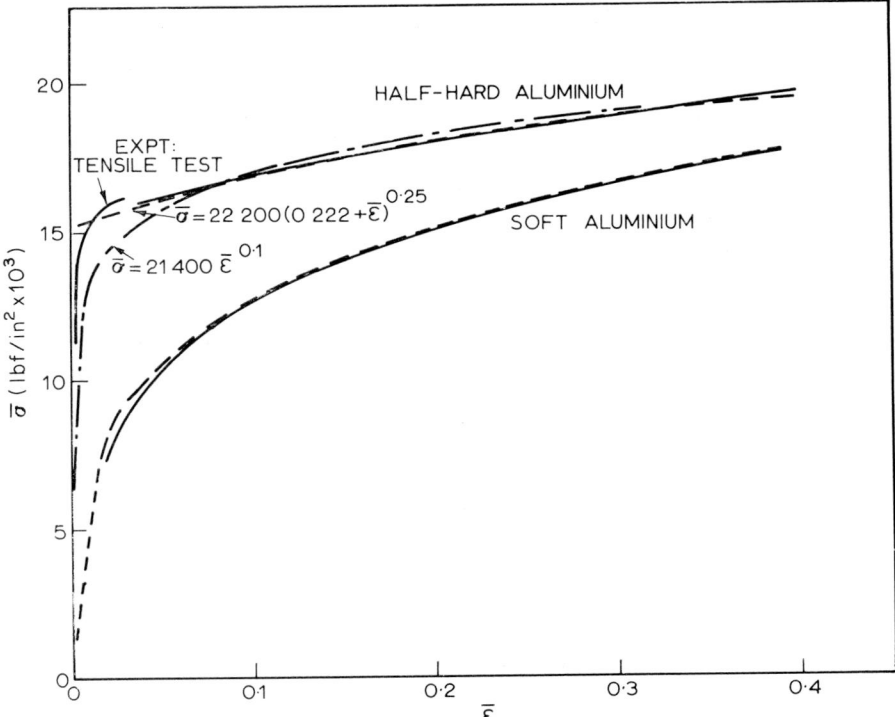

FIG. 10.23 Empirical equations to stress-strain curves of aluminium.

figure. The curve for half-hard aluminium to large strains can be fitted either by the equation

$$\bar{\sigma} = 22{,}200\,(0{\cdot}222 + \bar{\epsilon})^{0\cdot25} \quad \text{or} \quad \bar{\sigma} = 21{,}400\bar{\epsilon}^{0\cdot1}.$$

Taking the latter expression, the theoretical value of thickness ratio at instability of a circular diaphragm is 0·646, which is in very good agreement with the experimental value. However, the latter expression could not be used to predict the instability strain in tension, since it is seen that there is a large discrepancy between theoretical and experimental curves at low values of strain.

The experimental instability strain in tension for half-hard aluminium was found to be 0·03 and from the expression $\bar{\sigma} = 22{,}200(0{\cdot}222 + \bar{\epsilon})^{0\cdot25}$ the theoretical value is $(n - B) = 0{\cdot}028$. It was to be expected that the theoretical value would give the lower figure since elastic strains are neglected. If the equation $\bar{\sigma} = C\bar{\epsilon}^m$ is to be used the constants C and m must be re-determined.

10.12 Instability in Rotating Discs

When the angular velocity of a thin circular disc at room temperature is gradually increased the disc first of all behaves elastically then, at a certain velocity, plastic yielding begins and proceeds, with work-hardening, until all the material of the disc has reached the initial yield point. Further increase in speed causes unrestricted plastic flow to take place. The system is stable as long as an increase in speed is necessary to produce further plastic flow. The following analyses are restricted to conditions of large plastic strain and it will be assumed that the elastic strains are negligible in comparison with the plastic strain. It will also be assumed that the material is homogeneous and isotropic.

10.12.1 *Disc of Uniform Strength* (WALDREN, PERCY and MELLOR, 1966).

A disc of uniform strength is defined as one in which the radial stress σ_r and the tangential stress σ_0 are equal and have the same value at all radii. The axial stress σ_z is zero. This means that the disc must be solid and relatively thin for the assumption $\sigma_z = 0$ to be valid. Further, there must be radial loading at the periphery of the disc and this will be assumed to take the form of blades. The disc of uniform strength is usually thought of in elastic terms but we have here extended the definition to large plastic strains.

The effective stress is,

$$\bar{\sigma} = \sigma_r = \sigma_\theta \tag{10.36}$$

and the corresponding effective strain is,

$$\bar{\epsilon} = 2\epsilon_r = 2\epsilon_\theta = -\epsilon_z \tag{10.37}$$

where ϵ_r, ϵ_θ and ϵ_z are, respectively, the radial, tangential and axial (thickness) strains.

Figure 10.24 shows a model rotor which is assumed to have been so

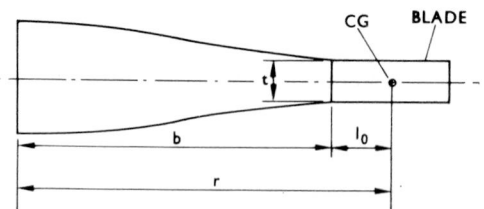

FIG. 10.24 Idealized disc of uniform strength.

designed that it is uniformly stressed in the plane of the disc. At a certain speed, ω, the current radius to the blade root is b, and the current radius to the centre of gravity of a blade is r. The current thickness at the blade root is t. As drawn, the smallest value of thickness is at the blade root. This would obviously not be the case for a practical rotor and the effect of thickening the disc at its join with the blades will be discussed later.

It will be assumed that the stress due to the centrifugal effect of the blading is distributed uniformly over the circumferential area at the root of the blades. We may then write

$$\sigma_r(2\pi bt) = C_1\omega^2 r. \tag{10.38}$$

The constant C_1 includes such factors as volume, density and number of blades. The choice of blade geometry to give a particular centrifugal effect is not unique. Though the multiple $C_1 r_0$, where r_0 is the initial radius to the centre of gravity of a blade, would have a certain value for a particular centrifugal effect, C_1 and r_0 can be varied (within certain limits) provided that their multiple does not change.

As the speed of the rotor is increased the current radius of the disc, b, increases and the centre of gravity of the blades also moves radially outwards. It will be shown later for the practical model rotor that the blades are only stressed elastically, even at burst, and that therefore the relative movement between the centre of gravity of a blade and the current radius b may be neglected. Assuming that this is true for all practical cases, then $r = b + l_0$ where l_0 is the constant distance between the blade root and the blade centre of gravity. Therefore,

$$\sigma_r(2\pi bt) = C_1\omega^2(b + l_0)$$

or

$$\sigma_r t = C_2\omega^2 \frac{(b + l_0)}{b} \tag{10.39}$$

where C_2 is a second constant.

At the instability condition the angular velocity is a maximum and the condition for this is, from equation (10.39),

$$\frac{d\sigma_r}{\sigma_r} + \frac{dt}{t} = \frac{db}{b + l_0} - \frac{db}{b}. \tag{10.40}$$

Now, $dt/t = d\epsilon_z$, $db/b = d\epsilon_\theta$ and since $2d\epsilon_\theta = -d\epsilon_z$ from equation (10.37) and $\bar{\sigma} = \bar{\sigma}_r$ from equation (10.36) the instability condition may be written as

$$\frac{d\bar{\sigma}}{\bar{\sigma}} = \frac{db}{b} + \frac{db}{b + l_0}. \tag{10.41}$$

This equation is open to particularly simple interpretations for values of $l_0 = 0$ and $l_0 = \infty$. The case $l_0 = 0$ means that theoretically the centrifugal loading is concentrated at radius b. Equation (10.41) then becomes

$$\frac{d\bar{\sigma}}{\bar{\sigma}} = 2\frac{db}{b} = 2d\epsilon_\theta$$

or

$$\frac{d\bar{\sigma}}{d\bar{\epsilon}} = \bar{\sigma}.$$

Thus for a material with a work-hardening characteristic $\bar{\sigma} = K\bar{\epsilon}^n$ the effective strain at instability is n. This is also the value of the thickness strain at instability. Also the stress components in the disc have values identical to the true stress at instability for the same material in simple tension. Note that the tangential strain in the disc at instability has a value $n/2$.

10—EP   *   *

As $l_0 \to \infty$ the term $db/(b + l_0) \to 0$ and equation (10.41) becomes

$$\frac{d\bar{\sigma}}{d\bar{\epsilon}} = \frac{\bar{\sigma}}{2}.$$

The thickness strain at instability has a value $2n$ and the tangential strain has a value n. The true stress at instability of the disc will be greater than the true stress at instability in the simple tensile test. The behaviour of the disc when $l_0 \to \infty$ is physically equivalent to the behaviour of a flat sheet of metal of uniform thickness subjected to a uniform tensile radial stress at the periphery.

The instability condition of a disc of uniform strength thus depends on the radial movement of the centre of gravity of the blades. Since the blades can be thought of as rigid, in this context, instability will depend on l_0 and on the value of the strain-hardening index n of the metal. The value of the tangential strain at instability for a practical rotor will lie between the extreme values $n/2$ and n. This variation in instability strain may be further examined by considering l_0 as a fraction, p, of the radius to blade root at instability b_m. If we put $l_0 = pb_m$ then equation (10.41) can be written

$$\frac{d\bar{\sigma}}{\bar{\sigma}} = \frac{2 + p}{1 + p} \frac{db_m}{b_m}$$

or

$$\frac{d\bar{\sigma}}{d\bar{\epsilon}} = \frac{1}{2} \frac{(2 + p)}{(1 + p)} \bar{\sigma}.$$

The effective instability strain is $\bar{\epsilon}_m = 2n(1 + p)/(2 + p)$ and the tangential strain $\epsilon_{\theta m} = n(1 + p)/(2 + p)$. The variation of $(\epsilon_{\theta m}/n)$ with (l_0/b_0) would be a more useful way of presenting the information and for a particular value of n the ratio (l_0/b_0) is easily obtained, since $\epsilon_{\theta m} = \ln (b_m/b_0)$. The only drawback is that the variation of $(\epsilon_{\theta m}/n)$ with (l_0/b_0) is different for each value of n. This method of representation has been adopted for a value of $n = 0.105$ and is shown in Fig. 10.25. This value of n was chosen since it is the value for the material used in experimental work on a disc of uniform strength at the National Gas Turbine Establishment. The difference between the two curves in Fig. 10.25 is small because of the low value of n chosen. Values of (l_0/b_m) and (l_0/b_0) up to unity are included in the figure but in practical cases these ratios will usually be less than 0.5. For a value of $n = 0.105$ and $l_0/b_0 = 0.244$ the tangential instability strain would be 0.058 or approximately 10 per cent greater than the minimum possible value of 0.0525. The chosen value of $l_0/b_0 = 0.244$ is the value for the model disc tested at the National Gas Turbine Establishment.

The plastic deformation of rotating discs, under plane stress conditions, has been analysed many times. The analyses have been based on either the Tresca yield criterion and its associated flow rule or on the Hencky deformation theory of plasticity. The Tresca and Hencky solutions have been compared by PERCY and MELLOR (1964). The Hencky theory would appear to be the most fruitful and the finite difference method of solution set down by MANSON (1951) has been adopted. This theory assumes a 1:1 relationship between stress and strain and will be the correct solution if it can be shown *a posteriori* that the stress ratios have remained constant throughout the straining. The use of an electronic digital computer has enabled many more results to be obtained, with greater accuracy, than has previously been possible.

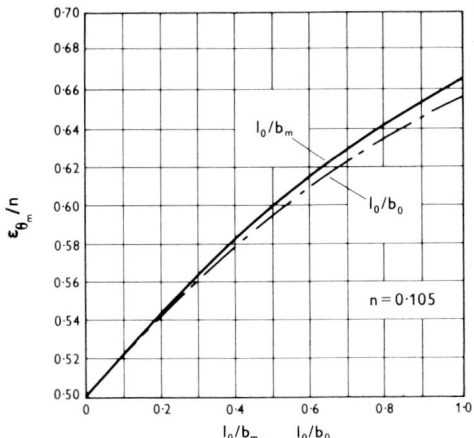

FIG. 10.25 Variation of instability strain with strain-hardening index *n* and geometry of disc. (*After* Waldren, Percy and Mellor, *Proc. Instn mech. Engrs.*)

The computational procedure was as follows. A strain at the centre of the disc and an angular velocity were chosen, and starting at the disc centre the stress and strain distributions were calculated by the finite difference method at a large number of stations across the disc. The stations were not uniformly distributed but were taken much closer together where the thickness of the disc changed rapidly or where rapid changes in stress and strain were anticipated. If the calculated radial stress at the periphery was correct, that is zero for a simple disc or equal to the radial stress calculated from the blade loading in the case of a model turbine disc, then the correct value of speed for the particular centre strain had been chosen. If the computed radial stress at the periphery was incorrect a different value of the angular velocity was taken keeping the same centre strain, and the computation repeated. Three or four estimates were usually required before an accurate value of speed was obtained. When the correct value of speed had been determined a further strain value was taken and the speed appropriate to this new value computed. Increasing values of strain were considered until the maximum speed, and instability strain, had been determined. The distribution of stress and strain across the disc was automatically tabulated for each value of centre strain. In this manner it was possible to study the effect of geometry and the effect of work-hardening on the stress and strain distribution across the disc and also on the instability condition.

The initial dimensions of a disc of uniform strength are as shown in Figure 10.26. The disc was designed and spun to burst at the National Gas Turbine Establishment. The following numerical analysis is based on the dimensions of the disc and on the strain-hardening characteristic of the disc material which could be represented by $\bar{\sigma} = 70 \cdot 5 \bar{\epsilon}^{0 \cdot 105}$. The actual material had an initial yield stress of 38·4 tonf/in², and the tensile strength was 50 tonf/in². It must be stressed that in solving this problem by numerical analysis

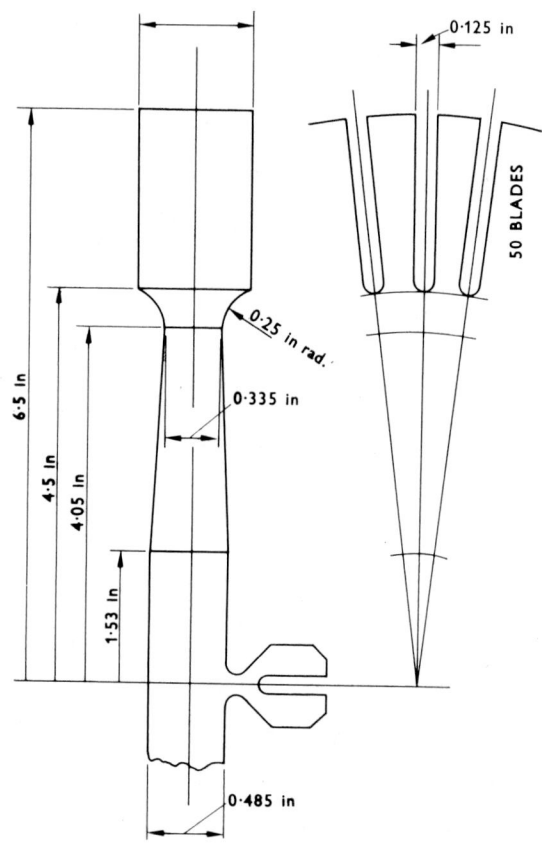

Fɪɢ. 10.26 Disc of uniform strength. (*After* Waldren, Percy and Mellor, *Proc. Inst. mech. Engrs.*)

it has nowhere been assumed that the disc is of uniform strength. If in fact the disc is of uniform strength this will be evident in the results. It is assumed that the centrifugal effect of the blades is to produce a uniform radial stress at the periphery, initial radius 2·25 in. The circumferential area at the periphery was assumed to be constant for the purposes of calculating the centrifugal effect. It is verifiable *a posteriori* that this assumption is justified.

The theoretical variation of disc speed with centre strain was graphed and the maximum speed found to be 54,307 rev/min. The associated natural tangential strain at the centre is estimated to be 0·06. The theoretical stress and strain distributions in the discs *just prior* to the instability point are shown in Figs. 10.27 and 10.28 when the speed was 54,305 rev/min. It will be seen from Fig. 10.27 that the tangential and radial stress components are almost identical and fairly uniform from the centre to the smallest section of the rotor. The deviation between the two stress components occurs because of the discontinuity of the disc profile at a radius of 0·765 in.

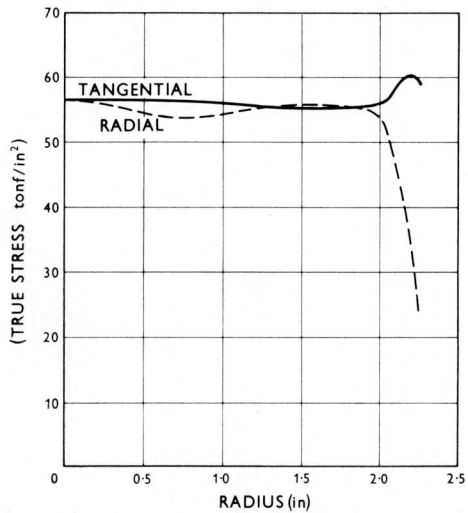

FIG. 10.27 Theoretical stress distribution for disc of uniform strength.
(*After* Waldren, Percy and Mellor, *Proc. Instn mech. Engrs.*)

The strain components Figure 10.28 show a similar deviation. Again, there is the greatest discrepancy at a radius of 0·765 in. It will be noticed that the tangential strain has a high value even at the rim, but that the radial strain drops off to a small negative value at the rim.

The stress and strain distributions at the join of the dummy blades to the main rotor are extremely complex. Except in a very narrow region close to the rim of the disc, the blades are subjected to a maximum radial stress of 24 tonf/in² which is well below the material yield point of 38 tonf/in² and this is accompanied by a tangential strain of 0·046. Considering the disc alone the radial strain at the rim is practically zero and therefore, from considerations of incompressibility, the thickness strain at the rim is numerically equal, but of opposite sign, to the tangential strain. Consequently there will be practically no change of circumferential areas at the rim and the radial stress at this radius may be based on the original area, as already stated.

The experimental burst speed was 54,900 rev/min. As is usual in this work there is thus good correlation with the theoretical instability speed of 54,307 rev/min. This is largely because the speed is sensibly constant for a large increase in strain. The computation had to be carried out with great accuracy in order to deal with speeds differing by only 5, 10 or at a maximum 200 rev/min from one another.

The experimental tangential strain at the centre of the disc measured after burst was approximately 0·05 while the theoretical value of the centre strain was 0·06. The simple theory predicted an instability strain of 0·058. Measurement of the actual disc after burst indicated that the strains were more uniformly distributed than predicted by theory.

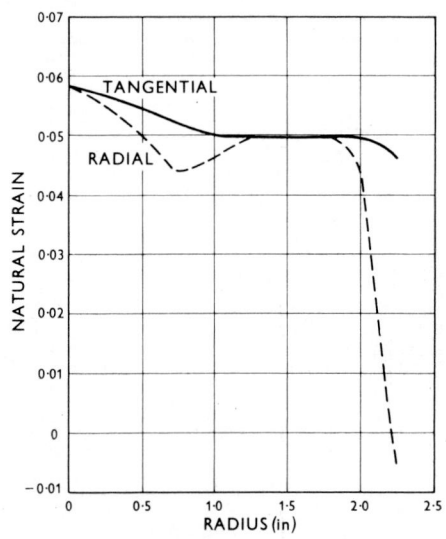

FIG 10.28 Theoretical strain distribution for disc uniform strength. (*After* Waldren, Percy and Mellor, *Proc. Instn mech. Engrs.*)

10.12.2 *Disc of Uniform Thickness*

The Hencky theory has also been applied to hollow discs of uniform thickness and the results compared with detailed experiments by PERCY (1967). The experimental work showed that at instability the bore of the disc becomes elliptical and the greatest thinning at the bore occurs along the minor axis of the ellipse. This is shown very clearly in Figure 10.29 where one face of a deformed disc has been successively ground down.

ROBINSON (1944) proposed that a disc fails when the nominal average tangential stress equals the tensile strength of the material. The nominal average tangential stress is defined as the centrifugal force on one half of the disc divided by the area of the original diametral cross-section. For a disc of uniform thickness this has a value

$$\frac{1}{3} \rho \omega_b^2 \left(\frac{b_0^3 - a_0^3}{b_0 - a_0} \right),$$

where a_0 is the internal radius and b_0 the external radius of the disc, ρ is the density of the material and ω_b the angular velocity at burst. A comparison

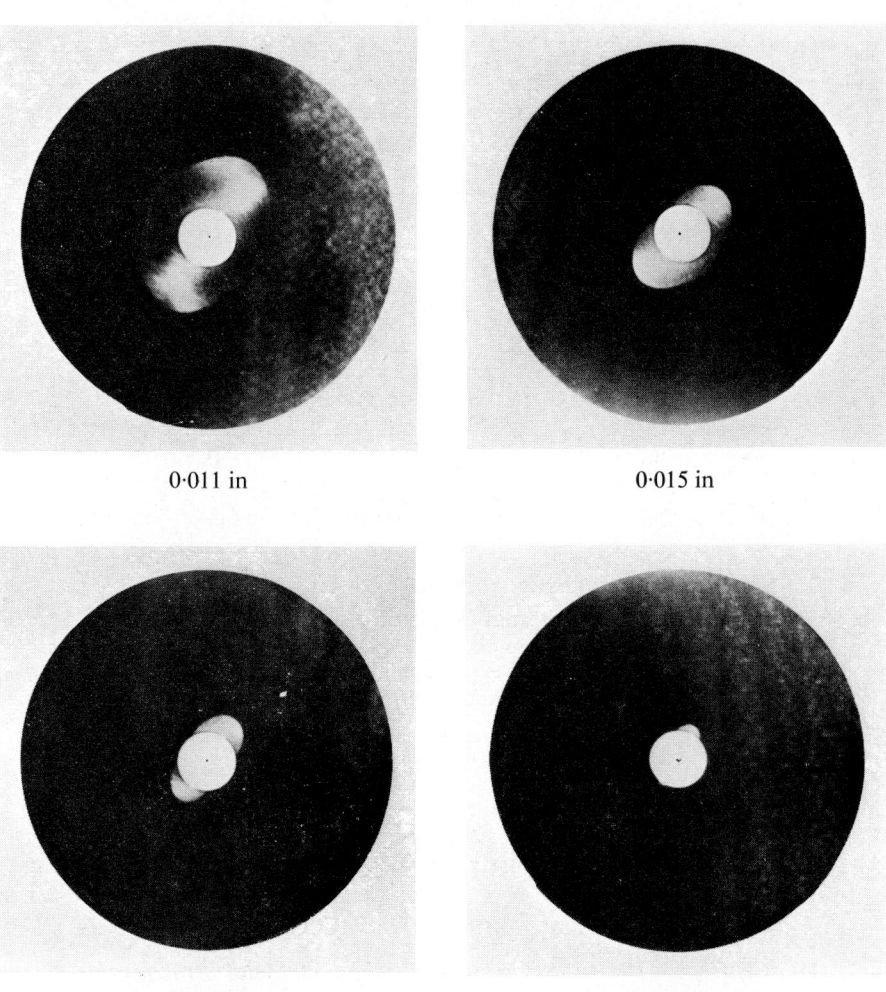

0·011 in   0·015 in

0·010 in   0·021 in

FIG. 10.29 Successive grinding operations in disc to locate point of
initiation of necking at the bore. Grinding depths as shown. (*After*
Percy, unpublished work.)

between predicted and experimental burst speeds is shown in Fig. 10.30. It
is seen that the Hencky theory predicts the burst speed with more accuracy
than the tensile strength criterion but for this particular material the latter
criterion is sufficiently accurate for design purposes. However, as explained
in Section 10.10, the accuracy of the tensile strength criterion depends on the
work-hardening characteristic of the material. In an extensive series of tests
WINNE and WUNDT (1958) showed that the criterion could overestimate the
burst speed of a disc by up to 5·5 per cent or underestimate it by up to 4 per
cent.

Radius Ratio $\dfrac{b_0}{a_0}$	Actual Burst Speed rev/min	Predicted Burst Speed (Hencky) rev/min	Predicted Burst Speed ($Av = T.S$) rev/min
3·5	89,100	86,874	86,573
5·0	93,850	92,188	90,903
7·5	98,280	96,550	94,400
10·0	99,270	98,420	96,051
15·0	101,170	100,200	97,850
20·0	102,120	100,968	98,680

b_0 is initial external radius of disc.
a_0 is initial internal radius of disc.

FIG. 10.30 Comparison of experimental and theoretical burst speeds for discs of uniform thickness. (*After* Percy, unpublished work.)

See Problems 40-42

REFERENCES

PLASTIC BUCKLING OF A COLUMN

BLEICH, F.   1952 *Buckling Strength of Metal Structures* McGraw-Hill, New York

HILL, R.   1957 'Stability of Rigid-plastic Solids' *J. Mech. Phys. Solids* **6**, 1

HILL, R. and SEWELL, M. J.   1959 'A General Theory of Inelastic Column Failure' I and II *J. Mech. Phys. Solids* **8**, 105

KARMAN, TH. V.   1910 *Untersuchungen über Knickfestigkeit, Mitteilungen über Forschungsarbeiten, Verein deutscher Ingenieure, Heft* 81 Springer, Berlin

SHANLEY, F. R.   1947 'Inelastic Column Theory' *J. Aero Sci.* **14**, 251

VAN DEN BROEK, J. A.   1945 'Column Formula for Materials of Variable Modulus' *Eng. J. (Canada)* **28**, 772

WANG, CHI-TEH   1948 'Inelastic Column Theories and an Analysis of Experimental Observation'. *J. Aero. Sci.*, **15**, 283

PLASTIC INSTABILITY IN TENSION

DAVIS, E. A.   1945 'Yielding and Fracture of Medium-carbon Steel under Combined Stress' *J. appl. Mech.* **12**, A–13

HILL, R.   1950 'A Theory of the Plastic Bulging of a Metal Diaphragm by Lateral Pressure' *Phil. Mag.* (Ser. 7), **41**, 1133

HILLIER, M. J.   1965 'Tensile Plastic Instability of Thin Tubes' *Int. J. mech. Sci.* **7**, 531

JONES, B. H. and MELLOR, P. B.   1967 'Plastic Flow and Instability Behaviour of Thin-walled Cylinders Subjected to Constant-ratio Tensile Stress' *J. Strain Analysis* **2**, 62

JOHNSON, W., DUNCAN, T. L., KORMI, K., SOWERBY, R. and TRAVIS, F. W.   1963 'Some Contributions to High Rate Sheet Metal Forming' *Proc. 4th Conf. Mach. Tool Des. and Res.*, Pergamon Press, 257 pp.

LANKFORD, W. T. and SAIBEL, E.   1947 'Some Problems in Unstable Plastic Flow under Biaxial Tension' *Metals Technol. tech. Publ.* 2238

MANSON, S. S.   1951 'Analysis of Rotating Discs of Arbitrary Contour and Radial Temperature Distribution in the Region of Plastic Deformation' *Proc. 1st U.S. Nat. Cong. of appl. Mech.* 595

MARCINIAK, Z.   1958 'Analysis of Plastic Instability of Thin-walled Tubes under Biaxial Stress' *Rozpr. inż.* **110**, 529

MARCINIAK, Z. and KUCZYNSKI, K.   1967 'Limit strains in the processes of stretch-forming sheet metal' *Int. J. mech. Sci.* **9**, 609

MELLOR, P. B.   1956 'Stretch-forming under Fluid Pressure' *J. Mech. Phys. Solids* **5**, 41

1960 'Plastic Instability in Tension' *The Engineer, London,* March 25

1960 'The Ultimate Strength of Thin-walled Shells and Circular Diaphragms subjected to Hydrostatic Pressure' *Int. J. mech. Sci.* **1**, 216

1962 'Tensile Instability in Thin-walled Tubes' *J. mech. Engng Sci.* **4**, 251

PERCY, M. J.   1967 'Instability in Rotating Discs with Large Plastic Strains' *Ph.D. Thesis,* University of Liverpool

PERCY, M. J. and
MELLOR, P. B.
1964 'Theoretical Prediction of Tensile Instability in Rotating Discs' *Int. J. mech. Sci.* **6**, 421

ROBINSON, E. L.
1944 'Bursting Tests on Steam Turbine Disc Wheels' *Trans. A.S.M.E.* **66**, 373

SACHS, G. and
LUBAHN, J. D.
1946 'Failure of Ductile Metals in Tension' *Trans. A.S.M.E.*, **68**, 271

SOWERBY, R.
and DUNCAN, J. L.
1971 'Failure in sheet metal in biaxial tension' *Int. J. mech. Sci.*, **13**, 217

STORAKERS, B.
1968 'Plastic and Visco-plastic Instability of a Thin Tube under Internal Pressure, Torsion and Axial Tension' *Int. J. mech. Sci.* **10**, 519

SVENSSON, N. L.
1958 'The Bursting Pressure of Cylindrical and Spherical Vessels' *J. appl. Mech.* **25**, 89

SWIFT, H. W.
1952 'Plastic Instability Under Plane Stress' *J. Mech. Phys. Solids* **1**, 1

THOMAS, D. G. B.
1954 'Calculations on the Plastic Bulging of an Annealed Copper Diaphragm' *B.I.S.R.A.*, Research Report, MW/B/4/54

VENTER, R.,
JOHNSON, W. and
DE MALHERBE, M. C.
1971 'The limit strains in inhomogeneous sheet metal in biaxial tension' *Int. J. mech. Sci.*, **13**, 299

WALDREN, N. E.,
PERCY, M. J. and
MELLOR, P. B.
1966 'Burst Strength of Rotating Discs' *Proc. Instn mech. Engrs* **180**, 111

WEIL, N. A.
1958 'Bursting Pressures and Safety Factors for Thin-walled Vessels' *J. Franklin Inst.* **265**, 97

1959 'Rupture Characteristics of Safety Diaphragms' *J. appl. mech.* **26**, 621

WINNE, D. W. and
WUNDT, B. M.
1958 'Application of the Griffith-Irwin Theory of Crack Propagation to the Bursting Behaviour of Discs, Including Analytical and Experimental Studies' *Trans. A.S.M.E.* **80**, 1643

Chapter 11

MECHANICS OF METAL FORMING I

11.1 Introduction

The degree of correlation between any theory and the physical situation it attempts to describe depends on how closely the theoretical model can be made to fit the real conditions. It is with this in mind that fairly full descriptive accounts are given of experimental investigations of a number of metal-forming processes. Wherever possible the accuracy of any theoretical solution will be checked against experimental work.

In this chapter the theoretical solutions are developed from an appreciation of homogeneous deformation, without friction between tools and deforming material. Therefore, if we consider a bar of non-hardening material of initial cross-sectional area A_0 deformed to a final cross-sectional area A, the work done per unit volume is $Y \ln A_0/A$ where Y is the yield stress of the material. This is an expression often met with in elementary treatments of metal-forming, though Y is sometimes looked on as a factor to be adjusted to give the best agreement with experiment. In practice, the work per unit volume required for a given deformation will be greater than that given in the above expression because of work-hardening of the material, friction between tools and material, and because of non-useful work of distortion.

In the following theoretical treatments of a number of metal-forming processes, elastic strains are neglected as being small compared with the plastic strains. This assumption of a rigid-plastic material means that residual stresses cannot be studied theoretically, though experimental determination of residual stresses will be discussed.

11.2 The Sinking of a Thin-walled Tube: General Considerations

Hollow sinking or tensile sinking consists in drawing a tube through a die such that its external diameter is reduced, Fig. 11.1(*a*). The operation modifies the mechanical properties of the material, gives a closer tolerance on the outside diameter and improves the degree of surface finish. In plug drawing, Fig. 11.1(*b*), a mandrel (or plug) is inserted in the die to govern the final thickness of the drawn tube. (See an investigation of the plug drawing process by BLAZYNSKI and COLE, 1960, also MOORE and WALLACE, 1967/68, and FLINN, 1969.) The hollow-sinking method generally allows larger reduction and is the process referred to below. In practice, the maximum reduction in diameter is approximately 30 per cent.

Equipment for the cold drawing of steel tubes is described at length by
Rozov (1968).

The drawing of a tube of uniform wall thickness through a straight-
tapered die, Fig. 11.2(*a*), will be examined, the tube wall being assumed
so thin in comparison with the diameter that the effects of plastic bending and
the variation of stress through the wall thickness can be neglected. As the tube
is drawn through the die, each part of the tube passes through exactly the
same operation, and if the tube is sufficiently long, a steady state is attained.

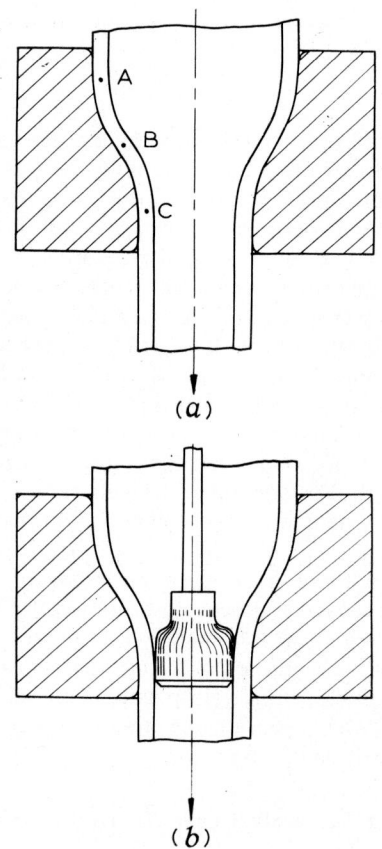

(*a*)

(*b*)

Fig. 11.1 Methods of cold-drawing tubes.
(*a*) Hollow sinking.
(*b*) Plug drawing.

The stresses in the tube are the tensile stress σ_1, parallel to the die face, the
compressive stress σ_2 due to contact with the die face and the compressive
hoop stress σ_3. If the tube wall is very thin, the compressive stress σ_3 may be
great enough to cause the tube to buckle, a fold forming along the length of
the tube, Sachs and Baldwin, (1946a). For convenience the stresses on the

element of tube in Fig. 11.2(b) are all shown as tensile, i.e., positive. The actual direction of the stresses will become apparent on solution.

There is also a frictional force between the die face and the tube wall dependent upon the degree of lubrication and the value of the pressure σ_2. Remembering that σ_2 is, in fact, compressive, the frictional force is $-\mu\sigma_2$ in the direction shown, where μ is the coefficient of friction. Strictly speaking, σ_2 is not therefore a principal stress. Also σ_1 on the perpendicular plane must be associated with a non-uniform shear stress, but for normal coefficients of friction this latter stress is negligible.

If a steady state condition has been reached, we can consider the equilibrium of an element of the tube, Fig. 11.2(b).

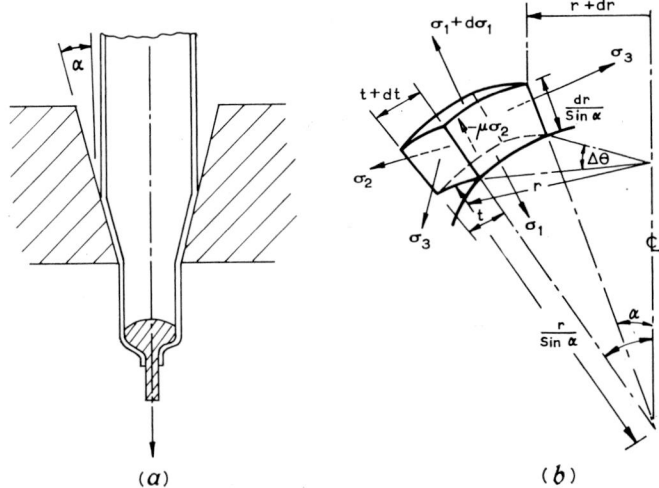

(a)   (b)

FIG. 11.2

(a) Hollow sinking through straight die.
(b) Stresses acting on small element of the wall.

Resolving forces perpendicular to the element of the die,

$$\sigma_2\left(r\Delta\theta\,\frac{dr}{\sin\alpha}\right) = \sigma_3\left(t\frac{dr}{\sin\alpha}\right)\Delta\theta\cos\alpha$$

or

$$\sigma_2 = \sigma_3\frac{t}{r}\cos\alpha. \tag{11.1}$$

σ_2 is small in comparison with σ_3 when t/r is small, as is assumed in this analysis.

Resolving forces parallel to the element of the die

$$(\sigma_1+d\sigma_1)(r+dr)\Delta\theta(t+dt)-\sigma_1 tr\Delta\theta-\sigma_3\left(t\frac{dr}{\sin\alpha}\right)\Delta\theta\sin\alpha-\mu\sigma_2\left(r\Delta\theta\frac{dr}{\sin\alpha}\right) = 0.$$

Neglecting second-order small quantities,

$$d(\sigma_1 tr)\,\Delta\theta - \sigma_3\,(t\,dr)\,\Delta\theta - \mu\sigma_2\left(r\Delta\theta\,\frac{dr}{\sin\alpha}\right) = 0$$

or
$$\frac{d}{dr}(\sigma_1 tr) - \sigma_3 t - \mu\,\frac{\sigma_2 r}{\sin\alpha} = 0. \tag{11.2}$$

Substituting the value of σ_2 from equation (11.1) into equation (11.2), we obtain

$$\frac{d}{dr}(\sigma_1\,tr) - \sigma_3 t\,(1 + \mu\cot\alpha) = 0. \tag{11.3}$$

It has been shown by SWIFT (1949) that little difference is made to the values of the calculated stresses and the thickness strains if t is assumed constant in the above equation and therefore the variation of t in the equilibrium equation will be neglected. Hence

$$\frac{d}{dr}(\sigma_1 r) - \sigma_3\,(1 + \mu\cot\alpha) = 0. \tag{11.4}$$

A second equation for the stresses is given by the yield criterion of the metal. Under the supposed and known stress conditions $\sigma_1 > 0 > \sigma_3$ and $\sigma_2 = 0$, which substituted into the von Mises' criterion gives

$$\sigma^2 - \sigma_1\sigma_3 + \sigma_3^2 = Y^2$$

where Y is the uniaxial yield stress.

Under the same stress conditions, the Tresca yield criterion gives

$$\sigma_1 - \sigma_3 = Y$$

remembering that σ_3 has been given a positive direction, Fig. 11.2(b).

While von Mises's criterion is more correct for non-ferrous metals, Tresca's criterion is mathematically simpler to apply when the magnitude and sense of the stresses are known. As a compromise, therefore, the yield criterion adopted will be

$$\sigma_1 - \sigma_3 = mY \tag{11.5}$$

where m is a constant obtained by the method of least squares to satisfy von Mises' criterion. The graphical representation of this modified criterion is shown in Fig. 11.3 and the value of m is $1\cdot1$.

The relations between stresses and strain increments for the assumed rigid-plastic material are given by the Lévy-Mises flow condition

$$\frac{d\epsilon_2}{d\epsilon_3} = \frac{dt/t}{dr/r} = \frac{\sigma_2 - \sigma_m}{\sigma_3 - \sigma_m}$$

where $\sigma_m = (\sigma_1 + \sigma_2 + \sigma_3)/3$, the hydrostatic component of the stress system.

Neglecting σ_2 in comparison with σ_1 and σ_3,

$$\frac{dt}{t}\bigg/\frac{dr}{r} = \frac{\sigma_1 + \sigma_3}{\sigma_1 - 2\sigma_3}. \tag{11.6}$$

In the following solutions, work-hardening will be neglected and the effect of this will be discussed later.

The final relation required for completing the solution is given by the equation of incompressibility

$$\frac{dt}{t} + \frac{dr}{r} + \frac{dl}{l} = 0 \tag{11.7}$$

where dl/l is the strain increment in the direction of σ_1.

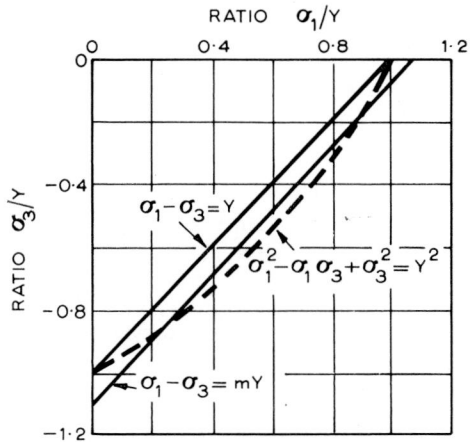

FIG. 11.3 Graphical representation of yield criteria.

11.3 Frictionless Tube-sinking

The simplest analytical solution is obtained for the hypothetical case of no friction between the tube and the die face, i.e., $\mu = 0$. The equilibrium equation (11.4) then reduces to

$$\frac{d}{dr}(\sigma_1 r) - \sigma_3 = 0. \tag{11.8}$$

Combining this equation with the yield criterion, equation (11.5),

$$d\sigma_1 = - mY \frac{dr}{r}$$

and, integrating $\qquad \sigma_1 = - mY \ln r + C$

where C is a constant ascertainable from the known boundary conditions. The longitudinal stress $\sigma_1 = 0$ at the entrance to the die, where $r = r_0$ provides $C = mY. \ln r_0$ and the stresses at any radius r are then

$$\sigma_1 = mY \ln \frac{r_0}{r} \text{ and } \sigma_3 = mY \left(\ln \frac{r_0}{r} - 1 \right). \tag{11.9}$$

In particular, the sinking stress $\sigma_1 = m Y \ln r_0/r_1$ where r_1 is the radius at exit.

The thickness strain is obtained directly by substituting the above stress values in the Lévy-Mises relation, equation (11.6). Thus,

$$\frac{dt}{t} = \frac{2 \ln (r_0/r) - 1}{2 - \ln (r_0/r)} \cdot \frac{dr}{r}.$$

Integrating and putting $t = t_0$, the initial wall thickness at entry to the die,

$$\ln \frac{t}{t_0} = 2 \ln \frac{r_0}{r} + 3 \ln \left[1 - \frac{\ln (r_0/r)}{2} \right].$$

Finally, the axial strain at any given die radius may be found from the incompressibility condition, equation (11.7).

11.4 Tube-sinking with Wall Friction

The stress, σ_3, is obtained from the assumed yield criterion (11.5) and substituted in the equilibrium equation (11.4), to give

$$\frac{d\sigma_1}{\sigma_1 \mu \cot \alpha - m Y(1 + \mu \cot \alpha)} = \frac{dr}{r}.$$

Integrating,

$$\frac{1}{\mu \cot \alpha} \ln \left\{ \sigma_1 \mu \cot \alpha - m Y(1 + \mu \cot \alpha) \right\} = \ln rK$$

or

$$\sigma_1 \mu \cot \alpha - m Y(1 + \mu \cot \alpha) = rK^{\mu \cot \alpha}$$

where K is a constant obtained by knowing that $\sigma_1 = 0$ at the entry where $r = r_0$. Hence σ_1 is given by

$$\frac{\sigma_1}{m Y} = \left(1 + \frac{1}{\mu \cot \alpha} \right) \left\{ 1 - \left(\frac{r}{r_0} \right)^{\mu \cot \alpha} \right\}. \tag{11.10}$$

By writing $(r/r_0)^{\mu \cot \alpha} = \exp [\mu \cot \alpha \ln (r/r_0] \simeq 1 - \mu \cot \alpha \ln (r_0/r)$

and allowing μ to tend to zero, equation (11.10) reduces to $\sigma_1 = m Y \ln r_0/r$ the stress in frictionless drawing.

The variation of sinking stress $\sigma_1/m Y$ with reduction r_1/r_0 is shown in Fig. 11.4 for different values of $\mu \cot \alpha$. The load required for hollow sinking is $2\pi r_1 t_1 \sigma_1/\cos \alpha$ where t_1 is the thickness at the exit side of the die. The thickness strain has already been found when $\mu = 0$ and the procedure is exactly the same when friction is considered.

Putting $a = \mu \cot \alpha$ and $b = (1 + a)/a$ the hoop stress σ_3 is, from the yield criterion

$$\frac{\sigma_3}{m Y} = b \left\{ 1 - \left(\frac{r}{r_0} \right)^a \right\} - 1$$

Substituting the values of the stresses into the Lévy-Mises equation (11.6)

$$\frac{dt}{t} = \frac{2b\{1 - (r/r_0)^a\} - 1}{2 - b\{1 - (r/r_0)^a\}} \cdot \frac{dr}{r}.$$

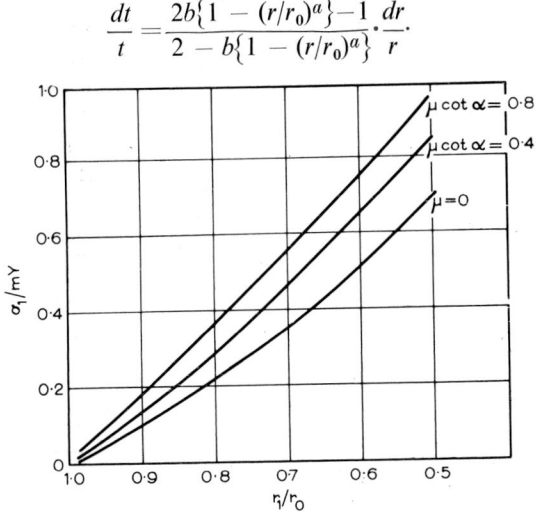

FIG. 11.4 Variation of sinking stress with reduction in radius.

Integrating and using the boundary condition $t = t_0$ where $r = r_0$

$$\ln \frac{t}{t_0} = \frac{3}{a - 1} \ln \left[\frac{2(r/r_0)^a}{2 - b\{1 - (r/r_0)^a\}} \right] - 2 \ln \frac{r}{r_0}. \quad (11.11)$$

The variation of the thickness ratio t_1/t_0 with reduction r_1/r_0 is shown in Fig. 11.5 for different values of $\mu \cot \alpha$. It is seen that, for the reductions and

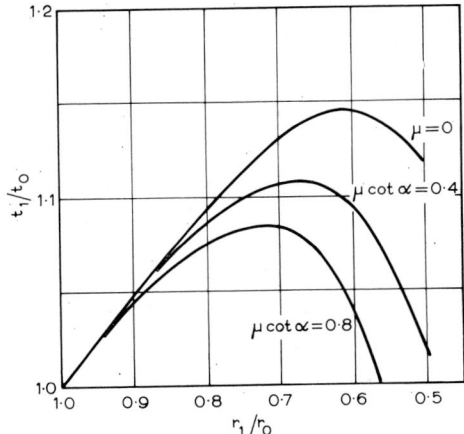

FIG. 11.5 Variation of thickness strain with reduction in radius.

coefficients of friction considered, there is a resultant thickening of the tube-wall.

The thickness, t_1, of the tube wall at exit can be found and the load required for hollow sinking evaluated.

It has already been stated that little difference is made to the values of the thickness strains if t is assumed constant in the equilibrium equation. It has also been shown again by Swift that strain-hardening has little effect on the strains; however, it may have a marked effect on sinking loads. If the tube is initially work-hardened, it may be sufficient to assume a mean yield stress, but for a tube that is initially annealed, a stress-strain relationship must be brought into the solution. (SWIFT, 1949; CHUNG, 1951; HILL, 1950.)

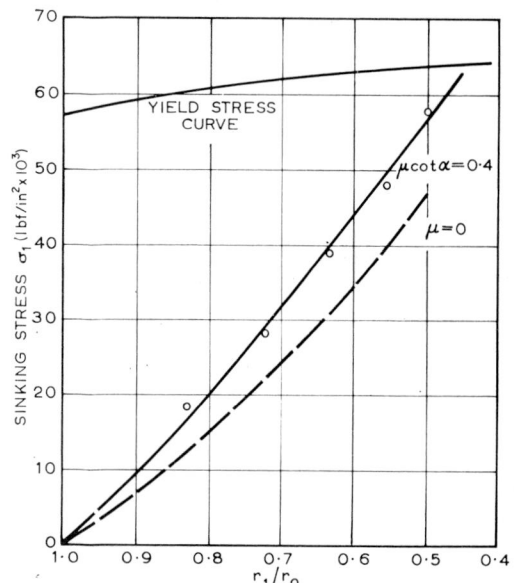

FIG. 11.6 Theoretical and experimental sinking stress for hard copper.

∘ Experimental values.

11.5 Comparison of Theoretical and Experimental Stresses in Tube–sinking

For a comparison of theoretical and experimental sinking stresses, some experimental results published by SACHS and BALDWIN (1946b) are reproduced in Fig. 11.6. In these experiments, the ratio t_0/r_0 was approximately 0·06, coefficient of friction $\mu = 0\cdot1$, $\alpha = 14°$; a mean yield stress $mY = 60,000$ lbf/in² was assumed. It is seen that when the frictional effect is taken into account, the correlation between theoretical and experimental results is good while neglecting the frictional effect may cause errors of up to 20 per cent.

The above analysis sets down the basic equations for the problem of hollow sinking. In industrial practice, the dies are curved and the tubes cannot always be considered as thin-walled (Fig. 11.1). This means that stresses due to bending and unbending must be taken into account and also the distribution of the stress, σ_2, across the thickness of the tube must be studied. CHUNG and SWIFT (1952) have put forward a theory which within the normal industrial range predicts satisfactorily the sinking loads, wall thicknesses and length increments. Descriptively, the mechanics of the process they investigated in a die composed of two oppositely directed radii is as follows, referring to Fig. 11.1(*a*).

When an element of material in the original tube enters the die, plastic deformation begins at point *A*, by bending the element to the profile radius; since there is no drawing tension at this section, no appreciable alteration in thickness accompanies the bending. From *A* to *B*, the point where the first profile radius ends, the tensile stress in the tube wall increases continuously owing to 'radial' drawing and interfacial die friction; the metal also continues to thicken under the compressive hoop stress. At *B* the metal unbends and then bends to the reverse curvature of the second part of the die-profile radius, a process involving thinning and a very rapid increase in the sinking stress. From *B* to *C*, both the sinking stress and the metal thickness increase continuously. At *C* unbending under considerable tensile stress occurs, causing a final decrease in the metal thickness.

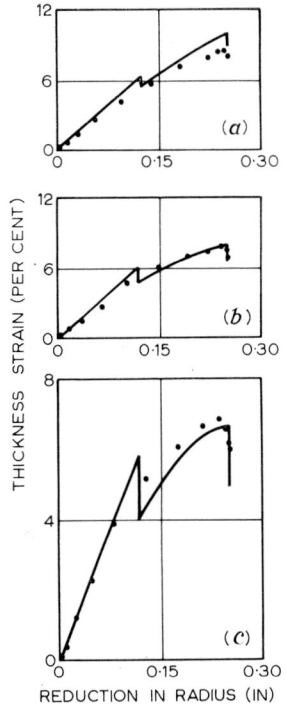

FIG. 11.7 Comparison of theoretical and experimental thickness strains during the sinking of mild-steel tubes.

        (*a*) 0·0635 in.
        (*b*) 0·1052 in.
        (*c*) 0·1305 in thick tubes.

(*After* Chung and Swift, *J. Iron Steel Inst.*)

The experiments by Chung and Swift consisted in reducing from $1\frac{3}{4}$ to $1\frac{1}{4}$ in diameter well-lubricated steel, brass, and copper tubes of different wall thicknesses, by drawing them through dies of total vertical conical angle 26° and profile radius 1·226 in, at the rate of 1 in/min. In Fig. 11.7 the results of thickness-strain/reduction-in-radius determinations, for mild-steel tubes, are compared with those predicted theoretically. The 'steps' in the predicted curves correspond to the (theoretically) sudden thinning due to plastic bending at the section contraflexure, B. The stress, σ_2, in the thickness direction is assumed to have a value equal to half the die face pressure.

A typical analysis of the constituents of the total sinking stress is given in Fig. 11.8.

Constituent	Wall Thickness, 0·0635 in		Wall Thickness, 0·1305 in	
	lbf/in^2	%	lbf/in^2	%
Pure radial deformation	24,800	75·6	27,320	62·2
Bending and unbending	3920	11·9	8280	20·7
Surface frication	5530	16·8	5970	15·0
Thickness variation	−1400	−4·3	−1560	−3·9
Total	32,850	100·0	39,990	100·0

FIG. 11.8 Constituents of total sinking stress.
(*After* Chung and Swift, *J. Iron Steel Inst.*)

11.6 The Detection and Measurement of Residual Stresses in Cold-drawn Tubes

The term residual or internal stress is applied to a stress or a stress system which is induced in an article during its manufacture and which does not disappear in the natural relaxation of the article when all external constraints are removed. Though the subject has received much attention, the measurement of residual stress has, until recently been only approximate. The object of much experimental work has been to improve the technique of measuring residual stresses, particularly as it applies to cold-drawn tubes and other tubular components.

Many methods have been proposed for detecting and measuring the internal strains in a residually stressed body, but generally these can be classified under three main headings:

1. Strain release and measurement by bending deflection.
2. Direct strain release and measurement.
3. X-ray and other non-destructive strain-detection methods.

The basic principles involved in the determination of residual stresses by bending-deflection methods are as follows: For circumferential stresses, a ring specimen is cut from a parent tube (the tube under test) and after careful measurement of all the leading dimensions, it is slit longitudinally along any one radial plane. The amount by which the outer diameter of the specimen in the plane perpendicular to that containing the slit alters from its original unslit diameter is a measure of the stress released. By successively reducing the tube-wall thickness unilaterally by machining or pickling, and by measuring the current 'sprung' diameter at each stage, the distribution of the residual circumferential stress throughout the tube wall can be deduced. For longitudinal stresses, a longitudinal tongue is cut from the tube wall and the amount of its deflection and curvature from the original straight line is a measure of the released stress. Further stress components are released by reducing the thickness of the tongue by pickling, and the longitudinal stress distribution throughout the walls of the tube can then be calculated.

Apart from approximate analyses for the calculation of the primary surface stresses released on slitting a tubular specimen or on cutting a longitudinal strip out of such a specimen, the only two analyses based on the bending-deflection method of strain release which attempt to provide means for measuring the stress distribution in cylindrical components are those of DAVIDENKOV (1932), and SACHS and ESPEY (1941). The Sachs and Espey analysis is found to be more convenient in application than the Davidenkov analysis, but it can be strictly applied only to relatively thin-walled tubes.

The measurement of residual stresses by direct strain release embodies in principle a reversal of the process by which the strains were produced, i.e., elements are removed from a stressed body and observations are made of the resultant deformations in the remaining material. The basic work in this field was published by Bauer and Heyn, but they considered only unidirectional stress systems, and in 1927 SACHS published a more general three-dimensional stress analysis for what is now widely known as the Sachs boring method. In this method, thin annular layers of material are successively removed from either the inner or the outer surface of a tubular specimen, and the consequent changes in length, diameter and wall thickness at each stage are measured.

Both bending-deflection and direct-strain-release methods have been considered, and in each case an acid pickling process has been used for removing material from the specimens. The measurement of the strains resulting from metal removal is generally made by micrometers in the bending deflection tests and by electric resistance strain-gauges in the direct-strain-release experiments.

The residual stresses present in a commercially produced tube are not generally the same at all points around the tube wall, and wide variations in derived-stress curves are obtained from identical specimens in the same tube that have been slit along different generators.

The residual-stress distribution in hollow-drawn steel tubes is found to be largely tensile in the outer 60 per cent of the tube wall (except very near the outer surface) and compressive over the remainder, as Fig. 11.9 shows. The rapidly changing compressive stresses at, and near to, the outer surface of the tube should be noted. This surface stress is affected by die friction and is reduced by the use of an efficient lubricant.

The residual stresses which develop in the wall of a hollow-drawn tube do not necessarily increase with the degree of reduction, and in fact, it has been shown experimentally that an inverse relationship may exist.

An extensive review of the status of some of the analyses referred to above and analytical analyses of residual stress distribution will be found in a paper by DENTON and ALEXANDER (1963).

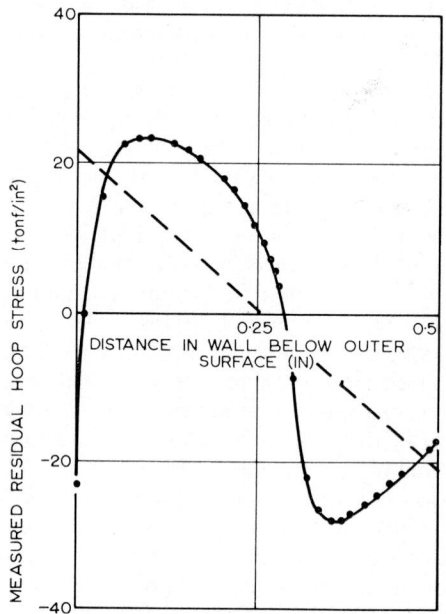

FIG. 11.9 Residual hoop-stress distribution in a 3·5 in outside diameter tube drawn down at 16 ft/min from a tube diameter of 4·25 in. Wall thickness of tube = 0·499 in.

11.7 Deep-drawing of a Circular Blank

11.7.1 *General Considerations*

We follow the approach used by HESSENBERG (1954) in the early part of the account given below. The simplest deep-drawing operation is the production of a cylindrical cup from a flat circular blank of sheet metal. However, compared with tube-drawing, where each particle undergoes the same deformation, the conditions of straining are very complex. The essentials of the tools are shown diagrammatically in Fig. 11.10(a) where the blank is shown divided into the three zones X, Y and Z. The outer annular zone X consists of material in contact with the die, the inner annular zone Y is initially not in contact with either the punch or the die and the circular zone Z is in contact with the punch.

Figure 11.10(c) shows the development of plastic strain in a blank drawn

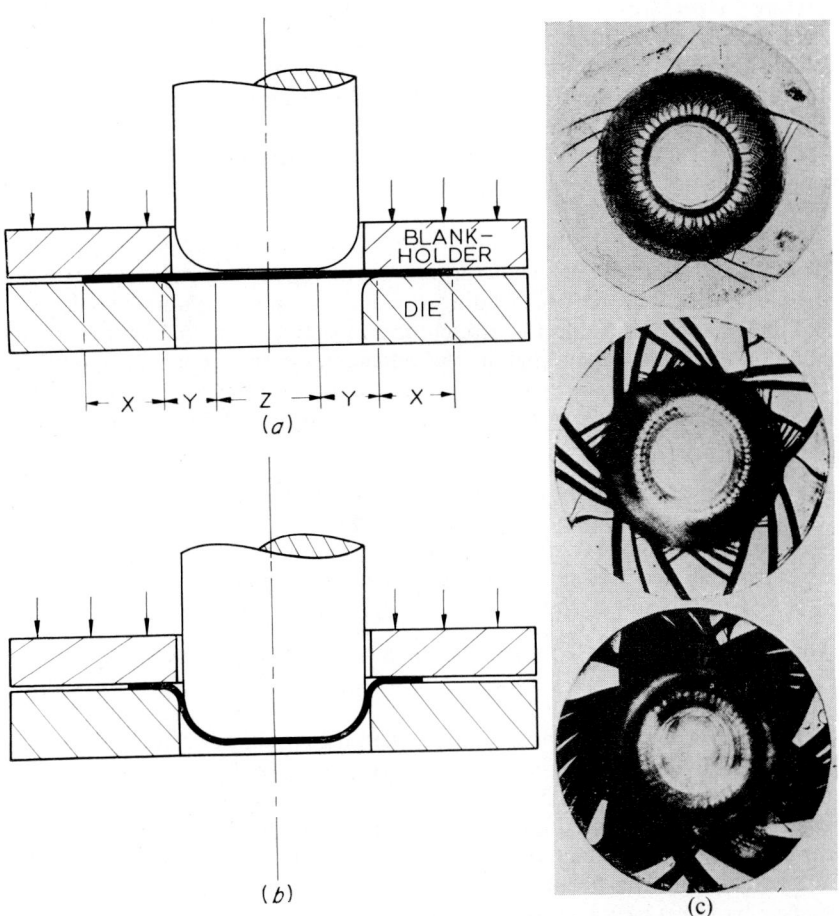

FIG. 11.10 First stage drawing.

with a flat bottomed punch during the first few millimetres of the draw. The black regions are the zones of plastic yielding in a mild steel blank, brought out by etching. The concentration of early straining in zone Y is evident and its progression into zone X also.

As the draw proceeds, the material in zone X is drawn progressively inwards towards the die profile under an applied radial tensile stress and the effect of continuously decreasing the radii in this zone is to induce a compressive hoop stress which causes a considerable increase in material thickness see Fig. 11.10(b). Unless a lateral pressure is applied by some form of blank-holding, this compressive hoop stress will, in general, cause the flange to fold or wrinkle. (See Section 11.10.) As the material in zone X passes over the die profile it is thinned by plastic bending under tensile stress. Finally, the inner parts of zone X are further thinned by tension between the die and punch. The net effect for the outer parts of zone X is an increase in thickness of the material.

Considering zone Y, it is easily seen from Fig. 11.10 that part is subject to bending and sliding over the die profile, part stretching in tension between die and punch, and part to bending and sliding over the punch profile.

Finally, zone Z is subject to stretching and sliding over the punch head, the amount of strain depending on the geometry of the punch head and the frictional conditions.

Summarizing the above remarks, we note that the following five processes take place:

1. Pure radial drawing between die and blank holder.
2. Bending and sliding over the die profile.
3. Stretching between die and punch.
4. Bending and sliding over the punch profile radius.
5. Stretching and sliding over the punch head.

Various parts of zone X may go through some or all of Processes 1, 2 and 3
,,　　,,　,,　,,　Y　,,　,,　　,,　　,,　,,　,,　,,　　,,　2, 3 and 4
,,　　,,　,,　,,　Z　,,　,,　　,,　　,,　,,　,,　,,　　,,　3, 4 and 5
The first process thickens the metal, the other processes thin it.

Between the fibres associated with drawing over the die and with stretching over the punch, there is a narrow band in zone Y which escapes plastic bending and is subject to predominantly simple tensile conditions throughout the drawing process. This material stretches and thins under tension to a markedly less extent than the fibres on either side, and consequently gives rise to a thicker band between two apparent 'necks'. Figure 11.11 shows the variation of metal thickness, exaggerated for emphasis, in cylindrical cups drawn with flat and hemispherical-headed punches, respectively. Fracture of the metal takes place at one or other of the necks, and usually at the one nearest to the punch head. Woo (1968) has predicted the strains in the stretch-forming region when drawing with a hemispherical punch. Bending effects at the die were neglected. The more general case of drawing with a flat-headed punch, making allowance for bending over the die and punch profiles, has not yet been solved. The punch load at any phase of drawing is determined by the radial drawing regime, and it in turn governs the strains in the stretch-forming region. If the blank is held rigidly at the die to prevent radial drawing, the process becomes one of pure stretch-forming.

Extensive and detailed experimental investigations of cup-drawing have been carried out by CHUNG and SWIFT (1951). Using tools producing cylindrical cups of 4 in internal diameter from blanks 0·039 in in thickness, they measured the punch loads and strains for various tool geometries, blank diameters and materials. Before attempting an analytical treatment of cup-drawing, the main points arising from these investigations will be summarized.

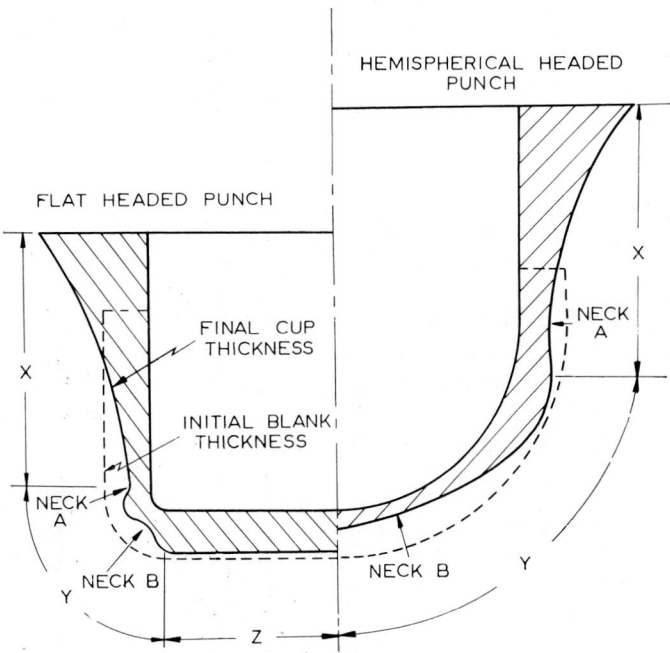

HEMISPHERICAL HEADED PUNCH

FLAT HEADED PUNCH

X

FINAL CUP THICKNESS

NECK A

X

INITIAL BLANK THICKNESS

NECK A

NECK B

NECK B   Y

Y

Z

FIG. 11.11 Sections through drawn cups. (Thickness changes greatly exaggerated.)

(*a*) BLANK-HOLDING. Two types of blank-holding are commonly used, clearance blank-holding and pressure blank-holding; the object in each case is to prevent wrinkling of the blank during radial drawing, but with the minimum of interference with free drawing. Early work (SWIFT, 1939–40) showed that with clearance blank-holding, an initial clearance of 5 per cent was sufficient for this purpose, though the blank would thicken much more than this amount, due to elastic deformation of the tools. With pressure blank-holding the minimum force necessary to prevent wrinkling was found, and it was established that increasing the force this amount had little effect on the maximum punch load or on the final thickness in the base and on the profile radius of the cups, though the walls were thinner with the higher loads. In tests on mild steel, a blank-holding pressure of 400 lbf/in² of blank contact area was adopted when this type of blank-holding was required and a clearance of 0·002 in when clearance blank-holding was used. The above pressure is well above the critical range without being excessive and it is also fairly representative of industrial practice.

(b) DRAWING RATIO. Drawing ratio is defined as the ratio of blank diameter to the throat diameter of the die. It was found that for any given drawing conditions, the punch load increases with the blank diameter in an approximately linear manner over the whole of the useful range with a slight tendency to droop near the limiting drawing ratio.

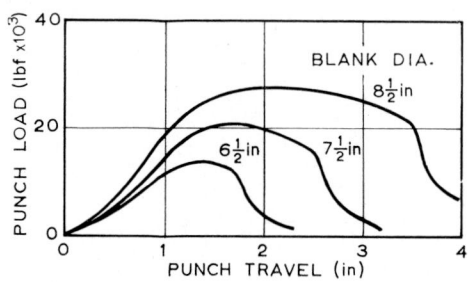

FIG. 11.12 Effect of blank diameter on punch load-travel.

Punch profile radius 1 in.
Die profile radius ⅜ in.
Positive pressure blank-holder.

(*After* Chung and Swift, *Proc. Instn mech. Engrs.*)

Figure 11.12 shows the variation of punch load against punch travel for different drawing ratios using annealed mild steel blanks. The profile radius of the punch was 1 in and the die profile radius was 3/8 in. The lubricant, which was applied to both sides of the blank, was graphite in tallow (1/3 by weight). The punch diameter was 4 in.

The measured thickness strains of the drawn cups are shown in Fig. 11.13. Two distinct zones of necking occur: one, A, at the lower end of the cup wall near its junction with the punch profile radius, the other, B, at some slightly lower radius. There is very little strain over the flat part of the punch head, except at the highest drawing ratio where the rapid development of neck B, leading to fracture at this point, should be noted.

(c) DIE PROFILE RADIUS. Tests with three blank diameters, covering a range of six die profiles of radii from 1/16 to 5/8 in showed little significant difference in maximum punch load as between die profiles, provided that the radius did not fall below 3/8 in (10 times the blank thickness). The sharper the die radius, the greater is the maximum punch load, because of the increased process work due to plastic bending under tension. Consequently, decreasing the die radius below a value 10 times the blank thickness lowers the limiting drawing ratio. The effect of die profile on the essentially practical problem of success or failure in drawing is shown in Fig. 11.14. Each point near the critical range represents at least five tests. The range of doubtful success is about 0·1 in in 8·5 in, an indication of the limit of consistency attainable in cup-drawing tests. Increasing the die radius above 10 times the blank thickness causes negligible improvement in drawability, while a greater tendency towards wrinkling is apparent on the drawn cups. This is due to the early removal of

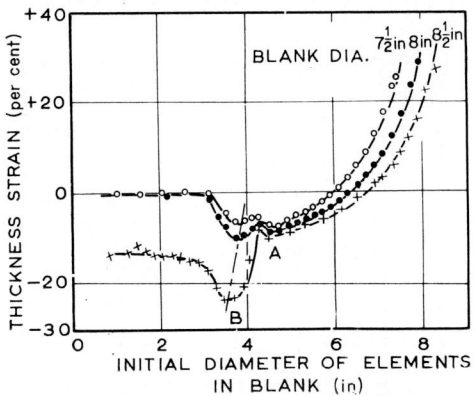

FIG. 11.13 Effect of blank diameter on thickness strain.

Punch profile radius: $\frac{1}{4}$ in.
Die profile radius: $\frac{1}{4}$ in.

(*After* Chung and Swift, *Proc. Instn mech. Engrs.*)

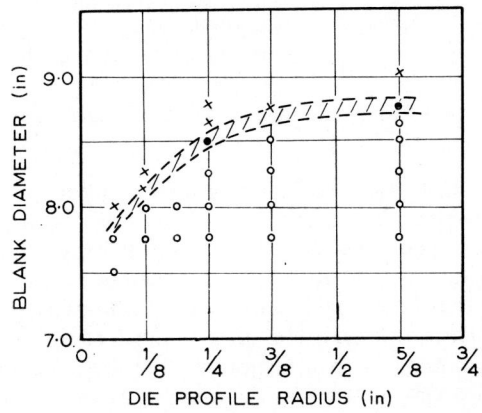

FIG. 11.14 Effect of die profile on drawing capacity.

   °   Success.
   ×   Failure.
   •   Some success and some failure.

(*After* Chung and Swift, *Proc. Instn mech. Engrs.*)

the guiding influence of the blank-holder, and it sets a practical upper limit to the die profile radius in terms of the blank thickness.

(d) PUNCH PROFILE RADIUS. The geometry of the punch profile is of utmost importance, since fracture usually occurs at neck *B*. A series of tests covering a full range of punch profile radii from 1/8 in to 2 in (hemispherical), but with a constant blank diameter of 8 in, showed that the more generous the punch radius, the more gradual is the rise of punch load and the longer the punch travel, but the maximum punch load is almost unaffected.

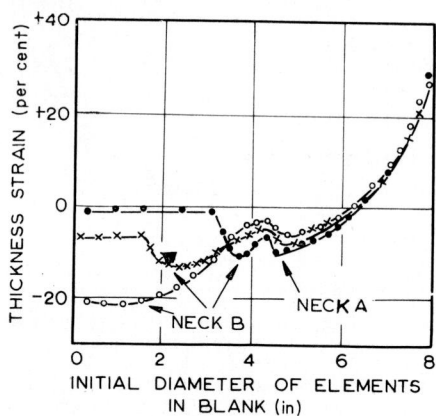

FIG. 11.15 Effect of punch profile on thickness strain.

$\bullet$ — $\bullet$ — $\bullet$ Punch profile radius $\frac{1}{4}$ in.
$\times$ — $\times$ — $\times$ Punch profile radius 1 in.
$\circ$ — $\circ$ — $\circ$ Punch profile radius 2 in.
(Hemispherical headed punch.)
Die profile radius: $\frac{1}{4}$ in.

(*After* Chung and Swift, *Proc. Instn mech. Engrs.*)

The distribution of thickness strain is shown in Fig. 11.15. The location and severity of profile neck *B* change systematically with punch form. The greatest thinning is seen to occur with the hemispherical punch. This does not indicate that rupture will occur first with a hemispherical punch, since the amount of strain that a material can withstand without instability and fracture depends very much on the existing stress conditions and also on the initial condition of the material. It will be noticed that the final strains in the walls of the cups are scarcely affected by the punch profile.

(e) RADIAL CLEARANCE. Radial clearance between punch and die throat may affect the drawing process directly by controlling the freedom of the walls either to thicken or to taper and pucker. It was found that a net radial clearance of about 30 per cent is suitable for general purposes, with free drawing and a reduction of, say, 50 per cent, and this has the sanction of practical experience. On the other hand, where ironing between punch and die can be tolerated, the net clearance may well be as low as 10 per cent. Too great a clearance may allow a bell-mouth to persist near the rim of the cup, which would be especially objectionable if the cup were required to undergo a re-drawing operation.

(f) MATERIAL. The majority of the work was carried out on annealed 0·08 per cent carbon rimming steel, but selected tests were made using aluminium, copper and 70/30 brass. To check the influence of material on punch load and strains, tests were made on blanks of 8 in in diameter, with pressure blank-holding and 1/4 in die profile radius. The blank-holding force was adjusted to suit a particular material. Tests with various punch profiles confirmed that with all the metals used, the maximum punch load and process work are almost independent of the profile radius. The thickness strains are shown in Fig. 11.16 when drawing with a punch of 1/4 in profile radius. It should be noted that the metals which do not thin in the base are those whose initial yield stresses are highest in comparison with their flow stresses under severe distortion, and that the results illustrated are specific to the 1/4 in punch profile radius and would be different with more generous profiles.

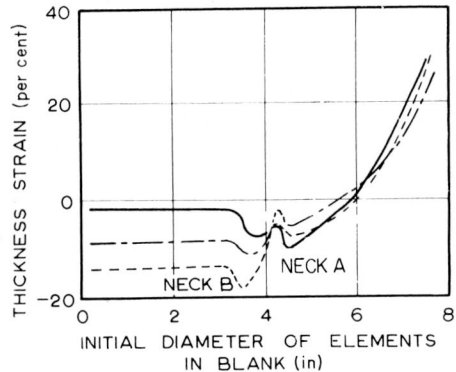

FIG. 11.16 Thickness strain distribution curves.

Punch profile radius and die profile radius: $\frac{1}{4}$ in.
Blank diameter: 8 in.
———————— Soft temper aluminium
— · — · — Half-hard temper brass
- - - - - - - Half-hard temper copper

(*After* Chung and Swift, *Proc. Instn mech. Engrs.*)

The formation of 'ears' at the free edges of a drawn cylindrical cup is caused by planar anisotropy and is undesirable in these circumstances since it represents a waste of material. It should be noted, however, that anisotropy can be usefully exploited in some asymmetric drawing operations, LANKFORD, SNYDER and BAUSCHER (1950), LLOYD (1962). In the drawing of a cylindrical cup, ears would be expected to develop from the rim of the blank at positions where the uniaxial yield strength has a minimum value in the circumferential direction. This is the case for some steels and aluminium where four ears are formed. The behaviour of some brasses, giving six ears, is more complex. The simple macroscopic theory of HILL (1950a) can explain the formation of four ears but an extension of the theory to cover the mechanical properties of six-eared brass was not successful (BOURNE and HILL, 1950).

WHITELEY (1960) showed experimentally that the limiting drawing ratio depends on the *normal anisotropy* in the sheet material as characterized by its *r*-value. With an *r*-value greater than unity the flow stress in biaxial tension is raised relative to the flow stress under a tensile-compression system. Thus the material over the punch head is strengthened relative to the material in the flange and a greater limiting drawing ratio can be achieved. When drawing with a high *r*-value material the change in thickness of the flange as it is drawn towards the die opening is less than for a low *r*-value material and the thickness of the resulting cup wall is more uniform. Thus for a given blank diameter a deeper cup will result for a high *r*-value material. WILSON (1966) has reviewed the effect of anisotropy in deep-drawing, and the effect of the *n*-value and *r*-value on the limiting drawing ratio has been discussed by EL-SEBAIE and MELLOR (1972).

(g) LUBRICATION. It was demonstrated by controlled experiments that lubrication has a very marked effect on the maximum size of blank that can be drawn successfully. This work has been reported by LOXLEY and FREEMAN (1954). The optimum conditions are obviously those which, for given blank size, give the smallest radial pull-back force and also restrict the straining in the region over the punch profile radius. This suggests good lubrication between blank and die face and no lubrication between blank and punch. This may not always be feasible in industrial practice, though measures are sometimes taken to restrict straining over the punch head.

PUNCH PROFILE	STATE OF LUBRICATION			
	FULL	NONE	DIE SIDE ONLY	PUNCH SIDE ONLY
FLAT HEADED PUNCH ($\frac{1}{8}$ in PROFILE RADIUS)	7·9 in	7·6 in	8·1 in	7·6 in
HEMISPHERICAL HEADED PUNCH	8·5 in	8·3 in*	8·9 in*	8·1 in

* FRACTURE OCCURED AT NECK A (SEE FIG. 11·11 OR FIG.11·13). IN ALL OTHER CASES FRACTURE OCCURED AT NECK B. PUNCH DIAMETER=4 in.

FIG. 11.17 Influence of punch form and lubrication on the limiting blank size when drawing 20 s.w.g. killed steel sheet with positive clearance (5·5 per cent) blank-holding. (*After* Loxley and Freeman, *J. Inst. Petrol.*)

Figure 11.17 shows the limiting blank diameters for mild steel drawn under different conditions of lubrication. In all cases, the drawing tools were degreased with trichlorethylene and the graphite in tallow lubricant applied to the blank as indicated. An increase of 0·1 in on any of the diameters caused fracture. In the case of the 'oversize' blanks fracture occurred at neck *B*, for the combination of hemispherical punch and no lubrication between punch and blank, when the flow over the punch head was restricted to such an extent that fracture took place at neck *A*.

WALLACE (1960) increased the friction between blank and punch still further by giving the punch a knurled finish. In this way he was able to increase the limiting drawing ratio of stabilized extra deep-drawing quality steel from 2·325 with a polished and lubricated punch to 2·550. KASUGA *et al.* (1961, 1965) have shown that even higher drawing ratios can be obtained by using high pressure fluid to hold the partly drawn cup hard against the punch. The same principle is also used to increase the depth of draw in the hydroform process.

We have discussed the main effects of a number of variables, but it must be remembered that they interact and cannot strictly be considered in isolation. One further important variable is the speed of drawing. This can affect the yield stress of the material and the efficiency of the lubricant. Systematic tests with drawing speeds of up to 30 ft/min on a set of dry blanks and on a set of blanks lubricated with graphite in tallow showed no speed effects. On the other hand, light oil exhibited pronounced speed effects, allowing a blank 8·7 in in diameter to be drawn successfully at the higher speeds against a limiting blank size of 8·3 in at very slow drawing speeds.

Further tests up to 90 ft/min have been carried out by COUPLAND and WILSON (1957). They concluded that at the higher drawing speeds partial fluid-film lubrication can develop.

The complex nature of cup-drawing has been emphasized in the above experimental details and in the following analysis of the problem we shall look only at a solution to pure radial drawing based on the work of CHUNG and SWIFT (1951). The complete analysis of radial drawing which includes bending, unbending and frictional effects over the die profile is beyond the scope of this book.

The monograph of WILLIS (1954) is well worth consulting; it was an attempt to rationalize and simplify Swift's work for presentation to non-mathematical readers. Much very useful work can be found reported in the collected papers of S. Fukui of Tokyo University.

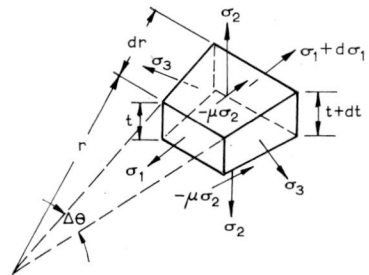

FIG. 11.18 Stresses on element in flange.

11.7.2 *Pure Radial Drawing*

For an isotropic material the principal stresses for plane radial drawing are the radial drawing stress σ_1, the stress σ_2 normal to the blank and the circumferential stress σ_3. The stresses will be regarded as positive when tensile. The stresses acting on an element in the flange, at current radius r, are shown in Fig. 11.18 and it can be shown that the equation of radial equilibrium is

$$\frac{d}{dr}(t\sigma_1) + \frac{t}{r}(\sigma_1 - \sigma_3) - 2\mu\sigma_2 = 0 \tag{11.12}$$

where μ is the coefficient of friction, assumed constant across the flange. The physical conditions show that σ_1 will be tensile, and σ_2 and σ_3 compressive. Also, since the blank thickens most at the rim, it will be assumed that all the blank-holding force is exerted at the rim and that elsewhere $\sigma_2 = 0$. Thus, considering frictional effects separately, the equilibrium equation becomes

$$\frac{d}{dr}(t\sigma) = \frac{t}{r}(\sigma_3 - \sigma) \tag{11.13}$$

where σ is the portion of the drawing stress due only to frictionless radial drawing.

The yield criterion may be written as

$$\sigma - \sigma_3 = m\bar{\sigma} \tag{11.14}$$

where $\bar{\sigma}$ is the representative stress and m has a value of $1 \cdot 1$ to give optimum agreement between the Tresca and von Mises criteria of yielding (see Fig. 11.3). The equilibrium equation then becomes

$$d\sigma = -m\bar{\sigma}\frac{dr}{r} - \sigma\frac{dt}{t}. \tag{11.15}$$

Further, since it is found that the actual difference in thickness across the flange at any moment is small, we may in the present theory neglect this difference in the equilibrium equation which then becomes

$$d\sigma = -m\bar{\sigma}\frac{dr}{r}. \tag{11.16}$$

Assuming that the strain-rate during radial drawing is constant and ignoring the effect of changing stress ratio, the current value of the representative stress $\bar{\sigma} = \sqrt{\Sigma(\sigma_1 - \sigma_2)^2}/\sqrt{2}$ depends only upon the representative strain $\bar{\epsilon} = \sqrt{2\Sigma(\epsilon_1 - \epsilon_2)^2}/3$. Also, HILL (1950b), has shown that $\bar{\epsilon}$ is never more than about 3 per cent greater than the numerical value of the hoop strain, ϵ_3.

Introducing strain-hardening, a suitable empirical expression is Ludwik's power law

$$\bar{\sigma} = \bar{\sigma}_0 + B\bar{\epsilon}^n \tag{11.17}$$

where $\bar{\sigma}_0$, B and n are constants. This can be made to fit the strain-hardening characteristics of a metal sufficiently well, and is not too difficult to handle in the following analysis. For practical purposes, we may therefore write

$$\bar{\sigma} = \bar{\sigma}_0 + B\left(\ln\frac{R}{r}\right)^n \tag{11.18}$$

where $\ln R/r$ is the hoop strain, R being the initial and r the current radius to a particle.

From Fig. 11.19, if t_m represents the current mean thickness of the metal between the rim and the element under consideration, the condition of incompressibility gives, equating the volume of metal before and during radial drawing

$$\pi(R_0^2 - R^2)t_0 = \pi(r_0^2 - r^2)t_m$$

where t_0 is the original thickness of the blank.

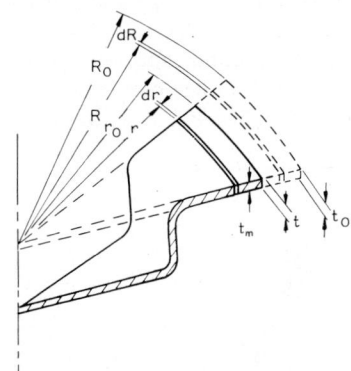

FIG. 11.19 Section of a partially drawn cup.

This may be written in the form

$$\left(\frac{R}{r}\right)^2\left(\frac{t_0}{t_m}\right) - 1 = \left(R_0^2 \cdot \frac{t_0}{t_m} - r_0^2\right)\frac{1}{r^2} = \frac{c}{r^2}, \quad \text{say.}$$

Hence, the hoop strain ϵ_3 is

$$\ln\frac{R}{r} = \frac{1}{2}\ln\left\{\frac{t_m}{t_0}\left(1 + \frac{c}{r^2}\right)\right\}. \tag{11.19}$$

Substituting in equation (11.18)

$$\bar{\sigma} = \bar{\sigma}_0 + B\left[\frac{1}{2}\ln\left\{\frac{t_m}{t_0}\left(1 + \frac{c}{r^2}\right)\right\}\right]^n. \tag{11.20}$$

If this is substituted into the equation of equilibrium (11.16),

$$d\sigma = -m\bar{\sigma}_0\frac{dr}{r} - mB\left[\frac{1}{2}\ln\left\{\frac{t_m}{t_0}\left(1 + \frac{c}{r^2}\right)\right\}\right]^n\frac{dr}{r}$$

and integrating, the portion of the drawing stress, at current radius r, due to radial drawing alone is

$$\sigma = -m\bar{\sigma}_0\ln\frac{r}{r_0} - mB\int_{r_0}^{r}\left[\frac{1}{2}\ln\left\{\frac{t_m}{t_0}\left(1 + \frac{c}{r^2}\right)\right\}\right]^n\frac{dr}{r}, \tag{11.21}$$

where r_0 is the current radius of the rim of the blank.

If H is the total blank-holding force, the portion of the drawing stress, σ'_1, due to blank-holding friction, is easily shown to be

$$\sigma'_1 = \mu H/\pi r_0 t \tag{11.22}$$

11—EP   *   *

where μ is the coefficient of friction, assumed constant and equal for both sides of the blank. The total radial drawing stress is then

$$\sigma_1 = \sigma + \sigma'_1.$$

The exact values of t_m and c in equation (11.21) and that of t in equation (11.22) can be approximated by the method of successive improvement, by alternate calculation of stress and strain. The volume of computation is greatly reduced by using the values of t_m, c and t obtained for the simple case of a non-hardening material ($B = 0$) and zero frictional stress, $\sigma'_1 = 0$. For practical purposes, the second approximation only is necessary in determining the value of σ_1. Once σ_1 has been evaluated, σ_3 can be determined from equations (11.14) and (11.20). The distribution of drawing stress across

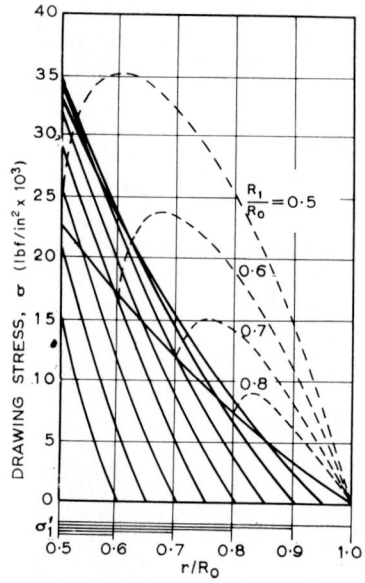

Fig. 11.20 Distribution of drawing stress across flange and variation at a given radius during drawing.

Stress-strain relation, $\bar{\sigma} = 30{,}000 + 46{,}500\bar{\epsilon}^{0.49}$.
Yield criterion $\sigma - \sigma_3 = 1 \cdot 1.\bar{\sigma}$
Strain rate, 3×10^{-4} per second.
————— Distribution of σ across flange at various stages.
- - - - - - - Variation of σ at a number of fixed radii R_1/R_0.

(*After* Chung and Swift, *Proc. Instn mech. Engrs.*)

the flange of an annealed mild-steel blank for a drawing ratio of 2 is given in Fig. 11.20. If there was no blank-holding friction, the base line of the curves would be the zero line. However, since σ'_1 is not zero, the base line has to be lowered to the appropriate position.

Once the stresses are known, the strains can be obtained from the Lévy-Mises relationship taking σ_2 to be zero.

Thus
$$\frac{dt}{t} = -\left(\frac{\sigma + \sigma_3}{2\sigma_3 - \sigma}\right)\frac{dr}{r}$$

and substituting $\sigma_3 = \sigma - m\bar{\sigma}$ from the yield criterion

$$\frac{dt}{t} = -\left(2 + \frac{3m\bar{\sigma}}{\sigma - 2m\bar{\sigma}}\right)\frac{dr}{r}. \tag{11.23}$$

If values of σ and $\bar{\sigma}$ are substituted into equation (11.23) it is possible to trace the thickness changes of any particular element of metal at initial radius R by numerical integration, provided that the relation between r and r_0 is known. This relation may be derived from the equation of incompressibility

$$(r_0^2 - r^2)\frac{t_m}{t_0} = (R_0^2 - R^2) = \text{constant} = c \quad \text{(say)}$$

and
$$r = \left(r_0^2 - c \cdot \frac{t_0}{t_m}\right)^{1/2} \tag{11.24}$$

where t_m is the current mean thickness of metal between the rim, radius r_0, and any radius r. Figure 11.21 shows the development of thickness strains in the flange for the same conditions as Fig. 11.20.

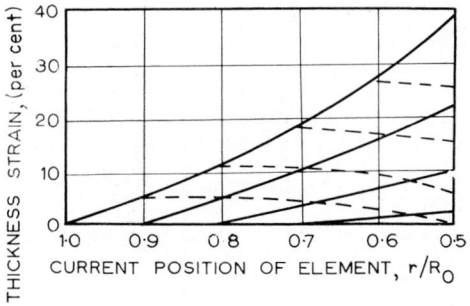

FIG. 11.21 Development of thickness strain due to radial drawing.

———————— Thickness strain history of a particular element.
- - - - - - - - - Thickness strain contour at a given stage.
Strain-hardening characteristic $\bar{\sigma} = 30{,}000 + 59{,}200\,\bar{\epsilon}^{0\cdot49}$.
Blank-holding force $= 15{,}500$ lb.
$\mu = 0{\cdot}06$

(*After* Chung and Swift, *Proc Instn mech. Engrs.*)

The final thickness strains in the walls of the drawn cup will be less than this due to the bending and unbending over the die. The stresses and strains involved in this further process have been computed by Chung and Swift. They were also able from these results to construct punch load against travel diagrams which gave good agreement with experimental results, as can be seen from Fig. 11.22. For the non-ferrous metals investigated, there was remarkably good agreement between the predicted and measured strains, but for mild steel, which exhibited marked earing effects, agreement was good only over the inner half of the blank annulus, corresponding to the lower half of the wall of the drawn cup (Fig. 11.23). They concluded that the computation involved in the application of the theory is necessarily complicated and not to be recommended for general use.

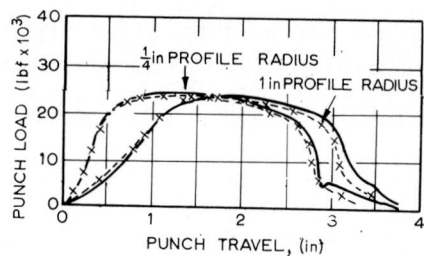

FIG. 11.22 Punch loads: comparison of theory and experiment.

Material: low carbon steel, 0·039 in in thickness.
Blank diameter: 8 in. Punch diameter: 4 in. Die diameter: 4·101 in.
Die profile radius: ¼ in.
×—×—× Theoretical.
———— Experimental.

(*After* Chung and Swift, *Proc. Instn mech. Engrs.*)

In view of the complexity in the analyses just described, an attempt has been made by DUNCAN and JOHNSON (1969) to give a simple analysis that will predict the load to draw a cup; this may be of value to plant engineers and for teaching purposes.

Other detailed accounts of deep-drawing mechanics will be found in the references asterisked on pp. 371-2.

11.8 Ironing

Ironing consists, principally, in reducing the wall thickness of a cup by restricting the clearance between punch and die, the reduction in internal diameter of the cup usually being sufficient to allow free entry of the punch only.

In early investigations at Sheffield University, ironing-load/reduction characteristics for various die angles and states of lubrication were obtained. These have been extended, and the two components of the ironing load, the punch-head load and the frictional or tractional load along the punch, have been measured separately as well as together, by FREEMAN and LEEMING (1953); the punch-head load is of primary importance, since it determines the tension in the cup walls and hence the maximum reduction possible for a given punch load.

Their experimental apparatus was based on a special 4 in diameter composite ironing punch, having a separate hemispherical head with electric resistance strain-gauges as load-sensitive elements. The recording was electronic and photographic. Their tests were made on hemispherical-headed cups, 0·039 in thick, to determine the ironing-load components for different conditions of wall thinning, die profile, and lubrication, and a typical set of load/reduction curves is shown in Fig. 11.24. A theoretical study of ironing has been made by FUKUI and HANSSON (1970).

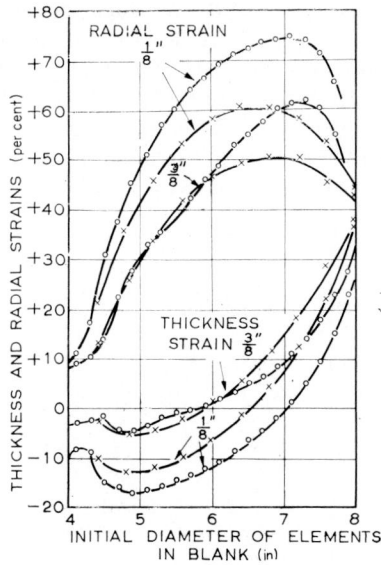

FIG. 11.23 Drawing strains for different die profile radii: comparison of theory and experiment.

Material: low carbon steel, 0·039 in in thickness.
Blank diameter: 8 in; Punch diameter: 4 in.
Die diameter: 4·101 in.
Punch profile radius: 1 in.
×—×—× Theoretical.
○—○—○ Experimental.

(*After* Chung and Swift, *Proc. Instn mech. Engrs.*)

11.9 The Re-drawing of Cups

When producing cups of the common engineering materials from flat circular blanks, the maximum first-stage drawing ratio seldom exceeds 2.2, corresponding to a cup height to diameter ratio of about unity, so that if the final proportions of a cup are to exceed this limiting value, re-drawing in one or more stages is necessary.

The methods of re-drawing fall into two main groups: (1) direct re-drawing and (2) reverse re-drawing. In the former, the external surface of the first-stage cup remains on the outside of the re-drawn cup, whereas in the latter, the cup is turned inside out, the inside surface of the first-stage cup becoming the outside surface of the resulting shell.

Diagrammatic representations of various methods of direct re-drawing are reproduced in Fig. 11.25. In methods *A* and *B* the metal in the wall of the first-stage cup has to bend and unbend twice, making two right-angled turns. Method *B* can in fact be regarded as a special case of method *A*, in which the die-profile runs directly into the blank-holder-profile radius, which is designed to fit the first-stage cup. In method *C*, the metal again has to bend and unbend twice, but on each occasion through an angle of less than 90°, so that an easier draw than in either *A* or *B* is to be expected, since there is less 'angular' friction. The first-stage cup, however, has to be prepared with a tapered corner, which is not very suitable for certain materials and tempers, owing to the danger of puckering. If the cup-diameter to wall-thickness ratio is not very large, the blank-holder may be dispensed with, as shown in method *D*.

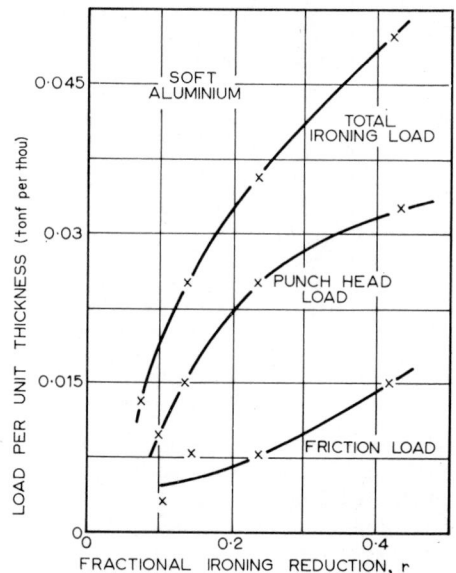

FIG. 11.24 Variation of ironing-load components with reduction.

Semi-angle of die $= 10°$.
Graphite-in-tallow lubrication.
(*After* Freeman and Leeming, *B.I.S.R.A. Report.*)

The essential difference between the two general methods of reverse re-drawing, as shown in Fig. 11.25 *E* and *F*, is that the die profile in *E* has two right-angled corners and a flat portion, which is essentially similar to method *A* of the direct re-drawing, whilst in *G* the die profile is semi-circular and so eliminates one bending and one unbending operation.

C is the most commonly used method in industrial practice. When methods *B* and *F* are used in practice, it is usual to modify the blank-holding technique as shown in Fig. 11.26. These modifications lessen the danger of wrinkling in the cup wall and hence allow a greater reduction (see communication by GRAINGER, J. A. to CHUNG and SWIFT (1952)).

Investigations have been carried out by CHUNG and SWIFT (1952) on brass, mild steel and aluminium in an experimental crank press based on a first-stage flat-bottomed cup of 4-in diameter. Their re-drawing apparatus consisted

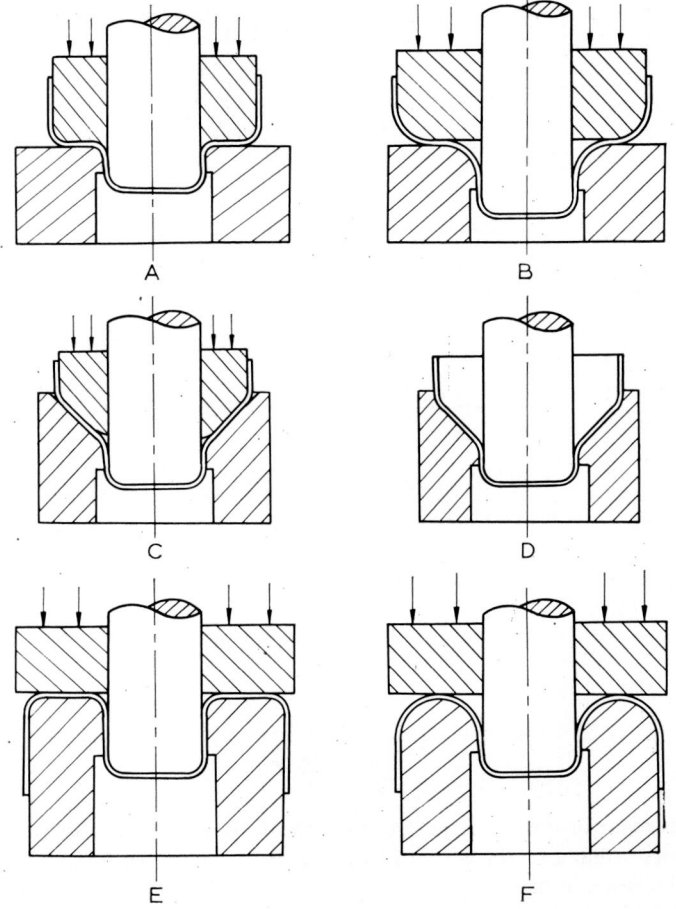

FIG. 11.25 Diagrammatic representation of direct and reverse re-drawing methods.

of three subpresses, having interchangeable parts, the blank-holding force being provided by a calibrated coil spring. The punch diameters covered the range 2·67–3·17 in and the nominal net clearance was 50 per cent with 0·04 in thick blanks.

Typical experimental results obtained in their work are shown in Figs. 11.27 and 11.28, and some of the conclusions reached are:

1. The significant difference between the direct and reverse methods of re-drawing lies in the number of bending operations and the degree of severity in bending. When this severity is the same for both methods, the reverse method definitely gives better results, since it can eliminate two bending operations. Unlike the reverse method, the direct method can, within limits

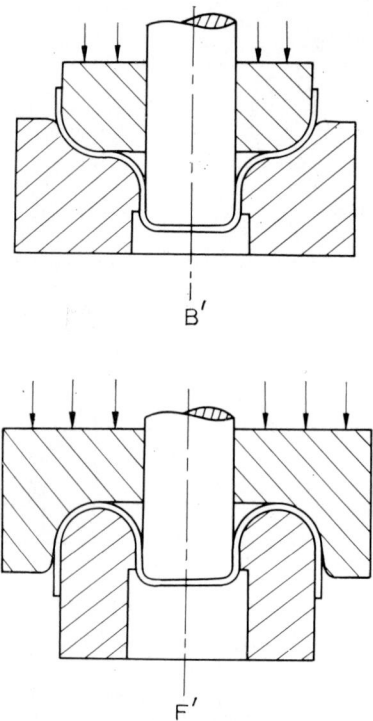

B′

F′

FIG. 11.26 Diagrammatic representation of two methods of re-drawing.

be designed to provide a large die-profile radius, and under favourable con-ditions may yield a better drawing capacity than the reverse method. Turning the shell inside out may permit a convenient arrangement of the drawing tool, but so far as stresses and strains are concerned, it has no significant advantage.

2. To obtain the maximum combined drawing capacity over two stages, the maximum first-stage blank size should be used, although the re-drawing capacity is greater with smaller initial blank diameters.

3. Interstage annealing greatly improves the re-drawing capacity, but from a given blank tends to produce a re-drawn cup of smaller depth because its walls and base are generally thicker.

4. Metals which give good results in drawing from flat blanks are not necessarily the best for re-drawing purposes. For first-stage drawing, materials with a high rate of work-hardening are advantageous, especially when using

hemispherical punch heads, but for re-drawing, materials less subject to strain-hardening are more suitable unless interstage annealing is adopted.

A theoretical analysis for the re-drawing of cups has been given recently by Fogg (1968).

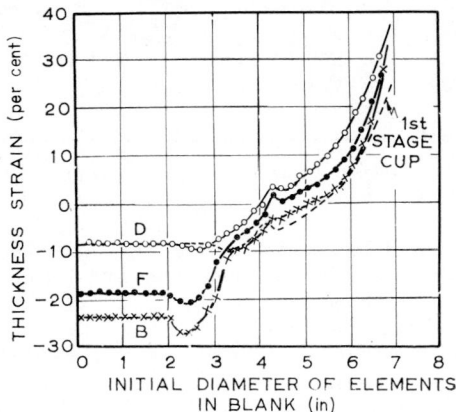

FIG. 11.27 Comparison of thickness-strain curves obtained by different re-drawing methods.

Re-drawing punch diameter—3·166 in.
o—o—o   Direct method D
●—●—●   Reverse method F.
×—×—×   Direct method B
- - - - - - - - - First stage flat-bottomed cup.

(After Chung and Swift, *Proc. Instn mech. Engrs.)*

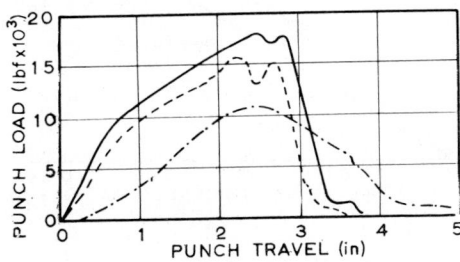

FIG. 11.28 Punch load/travel diagrams for three re-drawing methods.
——————— Direct method B.
- - - - - - - - - Reverse method F.
— · —— · — Direct method D.
(After Chung and Swift, *Proc. Instn mech. Engrs.)*

11.10 Flange Wrinkling in Deep-drawing

As a deep-drawing operation proceeds, the outer flange portion of the blank is subjected to a radial drawing stress and an induced compressive hoop stress ,and when the magnitude of these stresses exceeds a certain critical

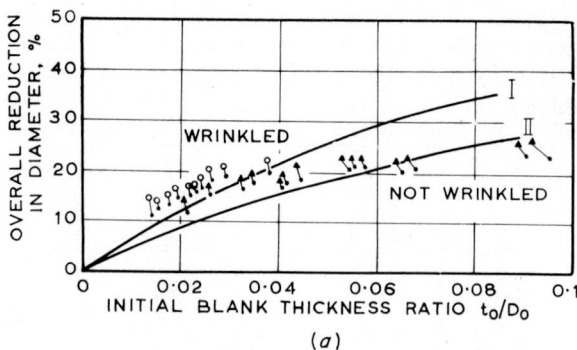

(a) Comparison of experimental results with theoretically predicted
stability limits for half-hard aluminium, with 0° and 30° dies.
Curve I: theoretical upper limit.
Curve II: theoretical lower limit.
Punch speed: 0·1 in/min.
Graphite-in-tallow lubricant.
▲—• 0·75 in diameter punch.
○—• 2·00 in diameter punch.

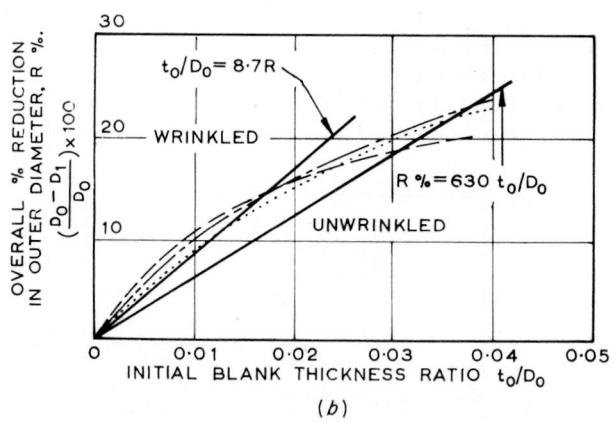

Fig. 11.29

(b) Comparison of experimentally observed stability limits for various materials.
$t_0/D_0 = 8\cdot7R$ is the suggested empirical curve.
$R = 630\, t_0/D_0$ is Esser and Arend's empirical curve.
D_i is the punch diameter.
D_0 is the original blank diameter.
- - - - - - - Half-hard aluminium.
—·—·— Killed steel.
· · · · · · · · · Soft 70 : 30 brass.

(*After* Senior, *J. Mech. Phys. Solids.*)

value, dependent on the current flange dimensions, lateral collapse into waves or wrinkles occurs.

If the flange is laterally unsupported, that is, no blank-holder is employed, it can be shown theoretically (GECKLER, 1928 and SENIOR, 1954) that instability sets in in the range

$$0.46 \left(\frac{t_0}{D_0}\right) \leqslant \frac{\sigma_\theta}{E_0} \leqslant 0.58 \left(\frac{t_0}{D_0}\right) \tag{11.25}$$

and that the number of waves formed, n, lies between the limits

$$1.65 \left(\frac{a}{b}\right) \leqslant n \leqslant 2.08 \left(\frac{a}{b}\right) \tag{11.26}$$

where σ_θ is the induced hoop stress, t_0 is the initial material thickness, D_0 the initial blank diameter, b is the flange breadth, a is the mean flange radius, and E_0 is the plastic buckling modulus $= 4EP/(\sqrt{E} + \sqrt{P})^2$, E being the elastic modulus, and P being the appropriate slope of the true-stress/natural-strain curve of the material. It was assumed in deriving these equations that the clamping of the inner periphery of the flange between the punch and die shoulder can be simulated by a lateral loading of the flange surface, proportional to the lateral deflection at any point. This factor may be evaluated in terms of the geometry of the flange and the mechanical constants of the material, using the accepted theory for the deflection of a flat annular plate (SENIOR, 1955).

Equation (11.25) can be evaluated on the basis of the true-stress/natural strain curve for the material, to give critical curves of the form shown in Fig. 11.29(a) for half-hard aluminium; experimental results from tests on tools of 0.75 in and 2.0 in nominal diameter have been inserted in the figure for comparison. Figure 11.29(b) shows collected mean experimental curves for a variety of strain-hardening materials. It seems that little difference exists between them in practice, a conclusion which is also suggested by the theoretical stability curves.

If blanks are used below the critical size, the depth of cup which can be obtained with thin material is severely limited. Thus the wrinkling of flanges is restrained in practice by using a blank-holder to provide lateral support. Then under quite moderate pressure the development of waves of large amplitude is prevented, though the number of waves is increased by a reverse buckling process, as shown diagrammatically in Fig. 11.30. However, the more numerous waves of small amplitude so formed are more easily ironed out to give a good finish on the cup rim without an unduly high drawing load. The types of blank-holder used in practice are divided into two main classes:

Type 1—The loading is provided by springs, or by the elastic deflection of the holding-down bolts of the blank-holder plate in the case of the so-called clearance-type blank-holder. Here, the load is a linear function of the blank-holder deflection, and therefore of the wave amplitude. It can be shown that under these conditions the number of waves formed is independent of the wave amplitude and is given by

$$n = a(3.8S/E_0 bat^3)^{\frac{1}{4}} \tag{11.27}$$

where S is the lateral stiffness of the blank-holder and t the current mean thickness of the flange. This equation can be evaluated from an idealized linear strain-hardening characteristic for a given material. Experimental results from the drawing of 2-in nominal diameter cups have given good confirmation of equation (11.27).

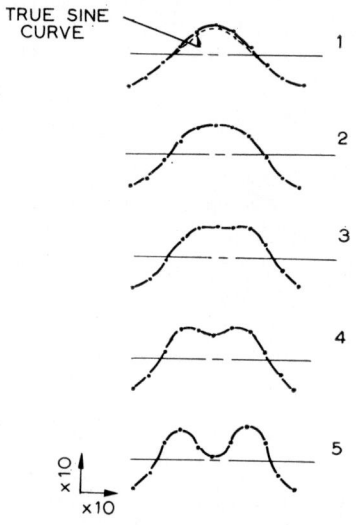

Fig. 11.30 Illustrating the manner in which the number of curves is increased in deep-drawing.

Figure 11.31 shows theoretical curves evaluated on the basis of an elastic and an idealized plastic modulus for 4·00 in diameter cups in 0·036 in half-hard 70 : 30 brass. Tests show that these are compatible with an initial elastic wrinkling in the early stages of the draw and with the final plastic wrinkling embodied in the finished cup. The fact that the final wrinkles are in general three times the number of the earlier ones can be explained by the mechanism of wave multiplication already referred to. It is clear that the theory, while by no means exact, indicates the correct trends.

Type 2 — Blank-holder load is provided by pneumatic or hydraulic cylinders and remains constant. In this case, theory indicates that at any stage of the draw the number of waves is dependent on the amplitude, and is given by

$$n = \left(\frac{1 \cdot 34 E_0 a^4 t^3 + 2 \cdot 5 a^3 b^3 Q / \gamma}{E_0 b^4 t^3} \right)^{\frac{1}{4}} \tag{11.28}$$

where γ is the wave amplitude and Q the total blank-holder load. A typical comparison between theory and experiment is shown in Fig. 11.32.

In the past, blank-holder-load requirements necessary to restrain undue wrinkling have often been quoted, but on comparison they are found to be somewhat at variance. Mean, minimum blank-holder-load curves for a range of materials are shown in Fig. 11.33. It may be added that the blank-holder load, above a certain minimum necessary to confine the wave amplitude to reasonable proportions, has little effect on the punch load required.

It appears, therefore, that in deep-drawing, a critical size of blank exists, above which wrinkling will occur in the absence of a blank-holder. To maintain a good finish on the cup rim and to prevent failure due to excessive

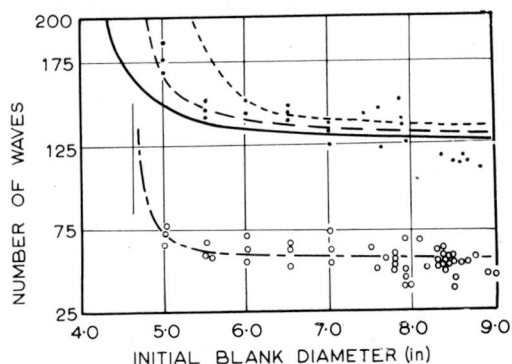

FIG. 11.31 Variation in number of waves with initial blank diameter, with clearance blank-holding.

Material: 0·36 in thick, half-hard 70 : 30 brass.
Die throat diameter 4·100 in.
Die profile radius: $\frac{1}{4}$ in.
Flat-headed punch with $\frac{1}{4}$ in profile radius.
Lubricant: graphite-in-tallow.
Punch speed: 6 ft/min.
The theoretical curves for the maximum number of waves are given by $n/a = (3\cdot8\ S/E_0\ \text{bat}^3)^{\frac{1}{4}}$ for extreme and mean shoulder diameters $E_0 = 4\cdot02 \times 10^5\ \text{lbf/in}^2$.
Blank-holder stiffness is $S = 88 \times 10^6\ \text{lbf/in}$.
- - - - - - - - $\phi = 4\cdot6$
— — — — $\phi = 4\cdot35$
————— $\phi = 4\cdot10$
Theoretical curve for minimum number of waves is given by $n/a = (3\cdot8\ S/E_0\ \text{bat}^3)^{\frac{1}{4}}$ with $E_0 = 12 \times 10^6\ \text{lbf/in}^2$.
— · — · — $= 4\cdot35$
ϕ is the effective flange-support diameter.
A blank stable without a blank-holder is equivalent to $n \to \infty$
o and • represent experimental points.
                    (*After* Senior, *J. Mech. Phys. Solids.*)

ironing, it is important to limit the amplitude of any waves which form. This involves certain minimum requirements for spring stiffness or total blank-holder load, depending on the blank-holding technique employed. It is better to err on the side of safety in blank-holder loads, as the effect on the punch load is slight.

11.11 Blanking

The terms associated with the study of the axisymmetric blanking process are adequately defined by reference to Fig. 11.34(*a*). If a slow-banking operation is carried out in a compression testing machine equipped with autographic recording apparatus, a typical punch force-punch displacement diagram as shown in Fig. 11.34(*b*) will be obtained.

The substantially proportional increase in punch force with punch displacement from O to A represents the predominantly elastic phase of the process. If the punch penetration stopped at the limit of this phase some elastic recovery would occur and no detectable penetration would be observed upon microscopical examination of the blank. From A to B, plastic shear deformation occurs and the manner in which the magnitude of the punch force

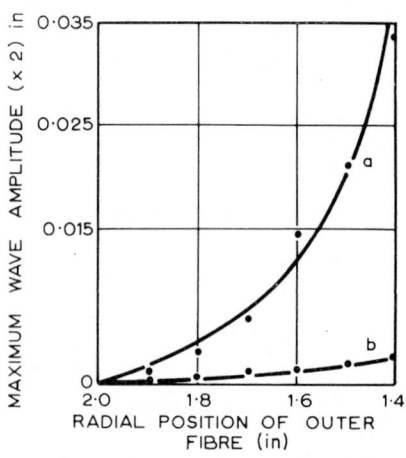

FIG. 11.32 Comparison of experimental and theoretical results for the variation of maximum wave amplitude with flange outer fibre radius.

(*a*) Theoretical curve, equation (11.28).
(*b*) Increase in flange thickness with flange outer fibre radius.
Material: 0·036 in rimming steel.
Blank-holder load: 0·2 ton.
Die diameter: 2·135 in; profile radius: ¼in.
● Experimental points.

(*After* Senior, *J. Mech. Phys. Solids.*)

changes with the punch displacement reflects the strain-hardening characteristics of the material which is used. The maximum blanking force is attained at B and from B to C although further strain-hardening of the material may occur, the punch force decreases in proportion to the decrease in the vertical area of shear. The shear stress thus increases from A to B while the shear area decreases.

It has been observed by Crasemann that cracks are formed after the blanking force reaches its maximum value at B. It is probable that cracks are formed as the punch is displaced from C to D. Beyond C no further plastic deformation occurs and the punch force decreases rapidly. From O to D the blanking force is thus determined by the current mean shear stress of the material and the frictional resistances present. However, beyond D the frictional conditions are the determining factor. From D to E there is frictional resistance between the blank and the hole, the punch and the hole and also between the blank and die. From E to F the blank passes further into the die and at F it may be assumed that the blank is completely separated from the sheet. The loop from F to G through H to O represents the return of the punch to its initial position, and the negative ordinates indicate the magnitude of the stripping force which is required to remove the strip from the punch.

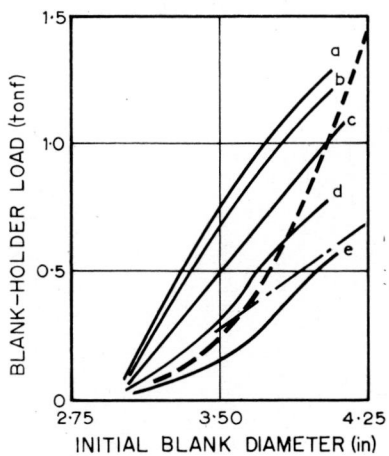

FIG. 11.33 Minimum blank-holder load curves required to prevent excessive wrinkling.

0·036 in thick materials.
Die diameter: 2·135 in; profile radius: $\frac{1}{4}$ in.
Curves a, b, c, d, and e are for hard, $\frac{3}{4}$H, $\frac{1}{2}$H, $\frac{1}{4}$H and soft aluminium respectively.
—— —— —— Rimming steel.
- —— - —— - Soft 70 : 30 brass
(*After* Senior, *J. Mech. Phys. Solids.*)

The area of the closed loop of the diagram represents to scale the energy required to effect the blanking operation and also strip the sheet from the punch. The earliest attempt to obtain records of punch pressure variation with punch displacement was due to Anthony and was reported in 1911. It has long been known that the maximum blanking force is greatest at no clearance and decreases but little with increase in clearance, see Fig. 11.34(*c*).

Useful fundamental analyses of the blanking process are not to be had and for this reason much reliance must be placed on the results of purely experimental investigation. It has been observed that the maximum quasi-static blanking force F_{max} for non-brittle materials is given by $F_{max} \simeq 0·6\,A\sigma_s,$

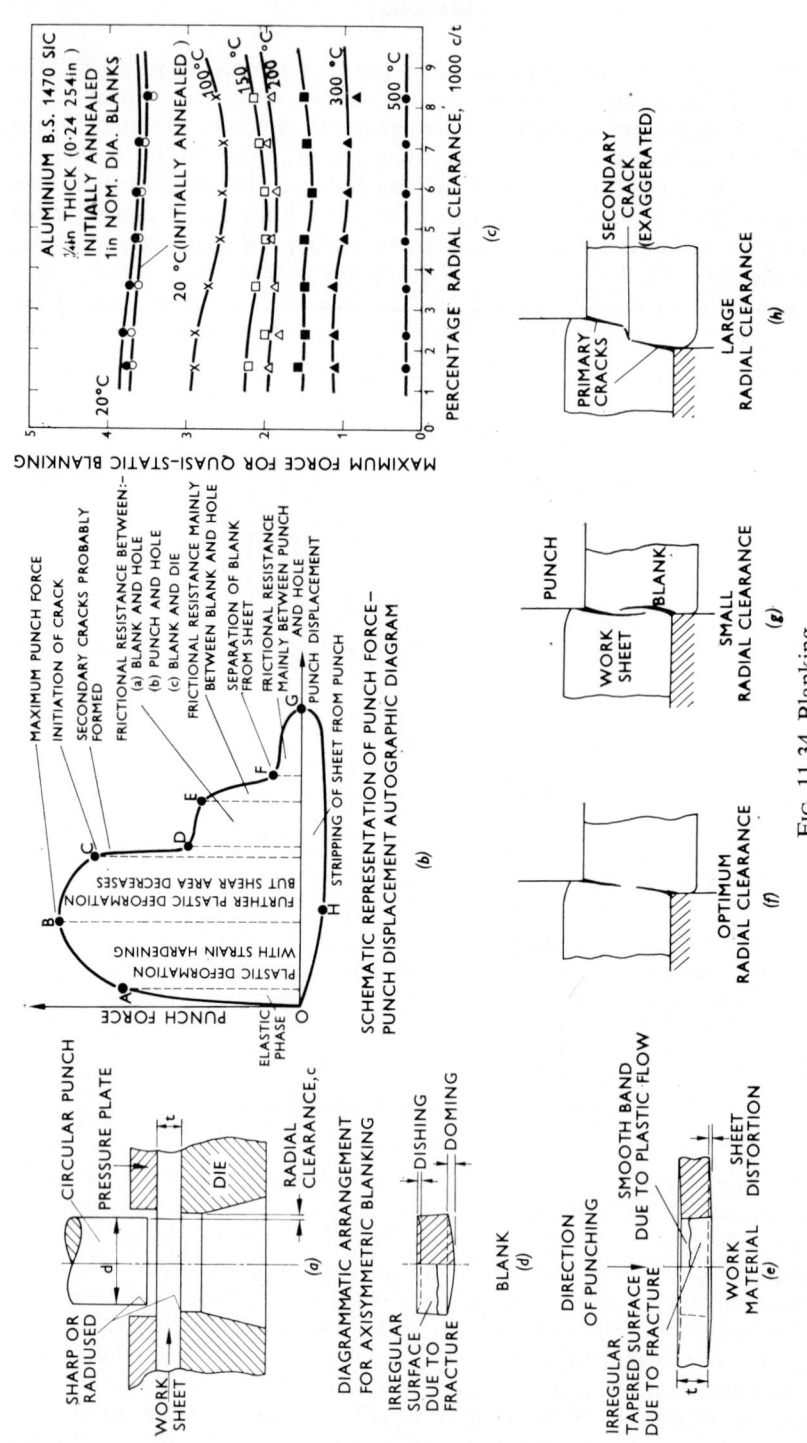

Fig. 11.34 Blanking.

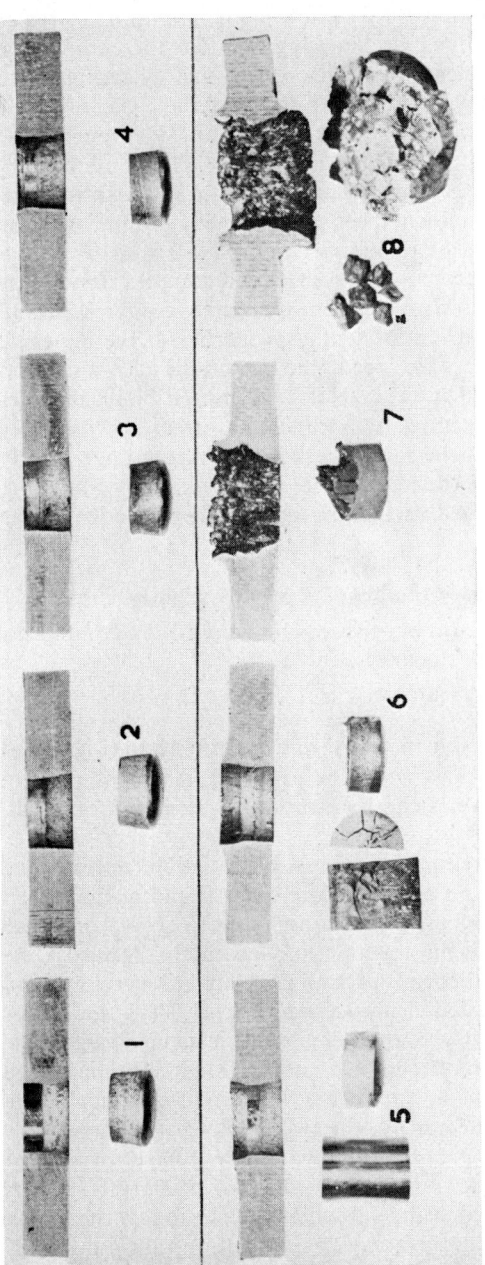

(i)

Fig. 11.34 Blanking.

(a) Diagrammatic arrangement for axisymmetric blanking.
(b) Schematic representation of punch force-punch displacement autographic diagram.
(c) Variation of maximum blanking force with radial clearance and temperature.
(d) Surface after blanking: blank.
(e) Surface after blanking: sheet.
(f) Crack formation and optimum radial clearance.
(g) Crack formation and small radial clearance.
(h) Crack formation and large radial clearance.
(i) Copper specimens blanked at increasing punch velocity. Velocities in ft/s, 1:0, 2:252, 3:438, 4:793, 5:930, 6:1500, 7:2100, 8:2350 (die removed). (In 5 and 6 the deformed punch is also shown).

(After Johnson and Slater, A.S.T.M.E.)

when $4 < d/t < 10$; A denotes the nominal shear area πdt and σ_s in the tensile strength of the material being blanked or the tensile strength of the material at a strain of 0·5, if blanking is taking place at elevated temperatures. Greater energy is generally required for dynamic blanking than for the comparable quasi-static process at all radial clearances. This can be mainly attributed to the effect of strain rate on the yield stress of the material. The order of magnitude of the plastic shear strain in a blank is unity and the time for the operation is that for the punch to penetrate the sheet. Thus elementary calculations show the order of the mean shear strain rate for a slow speed operation to be 10^{-4}/s when the punch speed is about 0·01 in/min, whereas for a dynamic operation in which the punch speed is 100 in/s it is 10^3/s.

The blanking of circular discs from strip or sheet is a complex non-steady state process involving a triaxial stress system, in some zones compressive and in others tensile; it involves the strain-hardening properties of the material, the tool geometry and such factors as speed, the punch-die clearance and the extent of punch penetration. The process of producing a blank involves both plastic deformation and fracture, and while the former is reasonably well understood, the same cannot be said of the latter. Accordingly, fairly precise predictions concerning the sheared edge of the blank are not to hand.

Three areas are usually identified as making up the sheared edge of the blank, see Fig. 11.34(d).

1. The periphery of the bottom of the blank is usually slightly rounded.

2. The next zone, is a smooth or burnished one which may comprise a considerable fraction of the blank thickness.

3. A fracture zone which usually tapers.

Zone 2, is due to plastic flow and the burnishing is caused during movement within the die throat; Zone 3 as would be expected, is irregular where fracture occurs. Corresponding zones can be identified in the work material, see Fig. 11.34(e).

Fracture, which consists of crack initiation and propagation, arises in the regions of high stress concentration and these are to be found at the punch and die corner or radii. Two cracks are initiated and start to grow from these two regions of high stress gradient, together or separately, towards one another. The sequence somewhat depends on tool geometry. General opinion is that at optimum percentage radial clearance, see Fig. 11.34(f), the lips of the two growing cracks approach one another along the same straight line and, on joining, a single fairly clean fracture occurs. At small clearances, see Fig. 11.34(g), the two cracks tend to embrace a very small quantity of the work sheet. For relatively large clearances, the two cracks grow more or less parallel but out-of-line, see Fig. 11.34(h); when they have proceeded sufficiently far, a secondary crack transverse to the two primary crack tips arises. This gives rise to a jagged edge and subsequent expulsion by the punch may cause some part of it to be burnished. Crack directions and initiations are often determined by inclusions in the work sheet.

When a large clearance between the punch and die is used (e.g., percentage radial clearances greater than about 8 per cent), the situation is similar to that of a small diameter but deep encastre circular plate uniformly loaded over its central portion. The work sheet or plate is now loaded in bending and shear; the larger the clearance the larger will be the bending stresses in relation to the shear and vice versa. However, also in blanking there is a kind of drawing operation—in part responsible for Zone (1)—in which there are large tensile bending stresses created near the punch and die radii. The extent of plastic deformation in these regions is limited by the tensile resilience of the metal with which fracture mechanics associates the onset of fracture.

Some of the effects of a high speed movement of the punch are

1. to give a blank edge at a punch speed of 30 ft/s which is better than would be the case at slow speeds when blanking mild steel, and

2. to enhance the tendency to dishing and doming, see Fig. 11.34(*d*).

At elevated temperatures the energy required to blank is greatly dependent on the speed of the punch. Punches fired at very high speed at plates at normal incidence, e.g., at about 600 ft/s, shatter the blank and create a hole much larger in diameter than that of the blank, see Fig. 11.34(*i*). This phenomenon is due to the punch moving faster than the plastic wave speed in the plate and thus material is forced out of the way of the punch transversely.

This section is a brief review of a survey paper by JOHNSON and SLATER (1967); a more ample treatment is provided by JOHNSON (1972).

11.12 Wire-drawing

11.12.1 *Theories and Experiment*

Wire-drawing is one of the most ancient crafts. The earliest wire production was probably entirely in non-ferrous metals, the manufacture of iron wire not being introduced until a few centuries ago. Until the beginning of the twentieth century single-die machines were used and the advances in the industry since then have been mainly due to the introduction of multi-die machines and tungsten-carbide dies. In the non-ferrous industry, tungsten-carbide dies are used down to a size of 0·055 in in diameter; below that diameter, diamond dies are generally used. (See CLEAVER and MILLER, 1950–51).)

In wire-drawing the metal is pulled through a tapered die, Fig. 11.35, the pull and the nip of the die stressing the metal so that it deforms plastically within the confines of the die. The main function of wire-drawing is to produce wire of specified size with a good surface finish. In industrial practice, the die has a trumpet-shaped bore, but since the curvature near the working surface is small, this may be simplified here into a conical portion which serves to deform the wire and a cylindrical portion which is meant to preserve the size of the bore in the face of wear. The total die angle can have values between 5° and 25° and the length of the cylindrical portion varies from nil to two wire diameters. The possible reduction at each die varies from about 10 to 45 per cent where percentage reduction is defined as $r = (1 - A_2/A_1)100$, A_1 being the initial and A_2 the final cross-sectional area of the wire.

In tube-sinking it is a good approximation to neglect the variation of stress through the wall thickness. In wire-drawing, however, the assumption of uniform stress throughout the wire leads to approximately correct values of drawing force for certain limited combinations of die angle and reduction. In general, the deformation is not homogeneous; plane sections do not remain plane on passing through the die, and extra work is expended in distortions which do not contribute to the final reduction in diameter, i.e., as redundant deformation; see ATKINS and CADDELL (1968).

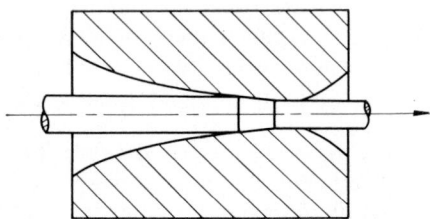

FIG. 11.35 Wire-drawing die.

The most detailed investigation into the mechanics of wire-drawing has been carried out by WISTREICH (1955). The investigation was limited to the slow drawing of round wire, without back-pull, through dies with bores in the shape of truncated cones, Fig. 11.36. The split-die technique suggested by MACLELLAN (1952–53) was used to ascertain the mean die pressure. The die was in two halves and the forces tending to separate the two halves was

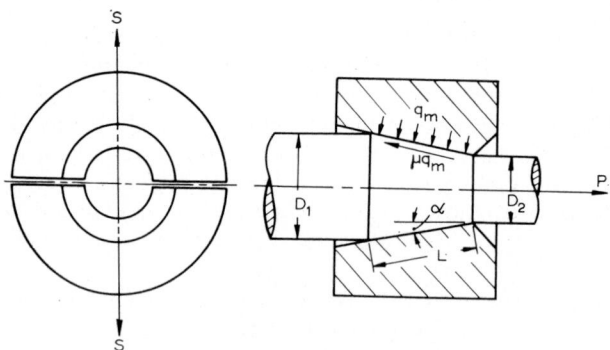

FIG. 11.36 Forces in wire-drawing.

measured concurrently with the drawing force. The external forces acting on the wire are shown in Fig. 11.36. The coefficient of friction, μ, between wire and die was assumed constant.

If α is the semi-angle of the die cone, the area of wire in contact with the die is $(A_1 - A_2)/\sin \alpha$ and for equilibrium, equating drawing force, P, and forces due to mean die pressure, q_m,

$$P = (A_1 - A_2)(1 + \mu \cot \alpha)q_m. \tag{11.29}$$

If D_1 is the initial diameter and D_2 the final diameter of the wire, the length of wire in contact with the die $L = (D_1 - D_2)/2 \sin \alpha$ and the splitting force S between the two halves of the die is given by

$$S = q_m \frac{(D_1 + D_2)}{2} L[\cos \alpha - \mu \sin \alpha]. \tag{11.30}$$

From equations (11.29) and (11.30)

$$\mu = \frac{1 - \pi(S/P) \tan \alpha}{\tan \alpha + \pi(S/P)} \tag{11.31}$$

and
$$q_m = \pi S/(A_1 - A_2)(\cot \alpha - \mu). \tag{11.32}$$

In the past the coefficient of friction has been treated as a parameter which could be adjusted to give the best correlation between theory and experiment. Its derivation from experiment is therefore an important development.

MAJORS (1955) has also obtained results for the mean die pressure by measuring the hoop strain on the outside of a complete die.

In his investigation, Wistreich used mainly light drawn electrolytic copper wire, since it was easy to lubricate well and did not cause rapid wear of the steel dies. Also the material in this condition does not work-harden rapidly and it is a good approximation in any theory to take account of the work-hardening by assuming a mean yield stress. As a result of other experiments, it was concluded that the deformation is independent of the properties of the metal except in the case of annealed wire drawn with very light reductions. The lubricant was sodium stearate, special care being taken to apply it evenly and so minimize frictional variations. The value of the coefficient of friction under these conditions averaged 0·02 to 0·03 and it did not vary significantly with die pressure. MAJORS (1955), in his tests, measured the coefficient of friction directly for the drawing of steel rods and obtained much higher values, ranging from 0·08 to 0·2.

The variation of die pressure with percentage reduction is shown in Fig. 11.37 for three different die angles. The mean yield stress of the material in the tests varied between approximately 22 and 27 tonf/in² and it can therefore be seen that for certain combinations of die angle and reduction, the die pressure greatly exceeded the values of yield stress. It was also established from further tests that, other things being equal, the greater the friction, the lower the die pressure.

The variation of drawing force with percentage reduction of area for the same three die angles is shown in Fig. 11.38. It will be noticed that the curves do not pass through the origin when extrapolated. The explanation of this apparent intercept on the ordinate axis lies in the fact that for very light reductions the wire bulges prior to its entry to the die.

If it is assumed that the deformation is homogeneous and that there is no friction between wire and die, then the work expended per unit volume in reducing the area of cross-section of the wire from A_1 to A_2 will be $Y_m \ln A_1/A_2$ where Y_m is the mean yield stress. The work done in producing unit length of wire is then σA_2 and the work per unit volume is simply the drawing stress σ. Therefore

$$\sigma = Y_m \ln A_1/A_2. \qquad (11.33.\text{i})$$

This is the most efficient, or ideal, means of reducing the wire diameter and it gives the unattainable minimum value of the drawing stress. The equation may be compared with the exactly similar case of the sinking of a thin-walled tube where the sinking stress $\sigma = Y_m \ln r_0/r_1$.

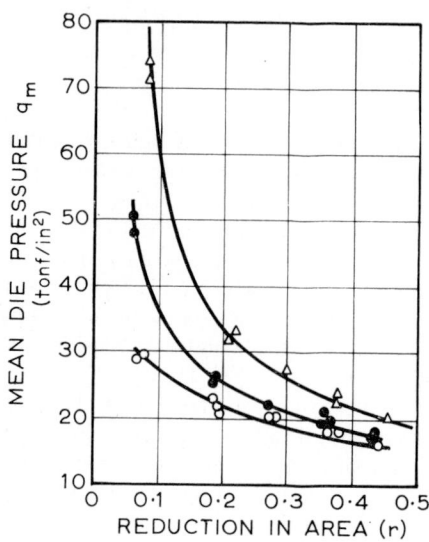

FIG. 11.37 Variation of mean die pressure with reduction.

	a(degrees)
o	2·29
•	8·02
△	15·47

(*After* Wistreich, *Proc. Instn mech. Engrs.*)

If the frictional force is taken into account, then

$$\sigma = Y_m(1 + \tan \alpha/\mu)[1 - (A_2/A_1)^{\mu \cot \alpha}]. \qquad (11.33.\text{ii})$$

This equation, derived by SACHS (1927), should be compared with the similar equation for tube-sinking.

So far, the non-useful work of distortion has not been taken into account. SIEBEL (1947) has proposed a theory of wire-drawing in which he assumes that the effects of homogeneous deformation, friction and non-useful distortion are additive. He assumes that the plastic region within the die is bounded by spherical caps with centres at the virtual apex of the cone. As the wire enters and leaves the die it is sheared instantaneously along these surfaces and within the die the metal moves towards the virtual apex of the cone. The equation for drawing force is given as

$$P = A_2 \left[Y_m \ln \frac{A_1}{A_2} + \frac{\mu}{\alpha} Y_m \ln \frac{A_1}{A_2} + \tfrac{2}{3} Y_m \alpha \right] \qquad (11.34)$$

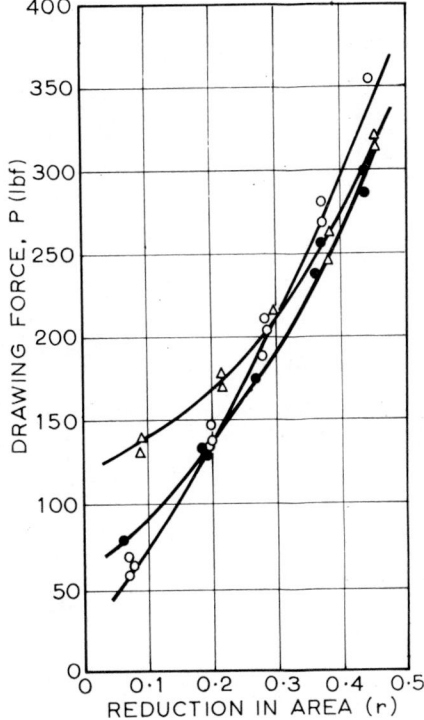

FIG. 11.38 Variation of drawing force with reduction.

	α(degrees)
○	2·29
●	8·02
△	15·47

(*After* Wistreich, *Proc. Instn mech. Engrs.*)

The first term accounts for the homogeneous deformation, the second the frictional component and the third the additional force required because of non-useful distortion. The mean die pressure in each case can be obtained from equation (11.29).

WHITTON (1958) has compared the drawing forces based on the above theories with experimental values obtained by Wistreich. A coefficient of friction of 0·025 was adopted and relevant values of mean yield stress were taken from the same experimental source. Such a comparison is made in Figs. 11.39 and 11.40 for two widely different die angles. The mean yield stress is taken as the average of the values before and after drawing. It will be noticed that the Sachs equation gives good correlation for a die of semi-angle 2·29° but that for a die of semi-angle 15·5° it underestimates the drawing force, especially at low reductions. This is, of course, because the Sachs equation does not allow for the non-useful distortion (redundant shearing),

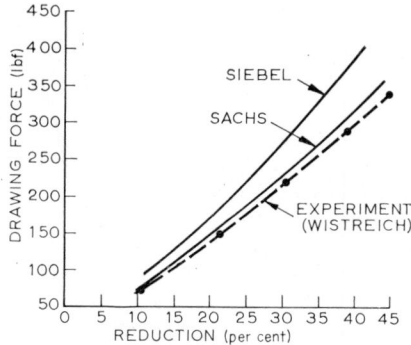

FIG. 11.39 Variation of drawing force with reduction; die semi-angle = 2·29 degrees. (*After* Wistreich, *Proc. Instn mech. Engrs.*)

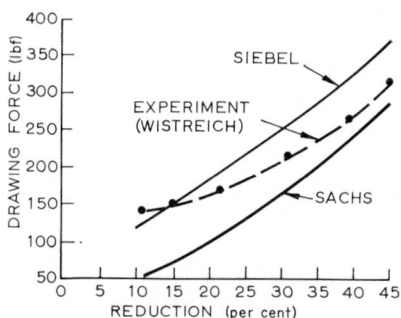

FIG. 11.40 Variation of drawing force with reduction; die semi-angle = 15·47 degrees. (*After* Wistreich, *Proc. Instn mech. Engrs.*)

the effect of which increases with increasing die angle, and which is attenuated by increasing reduction (Fig. 11.40). On the other hand, the Siebel equation overestimates the redundant shearing, and therefore the drawing force, and is only in good agreement with experiment at low values of reduction. In an attempt to get closer correlation with the experimental results, Whitton devised the following empirical formula

$$P = A_2 Y_m (1 + \cot \alpha / \mu)[1 - (A_2/A_1)^{\mu \cot \alpha}] + 2A_2 Y_m \alpha^2 (1 - r)/3r$$

where r is the reduction of area $(A_1 - A_2)/A_1$.

The final term is the extra drawing force for the non-useful distortion. The accuracy of the formula is within ± 10 per cent for the experimental results considered. Other empirical formulae have been suggested by WISTREICH (1965).

All the theories so far discussed make arbitrary assumptions about the deformation and the extent of the plastic region. In the complete solution of the problem the stresses and the mode of deformation must be compatible, As yet, no axisymmetric problems have been solved, but HILL and TUPPER

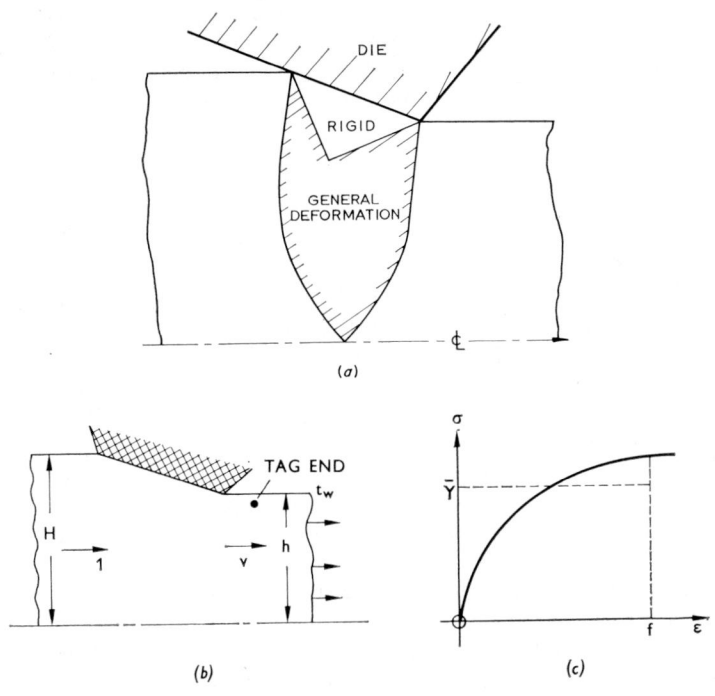

(a)

(b)   (c)

FIG.11.41

(a) Deformation zone in sheet drawing.
(b) Plane drawing defining tag end.
(c) True $\sigma/\bar{\epsilon}$ curve.

(1948) using slip-line field methods (see Chapter 12), have obtained a solution for the drawing of wide sheet which is the two-dimensional analogue of wire-drawing. Although, according to Wistreich, the calculations of forces in sheet-drawing cannot be applied to wire-drawing, the plane strain theory does account qualitatively for the salient features of the latter. For instance, the extent of the plastic region is sheet drawing as determined by theory (Fig. 11.41a) is in qualitative agreement with the actual deformation zone in wire-drawing, as determined by etching. Also, the theory predicts bulging of the sheet at light reductions. However, for a given die geometry, the deformation in wire-drawing is much less homogeneous than in sheet-drawing.

In the experimental work described above, the drawing speed was generally about 6 ft/min. In industrial practice, the drawing speeds range from 100 to 8000 ft/min and these high rates of deformation, and also friction, give rise to high temperatures and steep temperature gradients across the wire. High temperatures may hinder lubrication, may speed up age-hardening in ferrous and some other alloys, thereby lowering the ductility of the wire, and temperature gradients may cause residual stresses in the finished wire. Residual stresses are also caused by the inhomogeneous nature of the deformation. The reader is referred to a valuable review of the mechanical and physical aspects of deformation in wire-drawing contributed by WISTREICH (1958); this review gives emphasis to developments since 1948 whilst MACLELLAN (1948) has given a critical survey of wire-drawing theory before that date. The most up-to-date survey and bibliography on wire-drawing theory is that by JOHNSON and SOWERBY, (1969).

Discussions of wire-drawing in relation to lubrication will be found in the papers by CHRISTOPHERSON and NAYLOR, (1955) and THOMSON, HOGGART and SUITER (1967).

11.12.2 *Redundant Strain and Redundant Work Factors in Drawn Strain-hardening Materials*

The expenditure of redundant work and the creation of redundant strain in wire- and bar-drawing has been the subject of several recent papers and it is therefore useful to define and examine these quantities briefly: see CADDELL and ATKINS (1968).

Consider drawing through frictionless dies, see Fig. 11.41(b). If the necessary tension is t_w, the work done on unit volume of the drawn strip is $t_w.h.v/1.H$; v is the exit speed of the strip if the speed at entry to the die is unity. However, $h.v = 1.H$, and this is just the volume drawn in unit time. Hence, t_w is the work per unit volume of material drawn. If f denotes the mean strain imparted to the drawn, and therefore redundantly worked product, then for a strain-hardening material whose stress–strain is given by $\sigma = B\epsilon^m$, the necessary value of t_w is the product of f and the mean yield stress Y over the strain range 0 to f on the σ/ϵ diagram, see Fig. 11.41(c). (Note that f in some cases is calculable theoretically, as with slip-line field theory.) This quantity is just the area under the stress-strain curve from 0 to f and is the plastic work done on each unit volume drawn, i.e.,

$$t_w = \frac{Bf^{m+1}}{(1 + m)} .$$ (11.35.i)

For a strain-hardening material homogeneously strained in tension to level t_N, to give the same reduction, r, as is achieved by drawing, the strain which would be required is simply $\ln 1/(1 - r) = g$; the required work per unit volume, t_N, would be

$$\frac{Bg^{m+1}}{1 + m}.$$ (11.35.ii)

The redundant *work* factor ϕ_W, defined as the ratio t_W/t_N is

$$\phi_W = \frac{t_W}{t_N} = \left(\frac{f}{g}\right)^{m+1}. \tag{11.35.iii}$$

The redundant *strain* factor ϕ_s for both hardening and non-hardening materials is $\phi_s = f/g$.
The redundant *stress* factor ϕ_T for the hardening material is

$$\phi_T = Bf^m/Bg^m = (f/g)^m. \tag{11.35.iv}$$

For a non-hardening material, $m = 0$, and we obtain the obvious relationship that $\phi_W = \phi_s = f/g$ whilst by definition $\phi_T = 1$. For a hardening material $\phi_W/\phi_s = (f/g)^m$, so that in all real cases $\phi_W > \phi_s$.

In plane strain drawing through a $15°$ frictionless wedge-shaped die and using a slip-line field for a fractional reduction of 0.343 (see Chap. 12), $t_W/2k = 0.431$. However, for homogeneous straining we should have $t/2k = \ln(1/(1 - 0.343)) = 0.42$. Thus $\phi_W = 0.431/0.42 = 1.025$; and, of course, ϕ_s also is 1.025.

11.12.3 *The Maximum Reduction in Drawing: A Simple Analysis*

If homogeneous deformation is assumed then $t_W/2k = \ln(1/(1 - r))$, and if the material is non-hardening then the greatest permissible value of $t_W/2k$ is unity, and hence the greatest value of r is $(e - 1)/e \simeq 0.638$.

If a hardening material was considered then the maximum reduction that could be achieved in one pass would be different. We should have $t_W/\bar{Y} = \epsilon'$ where ϵ' is the mean strain imparted and $\bar{Y}$ the mean flow stress over the range 0 to ϵ' for the actual stress-strain curve. But,

$$\bar{Y} = \left(\int_0^{\epsilon'} B\epsilon^m \, d\epsilon\right)\bigg/\epsilon' = B\epsilon'^m\bigg/(1 + m)$$

and hence $t_W = B\epsilon'^{m+1}/(1 + m)$. Let the tensile strength of the material in simple tension be t_u, then (see p. 17), $t_u = Bm^m$. If the greatest permissible drawing stress is the tensile strength, then the greatest possible reduction should be given by,

$$Bm^m = B\epsilon'^{m+1}\big/(1 + m)$$

or $$\ln \epsilon' = \left\{m \ln m + \ln(1 + m)\right\}\bigg/(1 + m) \tag{11.36.i}$$

This analysis assumes that the tag end in drawing, when stressed to the tensile strength, is also of the same area as the die exit. The following table may be drawn up.

m	0	0·05	0·1	0·4	0·5	0·6
r	0·638	0·597	0·585	0·624	0·648	0·668

A different approach is to assert that the maximum reduction is defined by being able to apply to the drawn wire a stress which is equal to its yield stress after having been passed through the die; this stress is $B\epsilon'^m$. Thus in place of (11.36.i), we have

$$B\epsilon'^m = B\epsilon'^{m+1} \Big/ (1 + m)$$

and $\epsilon' = 1 + m.$ (11.36.ii)

Some typical values, on the assumption that $\epsilon' = \ln 1/(1 - r)$ are:

m	0	0·1	0·2	0·3	0·5
r	0·638	0·667	0·70	0·738	0·777

These results are very similar to those of ATKINS and CADDELL (1969).

11.13 Extrusion: General Considerations

Extrusion is one of the youngest of the metal-forming processes and was probably first developed at the end of the eighteenth century for the manufacture of lead pipes. The process is one whereby a slug or billet of metal is forced to flow, under high pressure, through a die shaped to give the required cross-section to the product. An extraordinary diversity of special sections, including many with re-entrant angles, can be produced by extrusion in a rapid and economical manner with high dimensional accuracy upon which finishing and machining operations are reduced to a minimum or dispensed with altogether. The versatility of the method has also been extended by combined forging and extrusion processes.

The two common methods of working, known as direct and reverse (or inverted) extrusion are illustrated by the sketches in Fig. 11.42. In direct extrusion, the punch and product travel in the same direction; in the reverse method, they travel in opposite directions. The form of die most frequently adopted is the square or 90° die shown in Fig. 11.42.

Until recent years, extrusion was almost exclusively a hot-working operation, though collapsible tubes and other such products were worked cold from the softer metals by 'impact' extrusion. The approximate temperature ranges for 'hot' extrusion billets at the beginning of extrusion are shown in Fig. 11.43. Above the upper temperature limit there is a tendency for the metal to crack or break up—called hot shortness—as it leaves the die. This phenomenon is dependent on the speed of extrusion; the faster the speed of extrusion, the quicker is the conversion of the work of extrusion into heat. The temperature of the billet then rises so far above the upper temperature limit as to cause phase changes. Below the lower end of the scale, the force required to deform the metal might become excessive and the billet becomes a 'sticker' in the container. The extrusion of the harder metals, such as steel, at temperatures over 1000 °C, means that the extrusion tools are subjected to very high thermal and mechanical stresses and the wear of the tools under these conditions will be appreciable. It is therefore necessary to provide for the heating or cooling, as may prove necessary, of the extrusion tools and also to have as good lubrication as possible between billet and tools. The glass-lubrication technique, known as the Ugine-Séjournet process, has greatly facilitated the hot extrusion of carbon and alloy steels, nickel and nickel alloys, as well as of other high-temperature alloys.

Such phenomena as unequal heating of the billet, scale formation on the billet, wear of the tools, etc., are of very first importance industrially but,

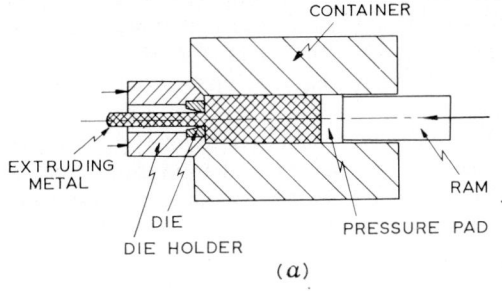

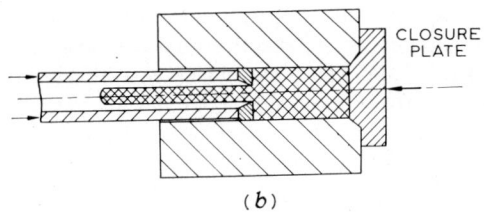

FIG. 11.42 Methods of extrusion.
(a) Direct extrusion.
(b) Reverse extrusion.

apart from the cost involved, these factors preclude the use of industrial plant for the derivation of systematic data on the mechanism of plastic flow. Controlled experiments in hot extrusion have been carried out by SIEBEL and FANGMEIER (1931), SACHS and EISBEIN (1931), PEARSON and SMYTHE (1931) and others.

Magnesium	280 – 320 °C
Aluminium	450 – 490 °C
68/32 Brass	700 – 750 °C
Copper	800 – 880 °C
Nickel	1110 – 1160 °C
Steels	1050 – 1250 °C

FIG. 11.43 Approximate extrusion temperature range.

The cold extrusion of steel was first successfully accomplished in Germany during World War II. The success of the process depends on the use of phosphating as a base for the lubrication film. The main advantages of cold extrusion are, economy of material, in many cases improved physical characteristics and a knowledge that the material in the resultant components is physically sound. The development of the cold extrusion of steel was surveyed by MORGAN (1959), and FELDMAN's monograph on the Cold Forging of Steel (1961) should be referred to. Friction and lubrication in metalworking are treated at great length in the book edited by SCHEY (1970).

The process calls for presses of extremely robust design, accurate guiding of the ram and very accurate tool setting. The billet must be very carefully prepared with removal of any scale, and it is preferable to anneal between each extrusion operation. In some cases, for example in the extrusion of gears, it is necessary to heat-treat the final product to give the required physical properties.

Most laboratory-controlled investigations into the mechanism of plastic flow have been carried out on lead, tin and aluminium at room temperatures. Attention has been restricted to the five fundamental factors: reduction, frictional and geometrical boundary conditions, the strain-hardening characteristic of the material, and the speed of operation. The fundamentals of the process will be more easily understood if initially we examine the cold extrusion of lubricated cylindrical rod from a solid cylindrical slug, of a non-work-hardening material, through square dies at slow speeds.

Autographic diagrams showing the variation of punch load with punch travel for direct and reverse extrusion under the latter conditions are of the form shown in Fig. 11.44. The diagrams consist of three principal and easily recognizable phases for non-hardening materials extruded through square dies.

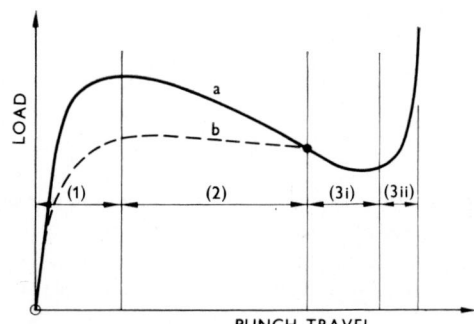

FIG. 11.44 Idealized autographic diagrams for (*a*) direct extrusion, and (*b*) reverse extrusion through square dies. (1) Coining phase. (2) Steady-state phase. (3) Unsteady-state phase.
- End of steady state.
(i) Non-piping.
(ii) With piping.

(1) THE COINING PHASE

There is a rapid build-up of pressure caused by the initial compression of the slug, which exactly fills and expands the container; there is also a small amount of extrusion of relatively unstrained material. The high initial pressure when extruding through conical dies has been discussed by DUFFILL and MELLOR (1969).

(2) THE STEADY-STATE PHASE

As the punch moves over this range, extrusion proceeds steadily, the total load in the case of direct extrusion continuously decreasing because the frictional load due to the relative motion of slug and container wall is decreasing. In reverse extrusion, no such steady decrease occurs, because there is no relative motion between the slug and the container wall; the top of the extrusion diagram is almost flat.

(3) THE UNSTEADY-STATE PHASE

The end of the steady-state phase is marked by a more rapid rate of decrease in load; this is the unsteady-state phase of deformation. The zone of plastic deformation fills the whole volume of the slug and the field of straining alters as the punch advances; this is shown as (3i) in Fig. 11.44. When the residual thickness of the slug is about half the diameter of the extruded rod product, a circular hole or pipe begins to form on the axis of the slug and a final phase, (3ii), of unsteady state deformation arises; the cavity which has been initiated continuing to grow. This is perhaps best understood as due to the residual disc buckling at the centre where it is unconstrained. The extrusion load is no longer carried over the whole of the pressure pad and the term 'extrusion pressure' as usually understood, ceases to be meaningful, since the area over which the extrusion load acts is no longer obvious. Ultimately, the load increases again at a rapid rate. The mechanics of this sequence were first arrived at by JOHNSON (1959) for plane strain: (see also Problem 23.) AVITZUR (1967) subsequently used the same ideas for the axisymmetric process; see also HOFFMANNER (1971).

In extrusion through square dies, some metal is lodged in the corner between the die and container wall and not extruded with the rest of the slug. This region is known as a dead-metal zone. The boundary of a dead-metal zone is marked by a narrow region of intense shear, and such zones effectively convert a square die into a curved conical die with a perfectly rough surface. A dead-metal zone can give rise to a bad surface finish on the product, and in practice it is avoided to a certain extent by using a radius ring, which fits on the square die, taking the place of the dead-metal zone that would normally be formed. The size and shape of a dead-metal zone depends on the reduction and frictional conditions between slug and extrusion tools.

Much experimental work has, in recent years, been devoted to hydrostatic extrusion. In this process, which was first studied by BRIDGMAN (1952), the billet is surrounded by fluid and extrusion usually takes place through a conical die. In simple hydrostatic extrusion the pressure of the surrounding fluid is increased until it is sufficiently high to extrude the billet through the die. The main advantages of hydrostatic extrusions are that there is no billet/container friction, and thus long billets can be used; the billet/die lubrication is also much improved. Lower die angles, giving more homogeneous deformation, are then feasible. The disadvantage of simple hydrostatic extrusion is that the product issues from the die in an uncontrolled manner. For practical applications it would seem necessary to control the speed of the product. This can be done by setting a pressure which is just insufficient to initiate extrusion and then applying a forward pull to the product. The main research into the process has been carried out at the National Engineering Laboratory, Scotland, at the Physics of Super Pressure Laboratory in Moscow and at the Battelle Memorial Institute, U.S.A. The work at these institutes has been reported by PUGH (1965), BERESNEV et al. (1963) and FIORENTINO et al. (1963), respectively. More recent developments have been reported by SLATER and GREEN (1967), by LENGYEL and ALEXANDER (1967), and by PUGH (1970).

11.14 Determination of Extrusion Pressure

The first rational relationship between extrusion pressure, p (ratio ram load/cross-sectional area of slug) and reduction in area, r, appears to have been put forward by SIEBEL and FANGMEIER (1931) and was based on the assumption of homogeneous deformation. It was assumed that

$$p = k' \ln \frac{A}{a} = k' \ln \frac{1}{1-r} \tag{11.37}$$

where A is the initial and a the final cross-sectional area of the material and k' the 'resistance to deformation . . . depends on temperature and . . . speed of extrusion'. This expression was not related to a particular length of slug, but for direct extrusion it was modified to include a term to allow for friction between the total length of the slug and the container wall. These authors also proposed the same expression for use in the case of reverse extrusion. The experimental values of extrusion pressure are always higher than the theoretical values based on the above formula because of neglect of the non-useful work of distortion. The correlation between experimental and theoretical values varies with the reduction and is best for high reductions.

HILL (1948), using the theory of slip-line fields, has calculated the plane strain extrusion pressure, in the case of extrusion of a non-work-hardening material through square dies, for reductions between 0·12 and 0·88 assuming dead-metal zones to cover the whole of the die face and no friction between the metal and the container wall. Calculations of the extrusion pressure have also been made (JOHNSON, 1955) for the case where there is friction between slug and container wall. These slip-line field solutions for plane strain extrusion are considered in detail in Chapter 12. At present no complete solutions are known even to the simplest problems of axisymmetric extrusion, although upper bound solutions can be obtained (e.g. KUDO, (1960), and see Chapter 16).

JOHNSON (1957) has suggested a useful empirical approach for estimating the pressure in axisymmetric extrusion. The method is developed from the known slip-line field solution for plane strain extrusion. The theoretical relations between extrusion pressure and reduction in plane strain extrusion can be fitted by the empirical equation

$$\frac{p}{2k} = a' + b' \ln 1/(1 - r) \tag{11.38.i}$$

where k is the yield stress in shear and a', b', are constants. JOHNSON (1956) has given results of plane strain experiments where, for comparison with slip-line field theory, the extrusion pressure was taken as the pressure at the end of the steady-state phase. Therefore, the variation of frictional load with slug length is not included in the correlation of theoretical and experimental results. (The degree of friction does, of course, affect the shape of the dead-metal zone.) The experimental results for the extrusion of pure lead and 0·065 per cent tellurium lead through square dies are shown in Fig. 11.45.

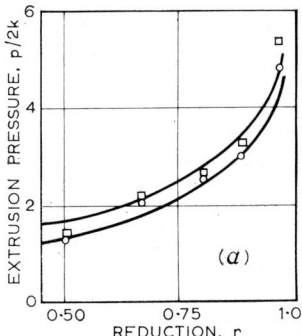 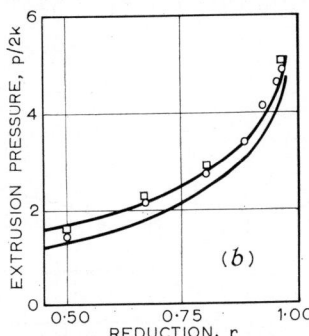

FIG. 11.45 Comparison of theoretical and experimental results for the plane strain extrusion of lead through square dies. The upper curve refers to a very rough container wall and the lower curve to a smooth container wall with a dead-metal zone on the die face.

○   Experimental point for lubricated extrusion.
□   Experimental point for unlubricated extrusion.
(a) Pure lead   (b) Tellurium lead
(*After* Johnson, *J. Mech. Phys. Solids.*)

Before commencing a test, the extrusion tools were cleansed with carbon tetrachloride, and graphite-in-tallow was used in tests on lubricated slugs. It is seen from Fig. 11.45 that there is reasonable agreement except at high reductions where the pressure is consistently underestimated. It was concluded that lubricated extrusions are accompanied by coefficients of friction that are not small, but that for extrusion through a square die, pressures will be predicted to better than about 15 per cent by the curves associated with the dead-metal zone hypotheses with a rough container wall. Friction coefficients at the container wall have been measured using the pin technique, see EL BEHERY *et al.* (1963).

For the axisymmetric extrusion of a non-work-hardening material, JOHNSON suggested that the experimental values of extrusion pressure and reduction could be fitted by the equation

$$\frac{p}{Y} = a + b \ln 1/(1 - r) \tag{11.38.ii}$$

Y is the uniaxial compressive yield stress and a and b are unspecified constants, different from a' and b' of equation (11.37). $Y = k\sqrt{3}$ or $2k$, depending on whether von Mises' or Tresca's yield criterion is adopted. *The right-hand side of equation* (11.38) *is to be regarded as the mean strain imparted to extruded lengths of slug.* (It slightly overestimates the mean strain due to the inclusion of friction.)

12—EP   *   *

Values of the constants a and b were obtained for the axisymmetric extrusion of lead, which simulates a non-work-hardening material. The true stress-strain curve, obtained in compression, is shown in Fig. 11.46(a) and experimental results for the ratio p/Y are set down in Fig. 11.47. When values of p/Y were plotted against $\ln 1/(1 - r)$ and the best straight line drawn through the points, it was found that

$$\frac{p}{Y} = 0{\cdot}8 + 1{\cdot}5 \ln 1/1 - r. \tag{11.39}$$

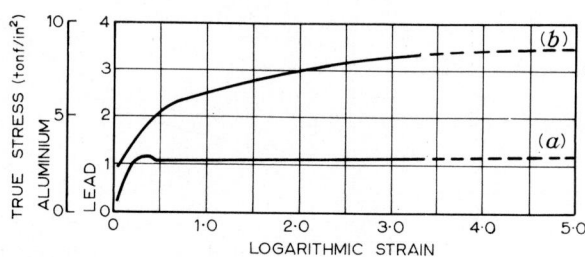

FIG. 11.46 True stress/logarithmic-strain curves for (a) pure lead, (b)
super-pure aluminium.
———————— Experimental
- - - - - - - - Extrapolation
(*After* Johnson, *J. Mech. Phys. Solids.*)

Fractional reduction, r	Load, lbf	Pressure, p, tonf/in²	Y tonf/in²	Experimental value, p/Y	Predicted value, p/Y
0·50	3,600	2·05	1·12	1·85	1·85
0·667	5,000	2·85	1·13	2·5	2·45
0·80	6,250	3·55	1·14	3·1	3·2
0·88	8,350	4·75	1·14	4·15	4·0
0·94	10,000	5·70	1·15	4·95	5·0

FIG. 11.47 Experimental results for extrusion of pure lead.
(*After* Johnson, *J. Mech. Phys. Solids*)

The extrusion of super-purity aluminium was next considered. When extruding a work-hardening material the individual elements are hardened to varying degrees, since the process is one of inhomogeneous deformation and no simple equation can be formulated to take account of this. It was suggested that the mean overall strain imparted to the slug may be expected to be closely approximated by that undergone by a non-work-hardening material under similar conditions, and hence that the differential hardening is taken account of by an average value, $\bar{Y}$, of the yield stress. $\bar{Y}$ is defined as the average value of the true stress over the range of logarithmic strain from 0 to $[0\cdot8 + 1\cdot5 \ln 1/(1 - r)]$ as given by the ordinates in Fig. 11.46(b), r being the value appropriate to the extrusion being performed. This method was applied to experimental values obtained for aluminium and the results of the calculation, $p/\bar{Y}$ are shown in Fig. 11.48. The values so derived are seen to be in remarkably close agreement with those predicted by

$$\frac{p}{\bar{Y}} = 0\cdot8 + 1\cdot5 \ln 1/(1 - r).$$ (11.40)

Fractional reduction, r	Load, lbf	Pressure, p, tonf/in²	$\bar{Y}$ tonf/in²	Experimental value, $p/\bar{Y}$	Predicted value, $p/\bar{Y}$
0·50	19,500	11·1	6·1	1·8	1·85
0·667	30,100	17·1	6·5	2·6	2·45
0·80	38,500	21·9	6·9	3·15	3·2
0·88	51,000	29·0	7·1	4·05	4·0
0·94	67,500	38·4	7·5	5·1	5·0

FIG. 11.48 Experimental results for extrusion of super-purity aluminium.

(*After* Johnson, *J. Mech. Phys. Solids.*)

Further experiments on 0·065 per cent tellurium lead and tin confirm the approximate correctness of the above expression and the method of evaluating the quantity $p/\bar{Y}$.

Implicit in the above approach is a general, inductive method for finding the load to effect extrusion, drawing or rolling, etc., if the process is a steady state one; it also applies no matter what the section shape. The procedure is to use a (nearly) non-hardening material and, on a model scale, find the pressure (or load) required; this is then divided by the yield stress of the material and the resulting number is an approximate measure of the mean strain imparted. If the stress-strain curve of the material which we are interested in working is then available, the mean yield stress over this mean strain obtained is then calculable and hence the required load for working.

WILCOX and WHITTON (1958) and (1959) have carried our measurements of extrusion pressure for a very wide range of reductions and die angles. They have also used an empirical equation having the same form as (11.38) but their definitions of extrusion pressure, p, and of $\bar{Y}$ are different and less fundamental than those above.

No mention has been made in this section of the geometry of deformation in extrusion. Some facets of this are considered theoretically and in detail for plane strain extrusion in Chapter 13. A detailed experimental investigation of the deformation during the extrusion of lead, has been carried out by YANG and THOMSEN (1953). Other important papers, books and reviews are asterisked in the list of References. The book by JOHNSON and KUDO (1962) is a detailed survey of extrusion metal mechanics up to 1960.

11.15 Rolling

Some useful and large collections of references, with detailed discussions of rolling theory and rolling mills will be found in the books by UNDERWOOD (1950), LARKE (1967), TSELIKOV (1967), SMIRNOV (1967), TSELIKOV and SMIRNOV (1965), and TARNOVSKII (1965). See Section 14.6 for reference to slip-line field treatments of rolling.

11.15.1 *Cold-rolling of Strip*

A theory of rolling aims at relating the externally applied forces to the mechanical strength properties of the material rolled. Any such theory will be of especial value in estimating the power requirements of a rolling mill for use on a newly developed metal or alloy.

Rolling theory may be divided into two main parts, that which applies to hot-rolling and that which applies to cold-rolling. In hot-rolling, the yield stress characteristic of the metal is strain-rate dependent and the frictional force between the rolls and stock is high, whereas in the cold-rolling of strip the yield stress characteristics of the metal is essentially independent of the rate of deformation, and the frictional force between rolls and stock is low.

We shall confine ourselves here to the cold-rolling of strip and in order to obtain a solution, we shall have to restrict further the scope of the theory. The assumptions that will be made are those set down by VON KARMAN (1925) and used by OROWAN (1943). BLAND and FORD (1948), basing their work on Orowan's theory, made further assumptions, and it is on their work that the following solution is based.

Consider what happens when a strip of material of initial thickness h_1 enters the rolls, Fig. 11.49(*a*). As the strip passes through the rolls, it is first compressed elastically until it yields, is then subjected to plastic deformation (work-hardening with increasing strain), and on leaving the roll gap there is elastic recovery to reduced thickness h_2.

In the theory, it will be assumed that the material is rigid plastic (work-hardening). This means that the contribution of the elastic arcs to roll force and torque, which occurs in the case of a real material, is ignored. When comparing the theory with experiment, this contribution may sometimes be allowed for in the theory by suitably adjusting the 'coefficient of friction' between the rolls and the strip. However, this method of adjustment is not applicable to very light passes (low reduction in thickness) or to passes on very hard strip where the elastic strains are not small compared with the plastic strains. This fact again reduces the scope of the theory.

As rolling takes place the rolls are themselves distorted elastically over the arc of contact with the material and this distortion must be allowed for in the calculation of the roll force and torque. It will be assumed that the arc of contact is circular and of radius R', greater than the radius R of the un-deformed roll. Again this assumption is not true during the rolling of thin hard strip such as occurs in the final rolling of stainless and razor-blade steels.

We now apply the further restriction that the strip width-to-thickness ratio must be sufficiently large to ensure that the rigid non-plastic material outside the roll gap prevents lateral spread. The deformation is then plane strain. In practice, the radii of the undeformed rolls are of the order of one hundred times the thickness of the strip and with a strip width-to-thickness ratio greater than ten, the lateral spread is seldom greater than 1 or 2 per cent.

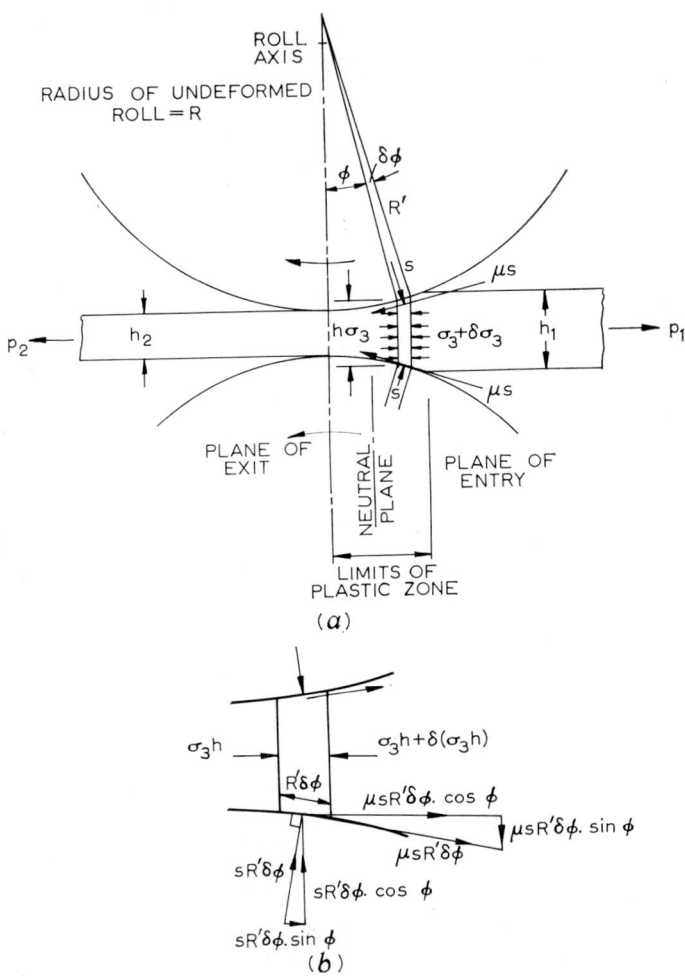

FIG. 11.49 Cold-rolling of strip: stresses in roll gap.

A most important simplifying assumption is that the horizontal stress and particle velocity do not vary through the sheet thickness. In the approach here it is also considered that there is slipping friction over the whole arc of contact, and it is further assumed that the coefficient of friction is constant over the arc of contact.

In passing through the roll gap, the strip, of constant width b, undergoes a thickness change from h_1 to h_2 and since the deformation is one of plane strain in an incompressible material

$$bh_1v_1 = bhv = bh_2v_2$$

where v_1 and v_2 are the velocities of the strip at entry to and at exit from the roll gap, and h and v refer to an intermediate position. Thus the velocity of the strip increases steadily from entry to exit and the velocity of the rolls must have some value between v_1 and v_2. On the entry side, the rolls move faster than the strip and the frictional forces draw the strip into the rolls; on the exit side, the strip moves faster than the rolls, and the frictional forces tend to oppose the delivery of the strip. At some intermediate plane (the neutral plane) the strip and rolls move with the same velocity. The position of the neutral plane is found by considering the equilibrium of the external forces, and it therefore depends on the values of the back and front tensions, p_1 and p_2, applied to the strip.

The assumptions that have been made may be summarized as:

1. The material is rigid plastic work-hardening.
2. The arc of contact is circular.
3. The deformation is one of plane strain.
4. The coefficient of friction is constant over the arc of contact.
5. Plane sections perpendicular to the direction of rolling remain plane.

(a) NORMAL ROLL PRESSURE

The stresses in the roll gap are shown in Fig. 11.49(a) and the forces on an element of material on the exit side are shown in the enlarged view, Fig. 11.49(b). It is convenient for this subject to adopt the convention that tensile stresses are negative and compressive stresses are positive. Considering the equilibrium of the longitudinal forces on the element of unit width

$$\sigma_3 h + 2sR' \, d\phi \sin \phi \pm 2\mu sR' \, d\phi \cos \phi = \sigma_3 h + d(\sigma_3 h)$$

where σ_3 is the stress in the longitudinal direction, s the normal roll pressure and μ the coefficient of friction.

Re-arranging

$$\frac{d(\sigma_3 h)}{d\phi} = 2sR'(\sin \phi \pm \mu \cos \phi). \tag{11.41}$$

The positive sign applies between the plane of exit and the neutral plane and the negative sign between the neutral plane and the plane of entry. In what follows, we shall adopt the convention that the upper sign always refers to the exit side and the lower sign to the entry side of the neutral plane.

From consideration of vertical equilibrium of the forces, the vertical stress, σ_1, in the material is given by

$$\sigma_1 R' \, d\phi \cos \phi = sR' \, d\phi \cos \phi \mp \mu sR' \, d\phi \sin \phi$$

or $$\sigma_1 = s(1 \mp \mu \tan \phi). \tag{11.42}$$

The relative magnitude of the stresses may be determined from the following reasoning. If the stress in the lateral direction, σ_2, (say) is equal to σ_3, then for reasons of symmetry, the spread in the lateral direction would be equal to the elongation in the direction of rolling. Since there is no lateral spread σ_2 must be greater than σ_3. Again, if σ_3 is equal to σ_1, then from symmetry the strip would suffer a lateral compression equal to the vertical compression. Therefore, σ_2 must be less than σ_1. Alternatively, it follows from the Lévy-Mises equations (putting $\epsilon_2 = 0$), that $\sigma_2 = (\sigma_1 + \sigma_3)/2$ and the von Mises yield criterion then has the form

$$\sigma_1 - \sigma_3 = 2k = \frac{2}{\sqrt{3}} Y \qquad (11.43)$$

where k is the yield stress in pure shear and Y is the yield stress in uniaxial compression. (The Tresca criterion requires $k = Y/2$.)

In cold-rolling, the angle of contact is small, rarely exceeding $6°$, and since the coefficient of friction is small, we can assume that the normal pressure approximately equals the vertical stress. That is, equation (11.42) may be written

$$\sigma_1 \simeq s \qquad (11.44)$$

with an error of less than 1 per cent.

We can also write $\sin \phi \simeq \phi$, $\cos \phi \simeq 1$ and equation (11.41) reduces to

$$\frac{d(\sigma_3 h)}{d\phi} = 2sR'(\phi \pm \mu). \qquad (11.45)$$

Finally, substituting the value of the roll pressure from equation (11.44) in the yield equation (11.43)

$$s - 2k = \sigma_3. \qquad (11.46)$$

Using these last three equations, we can estimate the roll pressures over the arc of contact.

From equations (11.45) and (11.46)

$$\frac{d[h(s - 2k)]}{d\phi} = 2sR'(\phi \pm \mu)$$

or

$$\frac{d}{d\phi}\left[2kh\left(\frac{s}{2k} - 1\right)\right] = 2sR'(\phi \pm \mu),$$

which may be written

$$2kh\frac{d}{d\phi}\left(\frac{s}{2k}\right) + \left(\frac{s}{2k} - 1\right)\frac{d}{d\phi}(2kh) = 2sR'(\phi \pm \mu).$$

The term $(s/2k - 1)\, d\,(2kh)/d\phi$ is very small compared with $2kh\, d(s/2k)/d\phi$ because the yield stress $2k$ increases as h decreases through the roll gap (and hence the product $2kh$ is almost constant) and the coefficient $(s/2k - 1)$ is small for practical cold-rolling.

Omitting this term, we can write, to give dimensionless quantities where possible

$$\frac{(d/d\phi)(s/2k)}{s/2k} = \frac{2R'}{h}(\phi \pm \mu).$$ (11.47)

The variation of strip thickness h through the roll gap is given by

$$h = h_2 + 2R'(1 - \cos \phi)$$

where h_2 is the strip thickness at exit, and writing $1 - \cos \phi = \phi^2/2$,

$$h = h_2 + R'\phi^2.$$

Substituting into equation (11.47)

$$\frac{(d/d\phi)(s/2k)}{s/2k} = \frac{2R'(\phi \pm \mu)}{h_2 + R'\phi^2},$$

and integrating

$$\ln\left(\frac{s}{2k}\right) = \ln\left(\frac{h}{R'}\right) \pm 2\mu \sqrt{\frac{R'}{h_2}} . \tan^{-1}\left(\sqrt{\frac{R'}{h_2}} . \phi\right) + \ln C$$

where C is a constant of integration.

The roll pressure is therefore

$$s = C\left(2k . \frac{h}{R'} . e^{\pm \mu H}\right)$$ (11.48)

where

$$H = 2\sqrt{\frac{R'}{h_2}} . \tan^{-1}\left(\sqrt{\frac{R'}{h_2}} . \phi\right)$$

which is zero at the exit where $\phi = 0$.

At exit, $\sigma_3 = -p_2$ the front tension and the roll pressure at exit $s_2 = 2k_2 - p_2$ using equation (11.46) where k_2 is the yield shear stress at exit.

From equation (11.48)

$$C = \frac{R'}{h_2}\left(1 - \frac{p_2}{2k_2}\right)$$

and the roll pressure on the exit side is

$$s^+ = \frac{2kh}{h_2}\left(1 - \frac{p_2}{2k_2}\right)e^{\mu H}$$ (11.49)

and on the entry side

$$s^- = \frac{2kh}{h_1}\left(1 - \frac{p_1}{2k_1}\right)e^{\mu(H_1 - H)}$$ (11.50)

where H_1 is the value of H at entry and k_1 is the yield shear stress.

(b) NEUTRAL PLANE

The neutral plane can be determined before the roll pressure curve s against ϕ is computed. From equations (11.49) and (11.50) at the neutral plane,

$$s_n^+ = s_n^-$$

giving the value of H_n. The angle ϕ_n at the neutral plane is then found to be

$$\phi_n = \sqrt{\frac{h_2}{R'}} . \tan \sqrt{\frac{h_2}{R'}} \times \frac{H_n}{2} .$$

The normal roll pressure rises from entry and exit sides to a maximum at the neutral plane. Its distribution over the arc of contact will be given therefore by a hill-shaped curve which is called the 'friction hill'. A similar curve represents the horizontal pressure σ_3.

Normal roll pressure curves derived from this theory for the rolling of high-conductivity copper are shown in Fig. 11.50, (BLAND and FORD, 1948). Two cases are illustrated: the lower curves show the distribution when front and back tensions are applied; the upper curves the distribution for the same pass but without tensions applied.

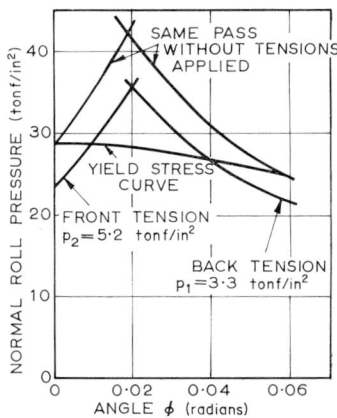

FIG. 11.50 Normal roll pressure curves showing effect of applying tensions to the strip. (*After* Bland and Ford, *Proc. Instn mech. Engrs.*)

(c) ROLL FORCE

As the strip passes through the rolls, the yield stress, $2k$, varies with the angle ϕ. Values of yield stress for the material for different reductions must be found experimentally. Since the deformation in rolling is one of plane strain, it is obviously best, if possible, to obtain a yield stress curve, using the same mode of deformation. Such a test, using smooth parallel dies for the compression of the strip has been devised by FORD (1948) and developed by

WATTS and FORD (1955). This method has already been described in Section 6.6.

Once the variation of yield stress through the roll gap and the coefficient of friction are known, then the roll force can be found by integrating the normal roll pressure over the arc of contact. Thus the roll force per unit width is given by

$$P = \int_0^{\phi_1} sR' \, d\phi. \tag{11.51}$$

See also HILL (1950).

(d) ROLL TORQUE

The power input to the mill is supplied by applying a torque to the rolls and by strip tension. The mean torque, G, per roll per unit width is the integral of the moment about the roll axis of the frictional force along the arc of contact. The contribution due to the moment of the normal forces may be neglected unless there is very large roll-flattening. The arm of the friction force is the distance from the roll axis to the roll surface, that is, approximately the undeformed roll radius R.

Therefore

$$G = \left[\int_{\phi_n}^{\phi_1} \mu s^- R' \, d\phi - \int_{\phi_2=0}^{\phi_n} \mu s^+ R' \, d\phi \right]$$

or

$$G = \mu RR' \left[\int_{\phi_n}^{\phi_1} s^- \, d\phi - \int_{\phi_2=0}^{\phi_n} s^+ \, d\phi \right] \tag{11.52}$$

Assuming that the coefficient of friction μ is constant over the arc of contact, this last equation is the difference of two quantities of the same order of magnitude and is not therefore suitable in this form for computation. BLAND and FORD (1948) have shown that an alternative formula, suitable for computation, is

$$G = RR' \left[\left(\int_{\phi_2=0}^{\phi_1} s\phi \, d\phi \right) + \frac{s_1 h_1 - s_2 h_2}{2R'} \right].$$

(e) DETERMINATION OF THE COEFFICIENT OF FRICTION

The problem here is to measure the coefficient of friction between two surfaces in contact, when one of the surfaces is undergoing considerable plastic deformation. If it is assumed that the coefficient of friction is constant over the arc of contact, a value for it may be found by direct measurement. The following method was proposed by BLAND and investigated by WHITTON and FORD (1955).

A strip of metal is rolled at a constant speed at any suitable pass reduction and the roll force and torque measured continuously. A gradually increasing back tension is applied to the strip until the neutral plane is forced to the exit. All the friction is then acting in the same direction, and equations (11.51) and (11.52) for roll force and torque reduce to

$$P = R' \int_0^{\phi_1} s \, d\phi \quad \text{and} \quad G = \mu R R' \int_0^{\phi_1} s \, d\phi. \tag{11.53}$$

It then follows from equation (11.53) that

$$\mu = G/PR, \tag{11.54}$$

a non-dimensional group independent of rolling theory.

(f) Roll Flattening

It has been shown how estimates can be made of the coefficient of friction and of the variation of yield stress through the roll gap. The only other unknown is the value of the radius of the deformed part of the roll R'. A formula devised by HITCHCOCK and put in the following form by BLAND and FORD (1948) gives

$$R' = R\left(1 + \frac{2cP}{h_1 - h_2}\right) \tag{11.55}$$

where $c = 1 \cdot 67 \times 10^{-4}$ in inch and ton units for steel rolls. This formula should not be applied to the rolling of thin hard strip.

A useful discussion of elastic effects in metal rolling and been given by WEINSTEIN (1963) and a treatment of transverse thickness variation in rolling is given by SAXL (1958).

(g) Some Further Comments

Within the restrictions applied, the above theory is not inconsistent with experimental results obtained for roll force and roll torque. It was found in some experiments that the yield stress curve, determined by a plane compression test, was too low for good agreement. It was suggested that redundant shearing in cold-rolling work-hardens the material a little more rapidly than in plane compression, the difference being from 5 to 7 per cent. Another factor, at high rolling speeds, is the raising of the yield stress above that measured in the plane compression test. Use in the theory of a yield stress value obtained by the latter method would in such cases underestimate the roll force. For comparisons between theory and experiment, the reader is referred to papers by HESSENBERG and SIMS (1951) and WHITTON and FORD (1955).

The above theory is applicable to quite a wide range of practical rolling. To facilitate computation, LIANIS and FORD (1956) have drawn up nomograms for determining roll force and torque. However, as has already been stated, the theory should not be applied to the rolling of thin hard strip or to rolling with high back tension.

Two interesting and useful papers on periodic surface finish and defects in cold-rolled products are by MOLLER and HOGGART (1967) and THOMSON and HOGGART (1967).

The roll force is decreased, and therefore performance increased, if the friction between the rolls and the strip is reduced to a minimum or if tensions are applied to the strip. WILCOX and WHITTON (1960) have investigated the effect of various lubricants on the value of the roll force when rolling thin titanium strip. Five different lubricants were used and it was shown that for the cold-rolling of hard, thin, titanium strip, reductions in load of up to 60 per cent are possible. The authors suggested that the same technique could be used for other high-strength materials such as nimonic alloys, high-strength steels and zirconium. The hydrodynamic effect of oil between strip and rolls, on roll force, torque and pressure has been analysed by BEDI and HILLER (1967). A recent important paper, especially as it touches on torque calculations, is that by ALEXANDER (1972).

11.15.2 Rolling Thin Hard Strip

An extensive bibliography on this subject has been given by AFONJA and SANSOME (1969).

As has been seen, there is generally a limit to the minimum gauge or thickness which may be achieved with a given material because of the elastic distortion of the rolls consequent on the high loading involved. For the mass-production of the thinner gauges specialized mills of the *Sendzimir planetary type*, with small, intricately backed work rolls and high tensions are used, the Sendzimir cluster mill or the Contact-Bend-Stretch mill. However, where demand does not justify the capital outlay on such specialized mills, ways must be sought to improve the performance of existing plant, such as in sandwich rolling where the hard metal to be rolled forms the middle layer of the sandwich between two layers of softer material. On rolling, there will be a tendency for the outer layers to elongate more than the central layer. This differential elongation is resisted by the frictional forces at the interface of the layers, adding to the existing longitudinal compressive stress in the softer material. In effect, this is the same as applying a tension in single-strip rolling, and consequently the roll force is reduced. The success of the technique depends on the selection and relative thicknesses of the soft and hard metals. It is important to get a good 'bite' between the rolls and the soft layers and to have as high a coefficient of friction as possible between the layers. The idea of using a sandwich technique to create requisite tensions or compressions to facilitate the working of hard metals in the various working processes, remains to be developed. See ARNOLD and WHITTON (1959).

11.15.3 Ring Rolling

Figure 11.51 illustrates one of the methods used to roll rings such as are used for railway tyres at one extreme and bearing races at the other. An initial ring is placed between the rolls P and D; the driving roll is fixed but free to rotate on its axis. The axis of the driven pressure roll, P, is caused to approach D and when the ring is gripped between the rolls it is caused to rotate, at the same time being continuously reduced. If there is no transverse spread the circumference of the ring steadily increases until the required diameter is obtained. In order to ensure that a circular ring is rolled a pair of guide rolls, correctly positioned, must be used.

The mechanics of ring rolling has received little attention, but some introductory discussion and a small number of references will be found in the papers by JOHNSON and NEEDHAM (1968a and b), JOHNSON, MACLEOD and NEEDHAM (1968) and CADDELL, NEEDHAM and JOHNSON (1968). One interesting feature is the creation of a plastic hinge in a ring about to be rolled, if it is first subjected to a specific amount of compression before rolling starts; the hinge lies diametrically opposite the nip.

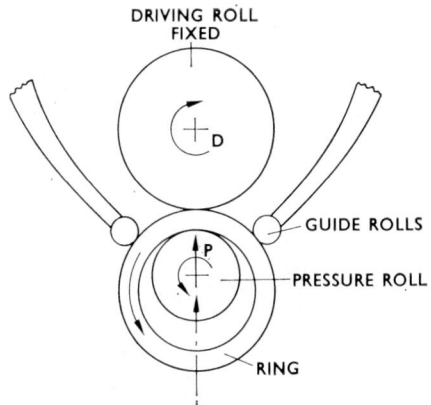

FIG. 11.51 Ring-rolling set-up.

11.15.4 *Asymmetrical Rolling*

In general, if strip is rolled under conditions where the two rolls are unequal in some respect, e.g., the roll diameters, surface roll speed, roll surface roughness, then the strip produced will be curved; this is asymmetrical rolling. In such cases the conditions in the roll gap are more complicated than they are for symmetrical rolling. Figure 11.52 shows three zones: the usual entry and exit zones exist, but a third additional zone arises—known as the cross shear zone; the shear and normal stresses are different and in particular the shear stresses on the same vertical section act in opposite directions. This situation is obviously difficult to analyse and clearly the Karman-type analysis, which only concerns itself with stresses, will be inadequate when attempting to predict final strip curvature. ZOROWSKI and SHUTT (1963) have attempted a detailed stress analysis of single-roll driven mills, i.e., with two rolls of equal radius, one driving and the other rotating freely. On the assumption that the surfaces of the strip are the same as those of the surfaces of the rolls which they contact, they are able to make predictions about the curvature of the emerging strip. An interesting introduction to this and related topics is to be found in the book by HOFFMAN and SACHS (1953). Some experimental results have been given by JOHNSON and NEEDHAM (1966).

11.15.5. *The Pendulum Mill*

The pendulum mill, developed by SAXL (1964–65) is shown, diagrammatically, in Fig. 11.53. The two work rolls oscillate as indicated while the metal is slowly propelled through them by feed rolls (not shown). The rolls oscillate under power at up to 2600 cycles per minute but do not rotate as they pass over the work-piece. The mill operates on cold strip and fractional reductions of up to 0·95 are claimed; obviously for such high reductions the metal will be greatly heated and thus the process is not one of cold-working.

The connection between this mill and the planetary mill will be obvious.

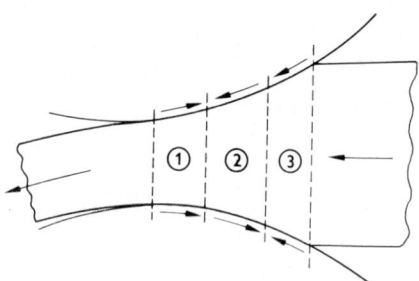

FIG. 11.52 Asymmetrical rolling.
① Normal exit zone.
② Cross shear zone.
③ Normal entry zone.

11.15.6 *Hot-rolling of Strip*

When strip is rolled hot the coefficient of friction between the rolls and the stock is higher than in cold-rolling and may even be high enough to cause shearing of the metal in contact with them. Also the yield stress depends on the temperature of the stock and on the strain rate, which vary through the pass. Slip-line field solutions for the case of hot-rolling with sticking friction over the whole arc of contact are referred to in Section 14.6; unfortunately these solutions can take no account of yield stress variations due to strain rate changes in the roll gap. JOHNSON and KUDO (1960) have used ALEXANDER's (1955) solution as an upper bound and derived the form to which an initially square grid of lines in the plane of rolling would be deformed after passing through the roll gap. LIPPMANN and JOHNSON (1960) have endeavoured to use an equation of state in conjunction with ordinary rolling theory to predict temperature changes in the roll gap during a hot-rolling operation.

The paper by SIMS (1954) is well known for its description of calculations for roll force and torque in hot-rolling, and the paper by SMITH, SCOTT and SYLWESTROWICZ (1952) describes the measurement of pressure distribution between stock and rolls in hot and cold flat-rolling.

11.15.7 *Section-rolling, Transverse Rolling, V-Groove Forming and Spread*

Pass design and associated calculations together with an outline of the equipment for section rolling will be found in the review by STEWARTSON (1959) and PARKINS (1968). KUDO and YOKAI (1968) have presented results concerning helical or transverse rolling. Ball and other kinds of rolling are referred to in an article in *The Engineer* (1967). KUDO and TAMURA (1968) have presented an analysis and some experiments in V-groove rolling, and

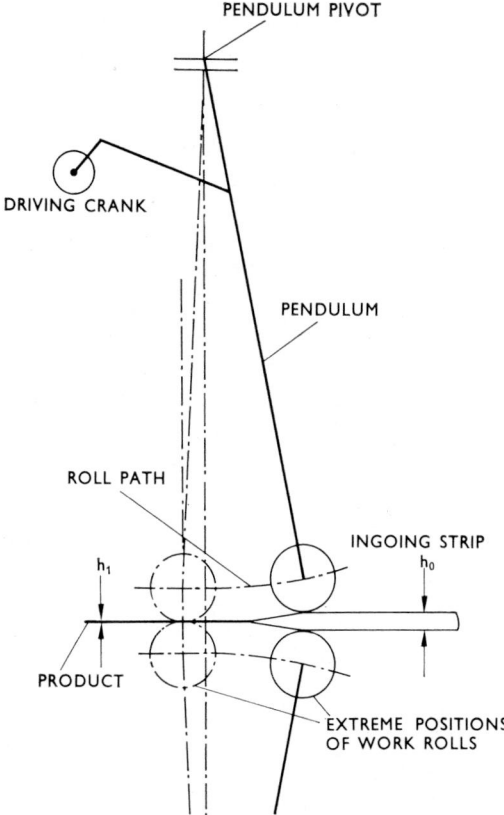

FIG. 11.53 The pendulum mill.

JOHANSSON (1965) a study of material flow between rolls. Amassed references and some experimental results concerning spread in rolling rectangular sections will be found in a paper by CHITKARA and JOHNSON (1966).

11.15.8 *Seamless Tube Making*

A readable account of the rotary piercing of tubes is to be found in the book by ALEXANDER and BREWER (1963), and other useful references are JUBB and BLAZYNSKI (1969), ZOROWSKI and HOLBROOK (1968) and BLAZYNSKI (1969).

11.16 Swaging a Cylindrical Rod

A cylindrical rod of current radius a and length $2l_0$ may be supposed to be compressed radially, see Fig. 11.54, so that material exudes at the ends. We require to find how the radial stress σ_r varies along the bar surface.

If mk denotes the constant frictional stress along the length of a die, the axial pressure variation may easily be found from the equation for axial force equilibrium. Supposing the axial stress σ_z to be independent of z, then for an element as shown,

$$\pi a^2 . d\sigma_z = 2\pi a . mk . dz.$$

Thus, $\qquad\qquad d\sigma_z = 2mkz + \text{constant.}$

If $\sigma_z = 0$ at $z = l_0$, then

$$\sigma_z = 2mk(l_0 - z)/a \qquad (11.56)$$

and $\qquad\qquad \sigma_{z,\text{max}} = 2mkl_0/a. \qquad (11.57)$

Using the Tresca yield criterion, $\sigma_r - \sigma_z = 2k$, where σ_r and σ_z are taken to be principal stresses, and substituting in (11.57)

$$\sigma_r = 2k \left(1 + \frac{(l_0 - z)m}{a} \right). \qquad (11.58.\text{i})$$

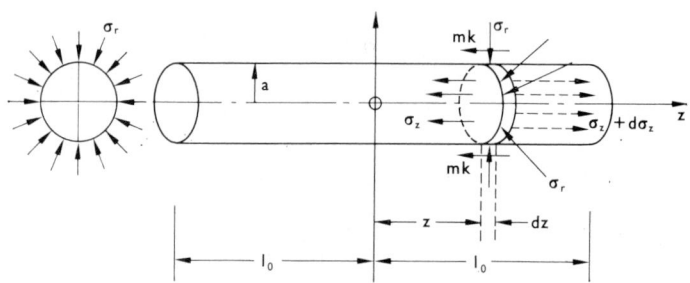

FIG. 11.54 Swaging of cylindrical rod.

If it had been assumed that Coulomb friction prevailed so that the frictional stress was not mk but $\mu\sigma_r$, then it would be found that

$$\sigma_r = Y \exp. [2\mu(l_0 - z)/a] \qquad (11.58.\text{ii})$$

11.17 The Simple Upsetting or Compressing of a Cylinder

In Section 1.6 simple, homogeneous, frictionless compression was discussed at length, and in Section 6.5, the effect of Coulomb friction between the cylinder and the dies. The latter analysis is reconsidered and extended below.

The distribution of normal pressure over the ends of a cylinder under compression may be found (approximately) in terms of a constant magnitude tangential frictional stress, τ, at the work-piece-platen interface. Denote this by mk where m is an empirical constant and k the yield shear stress of the material; obviously $0 \leqslant m \leqslant 1$. $m = 0$ denotes frictionless compression and $m = 1$, complete shearing at the interface. The equation of radial equilibrium for an element of the block is, as in Section 6.5,

$$\frac{d\sigma_r}{dr} + \frac{\sigma_r - \sigma_\theta}{r} = \frac{-2mk}{h}; \tag{11.59}$$

here, mk is substituted for μp. As before, the stresses are independent of z. Again assume (i) that $\sigma_r = \sigma_\theta$ and (ii) that σ_r, σ_θ and p are principal stresses. (This may be significantly in error, of course, when $mk = 1$.) Equation (11.59) reduces to,

$$\frac{d\sigma_r}{dr} = \frac{-2mk}{h}.$$

Integrating, and then noting that $\sigma_r = 0$ at $r = a$, we have

$$\sigma_r = \frac{-2mk}{h} \cdot r + \text{constant}$$

and thus,

$$0 = \frac{-2mk}{h} \cdot a + \text{constant}.$$

Hence,

$$\sigma_r = \frac{-2mk}{h}(a - r). \tag{11.60}$$

Using the Tresca yield criterion, where $2k$ denotes the compressive yield stress,

$$-(-p) + \sigma_r = 2k,$$

and then substituting in (11.60),

$$\frac{p}{2k} = 1 - \frac{\sigma}{2k} = 1 + m \cdot \frac{(a - r)}{h}. \tag{11.6.i}$$

(We could use the Mises yield criterion and write $p + \sigma_\theta = Y$.)

Figure 11.55(a) and (c) show the non-dimensional pressure, $p/2k$, and shear stress, τ/k, over the end face of the block. The total load, P, to enforce plastic compression is given by,

$$\frac{P}{2k} = \int_0^a 2\pi r \cdot \frac{p}{2k} \, dr = 2\pi \int_0^a \left[1 + \frac{m(a - r)}{h} \right] r \, dr$$

$$= \pi a^2 \left(1 + \frac{m}{3} \cdot \frac{a}{h} \right). \tag{11.61.ii}$$

The peak, or point, in the normal pressure distribution curve and the abrupt change in direction of the frictional shear stress, on the axis of the block, cannot be expected from a real compression test; experiment shows a 'smooth' change in both, see the dotted curve in Fig. 11.55(a). To derive these conditions analytically, assumptions other than the ones used here are introduced, e.g., that the frictional stress is proportional to the speed with which the metal slides across the platen.

In most practical situations neither the Coulomb coefficient of friction μ nor m are usually known and would require to be measured. Using an arrangement of relatively small pressure pins, see Fig. 11.55(a) Backofen and his students (see for instance van ROOYEN and BACKOFEN (1960),) have measured both the normal pressure distribution and determined the coefficient of friction. The calibrated pins are metal or plastic cylinders to which

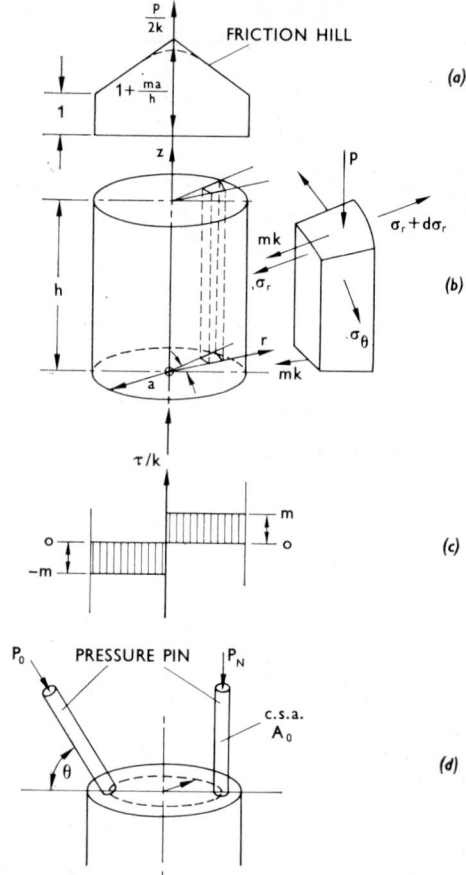

FIG. 11.55 Simple upsetting of a cylinder.

strain gauges are pasted, which operate (nominally frictionlessly) in a long thin cylindrical hole in the platen. The pins set normally to the platen-workpiece interface are used to determine the normal pressure, i.e., in Fig. 11.55(d), $p = P_N/A_0$. Radial oblique pins, in which the total force was measured, were used to deduce μ. In Fig. 11.55(d) if an oblique pressure pin is set at θ to the horizontal, and if P_0 denotes the measured force in the pin, then $P_0 = A_0 p \sqrt{(1 + \mu^2)}/\cos \theta$. However, with the aid of the normally set pin it follows that $\mu = [(P_0 \cos \theta/P_N)^2 - 1]^{1/2}$.

To determine the load-platen travel characteristic, let h be the current height of the block which has an initial height of H_0 and radius a_0. Assume that no significant amount of barrelling occurs and suppose the coefficient of friction remains constant during the compression. Then from Section 6.5,

$$P = Y.\pi a^2 \left(1 + \frac{2}{3}\mu.\frac{a}{h}\right);$$

but

$$\pi a_0^2 H_0 = \pi a^2 h,$$

so that

$$P = Y \pi a_0^2 \left(\frac{H_0}{h}\right)\left[1 + \frac{\frac{2}{3}\mu}{H_0/a_0}\left(\frac{H_0}{h}\right)^{\frac{3}{2}}\right]. \tag{11.62}$$

If the bottom platen is fixed, and denoting the upper platen travel by $z = H_0 - h$, then,

$$P = Y.\pi a_0^2 \left[\frac{H_0}{H_0 - z} + \frac{2}{3}\mu\frac{a_0}{H_0}\left(\frac{H_0}{H_0 - z}\right)^{\frac{5}{2}}\right]. \tag{11.63}$$

Figure 11.56 indicates how the load P and the maximum normal platen pressure p, vary with platen travel.

$\dfrac{Travel}{H_0} = \dfrac{z}{H_0}$		0	0·25	0·50	0·75
$\dfrac{h}{H_0} = \dfrac{H_0 - z}{H_0}$		1·00	0·75	0·50	0·25
a_0/H_0 = $\dfrac{1}{2}$	$\dfrac{P}{Y \pi a_0^2}$	1·03	1·40	2·17	4·96
	$\dfrac{p_{max}}{Y}$	1·09	1·17	1·29	2·05
a_0/H_0 = $\dfrac{1}{5}$	$\dfrac{P}{Y \pi a_0^2}$	2·46	4·46	13·00	
	$\dfrac{p_{max}}{Y}$	1·30	1·99	3·71	

FIG. 11.56 Variation of load, P, and maximum platen pressure, p_{max}, with platen travel, $\mu = 0.09$.

Alternatively, the characteristic curve of load versus current specimen height is described by

$$P = Y.\pi a_0^2 \left[\frac{1 + (m/3)(a_0/H_0)(1/y)^{3/2}}{y}\right], \tag{11.64}$$

where $y = h/H_0$.

The work done, W, in compressing a prismatic block from H_0 to H, when there is friction mk at the work-piece-platen interface, is

$$\frac{W}{Y} = \int_{H_0}^{H} -\frac{P}{Y} dh = \int_{H}^{H_0} \pi a^2 \left(1 + \frac{m}{3} \cdot \frac{a}{h} \right) dh$$

$$= \int_{H}^{H_0} \pi a_0^2 H_0 \left(\frac{1}{h} + \frac{m}{3} \frac{a_0 \sqrt{H_0}}{h^{5/2}} \right) dh$$

$$= \pi a_0^2 H_0 \left[\ln \frac{H_0}{H} + \frac{m\sqrt{a_0^2 H_0}}{3.(-3/2)} \left(\frac{1}{H_0^{3/2}} - \frac{1}{H^{3/2}} \right) \right.$$

i.e.,
$$\frac{W}{Y \pi a^2 H_0} = \ln \frac{H_0}{H} + \frac{2}{9} m \frac{a_0}{H_0} \left(\frac{H_0}{a_0} - 1 \right). \tag{11.65}$$

Equation (11.65) shows that a block for which $a_0/H_0 = \frac{1}{2}$ and $m = 1$, would require to be reduced to about 80 per cent of its original height in order that as much energy be used in overcoming interfacial friction as in securing homogeneous plastic deformation. The kind of analysis above, which leads to the idea of a friction-hill over the end surfaces of a compressed

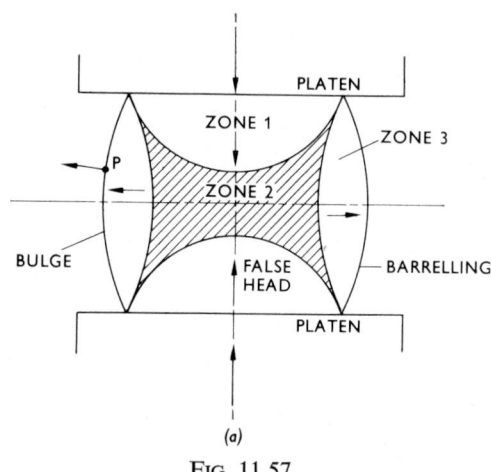

(a)

FIG. 11.57

(a) Compression of a cylinder with friction between platens and workpiece (Diagrammatic).

cylinder, is useful and reasonably reliable. It was developed on the tacit understanding that despite the presence of friction the compression proceeds homogeneously; this is untrue and leads to a gross over-simplification. Material adjacent to the platen tends *not* to slide transversely as was implied, but to act as if it was part of the platen. Cones of 'dead metal' or 'false heads' tend to be created and, as the platens approach, these force the remaining metal transversely outwards and cause barrelling, see Fig. 11.57(a). Three zones can be identified in an upsetting operation, see Fig. 11.57(a). Zone 1 identifies the cones, which are stationary relative to the dies, zone 3 is subject to a predominantly lateral movement and in zone 2 gross plastic deformation occurs as material is fed from the outside of the cones into zone 2 and thence into zone 3. The reader should better appreciate the situation after studying the plane strain homologue, see Fig. 12.13, in which the three zones just referred to can be identified.

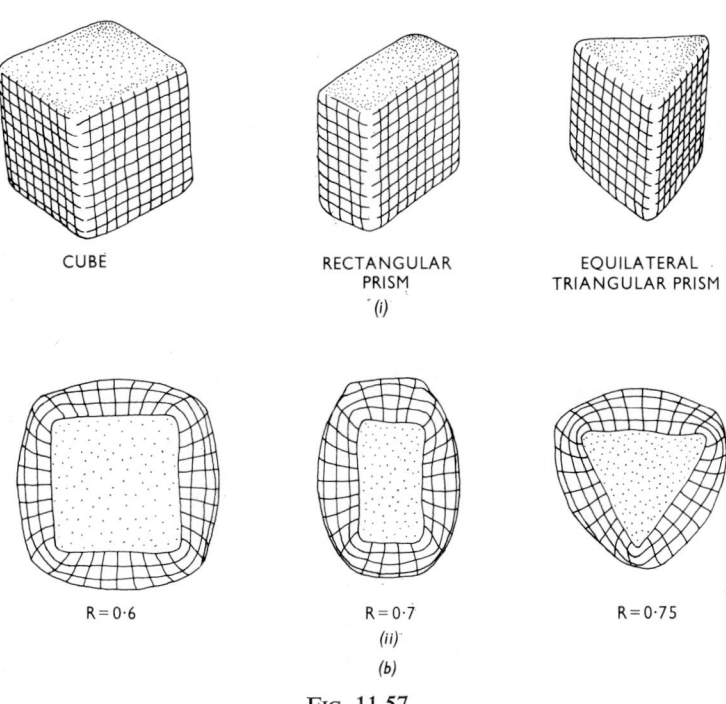

CUBE RECTANGULAR EQUILATERAL
PRISM TRIANGULAR PRISM
(i)

R = 0·6 R = 0·7 R = 0·75
(ii)
(b)

FIG. 11.57

(b) (i) Undeformed Plasticine cube, rectangular prism and equilateral triangular prism (ii) Corresponding dynamically upset specimens showing displacement, deformation and rotation of elements. These plan views show material which is substantially in the same horizontal plane as the ultimate material–platen interface (R denotes fractional reduction or compressive engineering strain). Note how after compression the original top surface is seen to be undeformed (*After* Aku, Slater and Johnson, *Int. J. mech. Sci.*)

One feature which confuses, but is easily understood in terms of these zones, is the observation that material on the cylindrical sides, after compression, may appear at the material-die interface, i.e., to have rotated through a right angle. A particle has lateral movement (predominantly) only until it is overtaken by the downward moving die, see Fig. 11.57(a). Examples of this phenomenon in relation to other shapes of prismatic block compressed dynamically under a hammer when shearing interfacial friction is present are seen in Fig. 11.57(b).

When a hollow cylinder is compressed between dies, see Fig. 11.58, (i) the inner radius expands if there is no friction and (ii) reduces if there is shearing friction. What precisely the inner cylinder does depends critically on the friction coefficient at the platen-work-piece interface and the geometrical characteristics of the block (see HAWKYARD and JOHNSON, 1967).

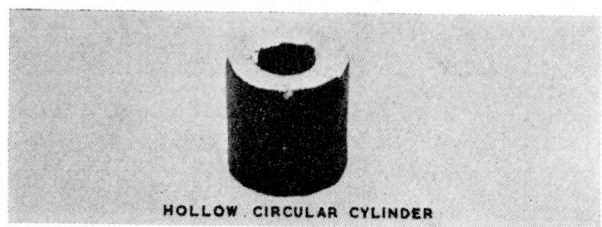

(*a*) Undeformed Plasticine hollow cylinder which has been coated with a 'flat' black paint on the external cylindrical surface and the internal surface of the hole.

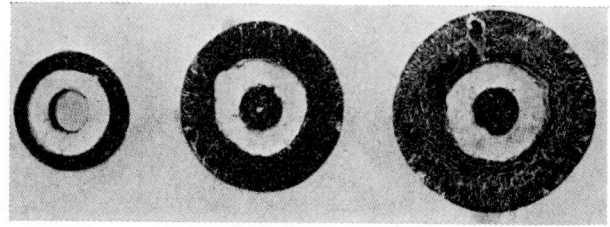

(*b*) Dynamically upset specimens corresponding to different fractional reductions. These plan views show material which is substantially in the same horizontal plane as the ultimate material–platen interface. (*After* Aku, Slater and Johnson, *Int. J. mech. Sci.*)

FIG. 11.58

11.18 The Compression of Non-Circular Prismatic Blocks

When prismatic blocks of section other than circular are upset under a falling weight, geometry changes take place; for $m = 1$ examples of shape change and rotation, (showing how material in one plane may appear after deformation in another), are shown in Figs. 11.59 and 11.60 for specimens of Plasticine, which of course simulate the behaviour of hot steel. When excessive bulging occurs, and therefore considerable tensile strain, cracking is likely to follow.

Homogeneous compression has been discussed above, entirely with reference to the unconstrained lateral flow of cylindrical blocks. However, the conception applies equally well to prismatic blocks of any cross-sectional shape. For homogeneous compression to be achieved it is only required that throughout the operation the plan cross-sectional shape shall remain geometrically similar with the original shape and be identical throughout the block or independent of vertical location. A rectangular prismatic block, originally $b_0 \times w_0$, see Fig. 11.61, which is homogeneously compressed takes on dimensions $b \times w$; $b_0/w_0 = b/w$ in order to preserve geometrical similarity. In plan view, see Fig. 11.61, since the original edge NQ must take up position $N'Q'$ in order to preserve geometrical similarity, it follows that general point P on NQ moves to P' on $N'Q'$, and that P' is the point of intersection of OP produced and $N'Q'$.

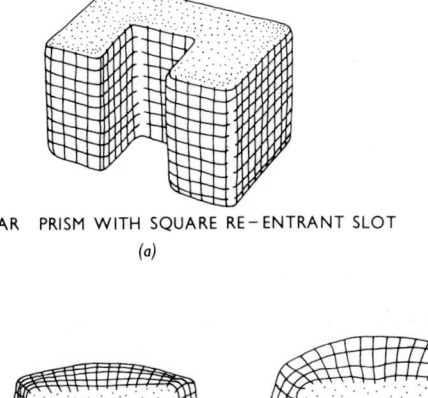

RECTANGULAR   PRISM   WITH   SQUARE   RE–ENTRANT   SLOT

(a)

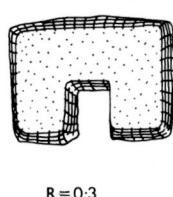
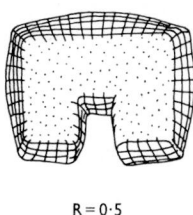
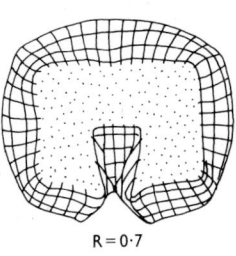

R = 0·3   R = 0·5   R = 0·7

(b)

FIG. 11.59

(a) Undeformed Plasticine rectangular prism having a square reentrant slot cut centrally in a longer side.

(b) Dynamically upset specimens corresponding to different fractional reductions. These plan views show material which is substantially in the same horizontal plane as the ultimate material–platen interface. (*After* Aku, Slater and Johnson, *Int. J. mech. Sci.*)

Figures 11.62 and 11.63 show rectangular blocks which have been inhomogeneously compressed; note the 'ears' at the corners and the suggestion of uniform plane strain deformation away from the smaller sides in blocks where the length is more than about twice the breadth. Some prismatic blocks of unusual section are shown compressed in Fig. 11.64; JOHNSON, SLATER and YU (1966) have suggested how the flow or bulging in cases such as these may be anticipated by a friction-hill concept, the hill shape being identified with a sand heap on the appropriate section. Some interesting forging studies have been described by WALLACE and SCHEY (1969) and JONES, DAVIES and SINGH (1969). The chapter on Compression by NADAI (1950), the book by UNKSOV

(1961) and the article by LIPPMANN on the dynamics of forging (1966) should be consulted. Figure 11.65 shows the movement of metal on planes of symmetry in a flat bar forging, Fig. 11.65(i). Spread is referred to for this latter case in the book by ALEXANDER and BREWER (1963). For an account of work on closed die forging, see the book by THOMSEN, YANG and KOBAYASHI (1965).

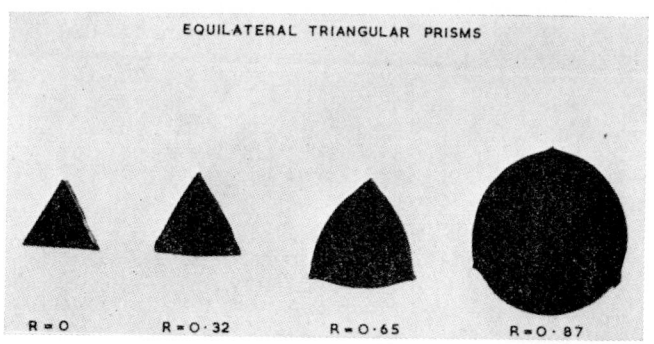

FIG. 11.60 Dynamically upset equilateral triangular prisms after increasing fractional reductions, R. (*After* Aku, Slater and Johnson, *Int. J. mech. Sci.*)

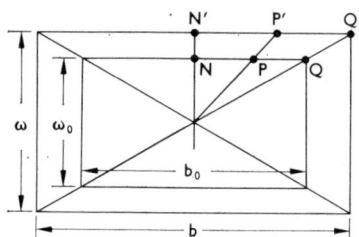

FIG. 11.61 Homogeneous compression of rectangular prismatic block.

11.19 Some Relationships between Engineering Strain Rate, Force, Time and Strain, in a Simple Upsetting Operation which Takes Place under a Drop Hammer

We may make some useful calculations relating to the above parameters, for a simple upsetting operation under a drop hammer if we assume that

1. frictionless homogeneous compression takes place,

2. that the rise in temperature in the compressed block may be neglected, and

3. that the current yield stress Y is related to the current engineering strain rate $\dot{e}$ by the equation $\sigma = Y_0 \dot{e}^n$, where Y_0 and n are constants. This equation neglects strain-hardening and may be closely identified with that used by ALDER and PHILIPS (1954).

Lead

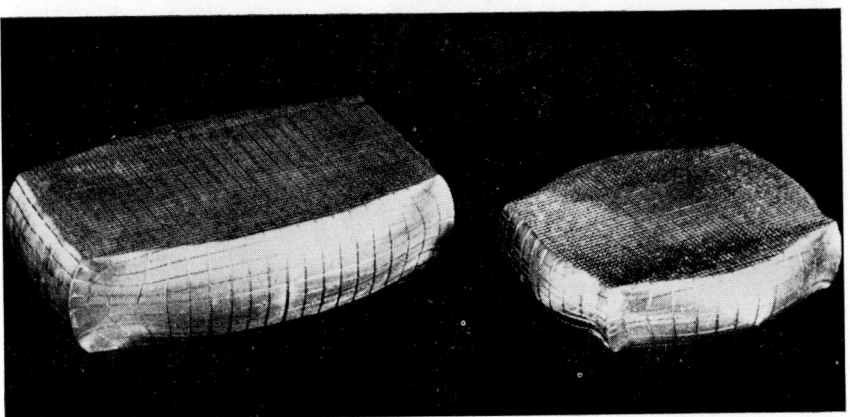

Aluminium

FIG. 11.62 Rectangular blocks 2 in $\times$ 1 in $\times$ $\frac{1}{2}$ in after 20 per cent compression. (*After* Johnson, Slater and Yu, *Int. J. mech. Sci.*)

The situation is shown in Fig. 11.66 where a rigid hammer of mass M is supposed to have compressed a cylindrical block from original height H_0 to current height H, during which the original speed of the hammer, v_0, is reduced to v. The equation of motion for the hammer is

$$- M \cdot \frac{dv}{dt} = A\sigma$$

$$= \frac{A_0 H_0}{H} \cdot Y_0 \dot{e}^n. \qquad (11.66)$$

t denotes time and A_0 and A are the initial and current cross-sectional areas of the block.

Now, see p. 23, $\dot{e} = v/H_0$

and thus $\dfrac{d\dot{e}}{dt} = \dfrac{dv/dt}{H_0}.$

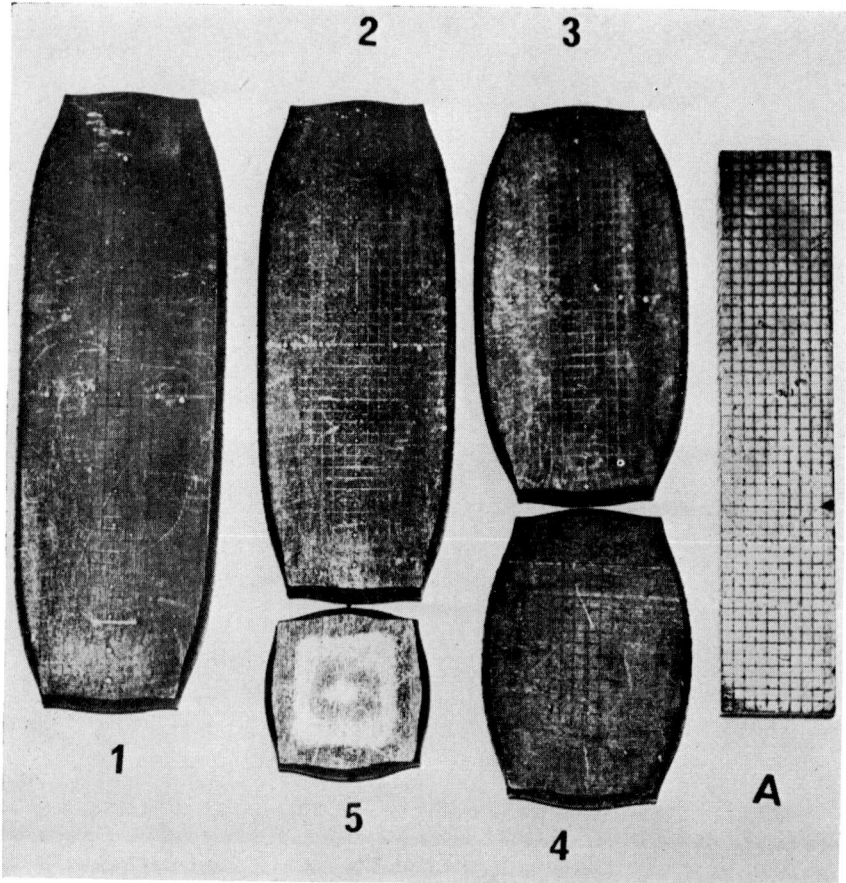

FIG. 11.63 Plan view of compressed rectangular blocks: note 'earing'.
(*After* Johnson, Slater and Yu, *Int. J. mech. Sci.*)

Substituting for dv/dt in equation (11.66) we find that

$$- M H_0 \frac{d\dot{e}}{dt} = A_0 Y_0 \frac{\dot{e}^n}{1 - e}, \qquad (11.67)$$

because, $H/H_0 = 1 - e$. Hence, also putting $de/\dot{e}$ for dt, (11.67) becomes

$$- \alpha \frac{d\dot{e}}{\dot{e}^{n-1}} = \frac{de}{1 - e},$$

where $\alpha = M H_0 / A_0 Y_0$.

Integrating, we see that

$$- \alpha \left[\frac{\dot{e}^{2-n}}{2 - n} \right]_{\dot{e}_0}^{\dot{e}} = \left[\ln \frac{1}{1 - e} \right]_0^e, \qquad (11.68)$$

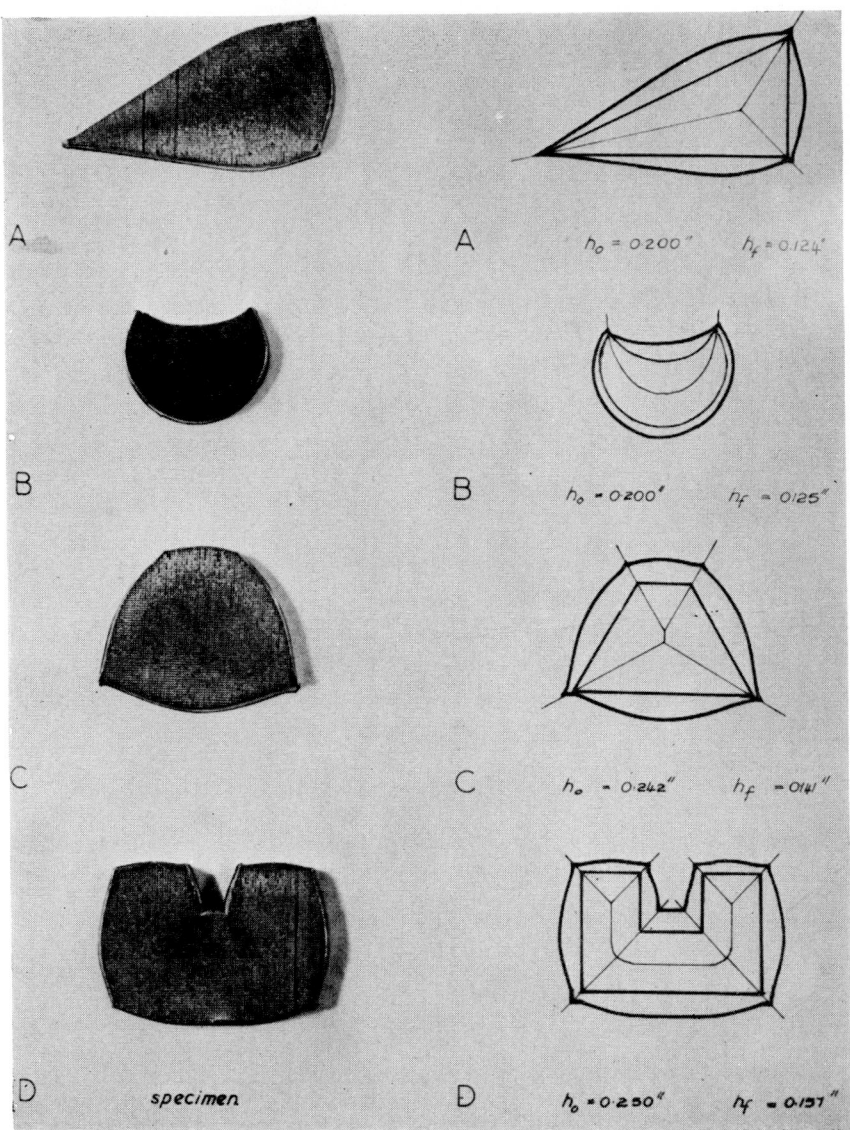

FIG. 11.64 On left: Compressed irregular shaped prismatic blocks.
On right: Outline in plan before and after compression. (*After*
Johnson, Slater and Yu, *Int. J. mech. Sci.*)

where $\dot{e}_0$ is the engineering strain rate at the start of compression when $e = 0$.
Equation (11.68) gives,

$$(\dot{e}/e_0)^{2-n} = 1 - \beta \ln (1 - e)^{-1},$$

where $\beta = (2 - n) A_0 Y_0/MH_0 \dot{e}_0^{2-n}$.

When $\dot{e} = 0$, i.e., at the end of compression, if e_F denotes the final engineering strain,

$$\ln \frac{1}{1 - e_F} = \frac{1}{\beta} = \frac{M H_0 \dot{e}_0^{\,2-n}}{(2-n) A_0 Y_0},$$ (11.69)

and equation (11.69) can be rewritten,

$$\left(\frac{\dot{e}}{\dot{e}_0}\right)^{2-n} = 1 - \frac{\ln[1/(1-e)]}{\ln[1/(1-e_F)]}.$$ (11.70)

Figure 11.67 shows how $\dot{e}/\dot{e}_0$ varies with e, for representative values of $n = 0.3$ and $e_F = 0.5$. The mean value of $\dot{e}/\dot{e}_0$ over the range $0 \leqslant e \leqslant e_F$ is about 0.69.

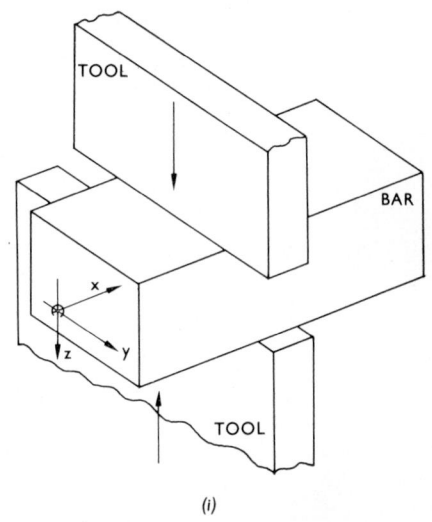

(i)

FIG. 11.65

(i) Flat bar forging.

If values of e which are not too large are considered, equation (11.70) can be written

$$\left(\frac{\dot{e}}{\dot{e}_0}\right)^{2-n} = 1 - \frac{e}{e_F}.$$ (11.71)

Figure 11.67 also shows equation (11.71) plotted for $n = 0$ and $n = 1$. When $n = 0$, (11.71) is a parabola and the mean value of $\dot{e}/\dot{e}_0$ is 2/3. If $n = 1$, (11.71) is linear and $\dot{e}/\dot{e}_0$ is 1/2. Since for real materials it turns out that $0 < n < 0.5$, and that $n = 0.3$ is a good typical value, it is evident that the mean engineering strain rate as based on the initial speed, i.e., $\dot{e} = (\dot{e}_0/2) = 0.5 \, v_a/H_0$, may considerably underestimate the actual mean rate of strain. It is seen that a good general figure to adopt is $\frac{2}{3} \, v_0/H_0$.

Relationships

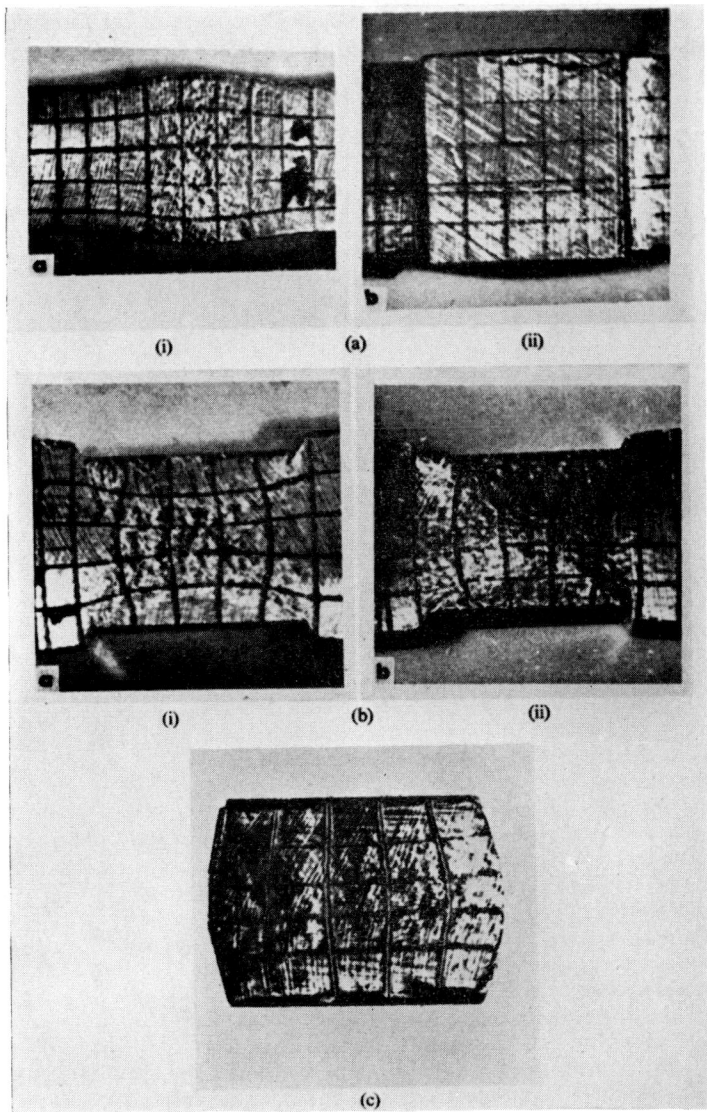

FIG. 11.65 (ii)

Grid deformation patterns in the three principal orthogonal directions
after 20 per cent reduction in flat bar forging.

(a) (i) Centre section; xOy plane   (ii) Contact surface

(b) (i) Centre section; xOz plane   (ii) Outside surface

(c) Centre section; zOy plane

(*After* Johnson and Baraya, *M.T.D.R. Conf.*, Pergamon.)

This approach may also be used to suggest the form of *the force-displacement curve* which may be expected during the compression. With equation (11.67), if F denotes force,

$$F = A Y = A_0 H_0 \frac{Y_0 \dot{e}^n}{H} = \frac{A_0 Y_0}{1 - e} \dot{e}^n. \tag{11.72}$$

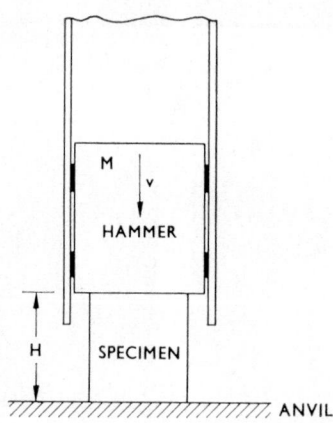

FIG. 11.66 Compression of a circular cylinder by a hammer blow.

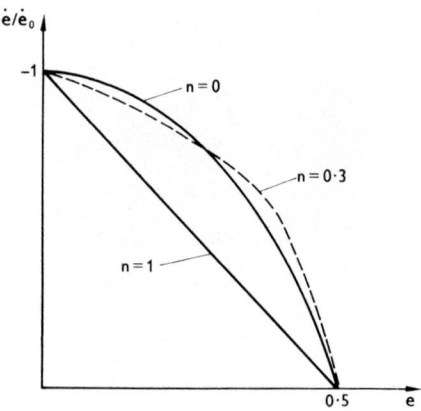

FIG. 11.67 Variation of strain rate with strain.

Introducing equation (11.70),

$$\frac{F}{\gamma} = \frac{1}{1 - e} \left[1 - \frac{\ln 1/(1 - e)}{\ln 1/(1 - e_F)} \right]^{n/(2-n)}, \tag{11.73}$$

or using alternatively, equation (11.71)

$$\frac{F}{\gamma} = \frac{[1 - (e/e_F)]^{n/(2-n)}}{1 - e} ,$$ (11.74)

where $\gamma = A_0 \, Y_0 \, \dot{e}_0^n$ is a constant. Putting $e_F = 0.5$, Fig. 11.68 shows the form of F/γ versus e using (11.74) for $n = 0, \frac{1}{4}, \frac{1}{2}$ and 1; also F/γ is a maximum when

$$e = \frac{n - e_F(2 - n)}{2(n - 1)}.$$ (11.75)

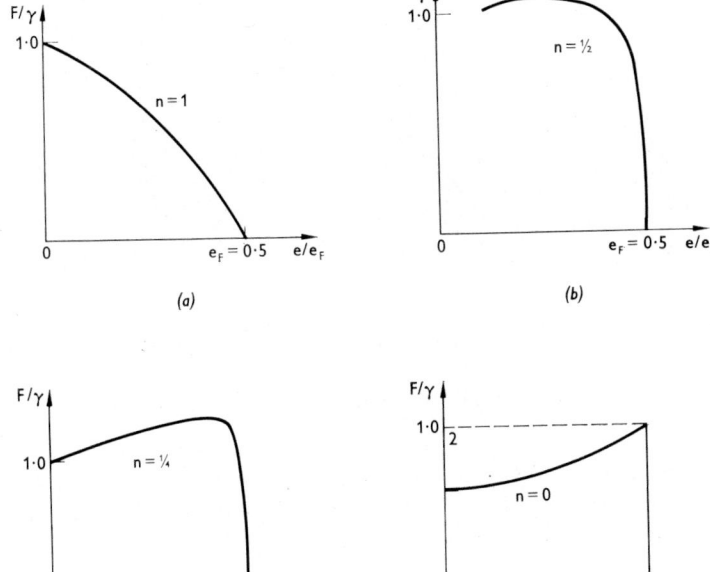

(a)   (b)   (c)   (d)

FIG. 11.68 Force-strain or displacement curves.

(11.75) is true of course only for $e_F < n/(2 - n)$. From equation (11.75) we see that the smaller the values of n the nearer the end of the compression operation does the maximum value of F occur.

The engineering strain versus time curve may be found approximately using (11.71) from which,

$$\dot{e} = \dot{e}_0 \left(1 - \frac{e}{e_F} \right)^{1/(2-n)}.$$ (11.76)

Rearranging equation (11.76) and integrating we find that,

$$1 - \left(1 - \frac{e}{e_F}\right)^{(1-n)/(2-n)} = \frac{\dot{e}_0}{e_F} \cdot \frac{1-n}{2-n}. \tag{11.77}$$

Equation (11.77) for $n = 1/4$ and $\dot{e}_0 = 200/s$ is plotted in Fig. 11.69, i.e.

$$1 - (1 - 2e)^{3/7} = 171t. \tag{11.78}$$

By combining the results from Fig. 11.68 and Fig. 11.69 we may arrive at the curve in Fig. 11.70 which shows the *form* of the F/t curve. This curve has the correct shape as comparison with experimental force-time records shows—apart from the small time region.

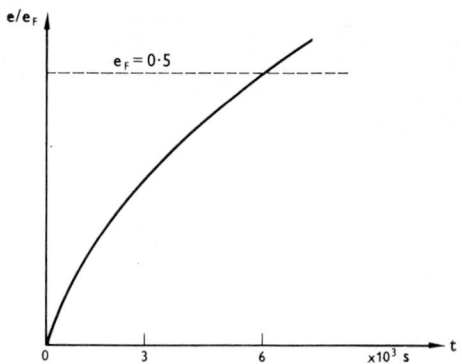

Fig. 11.69 Variation of strain with time.

It is sometimes useful to know how the *speed of the impinging mass* decreases with distance after impact has occurred. Combining equations (11.66) and (11.67) we see that,

$$- M \cdot \frac{dv}{dt} = \frac{A_0 H_0}{H} Y_0 \left(\frac{v}{H_0}\right)^n \tag{11.79}$$

but with $v = - dH/dt$, (11.79) becomes,

$$M v^{1-n} dv = A_0 Y_0 H_0^{1-n} dH/H$$

and with $v = v_0$ when $H = H_0$, after integrating,

$$M \left(\frac{v_0^{2-n} - v^{2-n}}{2 - n}\right) = A_0 Y_0 H_0^{1-n} \ln H_0/H. \tag{11.80.i}$$

Putting $n = 0$,

$$\tfrac{1}{2} M (v_0^2 - v^2) = A_0 H_0 Y_0 \ln H_0/H, \tag{11.80.ii}$$

and this could have been written down immediately from energy considerations assuming of course that all available kinetic energy in the hammer is dissipated as plastic work done.

If instead of $Y = Y_0 \, \dot{e}^n$, a relationship involving logarithmic strain rate is used, i.e., $Y = Y_0 \, \dot{\epsilon}^n$, then in place of equation (11.79) we have,

$$- M . \frac{dv}{dt} = \frac{A_0 \, H_0 \, Y_0}{H} . \left(\frac{v}{H}\right)^n \tag{11.81}$$

and thus the equation (11.82) in place of (11.80) is

$$M \frac{v_0^{2-n} - v^{2-n}}{2-n} = \frac{A_0 \, H_0 \, Y_0}{n} \left(\frac{1}{H^n} - \frac{1}{H_0^n}\right) \tag{11.82}$$

provided that $n \neq 0$. If $n = 0$, (11.81) gives rise simply to (11.80.*ii*).
The above analysis has also been carried out using a constitutive equation of the form $\sigma_D = (\sigma_s + B \dot{e}^n) . k e^m$; σ_D is the dynamic yield stress and σ_s the static yield stress. The results derived were tested experimentally and comparatively good correlations were obtained, see SLATER, JOHNSON and AKU (1967). The results of SAMANTA (1966) repay study.

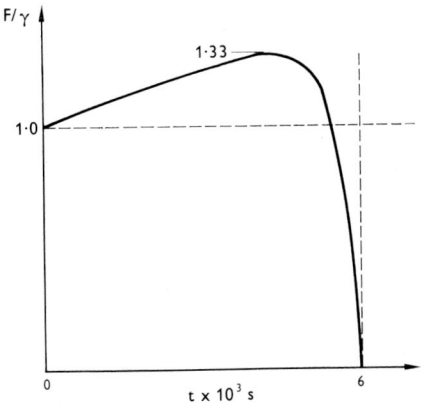

FIG. 11.70 Variation of force with time.

11.20 Superplasticity

Superplasticity is often taken to refer to large neck-free extensions when deforming certain metals and alloys; the materials which behave superplastically show low strength and high elongation before fracture, of the order of 1000 per cent in some cases. Such behaviour is obviously attractive for use in a forming process, especially as it may well be able to make use of other known pressure-forming techniques, e.g., those of the polymer and glass industries. Aluminium-zinc rods can be drawn into filaments as can glass, and very satisfactory deep-drawing-type products can be formed with superplastic materials, see JOHNSON, AL-NAIB and DUNCAN (1972).

There are two principal classes of superplastic process, (1) transformation plasticity and (2) micrograin plasticity.

13—EP * *

11.20.1 *Transformation Plasticity*

In this class, deformation occurs during a phase change whether it be martensitic or diffusion controlled, and because transformation reactions in steels are well understood many studies have concentrated on ferrous alloys. But allotropic transformations in iron, cobalt, titanium and zinc and various eutectoid and precipitation reactions, as well as the martensitic reactions, have also been investigated. It is observed that a threshold stress is necessary to initiate gross deformation; this is attributed to having to move dislocations out from their sources and to produce pile-ups at grain boundaries; in some systems the nucleation rate is then increased and thence the transformation rate. Coherency is then lost at the grain boundaries, the dislocations are freed and can move, and thus deformation proceeds. The following features, according to WEISS and KOT (1967), appear to characterize transformation plasticity,

(a) Large neck-free elongation.
(b) A linear relationship between applied stress and strain.
(c) Relative insensitivity to grain size.
(d) Plastic flow during a phase transformation even under external stresses much smaller than that required for normal yield or flow in the material.
(e) A sensitivity to the direction of the transformation and to heating and cooling rates.

GREENWOOD and JOHNSON (1965) have proposed a continuum plasticity model to account for this behaviour which is based on the volume differences that arise during transformation and which then create high internal stresses so that small external stresses thereafter can induce larger uniform elongations.

11.20.2 *Micrograin Plasticity*

Perhaps the best known example of superplasticity is that of the eutectoid system of approximately 80 per cent aluminium and 20 per cent zinc. Extremely high uniform elongations are found with this system at just below the eutectoid temperatures, for relatively small pressures. A lead–tin alloy of eutectic composition also shows superplastic behaviour. The distinguishing feature of these compositions is the stability and fineness of the microstructure.

The behaviour of superplastic alloys is highly strain-rate sensitive and Fig. 11.71 is typical, taken from a paper of BACKOFEN, TURNER and AVERY (1964). The strain-rate exponent in $\sigma = B\dot{\epsilon}^m$, is usually about 0·2 for many metals (see p. 27) but it may reach 0·7 for these special alloys. The consequence of a high value of m is easily demonstrated thus. In a tensile test, if P denotes the constant tensile force and A the current cross-sectional area then,

$$\sigma = \frac{P}{A} = B\dot{\epsilon}^m \quad \text{or} \quad A = \frac{P}{B}\dot{\epsilon}^{-m},$$

so that

$$-\frac{dA}{dt} = \frac{P}{B}m\dot{\epsilon}^{-m+1} = \frac{P^{1/m}}{B}\frac{1}{A^{(1-m)/m}},$$

on eliminating $\dot\varepsilon$. If $m = 1$, then the time rate of decrease of the cross-sectional area of the specimen is independent of the cross-sectional area; the load that could be carried would be independent of current cross-sectional area. For a low value of m, say 0·2, the rate of area reduction would be proportional to A^{-4} and thus it would reduce at an increasingly rapid rate. With this sort of reasoning in mind, it is usually stated that a high value of m inhibits neck development; in the act of forming a neck, the strain rate increases and thus the material strength, so that load bearing capacity increases.

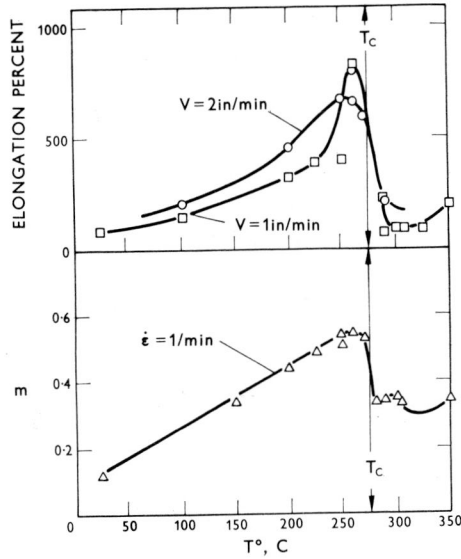

FIG. 11.71 The temperature dependence of elongation (upper) and strain-rate sensitivity (lower). T_c is the critical temperature, 80 per cent Zn — 2 per cent Al.
V: Straining head speed in tension test.
$\dot\varepsilon$: Strain rate.
(*After* Backofen, Turner and Avery, *Trans. A.S.M.*)

The characteristics of micrograin plasticity are:

1. large extensions due to a quasi-viscous behaviour associated with a relatively high strain rate sensitivity index, $m \simeq 0\cdot6$, and a test temperature in excess of 0·5 of the absolute melting point of the alloy or metal.

2. A stable microstructure and a small grain size in the micron range; the grain boundaries should permit grain boundary sliding. An eutectic or an eutectoid is not essential if a fine grain size can be produced and held.

3. Three mechanisms have been proposed to account for the high strain rate sensitivity, (a) vacancy creep, (b) creep by grain boundary diffusion and (c) grain boundary sliding.

The potentiality of superplasticity is not clear, but its applications are likely to be for non-load bearing (e.g. covers) and decorative components. Certainly the forming process innovations evoked will need to be carefully studied and developed. Forming times anyhow are slow, and there will be a critical need for optimizing forming pressures, stress, strain rate and deflection in sheet-forming. The paper of JOVANE (1968) is interesting in this respect. Papers of interest concerning superplastic forming will be found in the *International Journal of Mechanical Sciences* pp. 463–524, 1970 and pp. 63–76, 1971.

See Problems 43-44

REFERENCES

Asterisked articles have not been referred to in the text

TUBE-SINKING

BLAZYNSKI, T. Z. and COLE, I. M.	1960	'An Investigation of the Plug Drawing Process' *Proc. Instn mech. Engrs* 2/60
CHUNG, S. Y.	1951	'Theory of Hollow Sinking of Thin-walled Tubes' *Metallurgia* **43**, 215
CHUNG, S. Y. and SWIFT, H. W.	1952	'A Theory of Tube-sinking' *J. Iron Steel Inst.* **170**, 29
DAVIDENKOV, N. N.	1932	'Berechnung der Restspannungen in kaltgezogenen Rohren' *Z. Metallk.*, (Feb.)
DENTON, A. A. and ALEXANDER, J. M.	1963	'On the Determination of Residual Stresses in Tubes' *J. mech. Engng Sci.* **5**, 75
FLINN, J. E.	1969	'Wall thickness changes—in hollow drawn tubing' *Trans. A.S.M.E.* 792
HILL, R.	1950	'Tube-sinking' *The Mathematical Theory of Plasticity* O.U.P.
MOORE, G. G. and WALLACE, J. F.	1967 –68	'Theories and Experiments on Tube Sinking Through Conical Dies' *Proc. Instn mech. Engrs*, **182**, 19
ROZOV, N. V.	1968	'Cold Drawing of Steel Tube' (Translation of Russian book, Nat. Lending Lib. for Sci. and Tech., Boston Spa, U.K.)
SACHS, G. and BALDWIN, W. M.	1946a	'Folding in Tube-sinking' *Trans. A.S.M.E.* **68**, 647
	1946b	'Stress Analysis of Tube-sinking' *Trans. A.S.M.E.* **68**, 655

SACHS, G. and ESPEY, G. — 1941 'The Measurement of Residual Stresses in Metals' *Iron Age* **148** (12), 63 and (13), 36

SWIFT, H. W. — 1949 'Stresses and Strains in Tube-drawing' *Phil. Mag. Ser.* 7, **11**, 883

DEEP-DRAWING

*ALEXANDER, J. M. — 1960 'An Appraisal of the Theory of Deep Drawing' *Metall. Rev.* **5**, 349

BOURNE, L. and HILL, R. — 1950 'On the Correlation of the Directional Properties of Rolled Sheet in Tension and Cupping Tests' *Phil. Mag.* **41**, 671

CHUNG, S. Y. and SWIFT, H. W. — 1951 'Cup-drawing from a Flat Blank', Part I, Experimental Investigation Part 2, Analytical Investigation *Proc. Instn mech. Engrs* **165**, 199
1952 'An Experimental Investigation into the Re-drawing of Cylindrical Shells' *Proc. Instn mech. Engrs* **1B**, 437

COUPLAND, H. T. and WILSON, D. V. — 1957 'Speed Effects in Deep Drawing' Conf. Properties of Materials at High Rates of Strain *Instn mech. Engrs.*, 98

DUNCAN, J. L. and JOHNSON, W. — 1969 'Approximate Analyses of Loads in Axisymmetric Deep Drawing' *9th Int. M.T.D.R. Conf.*, Pergamon Press, Oxford, 303 pp.

EL-SEBAIE, M. G. and MELLOR, P. B. — 1972 'Plastic Instability Conditions in the Deep-drawing of a Circular Blank of Sheet Metal' *Int. J. mech. Sci.*, **14**, 535

FOGG, B. — 1968 'Theoretical Analysis for the Redrawing of Cylindrical Cups Through Conical Dies Without Pressure Sleeves' *J. Mech. Engng Sci.* **10**, 141

FREEMAN, P. and LEEMING, H. — 1953 'Ironing of Thin-walled Metal Cups—the Distribution of the Punch Load' *B.I.S.R.A. Report No. MW/E/46/53*

FUKUI, S. and HANSSON, A. — 1970 'Analytical Study of Wall Ironing, Considering Work-hardening' *Annals of C.I.R.P.*, **18**, 593

GECKLER, J. W. — 1928 'Plastic Folding of the Walls of Hollow Cylinders and Some Other Folding Phenomena in Bowls and Sheets' *Z. angew. Math. Mech.* **8**, 341

HESSENBERG, W. C. F.   1954 'A Simple Account of some of Professor Swift's Work on Deep-drawing' *B.I.S.R.A. Report MW/1954*

HILL, R.   1950 (a) 'The Earing of Deep-drawn Cups' (b) 'Deep-drawing' *The Mathematical Theory of Plasticity* O.U.P.

*JEVONS, J. D.   1949 *The Metallurgy of Deep-drawing and Pressing* 2nd edn., Chapman and Hall, London

*JOHNSON, W.   1956 'Research into some Metal-forming and Shaping Operations *J. Inst. Metals* **84**, 165

KASUGA, Y. and TSUTSUMI, S.   1965 Pressure Lubricated Deep Drawing *Bull. J.S.M.E.* **8**, 120

LANKFORD, W. T. SNYDER, S. O. and BAUSCHER, J. A.   1950 'New criteria for predicting the press performance of deep drawing sheets *Trans. A.S.M.* **42**, 1197

LLOYD, D. H.   1962 'Metallurgical engineering in the pressed metal industry' *Sheet Metal Industries* **39**, 82

LOXLEY, E. M. and FREEMAN, P.   1954 'Some Lubrication Effects in Deep-drawing Operations' *J. Inst. Petrol.* **40**, 299

*SACHS, G.   1934 'New Researches on the Drawing of –35 Cylindrical Shells' *Proc. Inst. Aut. Eng.* **29**, 588

SENIOR, B. W.   1956 'Flange Wrinkling in Deep-drawing Operations' *J. Mech. and Phys. Solids* **4**, 235

SWIFT, H. W.   1939 'Drawing Tests for Sheet Metal' –40 *Proc. Instn Auto. Engrs* **4**, 361 1954 'The Mechanism of a Simple Drawing Operation' *Engineering* **178**, 431

WALLACE, J. F.   1960 'Improvements in Punches for Cylindrical Deep Drawing' *Sheet Metal Industries* **37**, 901

WHITELEY, R. L.   1960 'The importance of Directionality in Drawing Quality Sheet Steel' *Trans. A.S.M.* **52**, 154

WILSON, D. V.   1966 'Plastic Anisotropy in Sheet Metals' *J. Inst. Metal* **94**, 84

*WILLIS, J.   1954 *Deep-drawing.* Butterworths, London

WOO, D. M.   1968 'On the Complete Solution of the Deep-drawing Problem' *Int. J. mech. Sci.* **10**, 83

BLANKING

JOHNSON, W. and   1967 'A Survey of the Slow and Fast Blanking of
SLATER, R. A. C.   Metals at Ambient and High Temperatures'
                           Int. Conf. Man Tech., *A.S.T.M.E.*, 825
JOHNSON, W.   1972 *Impact Strength of Materials*, Arnold, London

WIRE-DRAWING

ATKINS, A. G. and   1968 'The Incorporation of Work Hardening and
CADDELL, R. M.   Redundant Work in Rod-drawing Analyses'
                           Int. J. mech. Sci. **10**, 15
CADDELL, R. M. and   1968 'The influence of Redundant Work when
ATKINS, A. G.   Drawing Rods Through Conical Dies'
                           A.S.M.E., *Paper No.* 67–*WA*/*Prod.*–11
                      1969 'Optimum Die Angles and Maximum
                           Attainable Reductions in Rod-drawing'
                           A.S.M.E., *Paper No.* 68–*WA*/*Prod.*–11
CHRISTOPHERSON,   1955 'Promotion of Fluid Lubrication in Wire-
D. G.   drawing'
and NAYLOR, H.   *Proc. Instn mech. Engnrs* **169**, 643
CLEAVER, F. T. and   1950 'Wire-drawing Technique and Equipment'
MILLER, H. G.   *J. Inst. Metals* **78**, 537
HILL, R. and   1948 'A New Theory of Plastic Deformation in
TUPPER, S. J.   Wire-drawing'
                           J. Iron Steel Inst. **159**, 353
JOHNSON, W. and   1969 'Wire Drawing: A Survey of Theories'
SOWERBY, R.   *The Wire Industry*, pp. 137 and 249
LUNT, R. W. and   1946 'An Extension of Wire-drawing Theory with
MACLELLAN, G. D. S.   Special Reference to the Contributions of
                           K. B. Lewis'
                           J. Inst. Metals **72**, 65
MACLELLAN, G. D. S.   1952 'Some Friction Effects in Wire-drawing'
                      –53 *J. Inst. Metals* **81**, 1
MAJORS, H. J. R.   1955 'Studies in Cold-drawing' Part 3
                           'Determination of Coefficient of Friction'
                           Trans. A.S.M.E. **78**, 79
SACHS, G.   1927 'Zur Theories des Ziehvorgangs'
                           Z. angew. Math. Mech. **7**, 235
SIEBEL, E.   1947 'Der derzeitige Stand der Erkenntnisser über
                           die mechanischen Vorgänge bein Drahtsiehen'
                           Stahl und Eisen **66–67**, 171
THOMSON, P. F.   1967 'Drawing Copper Wire with a Lubricant
HOGGART, J. S. and   under Externally Generated Pressure'
SUITER, J.   *J. Inst. Metals* **95**, 152
WISTREICH, J. G.   1965 'Investigation of the Mechanics of Wire-
                           drawing'
                           Proc. Instn mech. Engrs **169**, 123
                      1958 'The Fundamentals of Wire-drawing'
                           Met. Rev. **3**, 97

EXTRUSION

AVITZUR, B.	1967	'Steady and Unsteady State Extrusion' *J. Engng Ind., A.S.M.E.* **89**, 175
BERESNEV, B. I., VERESCHAGIN, L. F., RYABININ, YU. N., and LIVSHITS, L. D.	1963	*Some Problems of Large Plastic Deformation of Metals at High Pressures* Pergamon Press, London
*BISHOP, J. F. W.	1957	'The Theory of Extrusion' *Metall. rev.* **2**
BRIDGMAN, P. W.	1952	*Studies in Large Plastic Flow and Fracture* McGraw-Hill, New York
DUFFILL, A. W. and MELLOR, P. B.	1969	'A Comparison between the Conventional and Hydrostatic Methods of Cold Extrusion through Conical Dies' *Annals of C.I.R.P.*, Vol xvii, **97**
EL-BEHERY, A. M., LAMBLE, J. H., and JOHNSON, W.	1963	'The Measurement of Container Wall Pressure and Friction Coefficient in Axisymmetric Extrusion', 4th Int. M.T.D.R. Conf., Pergamon Press, p. 319
FELDMAN, H. D.	1961	*Cold Forging of Steel* Hutchinson, London, 268 pp.
FIORENTINO, R. J., SABROFF, A. M., and BOULGER, F. W.	1963	*Hydrostatic Extrusion at Battelle* Machinery Lloyd (European Edition) 24th August
HILL, R.	1948	'A Theoretical Analysis of the Stresses and Strains and Extrusion and Piercing' *J. Iron Steel Inst.* **158**, 177
HOFFMANNER, A. L. (Ed.)	1971	*Metal Forming: Interrelation between Theory and Practice* Plenum Press, New York, 503 pp.
JOHNSON, W.	1955	'Extrusion Through Square Dies of Large Reduction' *J. Mech. Phys. Solids* **4**, 191
	1956	'Experiments in Plane-strain Extrusion' *J. Mech. Phy. Solids* **4**, 269
	1957	'The Pressure for the Cold Extrusion of Lubricated Rod through Square Dies of Moderate Reduction at Slow Speeds' *J. Inst. Metals* **85**, 403
	1959	'An Elementary Consideration of Some Extrusion Defects' *Appl. Sci. Res., Series A* **8**, 52
JOHNSON, W. and KUDO, H.	1962	*The Mechanics of Metal Extrusion* Manchester University Press
KUDO, H.	1961	'Axisymmetric Cold Forging and Extrusion' *Int. J. mech. Sci.* **2**, 102

LENGYEL, B and ALEXANDER, J. M.　1967 'Design of a Production Machine for Semi-continuous Hydrostatic Extrusion *Proc. Instn mech. Engrs* **182**, 207

MORGAN, R. A. P.　1959 'The Cold Extrusion of Steel' *J. Iron Steel Inst.* **193**, 285

*PEARSON, C. E. and PARKINS, R. N.　1960 *The Extrusion of Metals* Chapman and Hall, London

PEARSON, C. E. and SMYTHE, J. A.　1931 'The influence of Pressure and Temperature on the Extrusion of Metals' *J. Inst. Metals* **45**, 345

PUGH, H. Ll. D.　1965 *Recent Developments in Cold Forming* Bulleid Memorial Lectures, University of Nottingham, **IIIB**

PUGH, H. Ll. D. (Ed.)　1970 *Mechanical Behaviour of Metals under Pressure*, 785 pp. Elsevier, London

SACHS, G. and EISBEIN, W.　1931 'Power Consumption and Mechanism of Flow in the Extrusion Process' *Mitt. Mater. S* **16**, 67

SCHEY, J. A. (Ed)　1970 *Metal Deformation Processes*, 807 pp. Marcel Dekker, New York

*SEJOURNET, J.　1954 'The Hot Extrusion of Steel' *Engineering* 177, 463

SIEBEL, E. and FANGMEIER, E.　1931 'Researches on Power Consumption in Extrusion and Punching of Metal' *Mitt. K.-Wilhelm-Inst. Eisenforsch*

SLATER, H. K. and GREEN, D.　1967 'Augmented Hydrostatic Extrusion of Continuous Bar' *Proc. Instn mech. Engrs* **182**

WILCOX, R. J. and WHITTON, P. W.　1958 'The Cold Extrusion of Metals using Lubrication at Slow Speeds' *J. Inst. Metals* **87**, 289

1959 'Further Experiments on the Cold Extrusion of Metals using Lubrication at Slow Speeds' *J. Inst. Metals* **88**, 145

YANG, C. T. and THOMSEN, E. G.　1953 'Plastic Flow in a Lead Extrusion' *Trans. A.S.M.E.* **75**, 575

ROLLING

AFONJA, A. A. and SANSOME, D. H.　1969 'Review of Strip Rolling with Particular Reference to Rolling Thin Hard Strip' *9th Int. M.T.D.R. Conf.*, Pergamon Press

ALEXANDER, J. M.　1955 'A Slip-line Field for the Hot-rolling Process' *Proc. Instn mech. Engrs* **169**, 1021

1972 'On the Theory of Rolling' *Proc. R. Soc. Lond. A*, **326**, 535

ALEXANDER, J. M. and   1963   *Manufacturing Properties of Materials*
BREWER, R. C.   D. van Nostrand, 489 pp.

ARNOLD, R. R. and   1959   'Stress and Deformation Studies for
WHITTON, P. W.   Sandwich Rolling Hard Metals'
                                  Proc. Instn mech Engrs **173**, 241

BEDI, D. S. and   1967   'Hydrodynamic Model for Cold Strip Rolling'
HILLIER, M. J.   *Proc. Instn mech. Engrs* P7/68

BLAND, D. R.   1950   'A Theoretical Investigation of Roll
                                  Flattening'
                                  Proc. Instn mech. Engrs **163**, 141

BLAND, D. R. and   1948   'The Calculation of Roll Force and Torque
FORD, H.   in Cold Strip Rolling with Tensions'
                                  Proc. Instn mech. Engrs **159**, 144

BLAZYNSKI, T. Z.   1969   'Design of Tools for the Combined Piercing
                                  Elongating Tube Making Process',
                                  9th Int. M.T.D.R. Conf., Pergamon
                                  Press, Ms. No. 60.

CADDELL, R. M.,   1968   'Yield Strength Variation in Ring Rolled
NEEDHAM, G. and   Aluminium'
JOHNSON, W.   *Int. J. mech. Sci.* **10**, 749

CHITKARA, N. R. and   1966   'Some Experimental Results Concerning
JOHNSON, W.   Spread in the Rolling of Lead'
                                  J. basic Engng., A.S.M.E., *Paper No.*
                                  65–*WA*/Met.11

THE ENGINEER   1967   'Component Production by Transverse
                                  Rolling'
                                  10 Nov. 1967, p. 611

FORD, H.   1947   'The Effect of Speed of Rolling in the Cold-
                                  rolling Process'
                                  J. Iron Steel Inst. **156**, 380

                  1948   'Researches into the Deformation of Metals
                                  by Cold-rolling'
                                  Proc. Instn mech. Engrs **159**, 115

                  1957   'The Theory of Rolling'
                                  Metall. Rev. **2**, No. 5, 1

FORD, H. and   1960   'Rolling Hard Materials in Thin Gauges:
ALEXANDER, J. M.   Basic Considerations'
                                  J. Inst. Metals **88**, 193

                  1963   'Simplified Hot Rolling Calculations'
                  –64   *J. Inst. Metals* **92**, 397

HESSENBERG, W. C. F.   1951   'The Effect of Tension on Torque and Roll
and SIMS, R. B.   Force in Cold Strip Rolling'
                                  J. Iron Steel Inst. **168**, 155

*HILL, R.   1950   'Relations Between Roll-force, Torque and
                                  the Applied Tensions in Strip-rolling'
                                  Proc. Instn mech. Engrs **163**, 135

HOFFMAN, O. and SACHS, G. — 1953 *Introduction to the Theory of Plasticity for Engineers* McGraw-Hill, New York, Chapter 20

JOHNSON, W. and KUDO, H. — 1960 'The Use of Upper-Bound Solutions for the Determination of Temperature Distributions in Fast Hot Rolling' *Int. J. mech. Sci.* **1**, 175

JOHNSON, W. and NEEDHAM, G. — 1966 'Further Experiments in Asymmetrical Rolling' *Int. J. mech. Sci.* **8**, 443

1968a 'Experiments in Ring Rolling' *Int. J. mech. Sci.* **10**, 95

1968 'Plastic Hinges in Ring Indentation in Relation to Ring Rolling' *Int. J. mech. Sci.* **10**, 487

JOHNSON, W., MACLEOD, I. and NEEDHAM, G. — 1968 'An Experimental Investigation into the Process of Ring or Metal Tyre Rolling' *Int. J. mech. Sci.* **10**, 455

JOHANSSON, R. — 1965 'Flow of Material During Rolling Between Grooved Rolls' *R. Inst. tech. Rep.*, Stockholm

JUBB, C. and BLAZYNSKI, T. Z. — 1969 *Development of the Assel Tube Elongating Process into a Secondary Piercing Operation* 9th M.T.D.R. Conf., Pergamon Press, MS. No. 28

KARMAN, T. VON — 1925 'Beitrag zur Theorie des Walzvorganges' *Z. angew. Math. Mech.* **5**, 139

KUDO, H. and TAMURA, K. — 1968 'Analysis and Experiment in V-Groove Forming' *Annals of the C.I.R.P.*

KUDO, H. and YOKAI, M. — 1968 *Investigations into the Helical Rolling Process* 8th Int. M T.D.R. Conf., Pergamon Press, 1021

*LARKE, E. C. — 1957 *The Rolling of Strip, Sheet and Plate* Chapman and Hall, London

LIANIS, G. and FORD, H — 1956 'Graphical Solution of the Cold-rolling Problem when Tensions are Applied to the Strip' *J. Inst. Metals* **84**, 299

*LIPPMANN, H. and JOHNSON, W. — 1960 'Temperature Development Based on Technological Analysis: Fast Rolling as an Example' *Appl. Sci. Res.(A)* **9**, 345

MOLLER, R. H. and HOGGART, J. S. — 1967 *Periodic Surface Finish and Torque Effects During Cold Strip Rolling* 20th Ann. Conf. Jnl. Australian Inst. Metals

*NADAI, A. 1939 'The Forces Required for Rolling Steel
Strip Under Tension'
J. appl. Mech. **61**, A–54

OROWAN, E. 1943 'The Calculation of Roll Pressure in Hot and
Cold Flat Rolling'
Proc. Instn mech. Engrs **150**, 140

PARKINS, R. N. 1968 *Mechanical Treatment of Metals*
Allen & Unwin, 352 pp.

RUDISILL, C. S. and 1967 *A Three-Dimensional Theory of Hot Rolling*
ZOROWSKI, C. F. Int. Conf. on Manufacturing Technology,
A.S.T.M.E., p. 1083

SAXL, K. 1964 'The Pendulum Mill—A New Method of
–65 Rolling Metals'
Proc. Inst. mech. Engrs. **179**, 453

1958 'Transverse Gauge Variation in Strip and
Sheet Rolling'
Proc. Inst. mech. Engrs. **172**, 727

*SIMS, R. B. 1954 'Calculation of Roll Force and Torque in
Hot Rolling Mills'
Proc. Instn mech. Engrs **168**, 191

SMIRNOV, W. C. 1967 *Theory of Rolling*
In Russian, Moscow

SMITH, C. L., 1952 'Pressure Distribution Between Stock and
SCOTT, F. H. and Rolls in Hot and Cold Flat Rolling'
SYLWESTROWICZ, W. *J Iron Steel Inst.* **170**, 347

STEWARTSON, R. 1959 'The Rolling of Rods, Bars and Light
Sections'
Metall. Rev. **4**, No. 16, Inst. of Metals

TARNOVSKII, I. YA., 1965 *Deformation of Metals During Rolling*
POZDEYEV, A. A. and (English Translation), Pergamon Press, 340pp
LYASHKOV, V. B.

THOMSON, P. F. and 1967 'The Origin of Some Surface Defects on
HOGGART, J. S. Rolled and Drawn Products'
20th Ann. Conf., J. Australian Inst. Metals

TSELIKOV, A. I. 1967 *Stress and Strain in Metal Rolling*
M.I.R. Publishers, Moscow (In English)

TSELIKOV, A. I. and 1965 *Rolling Mills*
SMIRNOV, V. V. Translated from Russian and published by
Pergamon Press

UNDERWOOD, L. R. 1950 *The Rolling of Metals*
Chapman and Hall, Ltd., London

WATTS, A. B. and 1955 'On the Basic Yield Stress Curve for a Metal'
FORD, H. *Proc. Instn mech. Engrs* **169**, 1141

WEINSTEIN, A. S. 1963 'On Some Elastic Effects in Metal Rolling'
Int. Res. Prod. Engng, A.S.M.E. 374

WHITTON, P. W. 1955 'Surface Friction and Lubrication in Cold
and FORD, H. Strip Rolling'
Proc. Instn mech. Engrs **169**, 123

WILCOX, R. J. and   1960 'The Rolling of Thin Titanium Strip'
WHITTON, P. W.   *J. Inst. Metals* **88**, 200

———————   1960 'Research on the Rolling of Strip: a
Symposium of Selected Papers 1948–1958'
B.I.S.R.A.

ZOROWSKI, C. F. and   1963 'Analysis of Load and Torque
SHUTT, A.   Characteristics in Single Roll Drive Mills'
Int. Res. Prod. Engng, A.S.M.E. 380

ZOROWSKI, C. F. and   1968 *Influence of Mill Set-Up on Hollow Geometry*
HOLBROOK, R. L.   *Produced by Rotary Piercing*
8th Int. M.T.D.R. Conf., Pergamon Press, 1041

FORGING

AKU, S. Y.,   1967 The Use of Plasticene to Simulate the
SLATER, R. A. C. and   Dynamic Compression of Prismatic Blocks of
·JOHNSON, W.   Hot Metal
Int. J. mech. Sci., **9**, 495

ALDER, J. F. and   1954 'The Effect of Strain-Rate and
PHILIPS, K. A.   Temperature on the Resistance of
Aluminium, Copper and Steel to
Compression'
J. Inst. Metals **83**, 80

ALEXANDER, J. M. and 1963 *Manufacturing Properties of Materials*
BREWER, R. C.   D. Van Nostrand, 489 pp.

HAWKYARD, J. B. and 1967 'An Analysis of the Changes in Geometry
JOHNSON, W.   of a Short Hollow Cylinder During
Axial Compression'
Int. J. mech. Sci. **9**, 163

JOHNSON, W. and   1965 *Flat Bar Forging.*
BARAYA, G. L.   5th Int. M.T.D.R. Conf., Pergamon Press

JOHNSON, W.,   1966 The Quasi-Static Compression of Non-
SLATER, R. A. C. and   Circular Prismatic Blocks Between Very
YU, A. S.   Rough Platens using the 'Friction-hill'
Concept. *Int. J. mech. Sci.*, **8**, 731

JONES, M. G.,   1969 *Some High-Speed Cold Forging Operations*
DAVIES, R. and   9th Int. M.T.D.R. Conf., Pergamon Press
SINGH, A.

LIPPMANN, H.   1966 *On the Dynamics of Forging*
7th Int. M.T.D.R. Conf. Pergamon Press

*MALE, A. T. and   1964 'The Ring Test'
COCKROFT, M. G.   *J. Inst. Metals*, **93**, 38

NADAI, A.   1950 *Theory of Flow and Fracture of Solids*
McGraw-Hill, New York

VAN ROOYEN, G. T.   1960 'A Study of Interface Friction in
and BACKOFEN, W. A.   Plastic Compression'
Int. J. mech. Sci. **1**, 1.

SAMANTA, S. K.   1966 'Dynamic Compression of Steel'
                         R. Inst. Tech., Stockholm

SLATER, R. A. C.,   1967 'Experiments in the Fast Upsetting of
JOHNSON, W. and   Short Pure Lead Cylinders and a
AKU, S. Y.   Tentative Analysis'
                         Int. J. mech. Sci. **10**, 169

THOMSEN, E. G.   1965 *Mechanics of Plastic Deformation in*
YANG, C. T. and   *Metal Processing*
KOBAYASHI, S.   Macmillan, New York, 486 pp.

UNKSOV, E. P.   1961 *An Engineering Theory of Plasticity.*
                         Butterworths, London

WALLACE, P. W. and   1968 *Metal Flow in Forging: A Practical Study.*
SCHEY, J. A.   8th Int. M.T.D.R. Conf., Pergamon Press

SUPERPLASTICITY

JOHNSON, W., AL-NAIB, 1972 'Superplastic Forming Techniques and Strain,
T. Y. M., and   Distribution in a Zinc-Aluminium Alloy'
DUNCAN, J. L.   *J. Inst. Metals,* **100**, 45

JOVANE, F.   1968 'An Approximate Analysis of the Superplastic
                         Forming of a Thin Circular Diaphragm:
                         Theory and Experiments'
                         Int. J. mech. Sci. **10**, 403

WEISS, V. and   1967 *Superplasticity*
KOT, R.   Int. Conf. on Manufacturing Methods,
                         A.S.T.M.E., p. 1031

BACKOFEN, W. A.,   1966 'Superplasticity in an Al-Zn Alloy'
TURNER, I. and   *Trans. A.S.M.* **57**, 981
AVERY, H.

GREENWOOD, G. W.   1965 'The Deformation of Metals Under Small
and JOHNSON, R. H.   Stresses During Phase Transformations'
                         Proc. R. Soc., Series A **282**, 403

Chapter 12

THE SLIP-LINE FIELD: THEORY AND
EXAMPLES OF PLANE PLASTIC STRAIN

12.1 General Remarks

The tools of the metal-working processes discussed in Chapter 11 had simple geometric forms and were analysed in an empirical or approximate manner. In this chapter a general theory, known as the slip-line field theory, is developed to analyse non-homogeneous *plane strain deformation in a rigid-perfectly plastic isotropic solid*. It will be recalled that for such an ideal material the elastic strains developed are zero, while if plastic flow occurs, it does so without work-hardening. Clearly, these idealizations are not realized in industrial materials, but they do, however, give very good first approximations to loads required to perform operations and provide indications of the manner in which material deforms. The theory is valuable for investigating such industrial processes as sheet drawing, sheet extrusion, rolling and forging. Applications of the theory are discussed in this chapter and in Chapter 14.

The principal ways in which slip-line field theory fails to take account of the behaviour of real materials are:

1. It deals only with non-strain-hardening materials. (The stress aspect of slip-line theory as adapted for strain-hardening materials has been used by PALMER and OXLEY (1960) in connection with machining.) Whilst strain-hardening can be allowed for in calculations concerned with loads in an approximate way, the manner in which strain distribution is altered because of it is not always clear.

2. There is no allowance for creep or strain-rate effects. The rate of deformation at each given point in space and in the deforming body is generally different, and any effect this may have on the yield stress is ignored. Also, all inertia forces are neglected and the problems treated as quasi-static.

3. In the forming operations which impose heavy deformations, most of the work done is dissipated as heat; the temperatures attained may affect the material properties of the body or certain physical characteristics in the surroundings, e.g. lubrication. The thermal gradients of themselves create thermal stresses of which no account is taken.

Despite these shortcomings, the theory is extremely useful; it is very important, however, to remember its limitations and not to expect too high a degree of correlation between experimental and theoretical work.

Some readers will find it better to read Chapter 13 before continuing.

PLANE PLASTIC STRAIN

Deformation which proceeds under conditions of plane strain is such that
1. the flow or deformation is everywhere parallel to a given plane, say the (x, y) plane in a system of three mutually orthogonal planes and
2. the flow is independent of z. (Axisymmetric problems are two-dimensional ones but are not ones in plane strain because the flow pattern is not independent of z).

Since elastic strains are neglected, the plastic strain increments (or strain-rates) may be written in terms of the displacements (or velocities) $u_x(x, y)$, $v_y(x, y)$, $w_z = 0$, as below

$$\dot{\epsilon}_x = \frac{\partial u_x}{\partial x} \qquad \dot{\gamma}_{xy} = \frac{1}{2}\left(\frac{\partial u_x}{\partial y} + \frac{\partial v_y}{\partial x}\right)$$

$$\dot{\epsilon}_y = \frac{\partial v_y}{\partial y} \qquad \dot{\gamma}_{yz} = \frac{1}{2}\left(\frac{\partial v_y}{\partial z} + \frac{\partial w_z}{\partial y}\right) = 0$$

$$\dot{\epsilon}_z = \frac{\partial w_z}{\partial z} = 0 \qquad \dot{\gamma}_{zx} = \frac{1}{2}\left(\frac{\partial w_z}{\partial x} + \frac{\partial u_x}{\partial z}\right) = 0.$$

It follows from the Lévy-Mises relation (equations (5.10)) that τ_{xz} and τ_{yz} are zero and therefore that σ_z is a principal stress. Further, since $\dot{\epsilon}_z = 0$, then $\sigma_z' = 0$ and hence $\sigma_z = (\sigma_x + \sigma_y)/2 = p$, say. Again, because the material is incompressible $\dot{\epsilon}_x = -\dot{\epsilon}_y$ and each incremental distortion is thus a pure shear. The state of stress throughout the deforming material is represented by a constant yield shear stress k, and a hydrostatic stress $-p$ which in general varies from point to point throughout the material. k is the yield shear stress in plane strain and the yield criterion for this condition is

$$\tau_{xy}^2 + (\sigma_x - \sigma_y)^2/4 = k^2, \tag{12.1}$$

where $k = Y/2$ for the Tresca criterion and $k = Y/\sqrt{3}$ for the Mises criterion.

The state of stress at any point in the deforming material may be represented in the Mohr circle diagram, Fig. 12.1. Remember that for an isotropic material the directions of maximum shear strain-rate, represented by points A and B coincide with the directions of yield shear stress and that such directions are clearly directions of zero rate of extension or contraction. The loci of these directions of maximum shear stress and shear strain-rate form two orthogonal families of curves known as slip-lines.

The stresses on a small curvilinear element bounded by slip-lines are shown in Fig. 12.2(a). The slip-lines are labelled α and β as indicated. It is essential to distinguish between the two families of slip-lines, and the usual convention is that when the α- and β-lines form a right-handed co-ordinate system of axes, then the line of action of the algebraically greatest principal stress, σ_1, passes through the first and third quadrants. The *anti-clockwise* rotation, ϕ, of the α-line from the chosen x-direction is taken as positive.

In order to determine the load necessary for a particular plastic forming operation, we must first of all obtain the slip-line field pattern. This means that we must derive equations for the variation of p along both α- and β-lines. Also, we must check that all velocity conditions along α- and β-lines are satisfied. The method, adopted below, is not intended to be rigorous. The partial differential equations for the plane plastic flow of a plastic-rigid material are hyperbolic and it is therefore convenient to refer the equations to their characteristics (in this case the slip-lines). No further mention will be made of the theory of characteristics and readers who require a more rigorous treatment of the subject can refer to HILL (1950) or the monograph by JOHNSON, SOWERBY and HADDOW (1970).

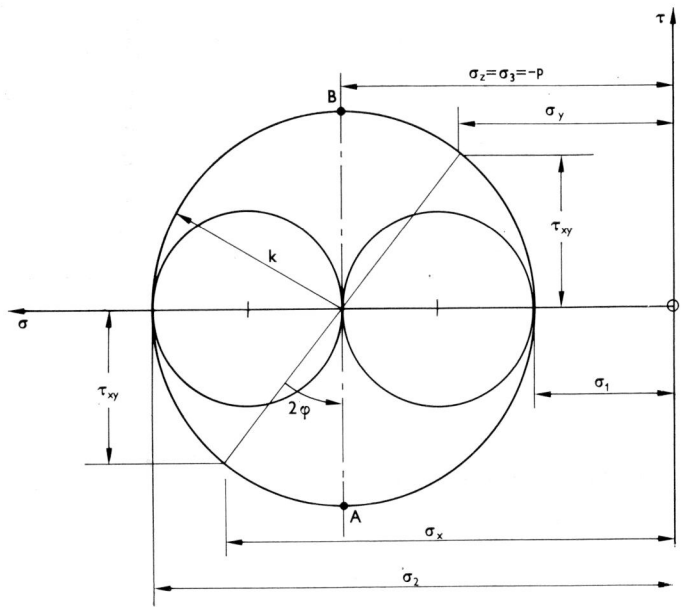

FIG. 12.1 Mohr circle diagram for stress in plane plastic strain. A and B represent the stress states $(-p, \pm k)$ at a point on planes parallel to the slip-lines through that point.

12.2 The Stress Equations

The equations of equilibrium for plane strain are, with neglect of body forces

$$\left.\begin{array}{c} \dfrac{\partial \sigma_x}{\partial x} + \dfrac{\partial \tau_{xy}}{\partial y} = 0 \\[3mm] \dfrac{\partial \tau_{xy}}{\partial x} + \dfrac{\partial \sigma_y}{\partial y} = 0. \end{array}\right\} \qquad (12.2)$$

The above stress components σ_x, σ_y, τ_{xy} expressed in terms of p and k are, see Fig. 12.1.

$$\left. \begin{array}{c} \sigma_x = -p - k \sin 2\phi \\ \sigma_y = -p + k \sin 2\phi \\ \tau_{xy} = k \cos 2\phi. \end{array} \right\} \qquad (12.3)$$

p is the normal or hydrostatic pressure on the two planes of yield shear stress.

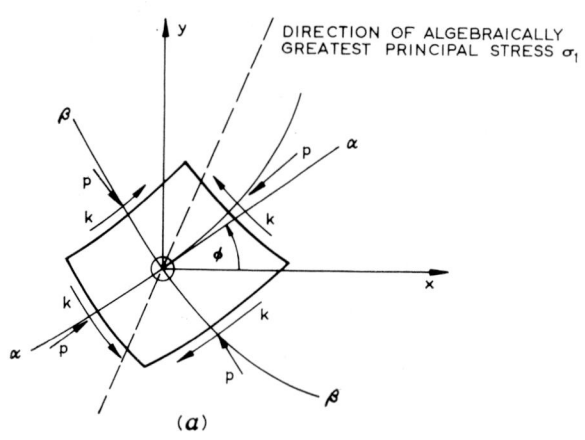

(a)

Fig. 12.2

(a) The stresses on a small curvilinear element.

Differentiating and substituting from equation (12.3) in equation (12.2) we have

$$\left. \begin{array}{c} -\dfrac{\partial p}{\partial x} - 2k \cos 2\phi \dfrac{\partial \phi}{\partial x} - 2k \sin 2\phi \dfrac{\partial \phi}{\partial y} = 0 \\[2mm] -2k \sin 2\phi \dfrac{\partial \phi}{\partial x} - \dfrac{\partial p}{\partial y} + 2k \cos 2\phi \dfrac{\partial \phi}{\partial y} = 0. \end{array} \right\} \qquad (12.4)$$

If now the α- and β-lines are taken to coincide with Ox and Oy at O, that is we take $\phi = 0$, equations (12.4) become

$$\left. \begin{array}{c} -\dfrac{\partial p}{\partial x} - 2k \dfrac{\partial \phi}{\partial x} = 0 \\[2mm] -\dfrac{\partial p}{\partial y} + 2k \dfrac{\partial \phi}{\partial y} = 0. \end{array} \right\} \qquad (12.5)$$

Thus, integrating

$$p + 2k\phi = f_1(y) + C_1,$$
$$p - 2k\phi = f_2(x) + C_2.$$

However, $f_1(y) = 0$ and $f_2(x) = 0$, because when ϕ is zero, p must have the same value whichever equation we take. Thus

$$p + 2k\,\phi = \text{constant along an } \alpha\text{-line}$$

and, $\qquad p - 2k\,\phi = \text{constant along a } \beta\text{-line.}$ $\qquad$ (12.6)

The equations (12.6) are known as the Hencky equations and are equivalent to the equilibrium equations for a fully plastic mass stressed in plane strain. In general, of course, the values of the constants C_1 and C_2 vary from one slip-line to another.

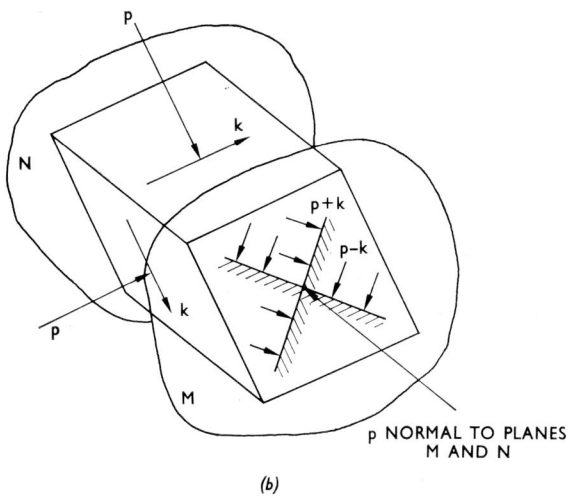

(b)

FIG. 12.2

(b) M and N are typical parallel planes in which the distortion occurs.

It should be noted that Ox and Oy are a general set of cartesian axes and once the α- and β-lines have been correctly designated, they may be taken to act in any chosen direction. In using the Hencky equations, it is often convenient to take the x-direction tangential to a point on the α-line; ϕ is then measured positively when it is understood to mean an anti-clockwise rotation when moving from one point to another along a given α- or β-line.

The equations of equilibrium (12.2) and the yield criterion (12.1) form a set of three equations in the three unknown stress components σ_x, σ_y, τ_{xy}. In some problems there will be sufficient boundary conditions, involving only the stresses, to enable a solution to these equations to be found. Such problems are known as *statically determinate* and the velocities can be calculated afterwards. In other problems the boundary conditions will involve both stresses and velocities and a solution can only be obtained by considering the stresses and velocities together. Such problems are known as *statically indeterminate*.

A major feature of the theory concerns the manner in which the solutions are arrived at. This is usually the result of a combination of experience and intuition; (instances are known in which careful experiments were performed with the aim of making evident the type of slip-line field in operation). The usual sense in which solutions to problems are obtained does not apply here where the practice is to propose the solution and then show that it fits all the boundary conditions. It does not follow, however, that the slip-line field solution obtained is a unique one and, in fact, a criterion of minimum work has to be invoked on some occasions to help choose the most probable mode of deformation from among several.

Figure 12.2(b) shows diagrammatically plane plastic flow.

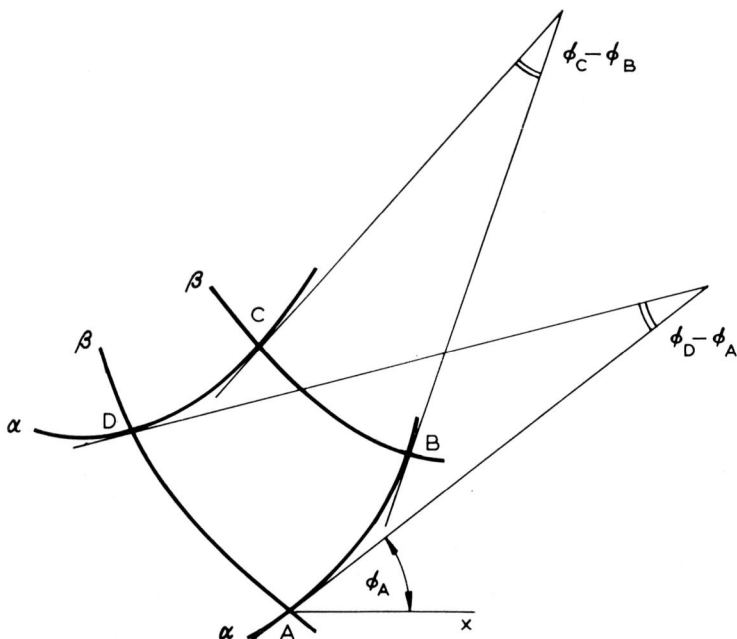

Fig. 12.3 Two pairs of α- and β-lines to demonstrate Hencky's Theorem.

HENCKY'S FIRST THEOREM

A theorem due to Hencky following from equations (12.6) and one which is repeatedly used in applications of the theory demonstrates the important slip-line field property that the angle between two slip-lines of, say, an α-family, where cut by a slip-line of the β-family, is constant along their lengths. For instance, in applying this to Fig. 12.3, it follows that the angle between the tangents at A and D is equal to that between the tangents at B and C. This theorem, see Fig. 12.3, is proved by arriving at the pressure difference between C and A in the two possible ways. If the x-direction is taken through A, as shown, then remembering that positive ϕ is defined as an anti-clockwise rotation of the α-line from the x-direction, we have, using the Hencky equations:

1. $A \to B$, α-line, $\quad p_B + 2k\phi_B = p_A + 2k\phi_A$

   $B \to C$, β-line, $\quad p_C - 2k\phi_C = p_B - 2k\phi_B.$

   Hence the pressure difference between C and A is

   $$p_C - p_A = 2k(\phi_A + \phi_C - 2\phi_B)$$

2. $A \to D$, β-line, $\quad p_D - 2k\phi_D = p_A - 2k\phi_A$

   $D \to C$, α-line, $\quad p_C + 2k\phi_C = p_D + 2k\phi_D.$

   Hence,

   $$p_C - p_A = 2k(2\phi_D - \phi_C - \phi_A)$$

Therefore, $\qquad\qquad\qquad \phi_C - \phi_B = \phi_D - \phi_A. \qquad\qquad\qquad$ (12.7)

12.3 The Velocity Equations

In Fig. 12.4, u, v are the component velocities of a particle at a point O along a pair of α- and β-slip-lines, the α-line being inclined at ϕ to the Ox

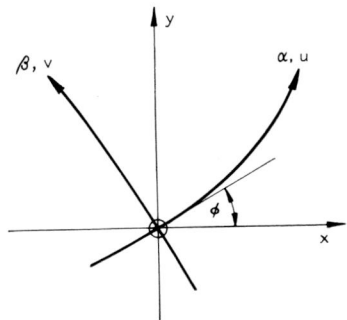

FIG. 12.4 Component velocities of a particle in plastically deforming body.

axis of a pair of orthogonal cartesian axes through O. The components of the velocity of the particle u_x and v_y parallel to Ox and Oy, respectively, are then

$$u_x = u \cos \phi - v \sin \phi$$
$$v_y = u \sin \phi + v \cos \phi.$$

Taking the x-direction at point O tangential to the α-line, i.e. $\phi = 0$,

$$\left(\frac{\partial u_x}{\partial x} \right)_{\phi=0} = \frac{\partial u}{\partial x} - v \frac{\partial \phi}{\partial x}.$$

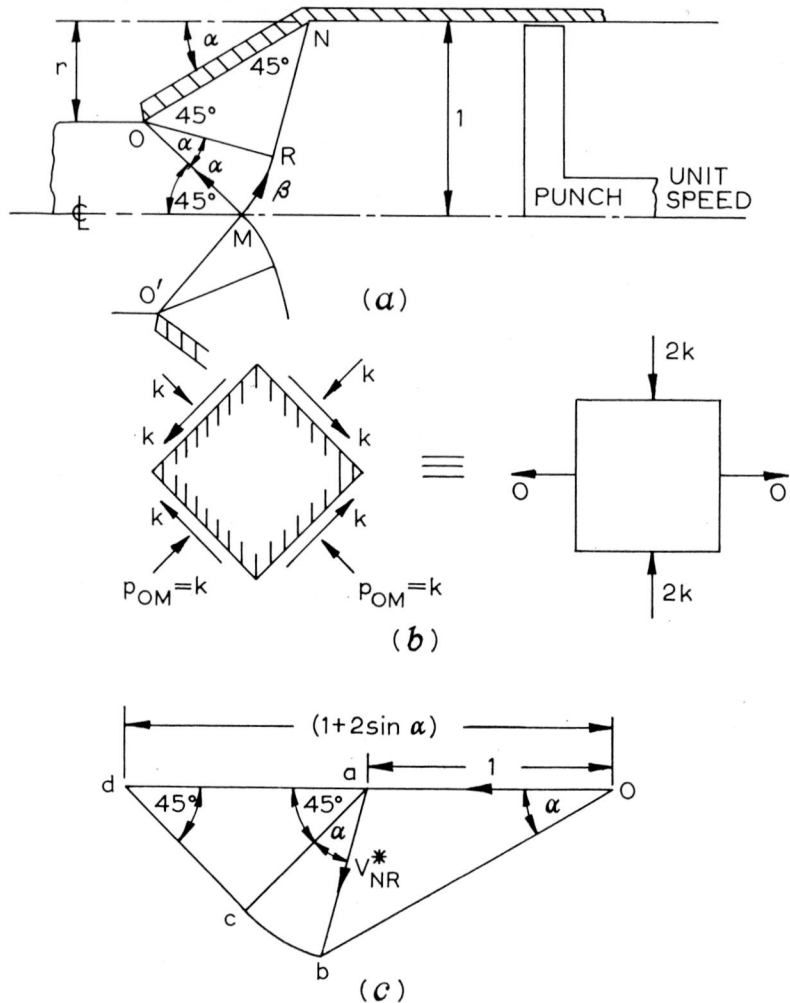

FIG. 12.5 A simple slip-line field solution for extrusion through a perfectly smooth wedge-shaped die of angle α.

(a) Top half of extrusion only is shown symmetrical about centre-line.

(b) Stress systems at M.

(c) Hodograph to (a).

Since $\epsilon_x = \partial u_x/\partial x$ is zero along a slip-line

$$\frac{\partial u}{\partial x} - v \frac{\partial \phi}{\partial x} = 0 \quad \text{along an } \alpha\text{-line}$$

or

$$du - v\, d\phi = 0 \quad \text{along an } \alpha\text{-line.} \tag{12.8}$$

Similarly, it can be shown that

$$dv + u \, d\phi = 0 \quad \text{along a } \beta\text{-line.} \tag{12.9}$$

Physically, it may be imagined that small rods lying on the slip-line directions at a point do not undergo extension or contraction.

Equations (12.8) and (12.9) are known as the velocity equations and are due to GEIRINGER (1930). These equations are essential for the purpose of calculating analytically the mode of deformation of metal; they have only been applied in cases where the slip-line fields are of the simplest kind and are therefore directly amenable to analytical manipulation. These are cases in which the slip-line field is composed of a single triangle and circular sector, see PRAGER and HODGE (1951) and JOHNSON, SOWERBY and HADDOW (1970).

12.4 A Simple Slip-line Field for Extrusion through a Frictionless Wedge-shaped Die of Semi-angle α, of reduction $r = 2 \sin \alpha/(1+2 \sin \alpha)$

Consider an extrusion for which the slip-line field is as shown in Fig. 12.5(a); it comprises an isosceles triangle ORN and a circular sector ORM. The die is assumed to be perfectly smooth and therefore the slip-lines OR and NR are both at 45° to the die face ON. For the simple slip-line field shown, OM meets the axis at 45° and OMR is a sector of angle α. For this particular case we choose OM so that it makes 45° with the axis; there is then symmetry in the slip-line field about the axis and the total force on the extruded material to the left of OM is zero. This implies that the *compressive* stress normal to the axis at M is $2k$, see Fig. 12.5(b). Thus in accordance with the convention introduced above, OM is an α-line and MRN a β-line because the direction of the *algebraically* greatest principal stress (zero in magnitude) is parallel to the axis.

For this slip-line field to be admissible, the velocity field must be compatible with the known flow of the material; before entering the field across NRM (Fig. 12.5(a)), it is rigid and has a speed of unity represented in Fig. 12.5(c) by Oa. Any particle crossing NR has its direction altered suddenly to proceed parallel to die face ON; this is brought about by a tangential velocity discontinuity V_{NR}^*. (The nature of such a discontinuity will be well understood after reading Section 13.5). From O a line is drawn at angle α to represent the direction of absolute movement in triangle ONR. From a, a vector ab is drawn parallel to NR: this triangle Oab is completed and the magnitude of the velocity vectors ab and ob determined. The velocity discontinuity is of constant magnitude along RM; it is represented in Fig. 12.5(c) by bc, an arc subtending angle α at a. ac is at 45° to Oa produced because the β slip-line is also at 45° to the axis at M. A particle on the axis is subject to two successive jumps in tangential velocity; first ac and secondly, one of equal magnitude in direction MO, i.e. also at 45° to Oa produced, and derived from the lower half of the extrusion; this is equal to cd. The value of do in Fig. 12.5(c) is found to be $(1 + 2 \sin \alpha)$; the material after emerging across OM is rigid. The completed diagram gives the velocity of any particle in the physical plane of Fig. 12.5(a) and is called the hodograph.

Notice that the area covered by the slip-line field is that in which large deformations are brought about; this deformation is accomplished in the die mouth.

The extruding slug is supposed to be long and the slip-line field once established remains fixed in space and does not alter with time; the extrusion is then said to be a steady-state one.

The reduction is r, i.e. the thickness is reduced by a fraction r and

$$r = 2 \sin \alpha / (1 + 2 \sin \alpha).$$

The normal pressure on the die is q, see Fig. 12.6(a), and

$$q = p_{OR} + k.$$

p_{OR} is the hydrostatic pressure normal to plane OR.
Using the Hencky equations to find p_{OR}, noting that $p_{OM} = k$ and that in

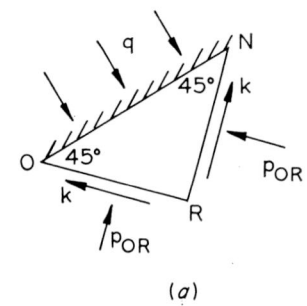

(a)

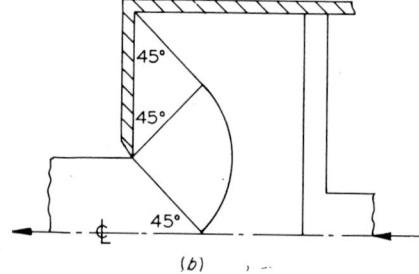

(b)

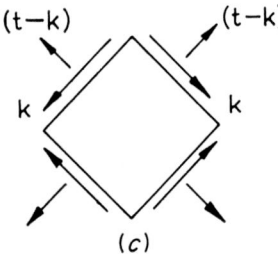

(c)

FIG. 12.6

(a) To calculate stress on die face.
(b) A square die; container wall and die face both perfectly smooth; $r = 2/3$.
(c) Stress system at M of Fig. 12.5 (a) for drawing.

moving from M to R along the β-line MRN, a positive rotation of angle α is described

$$p_{OR} - 2k\alpha = p_{OM} \quad \text{or} \quad p_{OR} = k(1 + 2\alpha).$$

Thus substituting in the expression for q above

$$q/2k = 1 + \alpha.$$

The uniform pressure applied by the ram is p, given by

$$p.1 = q \, ON \sin \alpha$$

or $\qquad p/2k = (q/2k).(r) = (1 + \alpha)r = \dfrac{2(1+\alpha)\sin\alpha}{1 + 2\sin\alpha}.$ $\qquad$ (12.10)

Note that in particular, when $\alpha = \pi/2$, $p/2k = 2(1 + \pi/2)/3 = 1\cdot71$. This is a much-discussed instance of extruding through a square or 90° die, the die face and the container wall *both* being perfectly smooth, see Fig. 12.6(b), when the reduction is 2/3 or 66·7 per cent.

Note that the tangential velocity discontinuity introduced at N, Fig. 12.5(c) is $\sqrt{2}\sin\alpha$; it is propagated along NRM to terminate at the other corner of the orifice opposite to O. The internal energy dissipated therefore along both lines of velocity discontinuity is

$$2k.V^*_{NM}.(\text{length } NRMO') = k(2 + \alpha)r.$$

Thus the fraction, f, of the total work done by the ram, dissipated along the two lines of velocity discontinuity is $(2 + \alpha)/2(1 + \alpha)$. The table below gives the fraction, f, of the total work done in the extrusion process which is dissipated in these velocity discontinuities for some angles and particular reductions.

$\alpha°$	15	30	45	60	75	90
r	0·343	0·50	0·58	0·63	0·66	0·67
f	0·89	0·83	0·78	0·745	0·72	0·695

The remaining fraction of work $(1 - f)$ is dissipated as a continuous deformation process in the sector.

To *draw* material through the die, suppose a uniform tension t is required. At M, see Fig. 12.5(a), the stress system is as shown in Fig. 12.6(c) but the hydrostatic component is $-(t - k) = -1 + k(1 + 2\alpha)$. Thus, $p_{OR} = -(t - k) + 2k\alpha = -t + k(1 + 2\alpha)$. Hence the die pressure is $q = [-t + 2k(1 + \alpha)]$. The drawing stress t is now found by equating the drawing load to the total horizontal component of the force on the dies

$$-r[t - 2k(1 + \alpha)] = (1 - r)t.$$

Hence $t/2k = r(1 + \alpha)$. The only difference between these two cases of extrusion and drawing is a hydrostatic component of p or t. Thus for smooth dies, drawing and extruding require equal drawing and extruding pressures if the reduction is the same in both cases. Clearly, however, the forward tension in drawing is restricted to being less than $2k$. A method investigated for the purpose of reducing die face pressure in wire-drawing practice—and hence wear—is to apply a back tension, say, t_0. The forward tension would then have to be increased to $(t + t_0)$ and the die face pressure would thereby be reduced from q to $(q - t_0)$.

12.5 The Compression of a Block between Rough Rigid Parallel Platens, the Platen Width Exceeding the Material Thickness

In Fig. 12.7, the platens AB and CD are rigid and perfectly rough, and when pressed together approach one another with a relative speed of two units. Approach cannot take place if it is possible to go from AB to CD through 'elastically' stressed material which is everywhere of finite thickness; since the stresses in the 'elastic' material induce strains of zero magnitude (remember that the material is rigid-perfectly plastic) the platens may only

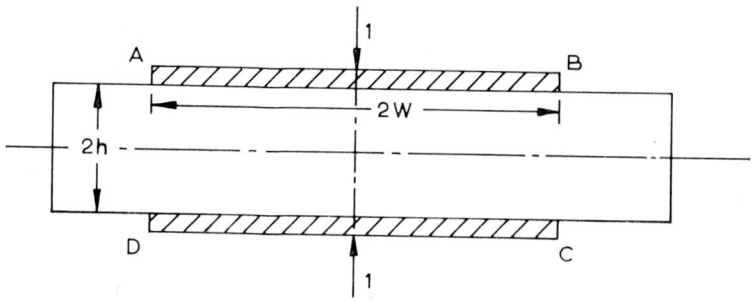

Fig. 12.7 Compression of a block between rough, rigid, parallel platens.

approach one another when a sufficient amount of the material between them has become plastic and which can thus be forced to flow outwards.

The load at which the platens approach one another is normally called the yield point load, P_0. The implication is that in a compression experiment of this kind, the ideal load versus platen movement would be somewhat as shown in Fig. 12.8. When once movement is induced, the ratio $2w/2h$ in-

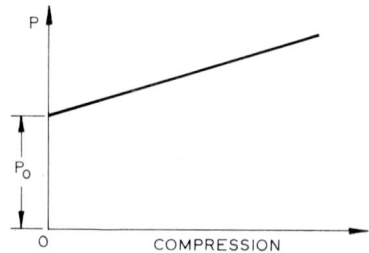

Fig. 12.8 Ideal load versus platen movement.

creases and the load to cause further compression must increase. In fact, in a typical experimental instance (see NYE, 1946) the graph of load versus amount of compression is as shown in Fig. 12.9 for a material having the compression stress-strain curve which is shown in Fig. 12.10.

It is clear that on applying compressive load, zones of plastic material will be initiated first at the corners of the plates and increasing load extends them until fusion occurs across from *A* to *D* and *B* to *C*. Further increase of load causes these latter two zones to extend inwards to the centre of the material and immediately final fusion occurs at the centre, the platens may then approach. The plastic zones initiated at *A*, *B*, *C* and *D* exclude two regions of material outside the platens, which remain sensibly rigid. (see Fig. 12.13.)

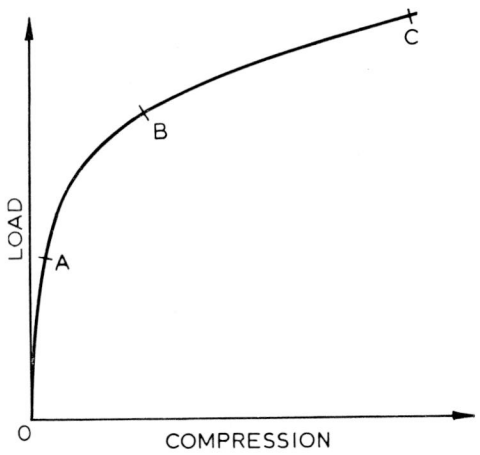

FIG. 12.9 Experimental graph of load versus amount of compression.

A: onset of yield.
B: yield point, contained plastic deformation.
C: load increase because of strain-hardening and because $2w/2h$ increases.

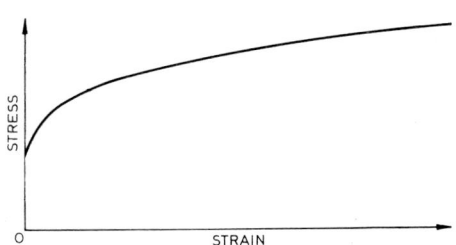

FIG. 12.10 Stress-strain curve in compression.

DRAWING THE SLIP-LINE FIELD

In Fig. 12.11, one half of a 15° net of slip-lines for solving the problem of compression between rough parallel platens for various ratios of $2w/2h$ is shown. The equiangular net is drawn with points *A*, *B*, *C* and *D* (Fig. 12.7) as centres of concentric circular arcs and radii of slip-lines, the radii meeting, (i) orthogonally on the axis of the block and (ii) the platens tangentially. (i) is required by symmetry considerations and is necessary in order that material to the left of DGA shall have no net force on it. (ii) is required because the notion of 'perfectly' rough means shearing of the material between it and the platen. Consequently arcs GG_3 and GG'_3 are orthogonal at G and to both platens at G_3 and G'_3. [It may be noted that if AD was not perpendicular to the block centre line, i.e., the platens were of unequal length, the slip-line field would be started from an isosceles triangle on AD, for reasons of equilibrium as before: see Fig. 12.12, (JOHNSON and KUDO, 1960).]

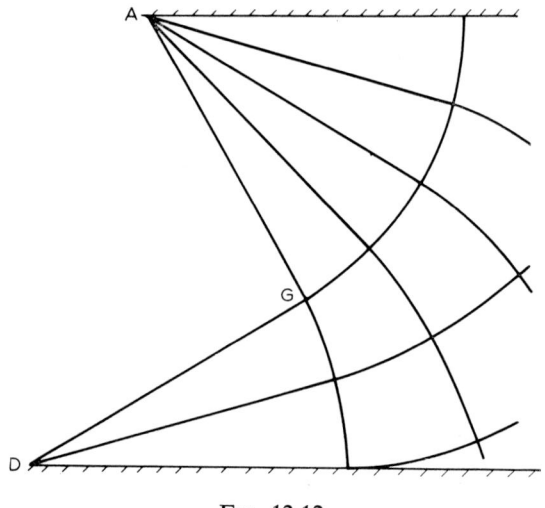

FIG. 12.12

To draw a 15° equiangular net, the angle GAG_3 is divided into three 15° angles, to locate points G_1 and G_2. Because all slip-lines intersect orthogonally at 45° on the horizontal centre-line of the compressed material, the slip-line leaving G_1 must rotate through 15° in passing from G_1 to H; this slip-line is approximated by a chord G_1H drawn from G_1 at $-37\frac{1}{2}°$ (i.e., $(45° + 30°)/2$) to cut the axis at H. To extend the field from G_2 draw a line at $-22\frac{1}{2}°$ to Ox (AG_2 is at $-15°$ to Ox at G_2 and at $-30°$ to Ox at H_1 and where it meets the slip-line of the other family). From H draw HH_1 at $52\frac{1}{2}°$ to Ox (this chord approximates the slip-line which rotates from being at 45° to Ox at H to 60° to Ox at H_1). Thus the chord from H at $52\frac{1}{2}°$ to Ox and from G_2 at $-22\frac{1}{2}°$ to Ox, intersect at H_1. As drawn here and due to this construction, the chords which approximate the slip-lines are not orthogonal. The field may now be continued in an obvious manner. From G_3 draw G_3H_2 at $-7\frac{1}{2}°$ to Ox, i.e., $(0 + 15°)/2$ and from H_1 draw H_1H_2 at $67\frac{1}{2}°$, i.e., $(60° + 75°)/2$ to Ox, to intersect at H_2. From H_2 draw H_2H_3 at $82\frac{1}{2}°$ to Ox, i.e., $(75° + 90°)/2$, to intersect platen surface AB at H_3. Next, start from H_1, and draw H_1J at $-37\frac{1}{2}°$ to Ox to determine J. The field may now be extended in the same manner to any degree required. An approximated slip-line field is thus obtained. Drawing pairs of orthogonal curved lines through the points of intersection of the chords, the true slip-line field is obtained to a fair degree of accuracy. The accuracy is evidently increased the smaller the equiangular mesh of slip-lines. Such a smoothed-out slip-line field is shown by Fig. 12.17, continued far enough to include a ratio, $2w/2h \simeq 6.60$. A typical picture for $2w/2h = 5.6$ is shown in Fig. 12.13, the two families of slip-lines intersecting at right angles at the block centre, M. The two shaded zones into which the slip-line field is not extended and contained between the platens and lines, $J_3 M J_3''$ and $J_3' M J_3'''$, are taken to be rigid zones which, in effect, behave as if they were rigidly attached to the plates.

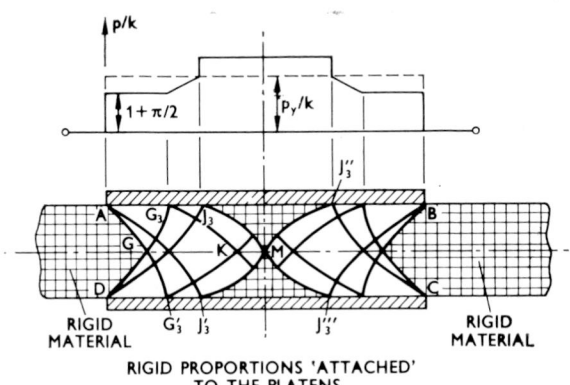

FIG. 12.13 Slip-line field and pressure distribution for compression between rough parallel platens when $2w/2h = 5.6$.

CALCULATION OF THE YIELD-POINT LOAD

To calculate the yield-point load, use is made of the Hencky equations. Continuing with Fig. 12.11, to the left of AGD, the total force is zero and the stresses on the material above AG are as shown in Fig. 12.14. The labelling

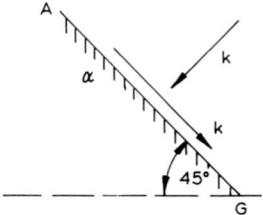

FIG. 12.14 Stresses along AG of Fig. 12.13.

of the α- and β-slip lines is in conformity with the convention that the direction of the algebraically greatest principal stress passes through the first and third quadrants of the local system. The zero principal stress parallel to Ox is algebraically greater than that parallel to Oy, which must therefore be

$-2k$ to cause shearing along AG and DG. The family of slip-lines radiating from A are α-lines and those from D, β-lines.

1. Normal Stress at G_3, p_{G3}

Move from G to G_3 via β-line GG_3, along which $p - 2k\phi =$ constant. At G take $\phi = 0$ and $p = k$, so that the constant for this line is k. Thus, $p - 2k\phi = k$ or $p = k(1 + 2\phi)$ along GG_3. Moving from G to G_3, a positive rotation of $\pi/4$ is made and hence, at G_3, $p = k(1 + 2\pi/4)$. Thus

$$p_{G_3} = k(1 + \pi/2) \doteqdot 2 \cdot 57\, k$$

2. Normal Stress at H_3, p_{H3}

(a) Moving from G_3 to H_2 via an α-line, the rotation is through $-15°$ or $-0 \cdot 262$ radians.

Now, $p + 2k\phi =$ constant and at G_3, taking $\phi = 0$, the constant is

$$p_{G3} = k(1 + \pi/2), \quad \text{from (i).}$$

Thus $p_{H2} = k(1 + \pi/2) - 2k\phi = k(1 + \pi/2 + 2 \times 0 \cdot 262) = 3 \cdot 09\, k$.

(b) Moving from H_2 to H_3 along a β-line, and rotating through $+15°$, $p - 2k\phi =$ constant $= p_{H2}$, taking $\phi = 0$ at H_2. Then

$$p_{H3} = k(3 \cdot 09 + 2 \times 0 \cdot 262) = 3 \cdot 62k$$

3. Similarly, $p_{J3} = 4 \cdot 66k$, etc., and thus the normal pressure at each

point on the platen may be obtained. This is plotted in Fig. 12.11 and shown by line $abc \ldots$ When the platens are wider than about $3 \cdot 6$ times the material thickness, the mode of deformation that is imagined to occur and for which there is good experimental evidence, is as referred to in connection with Fig. 12.13. There is a rigid block expelled at each end of the platen, and two regions of plastic material between them separated by two ostensibly rigid blocks which have point contact at the block centre. These two blocks effectively adhere to the platens and do not slide or shear outwards, relative to them. Their shape is defined by the pair of bounding slip-lines which intersect orthogonally at the block centre. When $2w/2h$ is less than $3 \cdot 61$ (which corresponds to taking any point between K and G as the block centre in Fig. 12.11), it will be observed as in Fig. 12.15 that the bounding slip-lines of these adhering blocks or false 'heads', stretch between the platen corners, i.e. the non-deforming material reaches across the *whole* of both platens.

Now since the stress distribution in the adhering rigid blocks is not calculable, only the total vertical force or average vertical stress, applied over the surface of contact with the platens is calculable. This is obtained thus:

Consider line NK_3, Fig. 12.11, to be one boundary of such a block and let δs be an elemental length; the stresses on δs are as shown in Fig. 12.16. The vertical force on δs is $\delta F = (k \sin \psi + p_s \cos \psi)\, \delta s$ and thus

$$F = k \int \sin \psi\, ds + \int p_s \cos \psi\, ds, \text{ taken over } \alpha\text{-line } NK_3,$$

$$= k \int dy + \int p_s dx.$$

Now $p_s = p_{K3} + 2k\psi$, where ψ is the magnitude of the rotation from K_3 to S

and hence $\quad F = kh + 2k \int \left(\dfrac{p_{K3}}{2k} + \psi \right) \cos \psi\, ds = kh + 2k \cdot \dfrac{p_{K3}}{2k} \int dx + \int \psi\, dx.$

Therefore, $\quad F/2k = h/2 + (p_{K3}/2k)\, X + \int \psi\, dx$

and $\qquad \bar{p}/2k = (2I + h)/2X + p_{K3}/2k \qquad\qquad (12.11)$

where $I = \Sigma \psi \Delta x$ which can be taken off from Fig. 12.11 for a particular case, and $\bar{p}$ is the *mean* vertical pressure acting on the false heads at the platens.

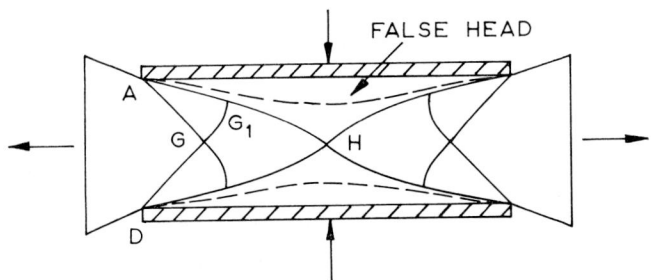

FIG. 12.15 Slip-line field for compression between rough parallel platens when $2w/2h < 3 \cdot 6$.

The mean platen pressure, p_y, is the mean value over the whole platen. Three skeleton calculations are given below, which the reader should check by reading off the quantities from Fig. 12.11 and calculating the summation term.

Fig. 12.13 shows a typical pressure distribution over a platen for $2w/2h \geqslant 5 \cdot 6$.

(a) *Case of* $2w/2h \simeq 6 \cdot 6$. (Rigid region over less than half the platen length.) Lengths are measured off from Fig. 12.11.

Line	NM_1	M_1L_2	L_2K_3
Δx	1·6	2·2	2·6
ψ radians	0·654	0·393	0·131
$\psi . \Delta x$	1·05	0·86	0·34

X	6·2
h	2·5
$p_{K3}/2k$	2·86

$\Sigma \psi \Delta x = 2 \cdot 25$

Using equation (12.10)

$$\bar{p}/2k = \frac{2 \cdot 5 + 2 \times 2 \cdot 25}{2 \times 6 \cdot 2} + 2 \cdot 85 = 3 \cdot 41.$$

The mean normal pressure on the platen, p_y, is the mean of

(i) a constant pressure over OG_3 (represented by ab)
(ii) a steadily rising pressure from G_3 to K_3 (pressure line bc) and
(iii) a mean pressure on the false head (pressure line de). Hence,

$$p_y/2k \doteqdot \frac{(1\cdot29 \times 3\cdot54) + 6\cdot78\,(1\cdot29 + 2\cdot87)/2 + 3\cdot41 \times 6\cdot2}{16\cdot5}$$

$$= \frac{4\cdot6 + 14\cdot1 + 21\cdot1}{16\cdot5} = \frac{39\cdot8}{16\cdot5} = 2\cdot41.$$

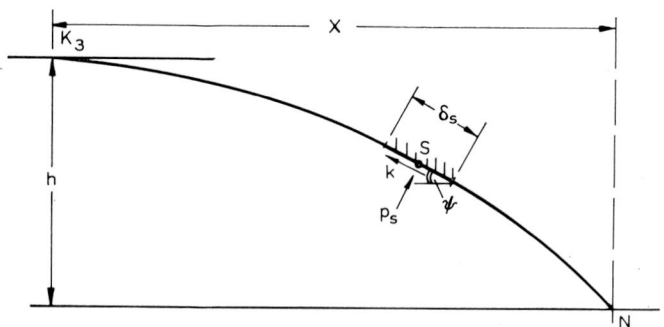

FIG. 12.16 Stresses on elemental length δs.

(b) *Case of* $2w/2h = 3\cdot6$ (Rigid region just covers whole of platen.)

Line	KJ_1	J_1H_2	H_2G_3
Δx	1·90	1·85	1·75
ψ radian	0·654	0·393	0·131
$\psi.\Delta x$	1·24	0·72	0·24

X	5·5
h	2·5
$p_{G3}/2k$	1·28

$\sum \psi \Delta x = 2\cdot2$

$$\bar{p}/2k = \frac{2\cdot5 + 2 \times 2\cdot2}{2 \times 5\cdot5} + 1\cdot28 = 1\cdot91$$

$$p_y/2k = \frac{(1\cdot29 \times 3\cdot54) + (1\cdot91 \times 5\cdot5)}{9\cdot04} = 1\cdot65.$$

(c) *Case of* $2w/2h = 1\cdot6$. (Rigid region covers whole of both platens, see Fig. 12.15.)

This instance is worked from first principles by reference to Fig. 12.11, H being taken to be the block centre.

Resolving the stresses over AG_1H parallel to Oy, or vertically,

$$w.p_y \doteq AG_1\{p_{AG_1}\sin 60° + k\cos 60°\} + G_1H\left\{\frac{p_H + p_{G_1}}{2}\sin 52\tfrac{1}{2}° + k\cos 52\tfrac{1}{2}°\right\}$$

Now
$$p_{AG_1} = k(1 + 0\cdot522) = 1\cdot522k,$$
$$p_H = k(1 + 1\cdot044) = 2\cdot044k,$$

therefore
$$p_y/k \doteq \frac{3\cdot54(1\cdot32 + 0\cdot5) + 1\cdot2(1\cdot42 + 0\cdot61)}{4\cdot0} = 1\cdot11.$$

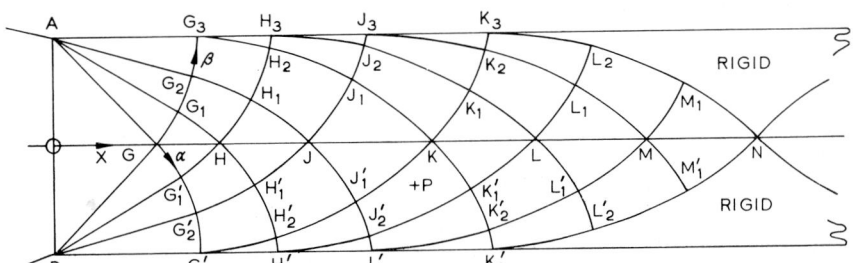

FIG. 12.17 Precise slip-line field for compression between rough parallel platens when $2w/2h \simeq 6\cdot6$.

The curve in Fig. 12·9 may be divided approximately into the three regions previously described in the caption, (i) predominantly elastic deformation OA, (ii) contained plastic deformation AB and (iii) unrestrained plastic flow flow with strain-hardening BC. The load at B is about the true yield point load. In many experiments, however, the definition of a yield load as in Fig. 12.9, is not easy to discern, and besides other things, much depends on the equipment used to identify it and the amount of pre-straining of the material. These remarks are necessary to ensure that the reader is not misled into believing that the mathematical precision of the methods of this chapter are accompanied by equally precise experimental predictions of load and the catastrophic flow behaviour envisaged by ideal models.

THE HODOGRAPH

A hodograph is a graphical representation of the speed or rate of displacement of every particle in the plane of deformation. To be able to draw such a diagram completely, is to apply a test that shows that all the elements of plastic deformation are compatible. This diagram is also immensely useful in helping to determine precisely how material elements deform. (The early sections of Chapter 13 may be usefully read before proceeding with this section.)

In Fig. 12.18, the hodograph is drawn for compression between perfectly rough plates when $2w/2h \simeq 6\cdot6$, to accord with the approximated 'slip-line' field of Fig. 12·11.

14—EP * *

N is the block centre and here a tangential velocity discontinuity of magnitude $\sqrt{2}$ at 45° to the axis is introduced, assuming the plates to approach one another at a relative speed of 2. In Fig. 12.18, Oa represents the speed of the top platen relative to the stationary centre line of the block. As material in the rigid zone adhering to the top platen crosses boundary K_3N at N, it is deflected and caused to move parallel to Ox, its velocity becoming On, Fig. 12.18. The remainder of the velocity field is easily constructed in a manner identical with that for constructing discontinuous velocity fields, see Chapter 13.

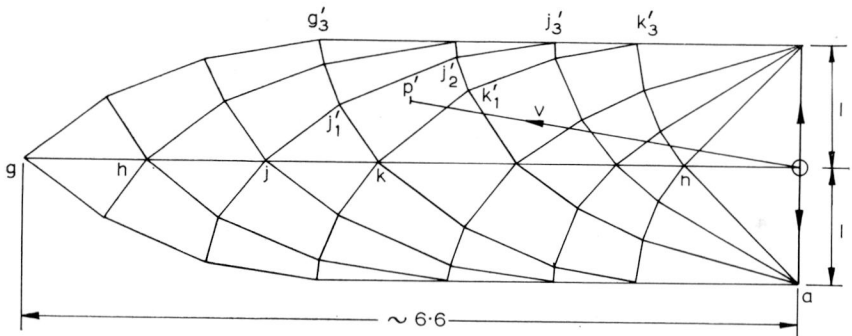

Fig. 12.18 Hodograph corresponding to approximate slip-line field of Fig. 12.11.

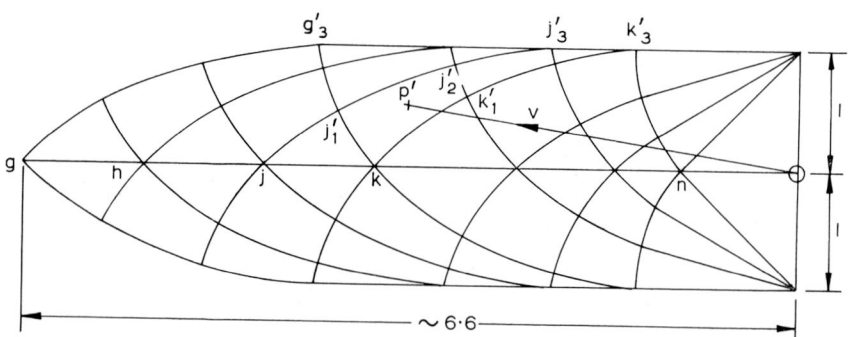

Fig. 12.19 Hodograph corresponding to slip-line field of Fig. 12.17.

Note that, strictly, *an* should be of magnitude $\sqrt{2}$; adjustment of the velocity field just constructed to become an orthogonal net that accords with all the known and assumed properties of the slip-line field is obvious; this has been done and the precise hodograph seen in Fig. 12.19 accords with the precise slip-line field of Fig. 12.17.

The mode of deformation of the material is easily obtained using either of the hodographs. If the displacement of a particle at P in Fig. 12.17 is required, corresponding to a bottom platen movement of Δh, it is necessary only to locate the corresponding point in the hodograph, i.e., p' in Figs. 12.18 or 12.19, and Op'. Δh then gives its magnitude. Op' gives the velocity of P.

It will be appreciated that as the platens approach, the slip-line field and the hodograph progressively change in shape; if a significant movement of the platens occurs, the final and precise position of P can only be obtained by taking these changes into account.

12.6 The Centred-fan Field

The slip-line field constructed from two circular arcs of equal radius whose centres are O and O' is a very frequently used form. Fig. 12.20(a) shows such

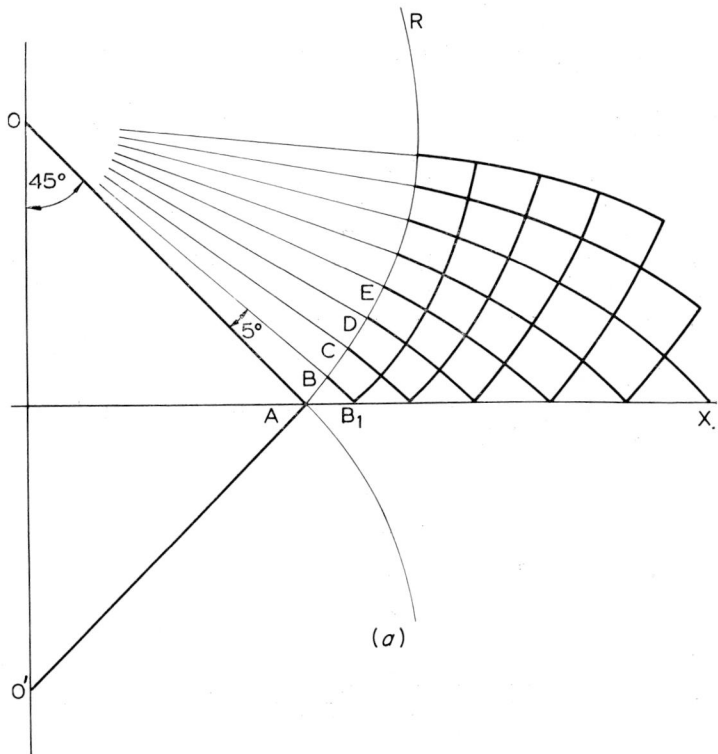

FIG. 12.20

(a) Centred-fan field: slip-line field, 5° net, defined by two equal circular arcs.

a field on a small scale. O and O' are singular points in a field from which the orthogonal net of lines spring. Let the arcs of equal radii intersect at A. It is clear that any field developed to the right of OO' will be symmetrical about the perpendicular bisector AX of OO'. Hence we will only concern ourselves with the half of the field above AX. Arc AR is divided so that arcs AB, BC, etc. subtend equal angles—5° in this instance—at O. The slip-line

OB when extended cuts AX at B_1 at 45°, which point can be found by drawing chord BB_1 at $42\frac{1}{2}$° to AX. Using the method of chords as described in the previous example, a net of considerable extent can be built up. Figure 12.20(b) is such a net—a 5° equiangular net—included here for the convenience of the reader to use in actual calculations.

12.7 Examples Using the Centred-fan Field

In this section a number of examples are given in shortened form, in which the reader should have no difficulty in filling in intermediate steps. Further details and cases, and the results of calculations will be found in papers by JOHNSON *et al.* (see References). Other examples, being applications to various forming processes, will be found in Chapter 14.

12.7.1 *Extrusion Through a Square Die Over a Smooth Container Wall;* $r > 0.5$, *Fig.* 12.21(a)

The starting slip-line AB is drawn at 45° to the centre line. ABC is a circular sector and the field is developed outwards from arc BC in the usual manner. A slip-line such as AC must ultimately meet the container wall at 45° because at a smooth wall there can be no shear stress. Also between B and E, all slip-lines must meet the axis at 45°. The general method of drawing the field by employing chords as in Fig. 12.17 may also be used here. The general field given in Fig. 12.20(b) contains, of course, the field of this figure. The region $ACDGA$ is taken to be a dead-metal zone; many experiments show that metal is deposited between the container wall and the die face.

Recognizing the force on the extruded material to the right of AB to be zero, at B the stresses must be as in Fig. 12.21(b) and the α- and β-lines must then be chosen as labelled. Thus

$$B \rightarrow C, \ \alpha\text{-line}, \ p_C + 2k(-\theta) = k$$

therefore

$$p_C = k(1 + 2\theta).$$

$$C \rightarrow P, \ \beta\text{-line}, \ p_P - 2k(\theta - \pi/4 - \psi) = p_C$$

therefore

$$p_P = k(1 + 2\theta + 2\theta - \pi/2 - 2\psi).$$

In particular, $p_D = k(1 + 4\theta - \pi)$.

On δs, an elemental length of the dead-metal zone, see Fig. 12.21(c) the horizontal thrust due to the flowing metal is

$$\Delta F = k \ \delta s \cos \psi + p_P \ \delta s . \sin \psi$$
$$= k \ \delta x + k(1 + 4\theta - \pi/2 - 2\psi) \ \delta y.$$

The total horizontal thrust exerted over AD is thus

$$F = \sum \Delta F = k \sum \delta x + k \sum (1 + 4\theta - \pi/2)\delta y - 2k \sum \psi \delta y$$
$$= kx + k(1 + 4\theta - \pi/2)y - 2k \sum \psi \delta y.$$

If the mean pressure exerted by the ram is p, then

$$\frac{p}{2k} = \frac{F}{2kH/2} = \frac{x + y(1 + 4\theta - \pi/2) - 2\sum \psi \delta y}{H}.$$

It is easy to evaluate $p/2k$ for a given reduction, $r = (H - h)/H$, by reading off the appropriate values of x, y, θ and H. The problem is only laborious in respect of the calculation of $\sum \psi . \delta y$.

12.7.2 *Extrusion Through a Smooth Wedge-shaped Die of Semi-angle α when $r \geqslant 2 \sin \alpha/(1 + 2 \sin \alpha)$, see Fig. 12.22.*

The starting line is AB, which is drawn at 45° to the slug axis. ABC is a circular sector where AC makes 45° with the die face. $C1$ also makes 45° with $A123 \ldots n$, i.e., the die face, at 1. The field is extended outwards from $BC1$ in the usual way.

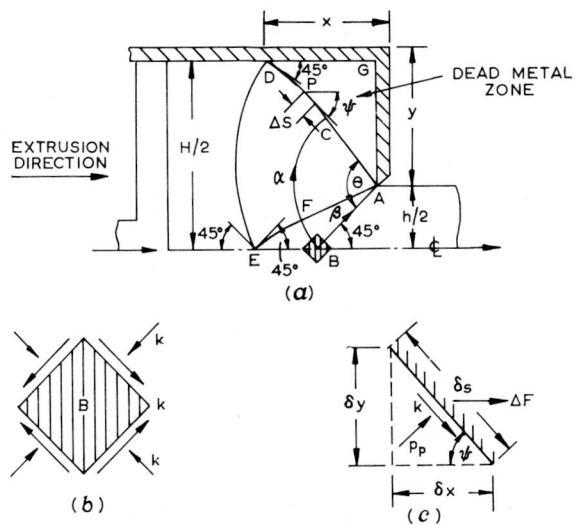

FIG. 12.21 Extrusion through a square die over a smooth container wall: $r > 0.5$.

The reader should have no difficulty in showing that the extrusion pressure, p, is given by

$$p/2k = (1 + \alpha)(h_1 - h_0) + \tfrac{1}{2}(q_1/2k + q_2/2k)(h_1 - h_2) + \ldots$$

$$+ \tfrac{1}{2}(q_{n-1}/2k + q_n/2k)(h_n - h_{n-1})$$

where the local die pressure, q, is

$$q_n/2k = 1 + \alpha + 2(n - 1)\Delta\phi,$$

$\Delta\phi$ being the angle of the slip-line field mesh employed, e.g., for a 5° mesh, $\Delta\phi = 0.0873^R$. Also, $r = (h_n - h)/h_n$.

12.7.3 *Extrusion Through a Smooth Wedge-shaped Die of Semi-angle α where $r \leqslant 2 \sin \alpha/(1 + 2 \sin \alpha)$, see Fig. 12.23(a).*

To draw this field, the isosceles triangle AED is first drawn and the arcs

CE and *EF*. These arcs consist of small equal arcs, e.g., *HE* and *EG*, subtending at their respective centres *D* and *A* equal angles; from *H* and *G*, *I* is found and the field is extended in the usual way. For a given reduction, the point *B* is chosen, being such that the slip-lines meet the axis at one point only and at angles of 45° to the axis. It will again be appreciated that the field, when drawn to scale, is implicit in Fig. 12.20(*b*).

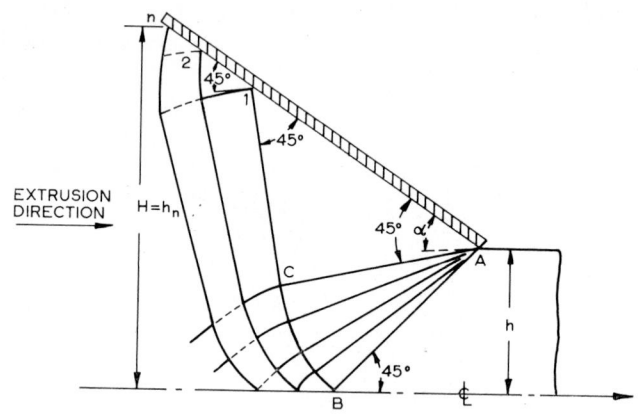

FIG. 12.22 Extrusion through a smooth wedge-shaped die of semi-angle $\alpha:r \geqslant 2 \sin \alpha/(1 + 2 \sin \alpha)$.

To determine the extrusion pressure, the pressure at a particular point in the field has first to be calculated. Let the pressure at *B* be p_B. Then $B \rightarrow P$, i.e., moving along a β-line gives

$$p_P - 2k(\psi - \pi/4) = p_B.$$

The resolved stresses on elemental length δs are as indicated in Fig. 12.23(*b*). The total force on the emerging sheet is zero, hence, from Fig. 12.23(*c*)

$$\int p_P \delta s \sin \psi - k \int \delta s \cos \psi = 0.$$

Thus,
$$\int_0^h \left(\frac{p_B}{2k} + \psi - \frac{\pi}{4}\right) dh = \frac{1}{2}\int_0^x dx$$

and
$$\frac{p_B h}{2k} = \frac{x}{2} + \frac{\pi h}{4} - \int_0^x \psi \, dh. \tag{12.12}$$

Using the Hencky equations again

$$B \rightarrow F, \ \beta\text{-line}, \ p_F = p_B + 2k\gamma$$

$$F \rightarrow E, \ \alpha\text{-line}, \ p_E = p_F + 2k\theta$$

$$= p_B + 2k(\theta + \gamma).$$

The stresses on block AED are shown in Fig. 12.23(d), and the die pressure q, is

$$q = p_E + k.$$

Therefore
$$\frac{q}{2k} = \frac{p_B}{2k} + (\theta + \gamma) + \frac{1}{2}$$

$$= \frac{x/2 - \int_0^x \psi\, dx}{h} + \frac{\pi}{4} + \theta + \gamma + \frac{1}{2}, \qquad \text{using (12.12)}$$

For a given geometrical configuration x, h, θ and γ are obvious and $\int_0^x \psi\, dx$ requires to be evaluated.

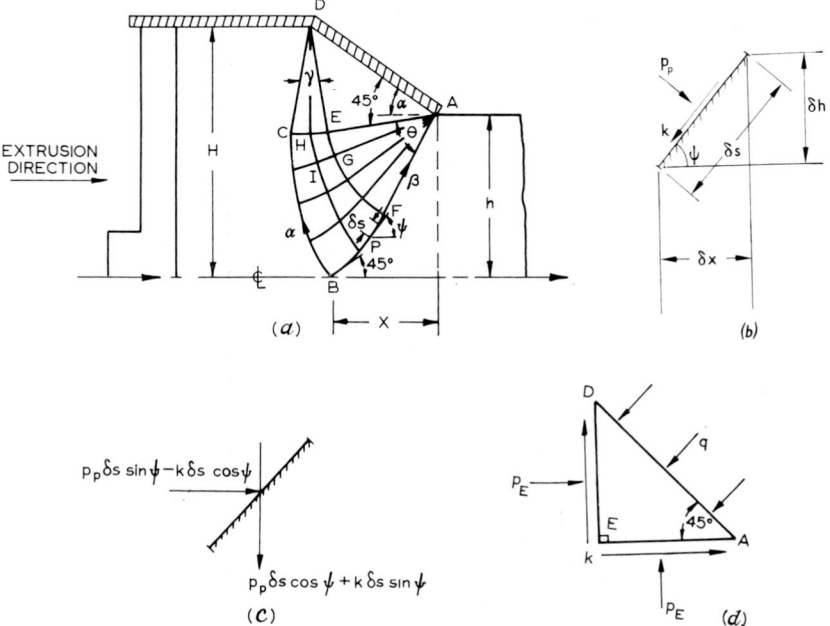

FIG. 12.23 Extrusion through a smooth wedge-shaped die of semi-angle $\alpha : r \leqslant 2 \sin \alpha/(1 + 2 \sin \alpha)$.

Hence, extrusion pressure p is $\dfrac{p}{2k} = r\left(\dfrac{q}{2k}\right)$ where $r = (H - h)/H$.

If the die face had been in some degree rough, the triangle DAE would cease to be isoceles. Whilst the angle at E still remains 90°, $D\hat{A}E$ becomes $< 45°$ say β. The coefficient of friction between the material and the die face, μ, is

$$\mu = \frac{\cos 2\beta}{q/k}.$$

It will be found then that

$$\frac{q}{2k} = \frac{\frac{x}{2} - \sum \psi \, \delta h}{h} + \frac{\pi}{4} + \theta + \gamma + \frac{\sin 2\beta}{2}$$

and
$$\frac{p}{2k} = \left(\frac{q}{2k} + \frac{\cos 2\beta}{2 \tan \alpha}\right) r.$$

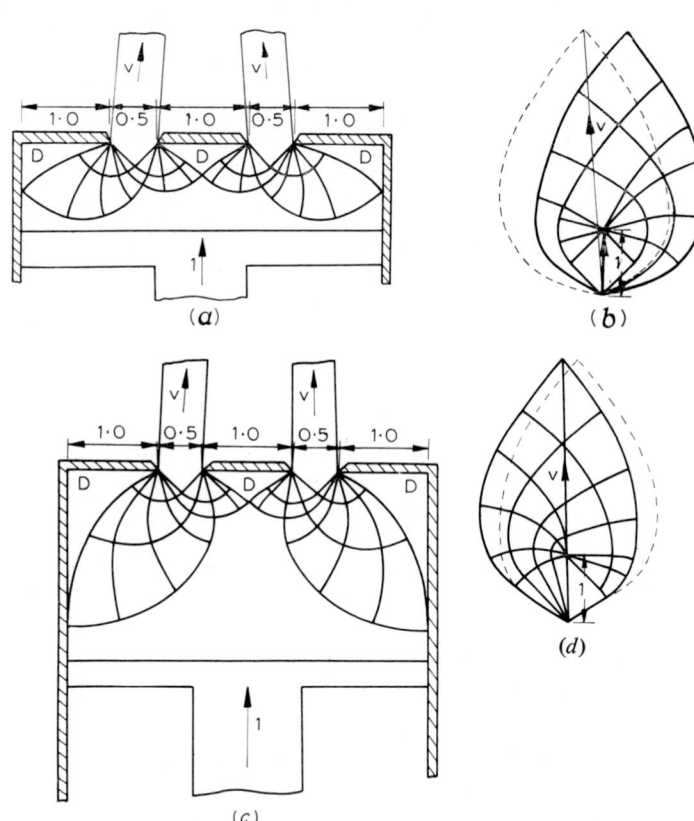

FIG. 12.24 Simultaneous extrusion through two orifices in a square die: (*a*) and (*b*), frictionless container wall, (*c*) and (*d*), shearing friction at container wall. *D*: dead metal zone.

Note that friction is similarly allowed for in Fig. 12.22; angle 1*AC* becomes some angle $\beta < 45°$. The orthogonal slip-lines, however, from point 1 up the die face, meet that face at angles which alter because the normal pressure on the die increases, if μ is constant.

12.7.4 *Simultaneous Extrusion Through Two Orifices in a Square Die*

In Figs. 12.24(*a*) and 12.24(*c*) are shown the slip-line fields for extrusion through orifices symmetrically situated about the container centre line, but asymmetrically disposed each in its own half of the die. In Fig. 12.24(*a*) the container wall is frictionless and in Fig. 12.24(c) so rough that metal shears along it; in this last case the slip-lines meet the wall at 0° and 90°. Note the three dead-metal zones over the three portions of the die face. The hodographs for each of these two cases are shown in Fig. 12.24(b) and 12.24(d). Results of calculations are given by DODEJA and JOHNSON (1957).

12.7.5 An Extrusion Involving Rotation

Figure 12.25(a) shows a *possible* slip-line field for an unusual form of extrusion, the extrusion through wedge-shaped dies of a bar of given curvature, which is introduced here simply as an example involving rotation which

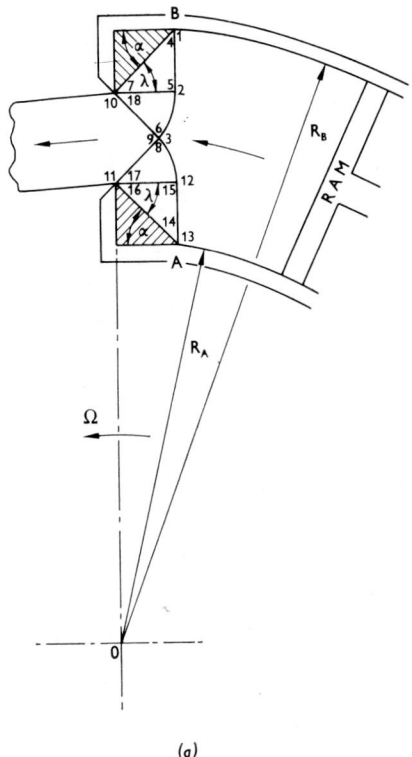

(a)

FIG. 12.25 Extrusion involving rotation.

thereby poses a problem of some complexity as regards the drawing of the hodograph. The hodograph is clearly labelled so that the velocity of points in the physical plane diagram, Fig. 12.25(a), are clearly identified in the hodograph. The situation is, essentially, a generalization of that discussed in Section 12.4 and it is not difficult to see the hodograph of Fig. 12.5(c) as a limiting case of that of Fig. 12.25(b). The pressure required for extrusion in Fig. 12.25(a) is identical with that calculated in Section 12.4, i.e., (12.10).

Note how the extrusion product emerges curved and that at exit from, as well as entry to, the die the velocity is *not* represented by a single point in the hodograph, as was the case in Fig. 12.5(*c*).

12.8 Stubby Cantilevers and Beams Carrying a Concentrated Load

In Chapter 7 the bending of cantilevers under end load and conditions of plane strain was presented, in which only a circular arc along which shearing occurred, was assumed. We refine the treatment in this section by giving a slip-line field solution which is valid over a certain range of length to thickness ratios. Confidence in the treatment here and that in Chapter 7 should reinforce one another.

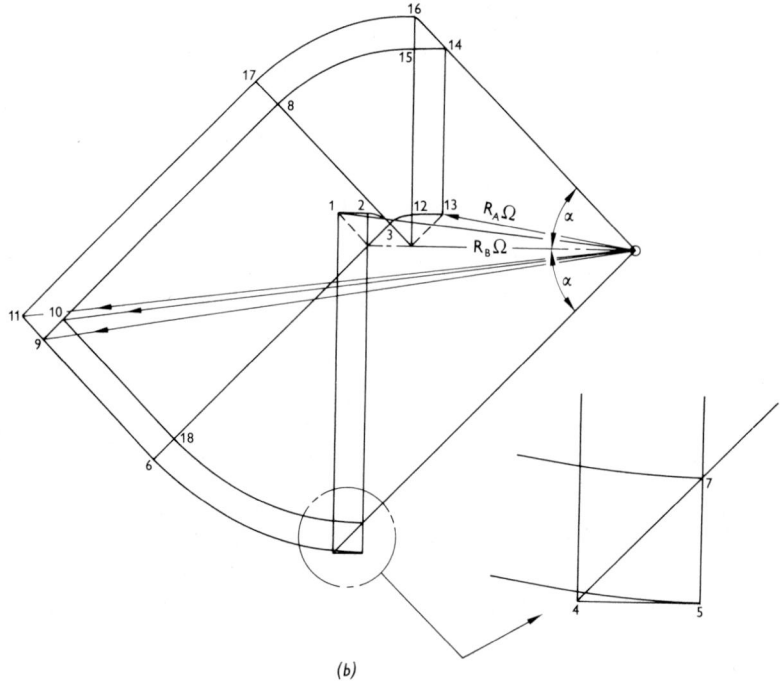

(*b*)

Fig. 12.25

This section is, further, presented because cantilevers are a prime concern of every engineer and because an example is provided of the unusual type of slip-line field usually associated with bending.

A stubby cantilever is perfectly built into a rigid wall, as shown in Fig. 12.26(*a*); it is one whose length to depth ratio is relatively small so that plastic deformation at the support occurs due to bending stresses and due to shear force in about equal degree.

The analysis of this situation, in which the cantilever carries a concentrated load *St* at its end, was first given by GREEN (1951).

Obviously yielding will be initiated at the points of stress concentration, A and B, and will grow therefrom. Choose A and B as points at which to attach isosceles triangles ACE and BDF and circular arcs ACG and BDH, respectively. The triangles will be such that the material parallel to the cantilever surfaces AE and BF are in tension and compression, respectively.

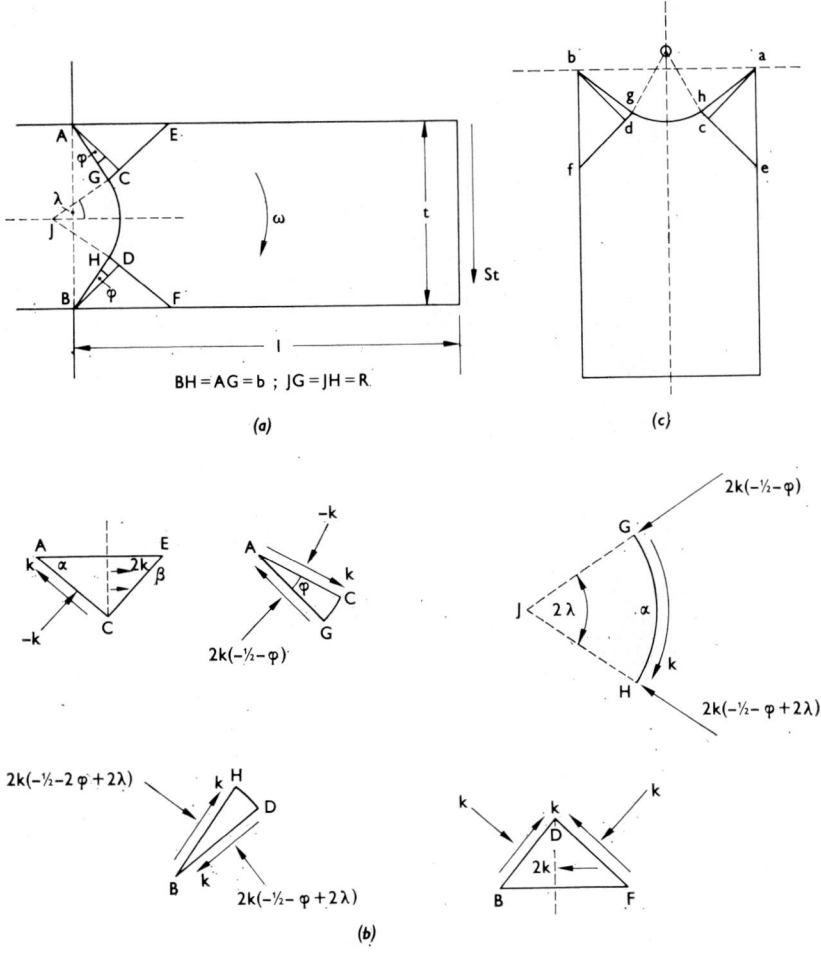

FIG. 12.26 Stubby cantilever with end load.

(a) Slip-line field.
(b) Equilibrium conditions.
(c) Hodograph.

Points G and H are joined by a circular arc of centre J; angles ϕ and λ are related because the Hencky equations, when used along EG, GH and HF, must lead from a tensile stress of $2k$ to a compressive stress of $2k$. The cantilever must (incipiently) rotate over surface GH about centre J; the material to the left of GH is stationary and that to the right rotates with angular speed ω, so that there is an angular velocity discontinuity across GH. The correctness of this field is well attested by the photograph in Fig. 12.27; etching has brought out clearly in a mild steel cantilever the region of plastic deformation.

The proportions of the slip-line field, i.e., ϕ, λ, b and R have to be found using the Hencky equations and the equations for force and moment equilibrium.

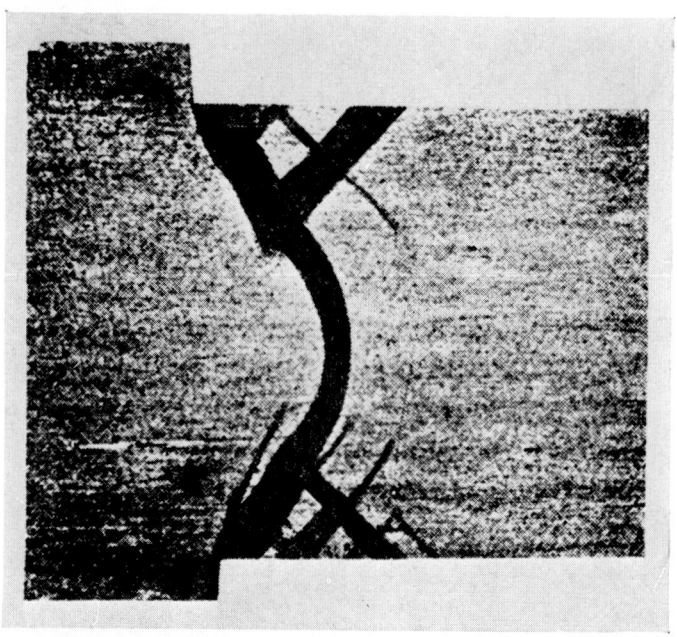

FIG. 12.27 Etched cross-section of stubby cantilever showing plastic yielding. (*After* Hundy, *Metallurgia*, 49, 109, (1954).)

For region ACE, the stress situation is as indicated in Fig. 12.26(b), and $p_c = -k$. Since ECG is a β-line, then

$$p_G - 2k\,(-\phi) = p_C = -k$$

or

$$p_G = 2k(-\tfrac{1}{2} - \phi).$$

AGH is an α-line, hence

$$p_H + 2k(-2\lambda) = p_G = 2k(-\tfrac{1}{2} - \phi)$$

or

$$p_H = 2k(-\tfrac{1}{2} - \phi + 2\lambda).$$

HDF is a β-line, hence

$$p_D - 2k(-\phi) = p_H = 2k(-\tfrac{1}{2} - \phi + 2\lambda)$$

or

$$p_D = 2k(-\tfrac{1}{2} - 2\phi + 2\lambda).$$

But, $p_D = k$, because triangle BDF is in compression and hence

$$k = 2k(-\tfrac{1}{2} - 2\phi + 2\lambda)$$

or $\qquad\qquad \lambda - \phi = \tfrac{1}{2}.$ (12.13)

But geometrically,

$$\frac{\pi}{4} + \phi + 2\lambda + \phi + \frac{\pi}{4} = \pi$$

or $\qquad\qquad \lambda + \phi = \frac{\pi}{4}.$ (12.14)

Thus, from equations (12.13) and (12.14),

$$\left.\begin{aligned}\phi &= \frac{\pi}{8} - \frac{1}{4} = 8° \ 10' \\[2mm]
\lambda &= \frac{\pi}{8} + \frac{1}{4} = 36° \ 50'.\end{aligned}\right\}$$

and (12.15)

By applying the Hencky equations and equating the resolved components of hydrostatic stress p and the shear stress k on $AGHB$ to St, it is found that,

$$\frac{St}{k} = t - 2b(1 - 2\lambda) \cos\left(\frac{\pi}{4} + \lambda\right) + 2R(2\phi \cos\phi - 2\sin\phi).$$ (12.16)

If the moment of the stresses on $AGHB$ about J is equated to that of St about J, it will be found that

$$St\left[l - b\cos\left(\frac{\pi}{4} + \lambda\right) + R\cos\phi\right] = k\left[2R^2\phi + b^2(1 - 2\lambda)\right].$$ (12.17)

A further equation is found for the depth of the cantilever as,

$$b\sin\left(\frac{\pi}{4} + \phi\right) + R\sin\lambda = \frac{t}{2}.$$ (12.18)

Equations (12.16), (12.17) and (12.18) may now be solved to give values for b and R, for given l/t ratios, and hence St. It may be verified, for example, that when $b/t = 0.4$ and $R/t = 0.3$ that $l/t = 2$ and $S/k = 0.28$.

The hodograph corresponding to Fig. 12.26(a) is shown in Fig. 12.26(c). This form of solution may be adapted for the case of a beam built-in at both ends and carrying a concentrated load W at its mid-point, see Fig. 12.28. (Note that there is no singularity, and hence no circular arc or fan, associated with bottom surface below W; there are just two triangles to which circular arcs attach.)

An interesting physical feature of this solution is shown up by experiment, see Fig. 12.29. After some degree of complete yielding of the cantilever, the material (mild steel) will have hardened; its extent is clear in Fig. 12.29 (in black). The load for further deformation in the same direction reaches a value so large that a different mode of deformation arises mainly in the form of a circular arc across the base of the cantilever, see the slip-line for an end loaded wedge in Fig. 12.30. The perceptive reader will be able to appreciate how the analysis of the cantilever, as above, may be extended to that of the pure bending of an asymmetrically notched bar, see Fig. 12.31. The predominance of the two circular arcs is clear; this feature is made use of in the next chapter.

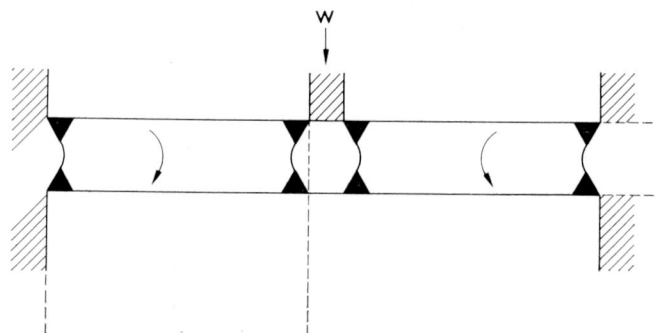

FIG. 12.28 Built-in beam with concentrated load at mid-point.

12.29 Etched cross-section of the stubby cantilever after further loading. (*After* Hundy, *Metallurgia*, **49**, 109 (1954).)

12.9 Concluding Note

The emphasis in this chapter and in Chapter 14 is placed on illustrations of how to apply basic equations and how to set about drawing a slip-line field. A complete and rigorous treatment of slip-line field theory has not been attempted. To do this it would be necessary, among other things, to establish tests to show that,

1. nowhere in the field of plastic deformation is negative work done, and
2. that outside the slip-line field in supposedly rigid material the yield criterion is nowhere violated.

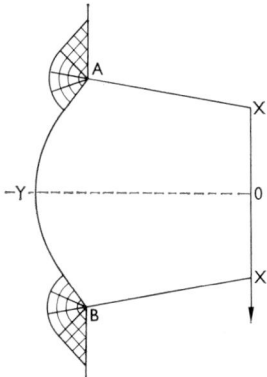

FIG. 12.30 Slip-line field for end loaded wedge.

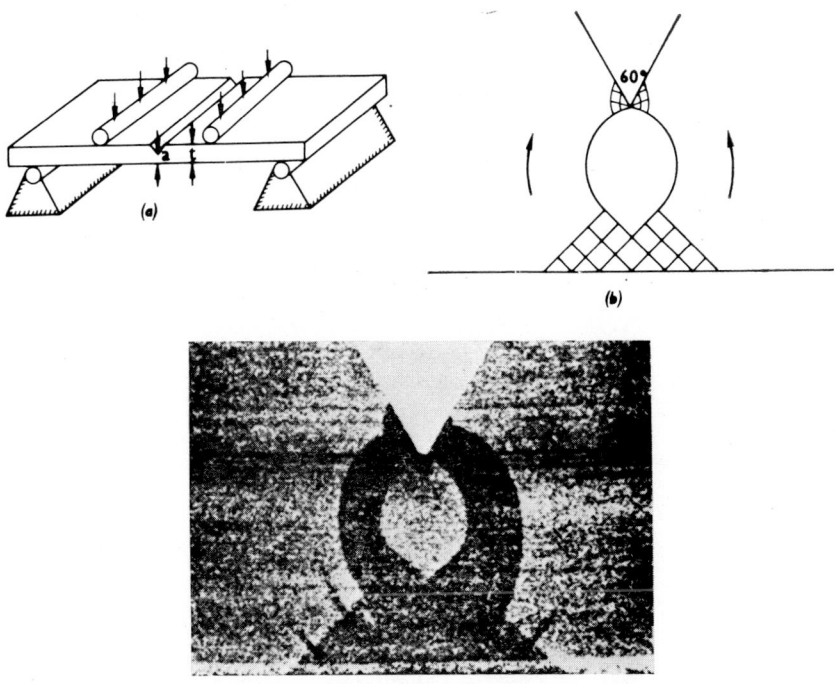

(c)

FIG. 12.31 Etched section of a notched bar bent by a couple. (*After* Green, *J. Mech. appl. Math.*)

An example of (1) is to be found in Section 14.10 but tests associated with (2) will not be given. The reader who requires to know these details will find them categorized, and ample references to the papers in which they are discussed, given in the monograph by JOHNSON, SOWERBY and HADDOW (1970).

See Problems 45, 56, 61, 63, 70, 72 and 73

REFERENCES

DODEJA, L. C. and JOHNSON, W.
1957 'On the Multiple Hole Extrusion of Sheets of Equal Thickness'
J. Mech. Phys. Solids **5**, 267

GEIRINGER, H.
1937 'Fondements mathématiques de la théorie des corps plastiques isotropes'
Mém. Sci. math. No. 86

GREEN, A. P.
1954 'A Theory of Plastic Yielding Due to the Bending of Cantilever and Fixed Ended Beams: Part I'
J. Mech. Phys. Solids **3**, 1

HENCKY, H.
1923 'Uber die einige statisch bestimmte Fälle des Gleichgewichts in plastischen Körpern'
Z. angew. Math. Mech. **3**, 241

HILL, R.
1950 *Mathematical Theory of Plasticity* Chap. 6. O.U.P.

JOHNSON, W.
1955 'Extrusion Through Wedge-shaped Dies,' Parts I and II
J. Mech. Phys. Solids **3**, 218–23 and 224–30

JOHNSON, W., MELLOR, P. B. and WOO, D. M.
1958 'Single-hole Staggered and Unequal Multi-hole Extrusions'
J. Mech. Phys. Solids 1958

JOHNSON, W. and KUDO, H.
1960 'Compression Between Inclined Plates'
Appl. Sci. Res. **9**, 206

JOHNSON, W., SOWERBY, R. S. and HADDOW, J. B.
1970 *Plane Strain Slip line Fields: Theory and Bibliography*
Arnold, London, 176 pp.

NYE, J. F.
1946 'Experiments on the Compression of a Body Between Rough Plates'
Ministry of Supply, Armament Research Dept. Rep. 39/47

PALMER, W. B. and OXLEY, P. B.
1959 'The Mechanics of Orthogonal Machining'
Proc. Instn mech. Engrs **173**, 623

PRAGER, W. and HODGE, P. G.
1951 *Theory of Perfectly Plastic Solids*
Wiley, New York, 264 pp.

Chapter 13

LOAD BOUNDING: INTRODUCTION AND APPLICATION TO PLANE STRAIN DEFORMATION PROBLEMS

13.1 Introduction

For many metal-working operations, no exact solutions for the load—which may be a force or a moment—to cause unconstrained plastic deformation are available, and accordingly means have been sought whereby this load can be established approximately. The methods that have been developed establish two values for the load, one of which is certainly an over-estimate (an *upper bound*) and the other, which is certainly an underestimate (a *lower bound*). Upper bounds are particularly valuable to mechanical or production engineers since generally they are required to estimate a load which will perform an operation—which an upper bound ensures—rather than one which will not—the lower bound. This facilitates calculations of the greatest possible load that a press or machine will be called upon to deliver or sustain.

There are several further reasons for bringing the notions of bounding techniques to the attention of engineers. They appeal to and make use of the physical intuition developed by designers; the methods, once appreciated, require little mathematical expertise. They are particularly valuable in plane-strain problems, since it is possible to estimate upper bound loads by completely graphical treatments. The results obtainable from the use of bounding techniques are usually approximate but they provide the engineer with a quantitative feel to his problem.

A warning must be given that the use of the simple plane strain upper bound, greatly exemplified in this chapter, is not always the quickest method of arriving at the best value for, say, the load in a given problem. Whilst *an* upper bound is generally easily obtainable, the least value of that bound may entail considerable algebraic and trigonometric manipulation. It may nullify the method's other advantages over the use of a slip-line field; this is often so when a general solution is sought. But if a particular solution is required, then it is probably best analysed by means of a graphical exploratory approach, as demonstrated below.

DRUCKER, GREENBERG and PRAGER (1951) stated three *limit theorems* from which the formulae for obtaining bounds can be obtained. The theorems were, however, deducible from certain work principles published by HILL (1950). The former workers and their associates stated these theorems in terms of the elastic-perfectly plastic material; Hill and his co-workers state them in terms of the rigid-perfectly plastic material. The material idealized by the former group is clearly the more realistic, the latter treatment the more rigorous.

The earliest 'proof' of the theorems of limit analysis appears to be due to GVOZDEV, and the keen student is well advised to examine the translation of Gvozdev's early paper of 1936 by HAYTHORNTHWAITE (1960).

Reasonably rigorous proofs of the limit theorems are given in the next two sections. This is, however, outside the scope of our proper intentions, and we therefore follow the section with an elementary justification of the particular form of the upper bound theorem as applied to plane strain problems.

13.1.1 *The Limit Theorems*

In Sections 13.2 and 13.3 a proof is given of two theorems which apply to rigid-perfectly plastic material. Before reading this section, however, it is suggested that the reader revises Sections 5.8 and 5.9.

In the proofs of the two theorems below, displacement increments du and plastic strain increments, $d\epsilon$ are referred to throughout. However, the theorems could as easily have been presented in terms of velocities and plastic strain rates and in terms of the rate at which work is done as opposed to an increment in work done, as was the case in Section 5.8.

13.2 **The Lower Bound Theorem**

On a body of volume V and total surface area S, let surface stresses or tractions T_i be specified over part of the surface, say S_T, and let the dis-

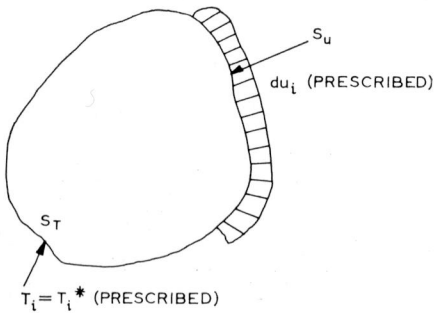

FIG. 13.1 Notation for use in proofs of theorems.

placement increments, du_i, also be specified over the portion of surface S_u, see Fig. 13.1. Further, let σ^*_{ij} denote any stress field in the body which satisfies the equilibrium equations, i.e., $\partial \sigma^*_{ij}/\partial x_i = 0$, the boundary conditions on S_T, and which nowhere violates the yield criterion. Suppose that σ_{ij} denotes an actual equilibrium stress field (which on S_T is in equilibrium with the specified T_i). The *principle of virtual work* provides the equation

$$\int_s T_i du_i.dS = \int_V \sigma_{ij} d\epsilon_{ij} dV = \int_V \sigma'_{ij} d\epsilon_{ij} dV. \tag{13.1}$$

σ'_{ij} is the deviatoric component of σ_{ij}, and the work done by the surface or external forces is equal to the plastic work dissipated in causing an increment of plastic strain, $d\epsilon_{ij}$. We may also write

$$\int_s T_i^* du_i dS = \int_V \sigma_{ij}^* d\epsilon_{ij} dV = \int_V \sigma'^*_{ij} d\epsilon_{ij} dV. \tag{13.2}$$

Associated with fictitious surface stresses, T_i^*, acting through the same incremental displacements as those undergone on the surface of the real body, is an amount of plastic or internal work equal to that associated with the fictitious equilibrium stress distribution.

Now,
$$\int_S T_i^* du_i dS = \int_{S_u} T_i^* du_i dS + \int_{S_T} T_i^* du_i dS$$

$$= \int_{S_u} T_i^* du_i dS + \int_{S_T} T_i du_i dS \tag{13.3}$$

because
$$T_i = T_i^* \quad \text{on} \quad S_T.$$

Equation (13.1) can be rewritten

$$\int_{S_T} T_i du_i dS + \int_{S_u} T_i du_i dS = \int_V \sigma'_{ij} d\epsilon_{ij} dV \tag{13.4}$$

and equation (13.2), using (13.3)

$$\int_{S_u} T_i^* du_i dS + \int_{S_T} T_i du_i dS = \int_V \sigma'^*_{ij} d\epsilon_{ij} dV. \tag{13.5}$$

Subtracting (13.5) from (13.4) gives

$$\int_{S_u} (T_i - T_i^*) du_i dS = \int_V (\sigma'_{ij} - \sigma'^*_{ij}) d\epsilon_{ij} dV \tag{13.6}$$

$$\geqslant 0, \quad \text{using equation (5.36).}$$

Thus,
$$\int_{S_u} T_i du_i dS \geqslant \int_{S_u} T_i^* du_i dS. \tag{13.7}$$

Equation (13.7), stated in words, provides the lower bound theorem. When a body is yielding and small incremental displacements are undergone, the increment of work done by the actual forces or surface tractions on S_u is greater than, or equal to, that done by the surface tractions of any other statically admissible stress field.

13.3 The Upper Bound Theorem

In the last section, the primary restriction was that the equilibrium equations should be satisfied at all points in the body. It was not essential to have regard to any restrictions that may be placed on the strain increments as derived from prescribed displacement increments. To derive this second theorem, this approach is reversed. Equilibrium equations are allowed to go unsatisfied and concern is primarily with strain increments and the conditions they have to fulfil in a fully plastic body.

Suppose the actual displacement increment field is denoted by du_i and any other different or assumed field by du_i^*, such that $du_i^* = du_i$ on S_u, i.e., du_i is prescribed over the part of the boundary S_u. Both fields are required to fulfil the incompressibility condition, i.e., $\partial u_i^*/\partial x_i = 0$, and $\partial u_i/\partial x_i = 0$—otherwise the spherical components of the stress would do work. Also this condition allows us to decide if the assumed field is a valid one, and if so then makes it possible to find the displacement increments at each point in it.

A kinematically admissible displacement increment field may have discontinuities in the tangential component along certain surfaces S_D, but normal components must be the same on either side of such surfaces in order that there be no plastic volume change.

Denote by $d\epsilon_{ij}^*$ the assumed plastic strain increments as derivable from du_i^* in the usual way.

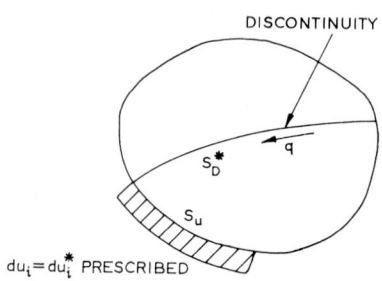

FIG. 13.2

If the principle of virtual work is applied to the kinematically admissible displacement increment field and the actual stress field, σ_{ij}, we have,

$$\int_S T_i\, du_i\, dS = \int_V \sigma_{ij}\, d\epsilon_{ij}^*\, dV + \sum \int_{S_D} q|du^*|dS_D^* \qquad (13.8)$$

where du^* denotes the discontinuity in tangential displacement increment on a surface S_D^* for the kinematically admissible displacement increment field and q is the shearing stress component of σ_{ij} in the direction of the displacement increment discontinuity. Now if σ_{ij} is a stress field, not necessarily statically admissible, derivable by way of the concept of the plastic potential from the strain increment field $d\epsilon_{ij}^*$,

$$\int_V (\sigma_{ij}^* - \sigma_{ij})\, d\epsilon_{ij}^*\, dV \geqslant 0$$

which is equation (5.36). Applying this to equation (13.8) we get

$$\int_S T_i du_i \, dS \leqslant \int_V \sigma_{ij}^* \, d\epsilon_{ij}^* \, dV + \sum \int_{S_D} k/du^*/dS_D^* \tag{13.9}$$

since $k > q$. Now

$$\int_S T_i du_i^* \, dS = \int_{S_u} T_i \, du_i \, dS_u + \int_{S_T} T_i \, du_i^* \, dS_T \tag{13.10}$$

and consequently,

$$\int_{S_u} T_i du_i \, dS_u \leqslant \int_V \sigma_{ij}^* \, d\epsilon_{ij}^* \, dV + \sum \int_{S_D} k|du^*|dS_D^* - \int_{S_T} T_i du_i^* dS_T. \tag{13.11}$$

The right-hand side of equation (13.11) gives an upper bound for the increment of work of the unknown surface tractions acting on S_u.

Since plane strain conditions only are examined in this chapter, in place of

$$\int_V \sigma_{ij}^* \, d\epsilon_{ij}^* \, dV$$

we can write

$$\int_V k \, d\gamma^* \, dV$$

where k is the yield shear stress in plane strain and $d\gamma^*$ is the maximum engineering shear strain increment. Further, we shall only treat of cases in which $d\gamma^* = 0$, i.e., modes of deformation comprised of rigid blocks of material separated by lines of tangential displacement discontinuity. Further, in every instance we examine, the term

$$\int_{S_T} T_i \, du_i^* \, dS_T = 0. \tag{13.12}$$

Thus the basic expression to be employed, from equation (13.11) is,

$$\int_{S_u} T_i \, du_i \, dS < \sum \int_{S_D^*} k|du^*|dS_D^* \tag{13.13}$$

13.4 The Upper Bound Theorem in Plane Strain: Elementary Justification

In the above sections the assumed conditions have been marked with asterisks. This differentiation between real and assumed conditions is relaxed in the following work. The problems to be analysed below are ones in plane strain and particular value attaches to the simple formula derived in this section. Consider a rigid parallelepiped of material of unit height defined by the particles at $ABCD$ in Fig. 13.3(a) which is also of unit thickness normal to the plane of the paper, moving to the left with unit speed. Since plane strain conditions are being considered, deformation is assumed to be proceeding equally and only in planes parallel to the paper. All material to the right of XX is rigid, but some time later, after crossing XX, parallelogram $ABCD$ is distorted to its new parallelogram form, $A'B'C'D'$ moving, say, in a new fixed direction at an angle α to its original direction; it will be assumed that AD is parallel to XX. The information regarding velocities is represented in the hodograph, or in a velocity plane, as Fig. 13.3(b). Choose an origin O and from it draw a vector to represent the unit velocity of the in-going material. This original unit speed may be resolved into components v_p and v_a, a velocity perpendicular to XX and a velocity parallel to XX. Next, from O draw a line at angle α to the unit vector as shown and terminate it where it is intersected by v_a produced (or by a line parallel to XX drawn

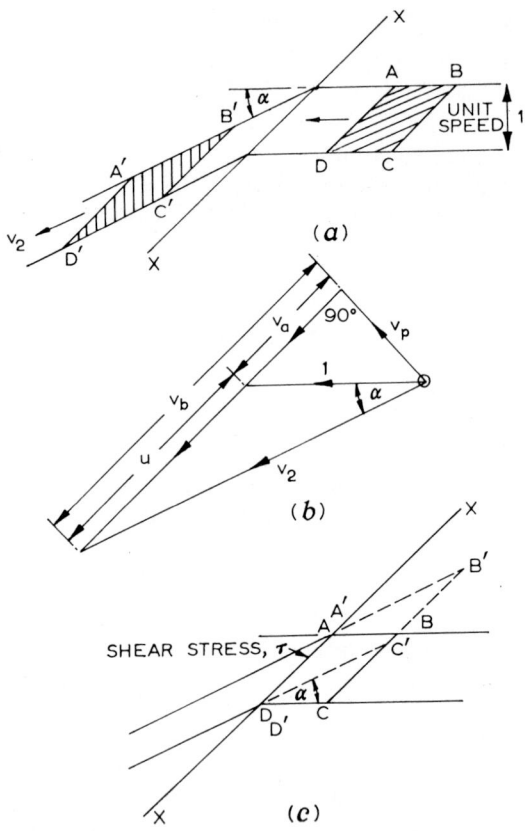

FIG. 13.3 Mode of deformation and hodograph for determining the rate of dissipation of internal energy for a plastic rigid material in plane strain.

from the end of the unit vector) to give vector v_2. The components of v_2 are now v_p and v_b; v_p is necessarily the same for both the unit speed and v_2 so that the condition of incompressibility or constancy of volume of the material, be adhered to. The difference between v_b and v_a, however, represents a 'jump' or 'velocity discontinuity' tangential to XX, i.e. u or say u_{xx}. It shows itself by changing the shape of $ABCD$ to $A'B'C'D'$; the 'jump' in reality is not, of course, instantaneous and infinitely thin across a line such as XX; however, velocity changes do occur in narrow bands, which, of course, in the limit become lines, see Section 13.5. (Here u and u_{xx} refer to velocity *change* parallel to XX but in Section 13.3, du denotes displacement increment.)

Now consider the work done in changing the block from shape $ABCD$ to $A'B'C'D'$. Since AD is initially parallel to XX, it and all parallel lines will remain so after traversing XX. Now, so place $ABCD$ and $A'B'C'D'$ that they share a common base AD, as shown in Fig. 13.3(c). If k is the yield shear stress on opposite sides of the block the work done is equal to $(k.BC).CC'$ or, expressed as a rate of dissipation of internal energy, $(k.BC).CC'/t$ where t is the time for DC to cross XX. Since the block is moving with unit speed, this may be written as $(k.BC).CC'/DC$.

Comparing triangles $C'CD$ and the hodograph, it is observed that they are similar so that $CC'/DC = u_{xx}$; the rate of energy dissipation, dW/dt, may now be written as

$$\frac{dW}{dt} = k.BC \: . \: \frac{CC'}{t} = k.BC \: . \: \frac{CC'}{DC}$$

$$= k.BC.u_{xx} = k.AD.u_{xx}.$$

If the line over which the discontinuity occurs is curved, then in place of AD we write ds and thus state

$$dW/dt = \int k u_{xx} ds \qquad (13.14)$$

the integration being performed along the line XX; note that u_{xx} would then be different at each point on XX. By dividing both sides of equation (13.13) by dt, we effectively obtain (13.14). If XX is a straight line, equation (13.14) can simply be written as

$$\dot{W} \equiv dW/dt = k.u_{xx}.s \qquad (13.15)$$

s being the length of line XX, and u_{xx} being the same at each point on it.

At no point in this discussion is there any question as to whether or not the conditions of stress equilibrium are fulfilled.

In many examples below we shall make continual use of (13.15). We shall employ systems or patterns of lines such as XX, calculate the rate at which work is done by material crossing each of them, sum these and equate the result to the rate at which work is done by some external body such as a punch. By this means we arrive at an estimate of the punch force.

The estimate which this technique provides always exceeds that which is actually required; this has been proved rigorously in Section 13.3 above. This method thus provides an upper bound or an over-estimate of the true load required to carry out the task investigated.

13.5 Examples of the Use of the Upper Bound Theorem

Before proceeding to direct applications of this theorem, it is appropriate to refer to its place in relation to slip-line fields. It may be appreciated that a slip-line field solution is incomplete in so far as no attempt is made to extend the stress field into the rigid zones; it is seldom shown that an equilibrium stress distribution satisfying the boundary conditions and not exceeding the yield-point exists in assumed rigid regions. Thus, as given, a slip-line field solution *usually* does not meet the requirements of the lower bound theorem. However, the velocity or displacement solution is always that required by the upper bound theorem, and, therefore, these kinds of incomplete solutions for plane problems are, strictly speaking, upper bound solutions.

Note: Some readers may find it easier to start with example 13.5.4.

13.5.1 *Bending of a Notched Bar* (GREEN, 1953)

Consider the bending of the notched bar, Fig. 13.4. Equal and opposite pure couples M are applied at the ends of the bar. The minimum thickness of the bar is a and the depth of the bar perpendicular to the plane of the paper is very large so that the bar bends in the plane strain.

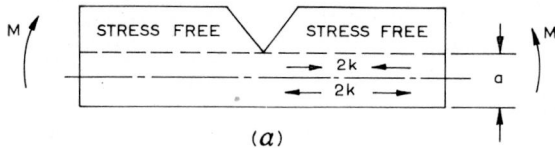

(a)

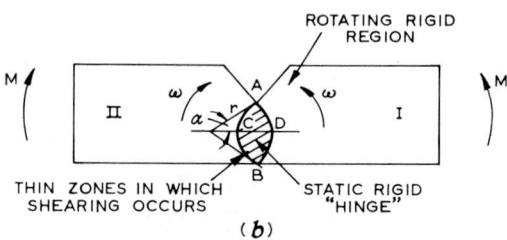

(b)

FIG. 13.4 Bending of a notched bar.

To determine the upper bound, Fig. 13.4(b), the mode of deformation is considered to be a shearing action along two circular arcs ACB and ADB of length l and radius r. The rigid outer portions of the bar rotate about the rigid hinge $ADBC$ with angular velocity ω, so that $u_{ACB} = u_{ADB} = r\omega$. The rigid arms I and II rotate about the stationary central pivot as if the joint was so 'solid' that the material actually sheared along the common planes; it will be clear that any pressure normal to the arcs of ACB and ADB would do no work. Thus the rate of internal energy dissipation from (13.15) is $2k.u_{ACB}.s = 2k.r\omega.l$; the external rate of work of the applied couples is $2M\omega$; the upper bound for the necessary load is, therefore, $M = klr$ where the product lr is chosen to make this upper bound as small as possible.

If the central angle of the circular arcs is 2α then $l = 2r\alpha$, $r = a$ cosec $\alpha/2$ and $lr = (a^2/2).\alpha/\sin^2 \alpha$ has a minimum value when $2\alpha =$ tan α or $\alpha \simeq 67°$. The upper bound is then $M = 0·69ka^2$. On p. 453 it is shown that a lower bound to the moment for this case is $M = 0·5ka^2$.

The bounds $0·5 \leqslant M/ka^2 \leqslant 0·69$ obtained by this approximate analysis may be sufficiently close for many circumstances. A slip-line field solution (GREEN, 1953) gives a value $0·63ka^2$ for an upper bound.

The reader's attention is called to the direct similarity of this case and that previously encountered in plane strain bending, see Section 7.6.

13.5.2 *Simple Indentation*

The slip-line field solution for the indentation of a semi-infinite material by a flat frictionless punch is considered in Chapter 14. At yield, material under the punch tends to move downwards and outwards causing 'piling-up' at the edges of the punch. An approximation to the mode of deformation is given in Fig. 13.5 where the triangles ABC, BCD and BDE are all equilateral. The lines AC, BC, CD, BD and DE are straight lines of tangential velocity discontinuity and the material below $ACDE$ is considered rigid.

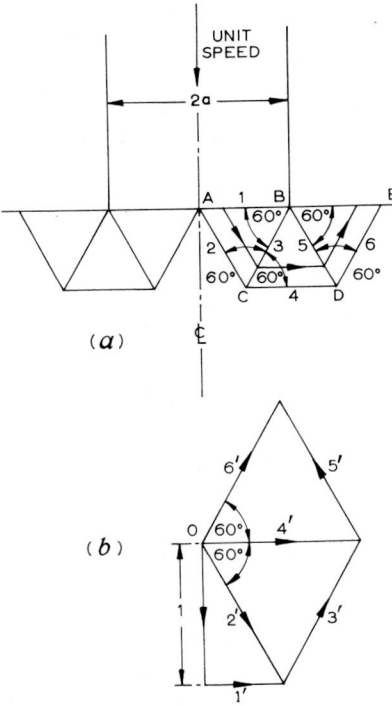

FIG. 13.5 Simple indentation.

Particles in triangle ABC are compelled to move downwards but also slide parallel to AC. As the particles reach BC they are 'sheared' and caused to move parallel to CD. Further change of direction takes place at BD when they are compelled to move parallel to DE. All material below $ACDE$ remains stationary.

If the punch is supposed to move downwards with unit speed the hodograph is as shown in Fig. 13.5(b) and the upper bound for the yield-point pressure is given after adding together $k.u_{xx}.s$ for each of the five lines.

$$p.1.a = (AC.u_{AC} + BC.u_{BC} + CD.u_{CD} + BD.u_{BD} + DE.u_{DE})k$$

where p is the load intensity, $2a$ the width of the punch, and terms such as u_{AC} denote the velocity discontinuity across AC.

Thus, $\qquad\qquad p.a.1 = 10ak/\sqrt{3}$ or $p/2k = 2.89.$ $\qquad\qquad$ (13.16)

If equal isosceles triangles had been chosen with $A\hat{C}B = C\hat{B}D = B\hat{D}E = \pi - 2\alpha$, then $p/2k = 3$ when $\alpha = \pi/4$ and the best value of $p/2k = 2\sqrt{2}$, when $\alpha = \tan^{-1}\sqrt{2}$.

13.5.3 *Compression Between Smooth Plates* (GREEN, 1951): *Heat Lines*

A rectangular block of metal, thickness $2h$, is compressed between rigid smooth parallel plates, width $2w$, moving with unit speed, Fig. 13.6(a). The overhang of metal at either side is supposed rigid.

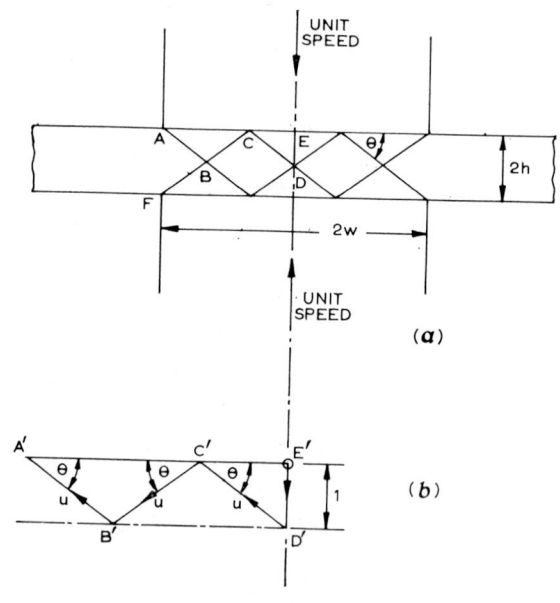

FIG. 13.6

(a) Compression between frictionless flat plates.
b) One quarter of complete hodograph.

The criss-cross of lines represent lines of discontinuity, making equal angles θ with the plates. Considering one-quarter of the system, then, from the hodograph Fig. 13.6(b), the velocity $u = \operatorname{cosec}\theta$ and the length AB is $s = w \sec \theta/n$ where n is the number of intersections of the lines of discontinuity on the horizontal centre-line of the block. The relation between $2w$ and $2h$ is given by $2h = 2w \tan \theta/n$.

From (13.15), the mean compressive pressure on the plates, p is

$$p.w.1 \leqslant k\sum[u_{xx}]s = \frac{k}{\sin \theta}.ns = \frac{k}{\sin \theta}.n.\frac{w}{n\cos \theta}$$

or $\quad p/2k \leqslant 1/\sin 2\theta \leqslant \dfrac{\cot \theta + \tan \theta}{2} \leqslant \dfrac{y/n + (n/y)}{2}$ $\qquad$ (13.17)

where $y = w/h$.

The variation of $p/2k$ with w/h is shown in Fig. 13.7 where it is compared with the slip-line field solution. The pressure oscillates and the solution is exact for the minimum value which occurs for integer values of w/h. For this latter case the lines of discontinuity meet the smooth parallel plates at angles of $\pi/4$ as would be expected. An obvious lower bound is $p = 2k$.

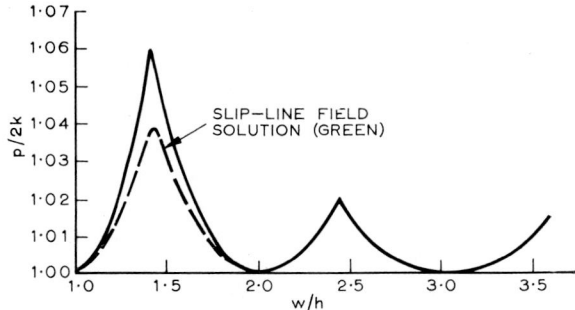

FIG. 13.7 Variation of $p/2k$ with w/h for compression between frictionless flat plates.

Consider a block in which $2w = 2h$, then from Fig. 13.7 we find that $p/2k = 1$; also this requires one simple cross. Evidently, when compression is occurring, all the work that is being done by the plates or dies is dissipated on the material passing through the lines of the cross; a rapid compression in which there is little time for heat loss due to conduction or radiation should thus cause a sharp rise in temperature in the region of the cross. Note that the heat release is confined to the immediate vicinity of the cross—it is *not* distributed uniformly. If the plates each descend with unit speed, the rate of displacement of material is $2.2w.1.1$ and this is also the rate of flow of material through the cross. Thus if ρ denotes density, J is the mechanical equivalent of heat, c is the specific heat and ΔT is the temperature rise, then

$$2k \cdot 2w \cdot 2 = \rho c \left(2 \cdot 2w \cdot 1 \cdot 1\right) \Delta T \cdot J$$

or $\qquad\qquad \Delta T = \dfrac{2k}{J\rho c}$

This assumes, of course, that all the work of compression reappears as heat, which is very nearly true. For steel at about 680 °C, ΔT lies between about

50 °C and 100 °C depending on the values taken for k, ρ and c. Thus, if a block of steel of the proportions chosen above is forged at about 680 °C (somewhat below normal forging temperature), a jump in temperature through the block should be visible and it should appear in the form of a cross. An experiment was carried out by JOHNSON, BARAYA and SLATER (1964) to test this prediction. A cross, using flat forging dies was evident and measuring the temperature rise with an optical pyrometer showed it to be about 100 °C.

These crosses or lines have been referred to as heat lines or lines of thermal discontinuity and they have been deliberately introduced at this point in the text to demonstrate the physical existence of lines of velocity discontinuity and to emphasize that they are not merely concepts. SLATER (1965/66) has written an interesting essay on these heat lines; they were apparently, first reported by MASSEY (1921).

13.5.4 *Extrusion Through Symmetrical Wedge-shaped Dies*

The slip-line field solution for the extrusion pressure for steady-state extrusion through a symmetrically placed single-hole die is, in certain cases, well established for all reductions. However, for extrusion through unsymmetrical or multi-hole dies, the slip-line field solutions involve very lengthy numerical procedures and the possible combinations of variables are too many to make a general tabulation of the results practicable. It is therefore worth while obtaining approximations to the extrusion pressure by assuming that the deformation zone is made up of straight-line discontinuities. This method gives values of extrusion pressures in excess of those predicted by slip-line fields, but it has the advantage of being of a straightforward graphical nature and is easy and quick to apply. It is therefore useful to engineers whose interest centres on the greatest loads likely to be encountered for a particular operation and set of conditions. Further, it gives a clear, if simplified, picture of the type of deformation taking place.

This section and Sections 13.5.5 to 13.5.8 deal with steady-state extrusion, but in Sections 13.5.9 and 13.5.10 it is shown that the method can easily be applied to a non-steady state.

The extrusion container and die for symmetrical deformation are shown in Fig. 13.8(*a*). The die is assumed to be perfectly smooth.

Consider the discontinuous velocity pattern shown, the velocity discontinuities occurring along AC, $A'C$, BC and $B'C$. Material to the right of BCB' is moving to the left at unit speed as a rigid block to which the uniform pressure p, is applied along FG. As the punch advances WX decreases in length and as each particle on WX crosses CB its path is instantaneously altered so that it proceeds parallel to the die face AB; at Y, the velocity of the particle is again instantaneously altered and it proceeds normal to the orifice, i.e., along YZ.

In Fig. 13.8(*b*) the hodograph appropriate to the field of Fig. 13.8(*a*) is shown. OP represents the velocity of particles in the rigid region adjacent to the punch, $CBFGB'$, assumed to be unity. When the particles from this mass encounter CB, they undergo a sudden change in velocity, u_{BC}, parallel to CB (so that from P a line is drawn at angle ϕ from OP produced) which causes them to move parallel to the die at angle α. Thus from O, a line OQ, u_{BA}, is drawn to complete the velocity triangle OPQ. This velocity u_{BA} is constant until CA is encountered, when it is altered, and the particles proceed parallel to their former direction. Thus from Q, Fig. 13.8(b), a line QP, u_{CA}, is drawn at angle θ to cut OP produced at R. OR is the speed, H/h, of the emerging sheet. The hodograph is completed by the reflection of triangle $OQ'R$ in OR for triangle $A'CB'$ of Fig. 13.8(a).

An upper bound to the extrusion force is, therefore, given by using (13.15) twice over as,

$$p.1.H = (BC.u_{BC} + CA.u_{CA})\, k, \tag{13.18}$$

there being an internal dissipation of work along the lines AC, $A'C$, BC and $B'C$ only, and none along the die faces since they are perfectly smooth.

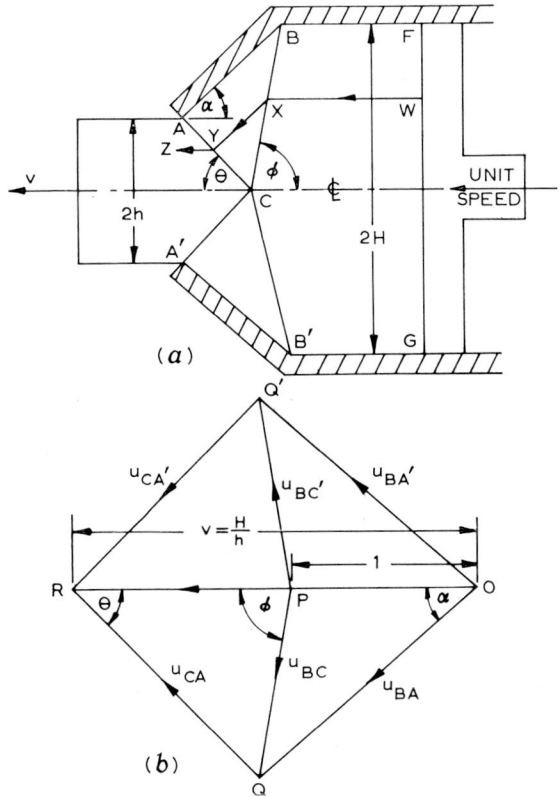

(a)

(b)

FIG. 13.8 Extrusion through symmetrical wedge-shaped dies.

The above equation yields a certain value of p depending on the value chosen for θ; see Fig. 13.9. The best estimate of $p/2k$ is the least value, so that by assuming various values of θ, the least value of p can be found.

Fig. 13.8(*a*) is a scale diagram for $\alpha = 40°$, $\theta \simeq 45°$ and fractional reduction $r = 0.57$. Fig. 13.8(*b*) is the hodograph of this system again drawn to scale. By taking off the various lengths from Fig. 13.8(*a*) and (*b*), the value of *p* can easily be calculated and is found to be 1·01. (The slip-line field solution gives the value of $p/2k$ as 0·95). The lower bound for this case is given on p. 456.

For small reductions the estimate for the pressure can be improved by choosing a velocity discontinuity pattern of the kind shown in Fig. 13.10(*a*) which has the hodograph of Fig. 13.10(*b*). Figs. 13.10(*a*) and (*b*) are drawn to scale for a smooth die of semi-angle 15° and reduction 0·1. The triangle on the die face has been taken to have angle $\beta = 45°$, the other positions and inclinations being chosen arbitrarily. The value of $p/2k$ is found to be 0·244 whereas the slip-line field solution is 0·209. It is possible to guess and find a different shape of field which will improve on the figure 0·244.

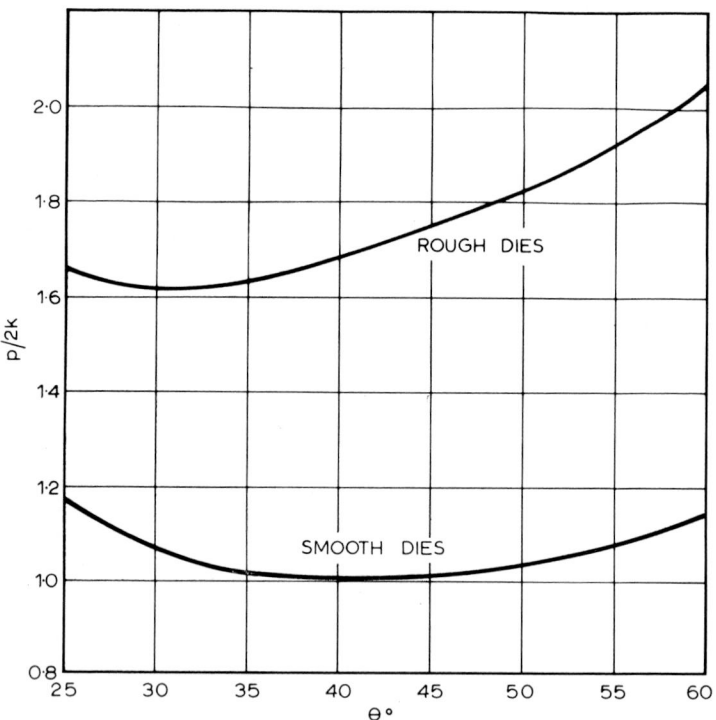

FIG. 13.9 Variation of $p/2k$ with θ with symmetrical wedge-shaped dies.

Suppose now that the die of Fig. 13.8(*a*) has a perfectly rough die face; that is one along which the material shears. Internal work is then dissipated along the die faces AB and $A'B'$ and another term, $AB.u_{AB}.k$, has to be added to the expression for total dissipation of internal energy. If this is done, it is found, Fig. 13.9, that the minimum value of $p/2k$ is 1·62. The slip-line field solution gives a value 1·48.

13.5.5 *Extrusion Through Unsymmetrical Wedge-shaped Dies*

The case of extrusion through smooth unsymmetrical wedge-shaped dies is shown in Fig. 13.11(*a*), in which the two unequal wedge angles are α and β. In this particular example, α = 40° and β = 20°, the reduction $(H - h)/H$ is 0·5 and eccentricity e, defined as the ratio $(NM - N'M)/(NM + N'M)$ where M is the mid-point of the orifice AA', is 0·161. The lines AC and $A'C$ have been

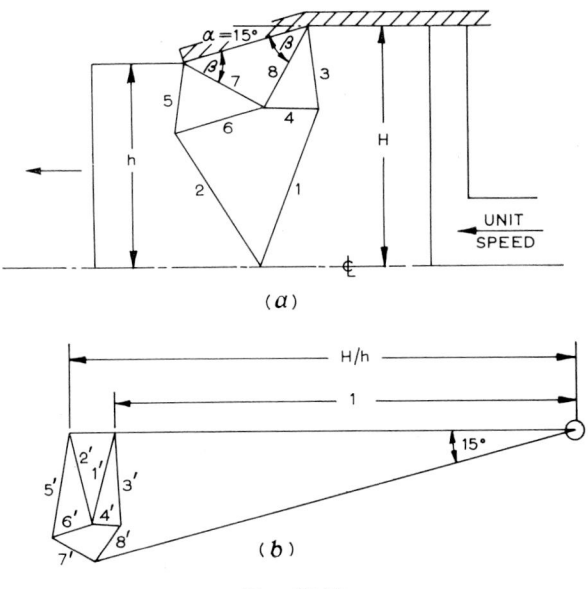

(*a*)

(*b*)

FIG. 13.10

(*a*) Pattern of velocity discontinuities for extrusion through a frictionless wedge-shaped die of small reduction.

(*b*) Hodograph.

chosen to be of equal length and are thus equally inclined to the direction of the punch travel; but this need not necessarily be the case. It will be observed from the hodograph, Fig. 13.11(*b*) that the predicted angle between the direction of emission of the sheet, and the direction of punch travel η is 8°. Fig. 13.12 shows the variation of $p/2k$ and η with θ; $p/2k$ is a minimum at 0·79 when $\theta = 42\frac{1}{2}°$. The slip-line field solution gives a value for $p/2k$ almost identical with this and $\eta = 9°$.

The process has been repeated for various combinations of die face friction and the results are also plotted in Fig. 13.12. The minimum values of $p/2k$ for all cases are shown on p. 430.

	$p/2k$	η°	θ°
1. Both die faces smooth	0·79	8	42·5
2. Top face smooth, bottom face rough	1·10	7	39
3. Top face rough, bottom face smooth	1·18	6	35
4. Both die faces rough	1·45	6	32

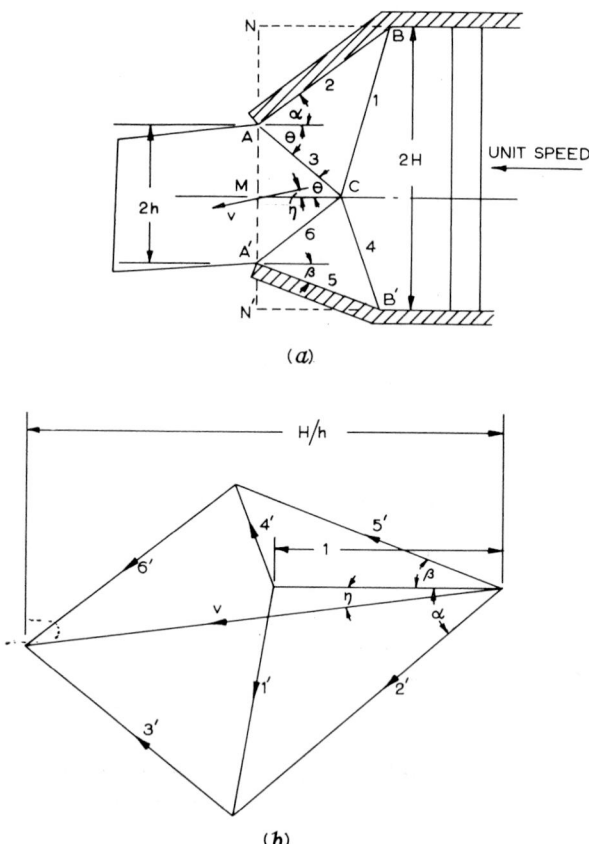

(a)

(b)

FIG. 13.11. Extrusion through unsymmetrical wedge-shaped dies.

It should be noted that these solutions only permit the consideration of interfaces which are perfectly smooth or perfectly rough, and also that it has been assumed in all cases above that there is no friction between the material and the container wall.

13.5.6 *Extrusion Through Square Dies*

The square die is the particular case of a wedge-shaped die when $\alpha = \pi/2$. Two cases will be considered: (a) where the die is frictionless and with free flow of material along it, and (b) where a dead-metal zone extends across the whole die face.

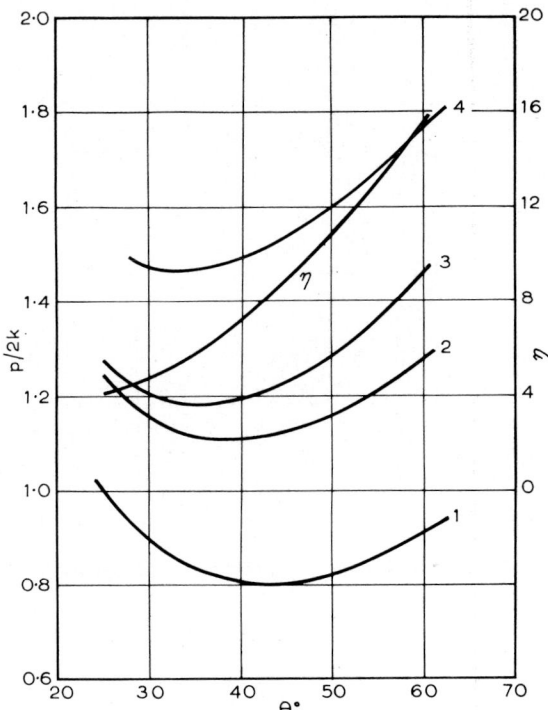

1. BOTH FACES SMOOTH.
2. TOP FACE SMOOTH, BOTTOM FACE ROUGH.
3. TOP FACE ROUGH, BOTTOM FACE SMOOTH.
4. BOTH FACES ROUGH.

FIG. 13.12 Variation of $p/2k$ and η with θ for unsymmetrical wedge-shaped dies.

When the face of the die is smooth, Fig. 13.13(*a*), the hodograph is that of Fig. 13.13(*b*). If $H/h = x$ is given, with θ the only independent variable, then

$$x.p.1 = (BC.u_{BC} + AC.u_{AC})k$$

$$= (x + 1)k \left[\frac{1 + x \tan^2 \theta}{x \tan \theta} \right].$$

The value of θ for p to be a minimum is $\tan^{-1}(1/\sqrt{x})$, which, substituted into the above equation, gives

$$p/2k = (x + 1)/\sqrt{x}. \tag{13.19}$$

15—EP * *

The variation of $p/2k$ with x is shown in Fig. 13.14 where it is compared with values given by slip-line field solutions. This simple triangle of velocity discontinuities over-estimates the pressure given by the slip-line field solution by about 30 per cent.

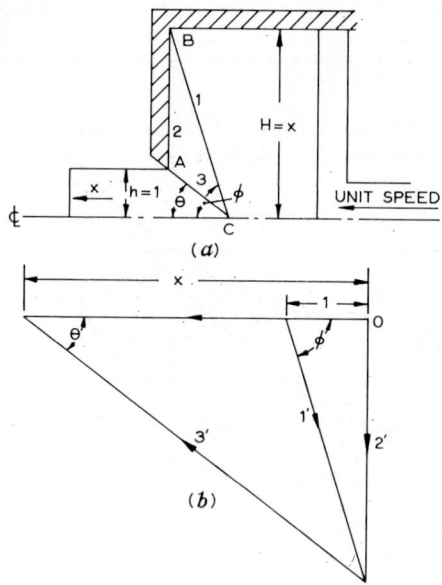

FIG. 13.13 Extrusion through a square die with sliding over the die face.

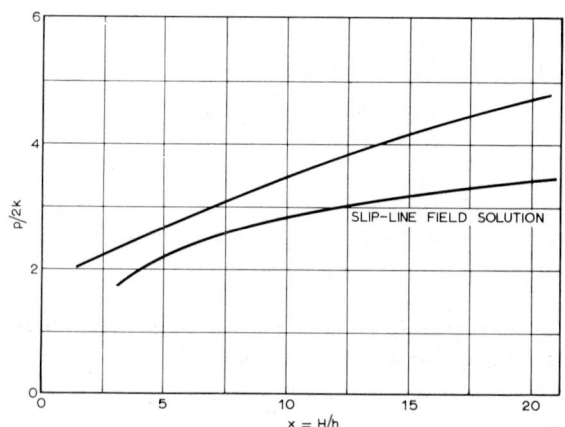

FIG. 13.14 Variation of $p/2k$ with x for extrusion through a square die with sliding over the die face.

The most important practical case arises when a dead-metal zone is formed on the die face. For moderate reductions, it is an experimental fact that dead-metal zones cover the whole of the die face and so, in effect convert the square die into a perfectly rough curved die. Let us begin by assuming that the dead-metal zone is simply a perfectly rough wedge-shaped die of unspecified angle. Suppose the reduction is given, then by assuming various values of wedge angle, the procedure outlined in sub-section 13.5.4. may be followed to determine the minimum value of $p/2k$ for each wedge angle selected. The least of these minima may then be accepted as giving an upper bound to the extrusion pressure for a square die of that particular reduction.

This procedure has been followed for the case $r = 0.75$ as shown in Fig. 13.15(a) for assumed wedge angles α having values 45°, 60° and 75° and

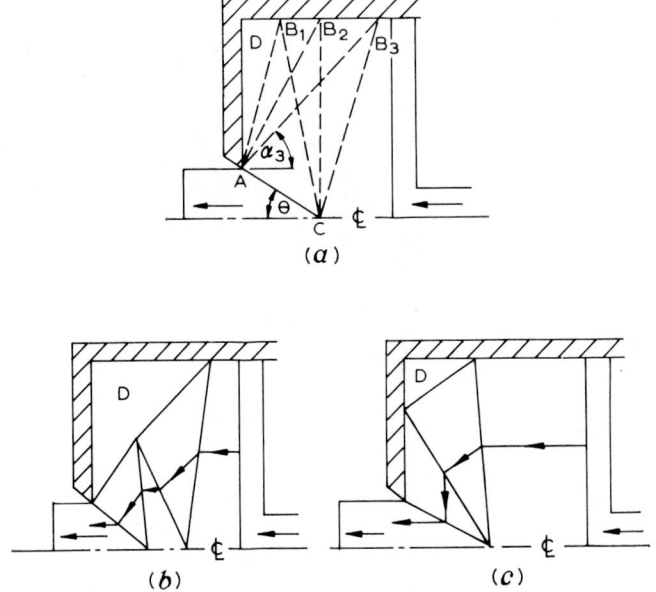

FIG. 13.15 Patterns of velocity discontinuities for square dies with a dead-metal zone on the die face.

for exit velocity discontinuity lines at θ to the slug axis such that $15° \leqslant \theta \leqslant 60°$. The variation of $p/2k$ with θ for these cases is shown in Fig. 13.16. The minimum value is about 2·55 which may be compared with the slip-line field solution of 2·48.

It is likely that more accurate estimates of the pressure, particularly for very large reductions may require modified velocity field patterns. Two such patterns are shown in Figs. 13.15(b) and (c). The latter pattern assumes what indeed appears to be the case for very large reductions, namely that the dead-metal zones lodge in the topmost corner at the junction of the face and container wall, the metal actually shearing along the lower part of the face.

13.5.7 *End Extrusion Through Three Holes* (JOHNSON, MELLOR and WOO, 1958).

End extrusion through three holes where the centre hole is larger than the two equal outer holes is shown in the half-section diagram, Fig. 13.17(*a*). Dead-metal zones, *D*, are assumed to be located in the topmost corner and on each intermediate portion of the die face. In Fig. 13.17(*a*) the symmetrical dead-metal zone assumed on the die face between *OA* and *O'A* is an isosceles triangle but an unsymmetrical triangle may, if so desired, be investigated.

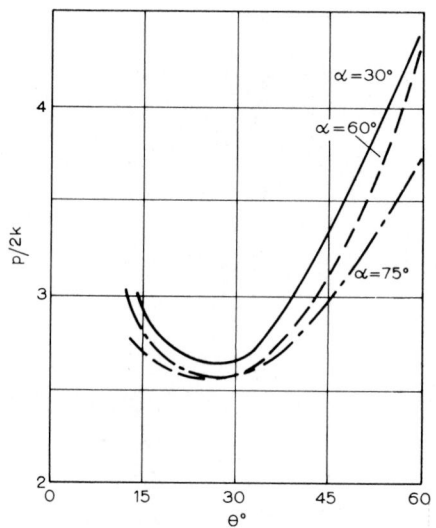

FIG. 13.16 Variation of $p/2k$ with θ for square dies with dead-metal zone on die face.

The value of $p/2k$ for the case drawn out in Fig. 13.17(*a*) is found to be 1·95 and, for an instance in which *OA* is perpendicular to *O'A* and *OA* less than *O'A*, 2·08. The total reduction in this example is 0·575 and if there had been one symmetrically placed orifice only, giving this particular reduction, and if the container wall was perfectly rough, $p/2k$ would be 1·83. Fig. 13.17(*b*) is the hodograph to Fig. 13.17(*a*).

13.5.8 *Sideways Extrusion from a Smooth Container*

In this case the pattern of velocity discontinuities and the hodograph are as shown in Fig. 13.18. The exit velocity discontinuities were taken to be at 45° to the plane of the orifice and those meeting the container wall also at 45°. The other lines were selected arbitrarily. The value of $p/2k$ for this particular example was found to be 1·26. Slip-line field calculations give a value of 1·20.

13.5.9 *Simultaneous Forward and Backward Extrusion of Sheets of Equal Thickness*

In Fig. 13.19(a) a slug of thickness x is subject to the compressive action of the punch and die face. The velocity discontinuities extend throughout the material between punch and die face, and since the thickness x is continually changing, so must the shape of the velocity pattern change; the state is an

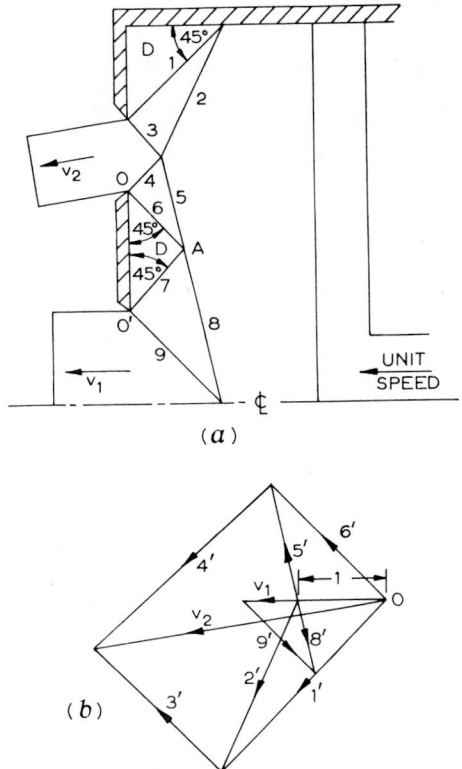

FIG. 13.17 End extrusion through three holes.

unsteady one. The resulting two extrusions of equal reduction, r, move with absolute speeds v_1 and v_2 in opposite directions. It is supposed that small zones of dead metal, D, are located in the corners as indicated, and that there is frictionless sliding of the material over the lower faces of the fixed and moving dies.

Only one 'cross' or intersection is shown in the discontinuous velocity pattern of Fig. 13.19(a) and in the corresponding hodograph of Fig. 13.19(b), but the analysis will be stated in terms of n equal crosses. If p denotes the uniform pressure over EF, then since the discontinuous field must obviously be symmetrical

$$p/2k = s_1 . u_1 + 2(n + \tfrac{1}{2})s_2 . u_2$$

where in Fig. 13.19(a), $s_1 = OA$, and $s_2 = OK$ and in Fig. 13.19(b) $u_1 = HJ$ and $u_2 = OG$.

The above equation may now be written

$$\frac{p}{2k} = \frac{r^2}{x}\left[\frac{(1-r)}{r} + \frac{1}{2n+1}\right] + \frac{x}{4}\left[\frac{r}{(1-r)} + (2n+1)\right] \quad (13.20)$$

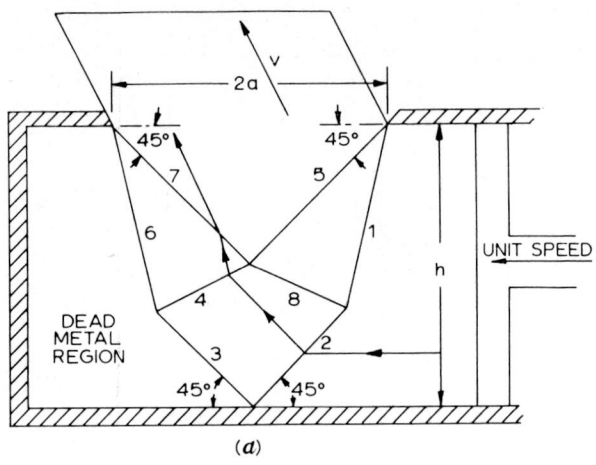

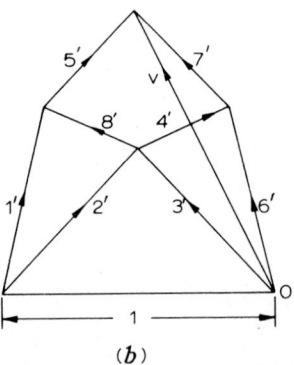

FIG. 13.18 Sideways extrusion from a smooth container.

and in particular if $n = 0$, i.e., when the dead-metal zones extend over the whole die face

$$p/2k = r/x + x/4\,(1-r). \quad (13.21)$$

13.5.10 *Simultaneous Piercing by Opposed Punches*

Figure 13.20 shows the cross-sectional view of the case in which an initially solid slug is subjected to a piercing operation by two equal size rams or punches moving in opposite directions at the same speed. (This type of industrial operation, where the slug is cylindrical, is commonly used for the production of tubular components, e.g., gudgeon pins.)

If the material forming the base to both residual common sheets is conceived to be divided-up into a number of equal size 'crosses' then, as in

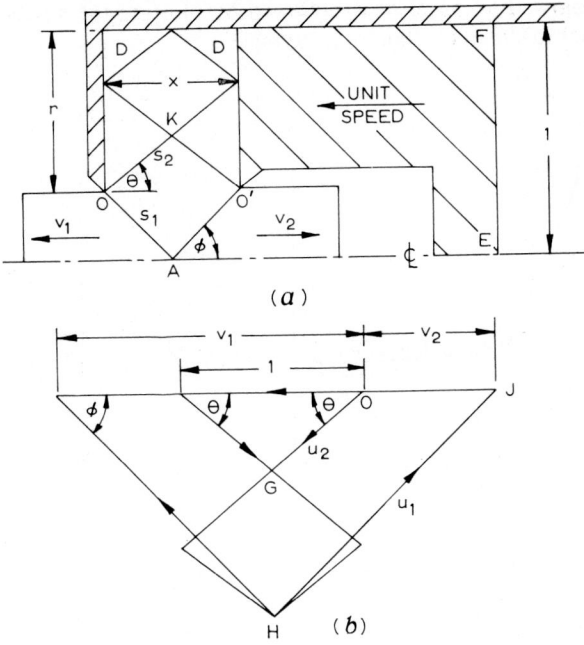

FIG. 13.19 Simultaneous forward and backward extrusion.

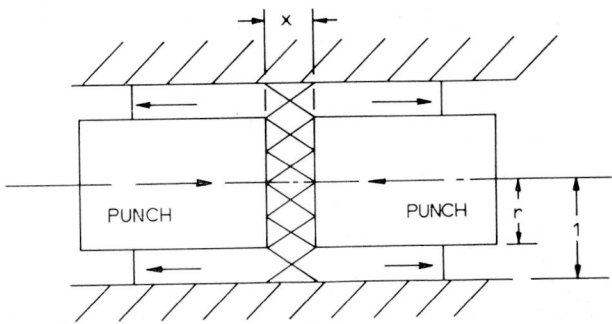

FIG. 13.20 Simultaneous piercing by opposed punches.

sub-section 13.5.9, it is easily shown that if the ram faces are both perfectly smooth

$$\frac{p}{2k} = \frac{r}{x}\left[\frac{(1-r)}{r} + \frac{1}{(2n+1)}\right] + \frac{x}{4r}\left[\frac{r}{(1-r)} + (2n+1)\right] \qquad (13.22)$$

where p is the average pressure required on the rams. To obtain the best estimate of p for a given value of r, that value of n is chosen which makes $p/2k$ least.

Figure 13.21(a) shows the variation of $p/2k$ with x when $r = 0.75$ for various values of n. The least value of $p/2k$ for each value of n chosen is tabulated below. Fig. 13.21(b) is obtained by plotting the smallest values of $p/2k$, for each particular value of x, against x.

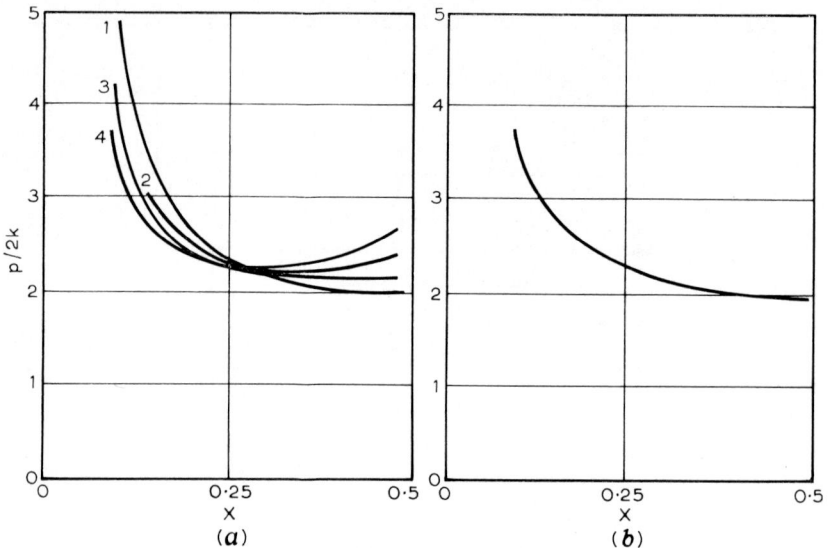

FIG. 13.21

(a) Variation of $p/2k$ with x for various crosses, $n = 1$ to 4.
(b) Variation of least values of $p/2k$, for any value of n, with x.

x	0·1	0·25	0·50
n	4	3	1
$p/2k$	3·74	2·26	2·0

If the punch faces are so rough that material sliding over them shears, then further terms would have to be added to the right-hand side of the above equation (13.22).

13.6 Friction

Throughout the examples discussed, it will have been observed that friction, other than shearing friction, has not been considered; indeed, it cannot be introduced directly, since the methods employed nowhere indicate how the pressure normal to a face over which material is sliding, may be calculated. Two methods may be noted for circumventing this. First, a load or pressure may be calculated as in the example shown in Fig. 13.8(a), for frictionless extrusion, and from this the pressure, q, on the die face may be deduced. Assuming q not to vary with the introduction of friction (having a coefficient μ) the extrusion pressure may then be recalculated; in this example the extrusion pressure with friction present would be $(1 + \mu \cot \alpha)$ times as great as that without friction. (This does not of course mean that an upper bound has been obtained.)

A second approach is to introduce a constant c, such that at the sliding interface, instead of k, ck is substituted. The calculations could then proceed in the usual way. The constant c can then be regarded as being analogous to μ. Its drawback in several cases would be the fact that the value of c varied along the interface, ie., the normal pressure on the interface varied from point to point along it. However, in many operations constant values for μ or k are assumed which cannot be realized due to the differential effects, throughout the body of temperature and speed or strain rate.

13.7 Closed Die Forging (JOHNSON, 1958a)

Figure 13.22(a) shows the forging of a block of metal of rectangular section, the vertical compression of the dies causing a sideways expulsion of the metal. A system of tangential velocity discontinuities is shown and the corresponding hodograph in Fig. 13.22(b). The reader should easily calculate the forging load.

13.8 Extrusion through Curved Dies (JOHNSON, 1958b)

Only a few slip-line fields have been given for extrusion through curved dies and some interest therefore attaches to the calculation of extrusion pressures by the method of upper bounds.

Consider an extrusion through a frictionless circular die which turns through 90°, i.e., from OA to OB; the case is shown in Fig. 13.23(a). The reduction is taken to be 0·5 and the die orifice is perpendicular to the direction of travel of the material. Assume a pattern of velocity discontinuities composed of the quadrant BC similar to BA, and the quadrant AC centre B. The hodograph is shown as Fig. 13.23(b); material is moved to the left by the punch with unit speed across line AC and into region ACB, where it rotates as part of the rigid mass ACB on arc AB and until, on crossing BC, it is subject to a second velocity discontinuity which leaves it with a translational speed of two. At A, the tangential velocity discontinuity administered, $\overrightarrow{ca}$ must cause particles to move at right angles to the slug axis along the die face with a speed of unity, $\overrightarrow{oa}$, in this instance. Since ABC is a rigid mass, the speed of the particle must remain constant, though its direction when it reaches B is parallel to the slug axis. The tangential velocity discontinuity applied at B must be such that the final speed is two, so that its magnitude is unity and its direction necessarily parallel to the axis, i.e., $\overrightarrow{cd}$. At C, the magnitude of the first tangential velocity discontinuity must again be $\sqrt{2}$ in order that the second tangential velocity discontinuity may be perpendicular to the axis and of magnitude unity, i.e., $\overrightarrow{bd}$. To each point in the region ABC there is a corresponding point in the region abc of the hodograph, and the angular velocity for all points is two radians per unit time.

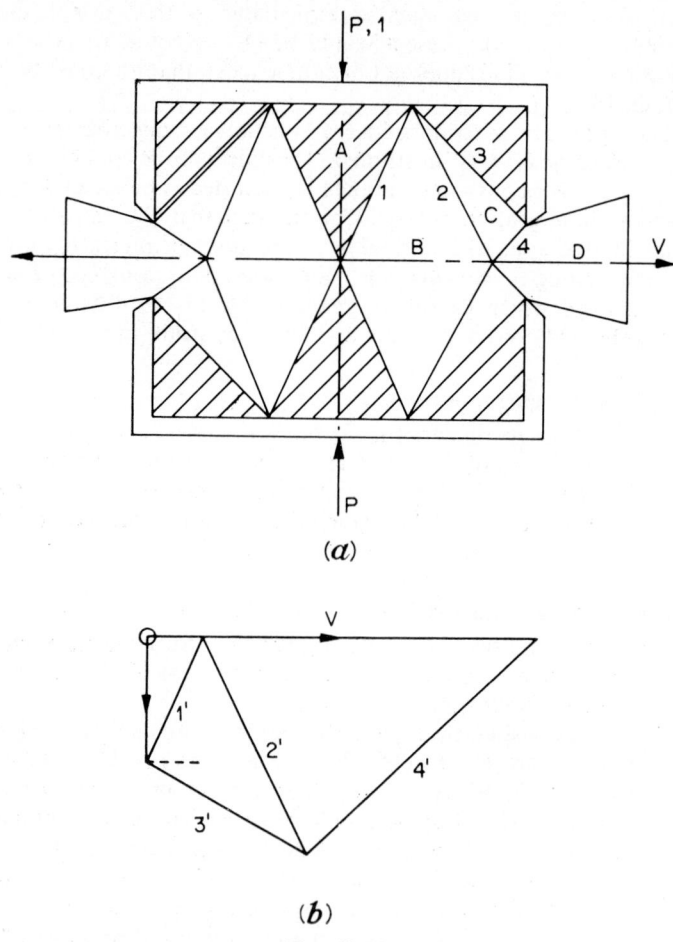

(a)

(b)

FIG. 13.22

(a) Forging of a rectangular block, showing pattern of velocity discontinuities; shaded regions denote dead metal.
(b) Hodograph for (a): quarter of complete pattern shown.

An upper bound to the extrusion pressure p is, from Figs. 13.23(a) and (b), given by

$$p/2k = [(\text{arc } CA)(ca) + (\text{arc } CB)(cd)]/2 = 3\pi/8 \simeq 1\cdot 18.$$

The circular die whose reduction is not 0·5 and whose exit and entry angles are α_1 and α_2 requires the diagram of velocity discontinuities and the hodograph to be constructed concurrently. After choosing arbitrarily angle ϕ, see Figs. 13.23(c) and (d), $\overrightarrow{ba}$ can be drawn to determine the speed of sliding over the die, i.e., oa, which is constant and only changes in direction to that of α_2 and is shown as $\overrightarrow{od}$. To have final rigid-body translational

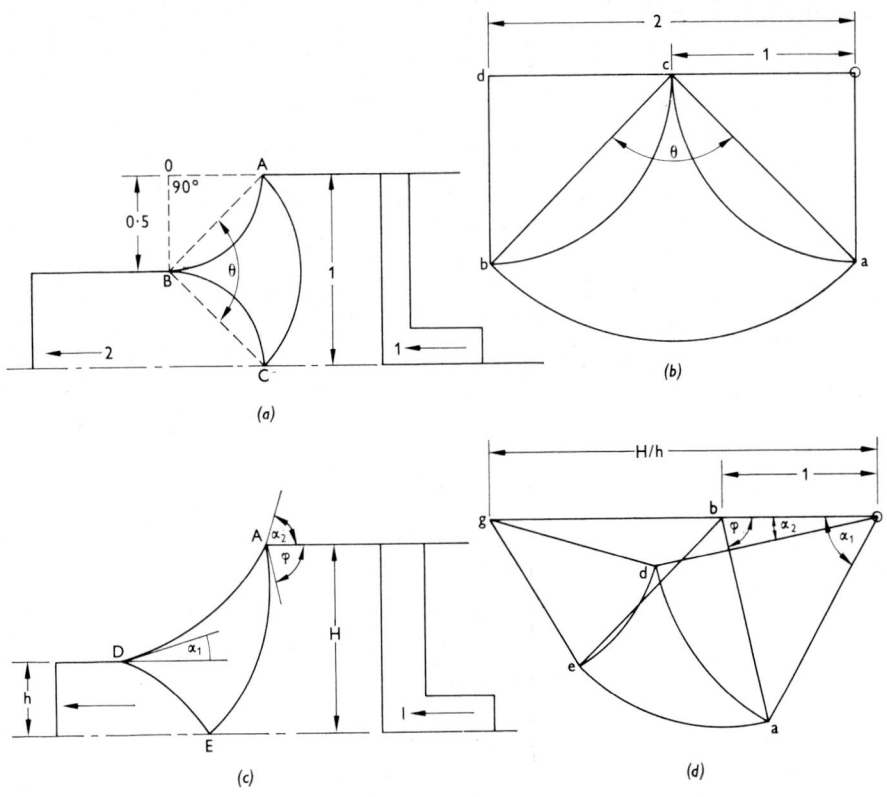

Fig. 13.23

(a) Half of pattern of velocity discontinuities, two circular arcs, for extrusion through a smooth circular die; $r = 0.5$.

(b) Hodograph for (a).

(c) Extrusion through a circular convex die having an entry angle α_2 and exit angle α_1.

(d) Hodograph for (c).

speed H/h, a tangential velocity discontinuity $\overrightarrow{gd}$ is required; this determines the direction at D. Point e is found, being the intersection of arc ae, centre b and radius ab, and arc de, centre g and radius dc. The two directions at E now being determined by eb and ag, circular arcs AE and DE can be constructed. The extrusion pressure can now be found, since the various arc lengths can be easily obtained and the absolute magnitude of the velocity discontinuities ab and dg read off from the hodograph. Several trials for various values of ϕ can be made and that value of ϕ which gives the least value of p selected.

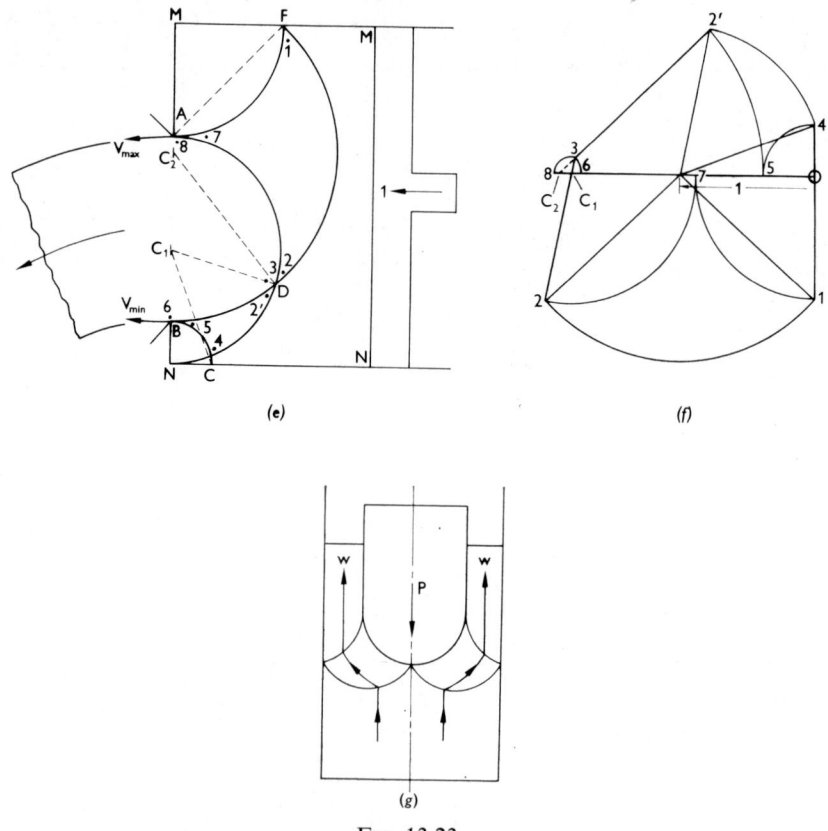

(e)   (f)

(g)

FIG. 13.23

(e) Convex dies of unequal curvature.
(f) Hodograph for (e).
(g) Piercing with a round-headed punch, P, to form walls, W.

An interesting case arises when convex dies are considered, which are such that their radii of curvature are unequal. Attention will be restricted to a particularly simple application of the methods previously described. The instance is shown in Fig. 13.23(e) which is for a reduction of 0·456, the radius of curvature of die FA being 8/3 times as great as that of die CB. Two arcs of velocity discontinuity intersecting at D will suffice; C_1 is the centre of the arc which passes through C and A and C_2 that of the arc passing through F

and B. The hodograph is drawn in Fig. 13.23(f), the procedure following that earlier described as far as lines $a2$ and $a2'$. 3 is the corner of the parallelogram opposite to a; 23 and $2'3$ produced cut the hodograph axis in c_1 and c_2 respectively. Arcs 83 and 63 have c_1 and c_2 as centres and $c_1 3$ and $c_2 3$ as radii, respectively. Thus the solution predicts that the emergent extrusion rotates. The mean extrusion pressure $\bar{p}$ is easily seen to be given by,

$$MN \ . \ \bar{p}/k = (\text{arc } ADC)\,(a4) + (\text{arc } BDF)\,(a1). \qquad (13.23)$$

For this instance, $\bar{p}/2k = 1.12$; when the dies are perfectly rough, $\bar{p}/2k = 1.39$, after adding to the right-hand side of (13.23) terms for energy dissipation at the two circular die interfaces.

By the simple expedient of interchanging the die wall and the centre line in Fig. 13.23(a), we arrive at a solution for the plane-strain piercing operation in which a semi-circular headed punch is used, see Fig. 13.23(g).

13.9 Rolling

There is much in common between the simple tangential velocity discontinuity patterns for providing an upper bound to a load for extrusion through curved dies and those for the rolling of sheet and plate. Indeed the patterns chosen can be identical; the rolling situation can be thought of in the same terms as an extrusion where the exit die angle is zero.

In Fig. 13.24 a simple two arc pattern of tangential velocity discontinuities is shown for the rolling of strip when there is no relative motion between the roll and the material. Zones A and B at the entry and exit from the roll gap are rigid whereas the zone between them, defined by the corners 1, 2, and 3 of the curvilinear triangle, rotates as if it were part of the roll. The change from translational to rotational motion and back to translation is effected in the circular arcs, 12 and 23, respectively. The reader should easily be able to verify the correctness of the hodograph shown as Fig. 13.24(b). The solution in (a), was obtained by starting with an arc tangential to the roll surface at exit and intersecting the strip axis at 45°; this provided point 2 and arc 12, centre C_2 was then inserted. This bound is, to some extent, framed in accordance with slip-line field solution shown as Fig. 14.30. The roll torque T is bounded as,

$$T \ . \ \omega \ . \ = k \, (\text{arc } 12 \ . \ \text{velocity discontinuity } a1 + \text{arc } 23 \ . \ \text{velocity discontinuity } b2).$$

ω is the angular velocity of the roll and is $O3/\text{roll radius}$. It may be verified that $T/2k \simeq 1{\cdot}4$ by measuring from Fig. 13.24 the various quantities.

It should be noted that the hodograph (b) is identical with the physical diagram (a), in shape and could therefore be usefully drawn on it directly. A square grid of lines is deformed in rolling as shown in Fig. 13.24(c) and this deformation is easily deduced with the help of Figs. 13.24(a) and (b).

This approach in some studies of rolling has been used by PIISPANEN, ERIKSSON and PIISPANEN, (1967).

AVITZUR (1963) has given an upper bound for use in calculations for the cold rolling of strip; the derivations involved are formidable.

Another interesting example of the use of this technique is that in which a rigid metal cylinder rolls over a rigid-perfectly plastic material. Consider a circular cylinder S of unit radius, see Fig. 13.25(a) which, is wide perpendicular to the plane of the paper and which rolls over a semi-infinite medium M, the centre of S moving with unit speed. Suppose that S is fitted into a preformed groove in M, of depth d, but that no plastic deformation takes place in S. To simplify the situation impose a uniform velocity of unity to the left on

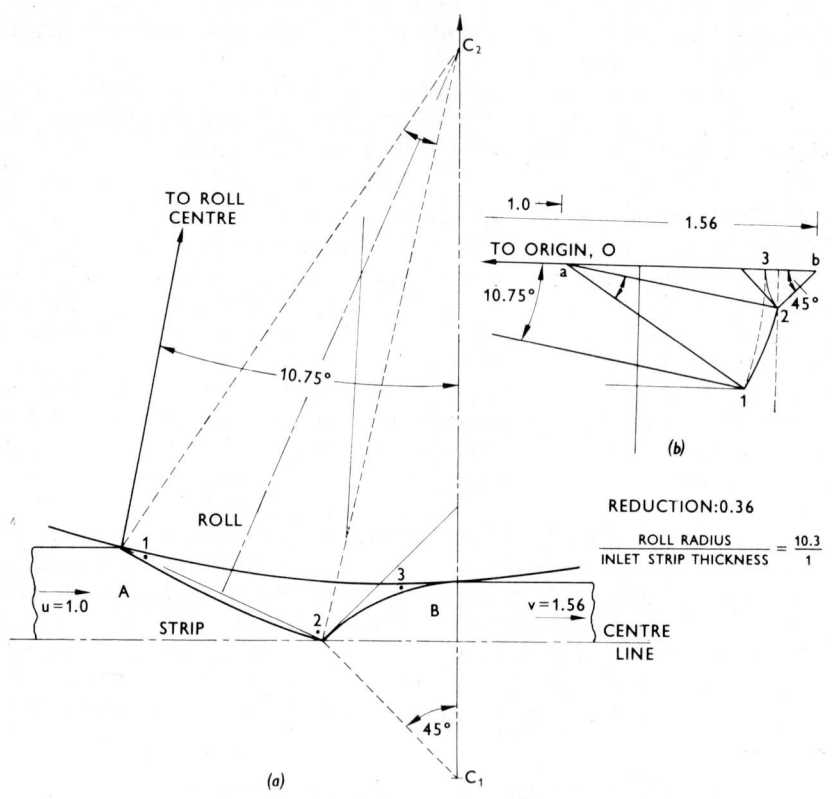

FIG. 13.24 Rolling of sheet or plate.

S and M; this brings the centre of S to rest and we then consider M to be moving towards S with unit speed. Since M is infinitely deep the surface of the material must be at the same level when S has been passed as it was before S was encountered. It follows that the surface material of M near S must move with a speed greater than unity; the possibility of density changes in the material is excluded.

Let DFD be a circular arc of tangential velocity discontinuity of radius r subtending an angle 2α which is on the perpendicular bisector of DD. Then

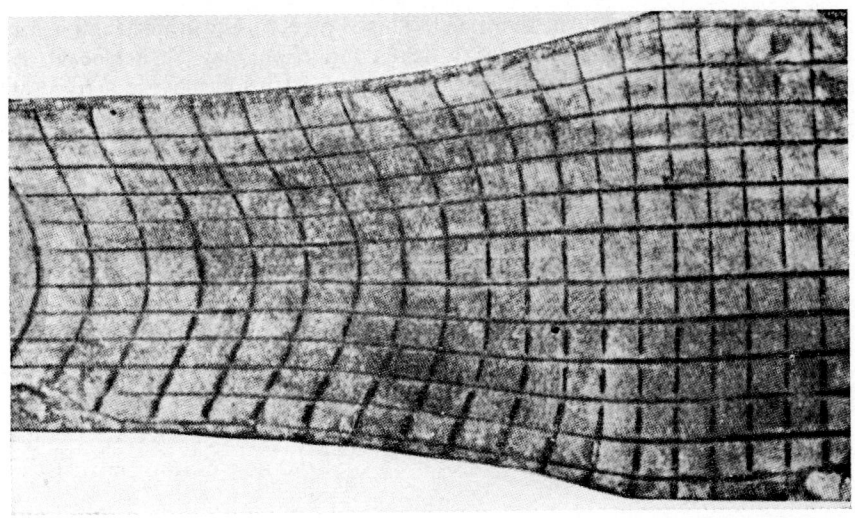

(c)

FIG. 13.24

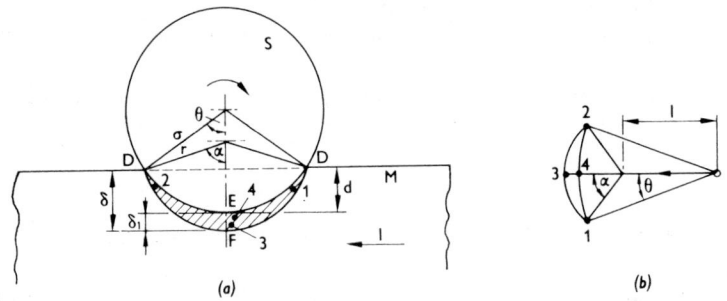

(a)   (b)

FIG. 13.25

(a) Rigid cylinder rolling over a semi-infinite plane; the cylinder is reduced to relative rest. The tangential velocity discontinuity is along arc *DFD*.

(b) Hodograph to (a).

all surface material to depth *d* passes through the shaded region where it rotates as part of a solid block about the centre of *S*. The hodograph is drawn in Fig. 13.25(*b*). This model explains some well-known features of rolling; it is discussed at length elsewhere and in relation to certain metal working processes. (JOHNSON, 1964; see also COLLINS, 1972).

13.10 Temperature Distribution in Swiftly Worked Metals

(TANNER and JOHNSON, 1960)

The plastic work done in a metal-working process is almost entirely dissipated as heat, and in some instances it is of value to be able to describe the temperature distribution in the metal. This is a relatively simple matter for *fast* plane strain processes because heat conduction may be neglected. A simple criterion of 'fast rate of working' for an extrusion may be arrived at as follows:

Consider a material moving with constant speed u up to line XX, where its temperature is θ_0, having been raised from zero at $x = \infty$. It may be imagined that XX is a line source of heat due to the existence of a tangential velocity discontinuity; the temperature endeavours to propagate heat upstream but the faster the speed of movement of the metal, i.e., u, the less successfully is this achieved, see Fig. 13.26. The steady-state heat conduction equation for the region to the right of XX is

$$\frac{\partial^2 \theta}{\partial x^2} + \frac{u\rho c}{k}\frac{\partial \theta}{\partial x} = 0 \qquad (13.24)$$

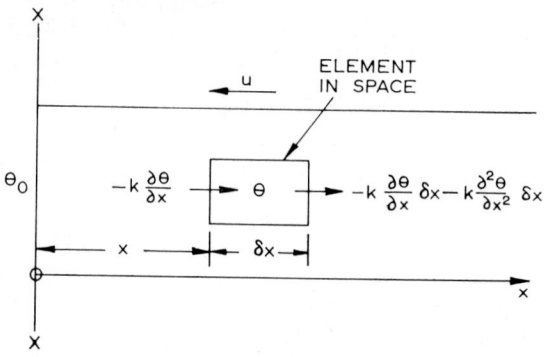

FIG. 13.26 Heat and mass flow through an elemental volume.

where θ is the temperature at distance x from XX, ρ is the density of the material, c its specific heat and k the coefficient of thermal conductivity. The solution to equation (13.24) is

$$\theta = \theta_0\, e^{(-u\rho c/k)x}.$$

For θ to fall, say to 2 per cent of θ_0 we require $x = k(\ln 50)/\rho c$. When u is 1 inch per second, the required values of x for the following metals, are

Material	$k/\rho c$ ft^2/hr	x in
Lead	0·94	0·15
Copper	4·4	0·70
Mild steel	0·48	0·07
Silver	6·6	1·03
Aluminium	3·7	0·58

The table shows that for all practical purposes, at speeds of 1 inch per second, the heat does not propagate upstream. With this kind of explanation in mind, thermal conduction effects are neglected in the following example. We now determine the temperature pattern in extruded metal about the mouth of the die in an adiabatic extrusion process.

13.11 Temperature Distribution in Extrusion Due to Fast Working

(TANNER and JOHNSON, 1960)

For simplicity, the method will be exemplified by reference to the simple case depicted in Fig. 13.27(*a*), extrusion through a smooth 90° die. First, the usual, well-known slip-line field is approximated using chords in place of circular arcs or triangles replacing sectors. The degree of coarseness acceptable in a replacement of the slip-line field by a system of chords is a matter of the degree of inaccuracy tolerable. In this instance, three 30° isosceles triangles replace a circular quadrant, see Fig. 13.27(*b*). This procedure gives a valid velocity field for this particular problem which can be used for estimating the ram pressure necessary to effect extrusion. Second, construct the hodograph to Fig. 13.27(*b*); this is shown in Fig. 13.27(*d*); it may be compared with that for Fig. 13.27(*a*) which is 13.27(*c*). This latter diagram enables the velocity of any particle which is part of the extrusion to be stated. Define a number of stream tubes—four in this case—by the 'horizontal' lines meeting the bounding entry line at its points of discontinuity, *A*, *B*, *C* and *D* in Fig. 13.28. The stream lines of particles passing through these points are easily obtained using the hodograph. All the material moving within a given tube undergoes a similar history of deformation; the progress of each of the four tubes through the deformation zone is now as indicated by the broken lines. (If there is concern with the *pattern* of temperature distribution only, all material constants, e.g. the yield stress of the material, may be omitted from the calculations). Consider stream tube II of thickness *t*. If material approaches AB at unit speed, the work done per unit time is $k.u_{AB}$; u_{AB} is the tangential velocity discontinuity across AB. The energy input per unit volume put through is $ku_{AB} AB/t$ and this reappears as heat and hence as a temperature rise, ΔT. Suppose that a fraction, f, of the energy dissipated manifests itself as temperature increase, then

$$fku_{AB} AB/t = \rho \cdot c \cdot \Delta T. \tag{13.25}$$

where ρ denotes density and c is the specific heat of the material. Hence,

$$\Delta T = \frac{k}{\rho c}.u_{AB}\frac{AB}{t}f = \left(\frac{k}{\rho c}.f\right)\frac{u}{v_p}. \tag{13.26}$$

Referring to Fig. 13.3 it is easily verified that $t/AB = v_p$. The quotient u/v_{AB} for AB is obtained from the hodograph and hence we may write $\Delta T = \delta.(u/v_p)_{AB}$ where δ is a constant.

The material in AA_1B is rigid and remains at constant temperature until it crosses A_1B where, for a second time, it undergoes a sudden change in direction. The temperature of the material moving into $BA_1A_2B_1$ jumps by an amount $\delta.(u/v_p)_{AB}$; the quantity u/v_p has to be found for the line A_1B from the hodograph. Further temperature jumps occur on crossing A_2B_1 and A_3B_2. The final temperature of the material is the sum of the separate jumps; for instance, the temperature in region $A_2B_1B_2A_3$ would be

$$\delta\left[\left(\frac{u}{v_p}\right)_{AB} + \left(\frac{u}{v_p}\right)_{A_1B} + \left(\frac{u}{v_p}\right)_{A_2B_1}\right]$$

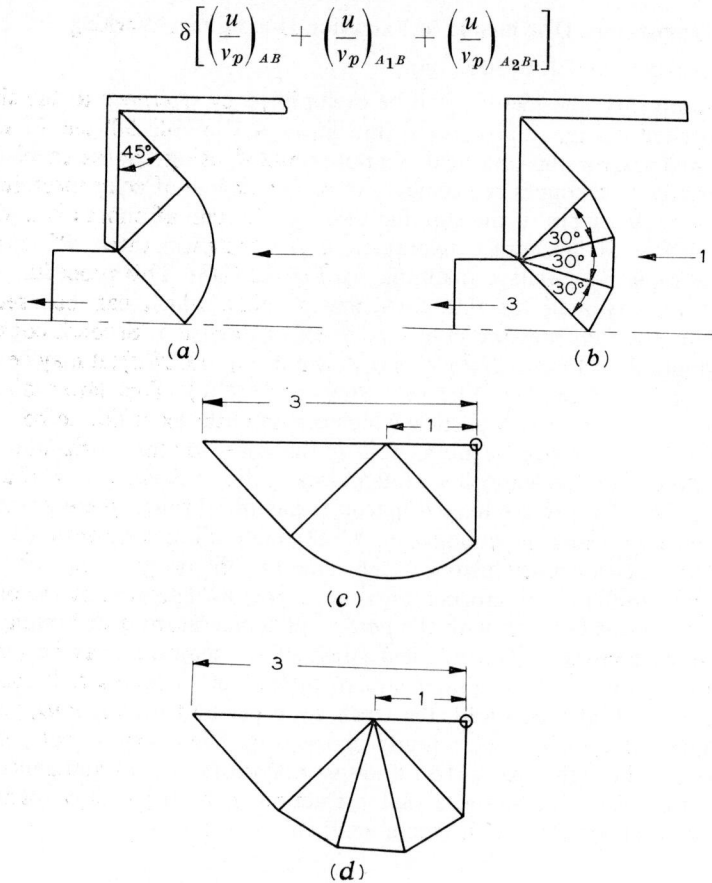

FIG. 13.27

(a) Half of slip-line field for extrusion through a 90° die, die face and wall perfectly smooth: $r = 2/3$.
(b) Pattern of velocity discontinuities substituted for (a).
(c) Hodograph corresponding to (a).
(d) Hodograph corresponding to (b).

and for the emerged stream tube we should have to add

$$\delta\left(\frac{u}{v}\right)_{A_3B_2}.$$

Similar calculations are made for each of the three remaining stream tubes. The largest value of the temperature has been converted to 100 and the value in each section of the diagram increased in proportion. The approximate temperature distribution is now clear and is depicted in Fig. 13.28.

It should be observed that above, u/v_P is nothing more than the shear strain imposed on an element as it crosses a line of discontinuity. Fraction $f \simeq 0.9$ and BISHOP (1956), (using a relaxation technique), has investigated one particular reduction allowing for conduction.

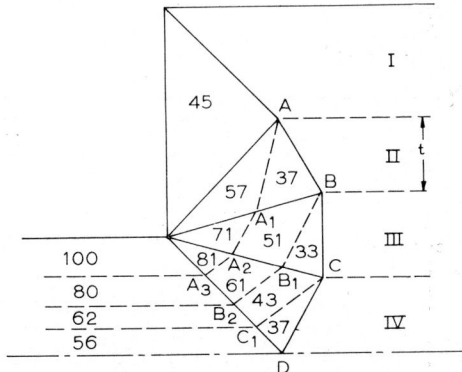

FIG. 13.28 Showing stream tubes and temperature distribution following Fig. 13.27 (b) and (d).

13.12 A Simple Notation for Constructing Hodographs

Reconsider the extrusion situation shown in Fig. 13.8; denote by a capital letter regions in the physical plane which are separated by lines of tangential velocity discontinuity see Fig. 13.29(a). Then the velocity of each region is represented by a single point in the hodograph and we may take advantage of this to draw the hodograph in a way which is already familiar to readers as either 'Bow's Notation' (or a Maxwell diagram). Region O is the origin in the velocity plane (the hodograph) and OA defines the tangential velocity discontinuity at the slug-extrusion chamber wall, $\beta\gamma$; this appears as $\overrightarrow{oa}$ in Fig. 13.29(b), where $\overrightarrow{oa} = u$. The velocity discontinuity across $\beta\delta$ is defined by AC and hence from a draw a line parallel to AC; the discontinuity at the die interface, i.e., across $\alpha\beta$, is defined by OC and hence from o draw oc parallel to OC to cut ac at c. Similarly BC defines the discontinuity across $\alpha\delta$ and is represented in the hodograph by bc.

It will be evident that we proceed by thinking of the lines of tangential velocity discontinuity as if they were rods or members in a structure.

A better upper bound to the extrusion pressure may be arrived at than that given by the simple two line discontinuity model of Fig. 13.29(a), by choosing the pattern shown in Fig. 13.29(c) which uses four lines encompassing the mass of the slug. To use this method, proceed thus:

Choose o and draw $\overrightarrow{oa} = u$, parallel to OA.
From o draw oc' parallel to OC.
From a draw ac'' parallel to AC; oc' and ac'' intersect at c.
Draw ad parallel to AD and cd parallel to CD, to intersect at d.

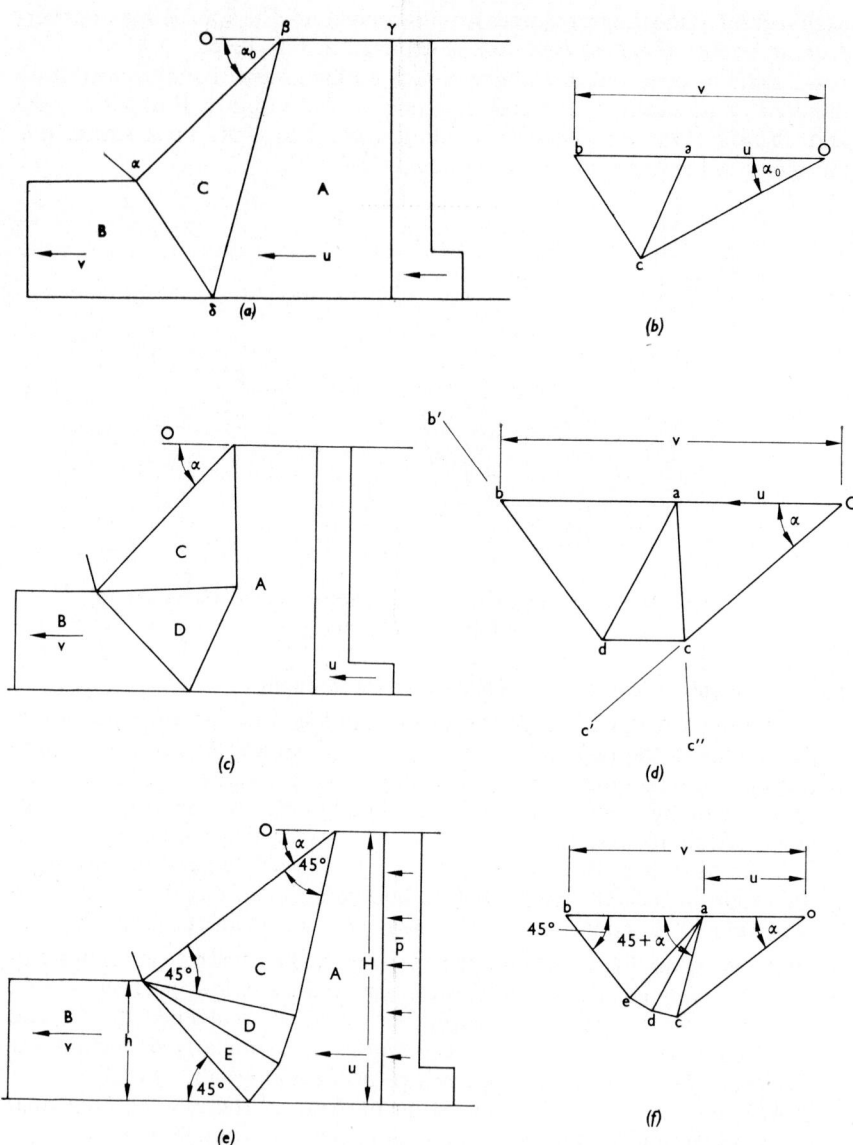

FIG. 13.29 Extrusion through symmetrical wedge-shaped dies.

Complete the hodograph by drawing db' parallel to DB, b being its point of intersection with oa produced, see Fig. 13.29(d) It will be seen that in choosing the more complex pattern of Fig. 13.29(c) as opposed to that of Fig. 13.29(a), use has been made of our knowledge of slip-line field patterns. If reference is made to Fig. 12.5(a), shown above as Fig. 13.29(e), with the circular arc OMR modified to two isosceles triangles, then the hodograph to its slip-line field will be easily drawn with the aid of this new simple procedure, see Fig. 13.29(f).

It should now be clear that the hodograph of Fig. 12.5(c) is just the limiting case of Fig. 13.29(f); and recognizing this we can now calculate the extrusion pressure $\bar{p}$ *without* the aid of the Hencky equations. Using Figs. 13.29(e) and (f), we should have,

$$(\bar{p}.H).u = k(\overline{AC.ac} + \overline{AD.ad} + \overline{AE.ae} + \overline{EB.eb} + \overline{CD.cd} + \overline{DE.de}).$$
(13.27.i)

In the limiting case, using the notation in Fig. 12.5(a) and (c), and following (13.23),

$$\bar{p}.1.1 = k(\text{length } \overline{NRM.v_{NR}^*} + \overline{OM.cd} + OR.\text{arc } cb)$$

and thus,

$$\frac{\bar{p}}{k} = (1 - r)\sqrt{2}(1 + \alpha).ab + (1 - r)\sqrt{2}.cd + (1 - r)\sqrt{2}.cd.\alpha$$

$$= (1 - r)\sqrt{2}.\frac{2 \sin \alpha}{\sqrt{2}} (1 + \alpha + 1 + \alpha).$$

Hence,

$$\frac{\bar{p}}{2k} = \left(1 - \frac{2 \sin \alpha}{1 + 2 \sin \alpha}\right).2 \sin \alpha \,(1 + \alpha)$$

$$= \frac{2 \sin \alpha}{1 + 2 \sin \alpha}.(1 + \alpha).$$
(13.27.ii)

13.13 Examples of the Application of the Lower Bound Theorem Using Stress Discontinuity Patterns

13.13.1 *The Mohr Circle in Plane Strain: The Pole of the Mohr Circle*

Figure 13.30(a) shows the stresses on planes through a point A. To draw the circle of stress for this system, supposing that (σ, τ) is given on AC and (σ', τ') on AC', we plot in the direct stress, σ,—shear stress, τ, plane, two points C and C', their co-ordinates being (σ, τ) and $(\sigma' \, \tau')$, respectively. If the perpendicular bisector of CC' is drawn to intersect the $O\sigma$ axis in C_0, then C_0 is the centre of the required stress circle and C_0C or C_0C' is the radius, see Fig. 13.30(b). It is easily shown that the angle contained between CC_0 and $C'C_0$ is twice the angle between CA and $C'A$.

If the stress circle is given in the first place, then C may be easily identified on the circle with the stress pair (σ, τ). If now the stress pair on any plane through A, say AC, at angle ϕ to the given plane is required, the following procedure may be justified. From C draw a line parallel to the plane on which (σ, τ) acts, i.e. CP, to cut the circle in P. From P draw PC'' parallel to AC'' to cut the circle in C''; then the co-ordinates of C'', (σ'', τ'') give the required stress pair acting on plane AC'', (σ'', τ''). The point P is known as the pole of the circle; it is of the greatest importance in calculating lower bounds by using the method of stress discontinuities described below.

13.13.2 *Stress Discontinuities, or Jumps, in Plane Strain*

The concept of a tangential velocity discontinuity has been used to great effect earlier in this chapter and we now introduce a similar idea, that of the stress discontinuity. We have already encountered one such example of this concept in the study of the full plastic bending of beams, see Fig. 7.2; on one side of the neutral axis the normal stress is $+Y$ and on the other it is $-Y$, so that there is a jump or discontinuity of amount $2Y$ (or $4k$) in crossing the line of stress discontinuity which is the neutral axis of the beam. Note that this discontinuity in normal stress is parallel to the line of discontinuity whilst the normal stress perpendicular to the neutral axis is continuous at zero magnitude.

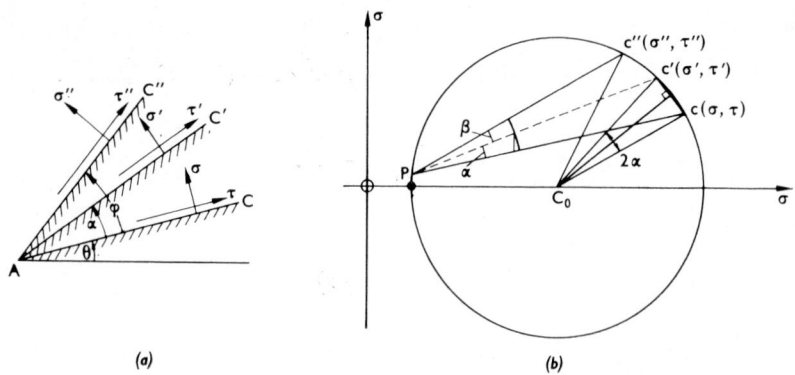

(a)   (b)

FIG. 13.30 The 'Pole of the Mohr circle'.
(a) Stresses on plane through point A.
(b) The Mohr circle.

The line XX in Fig. 13.31(a) is taken to be a line of stress discontinuity in the sense that the normal stress on an element parallel to XX, as XX is crossed, jumps from σ_t' to σ_t''. The other stress components remain unchanged and especially if $\sigma_1 = \sigma_2 = \sigma_n$ it follows that the equilibrium of an element such as AB is sustained. It is *not* necessary, however, that $\sigma_t' = \sigma_t''$ for equilibrium to be maintained. It is this feature of which we take advantage in cases where there is plastic deformation through a region, to build up a lower bound.

Two parts of the element straddling line XX are shown separated as A and B in Fig. 13.31(a). If the yield criterion is used in relation to the stresses on A and B and bearing in mind that the circumstances are those of plane strain, so that the yield shear stress is k, then

$$\left(\frac{\sigma_t' - \sigma_n}{2}\right)^2 + \tau^2 = \left(\frac{\sigma_n - \sigma_t''}{2}\right)^2 + \tau^2 = k^2.$$

These equations or conditions are represented in terms of Mohr circles of plastic stress, i.e., circles of radius k in the σ, τ plane, in Fig. 13.31(b). Points

A_0, B_0 and C_0 represent stress states (σ_n, τ), (σ_t', τ) and (σ_t'', τ), respectively. Evidently only one stress jump is possible since only two different circles of the same radius can pass through point A_0.

If (σ_n, τ) is given, then the jump is

$$\sigma_t' - \sigma_t'' = 4\sqrt{k^2 - \tau^2}. \tag{13.28}$$

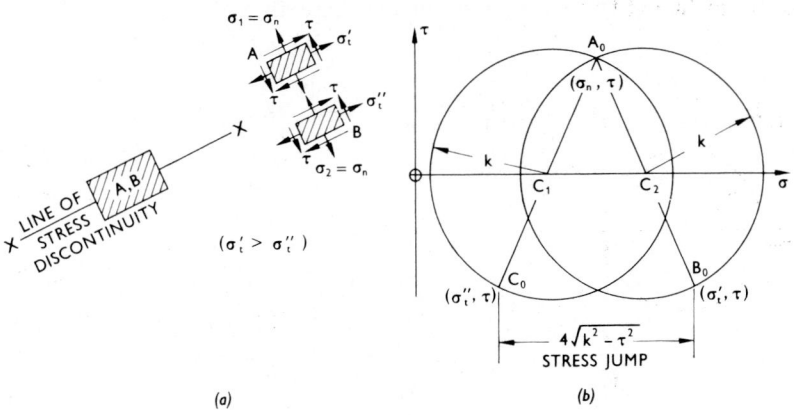

(a)

(b)

FIG. 13.31 Stress discontinuities in plane strain.

Examples in which lines of stress discontinuity are used with the aid of Mohr circles are given in subsections 13.13.4, 13.13.5 and 13.13.6. The approach used, is to devise a pattern of lines across which the normal stress may jump, but so that in every region thus bounded by these lines, the yield criterion is just fulfilled. Some region or regions at the boundary are usually only just stressed to the yield limit, so that we are thus enabled to invoke the lower bound theorem and to assert that equilibrium requirements are fulfilled throughout *the whole body* and that at no part of it is the requirement of the yield criterion violated.

13.13.3 *The Notched Bar in Bending*

See Fig. 13.4(*a*). A lower bound to the critical load is found by using the first limit theorem. The material to left and right of the notch is stress-free and the material below the notch is subject to simple bending, the neutral axis being half-way between the root of the notch and the lower surface of the bar. Thus at yield, there is a constant stress of $2k$, compressive above the neutral axis and tensile below. This gives rise to a resisting moment, M, and

$$M = 2k . \frac{a}{2} . \frac{a}{2} = 0.5ka^2.$$

This is a lower bound for the value of the bending couple, M.

The broken lines are lines of stress discontinuity; the stress normal to each line is the same on both sides of it, as is the shear stress tangential to these lines, so that stress equilibrium is maintained in progressing from the top face to the bottom face of the bar. The normal stress as a broken line is traversed, i.e., parallel to it, undergoes a jump; in moving from the top to the middle layer the jump is from 0 to $-2k$, and in moving from the middle to the lower layer it is from $-2k$ to $+2k$, the greatest that is permitted. These jumps are entirely permissible because the yield condition is everywhere fulfilled. The situation is trivial in terms of Mohr's stress circles, see Fig. 13.32(a), but we include it to connect with the next case discussed.

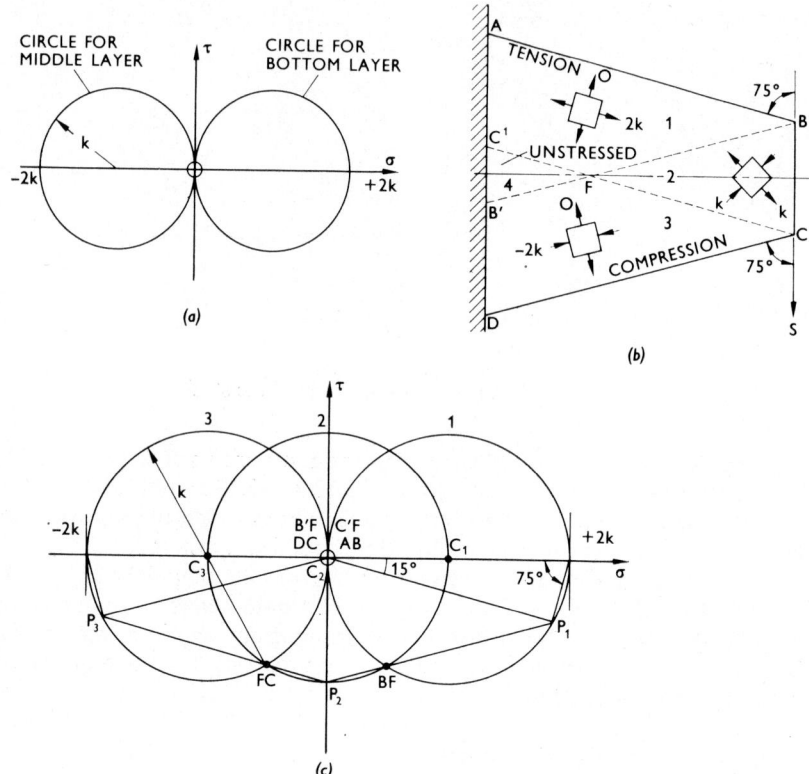

FIG. 13.32 Tapered cantilever under shear load.

13.13.4 *The Tapering Cantilever under Shear Loading*, see Fig. 13.32(b)

Consider a straight sided cantilever under an end shear load of S, the sides of the cantilever being at 75° to the line of action of S. This case has obvious affinities with that of sub-section 13.13.3. It is more complicated in that there are four zones of constant stress level, but the two regions of ($\pm 2k$, 0) are common to both. Regions 1 and 3 are evidently similar to those of sub-section 13.13.3 and as in Fig. 13.32(a), so in Fig. 13.32(c) we have circles 1 and 3 to represent the stress states. Pole P_1 is next located for stress circle 1 and the stress state on plane BF is found by drawing from P_1 a line parallel to plane *BF* to cut the circle in a point marked *BF*. The only other circle of radius *k* which can pass through point *BF* has its centre at the origin and thus circle 2 follows. Pole P_2 of circle 2 is next located by producing $P_1(BF)$. P_2 is equally well arrived at by starting from bottom layer 3 or circle 3, via pole P_3, point *FC* and line $P_3(FC)$. It is easily verified that zone 4 is completely unstressed and that zone 2 is in a state of simple shear; the stress on plane *BC* is zero normal stress and a shear stress of *k*. Thus a lower bound load for this cantilever is *k* times the area of the end of the cantilever.

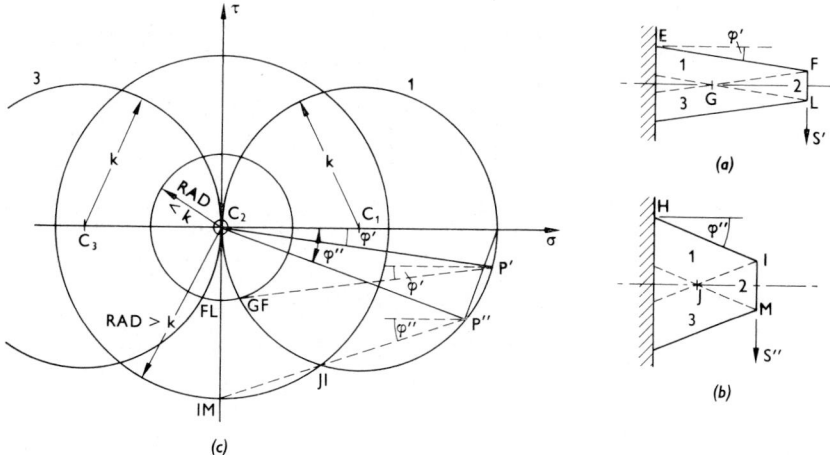

FIG. 13.33 Tapered cantilever; other geometries.

The cantilever tapered at 75° to the vertical is a special case. Fig. 13.33(*a*) and (*b*) show two cantilevers in which the taper of one is less and in the other more than 75°, respectively. The two stress circles of radius *k* for the top and bottom zones which are stressed in pure tension and pure compression, respectively, are identified again as circles 1 and 3. Proceeding as in the previous case, the poles *P'* and *P''* can be located and thence the stress points for planes *GF* and *JI*, respectively. These latter lines produced intersect the *O*τ axis in *FL* and *IM*. Circles through (i) *FL* and *GF* and (ii) *MI* and *JI*, have radii, respectively, less than *k* and greater than *k*. The shear stress on the end of each cantilever, $C_2(FL)$ and $C_2(IM)$, is $2k \sin 2\phi'$ and $2k \sin 2\phi$, see Fig. 13.33(*c*); the former $2k \sin 2\phi'$ is admissible but not the latter $2k \sin 2\phi''$. Region *JMI* is over-stressed; it implies that the stress along *IM* exceeds *k*, which is clearly unacceptable. The limiting case is clearly that in which the stress circle radius for a circle whose centre is at the origin, is *k*, i.e., $2\phi = 30°$; then the limiting taper angle of the cantilever is 75° (GREEN, 1954).

It should be pointed out before proceeding that for many metal-working operations an estimate of the minimum load or pressure that is not sufficient to cause the operation to happen can always be had by assuming the operation to be carried out without redundant deformation, i.e., a 100 per cent efficient process. The required change in shape brought about by homogeneous deformation is then $2k \ln [(1/1 - r)]$ for example for an extrusion process. This minimum load estimate should not, however, be confused with a lower bound as strictly defined in this chapter. Lower bounds as used here make no assumptions about operation efficiency or even ideal or hypothetical modes of straining. We are only concerned with the ability of the material to sustain a particular stress distribution *without* deformation and such that the stress boundary conditions are everywhere accommodated and the yield criterion nowhere exceeded.

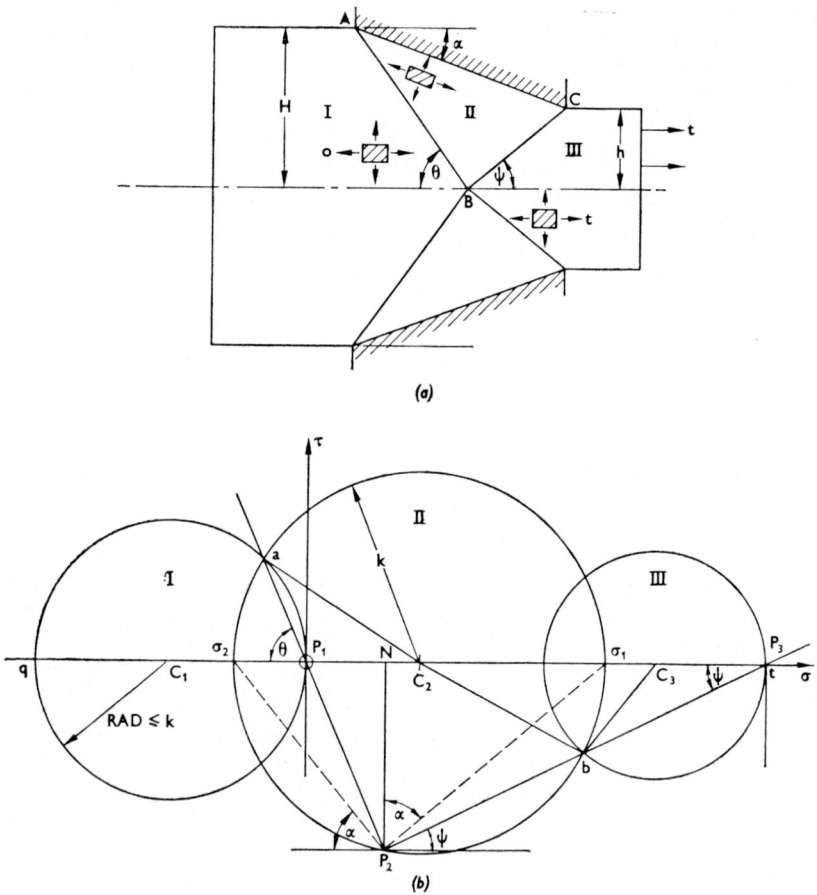

FIG. 13.34 Sheet drawing through frictionless wedge-shaped dies.

13.13.5 *Sheet Drawing and Sheet Extrusion for Frictionless Wedge-shaped Dies*

Figure 13.34 shows the drawing of sheet through a frictionless die of semi-angle α and we require to find a lower bound to the drawing stress, t. Choose lines of stress discontinuity BA and BC, thus dividing the sheet into three zones, I, II and III. For this particular pattern we shall find those values of θ and ψ which maximize t, i.e., give the best or greatest lower bound. Note that since the die is frictionless therefore the principal stresses in region II, which is everywhere plastic, must be parallel and perpendicular to the die face. Also because there is no pull back tension then the region I can be taken to be elastic and to carry stresses 0 and less than $2k$, as shown; the stress state for region I is the Mohr circle I centre C_1, as shown in Fig. 13.34(b). The stress on planes perpendicular to the axis in zone I is zero and thus the corresponding stress plane point on the Mohr circle is at the origin, O; the pole, P_1, for this circle is also at the origin, O. The stresses on plane AB are given by the co-ordinates of the point of intersection a of P_1a, which is parallel to BA, and in the circle, $a\hat{P_1}C_1 = \theta$.

Region II is obviously fully plastic and its stress state must therefore be represented by a circle of radius k. If C_2 is located where $C_2a = k$, then the circle of centre C_2 and radius k is the stress circle for region II; this construction ensures that in as far as point a is common to both circles I and II, then the normal and tangential stress on either side of line AB is the same, which is what is required for the maintenance of stress equilibrium. Of course, as indicated above, the normal stress on a plane perpendicular to AB undergoes a finite jump as AB is crossed.

Now point P_2 is the pole of circle II; a represents the stress on AB and to find the pole of the circle, line aP_2 must be drawn from a parallel to AB to intersect circle II in P_2. The principal stress normal to the die face σ_2 is found by drawing $P_2\sigma_2$ parallel to AC and the other principal stress σ_1 by drawing $P_2\sigma_1$ perpendicular to AC; $O\hat{\sigma_1}P_2 = \pi/2 - \alpha$. To find the stress on plane BC draw from P_2 a line parallel to BC to cut circle II in b.

Circle III for region III must pass through point b for the normal and tangential stress on BC to be the same, whether approached from zone II or zone III; also, the greater principal stress of circle III must be t, the required drawing stress. Point t is also the pole, P_3, for circle III and P_3b must be a continuation of P_2b, since both P_2b and P_3b refer to plane BC, from circle II and circle III, respectively.

As drawn in Fig. 13.34(b) the radius of circle III is less than k and thus it is implied that region III is elastically stressed. It is easy to see that if ψ was taken to be small enough, then the resulting circle would have a radius $R > k$ —and this is inadmissible; the region III would be overstressed. The diagram in Fig. 13.34(b) enables us to find a lower bound on t. We have

$$t = P_2N(\cot \theta + \cot \psi)$$

$$= k \sin 2\alpha(\cot \theta + \cot \psi). \tag{13.29}$$

Now from geometrical considerations,

$$\text{reduction, } r = \frac{(H \cot \theta + h \cot \psi) \tan \alpha}{H} \tag{13.30}$$

and also $r = (1 - h/H)$. \tag{13.31}

Thus,

$$r(\cot \alpha + \cot \psi) = \cot \theta + \cot \psi,$$

and substituting in (13.29),

$$t = k \sin 2\alpha(\cot \alpha + \cot \psi).r. \tag{13.32}$$

The greatest permissible value of t is found by taking the greatest permissible value of ψ consistent with the radius of circle III not exceeding k. When circle III is of radius k, $bC_3 = bC_2$ and hence

$$C_2\hat{C}_3b = b\hat{C}_2C_3,$$

that is,

$$\sigma_1\hat{P}_2b = C_3\hat{P}_3b,$$

or

$$\frac{\pi}{2} - \alpha - \psi = \psi$$

so that

$$\psi = \frac{\pi}{4} - \frac{\alpha}{2}. \tag{13.33}$$

Substituting for ψ in (13.32) from (13.33) it is found that,

$$t = 2k(1 + \sin \alpha).r. \tag{13.34}$$

Sheet extrusion is fundamentally identical with that of sheet drawing, especially as far as the frictionless conditions is concerned. Imposing an overall hydrostatic stress of t nullifies the drawing tension whilst it applies a pressure—an extrusion pressure—over the height, H; the die pressure increases, of course, by amount t also.

It is possible for long dies, where the angle will be small, to consider several 'crosses' instead of just one as dealt with above, and this may lead to a better lower bound; this procedure is analogous to using several crosses for certain cases in obtaining good upper bounds for extrusions of high reduction. The algebra is, however, lengthy and since the procedure introduces no substantially new idea we do not therefore pursue this example. The report by GREEN (1952) can be consulted for further discussion. It may be added that ANSOFF (1948) has apparently used a discontinuous stress field to obtain an approximate solution for drawing through a die with a circular contour, the contour being approximated by a polygon.

13.13.6 *Indentation*

In Section 13.5.2, Fig. 13.5(a), we considered the indentation of a semi-infinite block by a rigid punch and obtained a number of upper bounds to the indentation pressure. A lower bound of $5k$ for this pressure is obtained using the obvious pattern of stress discontinuities shown in Fig. 13.35, (DRUCKER and CHEN, 1968). In Fig. 13.35(b) the Mohr circles for each of the regions has been drawn. The poles for each stress circle for each stress region in the physical diagram are marked as P_1, P_2, etc., and the stresses on each plane, e.g., plane AH, are indicated by a point on the appropriate circle diagram against which the plane designation is marked. If the reader has assimilated the contents of the previous sections he should have little difficulty in checking the validity of Fig. 13.35(*b*).

The paper by DRUCKER and CHEN should be referred to in order to study how, instead of three 'supporting legs of stress', many more may be used and the Prandtl field solution (see p. 498) arrived at.

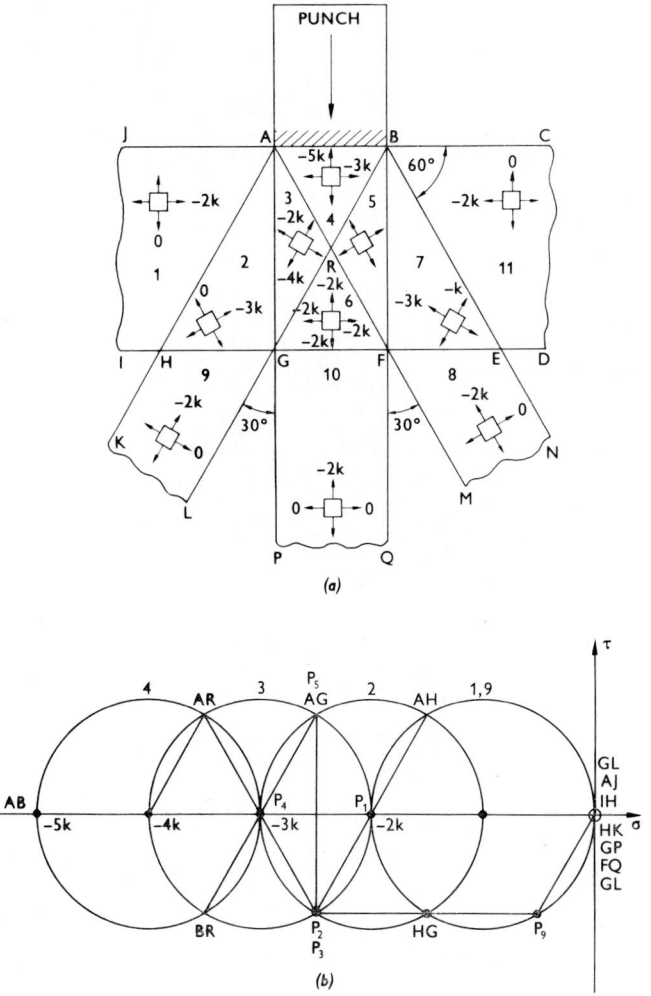

FIG. 13.35 Indentation of semi-infinite block by a rigid punch.

13.13.7 *Large Reduction Extrusion through Square Dies*

A lower bound for extrusion through square dies, given originally by ALEXANDER (1961), is based on the pattern of stress discontinuities defined by the broken lines in Fig. 13.36(*a*), which divide the whole region into eight zones. By starting with region 1 and choosing *AC* and *CI* at 30° and 60° to

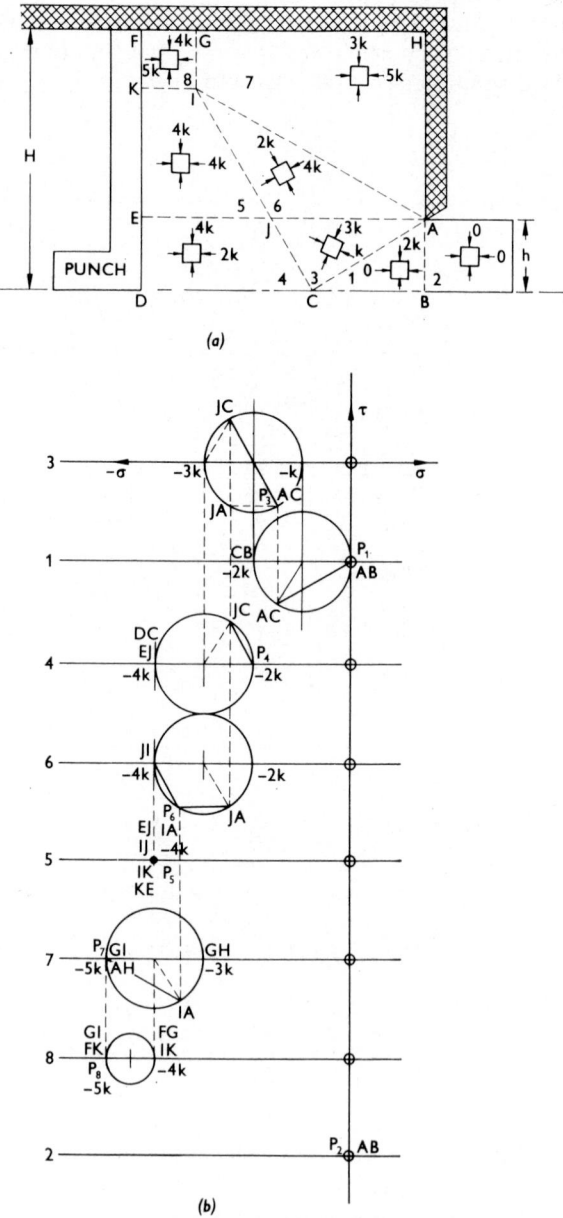

FIG. 13.36 Extrusion through square dies.

the slug axis, the Mohr circle for each region may be derived, once triangle *ABC* is assumed plastic due to principal stresses $-2k$ and 0, on planes parallel and perpendicular to the axis of the slug, respectively. The Mohr circle for triangle *ABC* is shown as 1 in Fig. 13.36(*b*); the stresses on plane *AC*, defined by letters *AC* are easily found. Since region *ACJ* is plastic, the circle for region 3, using stress co-ordinates *AC*, is easily arrived at. Each Mohr circle for each region is shown separately in Fig. 13.36(*b*) and the reader should easily be able to construct the successive circles by traversing the lines of stress discontinuity which become known as the problem is solved.

It transpires that the normal stress on *AH* is $5k$, so that the extrusion pressure $\bar{p}$ is given by

$$\bar{p}.H = 5k(H - h).$$

Hence,
$$\bar{p}/2k = 5r/2. \tag{13.35}$$

This lower bound is not greatly in deficit of the slip-line field value for values of r between 0.67 and about 0.8; it becomes increasingly unsatisfactory as the reduction increases. More useful patterns of stress discontinuity for the larger reductions have been given by ELLIS (1967).

13.13.8 *Hot-rolling*

A lower bound to the force and torque in one instance in the hot-rolling of strip has been given by ALEXANDER and FORD (1963), by adapting a solution originally given by WINZER (1951) and treating the roll as a polygon.

13.14 **An Analogy: Minimum Weight Frames: Michell Structures**

Consider a plane strain side extrusion process using a punch *A* and fitted bottom die *O*, both so rough that the extruded metal shears along a portion of the face of each (defined by *AH*, *AM*, *OI* and *ON*), and both equally inclined to the centre line of the orifice *XX*, see Fig. 13.37(*a*). If it is required that an upper bound to the mean vertical punch face pressure $\bar{p}$ be found, this may be done after constructing a valid system of tangential velocity discontinuities. (We follow the notation described in Section 13.12).

Regions *B* and *C* are dead metal zones and *R* refers to the extruded product. In terms of the plane strain yield shear stress, k, we have for the side extrusion process, to which the corresponding hodograph is that shown in Fig. 13.37(*b*).

$$\bar{p}.H./k = \sum s_{HK}v_{hk}. \tag{13.36}$$

s_{HK} means the length of line between regions *H* and *K* and v_{hk} is the velocity discontinuity across it; the $\sum$ sign refers to all such products. After measuring values of s_{HK} in the physical plane diagram and values of v_{hk} from the hodograph, $\bar{p}$ may be computed. The next step might be to try to improve on this estimate of $\bar{p}$ by selecting another valid pattern of discontinuities which will give rise to a lower value of $\bar{p}$.

Now let us consider the pin-jointed framework shown in Fig. 13.37(*c*), which carries a vertical load *P* at a fixed distance from a wall *H*. Further, let the framework be fixed with frictionless pins to the vertical wall at *X* and *Y*,

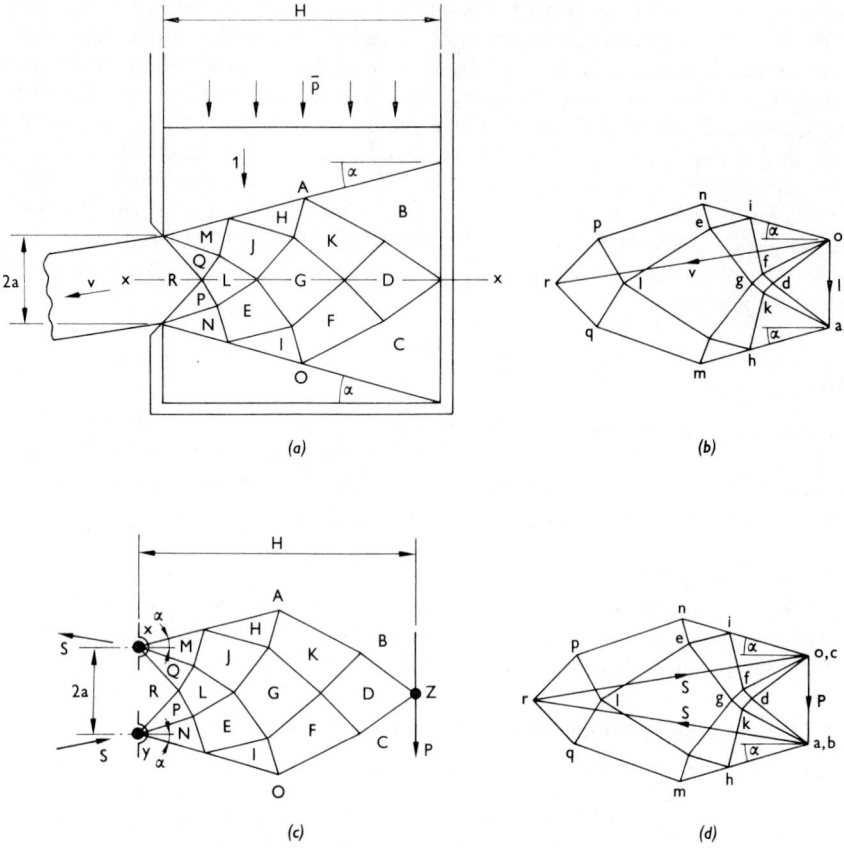

FIG. 13.37

(a) A system of tangential velocity discontinuities for the case of plane strain side extrusion using a punch, the rough face of which is not normal to the container axis.

(b) Hodograph to (a).

(c) A two-dimensional pin-jointed framework carrying a vertical load of P at Z and supported at X and Y.

(d) The force diagram to (c).

where the reactions are S and where XY is given as $2a$. Each member of the framework is denoted by the capital letters in the space on either side of it, e.g., HK. Now denote by A_{hk} the cross-sectional area of each member and by s_{HK} its length. If w_{hk} denotes the weight per unit volume of bar, then the total weight of the whole structure, W, is given by

$$W = \sum w_{hk} s_{HK} A_{hk}. \qquad (13.37)$$

The force diagram for the structure in Fig. 13.37(c) is shown in Fig. 13.37(d). Then if F_{hk} is the force in member HK, and σ_{hk} is the working stress (compressive or tensile) in the member, (13.37) can be rewritten

$$W = \sum w_{hk} s_{HK} \frac{F_{hk}}{\sigma_{hk}} = \sum \frac{w_{hk}}{\sigma_{hk}} s_{HK} F_{hk}.$$ (13.38)

Suppose now that w_{hk}/σ_{hk} is the same for every bar, say k', then (13.38) becomes

$$W/k' = \sum s_{HK} F_{hk}.$$ (13.39)

We observe that the structure in Fig. 13.37(c) is identical in form with the field of velocity discontinuities in Fig. 13.37(a), and that the hodograph of Fig. 13.37(b) is identical with the force diagram of Fig. 13.37(d). Further, (13.36) and (13.39) are formally identical. It follows then that to the extent to which $\bar{p}H/k$ of (13.36) provides an upper bound to the extrusion load, so W/k' is an upper bound to the weight of a structure of given k' value, restricted to points X, Y and Z being prescribed and the inclination of outer bars AM, AH, ON and OI also being prescribed.

An analogy is thus described which enables full use to be made of either known slip-line field solutions or systems of velocity discontinuities as used in analysing plane strain metal-working operations and the design or selection of nearly minimum weight pin-jointed frames having certain prescribed loading points. This, using other examples is described elsewhere, JOHNSON, 1962. A particular case is that in which all members of the frame are of the same material and working at the same stress level. Further solutions for this problem for different H/a values, when α is prescribed (i.e., is positive, zero or negative) could easily be obtained. A bibliography and formal proofs regarding this topic will be found in a recent paper by HEGEMEIER and PRAGER, 1968, and the paper by JOHNSON, et al. (1971) discusses the drawing of Williot diagrams for determining the deflection at each point in the frame. The Australian, Michell, in a paper in 1904 first discussed minimum weight frames.

13.15 Upper Bounds for Anisotropic Metals

The method of finding upper bounds to the load to work *isotropic* metals under conditions of plane strain using the idea of tangential velocity discontinuities can be easily extended to include metals which possess an anisotropic yield criterion, i.e. equation (4.16). An upper bound load is obtained as above by summing products $k.u.s.$ for each line element of a chosen pattern of velocity discontinuities, except that in the case of *anisotropic* metals it must be noted that the yield shear stress, k, varies with direction; a polar diagram in effect of, k versus direction, may easily be constructed for given anisotropic parameters to facilitate this operation. Such a diagram and examples of their use will be found in a paper by JOHNSON, MALHERBE and VENTER (1972).

See Problems 46–55, 59, 60 and 62.

16—EP   *   *

REFERENCES

ALEXANDER, J. M. — 1961 'On Complete Solutions for Frictionless Extrusion in Plane Strain' *Q. J. app. Math.* **19**, 31

ANSOFF, H. I. — 1948 Brown University Technical Report No. 23

AVITZUR, B. — 1963 'An Upper Bound Approach to Cold-strip Rolling' *J. Engng Ind. A.S.M.E.* Paper No. 63—Prod. 8

BISHOP, J. F. W. — 1956 'An Approximate Method for Determining the Temperatures Reached in Steady Motion Problems of Plane Plastic Strain' *Q. Jl. Mech. appl. Math.* **9**, 236

COLLINS, I. F. — 1972 'A Simplified Analysis of the Rolling of a Cylinder on a Rigid-Perfectly Plastic Half-Space' *Int. J. mech. Sci.* **14**, 1

DRUCKER, D. C. and CHEN, W. F. — 1968 *Engineering Plasticity* C.U.P., p. 129

DRUCKER, D. C., GREENBERG, W. and PRAGER, W. — 1951 'The Safety Factor of an Elastic Plastic Body in Plane Strain' *Trans. A.S.M.E.* **73**, *J. appl. Mech.* 371

ELLIS, F. — 1967 'Stress Discontinuities in Plane Plastic Flow' *J. Strain Analysis*, **2**, 52

FORD, H. and ALEXANDER, J. M. — 1963 *On the Limit Analysis of Hot Rolling* Prager Anniversary Volume Macmillan, New York

GREEN, A. P. — 1951 'The Compression of a Ductile Material Between Smooth Dies' *Phil. Mag.* **42**, 900

— 1952 'Calculations on the Theory of Sheet Drawing' *B.I.S.R.A. Report*, MW/B/7/52

— 1953 'The Plastic Yielding of Notched Bars Due to Bending' *Q. J. Mech. appl. Maths.* **6**, 223

— 1954 'A Theory of the Plastic Yielding Due to Bending of Cantilevers and Fixed Ended Beams' *B.I.S.R.A. Report* MW/B/9/54

GVOZDEV, A. A. (Trans. by HAYTHORNTHWAITE, R. M.) — 1960 'The Determination of the Value of the Collapse Load for Statically Indeterminate Systems Undergoing Plastic Deformation' *Int. J. mech. Sci.* **1**, 322

HEGEMEIER, W. and PRAGER, W. — 1969 'On Michell Trusses' *Int. J. mech. Sci.* **11**, 209

HILL, R.   1950   *The Mathematical Theory of Plasticity*
O.U.P.
1951   'On the State of Stress in a Plastic-rigid
Body at the Yield Point'
Phil. Mag. **42**, 868

JOHNSON, W.   1958a   'Over-estimates of Load for Some Two-
dimensional Forging Operations'
Proc. 3rd U.S. Cong. appl. Mech. 571
1958b   'Upper Bound Loads for Extrusion Through
Circular Shaped Dies'
Appl. Sci. Res. Series A, **7**, 437
1959   'Estimation of Upper Bound Loads for
Extrusion and Coining Operations'
Proc. Instn mech. Engrs **173**, 61
1962   'An Analogy Between Upper Bound
Solutions for Plane Strain Metal Working
and Minimum Weight Two-dimensional
Frames'
Int. J. mech. Sci. **3**, 239
1964   'An Approximate Treatment of Metal
Deformation in Rolling, Rolling Contact
and Rotary Forming'
Int. J. Prod. Res. **3**, 51

JOHNSON, W.   1972   'Upper Bounds to the Load for the Plane
DE MALHERBE, M. C.   Strain Working of Anisotropic Metals'
and VENTER, R.   *J.M.E.S.* **14**, 280

JOHNSON, W.,   1958   'Single-Hole Staggered and Multi-Hole
MELLOR, P. B. and   Extrusions'
WOO, D. M.   *J. Mech. Phys. Solids* **6**, 203

JOHNSON, W.   1964   'On Heat Lines or Lines of Thermal
BARAYA, G. L. and   Discontinuity'
SLATER, R. A. C.   *Int. J. mech. Sci.* **6**, 409.

JOHNSON, W.,   1971   'The displacement field and its significance
CHITKARA, N. R.,   for certain minimum weight two
REID, S. R. and   dimensional frames using the analogy with
COLLINS, I. F.   perfectly plastic flow in metal working'
Int. J. mech. Sci. **13**, 547

MASSEY, H. F.   1921   'The Flow of Metal During Forging'
Proc. Manchester Association of Engineers,
Nov. 1921

PIISPANEN, R.,   1967   'Plastic Processes During Rolling, II'
ERIKSSON, R. and   *Bänder, Bleche und Röhre* **8**, 819
PIISPANEN, O.

SLATER, R. A. C.   1965   'Velocity and Thermal Discontinuities
          –66   Encountered during the Forging of Steels'
Proc. Manchester Association of Engineers,
No. 5

TANNER, R. I. and   1960 'Temperature Distributions in Some Fast
  JOHNSON, W.   Metal-working Operations'
                                Int. J. mech. Sci. **1**, 28
WINZER, A.   1951 'Solution to the Rolling Problem for a
                                Strain-hardening Material by the Method of
                                Discontinuities'
                                J. appl. Mech. **18**, 90

Chapter 14

MECHANICS OF METAL FORMING II

14.1 Introduction

In Chapter 12 slip-line field theory was developed to analyse the deformation of a rigid-perfectly plastic solid under conditions of plane strain. The limitations of the theory were there stressed and extrusion was employed to demonstrate one application of slip-line field theory to the mechanics of metal forming. In this chapter, the theory is applied to machining, indenting forging, cutting and rolling under conditions of plane plastic strain. Ample use is also made of the methods described in Chapter 13. Our main aim is to use these techniques to predict the modes of deformation, to facilitate the calculation of the details of metal flow, and the loads and pressures which may be encountered.

14.2 Machining

14.2.1 General

The feature common to most machining operations is that of using a wedge-shaped tool to remove a thin strip from a large deep block. The action of the wedge is depicted in Figs. 14.1(*a*) and (*b*); the former shows the general case of oblique cutting and the latter orthogonal machining, the cutting plane being parallel to that of the work surface. For the latter case, plane strain or two-dimensional flow is assumed to occur. This is substantially the case provided that the tool width is large compared with the chip thickness.

The angle between the face of the tool—the rake face—and the vertical through the cutting point, α, is known as the rake angle.

Depending on whether the material cut is ductile or brittle, the strip formed by the tool is either continuous, or a series of discontinuous chips, see Figs. 14.2(*a*) and (*b*). Obviously, the behaviour of the material varies with strain rate and temperature.

In connection with continuous chip formation, two modes of deformation are usually identified:

1. Figure 14.3(*a*) depicts simple shear in a narrow plane which radiates from, and ahead of, the tip of the cutting tool to the undeformed surface; the angle between this plane and the direction of motion of the tool is denoted by ϕ and is called the shear plane angle.

2. Figure 14.3(*b*) shows a 'built-up' nose of dead metal attached to the tip of the tool; the chip cut rides over it and up the tool surface. From time to time, the nose breaks off and another is re-formed. The nose is responsible for roughness in the machined surface.

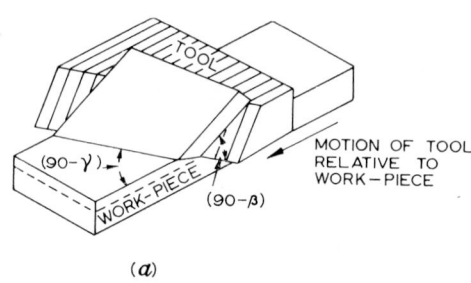

(*a*)

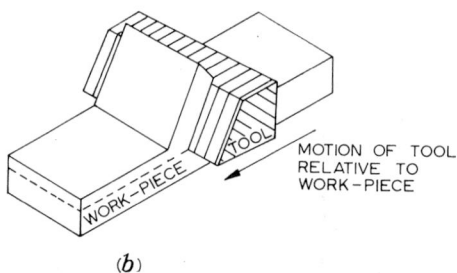

(*b*)

Fig. 14.1 Wedge-shaped cutting tools in action.

(*a*) Oblique cutting.
(*b*) Orthogonal cutting.

In discontinuous chip machining, see Fig. 14.3(*c*) the tool when at *LOM* is about to create a new chip, the previous chip having broken away to leave machined surface *MOBA*, line *AX* representing the original uncut surface. When the tool point has advanced to *F*, in effect it has penetrated the stationary surface *OBA* and the displaced material is piled up, forming a new surface, *BC*. Increase of *d* extends the zone of deformation *BCFED* until intersection with the free surface *AX* causes the chip to break away. The process of forming a new chip then commences again.

In a review of the mechanics of the cutting process, PUGH (1958), remarks that 'no theory yet produced is in quantitative agreement with experiment, even for a limited range of cutting conditions . . . where there is no built-up edge involved and no chatter occurs. This is perhaps not surprising, since none of the theories take into account the large strain rates, work-hardening and temperatures which undoubtedly exist in the process and their effects on the properties of the work-material'. Thus, whilst the major contributions to this topic are described below, the reader should not assume that results derived are immediately applicable with exactness.

14.2.2 *Ernst and Merchant's Theory* (1941)

These workers attempted an analytical solution to the cutting problem, other people having earlier suggested the physical model; it is based on the existence of a single shear plane. For simplicity, assume the tool to be stationary and the work-piece to be moving from left to right, see Fig. 14.3(*a*).

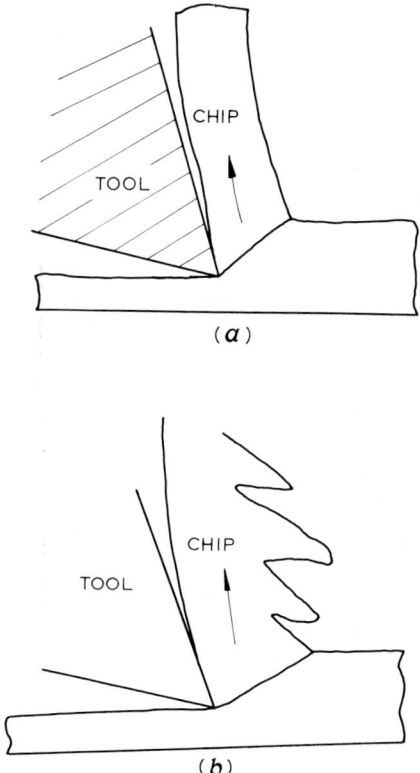

(*a*)

(*b*)

Fig. 14.2 The two kinds of chip formation.

(*a*) Continuous.
(*b*) Discontinuous.

Each layer, e.g., that shown shaded, is rigid until it encounters the shear plane *ST*, where it is sheared instantaneously and thus caused to slide again as a rigid body, parallel to the tool face *TR*.

Consider the relations necessary for force equilibrium per unit width of the chip. F is the frictional force and N the normal force at the chip-tool interface and F_N and F_s are the normal and tangential force components on plane *ST*.

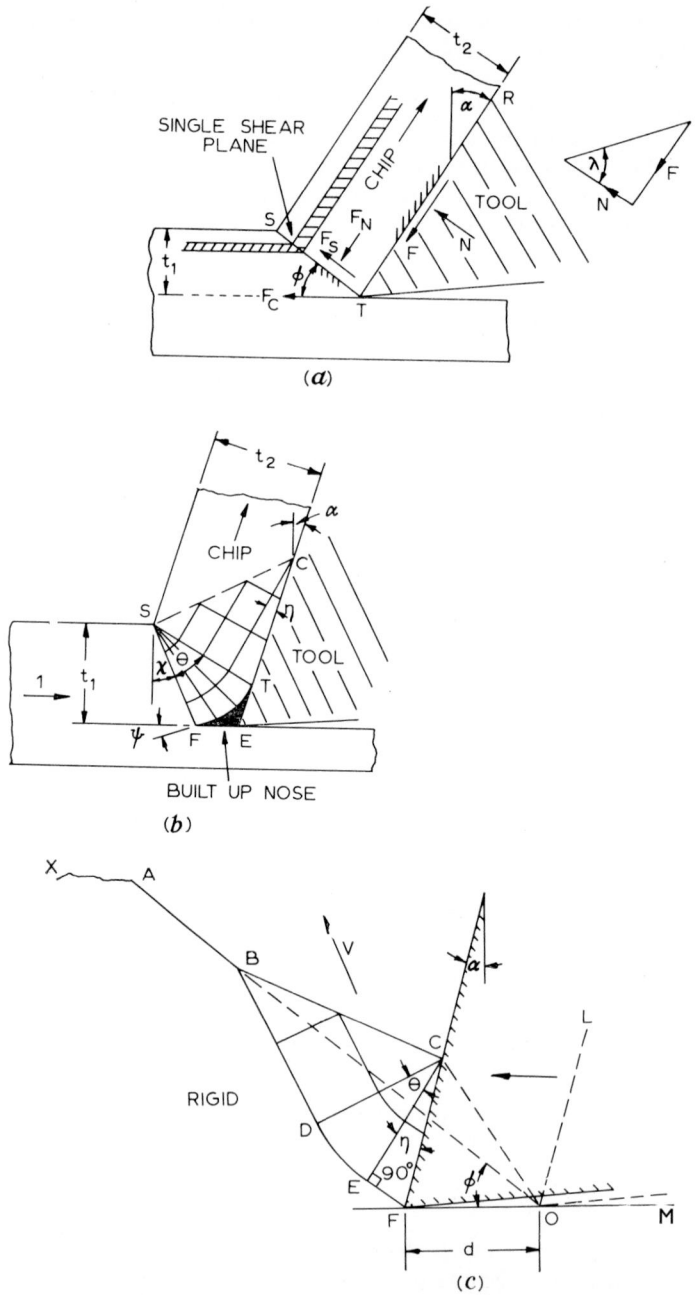

Fig. 14.3 The modes of deformation in chip formation.

(*a*) Single shear plane.

(*b*) Built-up nose.

(*c*) Slip-line field for discontinuous chip formation: this changes in scale only.

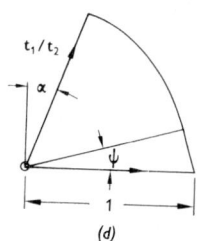

(d)

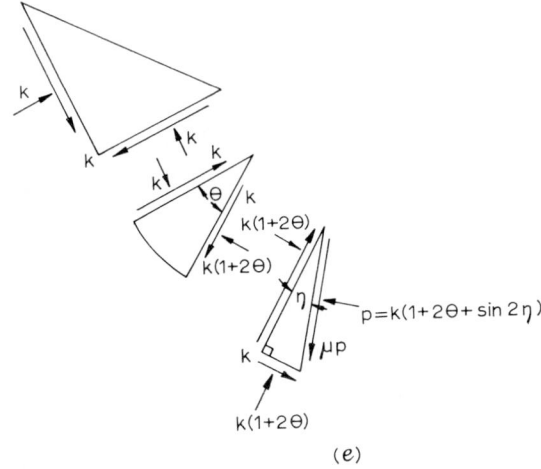

(e)

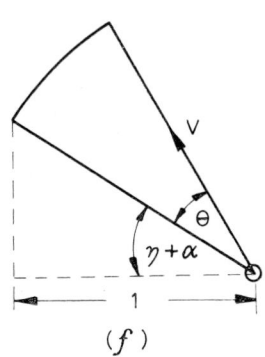

(f)

(d) Hodograph for case (b).
(e) Stress distribution for case (c).
(f) Hodograph for case (c).

Thus

$$F_s = \sqrt{F^2 + N^2} \cos(\phi + \lambda - \alpha). \tag{14.1}$$

Also, $$F_s = k(ST) = kt_1/\sin\phi \tag{14.2}$$

if k is the plane strain yield shear stress of the material.

Hence, $$\sqrt{F^2 + N^2} = \frac{kt_1}{\sin\phi \cos(\phi + \lambda - \alpha)}. \tag{14.3}$$

The cutting force F_c is the horizontal component of $\sqrt{F^2 + N^2}$, i.e., of the total force applied by the tool. Thus

$$F_c = \sqrt{F^2 + N^2} \cos(\lambda - \alpha)$$

$$= kt_1 \frac{\cos(\lambda - \alpha)}{\sin\phi \cos(\phi + \lambda - \alpha)}. \tag{14.4}$$

An assumption is now made that the shear plane takes up that value of ϕ which makes F_c a minimum; this is the same as saying that a moving cutting tool performs a minimum amount of work.

For F_c to be a minimum, $\sin\phi \cos(\phi + \lambda - \alpha)$ requires to be a maximum, and this occurs when

$$2\phi - \alpha + \lambda = \pi/2. \tag{14.5.i}$$

Hence, substituting in equation (14.4)

$$F_c = \frac{2kt_1 \cos(\lambda - \alpha)}{1 - \sin(\lambda - \alpha)} \tag{14.6}$$

$$= 2kt_1 \cot\phi. \tag{14.7}$$

The angle ϕ may be obtained experimentally from the chip thickness ratio r_c

since $$r_c = \frac{t_1}{t_2} = \frac{(ST)\sin\phi}{(ST)\sin\left(\dfrac{\pi}{2} - \phi + \alpha\right)} = \frac{\sin\phi}{\cos(\phi - \alpha)}$$

and hence, $$\tan\phi = \frac{r_c \cos\alpha}{1 - r_c \sin\alpha}. \tag{14.8}$$

MERCHANT (1945) conducted orthogonal machining tests on synthetic plastics and steel, but found the above expressions satisfactorily verified only for the plastics. The results for steel were only brought into agreement with the theoretical relations by assuming that the value of the yield stress was some function of the normal stress on the shear plane—a hypothesis generally regarded as inadmissible at ordinary metal-working levels. Merchant's results were well correlated by

$$2\phi° + \lambda° - \alpha° = 80°. \tag{14.5.ii}$$

14.2.3 *The Theory of Lee and Shaffer* (1951)

Ernst and Merchant's analysis deals only with the forces on the chip and makes no statement about stress distribution. Lee and Shaffer applied slip-line field theory and thereby ensured that the yield stress of the material was not exceeded in at least part of the chip.

Material is again imagined to move from left to right with, say, unit speed against a stationary tool, the coefficient of friction between the tool and material being $\mu = \tan \lambda$. On crossing shear plane *ST*, the material is supposed to move parallel to the tool face, the direction being altered instantaneously. The use in this process of the concept of the rigid-perfectly plastic material is said to be justified because:

1. for most metals the work-hardening rate falls to small values for large strains and so reaches a near-constant saturation stress,

2. the high strain rates which accompany the machining operation are said to raise the yield strength of the material and to make it approximate the idealized solid.

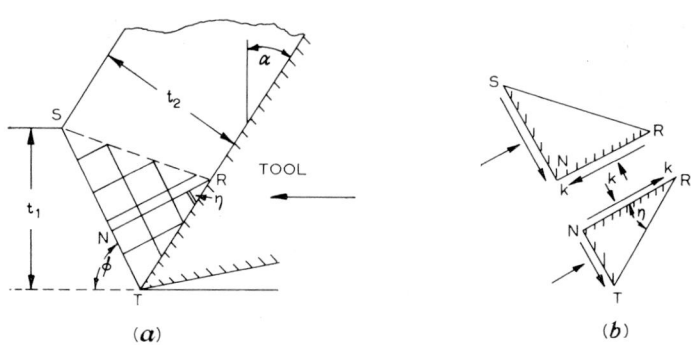

(a)   (b)

Fig. 14.4 Lee and Schaffer's slip-line field solution for a single shear plane.

(a) Slip-line field.
(b) Stress distribution.

The slip-line field proposed by Lee and Shaffer consists of a single triangle of orthogonal lines, *RST*, see Fig. 14.4, in which the whole of the material is raised to its yield point. It is easy to show that

$$\mu = \tan \lambda = \cos 2\eta/(1 + \sin 2\eta)$$

and thus

$$n = \pi/4 - \lambda. \tag{14.9}$$

But $\eta + \alpha = \phi$ and using equation (14.9)

$$\phi - \alpha + \lambda = \pi/4. \tag{14.10}$$

The machining force is, by considering the shear plane forces

$$F_c = ST (\cos \phi + \sin \phi) k$$
$$= k(1 + \cot \phi) t_1$$
$$= k \left[(1 + \cot\left(\frac{\pi}{4} + \alpha - \lambda\right)\right)\right]. \tag{14.11}$$

The chip thickness ratio, r_c is

$$r_c = \sin \left(\frac{\pi}{4} + \alpha - \lambda\right) \bigg/ \cos \left(\frac{\pi}{4} - \lambda\right). \tag{14.12}$$

If α is negative, say $-10°$, and with, say, $\lambda = 35°$, $\cot (\pi/4 + \alpha - \lambda)$ becomes infinite. It is obvious then that excessively high values of F_c can easily arise. Thus this solution is not tenable for certain values of the parameters α and λ. A solution which assumes a built-up nose on the tool tip may alternatively provide smaller and therefore more probable solutions. The slip-line field associated with the built-up nose with friction between the tool and sliding material is shown in Fig. 14.3(b) and the hodograph in Fig. 14.3(d). The cutting force is composed of the horizontal components of the normal and frictional shear stress along TC, F_c', the normal and yield shear stress components summed along the upper surface of the nose F_c'' and the frictional stress acting along FE, F_c'''. The normal stress along the plane ST is k because across SC there is no net force, and thus the first two components ($F_c' + F_c''$) of the cutting force are obtained as follows.

Note as before in equation (14.9) that $\eta = (\pi/4 - \lambda)$ and from simple geometry

$$\chi = \frac{\pi}{4} - \alpha + \lambda - \theta.$$

Thus if F is the horizontal cutting force acting on SF on which the normal pressure is p,

$$F_c' + F_c'' = F = SF (k \sin \chi + p \cos \chi)$$
$$= kt_1 (\tan \chi + p/k)$$
$$= kt_1 (1 + 2\theta + \tan \chi) \tag{14.13}$$

using the Hencky equations.

The value of χ is determined by the friction between the bottom of the nose and the cut surface. The force F_c''' acting along FE then follows.

14.2.4 *Remarks on the Comparison Between Theoretical and Experimental Results*

To distinguish between the three theoretical estimates for the cutting force, it is usual to choose that which requires the least force.

A test of the theories is to compare their predictions, see equations (14.5) and (14.10), with experiment by plotting ϕ against $(\lambda - \alpha)$. That the disagreement between them is quite marked has been confirmed by several independent workers. (See PUGH'S review.) These discrepancies have their origin principally in

1. theoretical over-idealizations, and
2. certain fundamental objections to the mechanics employed in analysing the process.

The first group comprises the assumption of achieving pure two-dimensional flow in experiments, the absence of strain-hardening, the neglect of any effects due to temperature on the material properties, and lubrication. The second group stems from the fact that any slip-line field solution is only an upper bound solution for the particular configuration chosen. (This implies that the single shear plane and the built-up nose solution are not mutually exclusive.) It is quite possible that the yield stress is exceeded at some particular singularity by assuming a solution and not checking this point. Using certain tests concerning stress singularities HILL (1954) has been able to determine ranges of single shear plane angle which are not valid. In all the analyses so far presented, it is assumed that a steady-state exists, but in fact, the whole machining process is one in which the mode of deformation is not clearly dictated by the tooling as in rolling and extrusion. The constraints appear to be insufficiently positive to admit of any well-defined steady-state process existing and being maintained. However, experiments by Low (1962) and Low and WILKINSON (1962) showed that initial conditions, which militate against the establishment of a steady-state, have no effect on the final deformation pattern.

Mention should be made of the work by CHRISTOPHERSON, PALMER and OXLEY (1956, 1959), in which account of strain-hardening is taken. Estimates of cutting force, etc., were made, using Hencky equations adapted to allow for strain-hardening, i.e., for values of k that vary with strain. In slow-speed experiments the flanks of the work machined, chip and tool were observed directly through a microscope and a film record taken. The flank of the work was polished and etched, and individual crystals were followed to ascertain the streamlines. It was possible to construct an orthogonal net of slip lines to which the modified Hencky equations could be applied, see Fig. 14.5(a). The usual correlation of theory and experiment was by this method improved, though discrepancies remained. The picture which emerges from their work is exemplified in Fig. 14.5 (c). Their conclusions were:

1. The plastic zone covers a considerable area as opposed to the plane required by the perfectly plastic solid.
2. The chip curls in the plastic zone.
3. Contact between chip and tool occurs in a zone some distance up the tool face, and there the deformation is elastic.
4. Tensile stresses occur near the tool tip and compressive stresses at the free surface.

14.2.5 *Bibliographical Note*

As the literature on the mechanics and physics of the machining process is immense, it must suffice to mention just a few references that any bibliography

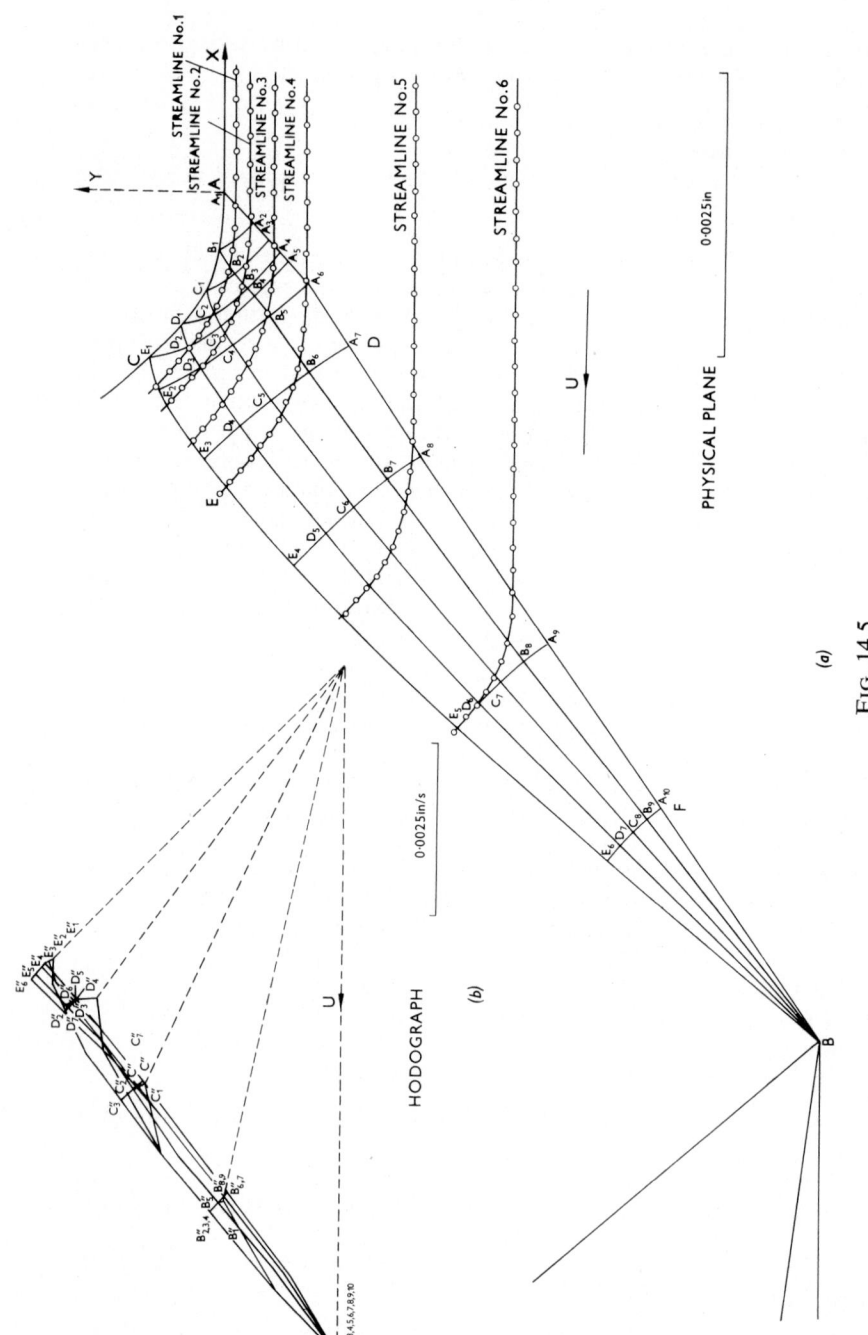

(a)

(b)

FIG. 14.5

(a) and (b) Slip-line field constructed for work-hardening theory and corresponding hodograph. (*After* Oxley and Palmer, *Proc. Instn mech. Engrs.*)

would include. The physics of the process includes items such as detailed discussions of diffusion-wear mechanisms on the tool face (LOLADZE, 1967) and the role of surface energy in cutting (RUBENSTEIN, 1967). The book by THOMSEN, YANG and KOBAYASHI (1965) deals with machining differently from that here, as does that by BOOTHROYD (1965). A very helpful supplement or alternative to this section is given by ALEXANDER and BREWER (1963).

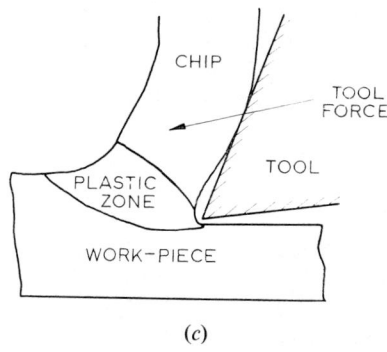

(c)

FIG. 14.5

(c) Showing a thickened shear plane.

Further useful papers are a very helpful survey of the mechanics of machining given by SHAW (1963) and PEKELHARING (1967), an extensive study of the built-up nose phenomenon by HEGINBOTHAM and GOGIA (1961), OXLEY, HUMPHREYS and LARIZADEH (1961) and OXLEY (1964) on rate effects in machining, FINDLEY and REED (1963) on extreme speeds and rake angles in metal cutting and a useful report by ENAHORO (1964). The most comprehensive summary of metal cutting mechanics yet produced is the book by ZOREV (1966). Two large sources of reference are the *Proceedings of the International Conference on Production and/or Manufacturing Engineering* (1963 A.S.M.E. and 1967 A.S.T.M.E.). A continuing and extensive source is the *Proceedings of the Machine Tool Design and Research Conferences* held bi-annually in Manchester and Birmingham, published by Pergamon Press.

14.2.6 *Discontinuous Machining*

The model whereby a chip is formed in Lee's theory (1954) has already been described. A slip-line field solution is framed, see Fig. 14.3(c), which is clearly an adaptation from an earlier solution in a paper by HILL, TUPPER and LEE (1947) entitled 'The Theory of Wedge Indentation in *Ductile* Materials'. To apply this same solution expressly to brittle materials is hardly consistent. It is briefly described for its exploratory value.

The reader should have no difficulty in appreciating Lee's slip-line field solution. Triangle BCD is an isosceles triangle, the material in it being subject to the compressive yield stress perpendicular to the surface BC. DCE is a centred-fan and triangle CEF is so defined as to give the required coefficient of friction, μ, over the length of tool face FC. The stress distribution is shown in Fig. 14.3(e). The following results can easily be derived. By resolving forces over length FC

$$\mu = \cos 2\eta/(1 + 2\theta + \sin 2\eta), \qquad (14.14)$$

the frictional stress along FC being $k \cos 2\eta$ and the normal stress,

$$k(1 + 2\theta + \sin 2\eta). \qquad (14.15)$$

The force applied by the tool is simply proportional to d and in principle is thus easily calculable from a knowledge of the rake angle and the inclination of the supposed surface of fracture ϕ.

From some particular calculations, Lee has shown that the apparent shear plane BF, Fig. 14.3(c), is little influenced by the value of μ.

Fig. 14.3(f) is the hodograph corresponding to Fig. 14.3(c) at a particular instant.

For an account of recent experimental work in this field and further bibliography see the paper by BANNERJEE and PALMER, (1966). The growth of discontinuous chips according to Merchant which is tacitly employed by LEE is shown in Fig. 14.6(i); but a different growth mechanism is proposed by Bannerjee and Palmer, basing their results on films taken of the process, see Fig. 14.6(ii). Relevant to this topic is the study of plastic deformation when cutting into an inclined plane, see PALMER, (1967).

14.2.7 Oblique Machining

The direction in which the chip flows over the rake face of the tool is inclined to the cutting edge at $(90 - \beta)$, see Fig. 14.1(a). It is found that approximately $\beta = \gamma$ for a wide range of cutting conditions, STABLER (1965), and this fact is used in the design of drills and milling cutters to ensure efficient chip disposal. Experiments by RAPIER and WILKINSON (1959) confirmed earlier work by SHAW, COOK and SMITH (1952) that a definite relationship between β and γ exists, but that a considerable deviation from equality can occur.

14.2.8 Machining Using a Restricted Contact Tool

If one examines the mechanics of the orthogonal machining operation using a tool which has a rake face of (theoretically) infinite length, it will be seen that the length of contact between the chip and the tool is unknown and ill-defined. It is common to divide this contact zone into two regions. Over the lower portion the shear stress is constant and equal to a fraction of the shear stress of the work-piece whereas over the remaining upper contact region the frictional stress decreases following an empirical $n - $ power law: WALLACE and BOOTHROYD, (1964), BHATTACHARYYA, (1965). In machining with a tool which only permits a fixed length of contact between chip and rake face, it can be seen that the definition of the physical situation prevailing is clearer, and it is reasonable to assume a constant frictional stress without the complicating presence of the second zone. For this reason, the restricted contact tool is a better object of study than the infinitely long rake face tool.

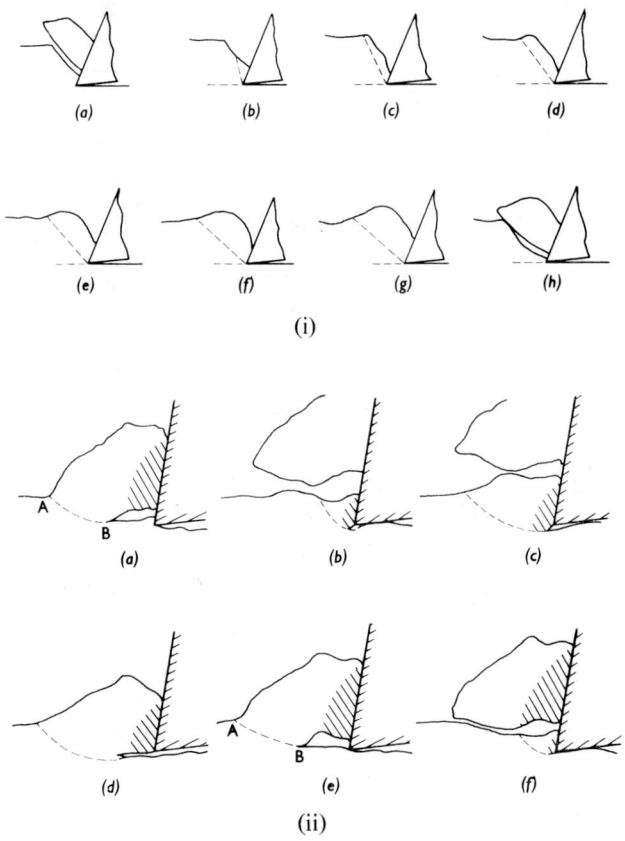

FIG. 14.6

(i) Discontinuous chip formation according to Merchant.
(ii) Discontinuous chip formation according to Bannerjee and Palmer.

Figure 14.7(a) shows a restricted contact tool, with the work-piece material deforming on passing through a single shear plane at some as yet unspecified shear plane angle ϕ. For simplicity assume the tool to be stationary and that the workpiece is moving from right to left at unit rectilinear speed. From the hodograph as shown in Fig. 14.7(b) we have,

$$\frac{u_{BC}}{\sin(90-\gamma)} = \frac{v}{\sin\phi} = \frac{1}{\sin[180-\phi-(90-\gamma)]}$$

or

$$\frac{u_{BC}}{\cos\gamma} = \frac{v}{\sin\phi} = \frac{1}{\cos(\phi-\gamma)}.$$

Now for a frictionless tool face the rate of energy dissipation is

$$\frac{dW}{dt} = \dot{W} = ksu_{xx} = k.\frac{t_1}{\sin\phi}.\frac{\cos\gamma}{\cos(\phi-\gamma)}$$

or

$$\frac{\dot{W}}{kt_1} = \frac{2\cos\gamma}{\sin(2\phi-\gamma)+\sin\gamma}. \tag{14.16}$$

Thus upper bounds for the rate of working $\dot{W}$ have been found and the actual value depends on the choice of ϕ. It is clearly most logical to assume within the framework of the assumptions made, that that value of ϕ will prevail, or is most appropriate, which makes $\dot{W}$ a minimum.

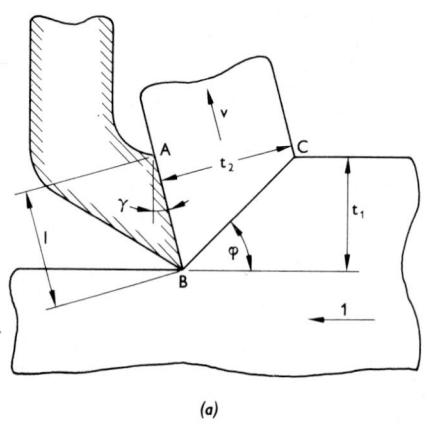

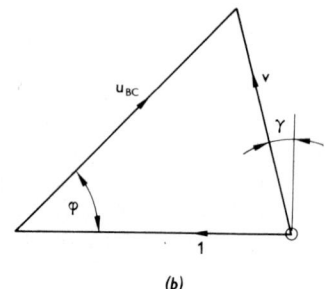

(a)   (b)

FIG. 14.7

(a) Movement in the physical plane with one line of tangential velocity discontinuity at an angle $= \phi$.

(b) Hodograph to (a).

Now $\dot{W}$ is least when

$$[\sin(2\phi-\gamma)+\sin\gamma]$$

is greatest, i.e., is equal to

$$(1+\sin\gamma)$$

or when,

$$2\phi-\gamma = \pi/2,$$

or

$$\phi = \pi/4 + \gamma/2. \tag{14.17}$$

Generally then,

$$\frac{\dot{W}}{2kt_1} = \frac{\cos\gamma}{1+\sin\gamma}. \tag{14.18}$$

When $\gamma = 0$, $\dot{W} = 2kt_1.$

The relationship $\phi = \pi/4 + \gamma/2$ is the same as Merchant's condition for frictionless tools.

When there is shearing friction at the tool face,

$$\dot{W} = ksu_{xx} + klv.$$

The second term on the right-hand side of the above equation is that due to shearing along the tool face.

Thus

$$\frac{\dot{W}}{kt_1} = \frac{\cos \gamma}{\sin \phi \cos (\phi - \gamma)} + \frac{l}{t_1} \sin \phi. \qquad (14.19)$$

If $\gamma = 0$,

$$\frac{\dot{W}}{kt_1} = \frac{1}{\sin \phi \cos \phi} + n \tan \phi,$$

where

$$n = l/t_1.$$

Differentiating $\dot{W}/kt_1$ in (14.19), the value of ϕ which makes it a minimum occurs when $n = 2 \cos (2\phi - \gamma)/\sin^2 \phi$. If $\gamma = 0$ it turns out that,

$$\phi = \sin^{-1} 1/\sqrt{(2 + n)}; \qquad (14.20)$$

substituting,

$$\frac{\dot{W}}{kt_1} = 2\sqrt{(1 + n)}. \qquad (14.21)$$

When

$$n = 1, \quad \phi \simeq 35° \quad \text{and} \quad \dot{W}/kt_1 = 2\sqrt{2}.$$

Generally also, the ratio of the undeformed to the final chip thickness,

$$t_1/t_2 = \sin \phi \cos (\phi - \gamma)$$

and for $\gamma = 0$,

$$t_1/t_2 = 1/\sqrt{1 + n} = 1/\sqrt{2}, \quad \text{when } n = 1.$$

When the optimized values of $\dot{W}$ are determined the cutting force is easily found.

It is not necessary, fortunately, to have to depend on the single shear plane solution described above, but it has been included to indicate the kinematical or the 'upper bound approach', which is seldom found in the general literature on machining. As has been emphasized several times earlier, the single shear plane assumption is best considered as a first approximation and indeed Briks (in 1896) pointed out that it may be improved upon by substituting 'a family of planes arranged fanwise and passing through the edge of the cutting tool', (ZOREV, 1966); this situation is studied and discussed at length by ZOREV and indeed the idea is well substantiated in the strain-hardening slip-line field of OXLEY and PALMER (1958). It is of interest to take Briks' model and to draw a hodograph for it. Note that with the fan of planes there is a progressive (i.e. in the limit) change in the slope of the work-piece surface until it becomes part of the chip; there is not one abrupt change as there is with the single shear plane model (see Section 13.12).

The Bows Notation technique (see Section 13.12) is used in Fig. 14.8(*b*) to give the hodograph corresponding to Fig. 14.8(*a*). The figure shows finite jumps in strain across each fan line but it is easy to progress to a continuous fan of lines—after some initial discontinuity—and a correspondingly continuous hodograph as in Figs. 14.8(*c*) and (*d*). As before, the total strain γ taken on by an element at any stage of its progress is

$$\gamma = \int d\gamma = \int (du_{xx}/v_p),$$

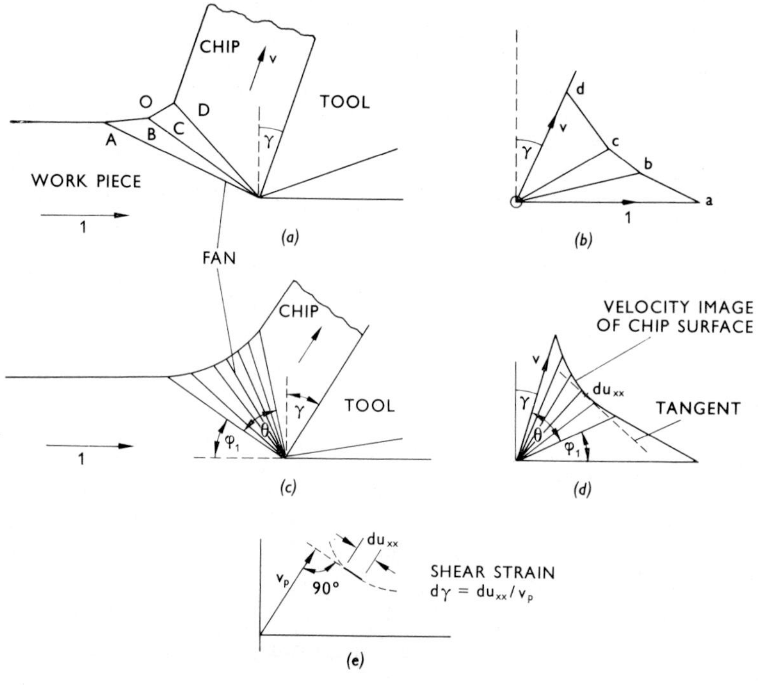

FIG. 14.8

see Fig. 14.8(*e*). The strength of an element is related to τ or σ through the stress-strain curve for the material at the appropriate temperature and rate of strain. It is pertinent to emphasize at this point, that when using the single shear plane theory, the value to be attributed to k should be at least the mean value of the yield shear stress for the work material taken from an appropriate stress-strain curve, over the strain range 0 to γ, i.e., the strain imposed in passing through the shear plane, i.e., (u/v_p); see THOMSON, *et al.* (1969). Slip-line field solutions for this particular kind of machining operation are easily arrived at and in Fig. 14.9 several of them are given. The normal uniform pressure, q, on the tool face for Fig. 14.9(*c*) is easily shown to be,

$$q = k(1 + \pi/2 - 2\gamma).$$

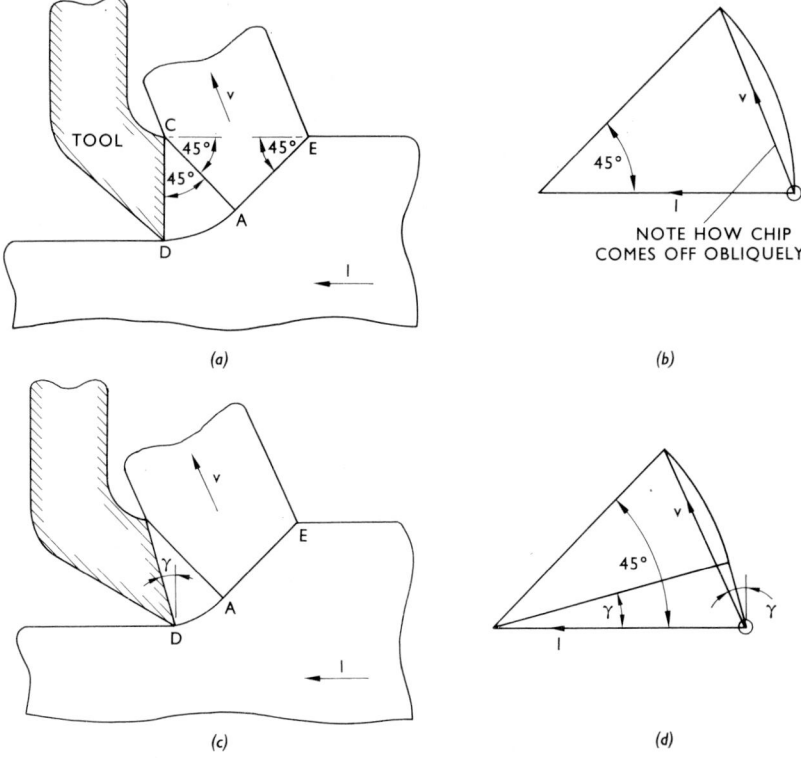

Fig. 14.9

(a) Movement in the physical plane. ACD is a circular sector.

(b) Hodograph to (a).

(c) Physical plane diagram for a tool with a positive rake angle γ along which the work material shears.

(d) Hodograph to (c).

Hence the minimum coefficient of friction required to cause shearing along the tool face is

$$\mu = \frac{k}{q} = \frac{1}{1 + \pi/2 - \gamma} \, . \tag{14.22}$$

Some typical values of $q/2k$ and μ are given in the table below.

$\gamma°$	0	15	30	45
μ	0·39	0·49	0·66	1
$q/2k$	1·28	1·02	0·71	0·5

In this connection Hsu's paper (1965) on normal and shear stresses on a tool face is of interest. It should be observed that the hodographs in Figs.

14.9 (*b*), (*d*) and (*f*), show the chip to be directed along a direction which is more oblique to the vertical than the rake face; this is consistent with what is observed in the actual cutting process. *Only the slip-line field or*, in some cases *the upper bound approach* of the previous chapter, *will make this kind of prediction*. The common force equilibrium approaches, being of a non-kinematical nature, cannot make a prediction of this character. A number of

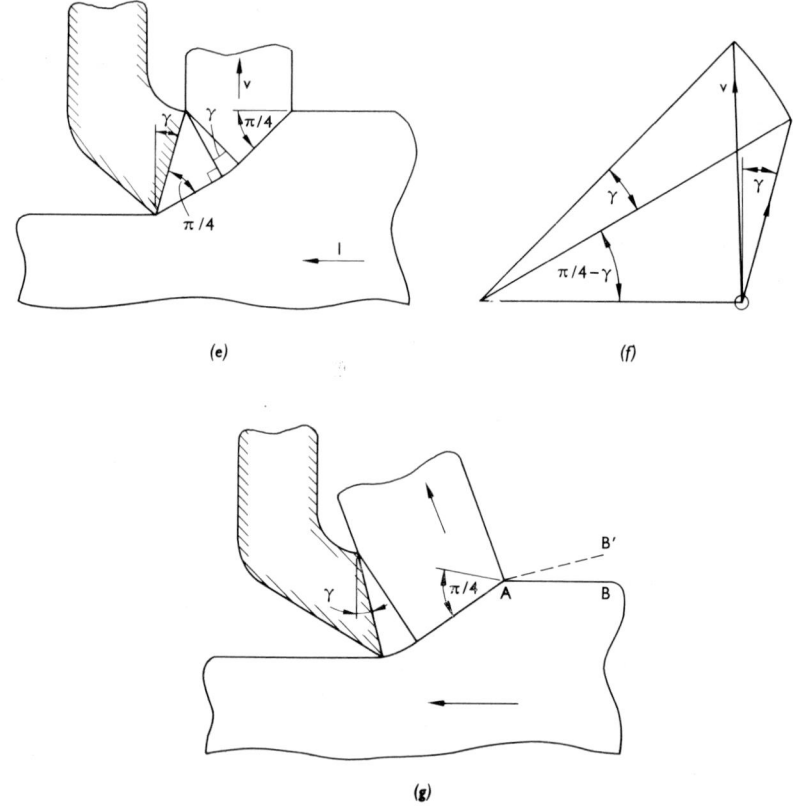

(*e*) (*f*)

(*g*)

FIG. 14.9

(*e*) Physical plane diagram for a frictionless tool of negative rake angle γ.

(*f*) Hodograph to (*e*).

(*g*) Work-piece surfaces AB or AB′ do not pass through top of the tool face.

other simple and useful predictions can easily be made with the help of the kind of slip-line field solutions shown in Fig. 14.9.

(*a*) THE ADIABATIC CHIP TEMPERATURE DISTRIBUTION

In the previous chapter (Section 13.10) it was demonstrated how, assuming that no heat was lost by conduction (we neglect effects at the tool-material interface) and provided that a constant fraction, *f*, of the plastic work done on an element manifests itself as a temperature rise, the distribution of temperature in a worked body may be approximated. For simplicity, we reduce the circular arc of Fig. 14.9(c) to an isosceles triangle see Fig. 14.10(a), and divide the uncut chip t_1 into two stream tubes at a stream line SS. If the material in stream tube I approaches the line of velocity discontinuity be-

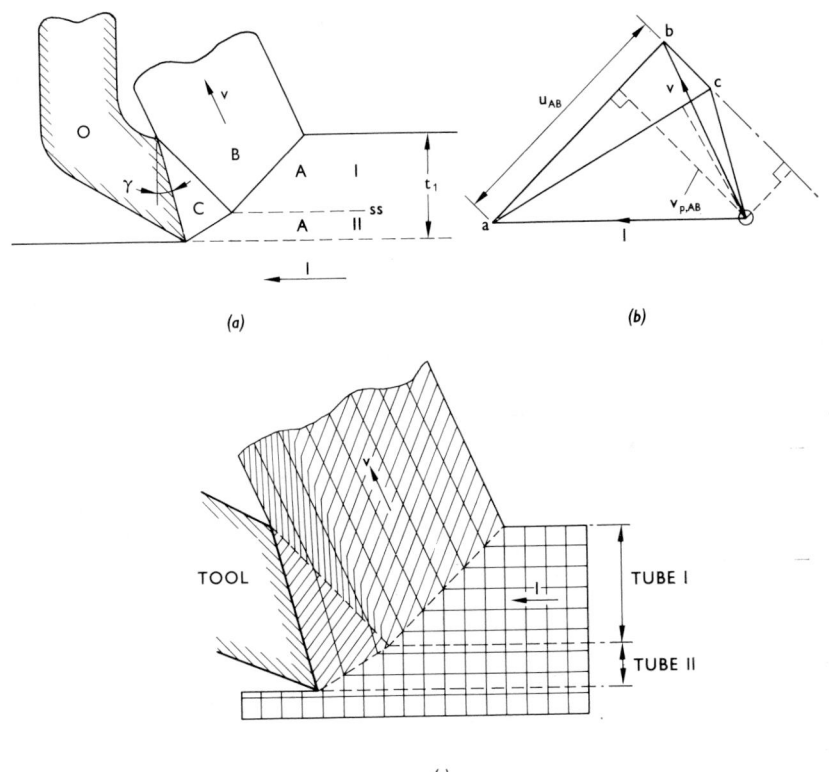

(a)   (b)

(c)

FIG. 14.10

(a) Showing division into two stream tubes for calculating grid deformation and temperature distribution.
(b) Hodograph to (a).
(c) Showing the shape of a square grid of lines in, and after passing through, the region of plastic deformation.

tween the zones B and A (tube I), at unit speed, the rise in temperature in all the particles in tube I, after machining due to crossing AB is,

$$\Delta T_{BA} = f \cdot \frac{k}{J\rho c} \cdot \left(\frac{u}{v_p}\right)_{BA}. \tag{14.23}$$

$(u/v_p)_{BA}$ is the tangential velocity discontinuity divided by the velocity perpendicular to the line BA, see Fig. 14.10(b); it is the shear strain imposed at BA.

Now the material in tube II has to cross *two* lines of tangential velocity discontinuity before it *finally* becomes part of the chip in zone *B*. Thus the rise in temperature in the left-hand portion of the chip *B* will be different from that of the right-hand side; it is given by

$$\Delta T = \Delta T_{AC} + \Delta T_{CB}$$

$$= \frac{f \cdot k}{J\rho c} \left[\left(\frac{u}{v_p} \right)_{AC} + \left(\frac{u}{v_p} \right)_{CB} \right]. \tag{14.24}$$

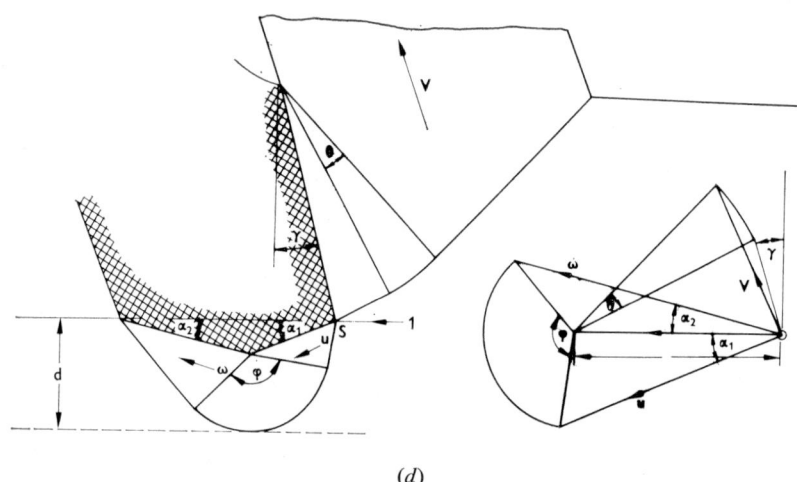

(*d*)

FIG. 14.10

(*d*) Slip-line field and hodograph for restricted tool contact.

(The notation for drawing and describing patterns of lines as referred to in Section 13.12 is used in all parts of this Section).

(b) GRID DEFORMATION PREDICTION

The 'horizontal' lines marked on the uncut work-piece can be traced through the deformation zone with the help of the hodograph; lines of particles which enter the chip across *AB* proceed parallel to $\overrightarrow{ob}$, and those traversing *AC* move parallel to the rake face or $\overrightarrow{oc}$ and then after crossing *CB*, parallel to $\overrightarrow{ob}$. The shape taken up by the 'vertical' lines can only be determined after keeping a careful log of the position to which individual points on them—the intersection of the vertical and horizontal lines, preferably—have moved at specific instants in time. Figure 14.10(*c*) shows the shape taken up by a square grid of lines after passing through the approximated deformation zone. This *simply* obtained distorted grid is a good approximation to what is actually observed.

(c) Strain and Mean Maximum Temperature Jump

The temperature plot of sub-section (a) is no more than a measure of the total shear strain $\gamma = \tan \xi$ imposed, at specific locations. For the solution shown in Fig. 14.7(a) the total shear strain, is

$$\tan \xi = \left(\frac{u}{v} \right)_{BC} = \frac{\cos \gamma}{\sin \phi \cos (\phi - \gamma)}. \tag{14.25}$$

With $\gamma = \phi \simeq 35°$, which is the case when there is shearing friction at the tool surface

$$\gamma = \tan \xi \doteqdot 2.5.$$

The temperature jump across the shear plane BC is thus

$$\Delta \theta = \frac{k}{J\rho c} \times 2.5. \tag{14.26}$$

For mild steel and aluminium at room temperature (and slow strain rate data) equation (14.26) predicts a temperature jump of approximately 300 °C and 100 °C respectively. If we had considered the case of a frictionless tool face, the temperature jump would only be about four-fifths of these figures.

(d) Hardness Variation in the Chip

For the solution as shown in Fig. 14.10(a) the materials in tubes I and II have different strain histories. Because material in tube II is subject to a larger shear strain than that in tube I, before it becomes part of the chip (two, in fact, consecutive increments) we would expect the left-hand side of the chip to be *harder* (for a real work-hardening material) than that on the right. (Annealing due to the general temperature increase in passing through the plastic zone and due to friction at the tool face, are neglected of course.) In making this last statement, we are implying that when machining real work-hardening materials, the strain distribution may be predicted to a first degree by use of the rigid-perfectly plastic model material. A continuous (probably the best) strain distribution pattern would be given by using the deformation mode instanced in Fig. 14.9(c). Hardness transverses across chips show the kind of hardness distribution here predicted: i.e., the hardness value increases as we approach the tool face.

(e) Strain Rate

Consider the solution in Fig. 14.7(a). If the shear zone thickness is assumed to be 2.5 mm, the work-piece speed 100 cm/s and $\tan \xi = 2.5$, then the strain rate $\dot\gamma$ is

$$\dot\gamma \simeq \frac{2.5}{2.5/10) \times (1/100)} = 10^4/\text{s } \dot\gamma$$

In an ordinary tensile test the strain rate is about 10^{-3} per second. Thus in machining the strain rate is about 7 orders of magnitude higher. Hence k the shear yield stress of the material will be much increased when machining occurs as against what it would be measured to be in a conventional tensile test.

(f) THE ACTIVE REGIONS IN METAL CUTTING

CASSIN and BOOTHROYD (1965) have remarked that it is usual to identify five active regions in metal cutting; see Fig. 14.11. It is evident that their region 1 is just the first slip-line encountered across which there is a finite velocity discontinuity; in machining real materials it appears, of course, as a 'line' of finite thickness. Zone 2 in Fig. 14.11 is the region approximately covered by the fan of Fig. 14.9(c); (or the triangle and fan of Fig. 14.9(e)). These two zones, referred to by students of metal cutting as zones of primary and secondary deformation, are in fact encountered in *all* metal-working opera-

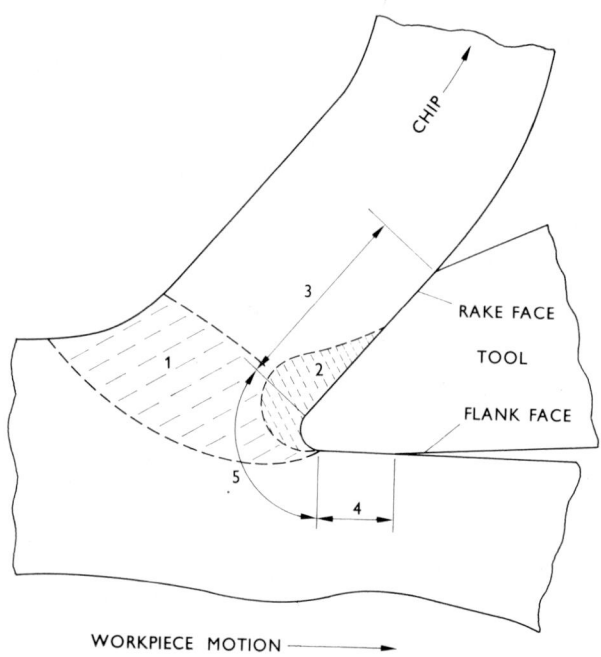

FIG. 14.11 Active regions in metal cutting.
1. Primary deformation zone.
2. Secondary deformation zone.
3. Chip-tool interface.
4. Work-tool interface.
5. Tool edge.

tions to some extent and there is nothing that is of special significance in metal cutting that is not similarly found elsewhere. 'Secondary deformation' in extrusion is the distortion brought about additionally at the boundary, imposed by the tools and the friction between the tools or dies and the work-piece.

If material shears about the tool or rake face, the hodographs of Fig. 14.9(b) and (d) show that the smaller the rake angle the slower, in relation to its original speed, is the passage of particles of the chip contiguous with the rake face across it. When the rake angle is zero, there is no relative velocity between the rake face and particles of the chip in contact with it; there would then appear to be adhesion of work material to the rake face.

It is easy to see that the slip-line fields shown in Figs. 14.9(a) and (c) imply very high rates of strain near the tool tip on crossing the first shear line EAD; the reader can see that $(u/v_p)_{DA}$—the shear strain imposed on crossing DA—is extremely large near D, and indeed in Fig. 14.9(a), at D, it is infinite. The implications of these remarks are,

1. since real materials cannot sustain indefinitely large, or simply, large strains, they must therefore fracture (or cut) at, or about, D;

2. the heat generation near the tool tip, in a given small element, because the shear strain is large, must be relatively great and hence the temperature rise must be large, indeed theoretically infinite, at D in Fig. 14.9(a);

3. the strain rates about D must be extremely large.

When there is shearing along a tool face, the theoretical implications are that the temperature rise at the tool face is infinite, in an infinitely thin layer, and the shear strain is infinite also; at small distances away from the rake face these quantities are still very large.

These latter considerations, or deductions, from a simple analysis using a slip-line field demonstrate that many features of the machining process may be understood as a consequence of simple mechanics, and much that passes for empirical knowledge is deducible and rational. Of course, in making the deductions above we have shown that the theory, on a basis of knowledge of the real behaviour of materials, is invalid—high temperatures, high strain rates and high shear strains in real materials combine to alter material properties—and the concept of the rigid-perfectly plastic material thought of as applying throughout the whole deforming region is certainly unsound at points. This applies also to all metal-working processes and it should therefore be kept in mind that the analysis we use is only intended as a model of a process, correct to a good first approximation.

Slip-line fields of the kind shown in Fig. 14.9 were originally given by Usui and Hoshi (1963), Johnson (1962, 1963), Usui, Kikuchi and Hoshi (1963) and Johnson and Mahtab (1965). Usui, et al., have given experimental results which show good agreement with theoretical predictions. Kudo (1965) has recently presented slip-line fields which give pressures and forces lower than those given by the fields described above; they are, however, of a more complicated character.

Figure 14.10(d) shows a slip-line field for a restricted contact tool in which the roundness of the tool edge is simplified to appear as three partly rough planes of angle α_1 and $-\alpha_2$, and the rake face. Point S can be regarded as a stagnation point; material above it is machined away and moves up the rake face while material below it to a depth d is distorted and this comprises a small worked surface layer. The field indicates infinite shear strain, etc., at depth d.

14.2.9 *Turning and Boring: Some Simple Upper Bounds*

The solutions discussed above referred to a machining operation in which the relative velocity between the tool and the work-piece is rectilinear. But for the operation where either tool or work-piece is rotating, the solutions have to be modified to satisfy the required velocity conditions.

Consider the solution as shown in Fig. 14.12(a) for a turning operation in which the tool is stationary and the work-piece is rotating; the shear

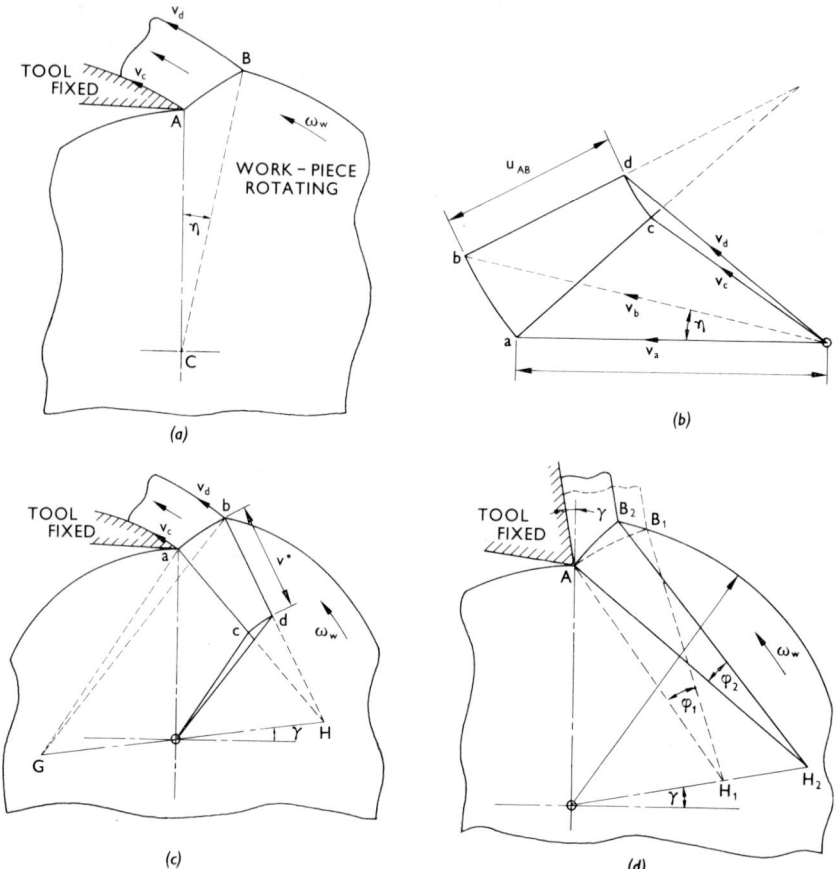

FIG. 14.12

(a) Showing a turning operation, in which the tool is stationary, and the work-piece.

(b) Hodograph for (a).

(c) The combined diagrams of the physical plane and the hodograph of (a) and of (b) to give the centre of chip rotation G.

(d) A special case of Fig. 14.11(a) where the magnitude of $u_{AB} =$ radius of curvature of arc AB.

plane AB must be either perpendicular to the tool face, or curved, in order to obtain a valid hodograph. Figure 14.12(a) shows such a solution and its hodograph in Fig. 14.12(b) for the case of a frictionless tool, and where we have assumed that the shear plane is an arc of a circle along which the tangential velocity discontinuity is constant, u_{AB}. It is interesting to note that the chip is predicted to rotate in the same direction as that of the work-piece. The centre of chip rotation G can be obtained by drawing lines from A and B perpendicular to the lines $\overrightarrow{oc}$ and $\overrightarrow{od}$ as shown in Fig. 14.12(c). It is also noted that the area Oab of the hodograph and CAB of the physical plane are similar so that the hodograph and the physical plane can very well be combined as shown in Fig. 14.12(c).

The rate of energy dissipation for this case is

$$\dot{W} = ku_{AB}\,AB.$$

Now, if for a given radius of work-piece and depth of cut, we keep say circular arc AB the same, but change u_{AB} then the corresponding rake angle and centre of chip rotation would also change. A special case arises when u_{AB} is just equal to the radius of curvature of AB, for then the direction of

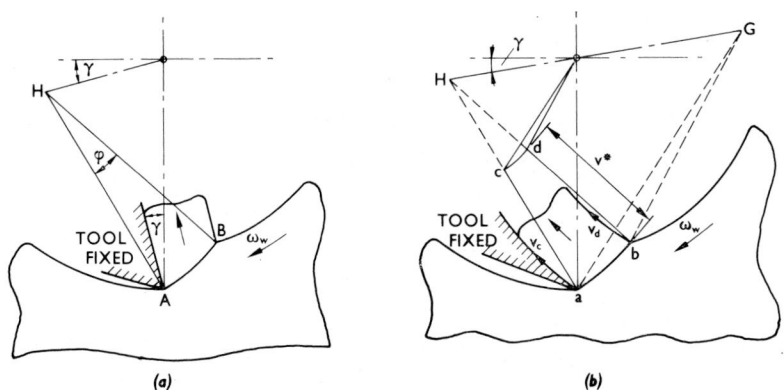

(a)   *(b)*

Fig. 14.13

    (*a*) Boring: the shear plane was so chosen that the chip emerges parallel to straight tool face.

    (*b*) Boring: Shear plane was so chosen that the curved chip rides up over the curved tool face.

chip emergence is perpendicular to CH as shown in Fig. 14.12(d); it has no rotation and the rake angle is γ. From the combined diagram of hodograph and physical plane, the rate of energy dissipation is

$$\frac{\dot{W}}{k} = r^2\theta = 2 \times \text{sector } AB_1H_1. \tag{14.27}$$

This suggests that for a given radius of work-piece R_2, radial depth of cut and rake angle γ, W/k is least when sector ABH_1 is least.

Similar solutions for boring can be obtained and are shown in Figs. 14.13(a) and (b).

If the analytical expressions which could be derived for the cases presented in this section were not too complex (unfortunately transcendental quantities would be involved) it would be possible to derive Merchant-type equations similar to those of equations (14.5.i) and (ii).

SHAFFER (1956) and (1958) has given a detailed analysis and some results for the orthogonal boring and turning operations using a circular shear plane similar to that used and described above, but on the assumption that the chip has just a rectilinear flow parallel to the tool face.

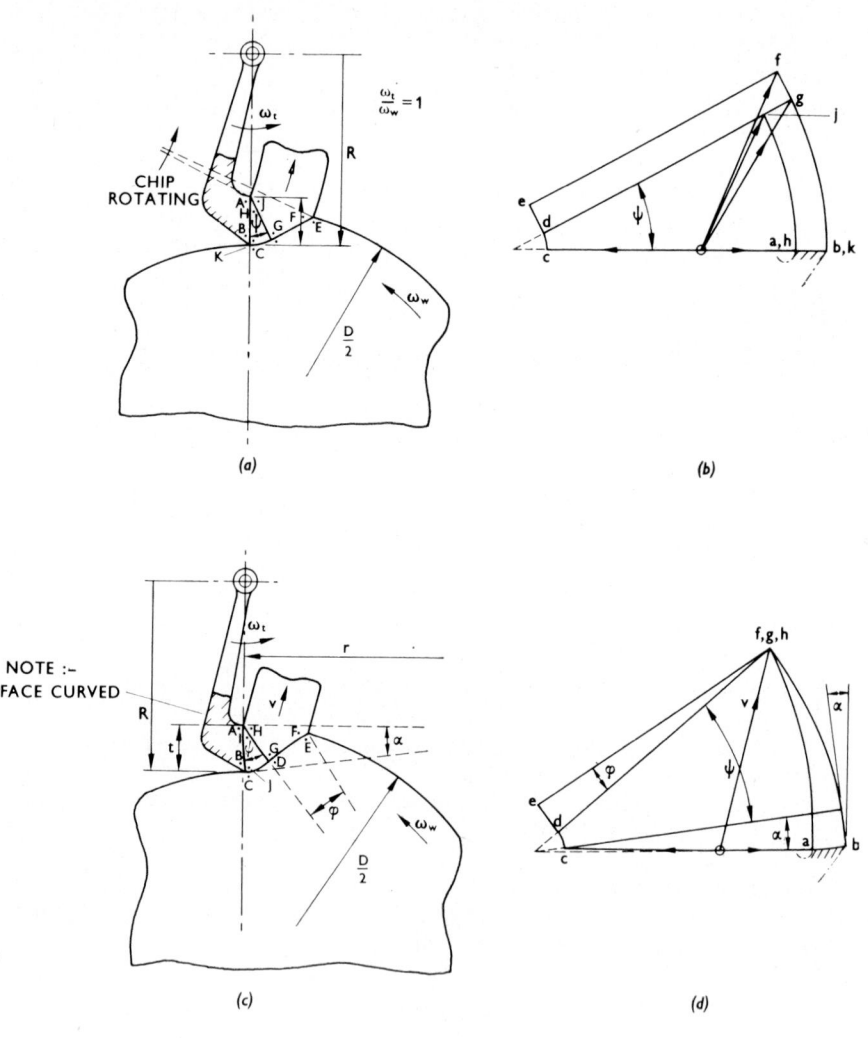

FIG. 14.14

It is possible to arrive at many solutions which should facilitate calculations of upper bounds to the force or torque necessary to cause cutting when either or both of the tool and the work-piece are rotating, see JOHNSON and MAHTAB (1965). However, a limited number of slip-line field solutions are available, and two of these are given in Fig. 14.14. Experimental results into the effects

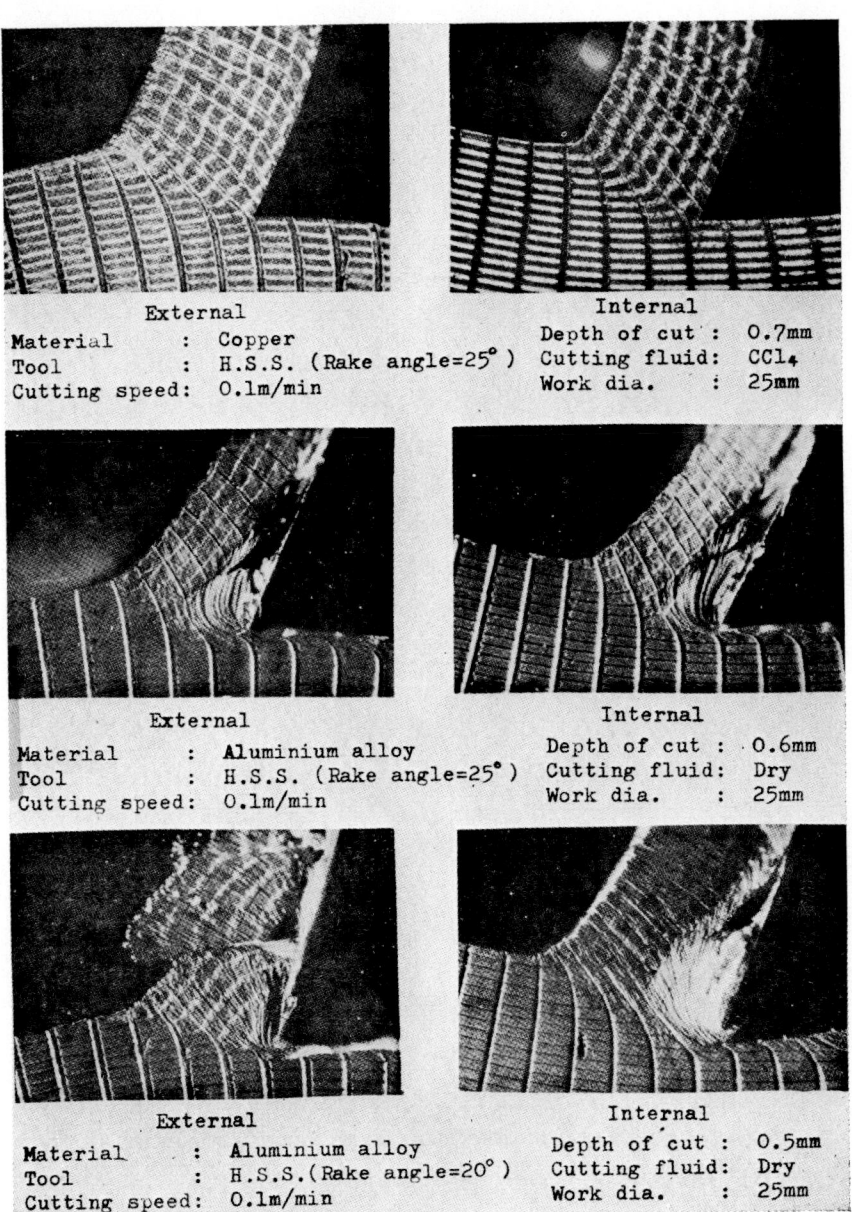

External	Internal
Material : Copper	Depth of cut : 0.7mm
Tool : H.S.S. (Rake angle=25°)	Cutting fluid: CCl_4
Cutting speed: 0.1m/min	Work dia. : 25mm

External	Internal
Material : Aluminium alloy	Depth of cut : 0.6mm
Tool : H.S.S. (Rake angle=25°)	Cutting fluid: Dry
Cutting speed: 0.1m/min	Work dia. : 25mm

External	Internal
Material : Aluminium alloy	Depth of cut : 0.5mm
Tool : H.S.S.(Rake angle=20°)	Cutting fluid: Dry
Cutting speed: 0.1m/min	Work dia. : 25mm

FIG. 14.15

of the curvature of the work-piece surface on cutting force in conjunction with speed have been given by OKOSHI and KAWATA (1966) and a selection of their grid distortion photographs are shown in Fig. 14.15; the types of flow should be easily identified in conjunction with Fig. 14.3.

14.3 Indenting with Straight-sided Dies

This subject has been dealt with at length by HILL (1950), there having been little theoretical development in it during the last decade. Consequently, the various problems and their solution will only be treated in outline.

Consider the frictionless indentation of a semi-infinite mass of rigid-perfectly plastic material by a rigid straight-sided, acute-angled indenter. About the indenting tip, three zones may be recognized, see Fig. 14.16(a). Zone 1 is

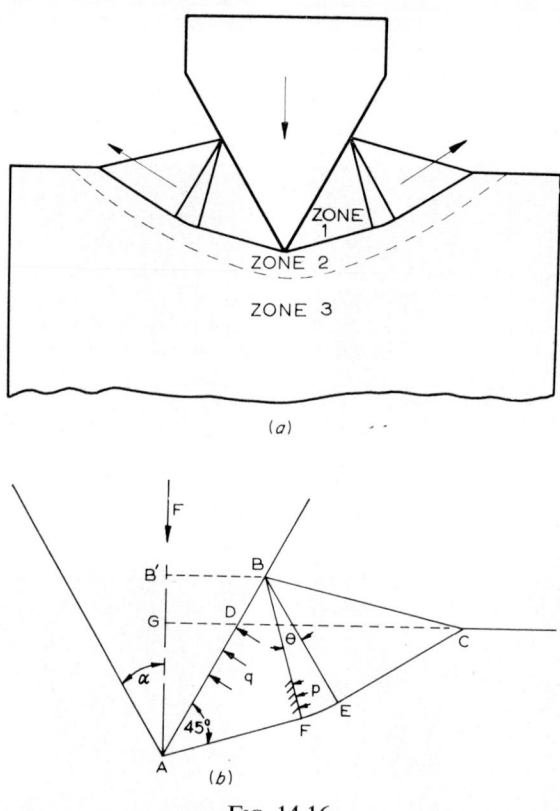

FIG. 14.16

(a) Indentation of semi-infinite mass of rigid-perfectly plastic material by a rigid straight-sided, acute angled frictionless indenter.
(b) Slip-line field for zone 1 of (a).

that of large plastic strains in which material is subject to large displacements and covered by a slip-line field; zone 3 is that in which the material is elastically stressed and therefore hypothetically rigid; zone 2 is ill-defined, but is contained between the other two zones where the material is plastic but constrained from moving. Our concern is only with zone 1. When the wedge tip has penetrated the original flat surface of the semi-infinite block from zero to a depth d, the slip-line field will appear as shown in Fig. 14.16(b). The coronet, or piled-up material CBD, is equal in volume to that displaced by the indenter, GDA. (A coronet is only clearly observed with work-hardened materials.) For all values of the penetration, the slip-line field configuration remains geometrically similar. It consists of an isosceles triangle BCE, in a state of uniform compression; BFE is a centred fan of angle θ, θ being determined in terms of α by use of the above fact of volumetric constancy and the assumption of geometrical similarity throughout triangle BEA is also a 45° isosceles triangle since the indenter side AB is perfectly smooth and the slip-lines must therefore meet it at 45°. The pressure along BF, p, is with the aid of the Hencky equations

$$p = k(1 + 2\theta)$$

and the pressure normal to AB, q, is

$$q = 2k(1 + \theta).$$

The necessary indentation force is

$$F = 2qAB \sin \alpha = 4k(1 + \theta).BB'. \tag{14.28}$$

If the coefficient of friction between the material and the die is μ the slip-line field triangle adjacent to AB is no longer isosceles, but has to be modified as shown in Fig. 14.17. A fairly complete presentation of results for various

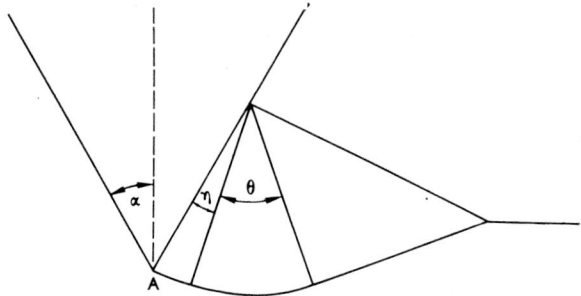

FIG. 14.17 Modification of the slip-line field when there is friction between material and indenter.

values of μ and α has been given by GRUNZWEIG, LONGMAN and PETCH (1954), with some experimental results.

When the indenter first enters a block of *finite* depth, the above mode of deformation prevails, the indentation load increasing linearly with depth. However, another or second mode of deformation is possible, in which the material indented no longer piles up, but in which the slip-line field passes through the whole thickness of the block, and indentation causes the two portions of the block to move oppositely outwards, horizontally. This applicable slip-line field is shown in Fig. 14.18(a), the block of material resting on a perfectly smooth foundation; Fig. 14.18(b) is the corresponding hodograph. To calculate the indenting force, it is necessary first to find the pressure at A, i.e., p_A. This is obtained by observing that the total horizontal force on

17—EP   *   *

the portion of the indented block to the right of bounding slip-line $PCBA$ is zero. p_A is then given by the equation (cf. equation (12.11)).

$$\left(\frac{p_A}{2k} - \frac{\pi}{4}\right)h + \int_0^h \psi \, dh = \frac{X}{2}. \tag{14.29}$$

It is simple then to use the Hencky equations and so find the pressure along PE and hence the indenting load.

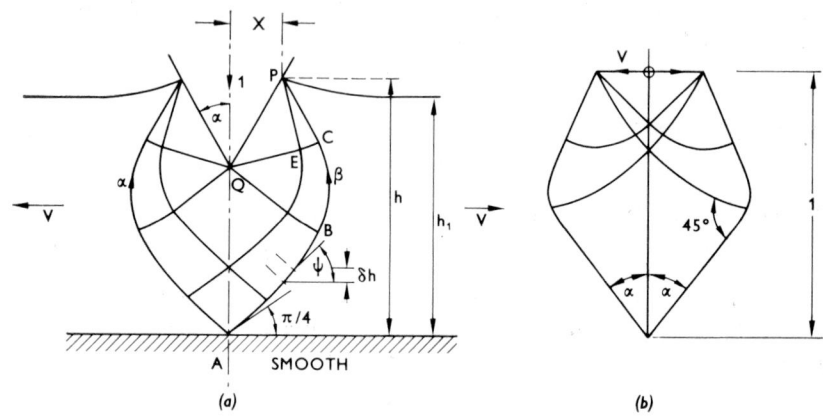

FIG. 14.18 Indentation of block of finite depth resting on smooth foundation.
(a) Slip-line field.
(b) Hodograph.

For each particular wedge angle α and value of μ along the interface PQ, a value of h_1/X can be found when this second mode of deformation will be preferred (on a basis of minimum indenting force) to a continuance of the first, piling-up mode. HILL (1953) has presented solutions and a full discussion of this problem.

The second mode of deformation holds, provided that the stress normal to the foundation at A is less than zero, i.e., is compressive. This slip-line field may, however, be extended beyond this latter configuration by adding an isosceles triangle at A as shown in Fig. 14.19(a). The stress normal to the foundation is then zero, and that parallel to it just the tensile yield stress $2k$. The hodograph for this instance, Fig. 14.19(b), shows that the material of this triangle rises vertically, causing contact with the foundation to cease over length XY.

The effect of interfacial friction along QP on the critical thickness at which the block is parted into two, has been examined by JOHNSON and KUDO (1961).

When μ reaches a certain value, shearing along the sides of the wedge occurs and the slip-line field appears as in Fig. 14.20. This solution and that shown in Fig. 14.17 are only valid as long as the slip-lines at A do not intersect at less than 90°. This requirement leads to the expectation that for some obtuse-angled wedges a false nose will extend over the wedge tip as shown in

Fig. 14.21. This false nose solution would also apply in cases where the value of μ is not high enough to cause shearing, but in which the indentation load consequent on a field such as that shown in Fig. 14.17, is higher than that required by a false nose solution.

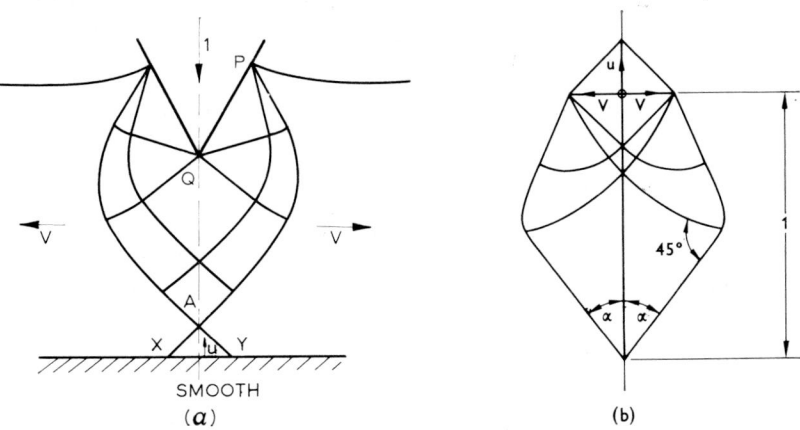

SMOOTH
(a)   (b)

Fɪɢ. 14.19 Possible extension of slip-line field given in Fig. 14.18(a).
(a) Slip-line field.   (b) Hodograph.

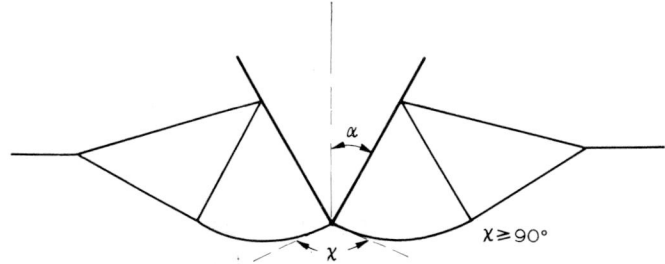

Fɪɢ. 14.20 Indentation with shearing of material along side of indenter, $\chi \geqslant 90°$.

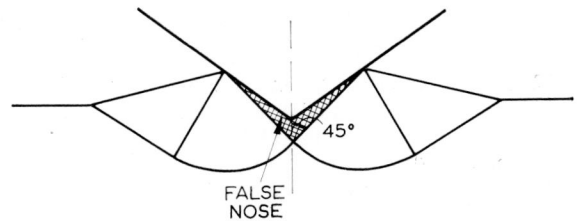

FALSE
NOSE

Fɪɢ. 14.21 Indentation with obtuse angled wedge showing false nose over wedge tip.

14.4 Indenting and Forging with a Flat Punch

Special mention must be made of the flat indenter, see Figs. 14.22 and 14.23. When indenting a semi-infinite medium, two solutions obtain according to whether or not the indenter is perfectly smooth. In the former case, the solution is that shown in Fig. 14.22(*a*) (HILL, 1950), the hodograph for which is given in Fig. 14.13(*b*); the material instantaneously displaced by, but adjacent to the punch, slides horizontally relative to the punch and the slip-lines must meet the punch at 45°. Another solution more applicable when the punch face is partly rough, is that due to Prandtl which requires a cap of dead

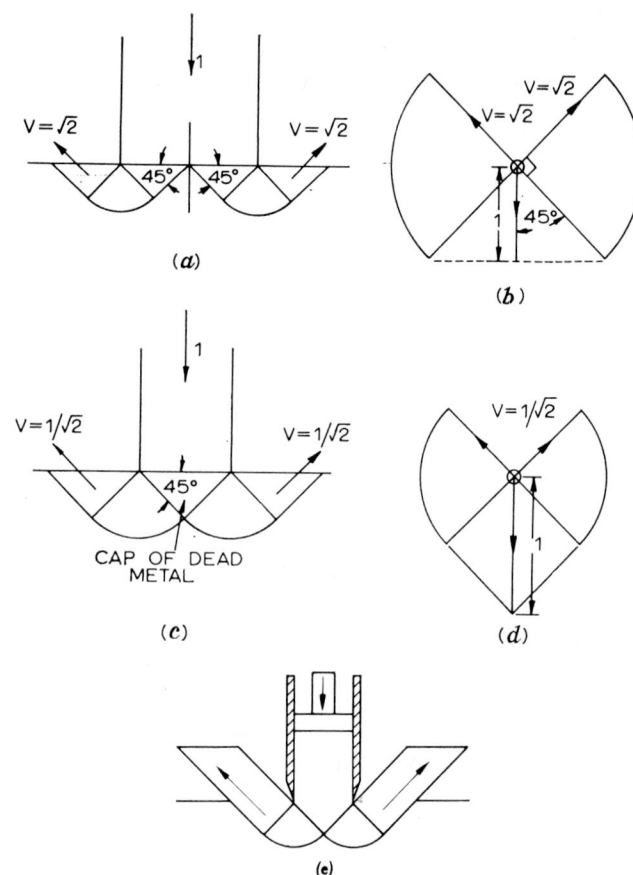

FIG. 14.22 Two solutions for the indentation of a semi-infinite mass
with flat punch.
(*a*) Slip-line field for perfectly smooth indenter.
(*b*) Hodograph for (*a*).
(*c*) Slip-line field when a cap of dead metal forms on the punch.
(*d*) Hodograph for (*c*).
(*e*) Steady-state indentation based on (*c*).

metal whose shape is that of an isosceles triangle, over the head of the punch; this shape is dictated by the requirement that the straight slip-lines radiating from the punch corners, see Fig. 14.22(*c*), must intersect at 90° on the axis of symmetry. The hodograph for this case is shown in Fig. 14.22(*d*). Both solutions give a punch indentation pressure of p where $p/2k = (1 + \pi/2)$.

These two solutions only hold momentarily; after a finite amount of penetration, a piling-up of uncertain shape will have occurred. This problem is obviously not one of geometrical similarity.

Figure 14.22(c) can, however, easily be converted to a steady state case by conceiving the punch to move in a filled container, see Fig. 14.22(e); the implication is that after some degree of indentation strips will have been extruded.

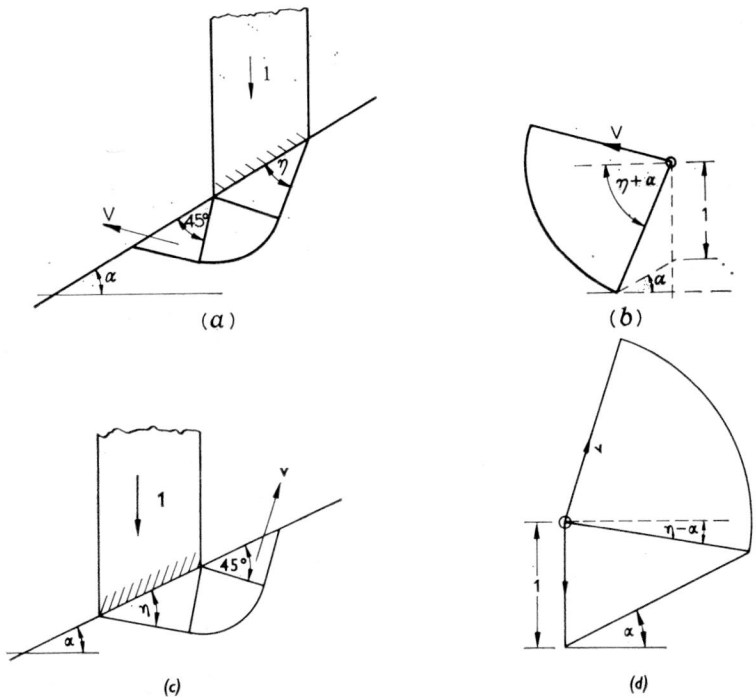

(a)
(b)
(c)
(d)

FIG. 14.23 Various forms of indentation.

(a) Oblique indentation when punch is in some degree rough.
(b) Hodograph to (a).
(c) Oblique indentation when punch is in some degree rough: alternative to (a).
(d) Hodograph to (c).

Some simple variants on these last two indentation examples are shown in Fig. 14.23, which the reader should have no difficulty in working; the cases referred to are as follows:

(a) and (c) present and leave for discussion two possible solutions for oblique indentation when the punch is in some degree rough.

(e) Shows indentation at the bottom of a flat trench; $p/2k = (1 + \pi/2 + \alpha)$.

(f) This is the same case as in (e) except that α is negative.

(g) Shows indentation at the foot of a very deep vertically sided groove; this mode of deformation is only possible if the material displaced by the foot of the punch can move upwards and out behind the punch;

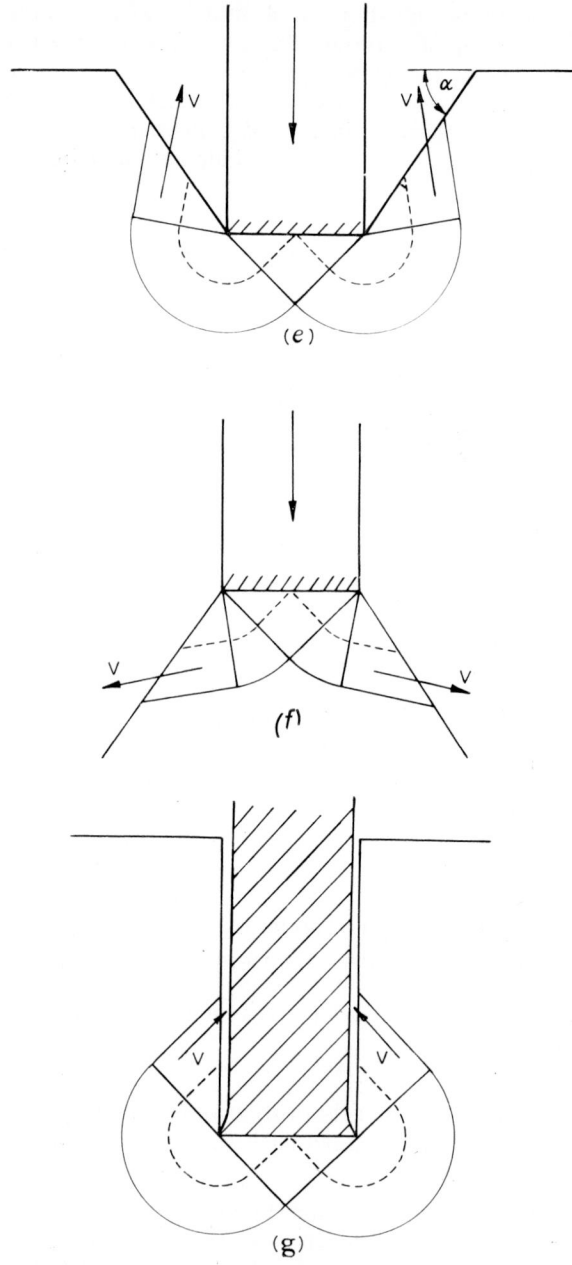

FIG. 14.23

(e) Indentation at the bottom of a flat trench.
(f) Same as (e) except that α is now negative.
(g) Indentation at the foot of a very deep vertically sided groove.

$p/2k = (1 + \pi)$. (Note that this result for $p/2k$ is of about the same magnitude as that obtained when considering the expansion of a hole deep in an infinite medium in Section 9.2.8., Chapter 9.).

In (e), (f), (g), alternative solutions for frictionless tools are shown by the broken line.

When the flat indenter is presented to a strip of finite thickness resting on a frictionless foundation, the slip-line field shown in Fig. 14.24 is applicable

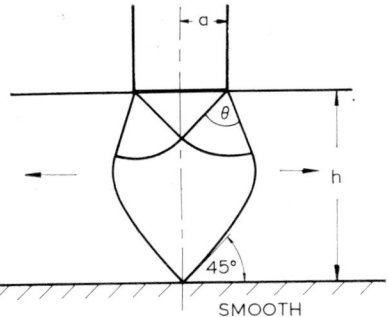

FIG. 14.24 Indentation of strip of finite thickness with flat indenter.

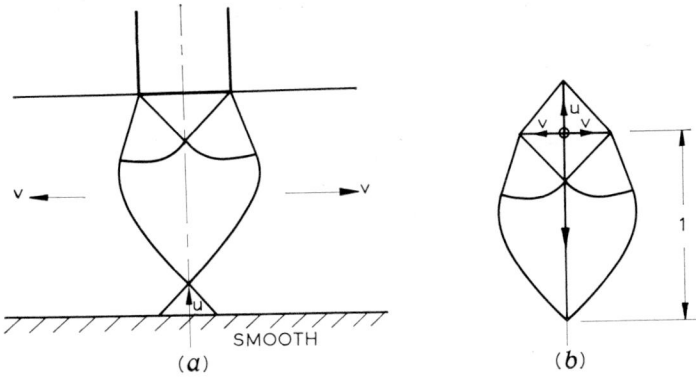

(a) (b)

FIG. 14.25 Indentation of strip of finite thickness with flat indenter when $h > 4.8\ a$.

(a) Slip-line field. (b) Hodograph.

for $h \leqslant 4.8a$ HILL (1953). A false head of metal attaches itself to the punch and indentation results in the strip being separated into two rigid portions, on either side of the slip-line field, which are caused to move oppositely outwards. Calculations of the indentation pressure proceed exactly as in the case of the wedge-shaped indenter described earlier. When $h > 4.8a$ this field gives rise to a tension on the foundation, and this is clearly inadmissible; hence a triangular block requires to be added to the tip of the slip-line field adjacent to the foundation (see Fig. 14.19), and as in Fig. 14.25(a). The hodograph (not to scale) shows the triangle moving upwards towards the indenting punch, Fig. 14.25(b).

The case shown in Fig. 14.26 is formally similar to that in Fig. 14.24, the foundation of the latter being identified with the centre-line of the former. In the forging case, however, tension normal to the centre-line can be withstood. When $h \simeq 8{\cdot}74a$, $\theta = 77{\cdot}3°$ and $p/2k = 1 + \pi/2$; it follows that for $h >$ $8{\cdot}74a$, the punch pressure is always $2k(1 + \pi/2)$ and indentation proceeds as in Fig. 14.22, there no longer being a 'through' penetration.

When the foundation is not smooth, friction may be accounted for in an approximate manner as suggested by HILL (1950) or exactly as given by JOHNSON and WOO (1956). The exact solution requires a false zone to attach itself to the foundation immediately beneath the punch.

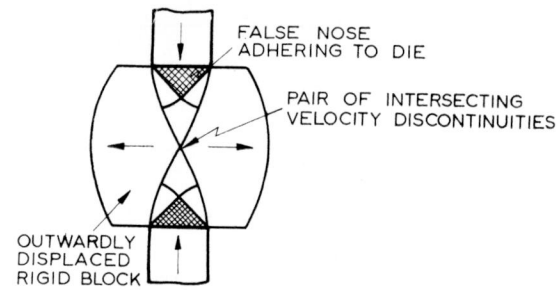

FALSE NOSE
ADHERING TO DIE

PAIR OF INTERSECTING
VELOCITY DISCONTINUITIES

OUTWARDLY
DISPLACED
RIGID BLOCK

FIG. 14.26 Slip-line field for indenting with two equal dies.

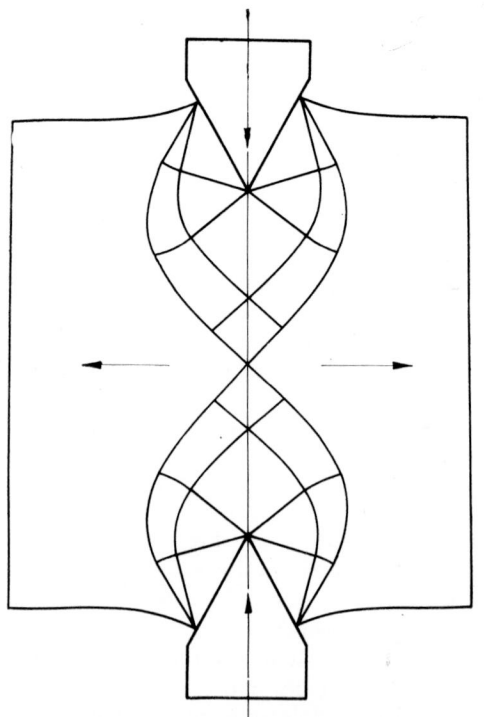

FIG. 14.27 Cutting with opposed wedge-shaped indenters.

14.5 Opposed Indenters: Forging and Cutting

Figures 14.26 and 14.27 show a pair of indenters oppositely situated. The first is typical of forging on a flat anvil whose width is equal to that of the indenter. When the two are unequal, the case is as discussed by JOHNSON (1958b). The second is typical of the cutting of strip with plier-like tools. The latter model may be used to explain the cutting of round wire with knife-edge pliers: first a coronet or piling-up occurs about each die, then one single plastic zone forms through the wire after some critical amount of penetration, further penetration causing the wire to extend laterally under tension. Eventually a tensile instability occurs with a transverse necking at the ends of the horizontal diameter between the dies. Wire-cutting has been discussed at length by JOHNSON (1957), JOHNSON and KUDO, (1961) and MAHTAB and JOHNSON, (1962).

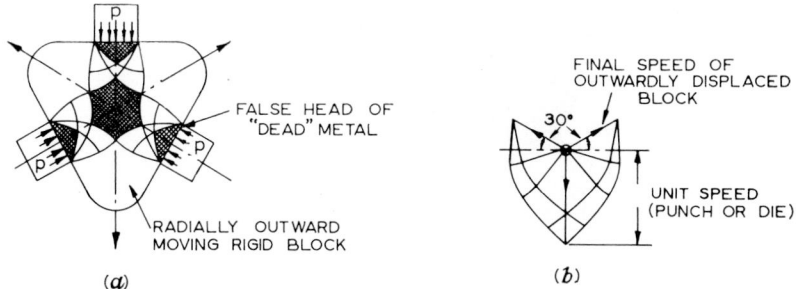

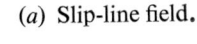

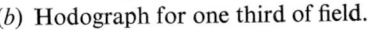

FIG. 14.28 Indentation by three equal size dies spaced at 120° to each other.
(a) Slip-line field.　　　　(b) Hodograph for one third of field.

FIG. 14.29 Defects in forging.
(a) Flat anvil forging.　　　　(b) V-anvil forging.

Figure 14.28 shows a block of material indented by three dies of equal size spaced at 120° to each other, all at the same distance from the block centre. On constructing the symmetrical slip-line field, it transpires that the core of the block remains rigid. This configuration of dies has obvious affinities with the problem of forging on a V-anvil (see Fig. 14.29(b)) and with Mannesman and Assel type tube-making operations. Nasmyth in 1839 claimed to have introduced the V-anvil as a device for producing internally sound forgings. Flat anvil forging requires a pair of intersecting velocity discontinuities in the block centre, which must make for unsoundness at the centre of any forging (cf. Figs. 14.26 and 14.29(a)). An unsoundness may, however, arise

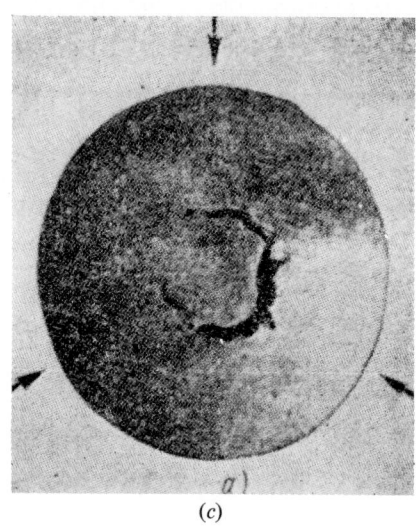

(c)

FIG. 14.29
(c) Central circular cavity.

with the V-anvil, or in the rolling of solid bar with three rolls, the weakness being created symmetrically at some distance from the centre along the lines of velocity discontinuity which bound the central 'sound' core. The photograph in Fig. 14.29(c) is taken from the book by TOMLENOV (1963) and shows a central circular cavity made by the use of three rolls applied at the points indicated by the arrows. The subject was discussed by JOHNSON (1958).

14.6 Slip-Line Fields for Hot- and Cold-rolling

The hot-rolling of strip presents a difficult problem for solution in terms of slip-line fields. Until recently only one solution was known and that was proposed by ALEXANDER (1955). The solution is complex and difficult to appreciate; FORD and ALEXANDER (1963–64) attempted to improve the situation by using a straight—rather than curved—entry slip-line and with the help of this approximation were able to adequately calculate roll loads and torques. A simple accurate solution, the elements of which will be easily appreciated is given in Fig. 14.30(a), with the hodograph in Fig. 14.30(b); this is due to CHITKARA and JOHNSON (1965). It uses a single circular arc as the exit slip-line and for the particular geometry shown, a dead metal cap extends across the whole of the roll where the strip and the roll are in contact. The exit circular arc is tangential to the roll at the exit—a necessary requirement in order that there should be shearing at exit between the strip and the roll. The hodograph, Fig. 14.30(b), has been shown turned through 90°, so as to facilitate an easy appreciation of the interplay between the slip-line field and the hodograph. The reader will also appreciate the common ground between Fig. 14.30 and the upper bound suggested in Fig. 13.24.

Fields similar to that just described and elaborations thereon, for a wide range of geometrical configurations, have recently been given by CRANE and

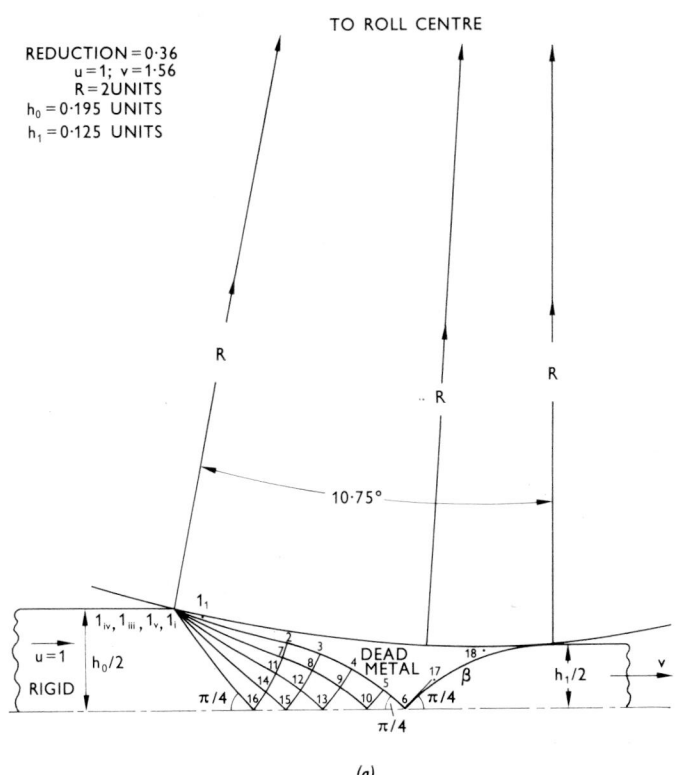

(a)

FIG. 14.30

(a) Hot rolling, physical plane: slip-line field.

ALEXANDER (1968). DRUYANOV (1965) has also proposed an exact *analytical* solution which assumes sticking contact between the strip and the rolls, for comparatively small reductions. One suggested solution for the cold-rolling of strip which is lubricated has been given by FIRBANK and LANCASTER (1965), and further discussion of that same problem is given by a later paper (1966).

14.7 Strip and Bar Drawing through a Perfect Die

A thick-walled cylinder of rigid-perfectly plastic material subject to an internal pressure p and of internal and external radii b and a respectively, is just fully plastic if $p = 2k \ln a/b$ and if the straining takes place under conditions of plane strain; this has been shown in Chapter 9. The hoop and radial stresses in the wall of the cylinder are everywhere principal stresses and hence the lines of maximum shear stress are everywhere at 45° to a radius These lines of yield shear stress are slip-lines and their equation is $r = be^\theta$, see Fig. 14.31; the radial stress at r is $\sigma_r = 2k \ln a/r$. (These features can be obtained directly from first principles if the reader is unable to appreciate them directly, see JOHNSON and SOWERBY (1967).)

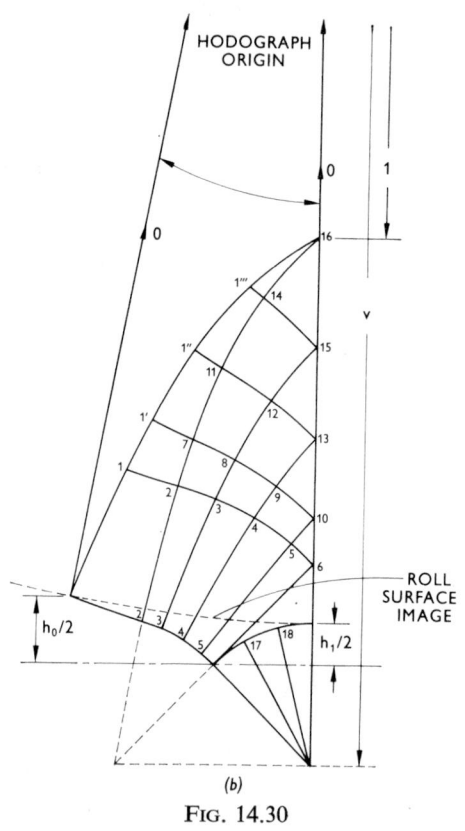

FIG. 14.30

(b) Hodograph.

(*After* Chitkara and Johnson, *M.T.D.R. Conf.*, Pergamon.)

Suppose that instead of a pressure $\sigma_r = p = 2k \ln a/b$, the cylinder was able to carry a radial tension, t, such that the whole cylinder became fully plastic and each particle in the tube wall had an incipient radially inwards motion, then $t = 2k \ln a/b$. Extract from Fig. 14.31 the curvilinear triangle *DCE*, where *DC* and *CE* are bounding slip-lines (logarithmic spirals), and from *D* and *E* draw straight lines at 45° to the radius, *DA* and *EB*, as in Fig. 14.32(a). The position of *B* is easily located since $EB = BO$, where O is the centre of the circle which gives rise to *DCE*. The height, *H*, which locates point *A* is chosen to be given by $H/h = a/b$. An identical triangle *dce*,

Fig. 14.32(*b*) is drawn and covered with an identical grid of lines; this will be used as a hodograph. At *F*, *G*, *H* and *I*, *J*, *K*, draw lines perpendicular to the slip-lines at those points, e.g., *FT*. Using Fig. 14.32(*b*), in Fig. 14.32(*a*) from *A* draw *AL* parallel to the direction which is the mean of *Od* and *Of*, to cut

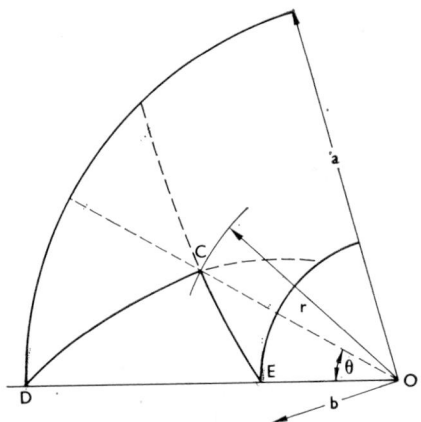

FIG. 14.31 Slip-line in the wall of a thick-walled cylinder.

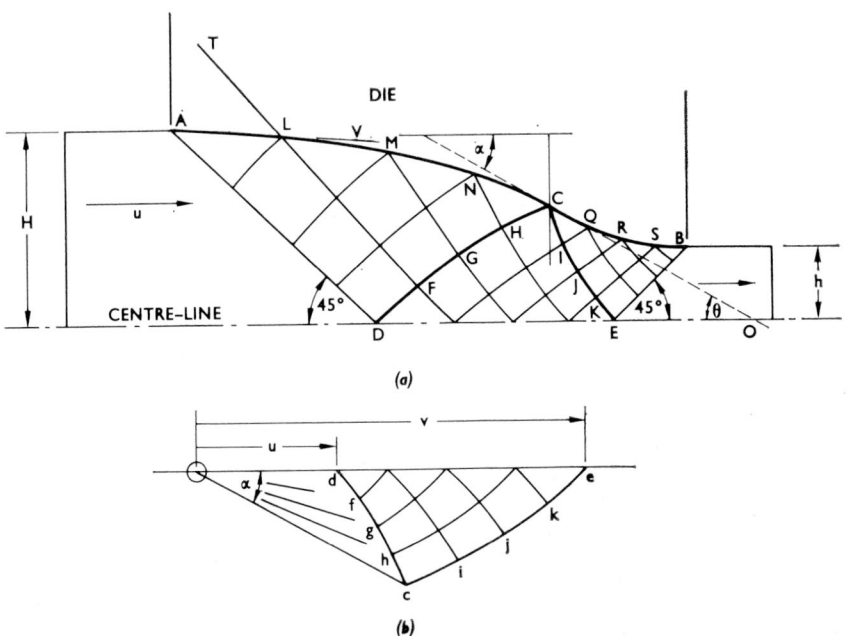

FIG. 14.32 Extrusion through sigmoidal die.

FT in *L*. Repeat this procedure to find points *M*, *N*, *Q*, *R* and *S*. Admitting some small degree of inaccuracy, the bell-mouth, or sigmoidal, die shape shown in Fig. 14.32(*a*) may be 'faired-in' through *A*, *L*, *M*, *N*, *C*, *Q*, *R*, *S* and

B. The die shown in Fig. 14.32(*a*) derived from the curvilinear triangle of the cylinder can be accurately specified and then constitutes a perfect die. If the die is frictionless it can be shown that strip of height *H* can be drawn through the die and reduced to height *h* by a tension of

$$t = 2k \ln H/h = 2k \ln a/b;$$

the strip may be drawn down and reduced homogeneously or with an efficiency of 100 per cent. Precise analysis of this remarkable die and its properties were originally given by RICHMOND and DAVENPECK (1962). In Fig. 14.32(*a*) the net of orthogonal slip-lines has been inserted; note that one family of slip-lines in the entry and exit zones are composed of straight lines meeting the die wall at 45°. The metal flowing through this die has a uniform radially inwards motion in the other zone, triangle *DCE*. The hodograph to the slip-line field is shown in Fig. 14.32(*b*); note that no tangential velocity discontinuity is called for with this die, and for this reason alone it would be expected to be of high efficiency. In real practical terms this die is not very attractive because some degree of friction is unavoidable which, taken together with the length of the die, would seriously increase the necessary drawing stress; this die would also be costly to manufacture.

The idea of plane radial flow which we have just used, immediately suggests that we might utilize the idea of spherical flow and thus come to the design of an axially symmetric perfect die. This has been achieved by RICHMOND and MORRISON (1967); interesting commentaries on this die are to be found in papers by HILL (1967) and SORTAIS and KOBAYASHI (1967). It is worth noting that the die shape is identical with a streamline through the die.

The above slip-line field solution for drawing applies only until the drawing tension *t* = 2*k*. Thus, the limiting fractional reduction,

$$r = (H - h)/H = (e - 1)/e \simeq 0\cdot63.$$

However, this die may be conceived of for use as an extrusion die; the maximum reduction may be shown to be 0·865 for this type of field. At this reduction the radius of curvature of slip-line *EC* at *E* is just equal to $h\sqrt{2}$ (TAKAHASHI, 1967).

14.8 Extrusion Through Bi-wedge Shaped and Curved Dies

We consider extrusion through a die comprised of two straight sides, see Fig. 14.33(*a*). Now the solution (i.e., the drawing of a valid slip-line field and hodograph) to extrusion through a straight partly rough wedge-shaped die of semi-angle $\alpha_1 = 70°$ is given in Fig. 14.33(*b*), the hodograph for which appears in Fig. 14.33(*c*). However, if we use the same hodograph, draw in the extra line *OB* and then interpret this diagram physically as an extrusion, we find it solves the problem depicted in Fig. 14.33(*a*), subject to there being a certain coefficient of friction on the upper die face. The same hodograph also accounts for extrusion through a die of different proportions though of the same total reduction and inlet and exit die angles; it may also be used to account for a 3-plane die face, see Fig. 14.33(*d*), and indeed a die face which curves continuously from $\alpha_1 = 70°$ to $\alpha_2 = 50°$, the friction coefficient along the die face also varying continuously, see Fig. 14.33(*e*).

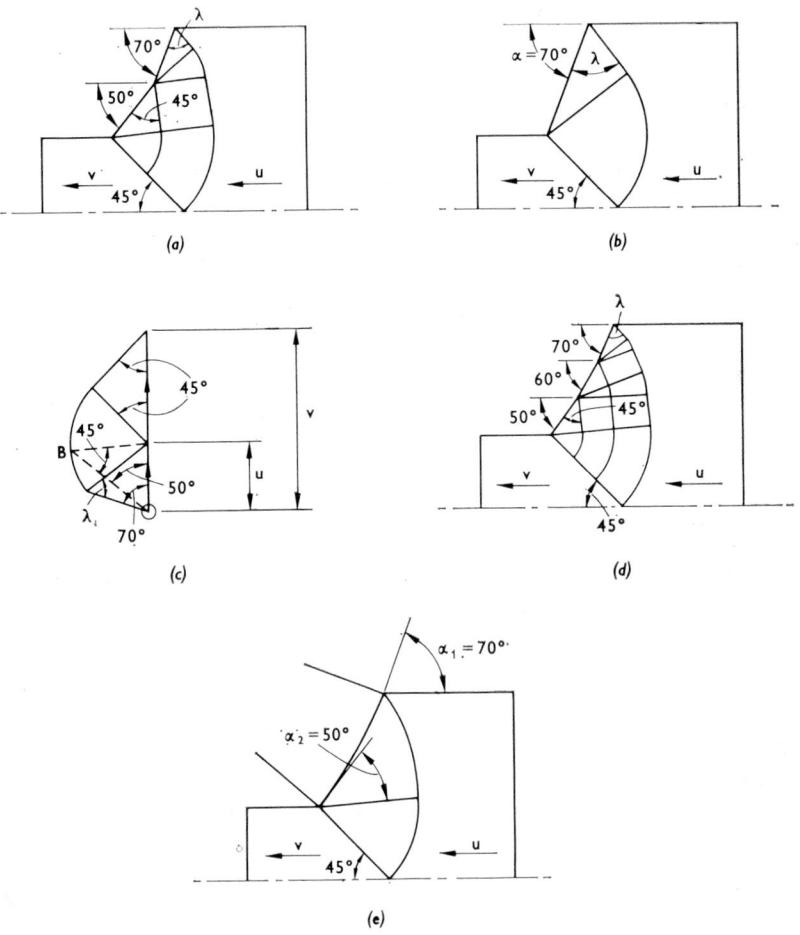

FIG. 14.33

(a) Slip-line field (SLF) for extrusion through a die composed of two straight sides; top half diagram only shown here (and in subsequent figures).

(b) A SLF for extrusion through a partly rough straight-sided die of semi-angle $\alpha_1 = 70$ degrees.

(c) Hodograph to (a), (b), (d) and (e), rotated through -90 degrees.

(d) A SLF for extrusion through a die composed of three straight sides; friction coefficient different on each face.

(e) A SLF for extrusion through a die continuously turning from $\alpha_1 = 70$ to $\alpha_2 = 50$ degrees; coefficient of friction varies along die face and is zero at exit.

It thus becomes clear that by re-interpreting many well-known hodographs we obtain, in fact, solutions to classes of problems of various geometry, see JOHNSON (1962, 1963). Following the same approach, solutions for a variety of curved dies can be found, see HILLIER and JOHNSON (1962).

14.9 Rotating Dies and Elements

Figure 14.34(a) shows a slip-line field for frictionless indentation using a rigid die and as originally given by HILL; this has already been discussed in Section 14.3. The hodograph for this situation is shown in Fig. 14.33(b).

The same field and hodograph may be employed to account for the forging action of a frictionless die, OA, rotating about fixed point O, with angular speed ω, see Fig. 14.34(c)). In this diagram $OBCD$ is stationary; triangle OAB undergoes a simple shear, particles on line BA rotating about B. Thus for a small rotation of die OA, particles on each radius through A are simply displaced hoopwise and finally AC is rotated, A moving to A^1; straight line AD, initially in the plane of the surface, is rotated about D to take up position $A'D$ and thus a small pile-up or coronet is formed. A particle of material at P, immediately under the die, Fig. 14.34(c), besides being compelled to move vertically downwards with speed $r\omega$, slides horizontally to the right with an equal speed. P thus moves parallel to OB and all particles along $PP_1P_2P_3$ are simply pushed along an equal amount; hence P_3 moves to P_3'. The hodograph is shown in Fig. 14.34(d) and the deformation imposed on an initially square grid of lines is shown in Fig. 14.34(e). To facilitate appreciation, the shear strain (or deformation) and yield shear stress are shown for the parts of the slip-line field in Fig. 14.34(f). Especially we observe that the sense of the shear strain is compatible with that of the shear stresses; this is to say that work done by the material is positive. If we had chosen the mode of deformation shown in Fig. 14.34(g), the work done on the material would require to be *negative* (see Fig. 14.34(h)), and the shear strain (or deformation), would be contrary to the direction of action of the shear stresses. The requirement that the work done should be positive, enables us to select the deformation mode in Fig. 14.34(c) as a valid one.

A slight modification to this last solution enables us to predict the incipient mode of deformation caused by a frictionless die rotating about an external point O in its own plane. The hodograph is shown in Fig. 14.34(i) and the physical picture in Fig. 14.34(j); pile-up occurs on one side only. The normal die pressure for both the previous cases above, i.e., Fig. 14.34(c) and Fig. 14.34(j), is uniform and just, $2k(1 + \pi/2)$.

Some straightforward examples that follow are:

1. When the die in Fig. 14.34(c) is not frictionless, the slip-line field would require to be modified to that shown in Fig. 14.34(k).

2. If, see Fig. 14.34(l), a pure couple (no net vertical force) could be applied to the die, i.e., if it was possible to exert a vertical pull on OB which permitted frictionless sliding of the material beneath it to right, and a vertical compression on OA in the manner shown in Fig. 14.34(c), the deformation mode would then be as shown by the broken line in Fig. 14.34(l). Over OB a tension of $2k(1 + \pi/2)$ would be required and over OA a corresponding compression. This incidentally suggests a solution for the plastic bending of a bar containing deep opposite cracks, see Fig. 14.34(q).

3. If the frictionless rotating die was located in a groove, see Fig. 14.34(m), the slip-line field and mode of deformation would be as depicted in that figure. The pressure here is $2k(1 + \alpha)$, where α is the semi-angle of the groove.

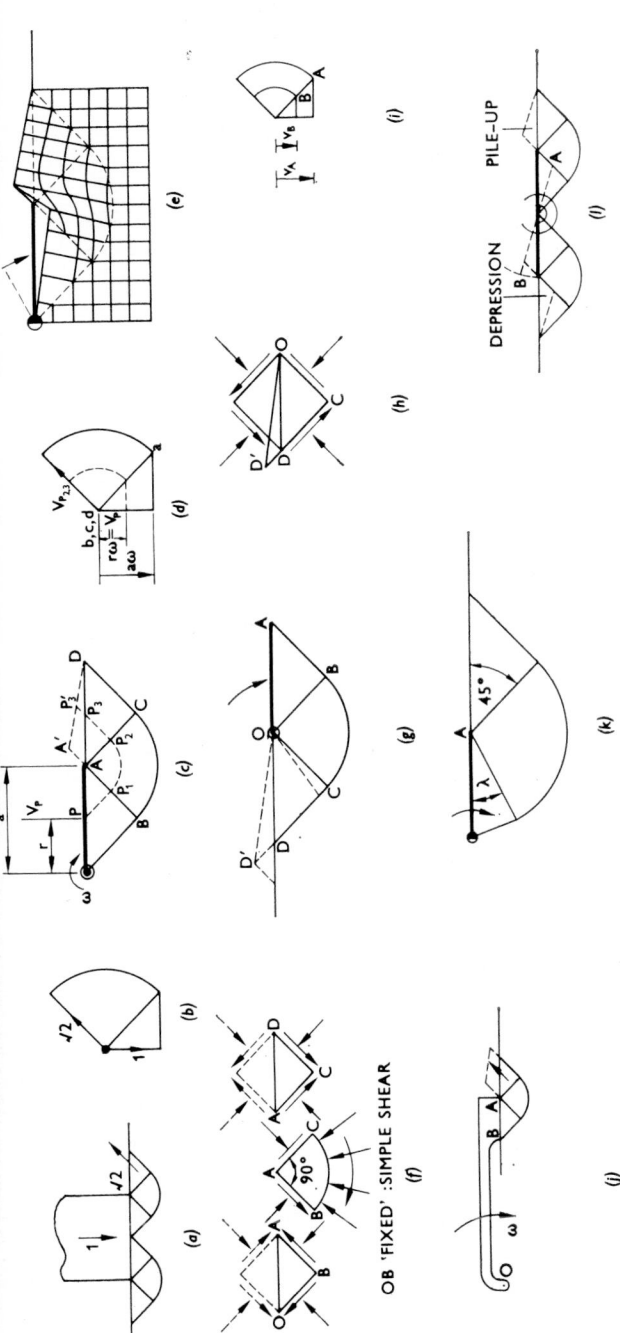

OB 'FIXED' :SIMPLE SHEAR

DEPRESSION   PILE-UP

FIG. 14.34

(a) A slip-line field (SLF) for indenting using a frictionless flat-ended rigid indenter (Hill-type field).
(b) Hodograph to (a).
(c) A SLF for forging with a frictionless rigid die rotating about O and indenting a semi-infinite medium.
(d) Hodograph to (c).
(e) Deformation of an initially square grid of lines, when the die in (c) rotates through a small angle.
(f) To show the relation of the shear stress for (c): positive work.
(g) Showing an incorrect mode of deformation.
(h) Negative work would require to be performed in the case shown in (g).
(i) Hodograph for (j)
(j) An extension of the solution in (c); rotation of die about external point O.
(k) Modification of (c) when the die is not frictionless. Frictionless die rotating about O.
(l) Pure couple applied to indenter.

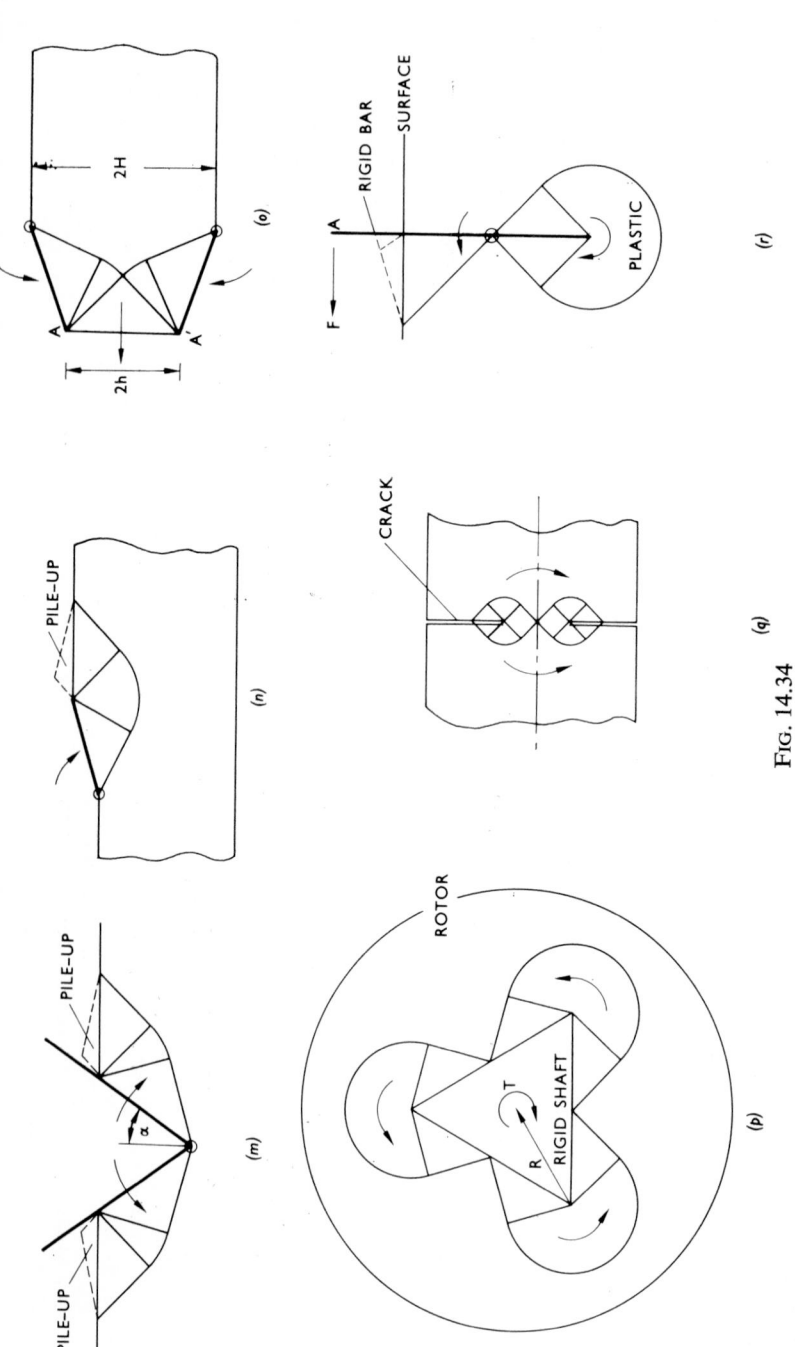

FIG. 14.34

(m) Die located in a groove and rotating about groove vertex.
(n) Edge forming with frictionless die.
(o) Crushing or swaging the end of a bar.
(p) To calculate the ultimate torque transmitting capacity of rotor R which just fits frictionlessly on to a triangular rigid shaft. The SLF for the incipient mode of deformation is shown. The rotor is stationary and the shaft rotates clockwise.
(q) Bending of a bar containing a crack (see (l)).
(r) A rigid bar partly imbedded in plastic mass and rotated

The last solution also solves the simple forging case shown in Fig. 14.34(*n*), and that of crushing in Fig. 14.34(*o*). (For the latter case appropriate slip-line fields are available if other values of *H*/*h* are given).

An interesting extension of this problem is the case in which it is required to calculate that torque which must be applied to a shaft whose section is that of a regular polygon of *n* sides, in order to cause full plastic yielding in the surrounding rotor; the width of the rotor perpendicular to the plane of the paper is large.

Fig. 14.34(*r*) shows a rigid lamina embedded in a plastic mass and subject to a lateral force *F*. When yield occurs the material piles up on one side whilst about some centre *O* deep in the mass, the lamina rotates creating a plastic 'vortex'.

These situations are discussed in papers by JOHNSON (1963) and JOHNSON and HILLIER (1963).

A comprehensive bibliography of slip-line field papers is to be found in the monograph by JOHNSON, SOWERBY and HADDOW (1969).

14.10 Plastic Deformation of Metals of Different Yield Strengths

Many practical problems involve the study of plastic deformation where two metals of different yield strengths are in contact. There is little or no theory of such a simplicity that it is of help to the ordinary engineer as in the case of the study of hot extrusion and therefore resort must be made to some artifice. One such suggestion is as follows and concerns a hot billet

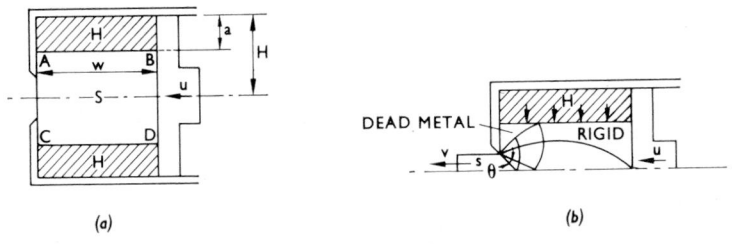

(a)   (b)

FIG. 14.35

(*a*) Block to be extruded which is composed of hard outer layers (*H*) and a soft inner layer (*S*).

(*b*) A SLF for accounting for extrusion of the block in (*a*) by assuming *AB* (and *CD*) advances towards the centre-line uniformly as punch moves to left. The punch is smooth. Half diagram only is shown.

slowly extruded which is supposed to be much harder on the outside than the inside. It is well known that the deformation pattern as between hot and cold extrusion can be radically different. An engineer's approach might be to idealize the situation and to conceive of the billet as made up of two hard homogeneous outer layers, *H*, and a single homogeneous thick inner layer, *S*, but relatively much softer, see Fig. 14.35(*a*). The compression of the outer layers by the punch could be supposed to cause the planes AB and CD to move towards the billet centre line *uniformly*, this causing the inner core to be extruded. (This can be interpreted in terms of slip-line fields, see Figs. 14.35(*b*) and (*c*). Alternatively, we could imagine as another mode of deformation that in which the triangle *EFG* is caused, because of punch advance, to slide towards the billet centre line across the die face but encouraging the rear central billet material to be extruded earlier than might otherwise be expected. Particular relative magnitudes for the lines of tangential velocity discontinuity in Fig. 14.35(*d*) would be determined by reference to a criterion

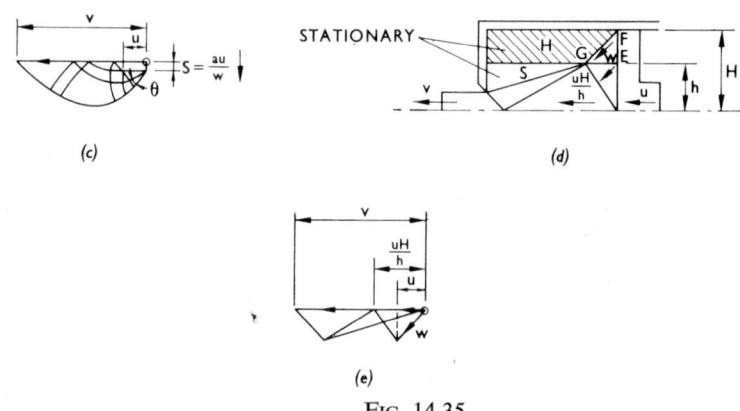

(c)

(d)

(e)

FIG. 14.35

(*c*) Hodograph to SLF in (*b*).
(*d*) A system of tangential velocity discontinuities to account for extrusion of block in (*a*). The triangular block of hard material slides transversely across the punch face.
(*e*) Hodograph to (*d*).

of minimum work done. It should well be possible to obtain coarse deformation modes from observations of extruded layered materials; compare SUKOLSKI, (1955) and see ADAMS (1966). Interesting cases of the use of slip-line fields being 'joined together' for different materials are, (i) when an indenter and the indented material deform concurrently, JOHNSON, MAHTAB and HADDOW (1964) and (ii) the extrusion of bimetals, ALEXANDER and LENGYEL (1966).

See Problems 64-69, 71 and 76

REFERENCES

MACHINING

ALEXANDER, J. M. and 1963 *Manufacturing Properties of Materials*
BREWER, R. C.   D. Van Nostrand, London, 489 pp.
BANNERJEE, J. and   1966 *Metal Cutting with a Discontinuous Chip*
PALMER, W. B.   Proc. 6th Int. M.T.D.R. Conf.,
                        Pergamon Press

BHATTACHARYYA, A. 1966 *On the Friction Process in Metal Cutting*
6th Int. M.T.D.R. Conf., Pergamon Press,
491

BOOTHROYD, G. 1965 *Fundamentals of Metal Machining*
Arnold, London, 144 pp.

CHRISTOPHERSON, 1956 Mech. Eng. Res. Lab., M.F./M.S.30.
D. G., Also, 1958 'Orthogonal Cutting of a Work-
OXLEY, P. L. B. and hardening Material'
PALMER, W. B. *Engineering* **186**, 113

CASSIN, C. and 1965 'Lubricating Action of Cutting Fluids'
BOOTHROYD, G. *J. Mech. Eng. Sci.* **7**, 67

ENAHORO, H. E. 1964 'Effect of Cold-Working on Chip Formation
in Metal Cutting'
Coll. of Aeronautics, M. and P., 1

ERNST, H. and 1941 'Chip Formation, Friction and High
MERCHANT, M. E. Quality Machined Surfaces'
Trans. A.S.M. **29**, 299

FINDLEY, W. N. and 1963 'The Influence of Extreme Speeds and Rake
REED, R. M. Angles in Metal Cutting'
J. of Engng for Ind. 49

HEGINBOTHAM, W. B. 1961 'Metal Cutting and the Built-up Nose'
GOGIA, S. L. *Proc. Instn Mech. Engrs.* Paper 40/60

HILL, R. 1954 'The Mechanics of Machining: A New
Approach'
J. Mech. Phys. Solids **3**, 47

HILL, R., 1947 'The Theory of Wedge Indentation of
TUPPER, S. J. and Ductile Metals'
LEE, E. H. *Proc. Roy. Soc.* A, **188**, 273

HSU, T. C. 1965 'A Study of the Normal and Shear Stresses
on a Cutting Tool'
A.S.M.E., Paper No. 64–WA/Prod–1

JOHNSON, W. 1962 'Some Slip-line Fields for . . . Machining . . .'
Int. J. Mech. Sci. **4**, 323
1962 'Extrusion, Forging, Machining . . .'
Int. Res. in Prod. Eng., *A.S.M.E.*, 342

JOHNSON, W. and 1965 *Slip-line Fields for . . . Turning and Boring*
MAHTAB, F. U. 6th Int. M.T.D.R. Conf., Pergamon Press,
436

KUDO, H. 1965 'Some New Slip-line Solutions for Two-
Dimensional Steady State Machining'
Int. J. Mech. Sci. **7**, 43

LEE, E. H. and 1951 'The Theory of Plasticity Applied to a
SHAFFER, B. W. Problem of Machining'
J. App. Mech., Trans. A.S.M.E. **73**, 405

LEE, E. H. 1954 'A Plastic-flow Problem Arising in the
Theory of Discontinuous Machining
Trans. A.S.M.E. **76**, 189

LOLADZE, T. N.   1967 'Problems of Wear and Strength in Cutting
                             Tools'
                             C.I.R.P. Annals, 1967/8, Pergamon Press
LOW, A. H.   1962 'Effects of Initial Conditions in Metal
                             Cutting'
                             N.E.L. Report No. 49
LOW, A. H. and   1962 'An Investigation of Non-Steady State
WILKINSON, P. T.   Cutting'
                             N.E.L. Report No. 65
MERCHANT, M. E.   1945 'Mechanics of Metal Cutting Process.
                             I—Orthogonal Cutting and a Type 2 Chip'
                             J. App. Phys. **16**, 267
OKOSHI, M. and   1966 'Effects of the Curvature of Work Surface
KAWATA, K.   in Metal Cutting'
                             C.I.R.P. Annals 1966., Pergamon Press
OXLEY, P. L. B.,   1961 'The Influence of Rate of Strain-Hardening
HUMPHREYS, A. G.   in Machining'
and LARIZADEH, A.   *Proc. Instn. Mech. Engrs* Paper 31/60
OXLEY, P. L. B.   1966 'Introducing Strain-Rate Dependent Work
                             Material Properties into the Analysis of
                             Orthogonal Cutting'
                             C.I.R.P Annals XIII, 127
PALMER, W. B.   1967 'Plastic Deformation When Cutting into an
                             Inclined Plane'
                             J. Mech. Eng. Sci. **9**, 1
PALMER, W. B. and   1960 'Mechanics of Orthogonal Machining'
OXLEY, P. B.   *Proc. Instn Mech. Engrs* **173**, 623
PEKELHARING, A. J.   1967 'Resumé and Critique on Chip Making
                             Processes'
                             Int. Conf. on Manu. Tech., A.S.T.M.E. 483
PUGH, H. LL. D.   1958 'Mechanics of the Cutting Process'
                             Proceedings of the conference on Technology
                             of Engineering Manufacture
                             Instn of Mech. Engrs. Paper 53, 237
RAPIER, A. C. and   1959 *Mech. Eng. Res. Lab.*, MF/MS, 1956
WILKINSON, P. T.
RUBENSTEIN, C.   1967 'The Role of Surface Energy in Metal
                             Cutting'
                             Iron and Steel Inst. Special Report Series,
                             No. 94, 49
SHAFFER, B. W.   1956 'An Analysis of the Orthogonal Boring
                             Operation'
                             A.S.M.E. Paper No. 55–A–67
                        1958 'Chip Formation During the Turning
                             Operation in the Presence of a Built-up Nose'
                             Proc. 3rd U.S. Cong. App. Mech., *A.S.M.E.* 6
SHAW, M. C.   1963 A Critique of Nine Papers
                             Int. Res. Prod. Eng., A.S.M.E. 3

SHAW, M. C.,
COOK, N. H. and
SMITH, P. A.

1952 'The Mechanics of Three-dimensional Cutting Operations'
Trans. A.S.M.E. **74**, 1055

STABLER, G. V.

1965 *The Chip Flow Law and its Consequences*
5th Int. M.T.D.R. Conf., Pergamon Press, 243

THOMSEN, E. G.,
YANG, C. T. and
KOBAYASHI, S.

1965 *Plastic Deformation in Metal Processing*
Macmillan, New York, Chapter 22, p. 486

THOMSON, P. J. *et al.*

1969 Proc. 9th Int. M.T.D.R. Conf.
Pergamon Press (In press)

USUI, E. and
HOSHI, K.

1963 'Slip-line Fields in Metal Machining which involve Centered Fans'
Int. Res. in Prod. Eng., A.S.M.E. 61

USUI, E.,
KIKUCHI, K. and
HOSHI, K.

1963 'The Theory of Plasticity Applied to Machining with Cut-away Tools'
A.S.M.E. Paper No. 63–Prod. 5

WALLACE, P. W. and
BOOTHROYD, G.

1964 'Tool Forces and Tool-chip Friction in Orthogonal Machining'
J. Mech. Eng. Sci. **6**, 74

ZOREV, N.

1966 *Metal Cutting Mechanics*
(Translation from Russian)
Pergamon Press, 526 pp.

EXTRUDING, DRAWING, INDENTING, FORGING, CUTTING AND ROLLING

ADAMS, D. J.

1966 'Inhomogeneous Extrusion'
Thesis for Ph.D., Univ. of Melbourne

ALEXANDER, J. M. and
LENGYEL, B.

1966 'Bi-metallic Extrusion'
Proc. Instn mech. Engrs, Appl. Mech. Conf., Cambridge

CHITKARA, N. R. and
JOHNSON, W.

1965 *Some Results for Rolling with Circular and Polygonal Rolls*
Proc. 5th N.T.D.R. Conf. (1964), p. 391, Pergamon Press

CRANE, F. A. A. and
ALEXANDER, J. M.

1968 'Slip-line Fields and Deformation in Hot Rolling of Strip'
J. Inst. Metals **96**, 289

DRUYANOV, A. B.

1965 'Kinematic Problems in the Theory of the Plane Plastic Flow of Ideally Plastic Bodies'
in the book *Investigation of Processes of Plastic Deformation of Metals* p. 134, Moscow

FIRBANK, T. C. and
LANCASTER, P. R.

1965 'A Suggested Slip-line Field for Cold Rolling with Slipping Friction'
Int. J. mech. Sci. **7**, 847

FIRBANK, T. C. and
LANCASTER, P. R.

1966 'Note: On some Aspects of the Cold Rolling Problem'
Int. J. mech. Sci. **8**, 653

GRUNZWEIG, J.,   1954 'Calculations and Measurements on
LONGMAN, I. M. and   Wedge-indentation'
PETCH, N. J.   *J. Mech. Phys. Solids* **2**, 81

HILL, R.   1950 *The Mathematical Theory of Plasticity*
O.U.P.
1953 'On the Mechanics of Cutting Metal Strips
with Knife-edged Tools'
J. Mech. Phys. Solids **1**, 265
1967 'Ideal Forming For Perfectly Plastic Solids'
J. Mech. Phys. Solids **15**, 223

HILLIER, M. J. and   1963 'Plane Strain Extrusion Through Partly
JOHNSON, W.   Rough Dies'
Int. J. mech. Sci. **5**, 191

JOHNSON, W.   1957 'The Cutting of Round Wire with Knife-edge
and Flat-edge Tools'
Appl. scient. Res. (A), **7**, 65
1958a 'Indentation and Forging and the Action of
Nasmyth's Anvil'
The Engineer **205**, 348
1958b 'Over-estimates of Load for Some Two-
Dimensional Forging Operations'
Proc. 3rd U.S. Cong. appl. Mech. 571
1962 'Slip-line Fields for Swaging and Indenting'
Int. J. mech. Sci. **4**, 323
1963 'Extrusion . . . and Indenting'
Int. Conf. Prod. Engng. Res., *A.S.M.E.*, p. 342

JOHNSON, W. and   1963 'Some Slip-line Fields for Indenting with
HILLIER, M. J.   Rotating Dies'
Int. J. mech. Sci. **5**, 203

JOHNSON, W.,   1964 'The Indentation of a Semi-Infinite Block by
MAHTAB, F. U. and   a Wedge of Comparable Hardness'
HADDOW, J. B.   *Int. J. mech. Sci.* **6**, 329

JOHNSON, W. and   1961 'The Cutting of Metal Strips Between Partly
KUDO, H.   Rough Knife-edge Tools'
Int. J. mech. Sci. **2**, 294

JOHNSON, W., and   1967 'An Analysis of a Rigid-Plastic Thick-Walled
SOWERBY, R.   Cylinder, Using Slip-line Field Theory'
B.M.E.E. **6**, 201

JOHNSON, W.,   1969 *A Bibliography of Slip-line Fields: Theory and*
SOWERBY, R. and   *Applications*
HADDOW, J. B.   Arnold, London, 176 pp

JOHNSON, W. and   1958 'The Pressure for Indenting Material resting
WOO, D. M.   on a Rough Foundation'
J. appl. Mech., *Trans. A.S.M.E.* **25**, 64

MAHTAB, F. U. and   1962 'The Cutting of Strip with Knife-edge and
JOHNSON, W.   Flat-edge Face Tools'
J. Mach. Tool Des. Res. **2**, 335

RICHMOND, O. and DAVENPECK, M. L. 1962 'A Die Profile for Maximum Efficiency in Strip Drawing'
4th U.S. Cong. appl. Mech., *A.S.M.E.*, 1053

RICHMOND, O. and MORRISON, H. L. 1967 'Streamlined Wire Drawing Dies of Minimum Length'
J. Mech. Phys. Solids **15**, 195

SORTAIS, H. C. and KOBAYASHI, S. 1967 'An Optimum Die Profile for Axisymmetric Extrusion'
Int. J. M.T.D.R. **8**, 61

SUKOLSKI, P. J. 1955 'Observations on the Flow of Metal with Particular Reference to Extrusion Defects'
B.I.S.R.A. Report MW/43/55

TAKAHASHI, H. 1967 'Some Slip-line Fields for Plane Strain Extrusion and Forging'
Jap. Soc. mech. Engrs. p. 9

TOMLENOV, A. 1963 *Mechanical Working of Metals*
Moscow, 195 pp.

Chapter 15

LOAD BOUNDING APPLIED TO THE PLASTIC BENDING OF PLATES

15.1 Introduction

Upper bounds to the load, P, required to enforce plastic flow in a body have been found previously, largely for various plane strain metal forming operations. A valid system of tangential velocity discontinuities in a physical plane diagram was proposed and the corresponding hodograph then drawn. With the aid of these two diagrams an upper bound to the load P at the boundary could be evaluated by equating the rate at which it does work to the rate of internal energy dissipated, which is usually designated as $\sum ksv^*$; s is the line length over which a tangential velocity discontinuity prevails and is known from the physical plane diagram, and v^* denotes the magnitude of the tangential velocity discontinuity, found from the hodograph. Homologously, the same method may be applied to find an upper bound to the carrying capacity of a plate which is transversely loaded. A valid velocity field is found which prescribes the angular-velocity of rotation for each element of the plate and then by equating the rate at which work is done by the external load to the rate at which energy is dissipated by the fully plastic bending moments in the plate and at its boundary, the load carrying capacity may be estimated.

The principles or justification for using the approach outlined below rests of course on the well-known limit theorems. We first consider in this chapter circular plates of uniform thickness of rigid-perfectly plastic material and then apply the same methods to rectangular and triangular plates; bending action only is considered and all membrane stresses and local shear stress distributions are neglected. It is also important to emphasize that unless the deflections undergone by a plate are large by comparison with possible elastic deflections, the results given by this work may be seriously in error.

The magnitude of the fully plastic bending moment per unit length of a plate is designed M_p and $M_\mathrm{p} = \sigma_0\, h^2_0/4$; σ_0 is the plane strain yield stress which is equal to the uniaxial yield stress of the material for a Tresca material, and h_0 is the plate thickness.

Because most of the cases investigated are treated as incipient problems, the deformation which starts to occur is related to initial configuration and no difficulties about the degree of deflection (i.e. infinitesimal or finite) arise; in effect incipient motion only is considered.

Some preparation for the material of the chapter is to be had from Chapter

7. The method given below was originally given by JOHNSON (1969). It has been rigorously justified by COLLINS (1971) who has also shown that loads to effect transverse bending may also be bounded from below by using the techniques of metal forming plasticity. The method may be made to take account of finite plastic deflection, see CALLADINE (1968) and JOHNSON (1972).

15.2 An Annular Plate Clamped at its Outer Edge

An annulus of outer radius a and inner radius b, is supposed firmly clamped at its outside and free along the inner circumference where, say, a uniformly distributed line load is applied of intensity P per unit length.

The velocity field we shall assume for the plate may be described as follows. Because of symmetry we first assume the plate divided into a very large number of sectors, 1, 2, 3, 4, 5, ..., each one of which subtends a small angle $\delta\theta$ at the centre of the plate, see Fig. 15.1(a). Let the inner circumference $A\,B\,C\,D$... descend with speed u, the outer circumference α, β, γ, δ... being stationary, and let each sectional element of the annulus, 1, 2, 3, 4 ... rotate as a rigid body with angular speed $u/(a - b)$, about $\alpha\beta$, $\beta\gamma$, $\gamma\delta$, etc.

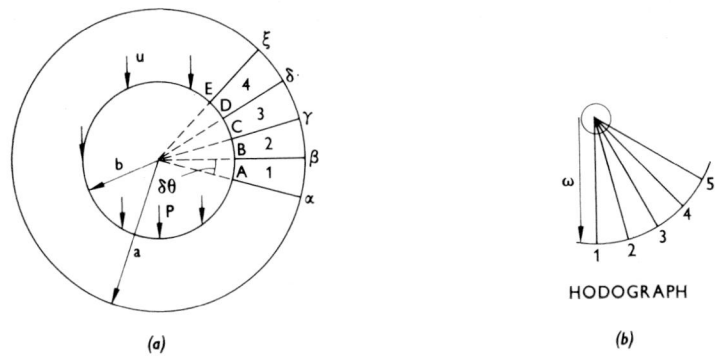

HODOGRAPH

(a)   (b)

FIG. 15.1 Annular plate clamped at its outer edge.

At the boundary $\alpha\beta$ there is full plastic bending—a plastic hinge—and additional plastic hinges are conceived to exist where the separate sectors are in contact along radii. The rotation of each sector may be represented by a vector (i) of magnitude $u/(a - b)$ and (ii), drawn in a clockwise direction perpendicular to its centre-line radius, in accordance with the conventional right-hand screw rule. The vector for sector 1 is represented by 01 and that for section 2 by 02, etc. Figure 15.1(b) shows the hodograph for the sectorially divided annulus; the hodograph in this case shows the angular velocity of all parts of the annulus. Note, in the hodograph, that 12, 23, 34, etc., represent the angular velocity discontinuities along radii A, B and C, etc.

In the limit $\delta\theta \rightarrow 0$, each sector becomes infinitely small and the hodograph becomes a circle of radius ω. With the limiting case in mind we now proceed to calculate an upper bound to the load.

The rate at which work will be done by the externally applied load is $2\pi b \cdot P \cdot u$. The rate of dissipation of energy in the outer circumferential plastic hinge is $2\pi a \cdot \omega \cdot M_p$ and the total for all the infinity of radial hinges is $(a - b) \cdot 2\pi\omega \cdot M_p$. Thus,

$$2\pi b \cdot P \cdot u = 2\pi a \omega M_p + (a - b) 2\pi\omega M_p \tag{15.1}$$

or

$$P = \frac{2a - b}{a - b}\frac{M_p}{b}. \tag{15.2}$$

The upper bound total applied load, P_0, is

$$P_0 = 2\pi b P = 2\pi \frac{2a - b}{a - b} M_p. \tag{15.3}$$

For a solid circular plate for which $b = 0$, an upper bound to the concentrated load required to cause collapse is

$$P_0 = 4\pi M_p. \tag{15.4}$$

If the same frustum-type mode of deformation was assumed when the annulus was carrying a uniform transverse pressure, p, then, by the same approach as above, the rate at which the work is done by the pressure, $\dot{W}$, is

$$\dot{W} = \int_b^a p \cdot 2\pi r \cdot dr \cdot u \cdot \frac{a - r}{a - b}$$

$$= \frac{\pi p u}{3}(a - b)(a + 2b), \tag{15.5}$$

and thus using the right-hand side of (15.1),

$$\frac{\pi p u}{3}(a - b)(a + 2b) = \frac{2\pi M_p u}{a - b}(2a - b)$$

or

$$p = \frac{6M_p \cdot (2a - b)}{(a - b)^2 (a + 2b)}. \tag{15.6}$$

For the solid circular disc, $b = 0$, and thus,

$$p = \frac{12 M_p}{a^2}. \tag{15.7}$$

The mode of deformation just used in connection with the solid plate is usually described as a 'conical'-type mode and it is frequently referred to as such below.

15.3 Plate Position Fixed (Zero Fixing Moment) at its Outer Periphery

The condition investigated is shown in Fig. 15.2. The approach, the method of analysis and the velocity field assumed are identical with those investigated in the clamped plate. We may arrive at an expression similar to (15.1); the first term on the right-hand side would be omitted since the outer edge of the plate is free. Thus,

$$2\pi b \cdot P \cdot u = (a - b) \, 2\pi\omega \, M_\mathrm{p} \tag{15.8}$$

or

$$P = \frac{M_\mathrm{p}}{b}. \tag{15.9}$$

The upper bound to the total applied load is

$$P_0 = 2\pi bP = 2\pi M_\mathrm{p}. \tag{15.10}$$

Obviously, this situation is identical to that in which the following conditions are prescribed for a plate, namely: position fixed at its inner radius, b, and carrying a uniform line load at its outer edge.

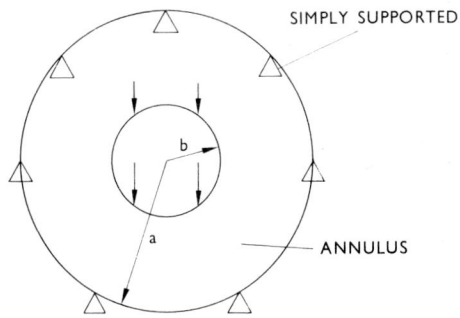

FIG. 15.2 Annular plate simply supported at its outer edge.

If the annulus was subjected to a uniformly distributed pressure, p, it is easily verified, with the help of (15.5) and (15.1) that

$$p = \frac{6M_\mathrm{p}}{(a - b)(a + 2b)}. \tag{15.11}$$

For the circular disc, $b = 0$, and

$$p = 6M_\mathrm{p}/a^2; \tag{15.12}$$

a conical-type mode of deformation has again been assumed.

15.4 Plate Perfectly Clamped along its Inner Boundary

In principle the analysis of this case, with the aid of the same assumptions as were used earlier, is identical with that used in the cases above, but the boundary conditions are slightly different.

It may be verified, see Fig. 15.3, first that

$$P = \frac{M_\mathrm{p}}{a - b}, \tag{15.13}$$

and

$$P_0 = 2\pi \frac{a}{a - b} M_\mathrm{p}; \tag{15.14}$$

and secondly, that $\qquad p = \dfrac{6M_p\,a}{(a-b)^2(2a+b)}.$ $\qquad\qquad$ (15.15)

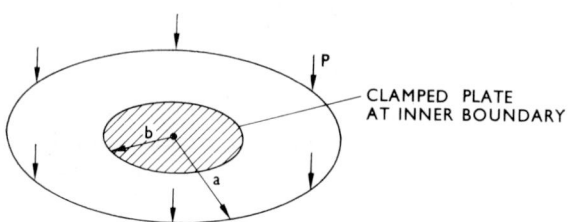

FIG. 15.3 Plate perfectly clamped along its inner boundary.

15.5 The Dynamic, or Inertia Loading of an Annular Plate Perfectly Clamped along its Inner Boundary

The previous case may be reconsidered in the following form which is useful in the study of situations which arise in dynamic plasticity.

A flat, uniformly thick annular plate, clamped at its inner edge, is subjected to an impulsive vertical load which causes the outer boundary to move with speed u_0; this initial kinetic energy is dissipated by enforcing full plastic bending throughout the plate (the possibility of buckling is neglected) and we require to estimate the plate deflection. Obviously on this occasion we do not consider an incipient problem, but assume that deflections will be not large, we may approach it in the same way as before.

The initial kinetic energy of the plate, E, is

$$E = \tfrac{1}{2}\int_b^a 2\pi r\,dr\,m\,h\,u^2 \qquad\qquad (15.16)$$

where u is the plate speed at radius r, h is the plate thickness and m the plate density. Now $u/u_0 = (r-b)/(a-b)$, and after substituting in (15.16), integrating and simplifying,

$$E = \frac{mhu_0^2}{12}\,\pi(3a^2 - 2ab - b^2). \qquad\qquad (15.17)$$

Instead of referring to ω, the angular rotation of sectors, we refer to ϕ, the rotation of a radius about the inner plate boundary. (Of course, $\phi = \omega \times$ time$/2$; sectors will be retarded at a uniform rate since the opposing bending moments are constant in magnitude.)

The total plastic work done on the plate, W, is

$$W = 2\pi b\,M_p\phi + (a-b)M_p\,2\pi\phi. \qquad\qquad (15.18)$$

Equating (15.17) and (15.18), we find

$$\phi = u_0^2\,\frac{m\,h\,(3a^2 - 2ab - b^2)}{24M_p a}. \qquad\qquad (15.19)$$

The peripheral deflection δ is simply

$$\delta = (a - b)\phi = mu_0^2 h(a - b)(3a^2 - 2ab - b^2)/24aM_p. \quad (15.20)$$

For dynamic investigations it should be kept in mind that the value of M_p will depend on the rate of straining of the material, and especially for steels this may be as much as three times the 'static' value.

We now proceed to apply the same methods for determining upper bounds to the load carrying capacity of non-circular plates as we have just used for circular plates.

15.6 Regular Polygonal Plates

Consider a uniformly thick polygonal plate having n sides, which circumscribes a circle of radius a, subject to, say, a uniform transverse pressure p, the plate being firmly clamped at its outer edge, see Fig. 15.4.

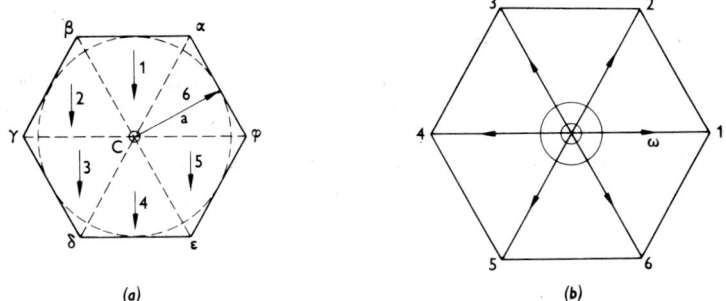

(a)   (b)

FIG. 15.4 Polygonal plate subjected to uniform pressure.

Assuming a pyramidal mode of deformation in which the centre of the plate descends with speed u, the lines of angular velocity discontinuity are $\alpha\beta$, $\beta\gamma$, etc., and $C\alpha$, and $C\beta$, etc. as in Fig. 15.4(a); the hodograph is shown in Fig. 15.4(b).

It is easy to verify that, (i) the rate at which work is done by the pressure is $[una^2 (\tan \pi/n)] p/3$ and (ii) that the rate of plastic work dissipation in the plate is $(2an M_p u \tan \pi/n)/a$ where the plate is perfectly clamped at the circumference, together with $[n . M_p a/(\cos \pi/n)] [2u \sin (\pi/n)/a]$ along lines such as $C\alpha$, $C\beta$ . . . in Fig. 15.4(a). Thus equating (i) and (ii) and simplifying, we find

$$p = 12M_p/a^2. \quad (15.21.i)$$

The upper bound to the total load carrying capacity of the plate is then the plate area times p.

The exercise may be repeated to determine an upper bound to the magnitude of the concentrated transverse load P_0 which may be applied at the plate centre to cause plastic collapse of the plate using the same pyramidal angular velocity field.

It may be proved that,

$$P_0 = (4n \tan \pi/n) \, M_{\mathrm{p}}. \qquad (15.22.\mathrm{i})$$

However, this estimate of P_0 can be improved upon, because a conical mode of deformation may be envisaged which involves a circular plastic hinge of radius a—shown by the broken line in Fig. 15.4(a)—and then $P_0 = 4\pi M_{\mathrm{p}}$; and 4π is always less than $4n \tan \pi/n$, where n is an integer.

Note, incidentally, that in the limit $n \to \infty$, and (15.22.i) gives $P_0 \to 4\pi M_{\mathrm{p}}$, i.e., the result for a circular plate.

If the plate was simply position-fixed at its boundary it is obvious that for a uniform loading of the plate, p,

$$p = 6M_{\mathrm{p}}/a^2 \qquad (15.21.\mathrm{ii})$$

and for a concentrated loading at the plate centroid,

$$P_0 = (2n \tan \pi/n) \, M_{\mathrm{p}}. \qquad (15.22.\mathrm{ii})$$

15.7 Rectangular Plate: Uniform Loading

The magnitude of an upper bound is dependent on the mode of deformation assumed and in this section an example of three quite different ones—all equally valid—is given.

A uniformly thick rectangular plate, sides of length $2a \times 2b$ ($a \leqslant b$) has its whole boundary $ABCD$ position-fixed, see Fig. 15.5, and a uniform trans-

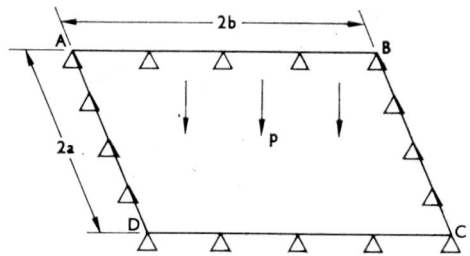

FIG. 15.5 Simply supported rectangular plate subjected to uniform pressure.

verse pressure p is applied. Three different upper bounds to p are given by using the modes of deformation indicated in Fig. 15.6(a), (b) and (c), in which, respectively, we have,

(a) the plastic hinges defined by the plate diagonals,
(b) a conical mode type of collapse and
(c) a symmetrical mode of deformation in which five hinges are formed.

For (a),

$$p \, 2a \, 2b \, \frac{u}{3} = M_{\mathrm{p}} \, 4 \, \sqrt{a^2 + b^2} \cdot \sqrt{\omega_1^2 + \omega_2^2},$$

but $u = a\omega_1 = b\omega_2$ and thus

$$p = 3 \cdot M_p\,(a^2 + b^2)/a^2\,b^2 = \frac{6M_p}{a^2} \cdot \frac{(1 + a^2/b^2)}{2}. \qquad (15.23)$$

For (b), we have as before, from (15.12)

$$p = 6M_p/a^2 \qquad (15.12)$$

For (c), we have

$$p\left\{\frac{2a.a.u}{2.3}.4 + \frac{2(b-a).2a.u}{2}\right\} = M_p\left\{(4a\,\sqrt{2}.\,\omega\sqrt{2} + 2(b-a).2\omega\right\}.$$

Now $u = a\omega$ and hence,

$$p = \frac{6\,M_p}{a^2} \cdot \frac{1}{3} \frac{a/b}{- a/b}. \qquad (15.24)$$

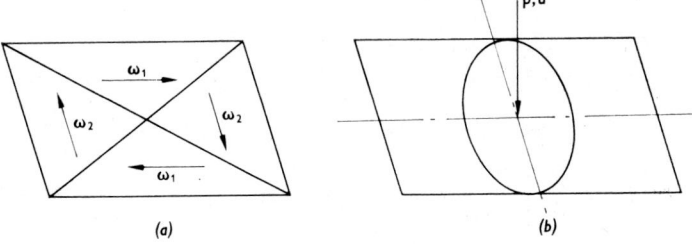

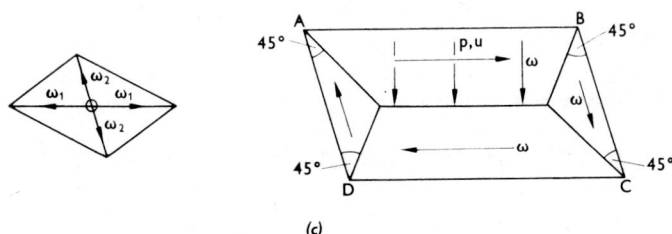

(c)

FIG. 15.6 Modes of deformation for rectangular plate.

Equation (15.24) provides a lower upper bound to p for all values of $0 \leqslant a/b \leqslant 1$ than either (15.23) or (15.12).

It is not difficult to show that if instead of assuming $D\hat{A}E$ to be 45° we denote it by ϕ, then in place of (15.24) we should derive,

$$p = \frac{6M}{a^2} \cdot \frac{1 + (a/b)\cot\phi}{3 - (a/b)\tan\phi}. \qquad (15.25)$$

Now p is least when,

$$\tan \phi = \sqrt{3 + (a/b)^2} - (a/b) \tag{15.26}$$

and hence using (15.26) in (15.25), we find

$$p = \frac{6M_{\mathrm{p}}}{a^2} \cdot \left[\sqrt{3 + (a/b)^2} - (a/b) \right]^{-2}. \tag{15.27}$$

Note that for a very long strip $a/b \to 0$ and then from both (15.24) and (15.27), $p \to 2\, M_{\mathrm{p}}/a^2$, which expression can, of course, be obtained from first principles very easily.

15.8 The Position-fixed Rectangular Plate Carrying a Concentrated Load

If the rectangular plate discussed in the previous section carries a concentrated load, P, at its centroid, several modes of deformation may again be envisaged. Consider that shown in Fig. 15.7, which has a fourfold symmetry

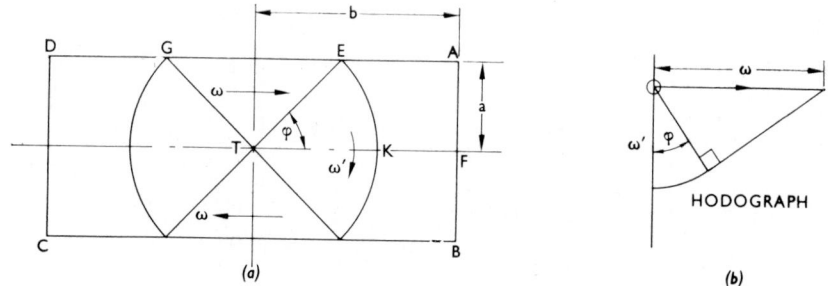

Fig. 15.7 Rectangular plate subjected to concentrated load at its centroid.

and which consists of the rigid rotation of plate portions such as *GTE*, stationary portion *EAFK* and the partly conical type of deformation *ETK*. Then with the help of the corresponding hodograph, see Fig. 15.7(*b*),

$$\frac{P_0 u}{4} = M_{\mathrm{p}} (TE \cdot \omega \cos \phi + TE \cdot \omega' \phi + TE \cdot \phi \cdot \omega')$$

$$= M_{\mathrm{p}} \cdot TE \cdot \omega (\cos \phi + 2\phi \sin \phi),$$

or
$$\frac{P_0}{4M_{\mathrm{p}}} = \cot \phi + 2\phi, \tag{15.28}$$

or
$$\frac{P_0}{4M_{\mathrm{p}}} = \sqrt{\left(\frac{b}{a}\right)^2 - 1} + 2 \sin^{-1}\left(\frac{1}{b/a}\right). \tag{15.29}$$

A table of values following (15.29) is,

$P_0/4M_{\mathrm{p}}$	π	$1 + \pi/2$	2·78	3·53
b/a	1	$\sqrt{2}$	2	3

$P_0/4M_p$ has its least value at $(1 + \pi/2)$ when $\phi = \pi/4$ and then $b/a = \sqrt{2}$. Thus with this mode of deformation, for $1 \leqslant b/a \leqslant \sqrt{2}$, the values of $P_0/4M_p$ lie between π and $1 + \pi/2$; for all $b/a > \sqrt{2}$, the same mode of deformation in which $\phi = \pi/4$ may be used, because all plates for $b/a \geqslant \sqrt{2}$ can accommodate the field proposed and thus the actual values given by (15.29) for $b/a > \sqrt{2}$ can be discarded. Values of $P_0/4M_p$ versus b/a are plotted in Fig. 15.8.

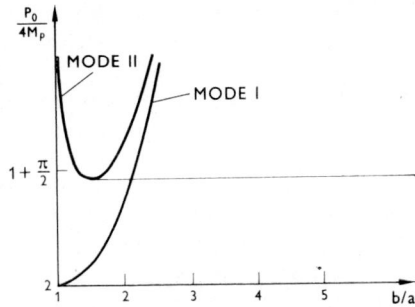

FIG. 15.8 Upper bound load for a rectangular plate subjected to concentrated load.

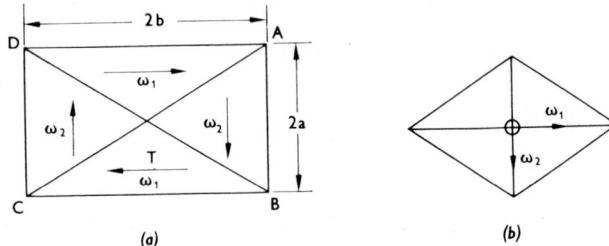

FIG. 15.9 Alternative deformation mode for a rectangular plate subjected to concentrated load.

Alternatively, a second mode of deformation which employs just four plastic hinges, is that of Fig. 15.9, and then

$$\frac{P_0 u}{4} = M_p \cdot TA \cdot \sqrt{\omega_1^2 + \omega_2^2}$$

$$= M_p \sqrt{a^2 + b^2} \cdot \omega_1 \cdot \frac{\sqrt{a^2 + b^2}}{b},$$

because $\omega_1/\omega_2 = b/a$, and thus,

$$\frac{P_0}{4M_p} = \frac{a^2 + b^2}{ab} = \frac{1 + (b/a)^2}{b/a}. \tag{15.30}$$

When $b/a = 1$, $P_0/4M_p = 2$ and for $b/a = 2$, $P_0/4M_p = 2\cdot5$. Values of $P_0/4M_p$ versus b/a following (15.30) are plotted on Fig. 15.8.

In the two solutions discussed and provided by (15.28) and (15.30) and plotted in Fig. 15.8, evidently the second mode is to be preferred over the range $1 \leqslant b/a < 2$ (approximately). For $b/a \geqslant 2$, the first mode of deformation restricted to $\phi = \pi/4$ gives the lower upper bound to $P_0/4M_p$.

15.9 A Plate Shape which is an Equilateral Triangle

An interesting case explored by Mansfield and Hodnestad (WOOD, 1961), and JOHNSON (1969) is that of a plate, transversely loaded by a concentrated load P_0 or a uniformly distributed pressure p, which is an equilateral triangle, see Fig. 15.10.

The degree of fixity along the sides of the plate is represented by μ; $\mu M_p = 0$ would mean that the plate was simply position-fixed at its boundary whilst $\mu M_p = M_p$ would represent a completely built-in or perfectly clamped plate. The mode of deformation envisaged, which has sixfold symmetry, see Fig. 15.11(a) is made up of the rigid body rotation of portions such as $MDNO$ together with plate portions which undergo a conical-type of deformation, such as $NHN'O$. The hodograph is drawn as Fig. 15.11(b); ω denotes the angular velocity of $MDNO$ and there is an angular velocity discontinuity along NO of $\omega \sin \phi$. Note that portions at the apices of the triangle, e.g., $NBN'HSN$, do not move or rotate.

The rate at which work is dissipated internally in the plastic hinges, i.e., along DN, NO and NHN' and along radii such as SO is $\dot{W}$.

Thus, $\dot{W} = (\mu\, M_p\, DN)\, \overrightarrow{oa} + (M_p\, NO)\overrightarrow{ab}$

$$+ (M_p\, NH)\, \overrightarrow{ob} + (M_p\, SO)\, (\overrightarrow{bsh})$$

$$= M_p \left[\mu \cdot \frac{r}{\tan \phi} \cdot \omega + \frac{r}{\sin \phi} \cdot \omega \cos \phi \right.$$

$$\left. + \frac{r}{\sin \phi} \cdot \left(\phi - \frac{\pi}{6} \right) \omega \sin \phi + \frac{r}{\sin \phi} \cdot \omega \cdot \left(\phi - \frac{\pi}{6} \right) \sin \phi \right]$$

$$= r\omega M_p \left[\frac{\mu + 1}{\tan \phi} + 2\left(\phi - \frac{\pi}{6} \right) \right]. \tag{15.31}$$

15.9.1 *The Concentrated Load*

If the external agency doing work is a concentrated load P_0 acting at the plate centroid 0 and moving with downward speed u, then

$$\dot{W} = \frac{P_0 u}{6} = \frac{P_0\, r\omega}{6}, \tag{15.32}$$

where r is the radius of the circle inscribed in the triangle. Equating (15.31) and (15.32) we find

$$P_0 = 6M_p \left[\frac{\mu + 1}{\tan \phi} + 2\left(\phi - \frac{\pi}{6} \right) \right]. \tag{15.33}$$

By differentiating (15.33) it can be shown that P_0 is least when

$$\cos 2\phi = -\mu. \tag{15.34}$$

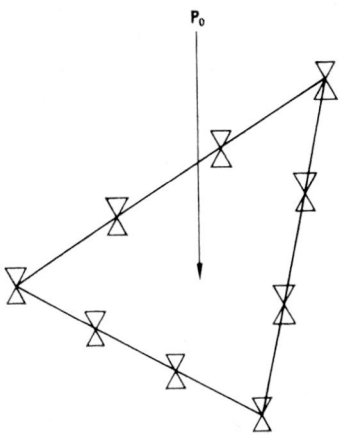

FIG. 15.10 Simply supported triangular plate subjected to concentrated load.

Thus, (a) for a simply supported plate for which $\mu = 0$, $\phi = \pi/4$ and $N\hat{O}H = 15°$. Then,

$$P_0 = 6M_\text{p} \cdot \left(1 + \frac{\pi}{6}\right) = 9 \cdot 1 M_\text{p}. \tag{15.35}$$

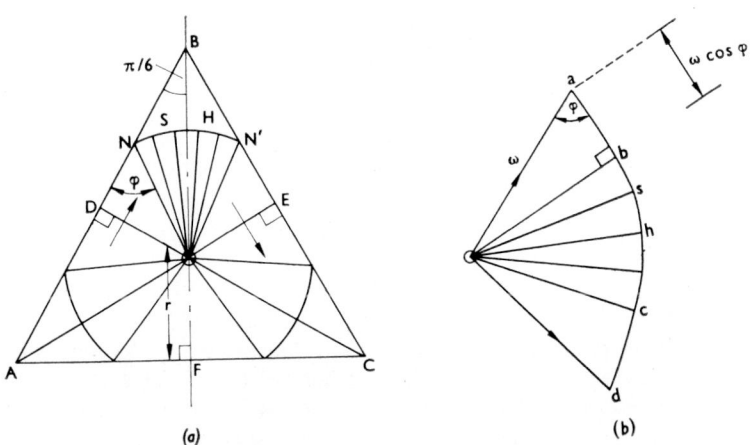

FIG. 15.11 Deformation mode and hodograph for triangular plate.

(Note that using (15.22.ii), $P_0 = 10 \cdot 4M_p$). And (b) for a perfectly clamped plate for which $\mu = 1$, $\phi = \pi/2$, and the usual complete conical pattern of deformation is obtained, for which

$$P_0 = 4\pi M_p. \tag{15.36}$$

15.9.2 The Uniformly Distributed Load

If the external agency doing work is a uniformly distributed pressure, p, which causes point O to move downwards with speed u, then for region $DOHN$,

$$\dot{W} = p\left\{\frac{r}{\tan\phi} \cdot \frac{r}{2} + \frac{1}{2} \cdot \frac{r^2}{\sin^2\phi}\left(\phi - \frac{\pi}{6}\right)\right\}\frac{u}{3}$$

$$= \frac{r^3\omega p}{6}\left[\cotan\phi + \left(\phi - \frac{\pi}{6}\right)\cosec^2\phi\right]. \tag{15.37}$$

Equating (15.31) and (15.37), and simplifying,

$$p = \frac{6M_p}{r^2} \cdot \frac{\mu + 1 + 2\left[\phi - (\pi/6)\right]\tan\phi}{1 + 2\left[\phi - (\pi/6)\right]\cosec 2\phi}. \tag{15.38}$$

Thus, (a) when $\mu = 0$, it can be shown that p is least when $\phi \simeq 37°$ and then

$$p \simeq 5 \cdot 7M_p/r^2, \tag{15.39}$$

(note that (14.2.ii) gives $p = 6M_p/r^2$); and (b) when $\mu = 1$, even when $\phi = 45°$, it is easily verified that $p = (6M_p/r^2) \cdot 1 \cdot 66 \simeq (10M_p/r^2)$; this is 17 per cent less than that predicted using a complete conical mode of deformation, using (15.21.i), i.e. $p = 12M_p/r^2$. MANSFIELD (1957) states that p is least when $\phi = 56 \cdot 4°$ and then $p = 9 \cdot 65 \, M_p/r^2$.

15.10 Simply Supported Plates

We may distinguish between plates which are position fixed along their entire boundary and those which are simply supported. In the latter case certain portions of the plate boundary may lift off the supports. For example, in Fig. 15.12(a), for a square plate simply supported at its boundary, and subject to a point load at O, we may suppose that there are plastic hinges along EG and HF. A quadrant such as $OEBF$ then rotates about a line EF; triangle OEF is depressed in the direction of action of the load while portion BEF is raised in the opposite direction to it. It can easily be shown that

$$\frac{P_0}{4M_p} = 2. \tag{15.40}$$

An alternative deformation mode is that of Fig. 15.12(c) in which elements such as $OIEJO$ rotate about side AB, while elements such as $OQAI$ rotate about QI. It is straightforward to calculate an upper bound load with the help of the hodograph, Fig. 15.12(d), either for P_0 or, indeed, if a uniform transverse pressure, p, was applied. In the latter case, however, account would have to be taken of the fact that triangles such as QAI, by virtue of the fact that they lift, *actually contribute negative work*.

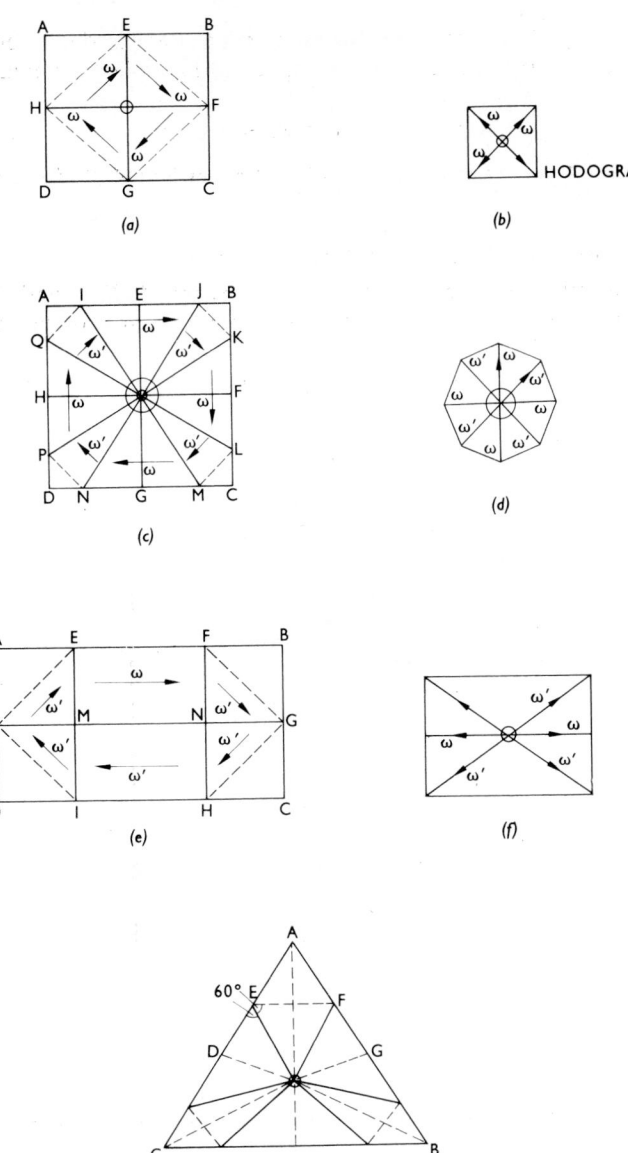

Fig. 15.12 Simply supported plates.

An example of where lifting could occur for a rectangular plate is given in Fig. 15.12(e). The plastic hinge lines would be JG, EI and FH. Lifting would take place for example in portions of the plate such as AEMJ, by rotation about JE, JEM being depressed and JAE being raised.

Lifting of the corner portions of a plate which has the shape of an equilateral triangle—compare Fig. 15.11(a)—is indicated in Fig. 15.12(g). Portions such as *EAF* rise, rotating about *EF*, whilst portion *EFO* is depressed. It can easily be shown that in this case, for a point load P_0 applied through centroid O,

$$P_0/4M_p = \sqrt{3} \quad \text{or} \quad P_0 \simeq 6\cdot93M_p. \tag{15.41}$$

This value for P_0 is about two thirds of that given by (14.35) for a position-fixed plate.

15.11 A Square Plate Supported only at its Corners

If a square plate is supported only at its corners, *A*, *B*, *C* and *D* in Fig. 15.13(*a*), modes of collapse for transverse loads (uniform or non-uniform)

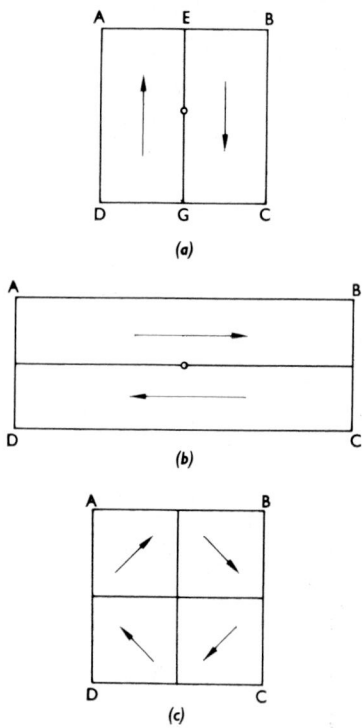

Fig. 15.13 Square and rectangular plates supported at their corners.

may be proposed which are slightly different from those already described for plates simply supported or position fixed at their boundary.

In Fig. 15.13(*a*), one hinge *EG* could be sufficient; *AD* and *BC* would remain fixed and the two halves of the plate *AEGD* and *BEGC* would rotate in counter directions: Fig. 15.13(*b*) for a rectangular plate is essentially the same as Fig. 15.13(*a*). Figure 15.13(*c*) shows another possible case which is attractive for reasons of symmetry.

Similarly, valid collapse modes are given by Figs. 15.12(*c*) and (*e*), if the broken lines are omitted therefrom.

Many problems of determining the collapse modes for plates supported only at specific points can easily be generated.

15.12 A Position-fixed Rectangular Plate, Unsymmetrically Loaded by a Concentrated Load

Figure 15.14(*a*) shows a rectangular plate carrying a concentrated transverse force P_0 which is loaded off one of the lines of symmetry, but lies on the

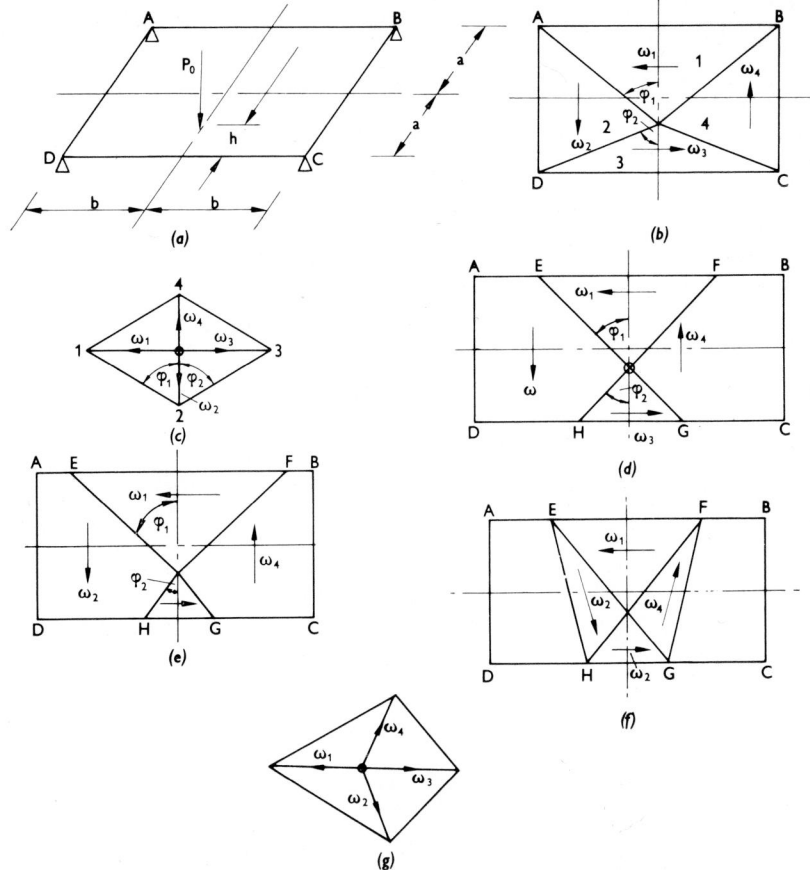

FIG. 15.14 Rectangular plate with offset loading.

other. Modes of deformation for calculating an upper bound to P_0 to cause plastic collapse of the plate and which immediately suggest themselves are shown in Fig. 15.14(*b*), (*d*) and (*e*); all the corresponding hodographs are indicated in principle as Fig. 15.14(*c*).

Figures 15.14(*b*) and (*d*) are just special cases of Fig. 15.14(*e*). Fig. 15.14(*d*) with *EOG* and *FOH* as straight lines will be easier for computation purposes than Fig. 15.14(*e*); Fig. 15.14(*b*) may be the best possible solution of this type to apply, where *E*, *F*, *G* and *H* of Fig. 15.14(*e*) are made to coincide with

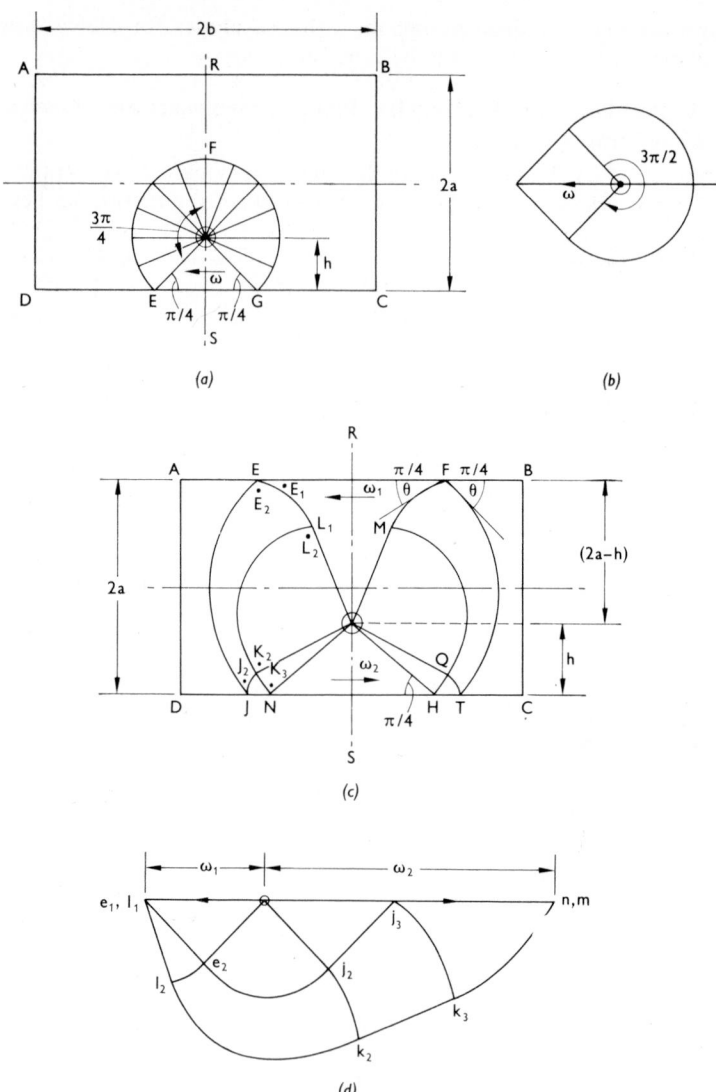

Fig. 15.15 Further deformation modes for a rectangular plate with offset loading.

A, B, C and D, respectively, for reasons of geometry. Yet another possibility is that shown in Fig. 15.14(f), where regions $AEHD$ and $BFGC$ remain rigid; the appropriate hodograph is Fig. 15.14(g). To find a value for P_0 using any of these deformation modes is straightforward.

Two modes of deformation more complicated than those just described are shown as Fig. 15.15(a) and Fig. 15.15(c), for a plate of large aspect ratio, i.e., when b/a is great. Figure 15.15(a) envisages a local mode of deformation, partly of the conical type, i.e., *EFG*, which is entirely contained within the plate; the hodograph is shown in Fig. 15.15(*b*). It may be easily proved, that

$$\frac{P_0}{4M_\text{p}} = \frac{1}{2} + \frac{3\pi}{4} \simeq 2 \cdot 86. \tag{15.42}$$

A more complicated net of lines, which is symmetrical about *RS* and which provides a valid hodograph, is indicated in Fig. 15.15(*c*) and Fig. 15.15(*d*), respectively. The form of the net of lines proposed in the plate is obviously identical with the slip-line field of Fig. 12.21(*a*) in Chapter 12; however, Fig. 15.15(*d*) gives the angular velocity at each point in the net. The end portions of the plate, *ADJEA* and *BCTFB* do not rotate, whereas portions *EL₁OMFE* and *NHON* rotate as rigid bodies. Now for the half portion of the plate to the left of *RS*,

$$\frac{P_0}{2}(2a - h)\,\omega_1 = M_\text{p}\sum l\omega^*,$$

or

$$\frac{P_0}{2M_\text{p}} \cdot \frac{2a - h}{2a} = \frac{\sum l\omega^*}{2a\omega_1}, \tag{15.43}$$

when $\sum l\omega^*$ stands for the product of a line length *l* and the angular velocity discontinuity across it, ω^*. Of course, across OL_1E, *OMF*, *EJ*, *FT*, *JKO* and *TQO* there are finite discontinuities of magnitude $\omega_1/\sqrt{2}$, but elsewhere through the whole field the discontinuities are infinitely small. It will be recalled from a knowledge of slip-line field theory for extrusion that using the well-known notation,

$$pHu = k\sum sv^*,$$

or

$$\frac{p}{2k} = \frac{\sum sv^*}{Hu}. \tag{15.44}$$

Thus if the slip-line field for a frictionless extrusion through a square die, where the reduction exceeds 0·5 and in which a dead metal zone is deposited on the die face, is used in a geometrically similar case to give an upper bound for the collapse load of a plate, and if we have $\sum l\omega^*/(2a)$. ω_1 and $\sum sv^*/Hu$, identical, then from (15.43) and (15.44) we may write,

$$\frac{P_0}{2M_\text{p}} \cdot \frac{2a - h}{2a} = \frac{p}{k}, \tag{15.45}$$

or

$$\frac{P_0}{4M_\text{p}} = \frac{p}{2k} \cdot \frac{1}{r}. \tag{15.46}$$

where $r = (2a - h)/2a$. Thus a useful upper bound to $P_0/4M_\text{p}$ is given by looking up the non-dimensionalized forward extrusion pressure $p/2k$ and multiplying by $1/r$ where the fractional reduction for the extrusion operation, *r*, also applies for the required plate geometry. Note that the number $(p/2k) \cdot (1/r)$ is just the non-dimensionalized pressure required in a piercing operation and is given as Fig. 46 in the book by HILL (1950).

At one extreme $h = a$, the equivalent value of $r = 0.5$ and $(p/2k)(1/r) =$

$1 + \pi/2$, so that $P_0/4M_p = 1 + \pi/2$; this is identical with the solution obtained previously, see p. 528.

When $h/a \simeq 0.48$, using the solution of Fig. 15.15(c), $P_0/4M_p \simeq 2.86$ and this value is the same as that derived using Fig. 15.15(a). Apparently the type of solution given in Fig. 15.15(c) is applicable for $0 \leqslant h/a \leqslant 0.48$ and thereafter that of Fig. 15.15(a).

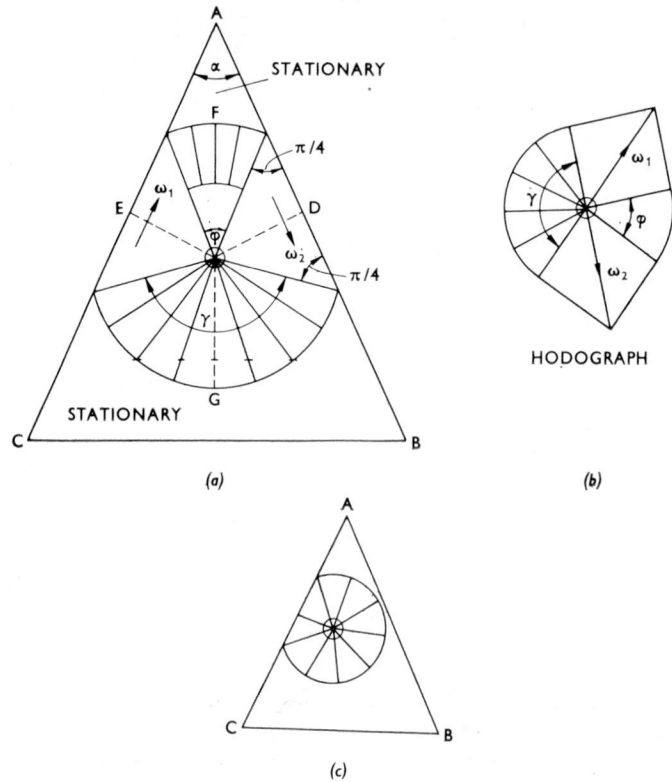

FIG. 15.16 Local collapse in a triangular plate.

15.13 Local Collapse in a Triangular Plate

It is obvious that a form of local collapse may occur in a position-fixed triangular plate, just as it does in a rectangular strip, see Fig. 15.15(a). Using the extrusion slip-line field analogy, the deformation pattern indicated in Fig. 15.16(a) is easily built up; the hodograph appears in Fig. 15.16(b). It is easy to show, by adding together the results for $OEFD$ and $ODGE$ that an upper bound to the concentrated load required to enforce plastic collapse is,

$$\frac{P_0}{M_p} = 2\left(1 + \frac{\pi}{2} - \alpha\right) + 2(1 + \gamma)$$

$$= 2(2 + \pi) = 10.28. \qquad (15.47)$$

Note that this result is independent of angle, α.

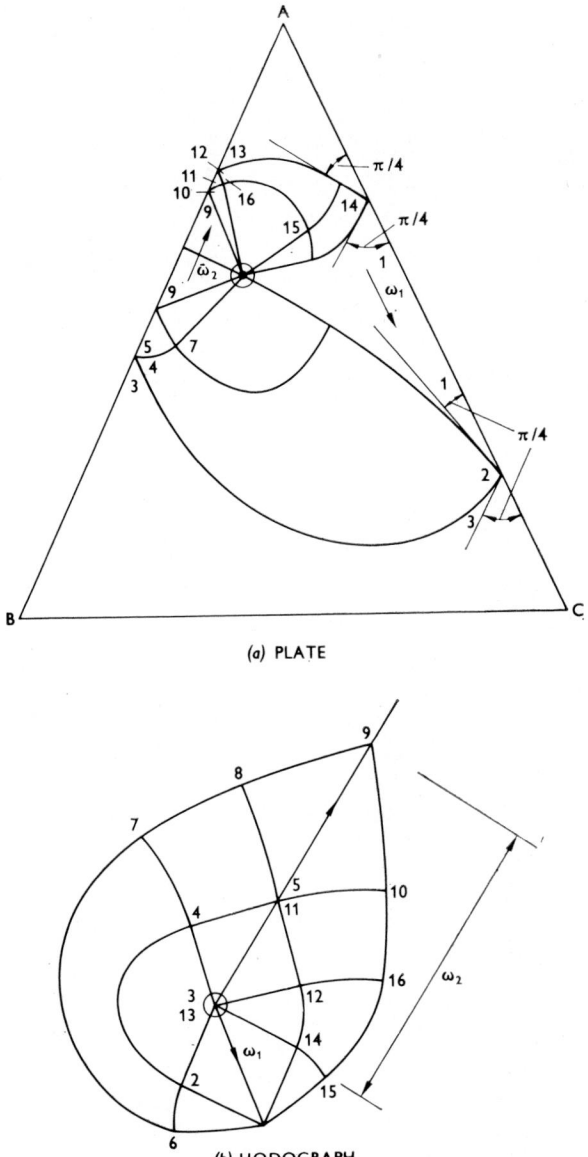

(a) PLATE

(b) HODOGRAPH

FIG. 15.17 Alternative mode of local collapse in a triangular plate.

An off-centre load, see Fig. 15.16(c), may perhaps be treated by using the same system as already referred to in Fig. 15.15(a). Just as Fig. 15.15(c) was an alternative to Fig. 15.15(a), so may Fig. 15.17(a) (with Fig. 15.17(b) as the valid hodograph to it) be an alternative to Fig. 15.16(c).

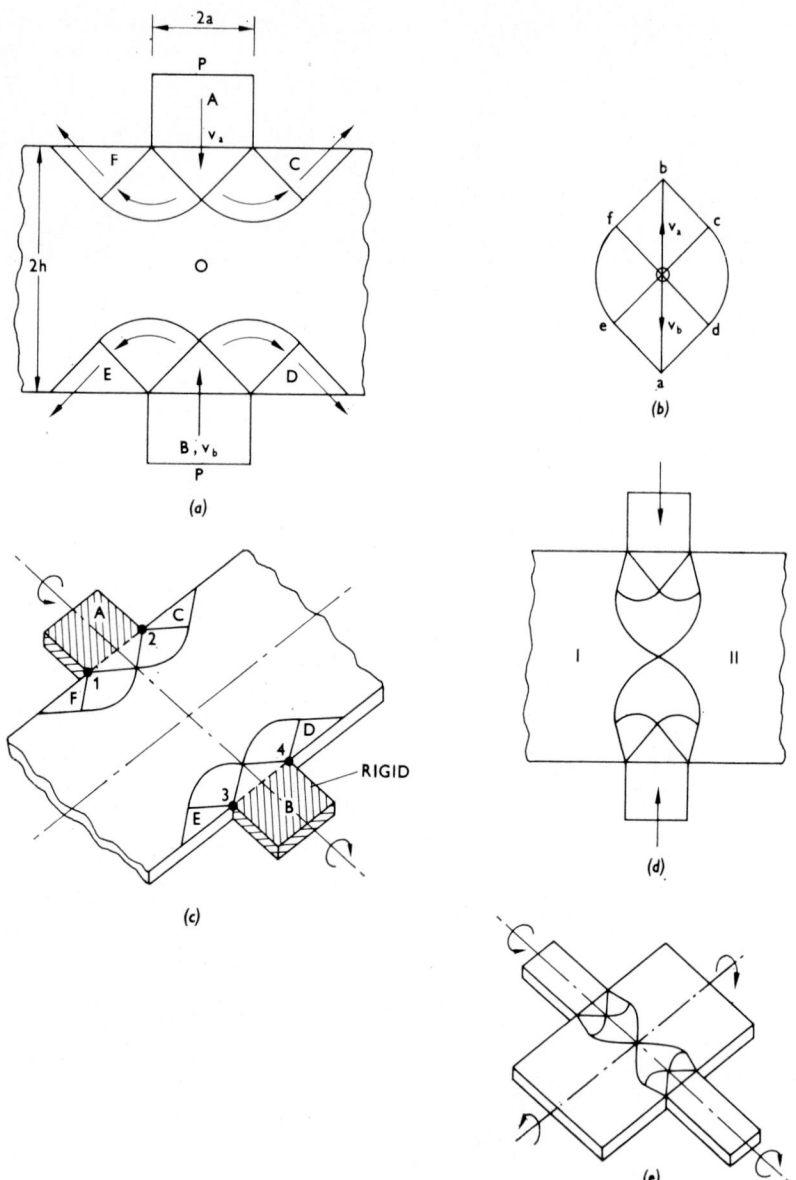

FIG. 15.18 Plate twisting analogy.

15.14 A Plate Twisting Analogy

If two rigid, flat ended punches indent, under conditions of plane strain, a relatively deep block, appropriate slip-line field nets of the Prandtl type, see Fig. 15.18(a), can be drawn and the load to cause plastic indentation calculated. The corresponding hodographs are as shown in Fig. 15.18(b).

Interpret the vectors in the hodograph as angular velocity rather than translational velocity vectors. When this is done we can associate with the hodograph a physical situation in which a wide flat rectangular plate has attached to it, at right angles, two short rigid plates A and B through which equal torques are applied to the main plate; a sketch of what is to be understood is given in Fig. 15.18(c). The hodograph of angular velocity vectors of Fig. 15.18(b) and of the slip-line fields of Fig. 15.18(a) then indicate the mode of plastic collapse due to twisting which takes place in the vicinity of the junction. It is not difficult to see, for instance, that it is implied that at points 2 and 3, in Fig. 15.18(c), the triangular portions C and E rise while portions D and F at 1 and 4 fall. The general manner in which distortion occurs will be intuitively appreciated.

To calculate the torque required to cause collapse, T, we have, in the well-known notation,

$$T\omega_A = M_p \sum l\omega^*.$$ (15.48)

ω_A is the angular velocity of A about its centre-line and is formally identical with v_A of the indentation problem. For the indentation problem we should have,

$$P \cdot v_A = k \sum sv^*$$ (15.49)

where P is the total punch load.

Thus, from (15.48) and (15.49),

$$\frac{T}{M_p} = \frac{P}{k},$$ (15.50)

when the torsion problem and the indentation problem possess identical geometries.

The indentation solution shown in Fig. 15.18(a), as is well known, holds only for $h/a \geqslant 8\cdot7$. When $h < 8\cdot7a$ another kind of slip-line field is most applicable, see Fig. 15.18(d); the outer portions I and II are not stationary but move outwards with equal speeds. This particular indentation problem, with the help of its hodograph, can also be interpreted as a plastic torsion, plate problem; in this case the whole of the main plate is everywhere subject to some degree of rotation, see the sketch in Fig. 15.18(e).

15.15 The Elastic-Plastic Bending of Circular Plates by Transverse Pressure

In a number of papers, OHASHI and MURAKAMI have presented the results of theoretical calculations for the elastic-plastic deflections undergone by solid and annular plates when transversely loaded by a uniform hydrostatic pressure. They have investigated simply supported and clamped mild steel plates and compared their results with those found by experiments (1964, 1966a and b and 1967). These authors found that at a hoop strain of about 0·1 (1964) at the plate centre the simple limit load was attained. Generally, however, it transpires that as distance from the plate centre increases it becomes increasingly difficult to identify a limit load by reference to a deflection or a strain, which increases rapidly with pressure, see Fig. 15.19(a).

Limit loads, due to Onat, which take deflection into account, and results which also allow for the development of membrane forces, give greatly improved correlation with experiment. A general conclusion is that after about 85 per cent of the limit load is reached, membrane forces become significant or, alternatively, when the central plate deflection is about half the plate

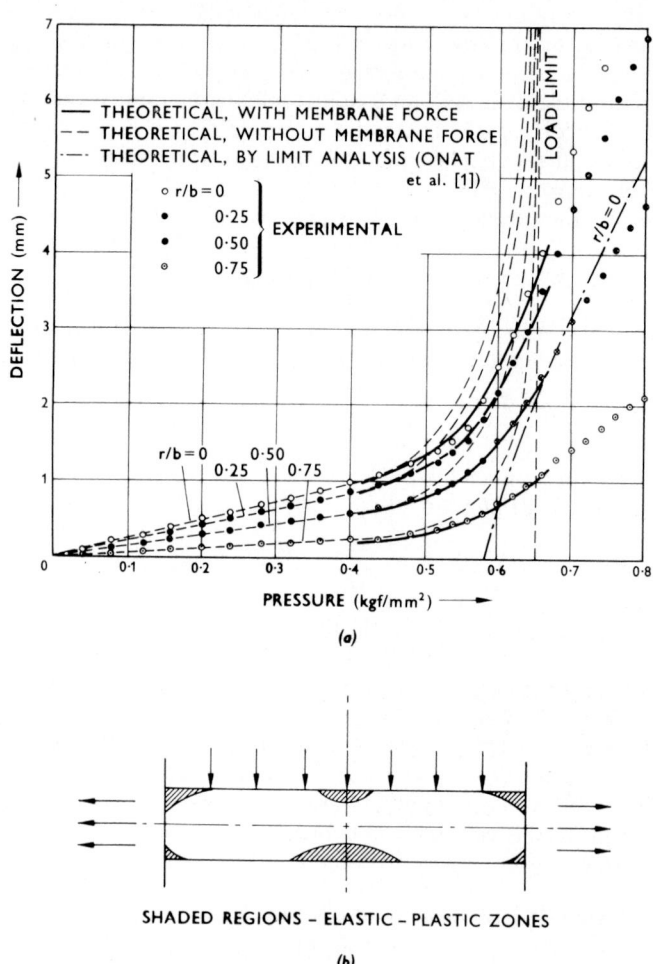

FIG. 15.19 Elastic-plastic bending of circular plates by transverse pressure.

(*After* Ohashi and Murakami, *J. appl. Mech. Trans. A.S.M.E.*)

thickness. The regions and the extent in plates in which material has become plastic, were calculated and then observed by sectioning and etching, see Fig. 7.19(*b*). Note that the extent of the region of plasticity is unsymmetrical due to the membrane tension imposed on the bending stresses; theory and experiment were in good agreement. The theoretical results of most of this work were numerically evaluated.

OHASHI and KAMIYA (1967b) have used a non-linear stress-strain curve for predicting deflection-transverse pressure curves in aluminium plates and obtained good agreement with experiment.

See Problems, 57 74, and 75.

REFERENCES

COX, A. D. and MORLAND, L. W.   1959   'Dynamic Plastic Deformations of Simply Supported Square Plates'
J. Mech. Phys. Solids **7**, 229

CALLADINE, C. R.   1968   'Large Deflection Theory of Plates'
Engineering Plasticity, p. 93
Cambridge University Press

COLLINS, I. F.   1971   'On the Analogy between Plane Strain and Plate Bending Solutions in Rigid Perfect Plasticity Theory'
Int. J. Solids Struct. **7**, 1037

JOHNSON, W.   1969   'Upper Bounds to the Load for the Transverse Bending of Flat Rigid-Perfectly Plastic Plates'
Int. J. mech. Sci. **11**, 913

1972   *Impact Strength of Materials*, 361 pp.
Arnold, London

MANSFIELD, E. H.   1957   'Studies in Collapse Analysis of Rigid-Plastic Plates with a Square Yield Criterion'
Proc. R. Soc., Series A, **241**, 311

OHASHI, Y. and MURAKAMI, S.   1964   'The Elasto-Plastic Bending of a Clamped Thin Circular Plate'
Proc. 11th Int. Conf. appl. Mech.,
Springer-Verlag, Berlin, 212

OHASHI, Y. and MURAKAMI, S.   1966a   'Elasto-Plastic Bending of Thin Annular Plates'
Bull. J.S.M.E. **9**, 271

1966b   'Large Deflection in Elasto-Plastic Bending of a Simply Supported Circular Plate Under a Uniform Load'
J. appl. Mech., Paper No. 66-APM-HH.

OHASHI, Y., MURAKAMI, S. and ENDO, A.   1967a   'Elasto-Plastic Bending of an Annular Plate at Large Deflection'
Ingenieur-Archiv **35**, 340

OHASHI, Y. and KAMIYA, N.   1967b   'Bending of Thin Plates of Material with a Non-Linear Stress-Strain Relation'
Int. J. mech. Sci. **9**, 183

WOOD, R. H.   1961   *Plastic and Elastic Design of Slabs and Plates*
Thames and Hudson, London, Chapter II

Chapter 16

LOAD BOUNDING APPLIED TO AXISYMMETRIC INDENTATION AND RELATED PROBLEMS

16.1 Introduction

The purpose of this chapter is to introduce the reader to the use of load bounding techniques for cases of axial symmetry. In previous chapters these techniques were applied almost exclusively to plane strain processes and these are inherently easier to deal with than ones in which there is some degree of axial symmetry. There has been relatively little exploration in this area as far as metal processing is concerned (we shall not consider the plastic analysis of shells) and the number of examples we can call upon is small. However, the reader who understands metal deformation through the study of plane strain processes is in a good position to appreciate these more-difficult-to-analyse situations and he will possess considerable insight into modes of deformation which are encountered in the more complicated circumstances; for this reason he probably does not now need many examples to aid his understanding. Further, most of the known axisymmetric upper bounds which have been found (mainly for axisymmetric extrusion) are described by analytical expressions and their application, unfortunately, usually leads to an immense amount of labour which thus diminishes their pedagogic value.

For these reasons we confine attention to a few relatively simple problems in indentation which demand no great amount of effort, and we shall thereby seek to demonstrate the principal methods of attack which have been developed.

16.2 Basic Equations

Recall equation (13.11) which states the Upper Bound Theorem in general, as applied to the rigid-perfectly plastic body, and rewrite it as,

$$\int_{s_u} T_i \, du_i \, dS_v + \int_{s_T} T_i \, du_i^* \, dS_T \leqslant \int_V \sigma_{ij}^* \, d\epsilon_{ij}^* \, dV + \sum \int_{s_D} k|dv^*| \, dS_D^*$$

(16.1)

Then the first term on the left-hand side of the inequality is the plastic work increment due to tractions acting through prescribed displacement increments; this work increment is usually the load or force whose value we seek, acting through a prescribed displacement increment which is a small distance moved by a tool or die, as in a compression operation. The second term on the left is the work increment supplied by specified tractions at the boundary; in the compression operation it might be specified that shearing occurs between the work-piece and the tools and therefore to find the work increment dissipated we should have to associate with this traction the relative velocity of sliding at the tool work-piece interface.

The first term on the right-hand side of the inequality is the plastic work increment dissipated in the body of the deforming metal; the asterisks attached to the symbols remind us that $d\epsilon_j^*$ is the plastic strain increment derived, or calculated, from the chosen, fictitious or supposed, acceptable velocity field. Likewise the second term on the right is the plastic work increment dissipated at the assumed velocity discontinuities which pass through the body of the given metal.

The first term on the right-hand side of the inequality may be simplified, (i) for material which obeys the Mises yield criterion and the Lévy-Mises flow rule, as

$$\int_V \sigma_{ij}^* \, d\epsilon_{ij}^* \, dV = \int_V \bar{\sigma}^* \, \overline{d\epsilon_{ij}^*} \, dV = Y \int \overline{d\epsilon_{ij}^*} \, dV, \qquad (16.2)$$

using first, equation (5.16) and then (4.5); and (ii), for a material which follows a Tresca yield criterion, i.e., equation (4.4), and the corresponding flow rule, equation (5.34), as

$$\int_V \sigma_{ij}^* \, d\epsilon_{ij}^* \, dV = 2k \int_V |d\epsilon^*| \max \, dV \qquad (16.3)$$

where $|d\epsilon^*|_{\max}$ represents the absolutely largest linear strain increment derived from the supposed velocity field.

As has been remarked before, *it is equally precise, and indeed now more frequent to operate with velocities rather than displacement increments and with strain rates rather than strain increments. In the examples presented below we shall in fact adopt this mode of description.*

There are two principal classes of approach to three-dimensional metal flow problems; they are (i) to make use of our knowledge of plane strain flows and (ii) to relay entirely on describing the supposed flow of metal by algebraic expressions, usually polynominals. Of course, what is usually done in connection with plane strain solutions subscribes to (16.1), (16.2) and (16.3), though in a highly formalized way; we have, however, made an extended statement about (16.1) and drawn special attention to the relationships in (16.2) and (16.3), in order that when we come to examine examples of (ii), we shall be prepared.

16.3 Simple Heading Operations with Frictionless Tools

16.3.1 *Relatively Narrow Bands: First Mode of Deformation*

Figure 16.1(a) shows a rigid flat-ended tool descending with unit speed on to a block of material, or work-piece, of width $2b$; the breadth of the tool is l and of the work-piece L; $L \gg l$. It is imagined that the tool displaces material by causing two blocks, e.g., $ABFEDC$, see Fig. 16.1(b), to slide over inclined planes, e.g., $ADCB$. Surface $ABFE$ of the block slides frictionlessly across the descending tool, transversely outwards. The hodograph is shown as Fig. 16.1(c). An upper bound to the mean forging or upsetting pressure applied by the die, $\bar{p}$, is given by

$$\bar{p}.l.b.1 = k\left(\text{Area } ABCD.\frac{1}{\sin\theta} + 2.\frac{\text{Area } AED}{\sin\theta}\right).$$

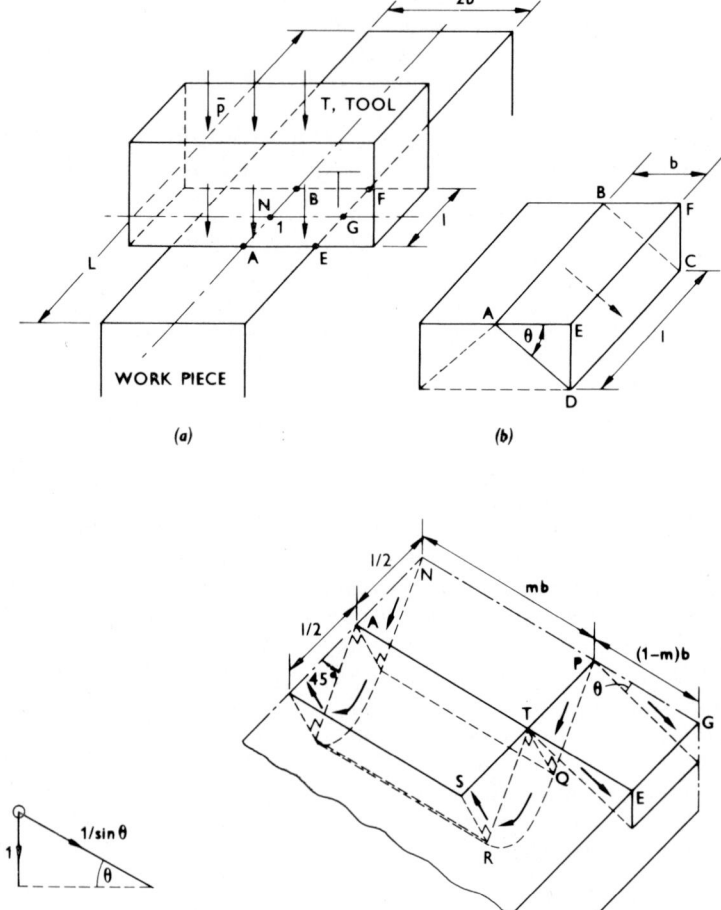

FIG. 16.1 Simple heading operation.

(a), (b) and (c) Relatively narrow band.
(d) Velocity field for moderately wide bands. (A–N–G–E is one quarter of loaded area. Heavy arrows indicate flow directions.)

After simplifying and writing $b/l = n$,

$$\frac{\bar{p}}{k} = \frac{2}{\sin 2\theta} + \frac{n}{\cos \theta}. \qquad (16.4)$$

The value of θ which makes $\bar{p}/k$ a minimum is given by

$$n \sin^3 \theta + 2 \sin^2 \theta - 1 = 0. \qquad (16.5)$$

Solving (16.5) for a given value of n, we may then calculate $\bar{p}/k$ using (16.4).

16.3.2 *Wide Bands: Second Mode of Deformation*

For wider bands or blocks a better upper bound than that just found may be obtained by combining the transverse flow mode just discussed, with the familiar indentation pattern of Fig. 14.22(*a*), see Fig. 16.1(*d*). An equation for $\bar{p}$ is,

$$\bar{p}.b.\frac{l}{2}.1 = \left(1 + \frac{\pi}{2}\right).2k.\frac{l}{2}.bm + \left[\frac{l^2}{8} + \frac{1}{2}\left(\frac{l}{2\sqrt{2}}\right)^2 \frac{\pi}{2}\right] k\sqrt{2}$$

$$+ \frac{(1-m)b}{\cos\theta \sin\theta} \cdot \frac{l}{2} \cdot k + \frac{(1-m)^2.b^2.\tan\theta.k}{2\sin\theta}. \qquad (16.6)$$

The first term on the right-hand side of (16.6) is the rate at which work is done inside the two triangles and sector; the second term is the rate at which energy is dissipated in the plane $PQRSTU$ and the third and fourth terms refer to the transverse sliding block as previously. Simplifying (16.6),

$$\frac{\bar{p}}{2k} = (2 + \pi)m + \frac{\sqrt{2}}{16n}(4 + \pi) + \frac{(1-m)}{\sin\theta\cos\theta} + \frac{n(1-m)^2}{\cos\theta}. \qquad (16.7)$$

We require to optimize (16.7) by choosing values of m and θ which will make $\bar{p}/k$ a minimum for a given value of n. This will be achieved when,

$$\left.\begin{array}{l} \theta \simeq 37 \cdot 2° \text{ (true for all values of } m) \\[2mm] \text{and} \quad m \simeq 1 - \dfrac{1 \cdot 22}{n}. \end{array}\right\} \qquad (16.8)$$

These two simple cases are taken from the paper by Ross (1955).

16.3.3 *A Third Mode of Deformation For a Very Deep Band*

If in the above case $l > L$, $2b$, i.e., if the compressing tool has a greater cross-sectional area than the work-piece to be squashed, then we may imagine a mode of deformation which has fourfold symmetry as indicated in Fig. 16.2, e.g., block $OCBD$ slides down plane ODB parallel to OD; the work-piece tends to 'mushroom' under compression. The mean die pressure $\bar{p}$ is given by

$$\bar{p}. \text{ Area } OAC = k. \text{ Area } OADC. |dv^*|. \qquad (16.9)$$

$|dv^*|$ is found from the hodograph and is cosec θ. Now area $OAC = b^2/2$ and area $OACD = b^2\sqrt{2 - \cos^2\theta}/2 \cos\theta$, so that

$$\frac{\bar{p}}{2k} = \frac{\sqrt{2 - \cos^2\theta}}{\sin 2\theta}. \tag{16.10}$$

$\bar{p}/2k$ is least when $\cos^2\theta = 2 - \sqrt{2}$ or $\cos\theta \simeq 0\cdot765$, i.e., $\theta \simeq 40°$. Then, $\bar{p}/2k \simeq 1\cdot20$. Note particularly, that if the asymmetric mode of deformation had been assumed in which there was plane strain, $\bar{p}/2k = 1$. The bound just obtained, based on a symmetric mode of deformation, is not therefore a good one and, indeed, it is clear that $\bar{p} = 2k$ must be a lower bound since the work-piece could not stand a load in excess of $4b^2.2k$.

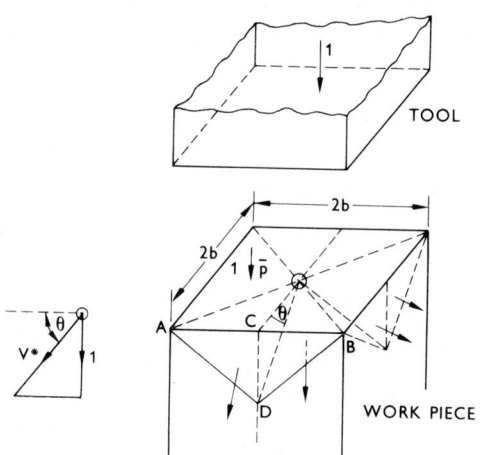

FIG. 16.2 Simple heading operation: very deep band.

16.4 Three-dimensional Punch Indentation Problem

A flat, rigid, frictionless, rectangular punch is caused to indent a semi-infinite block, as shown in Fig. 16.3(*a*). It is required to bound the true indentation load. Note that in obtaining upper and lower bounds below, the indications from the corresponding plane strain case are again used; the similarity of approach with Section 16.3.2 will be obvious.

16.4.1 *Lower Bound for a Rectangular Punch*

In Fig. 16.3(*b*), four 'legs' are first chosen to support the indentation pressure of $3k$; the material in the block under this system of stressing will nowhere be overstressed, i.e., the yield criterion will not be violated. It will be clear that viewed normal to either of the two planes of symmetry, the view is as in Fig. 16.3(*d*) and this has already been met as part of the valid lower bound plane strain case in Fig. 13.35. If an 'all-round hydrostatic compression of $2k$ is next imposed, see Fig. 16.3(*c*), the indentation pressure is put up to $5k$. However, this additional pressure would mean that a normal stress of $2k$

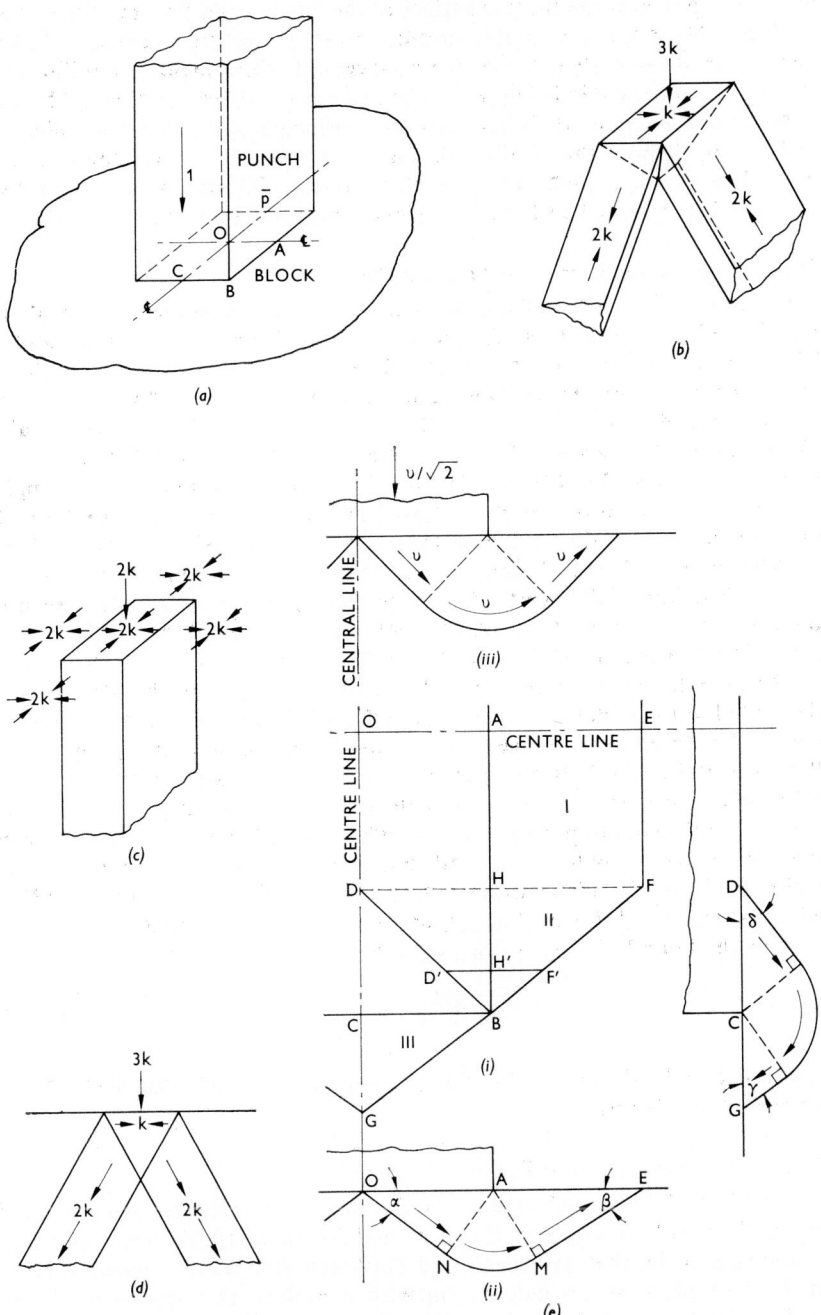

FIG. 16.3 The three-dimensional punch indentation problem.

would be applied along the free surface of the block—and this is not the stated problem. None the less, if this normal stress on the free surface is 'taken away', the stress system below the surface will still remain in equilibrium; the region immediately below would be in a permissible state of equal biaxial compression of amount $2k$ and the third principal stress has been allowed to be zero, without the yield condition being violated. Indeed, the situation arrived at is, on a plane of symmetry, almost identical with that in Fig. 13.35. A lower bound to the indentation pressure is thus $5k$.

16.4.2 *Upper Bound for a Rectangular Punch*

Consider one quarter of the block and the corresponding portion of the punch $OABC$, see Fig. 16.3(a); in plan $OABC$ is also shown in Fig. 16.3(e)(i). There are three zones in which metal flow is supposed to occur and all are conceived of as plane strain flow; material in zones I and II flows parallel to centre-line OA and that in zone III parallel to OC; recall the comparable plane strain case, shown as Fig. 16.3(e)(iii). As $OAHD$ descends the material is swept out parallel to OA, similar to Fig. 16.3(e)(iii), but choosing triangles and sectors of the proportions indicated in Fig. 16.3(e)(ii); regions I and II are different in that the surface of tangential velocity discontinuity, of shape $ENMO$, is conical and oblique, whereas for zone I it is cylindrical. As B is approached from DHF, vertical sections plane parallel to $DHFMN$, seen in plan at $D'H'F'$, give views similar to that of Fig. 16.3(e)(ii), but of decreasing scale. The flow in zone III is indicated in the sectional view of Fig. 16.3(e)(iv).

The indentation pressure is calculated after ascertaining the plastic work dissipated *in* the sector-type flows centred on AB and CB, together with the energy dissipated in the tangential velocity discontinuities arising due to the flow of metal at the bottom boundaries of the triangles and sectors where it takes place over stationary rigid material below. It is not difficult, after comparing the situation proposed with what might happen in a purely plane strain case, to anticipate that the indentation pressure for a rectangular punch will exceed that for plane strain indentation. SHIELD and DRUCKER (1953), who first examined this problem, found that for a square punch $p/k = 5 \cdot 71$.

Thus the bounds for p/k for a square punch are

or
$$\left. \begin{array}{c} 5 < p/k < 5 \cdot 71 \\ p/k = 5 \cdot 35 \pm 7\%. \end{array} \right\} \qquad (16.11)$$

The detailed calculations for this problem are not followed here since it involves much laborious algebra.

16.4.3 *Indentation with a Pyramid Indenter*

Some calculations and experimental results for indenting with pointed pyramids have been given by HADDOW and JOHNSON (1961). Their approach was suggested by that of SHIELD and DRUCKER (1953) and following Hill's analysis of plane strain indentation with a wedge. The approach will be evident from Fig. 16.4; it embodies Hill's original inclusion of pile-up in what is a problem of geometrical similarity. Again we eschew details and refer to the original paper.

16.4.4 *Comment*

These examples conclude our demonstration of the manner in which known plane strain fields can be easily adapted to facilitate calculations to bound the load necessary to perform an operation. We now turn to the second class of methods, in which metal flow is described in analytical terms and where no call is made on plane strain flow fields.

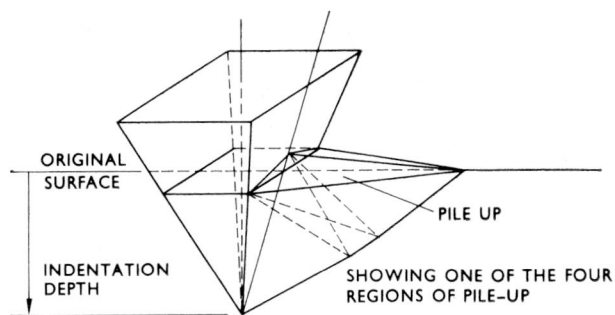

FIG. 16.4 Indentation with Pyramid Indenter.

16.5 An Upper Bound for the Load to Compress Square Discs between Rough, Parallel, Rigid, Overhanging Dies

Two upper bounds will be obtained for the load, or mean pressure, which must be exerted normally on parallel, rigid dies in order to plastically compress a thin square slab when the friction between the slab and dies is so large that the yield shear stress of the material is attained. The first bound is based on a very simple velocity field and the second on one which is more complicated though apparently more nearly describes the observed physical behaviour.

Figure 16.5 shows the square slab of material of height $2c$ and side length $2a$; an origin of co-ordinates is chosen at the block centroid. Suppose the dies approach the xOy plane, the top one with a speed of -1 and the bottom one with a speed of $+1$.

16.5.1 *Simple, Homogeneous Velocity Field*

We choose, or assume, the following simple velocity field, where u, v, w denote the velocity components of a particle parallel to Ox, Oy and Oz, respectively.

$$u = \frac{x}{2c}, \quad v = \frac{y}{2c} \quad \text{and} \quad w = -\frac{z}{c}. \tag{16.12}$$

Observe that at the die faces when $z = \pm c$, $w = \mp 1$, as required.

In order to apply (16.1) we need to calculate $\bar{\epsilon}^*$ and $|dv^*|$. Hence we require to calculate the strain rates. We have

$$\dot{\epsilon}_x^* = \frac{du}{dx} = \frac{1}{2c}; \quad \dot{\epsilon}_y^* = \frac{dv}{dy} = \frac{1}{2c}; \quad \dot{\epsilon}_z^* = \frac{dw}{dz} = -\frac{1}{c}. \tag{16.13}$$

$\dot{\epsilon}_x^*$, $\dot{\epsilon}_y^*$ and $\dot{\epsilon}_z^*$ are principal strain rates. Note that $\dot{\epsilon}_x + \dot{\epsilon}_y + \dot{\epsilon}_z = 0$; anticipation of the requirement that the sum of the three principal strain rates (or the three principal strain increments) should be zero, led us to choose the constants in (16.12) as $1/2c$, $1/2c$ and $-1/c$. Using (16.13) in (5.15), we find $\bar{\dot{\epsilon}}^* = 1/c$. Also, at the work-piece-die interface where the prescribed traction, i.e., yield shear stress k applies, the velocity discontinuity will be,

$$|dv|^* = \sqrt{u^2 + v^2} = \sqrt{x^2 + y^2}/2c.$$

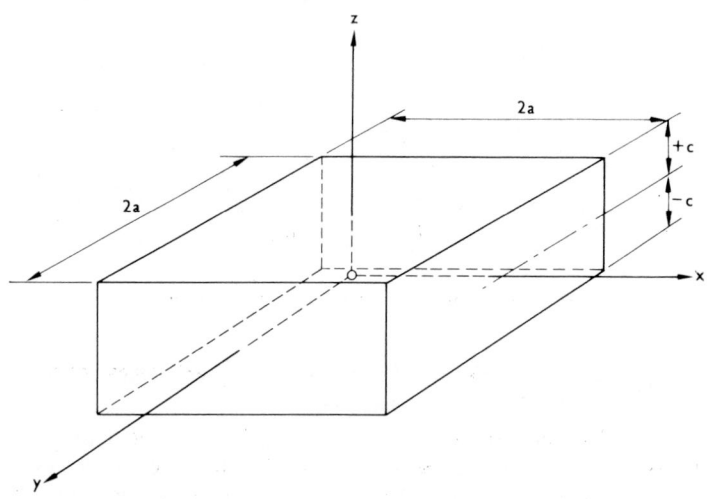

FIG. 16.5 Compression of thin square slab.

Thus denoting the mean normal pressure on the die by $\bar{p}_M$, for a Mises material we have, using (16.1) and (16.2),

$$2\bar{p}_M . 4a^2 . 1 = Y \int_V \bar{\dot{\epsilon}}^* . dV + 2 . \frac{Y}{\sqrt{3}} \int_A |dv|^* dA,$$

where A is the area of contact at one platen or die. Hence,

$$\bar{p}_M = \frac{Y}{8a^2} \left[\frac{1}{c} \int_V dV + \frac{2\sqrt{3}}{3} \int_{-a}^{+a} \int_{-a}^{+a} \frac{\sqrt{x^2 + y^2}}{2c} dx\, dy \right.$$
$$= Y(1 + 0 \cdot 43\, a/h), \tag{16.14}$$
where $h = 2c$.

If the Tresca yield criterion had been employed, then we should have, using (16.3),

$$\bar{p}_T\, 8a^2 . 1 = 2k \frac{8a^2 c}{c} + \frac{k}{c} \int_{-a}^{+a} \int_{-a}^{+a} \sqrt{x^2 + y^2}\, dx\, dy$$

or
$$\bar{p}_T = 2k(1 + 0 \cdot 375\, a/h). \tag{16.15}$$

16.5.2 *Velocity Field which Allows Bulging at Edges of Block*

We choose, or assume, a velocity field which is more complicated, algebraically, than (16.12); it has the advantage, however, of allowing bulging of the four free surfaces of the block, though no variation with z. Constant A is introduced and carried through the calculations and allows us to minimize the rate of energy dissipation in due course.

$$
\left.
\begin{aligned}
u &= Ax(x^2 - 3y^2) + Bx \quad \text{and} \quad \dot{\epsilon}_x^* = \frac{du}{dx} = 3Ax^2 - 3Ay^2 + B \\[2mm]
v &= Ay(y^2 - 3x^2) + By \quad \text{and} \quad \dot{\epsilon}_y^* = \frac{dv}{dy} = 3Ay^2 - 3Ax^2 + B \\[2mm]
w &= -z/c \qquad\qquad\quad \text{and} \quad \dot{\epsilon}_z^* = \frac{dw}{dz} = -\frac{1}{c}.
\end{aligned}
\right\} \quad (16.16)
$$

On this occasion $\dot{\epsilon}_x^*$ and $\dot{\epsilon}_y^*$ are not principal strain rates and we must therefore also determine

$$
\left.
\begin{aligned}
\dot{\gamma}_{xy} &= (\partial u/\partial y + \partial v/\partial x)/2 = -6Axy. \\
\dot{\gamma}_{yz}^* &= \dot{\gamma}_{zx}^* = 0.
\end{aligned}
\right\} \qquad (16.17)
$$

Also,

The incompressibility condition requires that

$$
\dot{\epsilon}_x^* + \dot{\epsilon}_y^* + \dot{\epsilon}_z^* = 0 = 2B - (1/c)
$$

and hence

$$
B = \tfrac{1}{2}c.
$$

The principal strain rates in the plane $z = $ constant, are denoted by $\dot{\epsilon}_1^*$ and $\dot{\epsilon}_2^*$ and are given by

$$
2\dot{\epsilon}_{1,2}^* = (\dot{\epsilon}_x^* + \dot{\epsilon}_y^*) \pm \sqrt{(\dot{\epsilon}_x^* - \dot{\epsilon}_y^*)^2 + 4(\dot{\gamma}_{xy}^*)^2} \qquad (16.18)
$$

Hence, substituting in (16.18) from (16.16),

$$
\dot{\epsilon}_1^* = 3A(x^2 + y^2) + \tfrac{1}{2}c
$$

and

$$
\dot{\epsilon}_2^* = -3A(x^2 + y^2) + \tfrac{1}{2}c.
$$

Thus,

$$
\bar{\dot{\epsilon}}^* = \sqrt{12A^2(x^2 + y^2)^2 + (1/c^2)}. \qquad (16.19)
$$

Ascertain the terms on the right-hand side in (16.2), thus

$$
Y . \int_V \bar{\dot{\epsilon}}^* dV = 2cY \int_{-a}^{+a}\int_{-a}^{+a}\left[12A^2(x^2 + y^2)^2 + (1/c^2)^2\right] dx \, dy
$$

and

$$
\frac{2Y}{\sqrt{3}} \int_{-a}^{+a}\int_{-a}^{+a} |dV^*| dx \, dy = 4Y \int_0^a \int_0^a \Big\{ (x^2+y^2) + 4Ac(x^4+y^4-6x^2y^2)
$$

$$
+ 4A^2c^2[y^6+x^6+3(x^4y^2+y^4x^2)]\Big\} dx \, dy.
$$

These integrals may be evaluated and that value of A chosen which gives the lowest upper bound to p_M; for a/h exceeding about 2·5 it transpires that,

$$\bar{p}_M/Y = 1{\cdot}093 + 0{\cdot}384\,a/h \qquad (16.20)$$

and for the Tresca criterion,

$$\bar{p}_T/Y = 1{\cdot}014 + 0{\cdot}332\,a/h.$$

Details of these calculations are given in the paper by HADDOW and JOHNSON (1962).

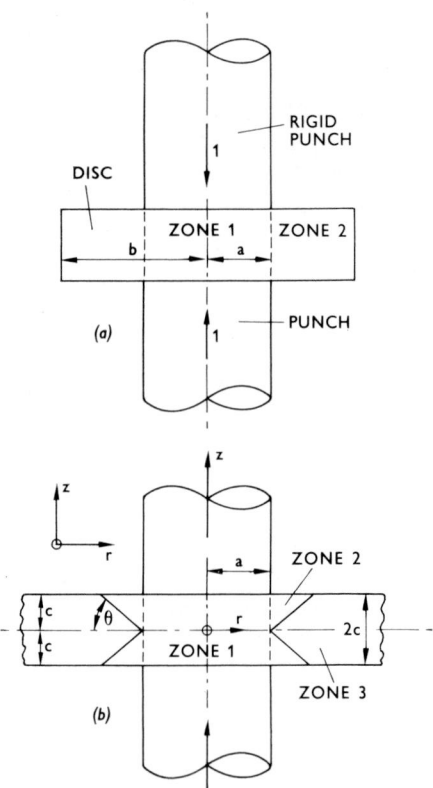

FIG. 16.6 Indentation of thin cylindrical slab by a pair of rough rigid punches.

16.6 The Indentation of a Thin Cylindrical Slab by a Pair of Rough Rigid Punches

Figure 16.6 shows a thin cylindrical slab being indented, or pierced, by a pair of rigid coaxial punches, the friction between the punches and the work material leading to shearing at the interface. We shall demonstrate how to apply two different modes of deformation, selecting the one which requires the smaller load. Both modes will be seen to be valid axisymmetric velocity fields. The first mode is that in which the whole of the disc becomes plastic and is radially expanded; it consists of two zones, I and II. Zone I is the cylindrical region between the punch faces and zone II the annular region surrounding it. The second mode is supposed to have three zones; zone I lies between the punch faces as before, zone II is annular in form and leads to incipient pile-up at the punch, while the remainder of the work-piece comprises zone III which is outside zone II and is supposed rigid.

16.6.1 *First Mode of Deformation: Disc Wholly Plastic*

Zone I: $0 \leqslant r \leqslant a$. Let the punches approach $Oz = 0$, with unit speed so that as in the previous example, $w = -z/c$ and thus $\dot{\epsilon}_z^* = -1/c$; see Fig. 16.6 (*a*). Any velocity chosen here, by virtue of the nature of the situation, has all particles possessing zero circumferential speed. The radial velocity u_r is chosen as $u_r = r/2c$, so that
$\dot{\epsilon}_\theta^* = u_r/r = 1/2c$ and $\dot{\epsilon}_r^* = du_r/dr = 1/2c$; note that $\dot{\epsilon}_\theta^* + \dot{\epsilon}_r^* + \dot{\epsilon}_z^* = 0$.
Then, (a) $\epsilon^* = 1/c$ and (b) $|\dot{\epsilon}^*|_{\max} = 1/c$.

There is a velocity discontinuity at the work-piece-die interface, of u_r. The rate at which plastic work is dissipated in zone 1, W_{M_1}, is then,

$$Y \int_V \bar{\epsilon}^* \, dV + 2 \frac{Y}{\sqrt{3}} \int_A |dv^*| \, dA$$

for a Mises material and

hence
$$\dot{W}_{M,I} = \frac{Y \pi a^2 2c}{c} + \frac{2Y}{\sqrt{3}} \int_0^a 2\pi r \frac{r}{2c} \, dr$$

$$= 2 \pi a^2 Y \left(1 + \frac{\sqrt{3}}{9} \frac{a}{c} \right).$$

For a Tresca material,

$$\dot{W}_{T,I} = 2k \int_V |\bar{\epsilon}^*|_{\max} \, dV + k \int_A |dv^*| \, dA$$

$$= 2\pi a^2 k \left(2 + \frac{1}{3} \frac{a}{c} \right)$$

Zone II: $a \leqslant r \leqslant b$. We choose, $w = 0$ and $u_r = a^2/2cr$; at $r = a, u_r = a^2/2c.a = a/2c$, which is also the magnitude of u_r as given by the expression for u_r for zone I. Thus

$$\dot{\epsilon}_\theta^* = u_r/r = a^2/2cr^2; \quad \dot{\epsilon}_r^* = du_r/dr = -a^2/2cr^2.$$

Again, because $\dot{\epsilon}_\theta^* + \dot{\epsilon}_r^* + \dot{\epsilon}_z^* = 0$, and since $\dot{\epsilon}_z^* = 0$, the two expressions for $\dot{\epsilon}_\theta^*$ and $\dot{\epsilon}_r^*$ from above, are compatible.

Thus, (a) $\bar{\epsilon}^* = \frac{1}{\sqrt{3}} \cdot \frac{a^2}{cr^2}$ and (b) $|\dot{\epsilon}^*|_{\max} = \frac{1}{2} \cdot \frac{a^2}{cr^2}$

Note that a velocity discontinuity exists at $r = a$, where in zone I, $w = -z/c$ and in zone II, $w = 0$.

Proceeding to calculate the rate of energy dissipation in zone II, we have,

$$\dot{W}_{M,II} = Y \int_a^b \frac{1}{\sqrt{3}} \frac{a^2}{cr^2} 2\pi r \, 2c \, dr + 2 \frac{Y}{\sqrt{3}} \int_c^0 2\pi a \left(-\frac{z}{c} \right) dz$$

$$= \frac{4 \pi a^2 Y}{\sqrt{3}} \ln \frac{b}{a} + \frac{4 \pi a^2 Y}{\sqrt{3}} \cdot \frac{c}{2a} = 2 \pi a^2 Y \left(\frac{2\sqrt{3}}{3} \ln \frac{b}{a} + \frac{c\sqrt{3}}{3a} \right)$$

and,

$$W_{T,II} = 2k \int_a^b \frac{1}{2} \frac{a^2}{cr^2} 2\pi r \, 2c \, dr + 2k \int_c^0 2\pi a \left(-\frac{z}{c} \right) dz$$

$$= 2\pi a^2 k \left(2 \ln \frac{b}{a} + \frac{c}{a} \right).$$

Thus, the mean die pressure $\bar{p}$ is

(a) for a Mises material,

$$\bar{p}_M \, \pi a^2. \, 2 = \dot{W}_{M,I} + \dot{W}_{M,II} = 2\pi a^2 \, Y \left(1 + \frac{\sqrt{3}}{9} \cdot \frac{a}{c} + \frac{2\sqrt{3}}{3} \ln \frac{b}{a} + \frac{c\sqrt{3}}{3a} \right)$$

or $\qquad\qquad \dfrac{\bar{p}_M}{Y} = 1 + \dfrac{\sqrt{3}}{9} \cdot \dfrac{a}{c} + \dfrac{2\sqrt{3}}{3} \ln \dfrac{b}{a} + \dfrac{\sqrt{3}}{3} \cdot \dfrac{c}{a}.$ $\qquad$ (16.21)

(b) For a Tresca material,

$$\bar{p}_T \, \pi \, a^2. \, 2 = 2 \, \pi \, a^2 \, k \left(2 + \frac{1}{3} \frac{a}{c} + 2 \ln \frac{b}{a} + \frac{1}{2} \frac{c}{a} \right)$$

or $\qquad\qquad \dfrac{\bar{p}_T}{2k} = 1 + \dfrac{1}{6} \dfrac{a}{c} + \ln \dfrac{b}{a} + \dfrac{1}{2} \dfrac{c}{a}.$ $\qquad$ (16.22)

16.6.2 *Second Mode of Deformation: Outer Annulus of Disc Remaining Rigid*

Figure 16.6(b) shows the three zones into which the disc is divided. Zone I is identical with that in the first mode of deformation, the assumed velocity field is the same and hence all calculations concerning it are the same. Zone III is rigid and no calculations follow therefrom. For zone II we assume, $u_r = a^2/2cr$ and thus $\dot{\epsilon}_r^* = du/dr = -a^2/2cr^2$ and $\dot{\epsilon}_\theta^* = u_r/r = a^2/2cr^2$. For w we choose $w = a^2 \tan \theta/2cr$, so that $\dot{\epsilon}_z^* = dw/dz = 0$ but $\dot{\gamma}_{rz}^* = (\partial w/\partial r + \partial u_r/\partial z)/2 = -a^2 \tan \theta/4cr^2$.

Now the principal strain rates are given by

$$2\dot{\epsilon}_{1,2}^* = (\dot{\epsilon}_r^* + \dot{\epsilon}_z^*) \pm \sqrt{(\dot{\epsilon}_r^* - \dot{\epsilon}_z^*)^2 + 4(\dot{\gamma}_{rz}^*)^2},$$

$$= \left(-\frac{a^2}{2cr^2} + 0 \right) \pm \sqrt{\frac{a^4}{4c^2 r^4} + \frac{a^4 \tan^2 \theta}{4c^2 r^4}},$$

that is $\qquad \dot{\epsilon}_{1,2}^* = -\dfrac{a^2}{4cr^2} (1 \mp \sec \theta).$

Thus, $\quad\quad\quad \bar{\dot\varepsilon} = 2\sqrt{1 + 3 \sec^2 \theta}/3 \cdot (a^2/4\ cr^2)$

and $\quad\quad |\dot\varepsilon^*|_{max} = a^2 (1 + \sec \theta)/4\ cr^2.$ $\quad\quad\quad\quad$ (16.23)

Now the magnitude of the vertical discontinuity between zones I and II is

$$|dv^*|_{1,2} = \frac{a \tan \theta}{2c} + \frac{z}{c},$$

and the sliding velocity discontinuity between zone II and zone III is $|dv^*|_{2,3}$ $= a^2 \sec \theta/2cr$.

We are now in a position to apply to equations (16.1), (16.3) and (16.23). It may be shown, see SAMANTA (1968), that

$$\frac{\bar{p}_T}{2k} = 1 + \frac{(1 + \sec \theta)}{2} \left[\left(1 + \frac{a}{c} \cdot \tan \theta \right) \ln \left(1 + \frac{c}{a} \cot \theta \right) - 1 \right]$$

$$+ \frac{a}{6c} + \frac{\tan \theta}{2} + \frac{c}{2a} + \frac{\sec \theta \operatorname{cosec} \theta}{2}. \quad\quad (16.24)$$

That value of θ is now determined which makes (16.24) least; according to Samanta, the best value of θ depends on a/c and lies between about 36° and 39°.

For $a/c \geqslant 3 \cdot 5$ and $b/a > 4$, this last, or second mode of deformation, gives a value for $p_T/2k$ lower than that of the first.

16.7 Velocity Fields for Axisymmetric Extrusion, Rolling and Indentation

Perhaps the first successful use of assumed, but valid velocity fields for predicting the mean pressure to extrude round bar, is that of JOHNSON and KUDO (1960). To extrude through square dies, two zones to the complete velocity field, with velocity discontinuities operating between them, were used; the two zones used were rather similar to zones I and II of the first mode of deformation discussed in the previous section—as comparison with the original paper will show. However, we shall not repeat those calculations for extrusion here. It will be appreciated from above that upper bounds based on velocity fields expressed analytically lead to much tedious calculation; this is especially the case in extrusion and rolling. The reader who wishes to gauge the quantity of effort in arriving at some solutions should examine the book by AVITZUR (1968). The principles applied in all the cases are basically identical and the reader who has grasped the methods employed above will be readily able to grasp the essence of many of the papers recently published in this field.

Attention should be drawn to some of the more outstanding papers that have been recently published. KUDO (e.g., 1961) has contributed immensely to this field and his work has helped to set the direction for many pieces of work in theoretical extrusion. HALLING and MITCHELL (1965) showed how to extend the idea behind the plane strain hodograph of Fig. 13.8 to axisymmetric extrusion through conical dies; Fig. 16.7(a) shows their physical plane diagram and Fig. 16.7(b) the hodograph. Material crossing AB undergoes a tangential velocity discontinuity parallel to AB (which is conical), i.e., $\overrightarrow{ab}$, and proceeds thereafter parallel to the die face. Its velocity increases as AC is approached, work being done on the metal as it flows conically inwards towards the cone apex O. On encountering AC, a velocity discontinuity $\overrightarrow{ca}$ is imposed and the material then emerges parallel to the initial axis of the slug.

ADIE and ALEXANDER (1967) extend the applicability and usefulness of the approach of HALLING and MITCHELL and deal with rod-can, tube-can and various other complicated forms of extruded product.

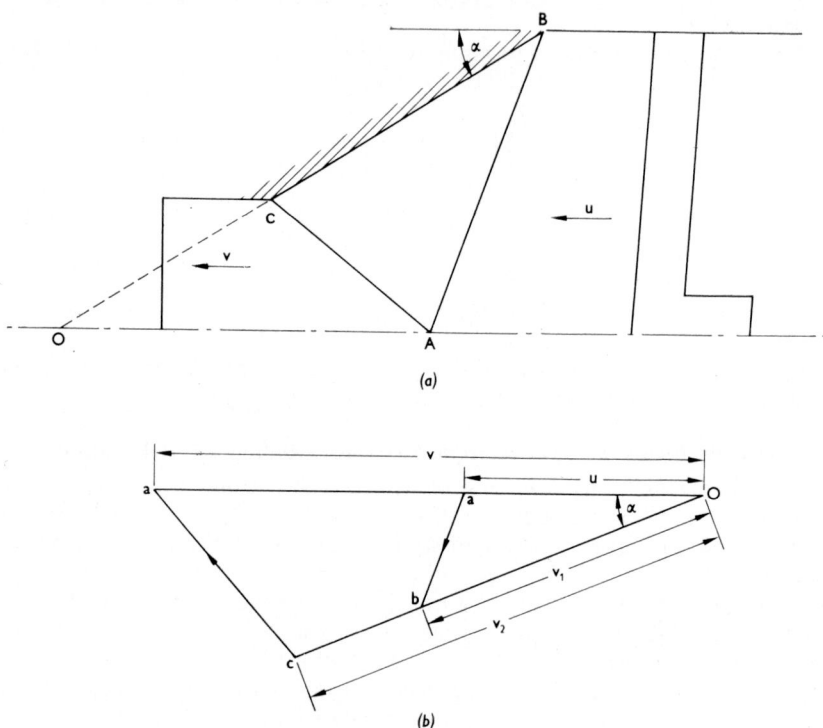

FIG. 16.7 Axisymmetric extrusion through conical dies.

The use of a computer, to great effect, in these sort of problems is well demonstrated by LAMBERT and KOBAYASHI (1969). A novel and very interesting approach, using potential theory, has been presented by SHABAIK and THOMSEN (1966).

The first application of the Limit Theorems to axisymmetric situations seems to be that of SHIELD (1955), who considered the indentation by a flat, circular, rigid punch of a plate of finite thickness; both upper and lower bounds to the indentation load were obtained. LEVIN (1955) solved the same sort of problem when the indenter entered a semi-infinite body. Some account of KUDO's approach to extrusion and forging will be found in JOHNSON and KUDO (1962).

16.8 Hill's General Method

HILL (1963) has given a general method of analysis for metal-working processes when the plastic flow is unconstrained and it applies whatever the friction or tool geometry. To date no applications of this method, except the ones discussed by Hill in his original paper, have been given. It would appear that the well-developed upper bound approach using a guessed velocity field, is very easily appreciated and applied, and for this reason engineers have tended to be well satisfied with it, especially when their experiments have confirmed their 'theoretical predictions'.

See Problem 77

REFERENCES

ADIE, J. F. and ALEXANDER, J. M.	1967	'A Graphical Method of Obtaining Hodographs to Axisymmetric Problems' *Int. J. mech. Sci.* **9**, 349
AVITZUR, B.	1968	*Metal-Forming: Processes and Analysis* McGraw-Hill Book Co., New York., 500 pp.
HADDOW, J. B. and JOHNSON, W.	1962	'Bounds for the Load to Compress Plastically a Square Disc Between Rough Dies' *Appl. Sci. Res., Series A*, **10**, 476
HALLING, J. and MITCHELL, L. A.	1965	'An Upper Bound for Axisymmetric Extrusion' *Int. J. mech. Sci.* **7**, 277
HILL, R.	1963	'A General Method of Analysis for Metal Working Processes' *J. Mech. Phys. Solids* **11**, 305
JOHNSON, W. and KUDO, H.	1962	*Mechanics of Metal Extrusion* Manchester University Press, 226 pp.
	1960	'Use of Upper Bound Solutions . . . For Axisymmetric Extrusion Processes' *Int. J. mech. Sci.* **1**, 175
KUDO, H.	1961	'Some Analytical and Experimental Studies of Axisymmetric Cold Forging' *Int. J. mech. Sci.* **3**, 91
LAMBERT, E. R. and KOBAYASHI, S.	1969	'An Approximate Solution for the Mechanics of Axisymmetric Extrusion' *9th Int. M.T.D.R. Conf.*, Pergamon Press
LEVIN, E.	1955	'Indentation Pressures of a Smooth Circular Punch' *Q. J. appl. Maths.* **13**, 133
ROSS, E. W.	1955	'On the Ideally Plastic Indentation of Inset Rectangular Bands' *J. appl. Mech., A.S.M.E.* Paper No. 55-A-52

SAMANTA, S. K.   1968 'The Application of the Upper Bound
                               Theorem to the Prediction of Indenting
                               and Compressing Loads'
                               Acta polytech. Scand. Me 38
SHABAIK, A. H. and   1966 'An application of Potential Theory
THOMSEN, E. G.   to the Solution of Metal Flow Problems'
                               6th Int. M.T.D.R. Conf. Pergamon Press
SHIELD, R. T.   1955 'The Plastic Indentation of a Layer
                               by a Flat Punch'
                               Q. J. appl. Maths **13**, 27
SHIELD, R. T. and   1953 'The Application of Limit Analysis
DRUCKER, D. C.   to Punch Indentation Problems'
                               Trans. A.S.M.E., J. appl. Mech. **20**, 453

PROBLEMS

1. A bar of material has a stress-strain curve given by $\sigma = 30\epsilon^{0\cdot4}$, σ is in tons per square inch. Such a bar which has already received an engineering tensile strain of 0·25, is to be pulled in tension until it begins to neck. What further amount of engineering strain may be expected?

2. If for a given material $\sigma = B\epsilon^n$ or $\sigma = Ce^m$, show that $m = 1 - \exp(-n)$.

3. If a material has a true stress-true strain curve expressed empirically by $\sigma = Y\tanh E\epsilon/Y$, show that the natural strain at maximum load in a simple tension test is $(Y/2E)\sinh^{-1}(2E/Y)$.

4. Find an expression for the tensile load-extension diagram for a material which has a constant natural strain-hardening rate and indicate its shape.

5. Calculate the tensile strength for a ductile metal for which $\sigma = 30.\ e^{0\cdot25}$; σ is in tons per square inch.

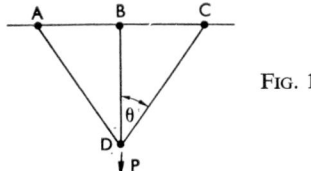

FIG. 1

6. Figure 1 shows a structure, sustaining a vertical load P, consisting of three bars AD, BD and CD, pin-jointed at A, B, C and D. AD and CD are of a metal whose true stress-engineering strain curve is expressed by $\sigma = B_2 e^{n_2}$ and BD is of a different metal for which $\sigma = B_1 e^{n_1}$. For optimum use of the material, it is considered necessary that all three bars should attain tensile instability simultaneously. At what angle θ should the bars be set initially? Calculate some representative values.

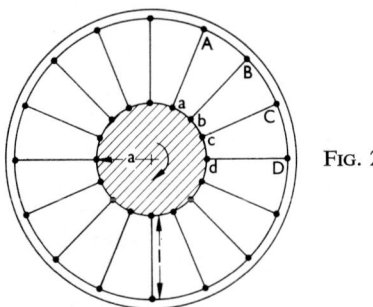

FIG. 2

7. Figure 2 shows the rim of a wheel, A, B, C ... and a perfectly smooth shaft a, b, c ... Aa, ... Bb represents n just taut wires. What torque would be transmitted at the instant instability sets in, in the wires when the shaft is rotated relative to a stationary rim? Assume that all wires are equally stressed and that the engineering strain in each wire is equal to or exceeds $[\sqrt{1 + (2a/l)} - 1]$. a is the shaft radius and l the length of the wires. Bending effects in the wires may be neglected.

8. A thin uniform hollow tube of mean internal diameter d and thickness t is pulled in tension. Find approximate expressions for the change in wall thickness and the mean tube diameter when tensile instability is about to occur. Assume the result of a tensile stress-strain curve on a solid bar to be given by $\sigma = Be^n$, e being engineering tensile strain and σ the true stress.

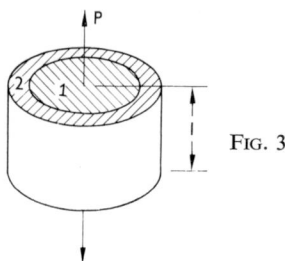

FIG. 3

9. A compound bar is made up of a solid cylindrical rod of material 1 surrounded by a uniformly thick tube of material 2 (see Fig. 3). Their respective cross-sectional areas are A_1 and A_2 and they are firmly fixed together at their common interface. Their true stress-engineering strain curves are $\sigma = B_1 e^{n_1}$ and $\sigma = B_2 e^{n_2}$. Set up an equation which, if solved, will give the engineering strain at which instability will occur for the compound bar acting as one unit.

What should be the ratio A_1/A_2 to give an instability strain of 0·6561 for a compound bar of soft brass, for which $\sigma = 100,000\ e^{0·5}$ and soft aluminium for which $\sigma = 20,000\ e^{0·25}$.

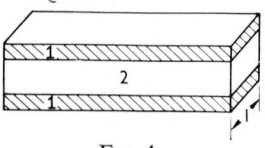

FIG. 4

10. A symmetrical straight bar is made of two different materials as shown in Fig. 4 and contains no residual stress. Its temperature is raised by amount ΔT and $\alpha_1 > \alpha_2$. Find the compressive stress induced in material 1 if the stress-strain behaviour of the bars may be represented by $\sigma_1 = Y_1 + P_1 e_1$ and $\sigma_2 = Y_2 + P_2 e_2$. If the bar cools to its original temperature, find the residual stress in material 1. It may be assumed that recovery in both bars is entirely elastic according to the relationships $\sigma_1 = E_1 e_1$ and $\sigma_2 = E_2 e_2$. α is the coefficient of linear expansion.

11. A frictionless conical drift perforates concentrically a plate of uniform thickness h_0, which contains a circular hole of diameter a_0. The drift of diameter $b_0(>a_0)$ moves along the axis of the hole and opens out a hole with raised lip of height H. Following the method described under Drifting, (see p. 6), show that

$$H = \frac{2}{3} b_0 \left[1 - \left(\frac{a_0}{b_0} \right)^{3/2} \right].$$

What is the thickness of the lip at its tip?

12. Find the plastic work done in Problem 11, when $a_0 = 0$ and if the plate material is linear strain-hardening, i.e., $\sigma = Y + P\epsilon$.

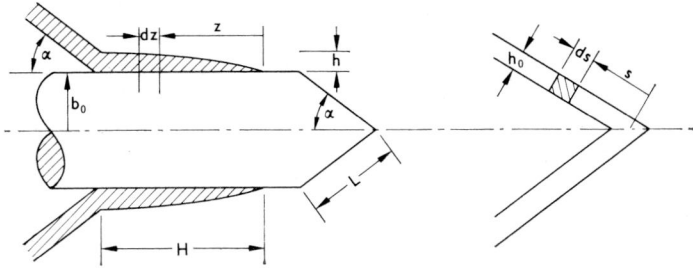

FIG. 5

13. If, instead of a plate of uniform thickness, a cone of uniform thickness was substituted, see Fig. 5, and if the circular drift moved along the axis of the cone, show that the horizontal length of the lip would be $(2/3)\,L$ where L is the slant length of the conical head.

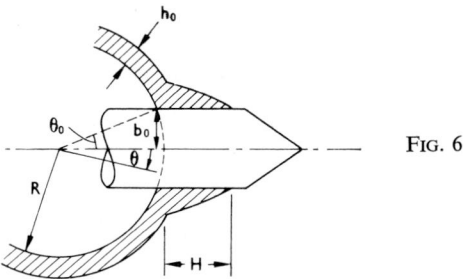

FIG. 6

14. If the conical drift of Problem 13, see Fig. 6, is moved radially outwards to develop a hole of its own diameter in a thin spherical shell of mean radius R, show that its height is given by

$$H = \left(\frac{R^3}{b_0}\right)^{1/2} \int_0^{\theta_0} \sqrt{\sin\theta} \; . \; d\theta.$$

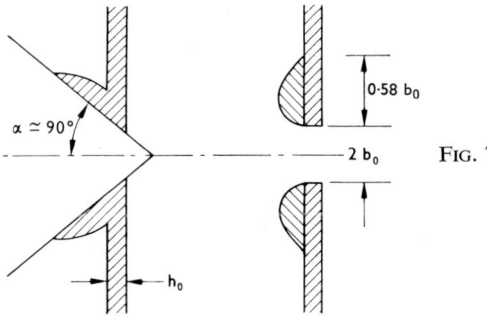

FIG. 7

15. If in Fig. 1.4, both the drift angle and a_0 are zero, the lip height is $\frac{2}{3}b_0$. Suppose another drift, whose angle $\alpha \simeq 90°$, is partially passed through the opening already made, to cause the lip to be folded back and to lie plane parallel to the original plate, see Fig. 7. Show that $H = (2^{\frac{2}{3}} - 1)b_0 \simeq 0.58b_0$.

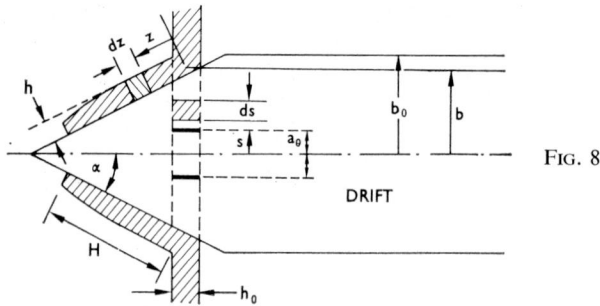

Fig. 8

16. Figure 8 shows a circular conical-ended, frictionless drift which has partially and symmetrically perforated a uniformly thick plate containing a circular orifice, radius a_0. Show, on the basis of the usual assumptions, that the current length of the lip H is given by,

$$H = \frac{b}{\sin \alpha} \left[1 - \left\{ 1 - \left(1 - \frac{a}{b} \right)^{3/2} \sin \alpha \right\}^{2/3} \right].$$

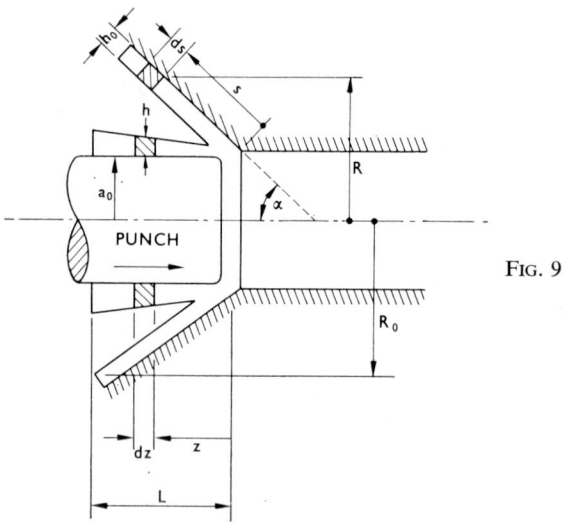

Fig. 9

17. Assuming that the approach of Problem 11 can be applied to the drawing of a flat-headed cylindrical cup (but neglecting bending, and assuming no ironing and buckling, etc.), see Fig. 9, show that the cup height is given by,

$$H = \frac{2 a_0}{3 \sin \alpha} \left[\left(\frac{R_0}{a_0} \right)^{3/2} - 1 \right]$$

What significant stress is overlooked by this analysis?

   (*a*) With reference to Problem 11 plot a curve to show how the cup wall thickness varies with cup height when $\alpha = 90°$, and
   (*b*) verify that the cup height $H \simeq a_0/2$ if $R_0/a_0 = 1.45$.

18. A thin-walled circular cylinder of wall thickness t, radius $D/2$, is subjected to internal pressure p and axial loading by the hydraulic ram arrangement shown in Fig. 10. For ratios of total ram area A_0 to cylinder area A, of (a) 2, (b) 5, determine

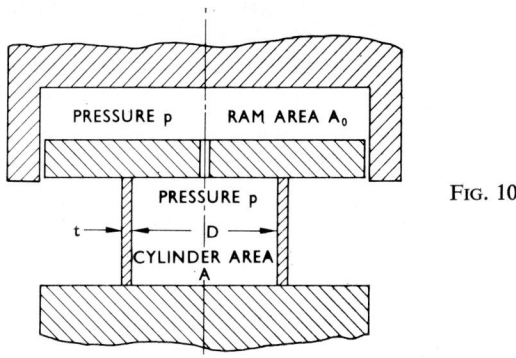

FIG. 10

the pressure p required to cause yielding of the cylinder if the yield stress in simple tension is Y. Use the Mises yield criterion. End constraints on the cylinder and radial stresses in the wall can be neglected. What ratio of A_0/A results in plane strain yield conditions in the cylinder?

19. Consider the plane stress situations which apply to sheet metal forming in general.

(a) Since one principal stress must be tensile, indicate this region by drawing the appropriate Mises yield locus in two-dimensional principal stress space.

(b) What is the magnitude of the maximum tensile stress that can be sustained by a sheet biaxially stressed only?

(c) Draw a graph to show how the mean normal stress varies with maximum shear stress.

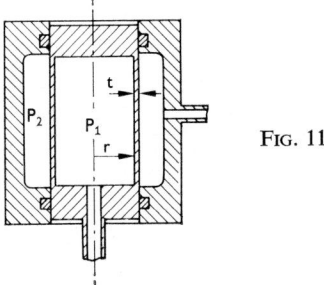

FIG. 11

20. Figure 11 shows a thin-walled circular cylinder of wall thickness t and radius r with closed ends and with the cylindrical surface enclosed by an outer cylinder. The inner cylinder has pressure p_1 acting internally and a lower pressure p_2 acts externally on the cylindrical surface. Sliding seals between the inner and outer cylinders ensure free relative axial movement.

If the inner cylinder has yield stress Y in uniaxial tension show that the pressure p_1 to cause yield according to the Mises criterion is

$$p_1 = \frac{2tY}{r} (4x^2 - 6x + 3)^{-\frac{1}{2}}$$

where $x = p_2/p_1$. Radial stresses can be neglected, and the radial constraint imposed on the cylinder walls by the ends,

21. A long closed ended thin-walled tube of current mean radius, r and current wall thickness t is subjected to an internal pressure p so that it deforms plastically. The material of which the tube is made possesses orthotropic anisotropy which is rotationally symmetric about a normal to the surface of the tube. Tensile tests on the material in its plane yielded an r-value of 2.

Determine the ratios of the circumferential, longitudinal and thickness strains. Compare the results with isotropic deformation. Elastic strains may be neglected.

22. A short stocky cylinder of internal radius a, and external radius b, is compressed between parallel platens, μ being the coefficient of friction between them and the material. Show how to obtain the 'friction hill' over the cylinder thickness, assuming that $\sigma_r = \sigma_\theta$ throughout the block.

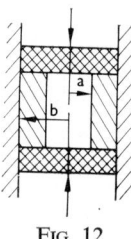

FIG. 12

23. Figure 12 shows a section through a container into which a short, hollow, thick cylinder just fits concentrically. Two pads sit on the ends of the cylinder as shown Find an expression in terms of Y, for the 'friction hill' along a radius of contact between the pads and the cylinder for incipient plastic flow. The coefficient of friction between pads and cylinder is μ and it may be assumed $\sigma_\theta = \sigma_r$ at all radii. The container is to be assumed to be rigid. What is the normal pressure on the container wall? What relationship between μ, a, b and h holds when the material touching the pad is about to shear at radius b? Criticize its validity. h is the cylinder height.

24. Consider the plane strain compression of a strip of metal of height h and width $2a$, between parallel platens when the coefficient of friction between metal and platens is μ, and derive an expression for the mean platen pressure to cause plastic flow, assuming $2\mu a/h$ is small. If the friction stress μp attains the yield shear stress, obtain expressions to take account of this. Draw diagrams to show the pressure variation over a, with μ, if $2a/h = 7$.

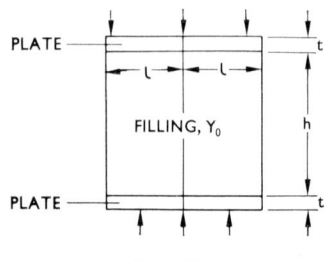

FIG. 13

25. Figure 13 shows two thin parallel plates or fibres of thickness t which are wide perpendicular to the plane of the paper; this material is substantially rigid but fractures in tension when the uniaxial tensile stress in it is Y_1. Sandwiched between the plates is non-hardening material of which the compressive yield stress is Y_0 and it is of thickness h. Bonding between the plate and filling is good but under frictionless compression at the two plates this filling may yield completely due to sideways flow; the traction at the plate-filling interface is everywhere equal to the yield shear stress $Y_0/2$. Show that, approximately, the maximum compressive load which can be carried by the sandwich is $[1 + (Y_1/Y_0) \cdot (t/h)]$ times as great as that which would be carried if the thin plates were not present.

If $Y_1 = 3Y_0$ and $t/h = 1/10$, show that due to the introduction of the plate fibres the increased load carrying capacity is 30 per cent.

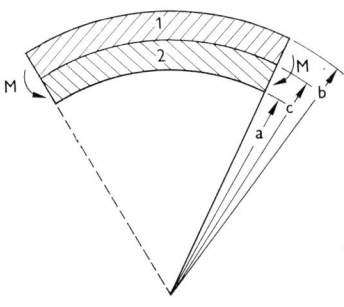

FIG. 14

26. Figure 14 shows two wide sheets of different materials; material 1 has a yield stress in uniaxial tension of Y_1 and material 2 a yield stress Y_2. The sheets have an initial curvature, having been formed to the radii shown, and before any external bending moment M, is applied, they are stress free. The two materials are firmly welded together at their common interface at radius c. Find an expression for the bending moment per unit width M_p, which when applied would bring the entire compound sheet to full plasticity, the hoop stress in sheet 1 being entirely tensile and that in sheet 2 entirely compressive. What relationship between a, b and c must apply? Plane strain bending is to be assumed What is the jump in hoop stress across $r = c$? Check that there are no net forces on the end faces and also verify your results by putting $Y_1 = Y_2$.

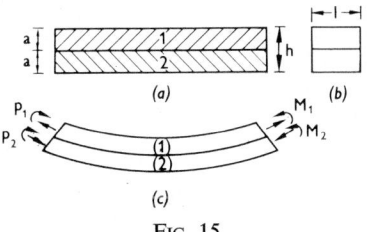

FIG. 15

27. A narrow strip consists of two metals of equal thickness welded together, see Fig. 15. They are straight at temperature t_0. When the temperature is raised to temperature t, the strip bends. Find the radius of curvature to which the bimetallic strip bends, assuming elastic deformation, and then verify that the greatest stress in each strip occurs at the interface. What is the rise in temperature if yielding starts simultaneously in both metals?

28. Figure 16 shows a cantilever of thin semi-circular section of uniform thickness supporting an end load W. Discuss how an approximate expression for the distance of the centre of shear from the centre of the circle, using a stress-strain relation $\sigma = Be^n$, may be obtained.

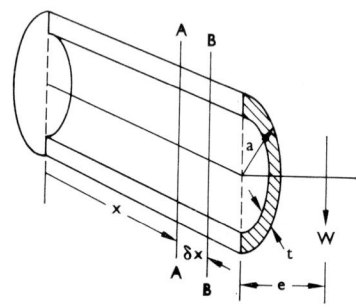

FIG. 16

29. Consider the bending of a wide straight rectangular cross-section beam to the ends of which is applied a bending moment which causes it to take up a curvature of $1/R$. Supposing the material to follow a stress engineering strain law $\sigma = Fe^n$, then the equations connecting σ, M and R are given by equation 7.3. Investigate what radial stresses are implied by the approach and results given in Section 7.2.

30. A cantilever AB is firmly built in at B and propped at its end A so that AB is horizontal. A downwards load W is applied at an intermediate position, C, distance a from A. If the total length of the cantilever is L, find the position of load W to give complete collapse of the cantilever for the minimum value of W.

31. A solid circular section bar of length l and radius r is held firmly at one end and twisted about its axis at the other, with an angular velocity $\dot{\theta}$. If the shear stress τ-shear strain rate $\dot{\gamma}$ relation is $\tau = k\dot{\gamma}^m$, where k and m are constants, find the necessary torque.

32. A hollow circular shaft of inner to outer radius ratio c is twisted until it is fully plastic and the torque then released. Show that the residual shear stress, τ_{RES}, at the outer radius is $k/3$ when $c = 0$.

Find the position of the radius at which the residual shear stress is zero, when $c = 0$ and when $c = 1/2$.

33. A hollow circular shaft in which $c = 1/2$ (see Section 8.3), is twisted until it has just become plastic at a radius of $3a/4$. Plot a diagram through the wall thickness showing the residual shear stress at salient points. What fraction of the original angle of twist remains when the shaft is unloaded?

34. Calculate the torque transmitted and the angle of twist per unit length for the double box beam whose dimensions are shown in Fig. 17, when the maximum shear stress is limited to 5000 lbf/in²; $G = 12 \cdot 10^6$ lbf/in².

35(a). The stress function ϕ to the membrane for the elastic torsion of a narrow rectangular bar of width c ($-\frac{1}{2}c \leqslant x \leqslant \frac{1}{2}c$) is given by

$$\phi = -G\theta\left(x^2 - \frac{c^2}{4}\right).$$

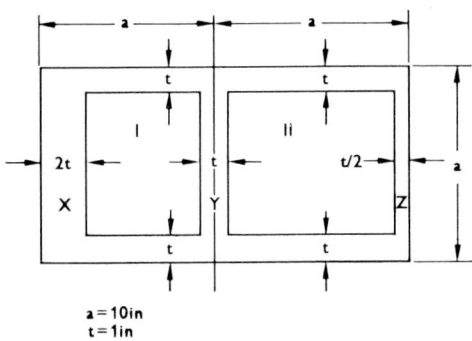

a = 10in
t = 1in

FIG. 17

Apply the above expression and derive an approximate formula for the torsional rigidity of the narrow tapered triangular section shown in Fig. 18.

(b). Explain briefly what you understand by the sand heap analogy in connection with the plastic torsion of non-cylindrical prismatic shafts.

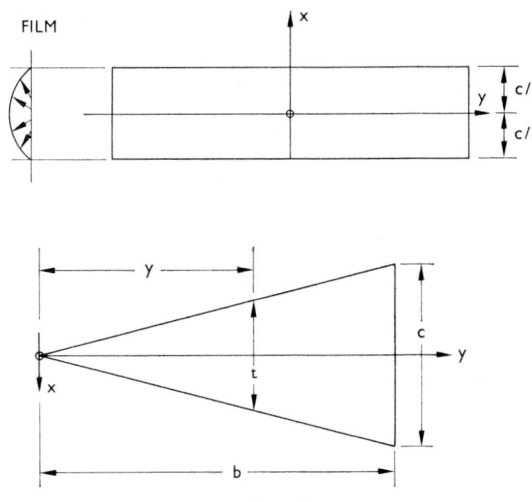

FIG. 18

Find the ratio of the torques required to produce full plastic sections of a circular and a square-section bar of equal areas.

36. A turbine, disc outer radius b, is shrunk on to a solid shaft of radius a, the pressure at the interface, assumed uniform, being p. If the thickness of the disc at the shaft is $2h_a$ and if in section it is hyperbolic (see Fig. 19), show that for the disc to become fully plastic,

$$p = -Ya(b - r)/rb.$$

Axial stresses in the disc may be neglected.

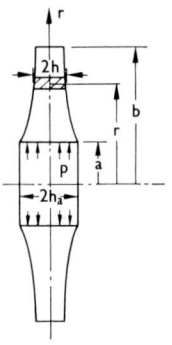

FIG. 19

37(a). A compound cylinder, inside radius r_0 and external radius r_n is made up of n long cylinders, all of the same material; the internal and external radii of the individual cylinders are (r_0, r_1), $(r_1, r_2) \ldots (r_m, r_n)$. The internal pressure carried is $-p$, i.e., at r_0, and the external pressure, i.e., at r_n, is zero (see Fig. 20). Show that if yielding is about to occur simultaneously at the inside surface of each cylinder, then the internal pressure is given by,

$$p = \frac{Y}{2}\left\{n - \left[\left(\frac{r_0}{r_1}\right)^2 + \left(\frac{r_1}{r_2}\right)^2 + \ldots + \left(\frac{r_m}{r_n}\right)^2\right]\right\}.$$

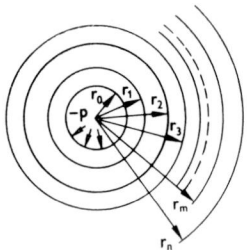

FIG. 20

If this pressure is to be a maximum, show that the ratio of the external to the internal radii of each cylinder is $(r_n/r_0)^{1/n}$, and then, that

$$p = \frac{Y}{2}n\left[1 - \left(\frac{r_0}{r_n}\right)^{2/n}\right].$$

Compare this value of p for various values of (r_0/r_n) and n with that which is required to bring the same compound cylinder to full plasticity. Use the Tresca criterion and assume that alterations in cylinder radii do not affect pressure calculations.

(b). Reconsider (a) when $n = 2$, if the yield stress of the two materials is different, i.e., Y_1 and Y_2.

38. A closed-ended thick-walled cylinder is subjected to internal pressure, p, and simultaneously a torque, T, about its longitudinal axis. If the diameter ratio is m, find the value of the ratio T/p to cause yielding at both the inner and outer surfaces. $[\sigma_z = p/(m^2 - 1)]$. Assume yielding to follow the Mises criterion.

39. If a thick-walled cylinder is made of a material which has an upper and lower yield shear stress of Y_U and Y_L, respectively, show that in the partially plastic state the internal pressure is given by p,

$$p = Y_L \ln n^2 + Y_U \left(\frac{m^2 - n^2}{m^2} \right).$$

m is the diameter ratio and n the ratio of the diameter of the plastic boundary to the bore diameter.

What is the expression for p, for the same material, for a thick-walled spherical shell?

40. A uniform thin-walled tube of outer diameter D just fits over a solid rod of diameter d (see Fig. 21). Find the load the tube can sustain before instability occurs. It is to be assumed that the tube can slide frictionlessly over the solid rod to which no load is applied. For the tube you may assume $\bar{\sigma} = A(B + \bar{\epsilon})^n$. Discuss the effect on the results of some initial clearance Δ between the tube and the rod.

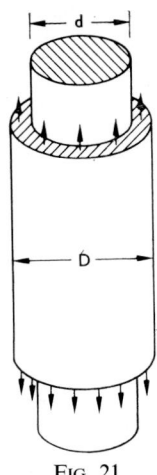

FIG. 21

41. A thin rotating ring, of initial radius $r_0 = 75$ mm, spins about an axis through its centre normal to its plane. If its density $\rho = 7840$ kg/m³ and if its work-hardening characteristics is represented by $\bar{\sigma} = 1400\bar{\epsilon}^{0.1}$ determine the angular velocity at instability.

42. Compare the effective strains and the surface strains at instability for a material having an isotropic work-hardening characteristic $\bar{\sigma} = A\bar{\epsilon}^n$ when subjected to the following stress systems.

(a) Uniform balanced biaxial tension applied to a flat sheet.

(b) A circular diaphragm clamped at its periphery and subjected on one side to a fluid pressure.

43. A wire is drawn through a die of cross-sectional area A, its initial cross-sectional area being A_0. If the drawing takes place when a back tension of f is applied and assuming that the shape change is brought about as one of homogeneous deformation, prove that the drawing stress, t, is given by $t = f + Y \ln A_0/A$. If the greatest reduction that can be achieved is 50 per cent. What is the back tension?

44. A long cylindrical rod of radius a and length l is encompassed over its curved surface by a tool which can be arranged to move radially inwards and thus cause axial extrusion of the bar at both ends. If μ denotes the coefficient of friction between the bar and the tool, find an expression for the variation of the radial pressure with distance along the bar. State your assumptions clearly.

For a given μ and beyond a certain value of l/a, the expression you derive would have to be modified. Why? Show that this modification would have to take place, if the yield shear stress of the bar is k when

$$\frac{l}{a} = \frac{1}{2\mu} \ln \frac{1}{2\mu}.$$

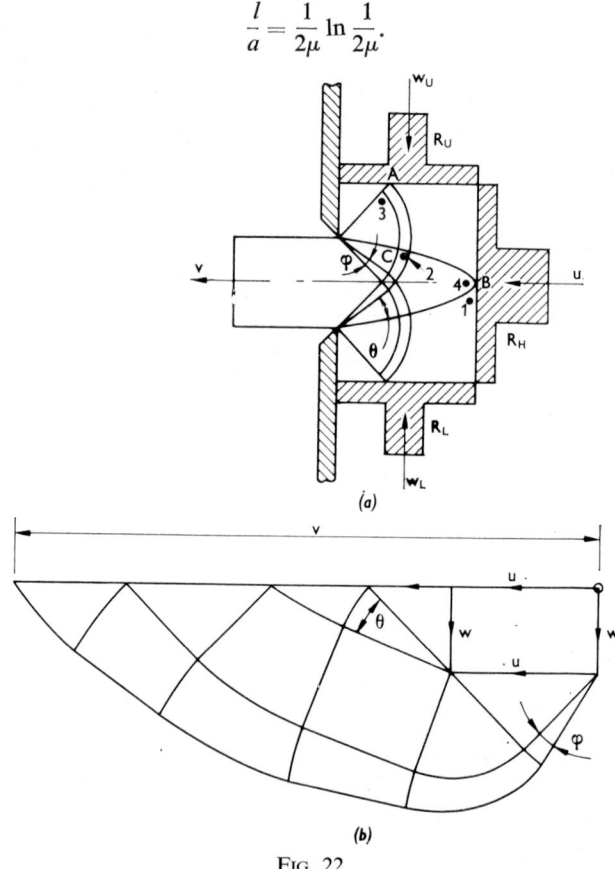

(a)

(b)

Fig. 22

45. Figure 22(a) shows an instantaneous situation in which material is caused to extrude normal to an orifice by three rams. The slip-line field (s.l.f.) and hodograph given in (b) is a general solution for combined end-extrusion and side-extrusion; the two rams R_U and R_L move in opposite directions with speeds w_U and w_L, which in the case shown in (b) are taken to be equal whilst ram R_H moves perpendicularly to the orifice with speed u. The s.l.f. is drawn for frictionless contact between the work material and the rams.

(a) In order to comply with the frictionless requirement referred to above, at what angles do the slip lines meet the rams?
If 'shearing friction' prevailed at the ram face for each of R_U and R_L, show how the s.l.f. should be modified.
(b) Mark on the slip-line field the α- and β-lines. In terms of θ and ϕ (in the given s.l.f.) and k, find the pressure on the rams at A and B showing clearly how you derive them.
(c) On the hodograph show the velocity vectors corresponding to the velocity at points 1, 2, 3 and 4 in the s.l.f. given.
(d) Show diagrammatically by re-drawing how the hodograph for the given s.l.f. changes when $w_u \neq w_L$ and $u = 0$.

46. Fig. 23 shows an instance of plane strain drawing through a smooth die of semi-angle 20°, the reduction being 0·19. A pattern of velocity discontinuities is also shown in the diagram. Draw the hodograph for the pattern shown and hence calculate an upper bound to the drawing stress.

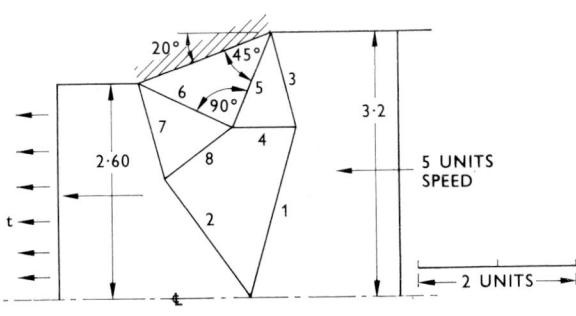

FIG. 23

47. Kunogi has described a method by which thick-walled tubes can be made using a stationary smooth punch of large conical angle ϕ, over which a given slug is caused to expand through a conical container, because of its compression between the ram and punch. The slug is pushed by a ram usually about the same diameter as the punch. The method will be clear from Fig. 24(*a*). The usual values of α and θ are 10° to 30°.
A pattern of velocity discontinuities for analysing the plane strain analogue to this method is shown dotted in the figure, in which $\alpha = \theta = 15°$, $D_1 = 1·0$ unit, $d_1 = 0·8$ and $D_2 = 1·3$ units. Draw the corresponding hodograph and hence calculate the mean ram pressure for the configuration given. Let the punch take up various positions to the right and left of that given, calculate the pressure associated with each and hence show that it may be presumed that the punch may be set in a position which will require a minimum ram pressure.
(KUNOGI (1956)* performed an extensive series of experiments and his conclusion about the best punch position was that it should be in something like the position found here.)

* 'A New Method of Cold Extrusion', *J. Sci. Res.* I, Tokyo 1956, **50**, 215.

48. Adapt Fig. 24 to deal with instances which require more than one triangle of velocity discontinuities on the punch face.

49. A process similar to those of the preceding problems is plug-drawing, diagrammatically represented in Fig. 24(c), with three lines of velocity discontinuity. The operation is one in which a tube is *drawn-down* over a fixed plug of angle α_1. The smaller the ratio of tube wall thickness to tube diameter, the more nearly the operation approximates to a plane strain process. Draw a hodograph for Fig. 24(c).

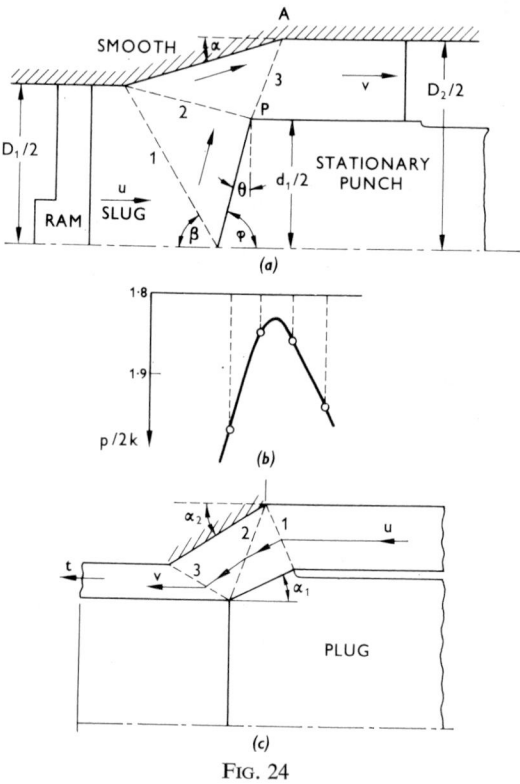

Fig. 24

50. In the manufacturing process shown in Fig. 66, as adapted from Fig. 24(a), it may be required that the point P on the punch enter the die throat. (See Fig. 69(a) where $\phi = 90°$ and P becomes G.) With this is associated a form of final product in which if, say, the die semi-angle $\alpha = 45°$, then the angle on the product, β, turns out not to be equal to α, and is less than 45° and probably nearer 30°. The punch causes the material to leave the die at its top corner and again meet with the container wall some distance further along. (There is no definite angle on the final product and it is found to be curved to some small degree.) Suggest some patterns of velocity discontinuity for analysing the plane strain analogue to this axisymmetric process and discuss the results they give.

51. Figure 25 shows the extrusion (unsteady motion problem) of a short slug. Show that a possible mode of deformation is that defined by lines of discontinuity OB

and OA. Show that if the punch be smooth, this mode prevails when $t < (h - x)$, point A being so chosen as to minimize the extrusion load. What is the physical implication attaching to points A and C not being coincident?

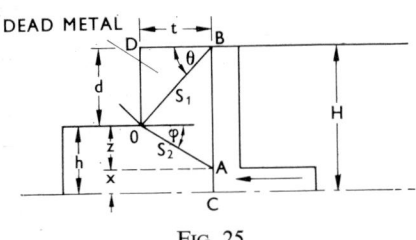

FIG. 25

52. Figure 26 shows the forging of an inset block of rigid, perfectly plastic material under conditions of plane strain (compare Problem 68) and lines of velocity discontinuity through it are to be assumed as indicated. Construct the corresponding hodograph and hence obtain an upper bound to the load necessary to effect forging, in terms of k. Attempt to improve this estimate of the load by making another trial solution of the same form but a different shape.

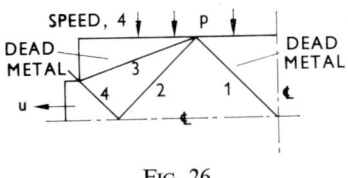

FIG. 26

53. Figure 27 presents a case similar to that of Problems 52 and 68, except that the vertical sides of the forging die have been rounded. If the radiused die is smooth, draw a possible field of velocity discontinuities, and the associated hodograph.

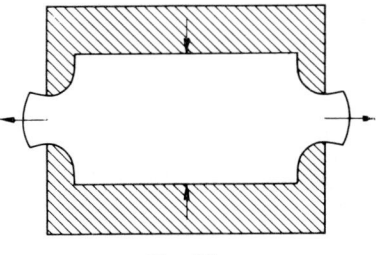

FIG. 27

54. Using Fig. 73, p. 617, which shows an extrusion through a perfectly smooth curved die with certain lines of constant magnitude velocity discontinuity, explain how, if two separate similar pieces of similar material in contact along the axis of the chamber are extruded, they may be expected to 'crocodile'.

55. Figure 28 shows an extrusion through smooth dies of semi-angle 15° for a simple and obvious slip-line field. Compare the extrusion pressure calculated using the slip-line field and that obtained by approximating it and calculating an upper bound.

Draw a diagram to show approximately the distribution of temperature in the vicinity of the die mouth, if the process proceeds adiabatically and the walls are non-conducting. All the heat generated is due to plastic work done; the highest temperature should be converted to a figure of 100.

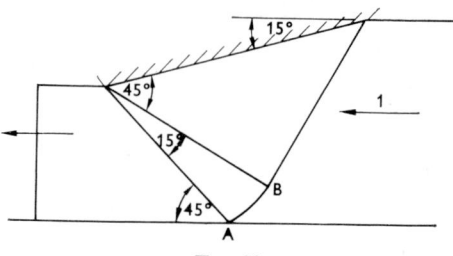

FIG. 28

56. From a simple solid cylindrical slug A (see Fig. 29(b)), it is required to make a component having the shape shown in Fig. 29(a) known as a 'boat-tail'. The container, die and punch are as shown in Fig. 29(b). The punch acting on A causes material to be forward-extruded through the smooth die of semi-angle α as in Fig. 29(c); this continues until the forward extrusion fills up the space available, when further metal is backward-extruded as a tube, flowing round the punch as it penetrates the block, see Fig. 29(d). Discuss the two 'mechanisms' necessary for explaining the metal flow, if this situation is conceived as a plane strain process.

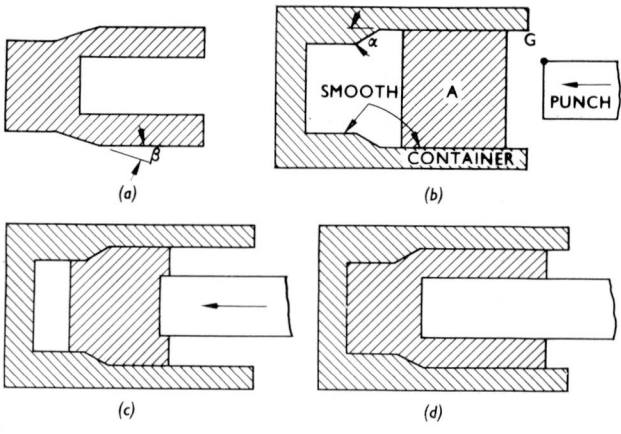

FIG. 29

57. A thin conical frustum of semi-angle α has its lower larger diameter $2a$ resting but fixed into a rigid foundation, and its upper smaller diameter is $2b$.

A rigid plate sits freely on the top of the frustum and carries a load P. At what value of P will there be plastic collapse of the frustum, if collapse due to bending only is considered?

58. A beam of span $2l$ has a uniform square section of thickness $2t$ over the two portions AC and DB; see Fig. 30. Over length CD the beam has a very heavy section which cannot yield—except where portions AC and CD are built-in to CD at C and D. The beam is built-in at A and B and parts AC and DB are of rigid-perfectly plastic material with a uniaxial yield stress of Y. From C and D hang uniform cables CC' and DD' of equal cross-sectional area A_0 in the unstressed state and the load $2W$ is equally divided between them. The true stress, σ, engineering strain, e, curve for the cables is represented by $\sigma = Be^m$ where B and m are constants. Find the value of A_0 if the beam is to be designed to undergo plastic collapse at the same time as the wires reach their (ultimate) tensile strength.

The beam and wires should be considered weightless and deflection due to shear is negligible.

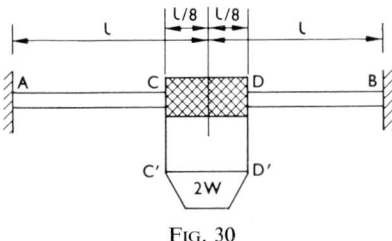

Fig. 30

59. Figure 31 shows extrusion through a square die. In order to obtain a lower bound the slug is divided into three regions. Put in the missing principal stresses q and s and hence find the lower bound to the extrusion pressure.

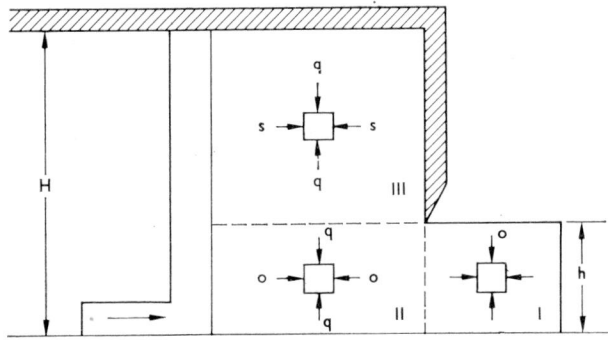

Fig. 31

60. Figure 32 shows a semi-infinite block of rigid-perfectly plastic metal through which passes a circular section hole of radius R and whose centre lies at a depth h below the surface. The hole carries an internal radial pressure p and it may be assumed that plastic deformation takes place by block $ABCD$ moving as shown, with shearing occurring along surfaces AD and BC. Find an upper bound to the pressure.

If the material was considered elastic-perfectly plastic then at a pressure of 9k, where k is the yield stress in shear, the expansion of the hole would be accommodated elastically (refer to Section 9.3.7). At what depth would this pressure be realized using the above upper bound?

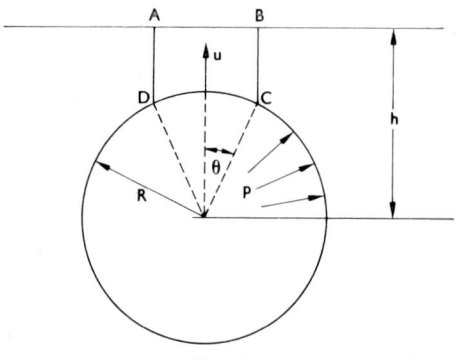

Fig. 32

61. The tool of a press brake, T, which moves vertically is shown in Fig. 33, where the plate it is bending, in plane strain, is partly formed to the radius of the tool nose, R; the female die is also indicated. The plate of rigid-perfectly plastic material is initially straight, but in Fig. 33 is presumed already formed or bent throughout the shaded region shown; the current point at which contact ceases between tool and work-piece is A and there is frictionless support of the plate at J. Draw an approximate slip-line field to show incipient plastic bending.

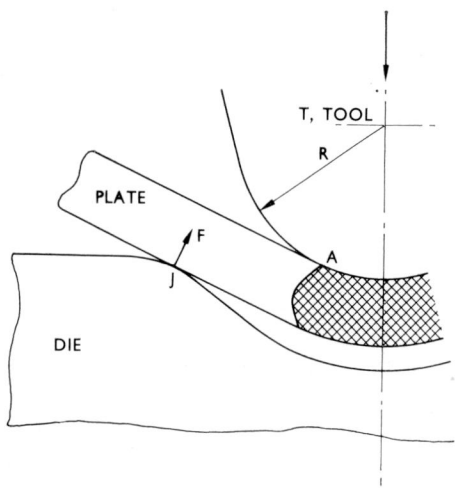

Fig. 33

62. Figure 34 shows how a straight, flat sheet of rigid-perfectly plastic material of thickness 2t can be bent, under conditions of plane strain, to a mean radius, ρ in a 3-roll pyramidal bending machine. The upper rigid working roll, W, has a radius R and rotates about its centre with angular speed ω; support rolls S_1 and S_2 rotate freely about their axes. The radial gap between W and S_1 and W and S_2 is just $2t$, so that when the sheet enters between W and S_1 it is immediately bent to mean radius $(R + t)$ and the inner surface of the sheet proceeds to move, without slipping, with the work roll surface speed $R\omega$. On passing out through the gap between W

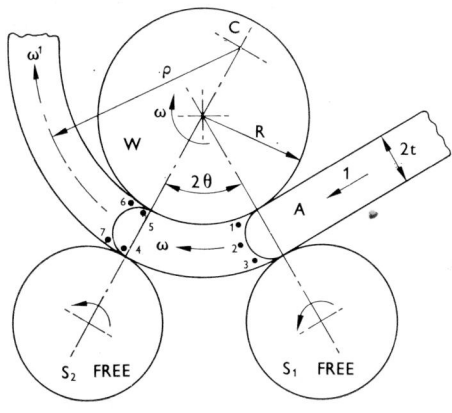

FIG. 34

and S_2, the curvature of the sheet is changed to the desired mean radius of curvature, ρ.

Draw the appropriate hodograph to represent this situation and show that the required working roll torque, T, is given by, $T = k\pi tR \cdot t(t - R)/\rho R$.

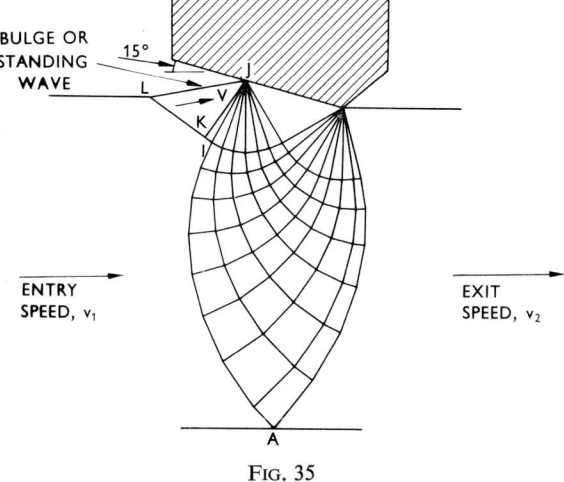

FIG. 35

63. Figure 35 shows one half of a strip being drawn through a 15° die. The reduction is small and due to the large die pressure a bulge (or standing wave) is created ahead of the die. Draw the corresponding hodograph.

64. Figure 36 shows material being extruded through a die of angle α. The reduction is *very small* and the slip-line field shown in the figure is proposed to account for it. Show that the field proposed is valid and that it implies the extrusion of a surface strip, or chip. Draw the hodograph and find the direction of the strip.

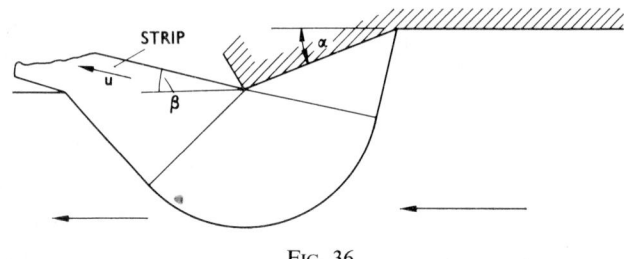

FIG. 36

65. The curved surface of a long frictionless, cylindrical indenter fits accurately into a pre-formed cavity machined in the surface of a semi-infinite solid and the situation is illustrated in Fig. 37. For incipient indentation a proposed slip-line field is inserted. The slip-lines everywhere meet the indenter surface and the free surface at 45°. Is the slip-line field proposed a valid one?

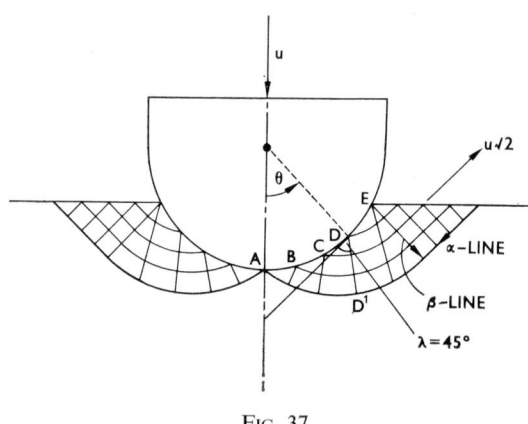

FIG. 37

66. Figure 38 shows a slip-line field for a rounded restricted contact tool, used for orthogonal machining, which has been simplified to appear as a tool composed of two planes. Draw the corresponding hodograph.

67. Figure 39 shows a block of material of which $D'C'B'$ $ABCD$ is a given stubby, straight-sided, symmetrical projection; it is required to find the force necessary to start compression by means of a smooth plate $F'B'$ ABF, using a slip-line field. Draw the hodograph and for a selected point in the projection, find its velocity. Take AB to be greater than BC.

Suggest the form of appropriate slip-line fields for the above problem when the interface is not frictionless and when a dead metal zone is wholly or in part in existence.

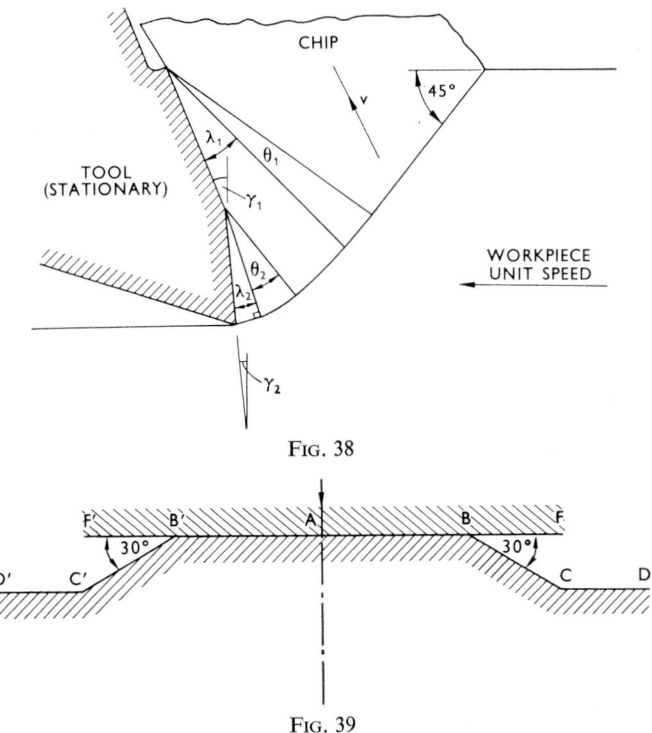

FIG. 38

FIG. 39

68. Consider the plane strain forging of an inset block of metal, of semi-width w and semi-height H, one quarter of which is shown in Fig. 40, by causing the top die to approach the horizontal centre-line with a vertical speed of unity and thus expel material horizontally with a speed u, where $u = w/h$. Assume that w/H is large. Let the surface of the upper die be perfectly rough, i.e., where there is relative motion

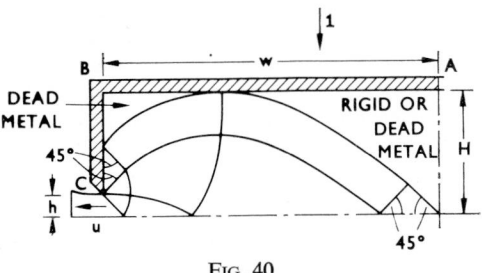

FIG. 40

of the metal and the top die, suppose shearing takes place. Suppose, too, that the vertical edge of the die BC is such that there is no friction between BC and the metal. Taking into account these boundary conditions, justify the slip-line field shown and draw a corresponding hodograph.

69. A smooth round bar of metal of diameter D and yield stress Y_b has a perfectly flexible and inextensible rectangular wire, $w \times t$, wrapped once round it as shown in Fig. 41. Estimate the tension required in the wire to start full plastic flow in the bar. w/D is small and coil friction is negligible.

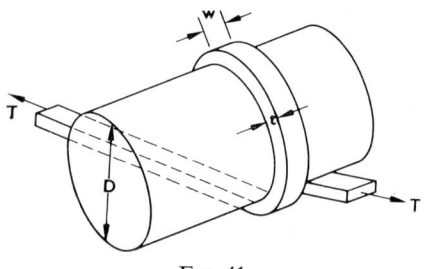

Fig. 41

70. Figure 42 shows a punch (plane strain) advancing to the left and acting on a slug of width X, causing a forward extrusion of thickness $2h_1$ and the rearward extrusion of two sheets each of thickness h_2. Assuming the container walls to be perfectly smooth, propose a slip-line field solution.

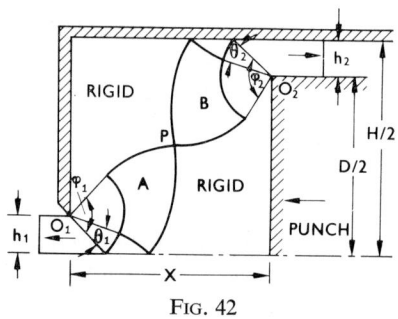

Fig. 42

71. Into a wide and relatively thin sheet of material two identical deep slots are machined, back-to-back, as shown in Fig. 43. A pair of rigid wedges just fit into the slots and pressure is applied through them just sufficient to cause unrestricted plastic flow. Discuss the possible behaviour using slip-line fields.

(This situation is not too far removed from the operation of cutting a wide bar or sheet with a pair of perfectly smooth pliers, the jaws of which may be assumed to approach the plane of the sheet at right angles since the angle of rotation of the jaws is small.)

72. Figure 44 shows a slip-line field for extruding material sideways from a rough container with a single punch (solid line). Calculate:

(a) the necessary pressure on the ram;
(b) the coefficient of friction between the material and the container wall;
(c) the true reduction.

If two punches moving with equal speed but in opposite directions were employed what does the new reduction become?

73. Figure 45 shows a particular slip-line field for an extrusion through smooth dies of unequal angles, the orifice being inclined to the line of punch travel. Calculate the mean punch pressure, draw the hodograph and obtain the true reduction.

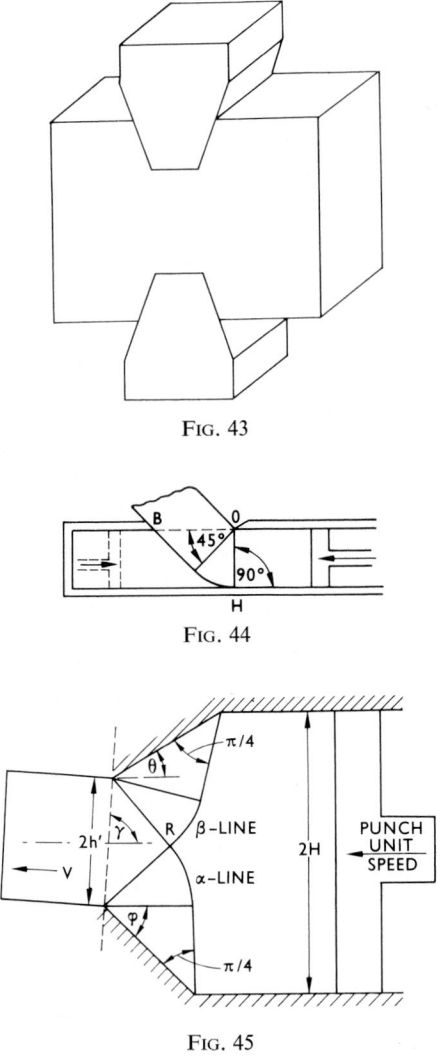

Fig. 43

Fig. 44

Fig. 45

74. A thin uniform plate of radius a is supported concentrically at radius b, $b < a$. What is the value of b/a for plastic collapse if the uniform pressure which may be applied is a maximum?

75. The diagram, Fig. 46, shows a portion of a semi-infinite plate in which the surface is BAB. Beneath the surface is a cylindrical hole of radius R whose axis is parallel to BAB, the minimum thickness of the material above the hole being h.

Assume a system of plastic hinges and show that the incipient plastic collapse load is given by

$$p/Y = (1 + \sqrt{2})h/R.$$

You should assume that any outer plastic hinges form at a location which makes the internal pressure p a minimum.

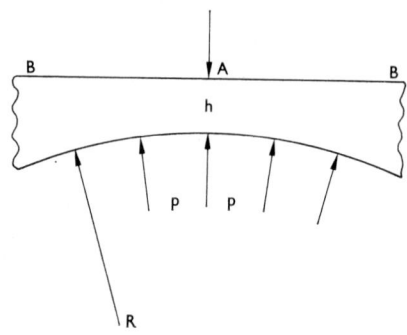

FIG. 46

76. The curved surface of a rigid cylinder is used as an indenter and slowly pressed into a semi-infinite rigid-perfectly plastic solid causing plane strain deformation. It is proposed to analyse the situation in an approximate way by using the slip-line field shown in Fig. 47. Beneath the indenter a cap of dead metal is always presumed to exist and its extent increases with the degree of penetration; all pile-up is neglected during indentation to depth y. Calculate the mean strain imposed on the volume of material encompassed by the slip-line field when the ratio of the width of the impression to the diameter of the indenter is 0·25.

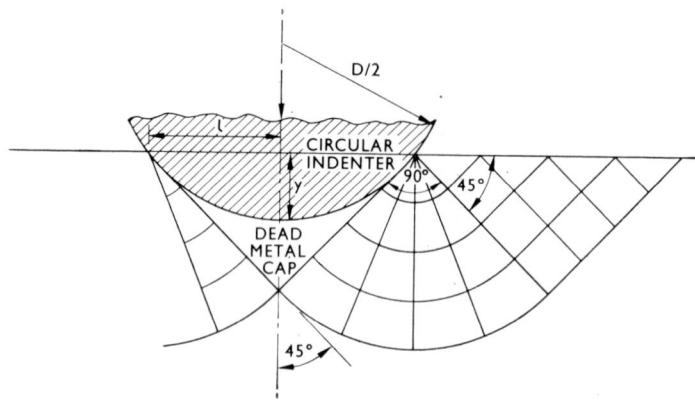

FIG. 47

77. Figure 48(a) shows a solid disc of material of height h and radius b which is being pierced by two rigid solid circular tools of radius a, both being driven by hydro-static pressure $\bar{p}$. The material of the disc between the tool faces when fully plastic is squeezed out radially causing the external radius of the disc to increase uniformly (plastically). If the friction between the disc and the tool face is such that the yield shear stress of the disc is obtained across the whole of their common interfaces, prove that, neglecting sealing ring friction, etc., $\bar{p}$ is given by

$$\frac{\bar{p}}{Y} = 1 + \tfrac{1}{3}\frac{a}{h} + \ln\frac{b}{a}.$$

For given values of Y, b and h what is the minimum value of $\bar{p}$?

Y denotes the uniaxial yield stress of the material and it is to be assumed that yielding follows the criterion of Tresca.

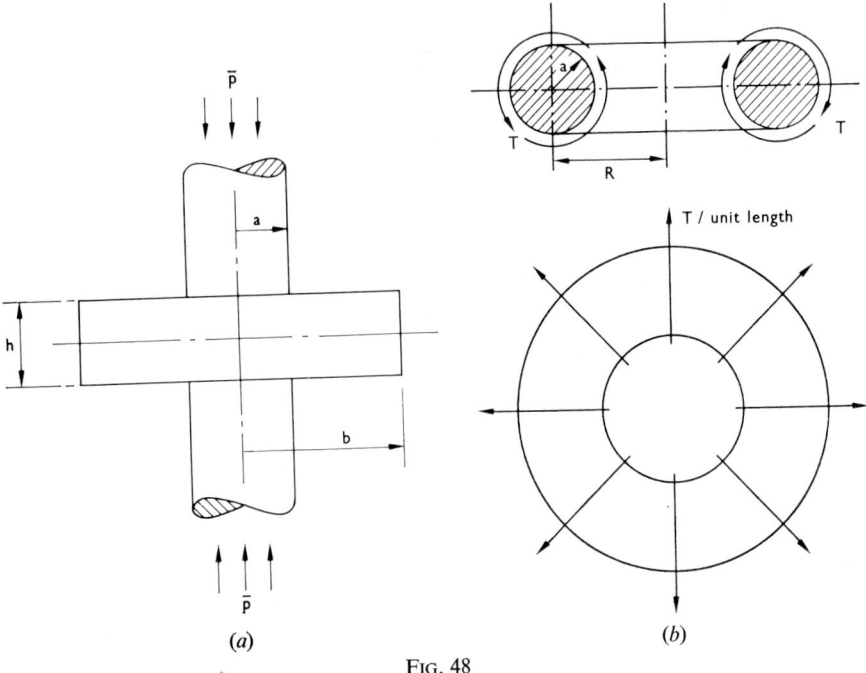

(a)   (b)

Fig. 48

78. A torus of rigid perfectly plastic material is shown in Fig. 48(b) and $R \gg a$. Find the torque per unit length which must be applied to the curved surface of the toroid to cause plastic collapse, i.e. rotation of each section (plastic) about its centre, C.

SOLUTIONS TO PROBLEMS

---

1. At necking, the natural strain is $\epsilon = 0{\cdot}4$.
Thus

$$\ln (l_1/l_0) + \ln (l_2/l_1) = \ln l_2/l_0 = \epsilon$$
$$\ln (1 + e_1) + \ln (1 + e_2) = 0{\cdot}4$$
$$\ln (1 + e_2) = 0{\cdot}4 - \ln 1{\cdot}25 = 0{\cdot}177$$

Hence, $\qquad\qquad\qquad\qquad e_2 = 0{\cdot}119$.

2. From $\epsilon = \ln (1 + e)$ and at necking,

(a) the natural strain $\epsilon = n$, and
(b) the engineering strain $e = m/(1 - m)$.

Hence, $\qquad\qquad n = \ln \left(1 + \dfrac{m}{1 - m}\right) = \ln \dfrac{1}{1 - m}$

or $\qquad\qquad\qquad m = 1 - \exp (-n)$.

3. Given $\qquad\qquad\qquad \sigma = Y \tanh \dfrac{E\epsilon}{Y}$,

then $\qquad\qquad\qquad \dfrac{d\sigma}{d\epsilon} = Y \operatorname{sech}^2 \left(\dfrac{E\epsilon}{Y}\right) \cdot \dfrac{E}{Y}$

$$= Y \tanh \dfrac{E\epsilon}{Y},$$

for the instability condition.

Thus $\qquad\qquad \dfrac{E}{\cosh^2(E\epsilon/Y)} = \dfrac{Y \sinh (E\epsilon/Y)}{\cosh (E\epsilon/Y)}$,

and $\qquad\qquad\qquad \dfrac{2E}{Y} = \sinh \dfrac{2E\epsilon}{Y}$

or $\qquad\qquad\qquad \epsilon = \dfrac{Y}{2E} \sinh^{-1} \dfrac{2E}{Y}$.

4. Let c denote the uniform strain-hardening rate, then

$$c = d\sigma/d\epsilon,$$

and thus $\sigma = c\epsilon$, assuming $\sigma = 0$ when $\epsilon = 0$.
Let x denote the extension. Then because

$$\epsilon = \ln \left(1 + \dfrac{x}{l_0}\right) \quad \text{and} \quad \sigma_0 = \dfrac{\sigma}{1 + e},$$

then
$$\sigma_0\left(1 + \frac{x}{l_0}\right) = c \ln\left(1 + \frac{x}{l_0}\right).$$

Hence, the tensile load F—extension diagram, x is $F = A_0\sigma_0$ and

$$F = A_0c\,\frac{\ln\left[1 + (x/l_0)\right]}{1 + (x/l_0)}.$$

A graph of $F/(x/l_0)$ is concave upwards as shown in Fig. 49.

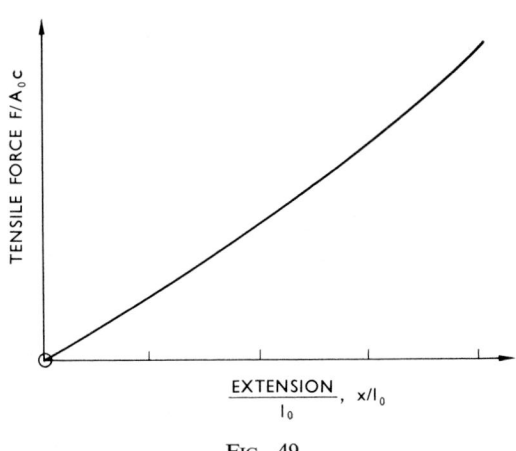

FIG. 49

5. At maximum load $e = m/1 - m$ and $m = 0.25$. Thus $e = \frac{1}{3}$.

Now $\sigma = 30(\frac{1}{3})^{0.25} = 22.9$.

Hence, $\sigma_0 = \sigma/(1 + e) = \frac{3}{4} \cdot 22.9 = 17.2$ tons/in².

6. At instability, $BD\ (=l_1)$ has a length $l_1/(1 - n_1)$ and $CD\ (=l_2)$, a length $l_2/(1 - n_2)$. Thus,

$$\left(\frac{l_1}{1 - n_1}\right)^2 + h^2 = \left(\frac{l_2}{1 - n_2}\right)^2,$$

where $h = AB = BC$.
Also, initially, $l_1^2 + h^2 = l_2^2$.
Eliminating h, it is found that:

$$\cos\theta = \frac{l_1}{l_2} = \left(\frac{1 - n_1}{1 - n_2}\right)\sqrt{\frac{n_2(2 - n_2)}{n_1(2 - n_1)}}.$$

7. When the strain exceeds $[\sqrt{1 + (2a/l)} - 1]$, the wires are tangential to the shaft. The T.S. may be developed in each wire, say P. The torque transmission T is then $T = Pna$.

8. The axial tensile strain e_L at instability is given by $d\sigma/de = \sigma/(1 + e)$ (see Section 1.3). If $\sigma = Be^n$ then at instability $e_L = n/(1 - n)$. The transverse strains are equal in magnitude to $-e_L/2$. The wall thickness decrease is thus

$$\Delta t = nt/2(1 - n).$$

The mean diameter of the tube becomes $dn/2(1 - n)$, because approximately $e_\theta = 2u/d$.

9. Load, $P = A_1\sigma_1 + A_2\sigma_2 = V_1 \cdot \dfrac{\sigma_1}{l_1} + V_2 \cdot \dfrac{\sigma_2}{l_2}$, where V denotes volume and

$l_1 = l_2 = $ length l. Assume no alteration in volume of either material; A is the current cross-sectional area of each component. At tensile instability, $dP = 0$. Thus:

$$V_1\left(\frac{d\sigma_1}{l} - \frac{dl}{l^2}\sigma_1\right) + V_2\left(\frac{d\sigma_2}{l} - \frac{dl}{l^2}\sigma_2\right) = 0.$$

Substituting,

$$\sigma_1 = B_1e^{n_1}, \; d\sigma_1 = n_1B_1e^{n_1-1}de, \; \sigma_2 = B_2e^{n_2} \text{ and } d\sigma_2 = n_2B_2e^{n_2-1}de,$$

in the above equation, after simplifying it is found that,

$$n_1 \times \frac{1 + (A_2B_2n_2/A_1B_1n_1)e^{n_2-n_1}}{1 + (A_2B_2/A_1B_1)e^{n_2-n_1}} = \frac{e}{1 + e}.$$

Taking suffix 1 to refer to aluminium and 2 to brass, $B_1 = 20,000$, $B_2 = 100,000$, $n_1 = 0\cdot25$ and $n_2 = 0\cdot5$.

Thus: $$0\cdot25 \times \frac{1 + (A_2/A_1) \cdot 5 \cdot 2e^{0\cdot25}}{1 + (A_2/A_1) \cdot 5e^{0\cdot25}} = \frac{e}{1 + e}$$

and $$A_1/A_2 = 3\cdot19.$$

Note that when $A_1/A_2 = 0$, i.e., the bar is entirely aluminium $e = 0\cdot333$ and when $A_1/A_2 \to \infty$, i.e., the bar entirely brass, $e = 1\cdot000$.

10. The net strain in 1 and 2 will be equal. Let $\alpha_1 > \alpha_2$. There will be a compressive stress in 1 and a tensile stress in 2 due to the temperature increase. Thus:

$$\alpha_1\Delta T - \left(\frac{\sigma_1 - Y_1}{P_1}\right) = \alpha_2\Delta T + \left(\frac{\sigma_2 - Y_2}{P_2}\right).$$

Also, if A denotes cross-sectional area, $A_1\sigma_2 = A_2\sigma_2$. Hence eliminating σ_2, it is found that

$$\frac{\sigma_1}{A_2} = \frac{(\alpha_1 - \alpha_2)\Delta T + Y_1/P_1 + Y_2/P_2}{A_2/P_1 + A_1/P_2}.$$

A temperature alteration of $-\Delta T$ from zero would, calculated elastically, produce a tensile stress in material 1 of

$$\sigma_1' = \frac{(\alpha_1 - \alpha_2)\Delta T}{1/E_1 + A_1/E_2A_2}.$$

The residual stress in material 1 is thus tensile and given by $(\sigma_1 - \sigma_1')$ or σ_1''; i.e.,

$$A_1\sigma_1'' = (\alpha_1 - \alpha_2)\Delta T \left\{\frac{1}{1/A_1E_1 + 1/A_2E_2} - \frac{1}{1/A_1P_1 + 1/A_2P_2}\right\} - \left\{\frac{Y_1/P_1}{1/A_1P_1} + \frac{Y_2/P_2}{1/A_2P_2}\right\}.$$

11. The procedure is similar to that set out in Section 1.4; the limits of integration for s are, however, a_0 to b_0, so that

$$\int_a^b \sqrt{z}\, dz = \int_c^H \sqrt{b_0}\, dz.$$

Thus,

$$\tfrac{2}{3}\left[b_0^{3/2} - a_0^{3/2}\right] = b_0^{\frac{1}{2}} H,$$

or

$$H = \tfrac{2}{3} b_0 \left[1 - \left(\frac{a_0}{b_0}\right)^{3/2}\right].$$

Thickness of the lip at its tip is $h_0(a_0/b_0)^{\frac{1}{2}}$.

12. The increment of plastic work done, dW, in forming an element of the lip, is equal to the volume of the element times the work done per unit volume in stretching the element in simple tension, i.e., through an increment of natural strain from ϵ to $\epsilon + d\epsilon$; thus if for natural strain ϵ the corresponding hoop stress is σ, then $dW = 2\pi s\, h_0 \cdot ds \cdot \sigma \cdot d\epsilon$. Hence, for the element,

$$W = 2\pi s\, h_0\, ds \int_0^\epsilon \sigma\, d\epsilon$$

Since, $\sigma = Y + P\epsilon$, where Y is the yield stress and P the plastic modulus, then

$$\int_0^\epsilon \sigma\, d\epsilon = Y\epsilon + P\epsilon^2/2.$$

Hence for the whole plate, the work done on each and every element must be

$$W = \int_0^b 2\pi\, h_0 \left[Y\ln\frac{b_0}{s} + \frac{P}{2}\left(\ln\frac{b_0}{s}\right)^2\right] s\, ds$$

since $s = \ln b_0/s$ and thus

$$W = \frac{\pi b_0^2\, h_0\, Y}{2}\left(1 + \frac{P/Y}{2}\right).$$

13. The equation of incompressibility gives

$$2\pi s \sin\alpha\, h_0\, ds = 2\pi b_0\, h\, dz.$$

And since

$$\ln\frac{dz}{ds} = \ln\frac{h}{h_0} = -\frac{1}{2}\ln\frac{b_0}{s \sin\alpha}$$

therefore

$$\int_0^H dz = \int_0^{b_0/\sin\alpha} \sqrt{s}\left(\frac{\sin\alpha}{b_0}\right)^{\frac{1}{2}} dz.$$

Therefore

$$H = \left(\frac{\sin\alpha}{b_0}\right)^{\frac{1}{2}} \frac{2}{3}\left(\frac{b_0}{\sin\alpha}\right)^{3/2} = \frac{2}{3} L.$$

14. Adapting the approach of Problem 13, $2\pi R \sin\theta\, R d\theta\, h_0 = 2\pi b_0\, h\, dz$. Again,

$$\ln\left(\frac{dz}{R d\theta}\right) = \ln\frac{h}{h_0} = -\frac{1}{2}\ln\frac{b_0}{R \sin\theta}$$

therefore

$$\int_0^H dz = \int_0^{\theta_0} \frac{\sqrt{R \sin\theta}}{\sqrt{b_0}} R\, d\theta_0.$$

Hence

$$H = \left(\frac{R^3}{b_0}\right)^{1/2} \int_0^{\theta_0} \sqrt{\sin\theta}\, d\theta$$

For small values of θ, $\sqrt{\sin\theta} \simeq \theta^{\frac{1}{2}}$ and $H \to \tfrac{2}{3} b_0$.

15. The solution is immediately found by letting $\alpha \to -90°$ and noting that $a/b \to 0$, using the equation derived in Problem 16,

Then,
$$\frac{H(-1)}{b_0} = 1 - (1+1)^{2/3}$$

or
$$H = (2^{2/3} - 1)b_0 \simeq 0.59\, b_0.$$

16. Refer to Fig. 8. The equation for no change in volume of an element at full penetration is

$$2\pi s \cdot ds \cdot h_0 = 2\pi (b_0 - z \sin \alpha)\, dz \cdot h.$$

But,
$$(ds/dz) = (h_0/h)$$

and thus
$$dz = \frac{h}{h_0} \cdot ds = \frac{s^{1/2}\, ds}{(b_0 - z \sin \alpha)^{1/2}}.$$

Thus,
$$\sqrt{b_0} \int_H^0 \left(1 - \frac{z \sin \alpha}{b_0}\right)^{1/2} dz = \int_{a_0}^b s^{1/2}\, ds,$$

and integrating,
$$\frac{b_0}{\sin \alpha}^{3/2} \left[1 - \left(1 - \frac{H \sin \alpha}{b_0}\right)^{3/2}\right] = b_0^{3/2} - a_0^{3/2}.$$

Hence the slant lip length, H is given by

$$\frac{H \sin \alpha}{b_0} = 1 - \left\{1 - \left[1 - \left(\frac{a_0}{b_0}\right)^{3/2}\right]\sin \alpha\right\}^{2/3}$$

Note, that when $\alpha \to 0$,

$$\frac{H \sin \alpha}{b_0} \simeq 1 - \left\{1 - \frac{2}{3}\sin \alpha \left[1 - \left(\frac{a_0}{b_0}\right)^{3/2}\right]\right\}.$$

$$= \frac{2}{3}\sin \alpha \left[1 - \left(\frac{a_0}{b_0}\right)^{3/2}\right].$$

and
$$H = \frac{2}{3}b_0 \left[1 - \left(\frac{a_0}{b_0}\right)^{3/2}\right].$$

17. As before, see Fig. 9,

$$\sqrt{\frac{R}{a_0}} = \frac{dz}{ds} = \frac{h}{h_0}$$

and hence,
$$\int_0^z a_0^{1/2}\, dz = \int_0^s \sqrt{R}\, ds = \int_0^s (a_0 + s \cdot \sin \alpha)^{1/2}\, ds.$$

Thus,
$$z a_0^{1/2} = \left[\frac{2 (a_0 + s \cdot \sin \alpha)^{3/2}}{3 \cdot \sin \alpha}\right]_0^s.$$

and
$$L = \frac{2}{3}a_0 \left[\left(\frac{R_0}{a_0}\right)^{3/2} - 1\right] \Big/ \sin \alpha.$$

It is assumed that the stress state throughout is one of simple uniaxial hoop compression; all radial stresses are neglected.

$$\text{If } \alpha = 90°, z = \frac{2}{3}\, a_0 \left[\left(\frac{R}{a_0} \right)^{3/2} - 1 \right]$$

and
$$h = h_0 \sqrt{\frac{R}{a_0}}.$$

Thus, eliminating R from the last two equations,

$$\frac{h}{h_0} = \left(1 + \frac{3}{2} \cdot \frac{z}{a_0} \right)^{\frac{1}{3}}$$

z/a_0	0	0·5	1	1·5	2
h/h_0	0	1·2	1·35	1·48	1·59

18. *Tensile* cylindrical hoop stress σ_0 is given by

$$\sigma_\theta\, 2t = p \cdot D \quad \text{or} \quad \sigma_\theta = \frac{pD}{2t}.$$

Compressive axial stress σ_z is given by,

$$\sigma_z\, 2\pi\, rt = (A_0 - A)\, p$$

or
$$\sigma_z = \frac{pD}{2t} \left(\frac{A_0 - A}{2A} \right) = \sigma_\theta \cdot \frac{\alpha}{2}.$$

Using the Mises yield criterion, assuming $\sigma_r = 0$ in,

$$(\sigma_\theta - \sigma_r)^2 + (\sigma_r - \sigma_z)^2 + (\sigma_z - \sigma_\theta)^2 = 2Y^2$$

and substituting,

$$\sigma_\theta{}^2 + \sigma_\theta{}^2 \cdot \frac{\alpha^2}{4} + \sigma_\theta{}^2 \left(1 + \frac{\alpha}{2} \right)^2 = 2Y^2$$

or
$$\sigma_\theta = \frac{2Y}{\sqrt{\alpha^2 + 2\alpha + 4}}.$$

$$(a) \quad \alpha = (A_0/A) - 1 = 2 - 1.$$

Thus,
$$\sigma_\theta = (2Y/\sqrt{7})$$

and
$$p = (4Yt/D\sqrt{7})$$

$$(b) \quad \alpha = 5 - 1 = 4.$$

Thus,
$$\sigma_\theta = (2Y/2\sqrt{6})$$

and
$$p = [(Yt\sqrt{6})/(D\sqrt{3})]$$

20—EP * *

The condition for plane strain yielding of the cylinder, that is $d\epsilon_z = 0$, is found using the Lévy-Mises equations.

Thus
$$\frac{d\epsilon_\theta}{\sigma_\theta'} = \frac{d\epsilon_r}{\sigma_r'}$$

Since
$$d\epsilon_\theta + d\epsilon_r = 0,$$

then
$$\sigma_\theta + \sigma_r = \frac{2}{3}\left(\sigma_\theta + \sigma_r + \sigma_z\right)$$

or
$$\sigma_\theta = 2\,\sigma_z.$$

Hence
$$\sigma_\theta = 2\,\sigma_\theta\frac{\alpha}{2} \quad\text{or}\quad \alpha = 1,$$

that is,
$$A_0/A = 2.$$

19. The Mises yield criterion is,

$$(\sigma_1 - \sigma_2)^2 + (\sigma_2 - \sigma_3)^2 + (\sigma_3 - \sigma_1)^2 = 2Y^2. \qquad (1)$$

For the biaxial stressing of sheet, $\sigma_3 = 0$, hence (1) reduces to

$$\sigma_1{}^2 - \sigma_1\sigma_2 + \sigma_2{}^2 = Y^2. \qquad (2)$$

Taking $O\sigma_1$ and $O\sigma_2$ as cartesian axes of principal stress, (2) appears as an ellipse and the portion in the third quadrant only does not have at least one principal stress positive.

After differentiating (2), we have

$$\frac{d\sigma_2}{d\sigma_1}(2\sigma_2 - \sigma_1) + 2\sigma_1 - \sigma_2 = 0.$$

Thus, when
$$d\sigma_2/d\sigma_1 = 0, \quad \sigma_1 = \sigma_2/2.$$

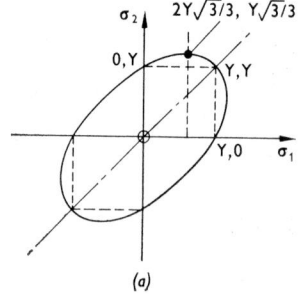

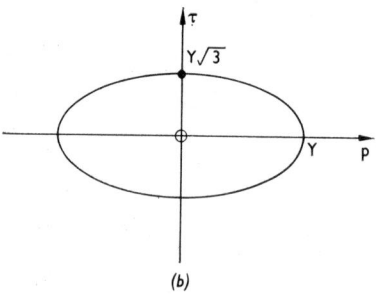

(a)   (b)

FIG. 50

Substituting in (2) we find that

$$\sigma_1 = Y\sqrt{3}/3 \quad\text{and}\quad \sigma_2 = 2Y\sqrt{3}/3.$$

Put
$$(\sigma_1 - \sigma_2) = 2\tau$$

and
$$(\sigma_1 + \sigma_2) = 2p,$$

so that equation (2) becomes

$$\frac{p^2}{Y^2} + \frac{\tau^2}{(Y/\sqrt{3})^2} = 1.$$

20. As given, radial stress $\sigma_r = 0$.

Axial stress, $$\sigma_z = \frac{p_1}{2} \cdot \frac{r}{t}.$$

Hoop stress, $$\sigma_\theta = (1 - x)\frac{rp_1}{t}.$$

Substituting in the Mises yield criterion, we see that

$$2Y^2 = \left[(1 - x)^2 + \left(\frac{1}{2}\right)^2 + \left(1 - x - \frac{1}{2}\right)^2\right]\left(\frac{rp_1}{t}\right)^2.$$

Simplifying and rearranging,

$$p_1 = \frac{2tY}{r}(4x^2 - 6x + 3)^{-1/2}.$$

21. From equilibrium conditions

$$\sigma_\theta = \frac{pr}{t}, \qquad \sigma_z = \frac{pr}{2t}, \qquad \sigma_r = 0.$$

Equations (5.37) give the following general relations between strain increment and stress components.

$$d\epsilon_z = d\lambda\left[H(\sigma_z - \sigma_\theta) + G(\sigma_z - \sigma_r)\right]$$
$$d\epsilon_\theta = d\lambda\left[F(\sigma_\theta - \sigma_r) + H(\sigma_\theta - \sigma_z)\right]$$
$$d\epsilon_r = d\lambda\left[G(\sigma_r - \sigma_z) + F(\sigma_r - \sigma_\theta)\right].$$

Writing in terms of σ_θ,

$$d\epsilon_z = \frac{\sigma_\theta}{2}d\lambda(G - H)$$

$$d\epsilon_\theta = \sigma_\theta d\lambda\left(F + \frac{H}{2}\right)$$

$$d\epsilon_r = -\sigma_\theta d\lambda\left(\frac{G}{2} + F\right).$$

For an isotropic material $F = G = H$, $d\epsilon_z = 0$ and $d\epsilon_\theta = -d\epsilon_r$.
When the anisotropy is rotationally symmetric, $F = G$ and,

$$d\epsilon_z = \frac{\sigma_\theta}{2}d\lambda(G - H)$$

$$d\epsilon_\theta = \sigma_\theta d\lambda\left(G + \frac{H}{2}\right)$$

$$d\epsilon_r = -\sigma_\theta d\lambda\left(\frac{3}{2}G\right).$$

The r-value is then equal to H/G and when $r = 2$

$$d\epsilon_z = -\frac{\sigma_\theta}{2} d\lambda\, G$$

$$d\epsilon_\theta = 2\, \sigma_\theta\, d\lambda\, G$$

and $\quad d\epsilon_r = -\frac{3}{2}\, \sigma_\theta\, d\lambda\, G.$

Therefore

$$d\epsilon_\theta \;:\; d\epsilon_z \;:\; d\epsilon_r = 1 \;:\; -\tfrac{1}{4} \;:\; -\tfrac{3}{4}.$$

22. We assume that radius c defines a circle of no slip; see Fig. 51. For $r > c$ the material moves outwards, and for $r < c$ inwards, on compression. The 'friction hill' for $r \geqslant c$, is $p = Ye^{+(2\mu/h)(b-r)}$ and for $r \leqslant c$, $p = Ye^{+(2\mu/h)(r-a)}$. At $r = c$, p must be the same by both expressions for p; this gives $b - c = c - a$, i.e., $c = (a + b)/2$. (It can be shown that the assumption $\sigma_\theta = \sigma_r$, used in Problems 22 and 23, is not a very good one.)

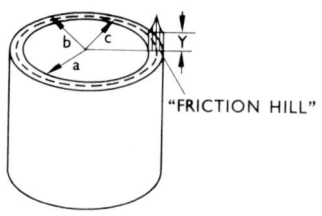

"FRICTION HILL"

FIG. 51

23. The method of solution (and with identical assumptions) proceeds as in Section 6.5. Set up an equation for radial equilibrium of forces; observe that $\sigma_r = 0$ at $r = a$, and find that the pressure on the cylinder at r is

$$p = Ye^{(2\mu/h)(r-a)}.$$

p at $r = b$ is $Ye^{(2\mu/h)(b-a)}$ Hence

$$\sigma_r = Y\{1 - e^{(2\mu/h)(b-a)}\}.$$

At $r = b$, $p = Ye^{(2\mu/h)(b-a)}$. Assume shearing occurs when $\mu p = Y/2$; this first occurs at $r = b$. Hence:

$$Y/2 = \mu Ye^{(2\mu/h)(b-a)}$$

or $\qquad \ln\dfrac{1}{2\mu} = \dfrac{2\mu}{h}(b - a).$

The yield criterion as usually applied in this kind of analysis assumes that p, σ_r, σ_θ are principal stresses; this is far from true when $\mu p = Y/2$.

24. The platen compressive stress at distance x from the centre-line is p, and q is the transverse stress assumed constant through the thickness h. (See Fig. 52(a).)

The equilibrium equation for the element shown, where μ is the coefficient of friction, is

$$h \cdot \delta q = 2\mu p \delta x$$

or
$$dq/dx = 2\mu p/h.$$

Assuming p and q are the greatest and least principal stresses, using the Tresca yield criterion,

$$p + q = 2k,$$

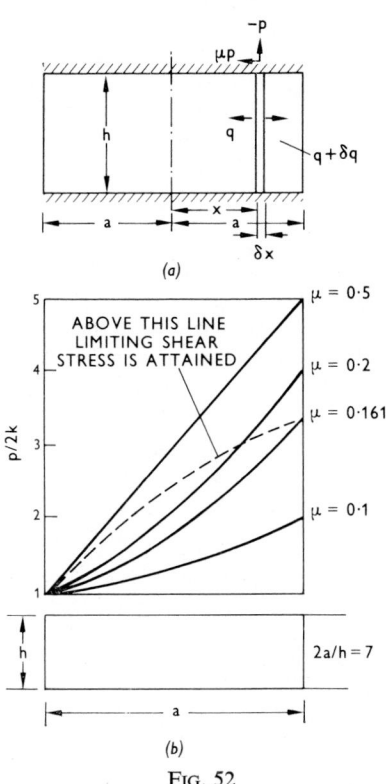

(a)

(b)

Fig. 52

where k is the yield shear stress in plane strain and hence $dp = -dq$. Thus, substituting, above

$$dp/p = -2\mu/h$$

and
$$\ln p = -2\mu x/h + \text{constant}.$$

At $x = +a$, $q = 0$ and thus $p = 2k$. The constant in the above equation can then be eliminated and

$$p/2k = \exp 2\mu(a - x)/h.$$

For $\mu = 0.1$ and $2a/h = 7$, the variation of $p/2k$ with x is shown across the top platen in Fig. 52(b). Note that there is a discontinuity in rate of change of p at the centre line and the general similarity in shape of this friction hill, to the normal pressure distribution between the rolls and stock in the rolling process. (In the rolling process and compression, the overall *relative* motion between metal and platens or rolls, is identical.) We may also point out that a slip-line field solution to this problem shows the pressure distribution on the plates to be made up of both steady increases in p with x, and lengths of constant pressure or pressure plateaux*.

The mean pressure on the platen $\bar{p}$ is:

$$\bar{p} = \frac{1}{a}\int_0^a p\,dx = \frac{2k}{a}\int_{\bullet}^a e^{2\mu(a-x)/h}\,dx$$

$$\frac{\bar{p}}{2k} = \frac{h}{2a\mu}(e^{2\mu a/h} - 1)$$

and if $2\mu a/h$ is small,

$$\frac{\bar{p}}{2k} \simeq \frac{h}{2a\mu}\left(1 + \frac{2\mu a}{h} + \frac{4\mu^2 a^2}{2h^2} - 1\right)$$

$$= 1 + \frac{\mu a}{h}.$$

We emphasize that this analysis is based on the assumptions that p and q are principal stresses and that q is constant through thickness h. Especially is the above result true only as long as $\mu p < k$.

We now find for what value of x, say x_1, $\mu p = k$. This is given by the equation

$$\mu p = k = 2\mu k e^{2\mu(a-x_1)/h}.$$

Hence,
$$\frac{x_1}{h} = \frac{a}{h} - \frac{1}{2\mu}\ln\frac{1}{2\mu}.$$

The initial equilibrium equation is only valid for $x \leqslant x_1 \leqslant a$; should $x_1 > a$ the equilibrium equation is

$$\frac{dq}{dx} = \frac{2k}{h} = -\frac{dp}{dx}$$

and thus,
$$p = -\frac{2k}{h}x + c_2.$$

c_2 can be eliminated because $p = p_1 = k/\mu$ at $x = x_1$.

Hence,
$$p = p_1 + \frac{2k}{h}(x_1 - x).$$

Now substitute for x_1 from the equation for x_1/h and find, for

$$0 < x < x_1 = a - \frac{h}{2\mu}\ln\frac{1}{2\mu}$$

$$\frac{p}{2k} = \left(\frac{a-x}{h}\right) + \frac{1}{2\mu}\left(1 - \ln\frac{1}{2\mu}\right).$$

*Alexander, J. M., *Mech. Phys. Solids* **3**, 223 (1955).

The mean pressure over the platen is now

$$\frac{\bar{p}}{2k} = \frac{a}{2h}\left(1 - \frac{\ln 1/2\mu}{2a\mu/h}\right)\left(1 + \frac{2 - \ln 1/2\mu}{2a\mu/h}\right) + \frac{1/2\mu - 1}{2\mu a/h}.$$

Summarizing the above results,

(a) $\qquad\qquad \dfrac{\bar{p}}{2k} = \dfrac{e^{2\mu a/h} - 1}{2a\mu/h}$ when $\dfrac{2a\mu}{h} < \ln\dfrac{1}{2\mu}$

(b) $\qquad \dfrac{\bar{p}}{2k} = \left[\dfrac{(1 + \beta)^2 + 1}{4\mu} - 1\right]\Big/\dfrac{2a\mu}{h}$ when $\dfrac{2a\mu}{h} < \ln\dfrac{1}{2\mu}$

where $\qquad\qquad\qquad \beta = \dfrac{2a\mu}{h} - \ln\dfrac{1}{2\mu}.$

Applying to the above expressions the value $2a/h = 7$, we have,

(a) $\qquad\qquad \dfrac{\bar{p}}{2k} = \dfrac{e^{7\mu} - 1}{7\mu}$ when $7\mu < \ln\dfrac{1}{2\mu}$

(b) $\qquad\qquad \dfrac{\bar{p}}{2k} = \left[\dfrac{(1 + 7\mu - \ln 1/2\mu)^2}{4\mu} - 1\right]\Big/7\mu$

when $\qquad\qquad\qquad 7\mu > \ln\dfrac{1}{2\mu}.$

$7\mu = \ln 1/2\mu$ when $\mu = 0\cdot161$. The pressure distribution curves given in Fig. 52 may now be verified.

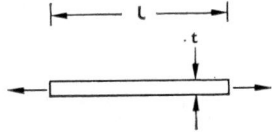

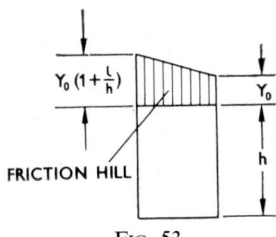

Fig. 53

25. The deformation is supposed to be one of plane strain.

(a) For the plate to fracture $l(Y_0/2) = Y_1 t$; therefore $l = (2Y_1 t)/(Y_0)$.
(b) Fully developed load on the filling is P_0 and $P_0 = 2Y_0 l[1 + (1/2)(l/h)]$.
(c) Without Y_1, $P'_0 = 2lY_0$.

Thus, $\qquad \dfrac{P_0}{P'_0} = \dfrac{Y_0 l[1 + (1/2)(l/h)]}{Y_0 l} = 1 + \dfrac{1}{2}\cdot\dfrac{2Y_1}{Y_0}\cdot\dfrac{t}{h}$

$$= \left(1 + \frac{Y_1}{Y_0}\cdot\frac{t}{h}\right).$$

If $t/h = 1/10$ and $Y_1/Y_0 = 3$, then $P_0/P'_0 = 1\cdot3$—an increase of 30 per cent.

26. The equation for radial equilibrium of an element is

$$\sigma_\theta - \sigma_r = r\,d\sigma_r/dr.$$

Material 1 $\qquad\qquad \sigma_\theta - \sigma_r = Y_1$

and thus $\qquad\qquad\quad \sigma_r = Y_1 \ln r + A.$

Because $\qquad \sigma_r = 0$ at $r = b, \qquad \sigma_r = -Y_1 \ln b/r.$

Hence at $r = c$, $\qquad\qquad\quad \sigma_r = -Y_1 \ln b/c.$

Material 2 $\qquad\qquad\quad \sigma_\theta - \sigma_r = -Y_2$

and thus $\qquad\qquad\quad \sigma_r = -Y_2 \ln r + B.$

Because $\qquad \sigma_r = 0$ at $r = a, \qquad \sigma_r = -Y_2 \ln r/a.$

Hence at $r = c$, $\qquad\qquad\quad \sigma_r = -Y_2 \ln c/a.$

σ_r, for equilibrium, must be the same on either side of the common surface or at $r = c$. Thus:

$$-Y_1 \ln b/c = -Y_2 \ln c/a \quad \text{or} \quad c^{Y_1 + Y_2} = a^{Y_2} b^{Y_1}.$$

The jump in σ_θ at $r = c$, is just $(Y_1 + Y_2)$.
The bending moment per unit width M_p is:

$$M_p = \int_c^b \sigma_\theta r \, dr + \int_a^c \sigma_\theta r \, dr$$

$$= Y_1 \int_c^b (1 - \ln b/r) r \, dr - Y_2 \int_a^c (1 + \ln r/a) r \, dr$$

$$= \frac{Y_1(b^2 - c^2) + Y_2(a^2 - c^2)}{4} + \frac{c^2}{2}\left[Y_1 \ln \frac{b}{c} - Y_2 \ln \frac{c}{a} \right].$$

Note that when $Y_2 = Y_1 = Y$, the results obtained in Section 7.8 are recovered, i.e., $c^2 = ab$ and $M_p = [Y(b - a)^2]/4$.
To check that there are no net forces over a section for sheet 1, we find

$$F_1 = \int_c^b \sigma_\theta . dr = \int_c^b Y_1\left(1 - \ln \frac{b}{r}\right) dr = Y_1 c \ln \frac{b}{c}$$

and for sheet 2,

$$F_2 = \int_a^c \sigma_\theta . dr = -Y_2 \int_a^c \left(1 + \ln \frac{r}{a}\right) dr = Y_2 c \ln \frac{a}{c}.$$

Clearly, $F_1 + F_2 = 0$.

27. Suffixes 1 and 2 refer to metals 1 and 2, respectively. We assume $\alpha_2 > \alpha_1$ and thus that bending occurs as shown in Fig. 15(c). P denotes the tensile or compressive force and M the bending moment on a metal. From equilibrium considerations:

$$P_1 = P_2 = P \quad \text{and} \quad M_1 + M_2 = Ph/2.$$

If R is the radius of curvature of the strip:

$$M_1 = E_1 I_1/R \quad \text{and} \quad M_2 = E_2 I_2/R.$$

E refers to Young's Modulus and I to the second moment of area of a section. At the interface the strain must be the same in both materials and thus:

$$\alpha_1(t - t_0) + \frac{P_1}{E_1 a} + \frac{a}{2R} = \alpha_2(t - t_0) - \frac{P_2}{E_2 a} - \frac{a}{2R}.$$

From these equations it is easily found that

$$\frac{1}{R} = \frac{24(\alpha_2 - \alpha_1)(t - t_0)}{h(14 + n + 1/n)} \quad \text{where} \quad E_1/E_2 = n.$$

It is next necessary to compare the magnitude of the direct stresses in the extreme fibres of both metals. It transpires that the maximum stresses σ occur on the interface. Then:

$$\sigma_1 = \frac{P}{a} + \frac{aE_1}{2R} = Y_1 \quad \text{and} \quad \sigma_2 = \frac{-P}{a} - \frac{aE_2}{2R} = -Y_2.$$

(Y denotes yield stress.)

Thus, adding, $Y_1 - Y_2 = \dfrac{a}{2R}(E_1 - E_2)$ and hence

$$\frac{Y_1 - Y_2}{E_1 - E_2} = \frac{a}{2} \times \frac{24(\alpha_2 - \alpha_1)(t - t_0)}{h(14 + n + 1/n)}$$

so that the temperature rise is given by

$$t - t_0 = \frac{(14 + n + 1/n)}{6(\alpha_2 - \alpha_1)} \times \left(\frac{Y_1 - Y_2}{E_1 - E_2}\right).$$

28. On a section such as BB, see Fig. 16, at a given value of y (see Fig. 54), the bending stress may be found. The total normal force on the section may next be obtained by integrating (bending stress $\times ad\theta t$) from 0 to θ. The same may be done

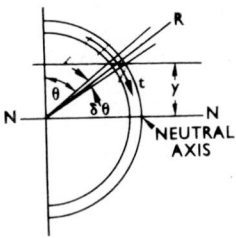

Fig. 54

on section AA, distant δx from BB, again over the same angle. The difference between these two normal forces must be accounted for by, say, a uniform shear stress q in the plane NR acting over area $t.\delta x$. The complementary shear stress acting tangentially at y is now calculable. The shear centre distance e, times the load W, must equilibrate the moment due to the complementary shear stress about section centre N. It may be shown that:

$$e/a = \int_0^{\pi/2} \int_0^\theta \cos^n \theta \, d\theta \, d\theta \Big/ \int_0^{\pi/2} \cos^{n+1} \theta \, d\theta.$$

(see Johnson and Mellor, *Applied Science Research*, Series A, Vol. 7, p. 467, 1957.)

29. Figure 55 shows a position of the beam. Let σ_r denote the radial stress over curved surface $ABCD$. Then the equation for radial equilibrium is, $2h$ being the height of the cross-section,

$$\sigma_r R = \int_y^h \sigma \, dz.$$

Hence
$$\sigma_r = \frac{1}{R} \int_y^h \sigma \, dz = \frac{F}{R^{n+1}} \int_y^h z^n \, dz \quad \text{using (7.3)}$$

$$= \frac{F}{R^{n+1}} \cdot \frac{(h^{n+1} - y^{n+1})}{(n+1)}.$$

If $n = 1$, i.e., for an elastic material, when $F = E$,

$$\sigma_r = \frac{E}{R^2} \cdot \frac{(h^2 - y^2)}{2}$$

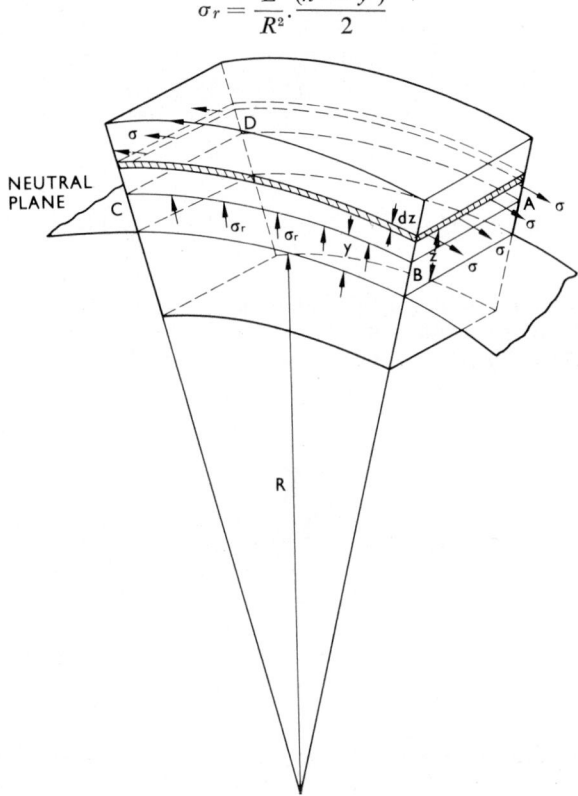

Fig. 55

and the radial stress distribution is parabolic. In particular $\sigma = \sigma_{\max}$ at $y = h/2$, and thus

$$\frac{\sigma_{\max}}{\sigma_{r \text{ at } y = 0}} = \frac{Eh}{R} \frac{2R^2}{Eh^2} = \frac{2R}{h}.$$

This result shows that the usual engineering theory implies radial stresses of negligible amount by comparison with the largest hoop bending stresses which can arise. Only when y is less than approximately $h^2/2R$, does σ_r exceed σ. For instance, if $R = 50$ in and $h = 1$ in, σ_r and σ are approximately equal at $y = 1/10$ in.

It is worth noting that σ in elementary beam theory implies mainly second order, negligible, radial stresses. These stresses are essential for the beam to form itself into the arc of a circle. The customary method, which is adopted in Section 7.3, of considering thin layers independently does not imply no radial or normal stress between them.

30. Refer to Fig. 56(i). The bending moment diagram will show maximum values at C and B, though they will be of opposite sign. Plastic hinges will then form at C and B.

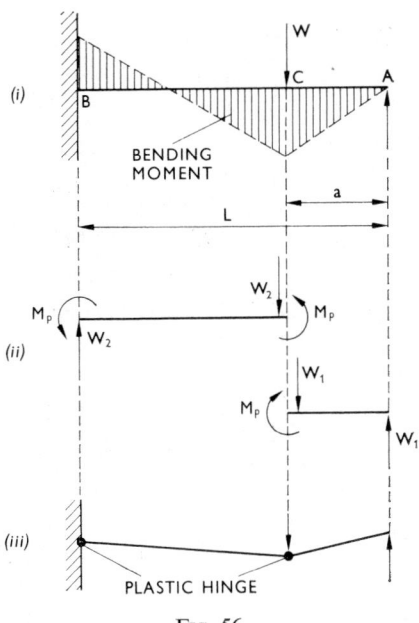

FIG. 56

(i) For moment equilibrium on AC, $M_p = W_1/a$, see Fig. 56(ii)

(ii) For moment equilibrium on CB, $W_2 b - M_p = M_p$, see Fig. 56(iii), and thus $W_2 = 2M_p/b$.

But,
$$W_1 + W_2 = W = M_p\left\{\frac{1}{a} + \frac{2}{b}\right\} = M_p\left\{\frac{1}{a} + \frac{2}{L-a}\right\}.$$

Thus,
$$\frac{\partial W}{\partial a} = M_p\left\{-\frac{1}{a^2} + \frac{2}{(L-a)^2}\right\} = 0,$$

when
$$\frac{L-a}{a} = \sqrt{2} \quad \text{or} \quad \frac{a}{L} = \frac{1}{1+\sqrt{2}}.$$

When W has a position such that $a/L = (\sqrt{2} - 1)$, W is also a minimum associated with complete collapse of the cantilever.

31. $T = 2\pi \int_0^a r \, dr . r . \tau = 2\pi \int_0^a r^2 . \dfrac{k . r^m . \theta^m}{l^m} \, dr$, since $\dot{\gamma} = r\dot{\theta}/l$.

Hence,

$$T = 2\pi k \left(\frac{\theta}{l}\right)^m \frac{a^{m+3}}{m+3}.$$

32. Putting $r = a$, and $\rho/a = b/a = c$ in equation (8.27.i), we find

$$\frac{\tau}{k} = 1 - \frac{4(1 - c^3)}{3(1 - c^4)}.$$

Hence when $c = 0$, $\qquad\qquad \tau/k = 1/3$

The position of zero residual shear stress is given by

$$\frac{r}{a} = \frac{3(1 - c^4)}{4(1 - c^3)}$$

and thus

c	0	1/2
r/a	0·75	0·80

33. The various results given in Section 8.3 may be used and the quantities given in Fig. 57 obtained.

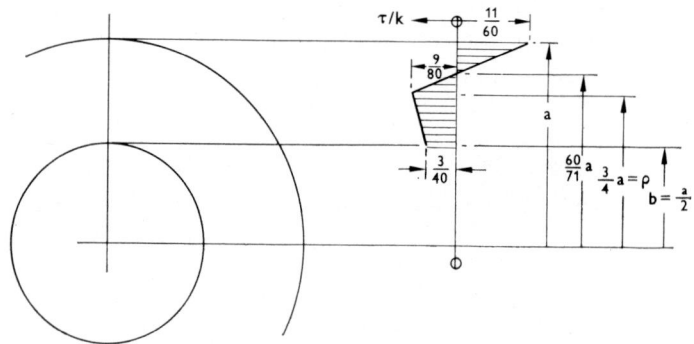

Fig. 57

Use the equation for θ_R given in Section 8.3.

$$\frac{\theta_R}{\theta_0} = 1 - \frac{\frac{4}{3}(\rho/a)[1 - \frac{1}{4}(\rho/a)^3 - \frac{3}{4}(a/\rho)c^4]}{1 - c^4} = \frac{9}{71}.$$

34. Using the soap film analogy, see Fig. 58, we have

Region I: $\qquad pa^2 \simeq T.a.\dfrac{h_1}{2t} + T.a.\dfrac{h_1}{t} + T.a.\dfrac{(h_1 - h_2)}{t} + T.a.\dfrac{h_1}{t}.$

Region II: $\qquad pa^2 \simeq T.a.\dfrac{h_2 - h_1}{t} + T.a.\dfrac{h_2}{t} + T.a.\dfrac{h_2}{t/2} + T.a.\dfrac{h_2}{t}.$

T denotes the membrane tension. (Remember that the membrane is very flat and, for instance, that the hypotenuse over thickness $2t$, is n $2t$.) Hence, for

Region I: $\qquad pa^2 = \dfrac{Ta}{t}\left(\dfrac{7}{2}h_1 - h_2\right)$ or $\dfrac{p.a.t}{T} = \dfrac{7}{2}h_1 - h_2.$

Region II: $\qquad pa^2 = \dfrac{Ta}{t}(5h_2 - h_1)$ or $\dfrac{p.a.t}{T} = 5h_2 - h_1.$

Thus, $\qquad\qquad h_1 = \dfrac{4}{11}\cdot\dfrac{p.a.t}{T}$ and $h_2 = \dfrac{3}{11}\cdot\dfrac{p.a.t}{T}.$

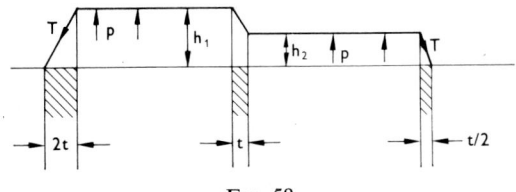

FIG. 58

The maximum shear stress in each of legs X, Y, Z, is,

in X: $\qquad\qquad \dfrac{h_1}{2t} = \dfrac{2}{11}\dfrac{p.a}{T}$

in Y: $\qquad\qquad \dfrac{h_1 - h_2}{t} = \dfrac{1}{11}\cdot\dfrac{p.a}{T}$

in Z: $\qquad\qquad \dfrac{h_2}{t/2} = \dfrac{6}{11}\cdot\dfrac{p.a}{T}.$

Hence,

$$\tau_{\max} = 5000 = \dfrac{6}{11}\cdot\left(\dfrac{p}{T}\right)\cdot a = \dfrac{6}{11}(2G\theta_0)10.$$

Thus, $\qquad\qquad \theta_0 = 0{\cdot}382 \ 10^{-4}$ radians per inch.

Transmitted torque is equivalent to twice the volume under membrane, i.e.,

$$T = 2a^2(h_1 + h_2)$$

$$= 2a^2\cdot\dfrac{7}{11}\cdot\dfrac{p.a.t}{T}$$

$$= 2a^2\cdot\dfrac{7}{11}\cdot(2G\theta)\cdot a.t$$

$$= 11{\cdot}65 \ . \ 10^5 \text{ in. lbf.}$$

35. In the rectangle, see Fig. 18, $\phi = -G\theta(x^2 - c^2/4)$.
In Fig. 18, $y/b = t/c$ or $t = yc/b$.

Assume that the membrane at y has a parabolic form in planes perpendicular to Oy.
Thus putting yc/b for t in the given stress function,

$$\phi = -G\theta\left(x^2 - \dfrac{c^2y^2}{4b^2}\right).$$

Note that along the boundary $\phi = 0$.

Hence, the elastic torque T required to cause a twist of θ per unit length of a shaft of the triangular section shown is given by

$$T = 2 \int_0^b \int_{-\frac{cy}{2b}}^{\frac{cy}{2b}} -G\theta\left(x^2 - \frac{c^2y^2}{4b^2}\right) dx\, dy$$

$$= -4G\theta \int_0^b \left[\frac{x^3}{3} - \frac{c^2y^2}{4b^2}x\right]_0^{\frac{cy}{2b}} dy$$

$$= -4G\theta \int_0^b \left[\frac{c^3y^3}{24b^3} - \frac{c^3y^3}{8b^3}\right] dy$$

$$= \frac{G\theta}{3} \cdot \frac{c^3}{b^3} \int_0^b y^3\, dy = \frac{G\theta\, c^3\, b}{12}.$$

Hence the torsional rigidity, $T/\theta = Gbc^3/12$.

36. Treat the disc as a short cylinder. The equilibrium equation for an element at radius r where the disc is of axial thickness $2h$ is

$$\frac{d}{dr}(h\sigma_r) = \frac{h(\sigma_\theta - \sigma_r)}{r}$$

The equation of the profile is $rh = ah_a$ and for full plasticity $\sigma_\theta - \sigma_r = Y$. Hence

$$h\sigma_r = \int \frac{ah_a}{r} \times \frac{Y}{r}\, dr = -\frac{Yah_a}{r} + A$$

Because $\sigma_r = 0$ at $r = b$, $A = \dfrac{Yah_a}{b}$,

therefore

$$\frac{p}{Y} = -a\left(\frac{1}{r} - \frac{1}{b}\right).$$

37(a). Writing down the condition for yielding at the inside surface of each cylinder, we have, using Tresca's criterion in the form $\sigma_\theta - \sigma_r = Y$, in conjunction with Lamé's equations:

$$p - p_1 = \frac{Y}{2} \times \left[1 - \left(\frac{r_0}{r_1}\right)^2\right]$$

$$p_1 - p_2 = \frac{Y}{2} \times \left[1 - \left(\frac{r_1}{r_2}\right)^2\right]$$

$$\cdots\cdots\cdots\cdots\cdots\cdots\cdots$$

$$\cdots\cdots\cdots\cdots\cdots\cdots\cdots$$

$$p_m - p_n = \frac{Y}{2} \times \left[1 - \left(\frac{r_m}{r_n}\right)^2\right]$$

$p_1, p_2 \ldots p_m$ are the pressures at the interface between the cylinders. We are given $p_n = 0$. Adding the above equations:

$$p = \frac{Y}{2}\left\{n - \left[\left(\frac{r_0}{r_1}\right)^2 + \left(\frac{r_1}{r_2}\right)^2 + \ldots + \left(\frac{r_m}{r_n}\right)^2\right]\right\}.$$

For p to be a maximum for given values of r_0 and r_n, we require

$$\frac{\partial p}{\partial r_1} = \frac{\partial p}{\partial r_2} = \ldots \frac{\partial p}{\partial r_m} = 0;$$

therefore, $\quad \dfrac{r_1}{r_0} = \dfrac{r_2}{r_1} = \dfrac{r_3}{r_2} = \ldots = \dfrac{r_n}{r_m} = \left(\dfrac{r_n}{r_0}\right)^{1/n}.$

The required ratio of the radii is then, $(r_n/r_0)^{1/n}$ and r_0, r_1, $r_2 \ldots r_n$ form a geometric progression. Hence:

$$p = \frac{Y}{2} n \left[1 - \left(\frac{r_0}{r_n}\right)^{2/n} \right]$$

For each cylinder to become fully plastic, we obtain from the usual equilibrium equation

$$\frac{d\sigma_r}{dr} = \frac{\sigma_\theta - \sigma_r}{r}, \quad \text{therefore} \quad \sigma_r = \int Y \frac{dr}{r} = Y \ln r + C.$$

Thus for each cylinder we have

$$-p + p_1 = Y \ln r_0/r_1$$

$$-p + p_2 = Y \ln r_1/r_2$$

$$\ldots\ldots\ldots\ldots\ldots\ldots\ldots\ldots$$

$$-p_m + p_n = Y \ln r_m/r_n$$

and adding and putting $p_n = 0$

$$-p = Y \ln r_0/r_n$$

or $\qquad\qquad p = Y/2 \ln (r_n/r_0)^2.$

$$\textit{Values of } \frac{p}{Y/2}$$

		\multicolumn{8}{c}{$n\left\{ 1 - \left(\dfrac{r_0}{r_n}\right)^{2/n} \right\}$}
n	r_n/r_0	1·2	1·5	2	3	4	5	7	10
1		0·305	0·555	0·750	0·790	0·938	0·960	0·980	0·990
2		0·334	0·660	1·000	1·333	1·50	1·60	1·71	1·80
3		0·345	0·711	1·113	1·545	1·803	1·971	2·181	2·352
4		0·348	0·740	1·172	1·696	2·00	2·220	2·524	2·736
5		0·352	0·748	1·210	1·779	2·128	2·373	2·704	3·010
10		0·36	0·78	1·29	1·98	2·42	2·75	3·22	3·69
Fully plastic		0·365	0·811	1·386	2·20	2·77	3·22	3·89	4·61

37b. Let the interfacial pressure between the cylinders be $-\sigma_{r_1}$. By the usual methods, we can show that for cylinder 1 to be fully plastic:

$$Y_1 \ln \frac{r_0}{r_1} = -p + \sigma_{r_1}$$

and for cylinder 2,

$$Y_2 \ln \frac{r_1}{r_2} = -\sigma_{r1}$$

Adding these two equations:

$$p = Y_1 \ln \frac{r_1}{r_0} + Y_2 \ln \frac{r_2}{r_1}$$

Thus for a given p, r_0 and r_2

$$r_1 = \left[\frac{r_0^{Y_1}}{r_2^{Y_2}} \exp(p) \right]^{1/(Y_1 - Y_2)}$$

38. The stresses on an element of the tube are shown in Fig. 59. Recall that the elastic shear stress at radius r is $\tau = 32Tr/\pi(R_0^4 - R_i^4)$.
The general equation for applying the Mises criterion is

$$[(\sigma_\theta - \sigma_r)^2 + (\sigma_r - \sigma_z)^2 + (\sigma_z - \sigma_\theta)^2 + 6\tau_{\theta r}^2 + 6\tau_{rz}^2 + 6\tau_{z\theta}^2]^{\frac{1}{2}}$$

$$= \left[\left(\frac{2p}{m^2 - 1} \cdot \frac{R_0^2}{r^2} \right)^2 + \frac{p}{m^2 - 1} \cdot \frac{R_0^2}{r^2} + \frac{p}{m^2 - 1} \cdot \frac{R_0^2}{r^2} + 6\tau^2 \right]^{\frac{1}{2}}$$

$$= \left[\frac{6p^2}{(m^2 - 1)^2} \cdot \frac{R_0^4}{r^4} + 6\tau^2 \right]^{\frac{1}{2}}.$$

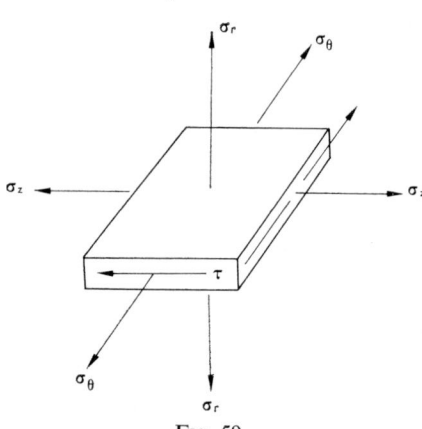

FIG. 59

The condition that yield occurs simultaneously at $r = R_0$ or $r = R_i$, is met by

$$\frac{6p^2 R_0^4}{(m^2 - 1)^2 \cdot R_0^4} + 6 \left(\frac{32 \, TR_0 \, m^4}{\pi R_0^4 \, (m^4 - 1)} \right)^2 = \frac{6p^2}{(m^2 - 1)^2} \cdot \frac{R_0^4}{R_i^4} + 6 \left(\frac{32 \cdot TR_i \cdot m^4}{\pi R_0^4 \, (m^4 - 1)} \right)^2.$$

Simplifying,

$$\frac{T}{p} = \frac{(m^2 + 1)^{3/2} \pi R_0^3}{32m^6}.$$

39. The equilibrium equation in the plastic zone using (9.40), is

$$\sigma_\theta - \sigma_r = r \frac{d\sigma_r}{dr}$$

and throughout it, $\sigma_\theta - \sigma_r = 2Y_L$, so that

$$2Y_L \int_{R_i}^{nR_i} \frac{dr}{r} = \int_p^{\sigma_r} d\sigma_r$$

or

$$2Y_L \ln n = \sigma_r - p.$$

The outer elastic cylinder is just on the point of yield at its inner radius so that using,

$$2Y_n = -\sigma_r \frac{2m^2 R_i^2}{m^2 R_i^2 - n^2 R_i^2},$$

Eliminating σ_r between the last two equations, gives

$$p = \sigma_r - 2Y_L \ln n$$

$$-p = \frac{m^2 - n^2}{m^2} \cdot Y_u + Y_L \ln n^2.$$

For a thick-walled spherical shell we have, using equation (9.20),

$$p = 2Y \ln \frac{c}{a} + 2Y \frac{b^3 - c^3}{3b^3}.$$

Therefore, in the present notation,

$$p = Y_L \ln n^4 + Y_U \frac{4}{3} \left(\frac{m^3 - n^3}{m^3} \right)$$

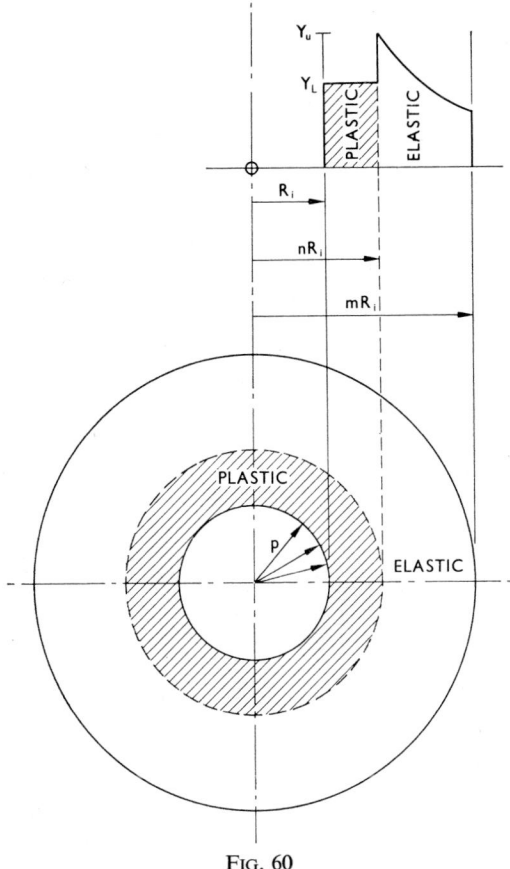

Fig. 60

40. No change in mean radius of thin-walled tube, i.e., $d\epsilon_\theta = 0$, is assumed.

Also, $\sigma_t = 0$

By constancy of volume, $d\epsilon_l = -d\epsilon_t$ or $\epsilon_l = -\epsilon_t$ (constant ratio). The Lévy-Mises flow rule gives

$$\frac{d\epsilon_l}{\sigma_l - \sigma_\theta} = \frac{d\epsilon_l}{\sigma_\theta} \quad \text{or} \quad \sigma_l = 2\sigma_\theta$$

and hence
$$\bar{\sigma} = \frac{\sqrt{3}}{2}\sigma_l \quad \text{and} \quad d\bar{\epsilon} = \frac{2}{3}d\epsilon_l = -\frac{2}{\sqrt{3}}d\epsilon_t.$$

The radius r is constant and thus:
$$P = (2\pi r)t\sigma_l$$

Instability occurs when $dP = 0$, i.e., when $\dfrac{d\sigma_l}{\sigma_l} = -\dfrac{dt}{t} = -d\epsilon_t = d\epsilon_l.$

Hence
$$\bar{\epsilon} = \left(\frac{2}{\sqrt{3}}n - B\right) = \frac{2}{\sqrt{3}}\epsilon_l,$$

$$\bar{\sigma} = A\left(\frac{2}{\sqrt{3}}\right)^n = \frac{\sqrt{3}}{2}\sigma_l,$$

$$t = t_0 \exp\left(\frac{\sqrt{3}}{2}B - n\right).$$

Load at instability
$$P = 2\pi r t_0 \exp\left(\frac{\sqrt{3}}{2}B - n\right)\frac{2}{\sqrt{3}}A\left(\frac{2}{\sqrt{3}}n\right)^n$$

$$P = X_0 \frac{2}{\sqrt{3}}A\left(\frac{2}{\sqrt{3}}n\right)^n \exp\left(\frac{\sqrt{3}}{2}B - n\right)$$

where X_0 is the original cross-sectional area.
For *simple tension*, T.S. $= X_0 . A(n)^n \exp(B - n)$.

41. For a thin rotating ring considerations of equilibrium give the circumferential stress as
$$\sigma_\theta = \rho r^2 \omega^2.$$

The stress components in the radial and thickness directions can be considered to be negligible. Instability will occur when the angular velocity reaches a maximum, that is when
$$\frac{d\sigma_\theta}{\sigma_\theta} = 2\frac{dr}{r} = 2\,d\epsilon_\theta$$

or
$$\frac{d\sigma_\theta}{d\epsilon_\theta} = 2\sigma_\theta$$

Consider now the general expression $\sigma_\theta = A\epsilon_\theta{}^n$.
From this
$$\frac{d\sigma_\theta}{d\epsilon_\theta} = \frac{n}{2}\sigma_\theta.$$

Comparing the last two equations, the strain at instability is given by $\epsilon_\theta = n/2$.
The radius of the ring at instability is
$$r = r_0 \exp\frac{n}{2}$$
$$= 78 \cdot 85 \text{ mm}.$$

The circumferential stress at instability is

$$\sigma_\theta = 1400(0 \cdot 05)^{0 \cdot 1}$$
$$= 1012 \ \text{MN}/m^2.$$

Therefore, at instability

$$\omega^2 = \frac{\sigma_\theta}{\rho r^2} = 20 \cdot 8 \times 6$$

giving

$$\omega = 4560 \ \text{rad/s (or 43,600 rev/min)}.$$

42. The work-hardening characteristic is represented by $\bar{\sigma} = A\bar{\epsilon}^n$. Differentiating,

$$\frac{d\bar{\sigma}}{d\bar{\epsilon}} = \frac{n}{\bar{\epsilon}} \cdot \bar{\sigma}.$$

(a) Balanced biaxial tension applied to a flat sheet.
The stress system is $\sigma_1 = \sigma_2$, $\sigma_3 = 0$.

At instability,

$$\frac{d\bar{\sigma}}{d\bar{\epsilon}} = \frac{\bar{\sigma}}{2}$$

where

$$\bar{\sigma} = \sigma_1 = \sigma_2$$

and

$$\bar{\epsilon} = 2\epsilon_1 = 2\epsilon_2 = -\epsilon_3.$$

The effective strain at instability is

$$\bar{\epsilon} = 2n$$

and therefore

$$\epsilon_1 = \epsilon_2 = n; \ \epsilon_3 = -2n.$$

(b) Circular diaphragm (see Section 10.9).
The effective strain at the pole when instability occurs is

$$\bar{\epsilon} = \frac{4}{11} (2n + 1)$$

and the surface strains

$$\epsilon_1 = \epsilon_2 = \frac{2}{11} (2n + 1).$$

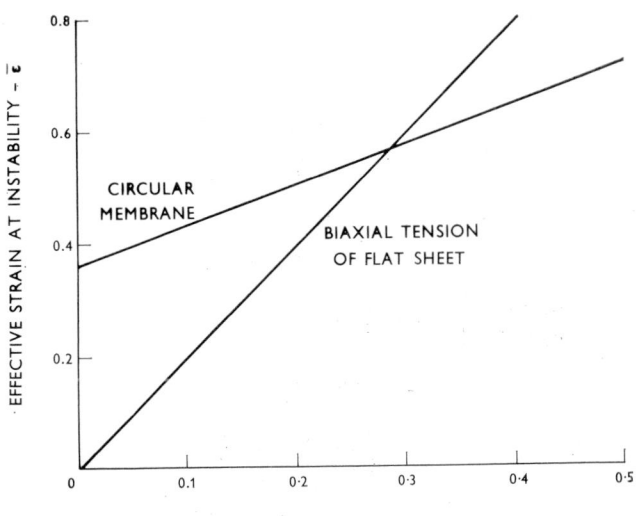

WORK-HARDENING INDEX – n

FIG. 61

The variation of the effective strain with the index n is shown in Fig. 61. Note that the effective instability strains for the two systems are equal only when $n = 0.286$.

43. The drawing tension t in pulling unit volume of the wire through the die, does an amount of plastic work $A_0.uY \ln A_0/A$ where uY is the speed of entry of the wire into the die mouth, if homogeneous deformation only, occurs. However, tension t always does work against a back tension f, so that

$$t.A.v = fA_0u + A_0uY \ln A_0/A.$$

v is the speed of emergence of the wire from the die mouth.

Hence, $t = f + Y \ln A_0/A.$

For a given greatest value of A_0/A and since $t/Y = 1$ at most, f/Y can be found. If $(A_0 - A)/A_0 = 0.5$, $A_0/A = 2$ and hence $t/Y = 1 = f/Y + \ln 2$. Thus $f/Y \simeq 0.31$.

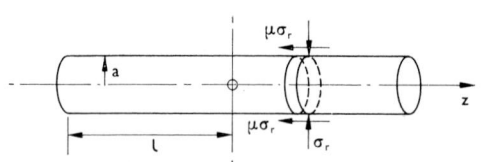   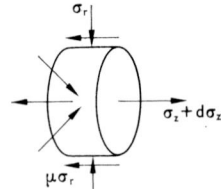

FIG. 62

44. The equilibrium equation parallel to the axis is,

$$\pi a^2 . d\sigma_z = 2\pi a\, dz\, \mu\sigma_r$$

or $a d\sigma_z = 2\mu\sigma_r\, dz.$ (1)

The Tresca yield criterion is,

$$\sigma_r + \sigma_z = 2k$$

so that, $d\sigma_r = -d\sigma_z.$ (2)

Put (1) into (2), $-a d\sigma_r = 2\mu\sigma_r\, dz.$

Integrate, $-a \ln \sigma_r = 2\mu z = C.$

Now $\left.\begin{matrix} \sigma_z = 0 \\ \sigma_r = 2k \end{matrix}\right\}$ at $z = \pm l$:

$$-a \ln 2k = 2\mu\, l + C$$

$$a \ln \frac{\sigma_r}{2k} = 2\mu(l - z)$$

or $\sigma_r/2k = \exp 2\mu \dfrac{l}{a}\left(1 - \dfrac{z}{l}\right).$

The limitation to the above approach lies in fact thus $\mu\sigma_r < k$. The limit is set when $\mu\sigma_r = k$.

Thus
$$\mu\sigma_r = k = 2k \cdot \mu. \exp 2\mu\frac{l}{a}\left(1 - \frac{z}{l}\right).$$

Putting $z = 0$.

Therefore,
$$\frac{l}{a} = \frac{1}{2\mu}\ln\frac{1}{2\mu}.$$

45. Refer to Figs. 22 and 63.

(a) The angles in Fig. 63 (a) marked with $\times$ must be 45° for frictionless extrusion.

The modification to the slip-line field for the shearing friction condition is shown in Fig. 63 (b).

(b) The α- and β-lines are as shown in Fig. 63 (a), the convention being that used in Chapter 12.

Using the Hencky equations, $p_D = k$, see Fig. 63 (cii).

Figure 63 (ci), $D \rightarrow F$, β-line, $p_F - 2k\,_D(\Delta\phi)_F = k$

$$p_F = 2k\phi\,_D + k$$

$$F \rightarrow G, \quad \alpha\text{-line}$$

$$p_G + 2k\,_F(\Delta\phi)_G = p_F$$

$$p_G = 2k\phi + k - 2k(-\phi)$$

$$= k(1 + 4\phi).$$

$G \rightarrow A$, β-line, and thus

$$p_A - 2k\,_G(\Delta\phi)_A = p_G$$

$$p_A - 2k\,(\pi/2) = p_G$$

$$p_A = k(\pi + 1 + 4\phi)$$

$$q_A = p_A + k = 2k[1 + \pi/2 + 2\phi].$$

See Fig. 63(c iv), $D \rightarrow C$, β-line

$$p_C - 2k\,_D(\Delta\phi)_C = p_D$$

$$p_C = 2k(\theta + \phi) + k.$$

And for $C \rightarrow B$, α-line

$$p_B + 2k_C(\Delta\phi)_B = p_C$$

Thus,
$$p_B = p_C + 2k(\theta + \phi)$$

$$= k[4(\theta + \phi) + 1]$$

Hence, $q_B = p_B - k = 4k(\theta + \phi)$, see Fig. 63 (cv)

(c) See Fig. 63(d).

(d) Product emerges from orifice at an angle $\pm 90°$ to orifice, see Fig. 63(e) for hodograph.

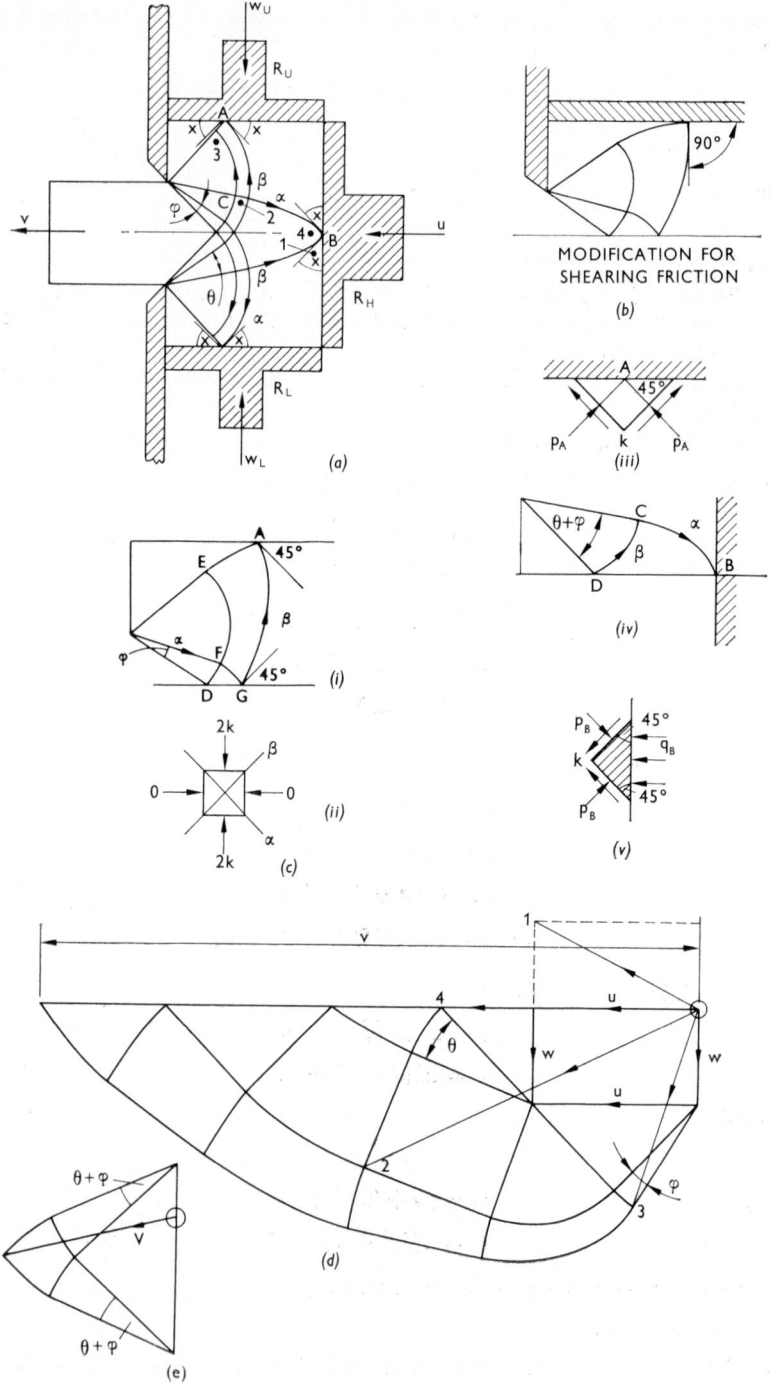

Fig. 63

46. $u.2·6 = 5.3·2$ ∴ $u = 6·17$.

The values of s and v^* from physical plane diagram Fig. 23 and hodograph Fig. 64, respectively, are tabulated below.

Line	s	v^*	sv^*
1	2·12	1·2	2·54
2	1·75	1·4	2·45
3	1·2	1·1	1·32
4	0·80	0·65	0·52
5	1·2	0·85	1·02
6	1·2	0·9	1·08
7	1·2	1·5	1·80
8	1·1	0·55	0·61
$\sum sv^*$		Total	11·34

Drawing stress t is thus given by $h.t.u = \sum ksv^*$,

$$\therefore \frac{t}{2k} = \frac{11·34}{2} \times \frac{1}{6·17 \times 2·6} = 0·34.$$

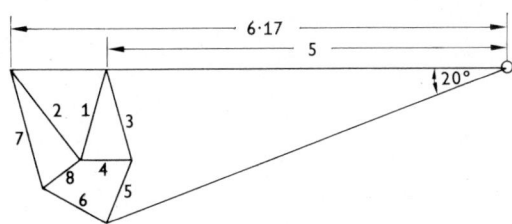

FIG. 64

47. Figure 65 shows the hodograph for Fig. 24(*a*). Combining values of s and v^* from those two figures, we may arrive at a value of $p/2k$, p being the ram pressure for the punch position selected. Fig. 24(*b*) shows how $p/2k$ varies with the position of the point P on the punch.

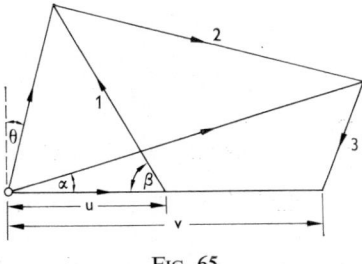

FIG. 65

If a simple piercing operation had been analysed, as in Fig. 66, by the same analysis as that described above, one would observe how the punch pressure varied with punch position and would conclude that if a method could be found whereby the punch could be fixed so that it would have to withstand a minimum stress

(= ram stress × D_1/d_1), then improvement in process performance would result. Indeed, the limit to the cold extrusion of steel tubes is set by the strength of the tool steels now available for the manufacture of punches. In effect, then, this problem was solved by Kunogi as demonstrated in Fig. 24(*a*).

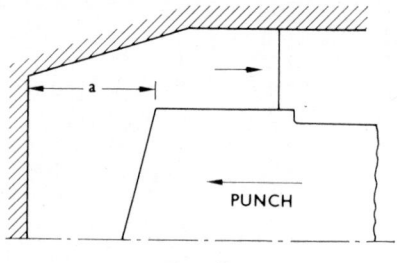

FIG. 66

48. Figure 67(*a*) shows a possible system of velocity discontinuities and Fig. 67(*b*) the corresponding hodograph.

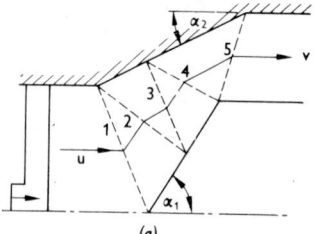

FIG. 67

49. Figure 68 shows the hodograph required for Fig. 24(*c*).

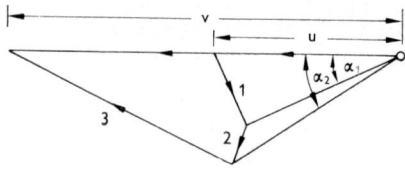

FIG. 68

50. See Fig. 69(*a*). We shall assume that:
    (a) Material slides transversely over the punch head, frictionlessly, parallel to *CG*, and
    (b) there is a line of discontinuity from *D* to *G*, so that after crossing *DG*, it moves parallel to die face *DA*.
We now assume that it follows either of two courses:
    (a) It continues to slide over the whole die face from *D* to *A* and is finally discharged parallel to the punch across line *AG*, or
    (b) at a particular point on *DA*, say *C*, contact ceases and the material proceeds in a direction such as *C*3, contact with the container wall being made at 3 and after which it moves parallel to the punch; these changes of direction are brought about by the 'shock' lines *CG* and 3*G*. (*β* is then angle *A*3*C*.)

Now it is reasonable to suppose the deformation mode taken up will be that which requires the punch to do least work and this is determined by selecting various combinations of a point on AD and one on the container wall, and calculating a number indicative of the amount of work done. For the 'shock line' configuration shown in Fig. 69(a), the hodograph is given in Fig. 69(b). The following table gives a magnitude representative of the work done associated with possible various surface boundaries obtained by joining together a point on AD and a point on AN.

A	5100	$C, 2$	5020	$D, 2$	5140
$B, 2$	5000	$C, 3$	5070	$D, 3$	5330
$B, 3$	4980	$C, 4$	5165	$D, 4$	5455
$B, 4$	5010	$C, 5$	5260	$D, 5$	5710
$B, 5$	5065				

Clearly $B3$ yields the lowest value, and we presume that the product shape appropriate to the punch position shown has exterior surface $DB3N$. Joining B to 3 the effective angle, i.e. $A3B = 30°$. This is typical of results obtained in actual experimental work, an original angle of 45° being set for the work and a 'rounded angle' of about 30° resulting.

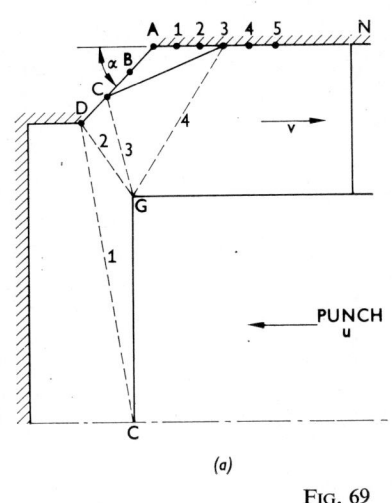
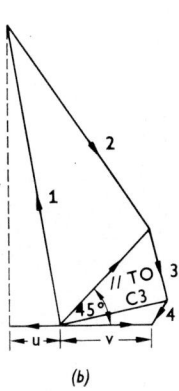

(a)   (b)

FIG. 69

51. See Fig. 70 for hodograph.

$$u = (H - x)/(h - x).$$

If P is the punch load, then

$$P/2k = s_1v_1{}^* + s_2v_2{}^*$$

$$\therefore \ Pt/2k = (d^2 + t^2) + (H - h)(z + t^2/z)$$

For given H, h, d and t, P is least when $z = t$ and hence

$$(P/2k)_{\min} = d^2/t + t + 2d; \quad \text{also} \quad u = 1 + d/t.$$

When A and C are not coincident, the speed of the emerging product, u, is such that the metal ceases to have contact with the pressure pad over length AC. This 'model' can be used to explain the reason for the formation of a hole on the back face of an extrusion.

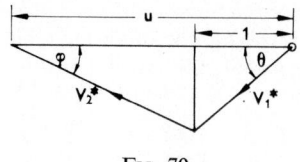

FIG. 70

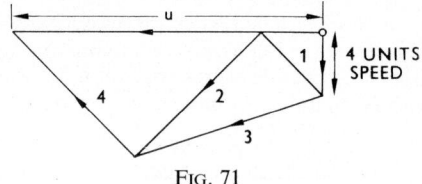

FIG. 71

52. Refer to the hodograph in Fig. 71.

Line No.	1	2	3	4
s	$4\sqrt{2}$	$4\sqrt{2}$	$\sqrt{40}$	$\sqrt{8}$
v^*	$4\sqrt{2}$	$8\sqrt{2}$	$2\sqrt{40}$	$8\sqrt{2}$
sv^*	32	64	80	32

$$\therefore\ p/2k = \sum sv^*/2 \times 10 \times 4 = 208/80 = 2 \cdot 6$$

53. Figure 72(a) shows a possible system of lines of tangential velocity discontinuity. Lines 6 and 7 are circular arcs. Fig. 72(b) is the hodograph to Fig. 72(a).

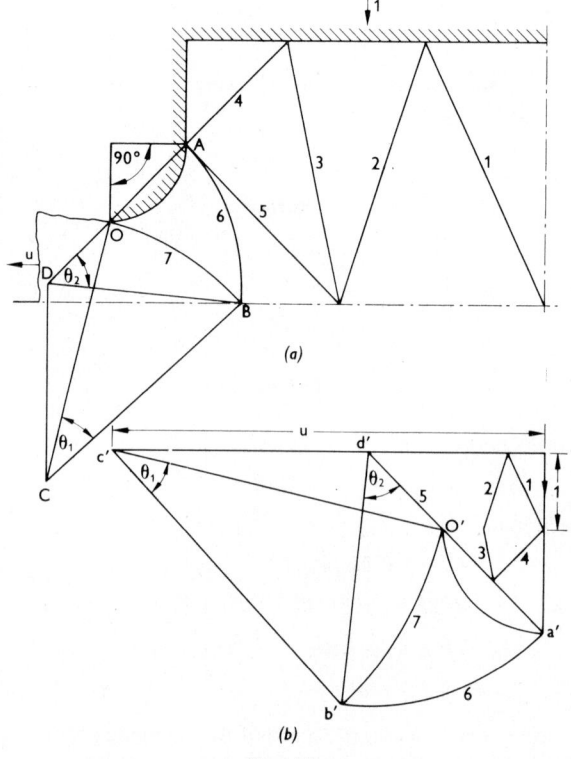

(a)

(b)

FIG. 72

54. See Fig. 73. The two layers are in contact along surface NM. Assume lines of velocity discontinuity A_1N and A_2N; these are circular arcs having centres C_1 and C_2. On crossing one of these arcs, the material then proceeds to slide over the surfaces of the die rotating about their centres of curvature. The two strips part company at N, each continuing to rotate. This is 'crocodiling'. (A phenomenon similar to this is encountered in rolling.) The energy required for this mode of deformation, when the arcs A_1N and A_2N are well chosen, is less than that required when dealing with a single solid sheet when other things are the same.

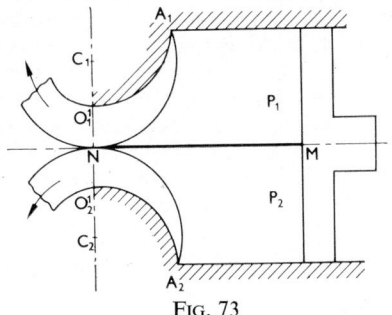

FIG. 73

55. For the given slip-line field, reduction $r = 2 \sin 15°/(1 + 2 \sin 15°) = 0{\cdot}336$ and the pressure is $p/2k = (1 + \alpha)\, r = 0{\cdot}423$.

Approximate the slip-line field by substituting a chord for the arc AB (Fig. 28), and using an obvious pattern of lines as lines of velocity discontinuity, see Fig. 74(a), draw a hodograph, see Fig. 74(b). Measure values of s and v^* and compile a table as shown. Hence find $p/2k$. From calculations of v^*/v_P the temperature distribution is approximated.

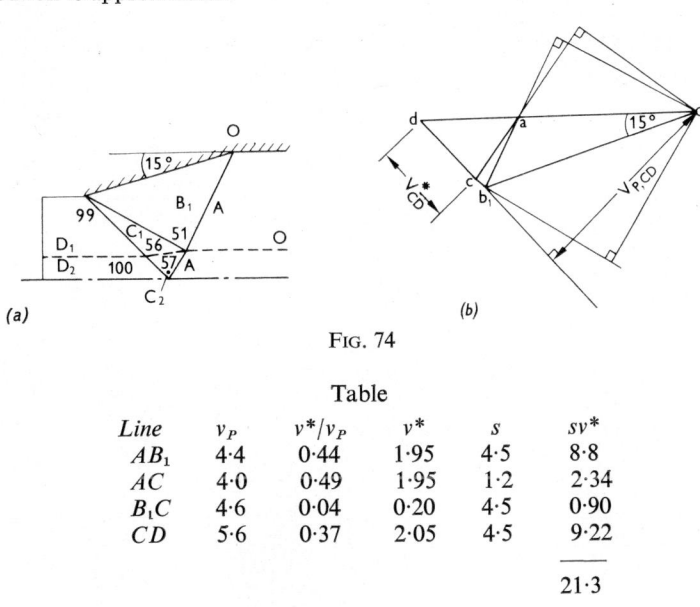

FIG. 74

Table

Line	v_P	v^*/v_P	v^*	s	sv^*
AB_1	4·4	0·44	1·95	4·5	8·8
AC	4·0	0·49	1·95	1·2	2·34
B_1C	4·6	0·04	0·20	4·5	0·90
CD	5·6	0·37	2·05	4·5	9·22

$$21{\cdot}3$$

$$\therefore\ p/2k = 21{\cdot}3/2 \times 5{\cdot}05 \times 4{\cdot}95 = 0{\cdot}43$$

Temperatures: see obvious construction lines in Fig. 74(b)

Region	B_1	C_1	D_1	C_2	D_2
	0·44	0·48	0·85	0·49	0·86
	51	56	99	57	100

See Fig. 74(a).

56. We first observe that the tools must be so proportioned that the forward extrusion involves a distinctly lower pressure on the punch than does that necessary to cause backward extrusion. A diagrammatic representation of a typical punch load punch travel card for a mild steel product, the forward and backward reductions both being 0·36, is indicated in Fig. 75. The analysis for the forward extrusion, the first 'mechanism', might be that given on p. 405. (If the diameter of the completed forward extrusion is the same as that of the chamber into which it emerges, in practice there would be considerable frictional interference to be overcome by the punch because the extruded block will be elastically reasserting itself.) The backward tube extrusion analysis, the second 'mechanism', will be seen to be susceptible to two analyses: (a) When the punch is well away from the die, it is similar to an extrusion through a square die (see p. 403), and (b) when it is about to penetrate the throat of the die it is then similar, partly, to Kunogi's process, see Problem 50.

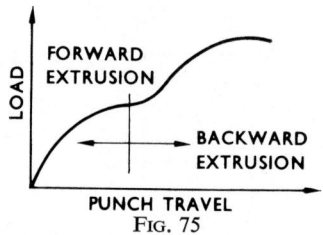

FIG. 75

57. W is the rate at which plastic work is done per element (e.g. $ABCD$).

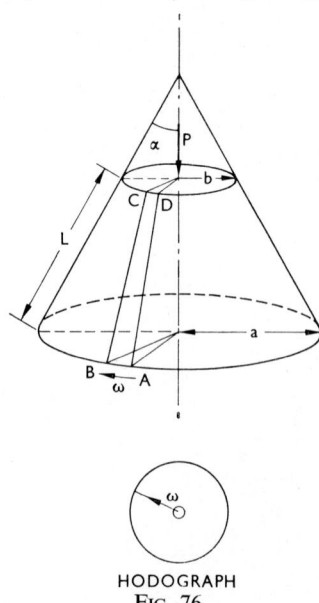

HODOGRAPH
FIG. 76

Thus $$\dot{W} = 2\pi\omega . L . M_p + 2\pi a\, M_p\, \omega.$$

(e.g. *BC*, *AD*)   (e.g. *AB*).

But if load *P* descends with speed u_0,

$$L\,\omega \sin \alpha = u_0$$

Thus, $$P . L\omega \sin \alpha = 2\pi\omega\, M_p\,(L + a)$$

Therefore, $$P = 2\pi\, M_p \left(1 + \frac{a}{L}\right) \operatorname{cosec} \alpha$$

or $$P = 2\pi\, M_p \left(\frac{L + a}{a - b}\right)$$

58. For the beam, at collapse, if the fully plastic moment is M_p, with plastic hinges at *A*, *B*, *C* and *D*, assuming beam portion *CD* to descend with unit speed,

$$2 . W . 1 = 4 . M_p . \theta,$$

and $$1 = \frac{7}{8} . l\,\theta,$$

so that $$2W = \frac{32}{7} . \frac{M_p}{l} .$$

Now, $$M_p = 2t^2 Y . t = 2 Y t^3,$$

and thus, $$W = \frac{32}{7} . \frac{t^3}{l} . Y.$$

For the wire, at T.S., $e = m/(1 - m)$ because $\sigma = Be^m$. Hence,

$$W = A\sigma = \frac{A_0 l_0}{l} . Be^m = \frac{A_0 B}{1 + e} . e^m$$

$$= A_0 B m^m . (1 - m)^{1-m}.$$

Combining the two expressions for *W*, gives

$$A_0 = \frac{32}{7} . \frac{t^3}{l} . \frac{Y}{B} . \frac{(1 - m)^{m-1}}{m^m} .$$

59. We may put $q = 2k$ so that the yield criterion is not violated in region II. For similar reasons we may put $s = 4k$. There is continuity of normal stress across the

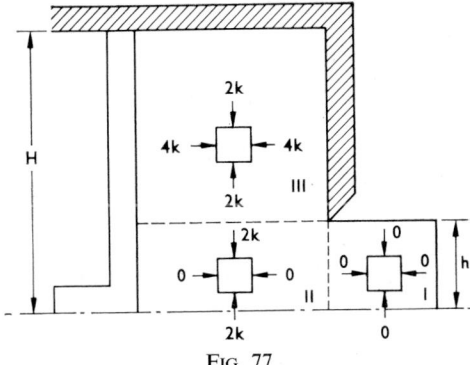

Fig. 77

I–II and II–III boundaries; thus a lower bound to the extrusion pressure $\bar{p}$ is given by

$$\bar{p}.H = 4k(H - h)$$

or

$$\frac{p}{2k} = 2\left(1 - \frac{h}{H}\right)$$

$$= 2r,$$

where r is the fractional reduction.

60. The rate at which work is done by the pressure p on the block $ABCD$ is $p.2\,R\sin\theta.u.$; this is equal to the rate of energy dissipation along AD and BC which is $2(h - R\cos\theta).u.k.$ Hence, $p/k = (h - R\cos\theta)/R\sin\theta$. Now minimize p/k by putting $d(p/k)/d\theta = 0$.
It is then found that $\sin\theta = \sqrt{h/R}$ and thus that,

$$p/k = [(h/R)^{3/2} - 1]\Big/\left[\frac{h}{R} - 1\right]^{1/2}.$$

When $h/R \simeq 7{\cdot}4$, $p/k = 9$.

61. It may be imagined that a slip-line field of the kind shown in Fig. 78 is applicable. A is made a point of singularity; on the lower surface of the plate there is no similar singularity; triangle EFG is subject to the tensile yield stress parallel to FG. The Hencky equations require, $1 + 2.C\hat{A}D - 2.D\hat{H}E = -1$, obtained by applying them along 'path' $BCDEF$.

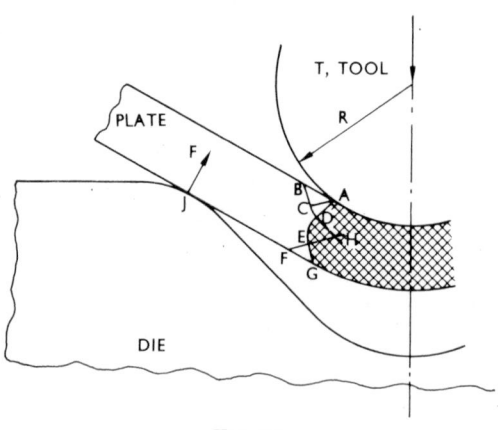

FIG. 78

62. If ω' denotes the angular speed of the emerging sheet about C, its centre of curvature, then

$$(\rho + t)\,\omega' = (R + 2t)\,\omega - v^*$$

and

$$(\rho - t)\,\omega' = R\omega + v^*,$$

where v^* is the magnitude of the tangential velocity discontinuity along the semi-circular arc in the exit roll gap. (There is a velocity discontinuity at the entrance to the roll gap of $t\omega$.)

$$v^* = R \, \omega \, [(1 + t/R)(1 - t/\rho) - 1].$$

As all the work is supplied from roll W, which applies torque T, and denoting by k the plane strain yield shear stress,

$$T\omega = k \, (\pi t \, . \, t\omega + \pi t \, . \, v^*).$$

Hence substituting for v^*,

$$T = k\pi R t \left[(1 + t/R)\left(1 - \frac{t}{\rho}\right) - 1 + \frac{t}{R} \right] = k\pi t R \frac{t}{\rho}\left(\frac{t}{R} - 1\right).$$

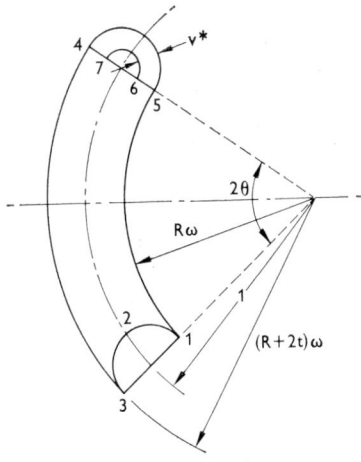

FIG. 79

63. See Fig. 80 for hodograph.

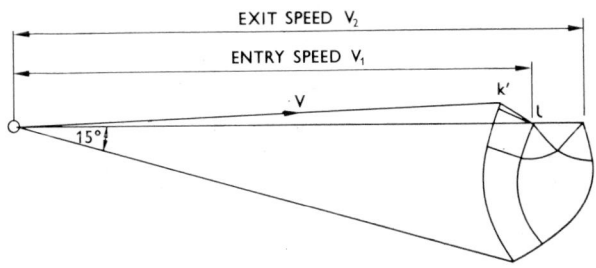

FIG. 80

64. See Fig. 81 for hodograph and chip direction.

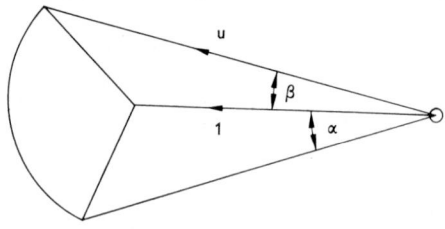

FIG. 81

65. The hodograph corresponding to the slip-line field of Fig. 37, drawn such that the tangential velocity discontinuity along the rigid-deforming boundary is a constant, is shown in Fig. 82. However, this is unacceptable because the constant angle of 45° between the slip-lines and the tool is not found in the hodograph. The motion of a particle at D, Fig. 37, is vertically downward with speed u and tangential to the tool at D, i.e., in the hodograph $\overrightarrow{OT}$ plus a component velocity parallel to TD', i.e. at θ to TA. The direction of the particle movement should, however, be the same at all points on straight slip-line DD' and in particular be perpendicular to it at D' where the velocity has been represented by OD'', and where the rotation from A to D' is θ. According to Fig. 82 the angle between the α slip-line at D', or D on the tool surface, should be ϕ, but Fig. 37 presumes it is 45° which is not so except at A.

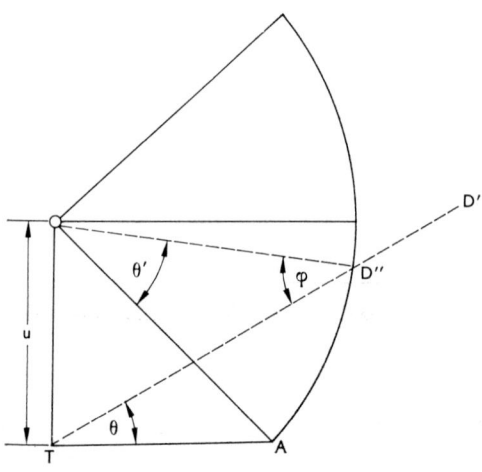

FIG. 82

66. The hodograph is shown in Fig. 83.

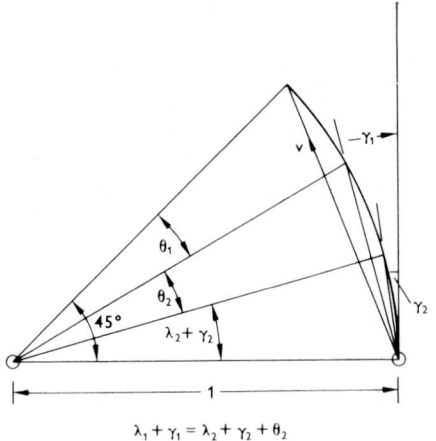

$$\lambda_1 + \gamma_1 = \lambda_2 + \gamma_2 + \theta_2$$

FIG. 83

67. To draw the slip-line field from B and C (see Figs. 39 and 84(a)) as singular points, equal radii are hinged, and triangle BEC is isosceles. The slip-line field is extended such that the slip-lines meet the plate at 45°. The method of construction will be clear from the figure. To compute the force on the plate, first observe that there is zero force in the direction perpendicular to BC so that for triangle BEC to be plastically deformed, there must be a compressive stress parallel to BC of $2k$.

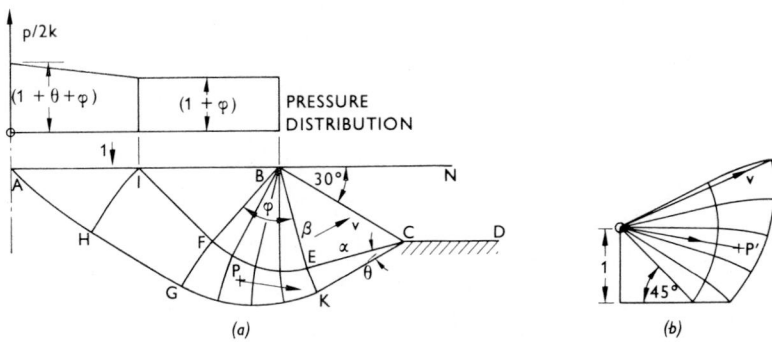

FIG. 84

We are able thus to identify the α- and β-lines at E. To find the pressure on the plate, consider a movement along the α-line from E to F and use the Hencky equations. $N\hat{B}C$ is given as 30°; then since $I\hat{B}F$ is 45°, $E\hat{B}F = 60°$. Now, at E we have

$$p + 2k\phi = \text{constant} = C_1.$$

Choosing ϕ to be zero at E, then $p = k = C_1$. The rotation from E to F is $-60°$ or $-\pi/3$ radians and hence, the hydrostatic pressure at F denoted by p_F is given by

$$p_F + 2k(-\pi/3) = k \quad \text{or} \quad p_F = k(1 + 2\pi/3).$$

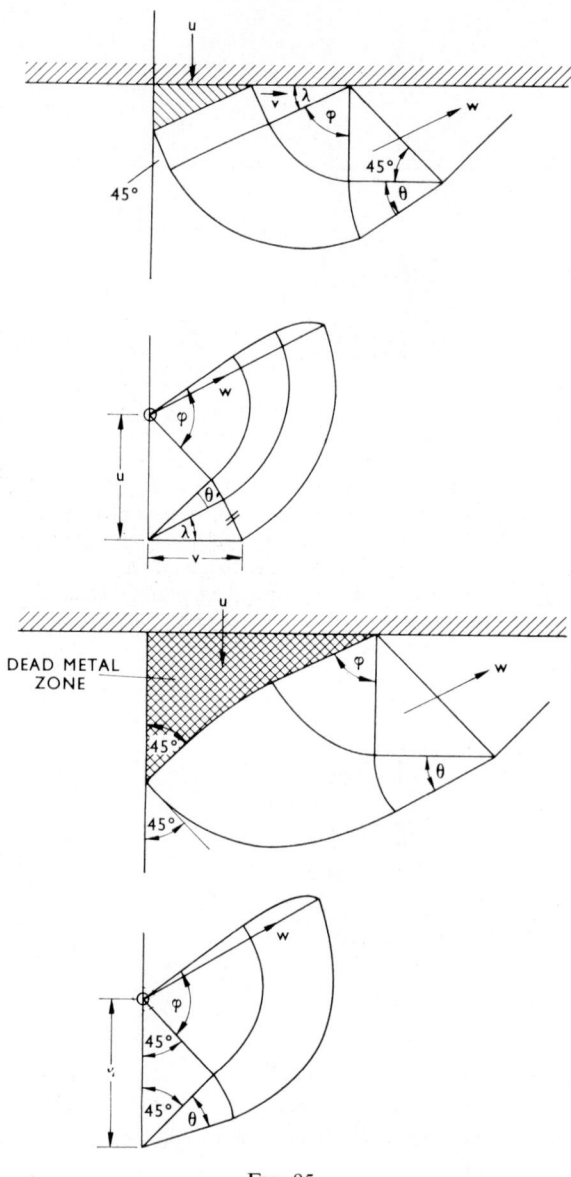

Fig. 85

The stress normal to IB is thus

$$p_F + k = 2k(1 + \pi/3).$$

Moving from I to H, i.e. along a β-line, and thus rotating through $+15°$ or $+\pi/12$ radians, the hydrostatic pressure at H is p_H and

$$p_H - 2k(\pi/12) = p_I \quad \text{therefore} \quad p_H = k(1 + 2\pi/3 + \pi/6) = k(1 + 5\pi/6).$$

Similarly, moving from H to A along an α-line, gives the hydrostatic pressure at A, p_A. Hence

$$p_A + 2k(-\pi/12) = k(1 + 5\pi/6 + \pi/6)$$

therefore, $p_A = k(1 + \pi).$

The pressure on the plate at A is

$$p_A + k = 2k(1 + \pi/2).$$

The pressure over the section BI is constant, but from I to A it rises continuously. In the upper part of Fig. 84(a), this pressure distribution is indicated. The length of contact between the material and the plate, i.e., $2AB \simeq 4$ times the projection length BC in Fig. 39; it has been chosen so that the slip-line field drawn here, which is symmetrical about a centre-line through A, fits these proportions. If the contact area were greater, it would be necessary to extend the field by increasing angle θ.

The mean pressure on the plate $\bar{p}$ is easily found to be $\bar{p}/2k = 2.2$.

To examine the metal movement, a hodograph must be drawn. This is shown in Fig. 84(b). To each point in the slip-line field there is an identifiable point in the hodograph, e.g., to point P, the point P'; the velocity of P' is given by OP' in Fig. 84(b).

Two suitable slip-line fields are shown as Fig. 85.

68. This problem has previously been discussed by Ross (1957)*, assuming that a dead metal zone covers the whole of die face BC, and not only partly so as here. (Johnson† (1958) has also discussed this problem with the aid of velocity discontinuities.) Starting from the die mouth and using the Hencky equations, it is a straightforward task to determine the pressure along the die and dead metal surfaces, to resolve the local pressures vertically and horizontally, and so obtain the total downward force required on AB and the sideways thrust on BC.

The hodograph is shown in Fig. 86.

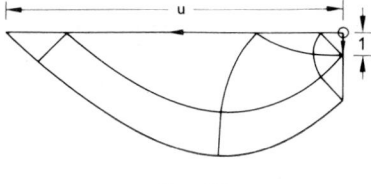

FIG. 86

69. This situation is similar to that in which a wire is used to cut butter and cheese! Since w/D is small we can imagine the wire to cut into the bar radially; the indentation may (approximately) be envisaged as one of plane strain indentation

* On the plane plastic flow of an Inset Block', *J. App. Mech.* 1957 **24**, 457.
† Some two dimensional forging operations,' *3rd U.S. Congress App. Mech.* 1958, 571.

using a flat-ended punch (see p. 498 and Fig. 14.22). The pressure to effect this is $2(1 + \pi/2)k$. (k is related to Y_b through the Mises or Tresca yield criterion). In the usual way the tension T to give the required normal radial stress on the bar surface is found to be

$$T = D.(1 + \pi/2)kw.$$

If the stress in the wire is denoted by σ_w and we assume $k = Y_b/2$, then

$$\sigma_w = \frac{D(1 + \pi/2) Y_b}{2t} \simeq 1\cdot 28 \frac{D Y_b}{t}.$$

70. We require to construct two slip-line fields, A and B, see Fig. 42, started, respectively, from the orifices at O_1 and O_2; A and B meet at a single point P, which is so chosen that, (a) the total axial extent of the fields is equal to the current slug length X and (b) the hydrostatic pressure at P, as determined by using the Hencky equations and working from each of the two orifices, shall be the same, i.e.,

$$\theta_1 + \phi_1 = \theta_2 + \phi_2.$$

At its simplest, the problem may be envisaged as shown in Fig. 87(a), for which $X/D = 1/3$ and $h_1 = h_2 = h$. The reduction is $r = (H - 2h)/H = 0\cdot 5$, and the slip-line fields are simply a pair of back-to-back quadrants. The punch pressure is easily shown to be $0\cdot 86 \times 2k$; the top extruded sheet moves rearward with a speed of unity relative to the container wall, and the axial sheet forward with an absolute speed of two. The hodograph is shown in Fig. 87(b).

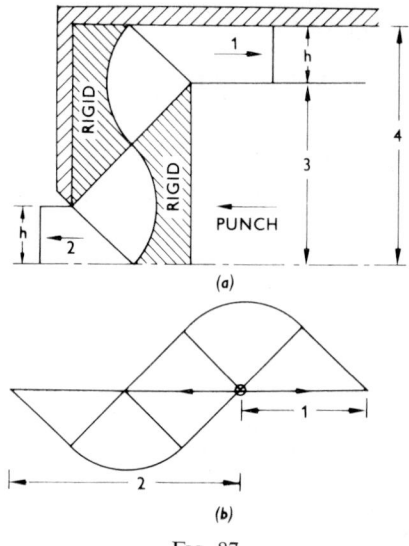

(a)

(b)

Fig. 87

71. When pressure is applied to the jaw or wedge, the material immediately beneath it will be (a) compressed vertically mainly by the flat end of the wedge, and (b) subject to a horizontal tension due to the sloping sides of the wedge. One quarter of the slip-line field proposed to 'explain' this action is shown in Fig. 88. It consists in effect of two separated slip-line fields, both of which are well known; the two triangles and one circular sector field $BCGHIB$ on the sloping side BC was apparently first proposed by Sokolovsky and developed by Hill, Lee and Tupper for indentation with a wedge-shaped die, allowing also for the formation of a lip. The second field beneath AB, an extension of the sector centred on B and giving $ABDEFA$, and in which ABF constitutes a false head on AB, was originally proposed by Prandtl; throughout region $BDEFB$, the material is plastic. This solution is valid, provided that angle IBD exceeds $90°$—otherwise the rigid material (i.e. that to the right of $EDBIHG$) at this corner would yield—and provided that a hodograph can be drawn. Suppose the wedge approaches the stationary sheet centre line with a vertical speed of unity. Then the material to the right of $EDBIHG$, which is rigid, moves instantaneously with a horizontal speed of u and, using the fact of volumetric constancy,

$$1.a = uh$$

where a is half the plier width and $2h$ the minimum sheet thickness. The hodograph is drawn in Fig. 88(b), and slip-line field region $ABDEF$ is represented in the hodograph by $okdfeo$; $\overrightarrow{ok} = u$. The slip-line field region $CGHIBC$ is represented by klm in the hodograph: radius kl is determined by l being the point of intersection t of kl

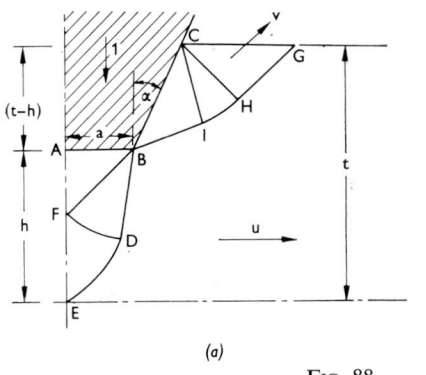
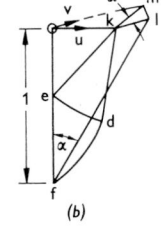

(a)　　　　　　　　　(b)

FIG. 88

which is parallel to BI drawn from k, and fl which is parallel to side BC. The force to initiate cutting is the sum of that due to pressure on BC, i.e., $2k(1 + \alpha)$, and the pressure on AB. This latter pressure is less than that required to indent or start the cutting of a strip of simple thickness $2h$, by the horizontal tension q, applied by means of BC; this is of amount $2k(1 + \alpha)$. $(t - h)/h$. We may observe that (i) for AB to be effective in applying pressure, $q \leqslant 2k$ and (ii) for BC to be effective in creating tension $|\overrightarrow{kl}| > 0$ (in Fig. 88(b)), i.e. $\tan \alpha > u = a/h$. At some stage (if the material possesses sufficient ductility) u would be large enough to create a gap between side BC and the material. Alternatively, we may say that wedge action is maintained as long as $A\hat{E}B$ is less than α; the possibility of superposing parts of the hodograph on the slip-line field of Fig. 88(a) is obvious.

The above discussion is not completely applicable to that of cutting with pliers the processes being envisaged as a succession of changing slip-line fields of the form described. First, such a solution ignores the pile-up of material at the wedge sides, —as occurs even in a steady state problem. Secondly, this problem is *not* a steady state one. Here there is no geometrical similarity between successive states; the ratio of AB to BC alters with penetration. The lip made near C will be curved; $\overrightarrow{v}$ alters as does u, i.e., as h decreases.

21*—EP　＊　＊

Intuitively it seems doubtful that by ignoring the lip formation, much serious error would in fact be made in any calculation, since the tendency to lip formation is the smaller, (i.e., $|\overrightarrow{kl}|$ in Fig. 88(b)) the larger is CB. (This form of solution was arrived at by W. Johnson and H. Kudo.)

72(a). Using the Hencky equations, and because there is zero force at right angles to OB, in Fig. 44,
$$p_{OH} = k(1 + 2\pi/4) \quad \text{or} \quad p_{OH}/2k = 1\cdot285$$
This neglects the frictional drag due to the motion of the slug relative to the container wall.

(b) $\mu = k/k(1 + \pi/2) \simeq 0\cdot39$

(c) $r = 0$

'Expansion' from unit thickness to $\sqrt{\ }$ with emergence at right angles to orifice occurs.

73. The hodograph to Fig. 45 is shown in Fig. 89.
r' is the nominal reduction, $= (2H - 2h')/2H = 1 - 2h'/2H.$

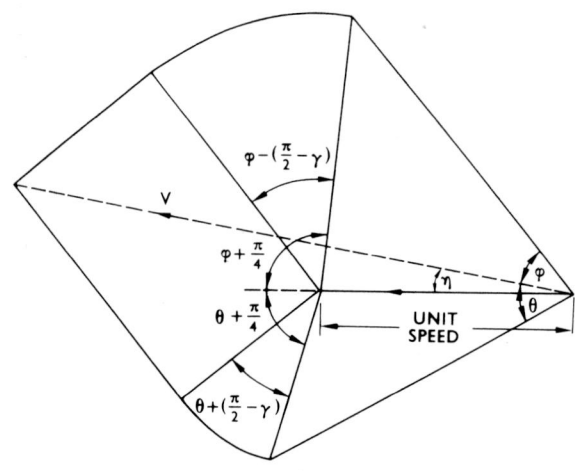

FIG. 89

If $2h' = \sqrt{2}, r' = 1 - \sqrt{2}/\sqrt{2}(\sin \gamma + \sin \theta + \sin \phi)$ or
$$r' = \frac{\sin \gamma + \sin \theta + \sin \phi - 1}{\sin \gamma + \sin \theta + \sin \phi}.$$

Using the Hencky equations, it is easily shown that q, the die pressures, are given by
$$q_1 = 2k(1 + \theta + \pi/2 - \gamma)$$
$$q_2 = 2k(1 + \phi - \pi/2 + \gamma).$$

and

Extrusion pressure p is thus,
$$\frac{p}{2k} \times 2H = \frac{q_1}{2k} \sqrt{2} \sin \theta + \frac{q_2}{2k} \sqrt{2} \sin \phi$$

or $$\frac{p}{2k} = \frac{(1 + \theta) \sin \theta + (1 + \phi) \sin \phi + (\pi/2 - \gamma)(\sin \theta - \sin \phi).}{\sin \gamma + \sin \theta + \sin \phi}$$

The actual emission of the strip is inclined to the normal to the orifice and the true reduction is given by:

$$r = 1 - \frac{2t}{2H}$$

where $2t$ is the strip thickness and $2t = 2H/v$.

74. The most general symmetrical mode of deformation is indicated in Fig. 90(a), the hodograph consisting of two concentric circles as in Fig. 90 (b). If p denotes the uniform transverse pressure, then following the approach of Chapter 15, p. 523, we have

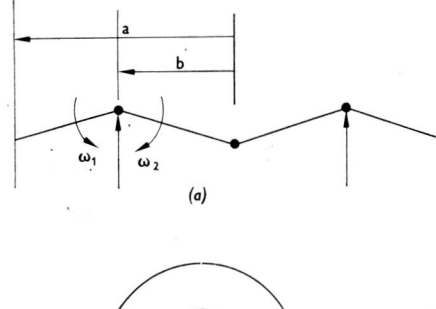

(a)

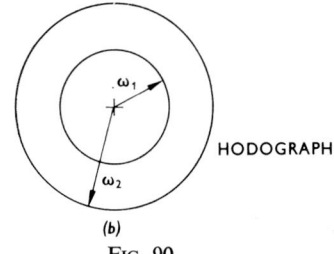

HODOGRAPH

(b)

FIG. 90

$$p \cdot 2\pi\, \omega_1 \left[b + \frac{2(a - b)}{3} \right] \left[\frac{(a - b)^2}{2} \right] + p \cdot \frac{\pi b^2\, b \omega_2}{3}$$

$$= M_p \left[(a - b) \cdot 2\pi + 2\pi b \right] \omega_1 + M_p \left[2\pi b + b \cdot 2\pi \right] \omega_2$$

$$p \frac{\pi}{3} \left[\omega_1 (2a + b)(a - b)^2 + b^3 \omega_2 \right] = M_p \left[\omega_1 \cdot a + 2b\, \omega_2 \right] 2\pi$$

Put $\omega_2/\omega_1 = y$ and $b/a = x$,

then $$\frac{p}{6M_p/a^2} = \frac{1 + 2xy}{(2 + x)(1 - x)^2 + x^3 y} = \frac{1 + 2xy}{D}.$$ (1)

For p' to be a maximum,

where $$p' = \frac{p}{6M_p/a^2},$$

$$\frac{\partial p'}{\partial y} = 0 \text{ and } \frac{\partial p'}{\partial x} = 0.$$

$$\frac{\partial p'}{\partial y} = \frac{2x}{D} - \frac{(1 + 2xy) x^3}{D^2} = 0$$

when, $2xD = (1 + 2xy) x^3.$ (2)

(b) $\frac{\partial p'}{\partial x} = \frac{2y}{D} - \frac{(1 + 2xy)}{D^2} [(1 - x)^2 + (- 2)(1 - x)(2 + x) + 3x^2y] = 0$

when $2yD = 3(1 + 2xy) [x^2y + x^2 - 1].$ (3)

Dividing (2) by (3), gives

$$y = \frac{3(1 - x^2)}{2x^2}.$$ (4)

Putting (4) into (2), we find

$$2x^3 - x^2 - 6x + 4 = 0.$$

A real root is $x = 0.7$, which, used in (1) provides $p = 24.5 \, M_p/a^2$.

75. Suppose hinges form at sections AA and BB and that when collapse takes place portions AB rotate about the hinges at BB. Assume the plastic bending moment at

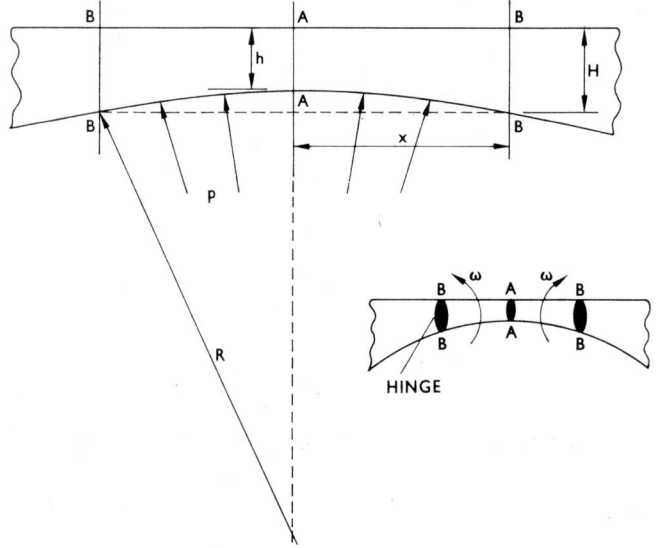

FIG. 91

AA is $h^2 Y/4$ and that at BB is $H^2 Y/4$, per unit length normal to the plane of the diagram.

Then, $p \cdot \frac{\omega x^2}{2} = (h^2 \, Y/4 + H^2 \, Y/4) \, \omega$ (1)

where ω is the rate of rotation of the (nominally) rigid material AB; p is the internal pressure applied and x is the distance between sections AA and BB, which is to be determined from (1),

$$\frac{2p}{h^2Y} = \frac{h^2 + H^2}{h^2x^2} = \frac{h^2 + (h + \Delta)^2}{h^2x^2}$$

where $\Delta \simeq x^2/2R$.

Hence,

$$\frac{2p}{h^2Y} = \frac{1}{x^2} + \left(\frac{1}{x} + \frac{x}{2Rh}\right)^2.$$

Assuming that the hinges at BB form at a location which requires a minimum p, $dp/dx = 0$, when

$$\frac{2}{x^3} = 2\left[\frac{1}{x} + \frac{x}{2Rh}\right]\left[-\frac{1}{x^2} + \frac{1}{2Rh}\right].$$

And then
i.e.

$$(p/Y) = (1 + \sqrt{2})(h/R).$$

$$x = 8^{1/4}\sqrt{Rh}; \quad \text{or} \quad x \simeq 1{\cdot}7\sqrt{Rh}.$$

76. Refer to Fig. 47. The indentation pressure is constant at $2k(1 + \pi/2) \simeq 2{\cdot}6.2k$. When the penetration is y, $l^2 \simeq y$. D, and the penetration force, F, is $2{\cdot}6. 2k. 2l$. The plastic work done in securing this degree of penetration is

$$\int_0^y F \, . \, dy = 4k \int_0^y 2{\cdot}6 \sqrt{yD} \, dy = 4k \, . \, \frac{2}{3} \, . \, 2{\cdot}6y \sqrt{yD}.$$

The volume of material encompassed by the slip-line field and dead metal zone when the indentation has proceeded to depth y, is,

$$\frac{l^2}{2} + \frac{\pi}{2}(l\sqrt{2})^2/2 + l^2 - \frac{2}{3}\frac{l^3}{D} - l^2\left(3 \, . \, 1 - \frac{1}{4} \, . \, \frac{2l}{D}\right)$$

Now, if ϵ denotes the mean strain, then
Volume of material . $2k$. $\bar{\epsilon}$ = Plastic work done and

hence,

$$\bar{\epsilon} = \frac{2 \, . \, k \, . \, 2/3 \, . \, 2{\cdot}6 \, . \, y \, . \, \sqrt{yD}}{2k \, . \, y \, . \, D[3.1 - 1/3 \, . \, (2l/D)]} = \frac{1{\cdot}73 \, . \, (2l/D) \, . \, 1/2}{3{\cdot}1 - 1/4 \, . \, (2l/D)}$$

$$= \frac{1{\cdot}73 \, . \, 0{\cdot}25 \, . \, 0{\cdot}5}{3{\cdot}1 - 1/3 \, . \, 0{\cdot}25} = 0{\cdot}10.$$

77. The radial pressure to plastically expand the block is taken to be $p = Y \ln b/a$, see Fig. 92(i).
The mean normal platen pressure p is, using the friction-hill approach, easily found to be,

$$\frac{\bar{p}}{Y} = 1 + \frac{f}{Y} + \frac{1}{3} \, . \, \frac{a}{h}$$

where f is the radial stress in the cylindrical block of radius a, see Fig. 92(ii). However, $f = p$

and hence

$$\frac{\bar{p}}{Y} = 1 + \frac{1}{3} \, . \, \frac{a}{h} + \ln\frac{b}{a}.$$

Thus $\dfrac{d}{da}\left(\dfrac{\bar{p}}{Y}\right) = \dfrac{1}{3} \cdot \dfrac{1}{h} - \dfrac{1}{a}$ and hence $h = \dfrac{a}{3}$.

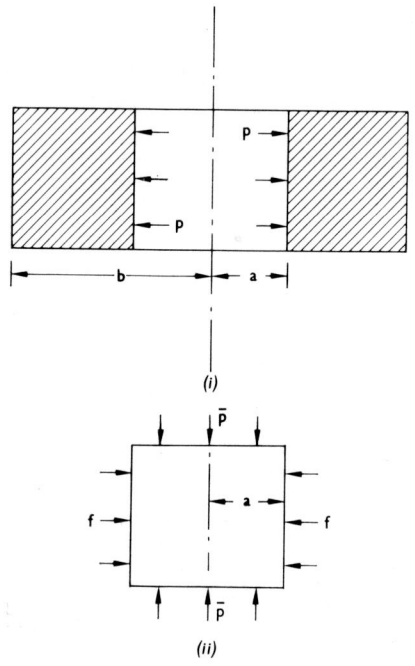

(i)

(ii)

FIG. 92

78. At the instant when rotation of a whole section occurs, (see Fig. 93) all hoop elements above XX begin to be stretched in simple hoop tension and those below XX put into simple compression. In the sectional view in Fig. 93, the tension and compression give rise to bending moments M_p which equilibrate the externally applied torque πRT, see the moment diagram of Fig. 93. Thus $2M_p = 2TR$ where $M_p = 4a^3 Y/3$ and hence, $T = 4a^3 Y/3R$.

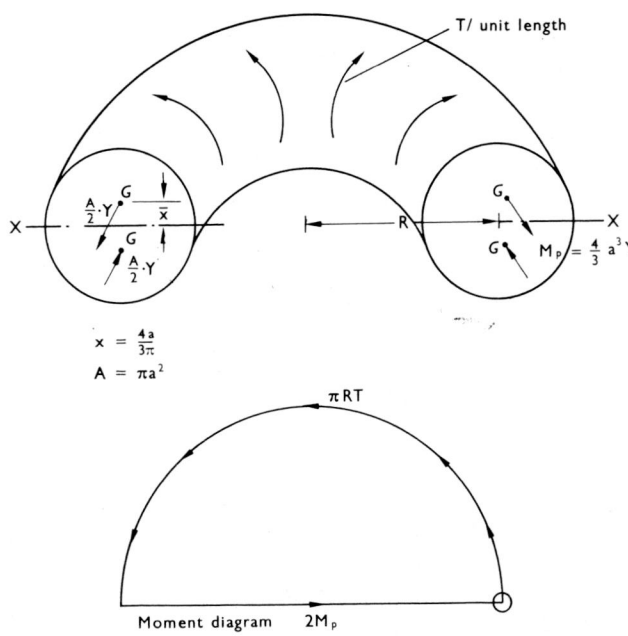

$$x = \frac{4a}{3\pi}$$

$$A = \pi a^2$$

Moment diagram $2M_p$

FIG 93.

AUTHOR INDEX

SUBJECT INDEX